Advances in Intelligent Systems and Computing

Volume 1398

The series "Advances in Intelligent Systems and Computing" contains publications on theory, applications, and design methods of Intelligent Systems and Intelligent Computing. Virtually all disciplines such as engineering, natural sciences, computer and information science, ICT, economics, business, e-commerce, environment, healthcare, life science are covered. The list of topics spans all the areas of modern intelligent systems and computing such as: computational intelligence, soft computing including neural networks, fuzzy systems, evolutionary computing and the fusion of these paradigms, social intelligence, ambient intelligence, computational neuroscience, artificial life, virtual worlds and society, cognitive science and systems, Perception and Vision, DNA and immune based systems, self-organizing and adaptive systems, e-Learning and teaching, human-centered and human-centric computing, recommender systems, intelligent control, robotics and mechatronics including human-machine teaming, knowledge-based paradigms, learning paradigms, machine ethics, intelligent data analysis, knowledge management, intelligent agents, intelligent decision making and support, intelligent network security, trust management, interactive entertainment, Web intelligence and multimedia.

The publications within "Advances in Intelligent Systems and Computing" are primarily proceedings of important conferences, symposia and congresses. They cover significant recent developments in the field, both of a foundational and applicable character. An important characteristic feature of the series is the short publication time and world-wide distribution. This permits a rapid and broad dissemination of research results.

Indexed by DBLP, EI Compendex, INSPEC, WTI Frankfurt eG, zbMATH, Japanese Science and Technology Agency (JST).

All books published in the series are submitted for consideration in Web of Science.

More information about this series at http://www.springer.com/11156

Jemal Abawajy · Zheng Xu ·
Mohammed Atiquzzaman ·
Xiaolu Zhang

Editors

2021 International Conference on Applications and Techniques in Cyber Intelligence

Applications and Techniques in Cyber
Intelligence (ATCI 2021) Volume 1

 Springer

Editors
Jemal Abawajy
Faculty of Science, Engineering
and Buil
Deakin University
Geelong, VIC, Australia

Mohammed Atiquzzaman
School of Computer Science
University of Oklahoma
Norman, OK, USA

Zheng Xu
School of Computer Engineering
and Sciences
Shanghai University
Shanghai, China

Xiaolu Zhang
Department of Information Systems
and Cyber Security
University of Texas at San Antonio
San Antonio, TX, USA

ISSN 2194-5357   ISSN 2194-5365 (electronic)
Advances in Intelligent Systems and Computing
ISBN 978-3-030-79199-5   ISBN 978-3-030-79200-8 (eBook)
https://doi.org/10.1007/978-3-030-79200-8

This Springer imprint is published by the registered company Springer Nature Switzerland AG
The registered company address is: Gewerbestrasse 11, 6330 Cham, Switzerland

Foreword

The 2021 International Conference on Applications and Techniques in Cyber Intelligence (ATCI 2021), building on the previous successes in Fuyang, China (2020), Huainan, China (2019), Shanghai, China (2018), Ningbo, China (2017), Guangzhou, China (2016), Dallas, USA (2015), Beijing, China (2014), and Sydney, Australia (2013), is proud to be in the eighth consecutive conference year. ATCI 2021 has moved online due to COVID-19.

The purpose of ATCI 2021 is to provide a forum for presentation and discussion of innovative theory, methodology and applied ideas, cutting-edge research results, and novel techniques, methods, and applications on all aspects of cyber and electronics security and intelligence. This conference establishes an international forum and aims to bring recent advances in the ever-expanding cybersecurity area including its fundamentals, algorithmic developments, and applications.

Each paper was reviewed by at least two independent experts. This conference would not have been a reality without the contributions of the authors. We sincerely thank all the authors for their valuable contributions. We would like to express our appreciation to all members of the Program Committee for their valuable efforts in the review process that helped us to guarantee the highest quality of the selected papers for the conference.

We would like to express our thanks to the strong support of the General Chairs, Publication Chairs, Organizing Chairs, Program Committee Members, and all volunteers.

Our special thanks are due also to the editors of Springer book series "Advances in Intelligent Systems and Computing", Thomas Ditzinger, Holger Schaepe, Beate Siek, and Gowrishankar Ayyasamy, for their assistance throughout the publication process.

Jemal Abawajy
Zheng Xu
Mohammed Atiquzzaman
Xiaolu Zhang

Organization

General Chairs

Hui Zhang   Tsinghua University, China
Liang Wang   Chinese Academy of Sciences, China

Online Conference Organizing Chairs

Bingkai Zhang   Fuyang Normal University, Director of scientific
                         research, China
Shibing Wang   Fuyang Normal University, Dean of School
                         of Computer Information Engineering, China

Program Chairs

Jemal Abawajy   Deakin University, Australia
Zheng Xu   Shanghai Dianji University, China
Mohammed Atiquzzaman   University of Oklahoma, USA
Xiaolu Zhang   The University of Texas at San Antonio, USA

Publication Chairs

Mazin Yousif   T-Systems International, USA
Vijayan Sugumaran   Oakland University, USA

Publicity Chairs

Kewei Sha   University of Houston, USA
Neil Y. Yen   University of Aizu, Japan
Shunxiang Zhang   Anhui University of Science and Technology,
                         China

Website and Local Service Chairs

Xianchao Wang	Fuyang Normal University, China
Jia Zhao	Fuyang Normal University, China

Program Committee Members

William Bradley Glisson	Sam Houston State University, USA
George Grispos	University of Nebraska at Omaha, USA
V. Vijayakumar	VIT Chennai, India
Aniello Castiglione	Universit di Salerno, Italy
Florin Pop	University Politehnica of Bucharest, Romania
Neil Yen	University of Aizu, Japan
Xianchao Wang	Fuyang Normal University & Tech., China
Feng Wang	Fuyang Normal University & Tech., China
Jia Zhao	Fuyang Normal University & Tech., China
Xiuyou Wang	Fuyang Normal University & Tech., China
Gang Sun	Fuyang Normal University & Tech., China
Ya Wang	Fuyang Normal University & Tech., China
Bo Han	Fuyang Normal University & Tech., China
Xiuming Chen	Fuyang Normal University & Tech., China
Xiangfeng Luo	Shanghai Univ., China
Xiao Wei	Shanghai Univ., China
Huan Du	Shanghai Univ., China
Zhiguo Yan	Fudan University, China
Abdulbasit Darem	Northern Boarder University, Saudi Arabia
Hairulnizam Mahdin	Universiti Tun Hussein Onn, Malaysia
Anil Kumar K. M.	JSS Science & Technology University, Mysore, Karnataka, India
Haruna Chiroma	Abubakar Tafawa Balewa University, Bauchi, Nigeria
Yong Ge	University of North Carolina at Charlotte, USA
Yi Liu	Tsinghua University, China
Foluso Ladeinde	SUNU, Korea
Kuien Liu	Pivotal Inc., USA
Feng Lu	Institute of Geographic Science and Natural Resources Research, Chinese Academy of Sciences, China
Ricardo J. Soares Magalhaes	University of Queensland, Australia
Alan Murray	Drexel University, USA
Yasuhide Okuyama	University of Kitakyushu, Japan
Wei Xu	Renmin University of China, China
Chaowei Phil Yang	George Mason University, USA

Hengshu Zhu	Aidu Inc., China
Morshed Chowdhury	Deakin University, Australia
Elfizar	University of Riau, Indonesia
Rohaya Latip	Universiti Putra, Malaysia

Welcome Message

The 2021 International Conference on Applications and Techniques in Cyber Intelligence (ATCI 2021), building on the previous successes in Fuyang, China (2020), Huainan, China (2019), Shanghai, China (2018), Ningbo, China (2017), Guangzhou, China (2016), Dallas, USA (2015), Beijing, China (2014), and Sydney, Australia (2013), is proud to be in the eighth consecutive conference year. ATCI 2021 has moved online due to COVID-19.

The purpose of ATCI 2021 is to provide a forum for presentation and discussion of innovative theory, methodology and applied ideas, cutting-edge research results, and novel techniques, methods, and applications on all aspects of cyber and electronics security and intelligence. This conference establishes an international forum and aims to bring recent advances in the ever-expanding cybersecurity area including its fundamentals, algorithmic developments, and applications.

Each paper was reviewed by at least two independent experts. This conference would not have been a reality without the contributions of the authors. We sincerely thank all the authors for their valuable contributions. We would like to express our appreciation to all members of the Program Committee for their valuable efforts in the review process that helped us to guarantee the highest quality of the selected papers for the conference.

We would like to express our thanks to the strong support of the General Chairs, Publication Chairs, Organizing Chairs, Program Committee Members, and all volunteers.

Our special thanks are due also to the editors of Springer book series "Advances in Intelligent Systems and Computing", Thomas Ditzinger, Holger Schaepe, Beate Siek, and Gowrishankar Ayyasamy, for their assistance throughout the publication process.

Jemal Abawajy
Zheng Xu
Mohammed Atiquzzaman
Xiaolu Zhang

Conference Program at a Glance

Saturday, June 19, 2021, Tencent Meeting Online Link		
9:00–9:30	Opening ceremony by conference PC Chair	Tencent Meeting
9:30–10:10	Keynote 1: Vijayan Sugumaran	Tencent Meeting
10:10–10:50	Keynote 2: Jemal Abawajy	Tencent Meeting
Saturday, June 19, 2021, Tencent Meeting Online Link		
13:00–18:00	Session 1	Tencent Meeting
	Session 2	Tencent Meeting
	Session 3	Tencent Meeting
	Session 4	Tencent Meeting
	Session 5	Tencent Meeting
	Session 6	Tencent Meeting
	Session 7	Tencent Meeting

ATCI 2021 Keynotes

Vijayan Sugumaran

Vijayan Sugumaran is Professor of Management Information Systems and Chair of the Department of Decision and Information Sciences at Oakland University, Rochester, Michigan, USA. He is also Co-Director of the Center for Data Science and Big Data Analytics at Oakland University. He received his Ph.D. in Information Technology from George Mason University, Fairfax, Virginia, USA. His research interests are in the areas of big data management and analytics, ontologies and semantic web, intelligent agent and multi-agent systems. He has published over 200 peer-reviewed articles in journals, conferences, and books. He has edited 12 books and serves on the Editorial Board of eight journals. He has published in top-tier journals such as Information Systems Research, ACM Transactions on Database Systems, Communications of the ACM, IEEE Transactions on Big Data, IEEE Transactions on Engineering Management, IEEE Transactions on Education, and IEEE Software. Dr. Sugumaran is Editor-in-Chief of the International Journal of Intelligent Information Technologies. He is Chair of the Intelligent Agent and Multi-Agent Systems mini-track for Americas Conference on Information Systems (AMCIS 1999–2019). Dr. Sugumaran has served as Program Chair for the 14th Workshop on E-Business (WeB2015), the International Conference on Applications of Natural Language to Information Systems (NLDB 2008, NLDB 2013, NLDB 2016, and NLDB 2019), the 29th

Australasian Conference on Information Systems (ACIS 2018), the 14th Annual Conference of Midwest Association for Information Systems, 2019, and the 5th IEEE International Conference on Big Data Service and Applications, 2019. He also regularly serves as Program Committee Member for numerous national and international conferences.

Jemal Abawajy

Jemal Abawajy is a faculty member at Deakin University and has published more than 100 articles in refereed journals and conferences as well as a number of technical reports. He is on the editorial board of several international journals and edited several international journals and conference proceedings. He has also been a member of the organizing committee for over 60 international conferences and workshops serving in various capacities including best paper award chair, general co-chair, publication chair, vice-chair, and Program Committee. He is actively involved in funded research in building secure, efficient, and reliable infrastructures for large-scale distributed systems. Toward this vision, he is working in several areas including pervasive and networked systems (mobile, wireless network, sensor networks, grid, cluster, and P2P), e-science and e-business technologies and applications, and performance analysis and evaluation.

Oral Presentation Instruction

1. Timing: a maximum of 10 minutes total, including speaking time and discussion. Please make sure your presentation is well timed. Please keep in mind that the program is full and that the speaker after you would like their allocated time available to them.
2. You can use CD or USB flash drive (memory stick), and make sure you scanned viruses in your own computer. Each speaker is required to meet her/his session chair in the corresponding session rooms 10 minutes before the session starts and copy the slide file (PPT or PDF) to the computer.
3. It is suggested that you email a copy of your presentation to your personal inbox as a backup. If for some reason the files can't be accessed from your flash drive, you will be able to download them to the computer from your email.
4. Please note that each session room will be equipped with a LCD projector, screen, point device, microphone, and a laptop with general presentation software such as Microsoft PowerPoint and Adobe Reader. Please make sure that your files are compatible and readable with our operation system by using commonly used fronts and symbols. If you plan to use your own computer, please try the connection and make sure it works before your presentation.
5. Movies: If your PowerPoint files contain movies, please make sure that they are well formatted and connected to the main files.

Registration

Since the COVID-19 problem, ATCI 2021 will use online conference way. Authors do not need pay registration fee.

Contents

Cyber Intelligence for Business and Management Innovations

Prediction of Total Investment in Fixed Assets Based on the SARIMA
Model and X_12_ARIMA Seasonal Adjustment Method   3
Canyi Yang

Analysis of the Impact of Green Credit on the Operating Efficiency
of Listed Commercial Banks Based on Python.   13
Jiahan Song

The Influence of Computer Network Technology on Trade
Development.   22
Ning Wang

The Application of Knowledge Map and the Construction
of Enterprise Knowledge Map   28
Hua Xin

Interaction Mechanism of Regional Tourism Industry
and Information Industry.   35
Xincheng Zhang

Service Quality Evaluation of Office Property Management Based
on SERVQUAL Model.   42
Lin Zhu

Characteristics and Trends of Tourism Informatization Research -
Based on Citespace Knowledge Map Analysis   50
Weitong Qiao, Nan Gao, and Xincheng Zhang

Policy Guarantee for the Development of FinTech Under the Flow
of Information Technology in Guangdong-Hong Kong-Macao
Greater Bay Area.   57
Baoxin Chen, Kan Chen, and Xi Wang

Training Mode of Applied Talents in Tourism Management Specialty
Under Artificial Intelligence   64
Yan Li

Risk Analysis and Development Research of Cross-Border
E-commerce Payment Under the Background of Internet Finance   72
Rui Lin

The Construction and Application of Financial Sharing Service
Center of Enterprise Group   80
Zijian Han

Analysis on Employment Situation of Applied Universities
in Transformation and Development   87
Peng Li, Hongjia Yu, Haitao Chang, and You Tang

Corpus-Based Comparative Study on Chinese
and English Subjectivity   93
Weiyan Ma

Analysis on the Development of Regional Innovation Ability Driven
by Internet Under the Background of "Internet Plus"   99
Rongting Long

Influence Mechanism of Mobile Social Network Users' Product
Recommendation Information on Consumers' Intention to Participate
Sharing Economy 106
Ying Dong and Li Dong

Optimizing System Management Based on Trust of Domain 113
Xiaozhu Wang and Ping Song

Service Clustering Based on GSDMM Topic Model 120
Bing Wang

Opportunities and Challenges of Hotel Marketing Under
the Background of Big Data 128
Jingfeng Jiang and Ziwei Yu

The Impact of Technological Innovation on Regional Economic
Development Under the Background of Internet 134
Xia Zhong

Digital Inclusive Finance and Innovation of Small and Micro
Enterprises Under Mobile Internet 141
Qun Cao and Jinyuan Zhang

Artificial Intelligence Technology in Enterprise
Economic Management 151
Tingting Li

Face Detection of Innovation Base Based on Faster RCNN 158
Haixing Guan, Hongliang Li, Rongqiang Li, and Mingyang Qi

**Application of Computer Information Technology and Liang Ping's
New Year's Paintings in the Field of Modern Design** 166
Yao Lu

**Exploration of Innovative Development of Logistics Management
in the Internet Age** 173
Jingde Weng

**Optimizing Short Video Platform Based on User Portrait
Theory——Take KuaiShou APP as an Example** 180
Han Li

**Fault Analysis and Diagnosis Method for Intelligent Maintenance
of OPGW Optical Cable in Power Systems** 186
Jing Song, Ke Sun, Di Wu, Fei Cheng, and Zan Xu

**Impact of Margin Trading on the Liquidity of China's Stock Market:
Based on VAR Model** 192
Yiting Shen

**Computer Information Processing Technology in Financial
Accounting Information System** 200
Ran Lv

**The Improvement Path of Police's Scientific and Technological
Literacy Based on Multi-objective Optimization Algorithm** 206
Yujie Wu and Lan Sun

**Design and Application of Tourism Marketing Platform Based
on Cloud Computing** 211
Yuhua Jiang, Haizhi Yu, and Jing Wang

**Analysis on Application Framework Construction of Transportation
and Tourism Integration Under Big Data Platform** 216
Jing Wang, Yuhua Jiang, and Haizhi Yu

Data Mining Method of Intelligent Civil Aviation Cabin Sales 221
Feng Guo

**BP Neural Network Based on Genetic Algorithm in Economic
and Industrial Structure Analysis and Prediction** 226
Yajun Yang

Implementation of ID3 Algorithm in Asset Inventory Model 231
Liuqing Ye, Canhui Zhang, Dan Zhao, Longmin Bu, and Huiting Xu

Computerized Accounting System of E-commerce Platform 236
Yuehui Hu

Big Data in the Network Live Platform of Enterprise
Brand Communication 241
Ou Yangli and Yue Wang

Machine Learning Differences in Machine Translation of Urban
Publicity Texts 246
Zhang Hong

Cyber Intelligence for Network and Cloud Technologies

Analysis on the Construction of Cultivation Mechanism of the College
Counselors' Scientific Research Ability with the Computer Technique
Support in the Internet Age 253
Chunsheng Deng, Yuebing Huang, and Jing Yu

Optimal Strategy of Front-End Development Technology
of Computer Website 260
Rong Wu

SAR Image Change Detection Based on Complex Neural Network 268
Xia Wen, Huiyong Jiang, Yanghong Mao, and Hongqin Wang

Time Series Forecasting Model Method Based on Neural Network 277
Peng Hua

Improve the Application of EMD and SVM in the Recognition
of Emotional EEG Signals 285
Zhiyi Li

College Students' Internet Altruistic Behavior from the Perspective
of Big Data 294
Jing Lin

Monitoring and Prediction Mechanism of Artificial Intelligence
Technology in Online Language Violence 300
Zhihui Yu and Weimin Wang

Medium Access Control Protocol Based on Continuous Listening
and Hierarchical Scheduling 308
Tongfei Shang, Liqiong Yang, and Jing Liu

Analysis of Network Public Opinion Based on BiLSTM
and Self-attention Fusion Mechanism 315
Jianming Sun and Yibo Sun

Prediction and Analysis of Saturated Electricity Consumption Based
on Logistic - BP Neural Network 322
Xiang Cui, Zhenghao Jia, Ping Xue, Qidan Xu, Shuang Li,
and Liankai Zhou

The Sustainable Development of the Internet Economy Under
the Background of Big Data 328
Da Kuang and Mo Kuang

Modeling and Simulation of Gansu Province Emergency Logistics
System Based on Petri Net 334
Xueqin Zhang

Overview of Cyber Threat Intelligence Description 343
Liang Guo, Senhao Wen, Dewei Wang, Shanbiao Wang, Qianxun Wang,
and Hualin Liu

Snooker Match Outcome Prediction Using ANN with Inception
Structure 351
Shanglin Li, Bo Li, Haiyang Lu, and Juan Xiao

Construction of Secure and Stable Communities in Higher Vocational
Colleges in the "Internet+" Era 360
Yunshan Liu

Competitiveness Evaluation of Port Based on BP Neural Network 367
Chongkai Zhang, Jianmin Li, Zhipeng Zhang, and Wan Zhang

Target Simulation of Bucket Reflector Based on Complex Structure . . . 373
Jiaxing Hao, Xuetian Wang, Hongmin Gao, Sen Yang, and Cuicui Yu

Computer Molecular Biology Technology in the Field of Clinical
Microbiological Examination 380
Yu Liu, Huizi Sun, and Xiaoming Dong

Cloud Computing Network Security Technology Based
on Big Data Analysis 387
Fanxing Meng

Balance Detection System Based on the Entropy Weight Method
and Decision Tree Classification 395
Bocheng Liu and Boxiang Ji

Early Warning Mechanism of Network Public Opinion Crisis in Big
Data Environment 401
Qing Liu

Mobile Application Behavior Recognition Based on Dual-Domain
Attention and Meta-learning 407
Wenjun Zhang

"Internet + Government Service" Optimization Based on Big Data 414
Xin Jin and Yiheng Yao

Application Analysis of User Portrait in Library Field 422
Jie Dong and Xichen Xing

The Application of AutoCAD in the Drawing of Archaeological
Artifacts—Line Plot Surveying and Mapping of Standard and
Complete Artifacts   429
Qiwang Zhao and Qianyun Lyu

Application of Analytic Network Process in Power Grid
Development-Diagnosis Management   437
Shiyuan Lin, Yingjin Ye, Yaru Han, Yafang Zhu, Qingxian Li,
and Jing Wu

Time Synchronization Algorithm of Airborne Network Data   444
Jian Li and Kun Zhao

Reservoir Storage Rainfall Dynamic Monitoring System Based
on Data Mining Algorithm   449
Chenchen Yao, Junlong Tang, Jinhua Liu, and Leilei Zhang

Numerical Calculation of Surface Plasmon Polariton Lens Based
on 3D Model of Cloud System   454
Min Li, Hairong Wang, and Jingmei Zhao

Cloud Computing Platform for Solar Semiconductor
Lighting System   459
Hairong Wang, Min Li, and Jingmei Zhao

Evaluation System of Scientific and Technological Innovation Talents
Benefit Based on Cloud Operation Management and Optimization   463
Yejun Wang

Enterprise Management Simulation Training System Based on Cloud
Information System Architecture   468
Wen-Jing Hu

Genetic Algorithm and Cloud Computing Platform for SaO2   473
Zhenwu Zhou and Weizheng Sun

Landscape Design Virtual Platform Based on iPad
Operating Environment   478
Shihan Hu

Cloud Based Information System Architecture in Construction Site . . .   483
Jun Zhao

BIM Technology Based on Cloud Computing in Urban Design   488
Qianyi Zhu and Rui Wang

Cyber Intelligence for AI, VR, Blockchain Applications and Innovations

Application of Robot-Assisted Percutaneous Spinal Endoscopy
in the Treatment of Lumbar Fractures 495
Weiguo Chen, Xiangfu Wang, Wanqian Zhang, Heng Zhao, Xiangyu You,
Huanying Yang, Gang Zheng, Tingrui Sang, and Chao Zhang

Application of Robot-Assisted Percutaneous Endoscopic Spine
Surgery for Thoracolumbar Tractures 504
Wanqian Zhang, Xiangfu Wang, Huihui Zheng, Weiguo Chen,
Xiangyu You, Huanying Yang, Gang Zheng, Tingrui Sang,
and Chao Zhang

Development and Application of Artificial Intelligence Technology
to Unmanned Driving Under the Background of Wireless
Communication 511
Zhenxing Bian

Key Technologies of Autonomous Driving System of Rice and Wheat
Agricultural Robots Based on High Precision Temporal and Spatial
Information 519
Wei He, Lunning Zhang, Liankai Song, and Guoxin Yu

Protection Method of Network Data Privacy Security Issues Based
on Blockchain Technology 526
Huikui Zhou and Mudan Gu

Analyses on the Monitoring Technology of CNC Machiner Under
Visual System 534
Gan Lu

Athlete's Temperature Characteristics Based on Artificial Intelligence
Perspective Adapt to Infrared Thermal Image 541
Wumei Li

Artificial Intelligence-Based Tennis Match Technique and Tactics
Evaluation System 548
Shanshan Yin

Digital Realization Technology of Qiannan Ethnic Pattern 556
Jun Yuan

Application of Artificial Intelligence Technology in Lacquerware
Process Restoration 563
Chao Deng and Ting Zhang

Digital Protection Platform of Pingxiang Nuo Mask Based
on AR Technology 572
Wei Yu

Discussion on Application of Artificial Intelligence in the Construction
of the Model for Senile Intertrochanteric Fracture After Hidden
Hemorrhage During Treatment   581
Yi Liu and Difeng Ke

Auxiliary Role of Artificial Intelligence Technology in Landscape
Architecture Design   588
Yinglin Xiao

Application of Computer Environment Simulation Technology
in Explosion Proof Equipment Experiment   596
Ruotong Shi and Weibin Zhang

Rapid Real-Time Collision Detection for Large-Scale Complex Scene
Based on Virtual Reality   605
Sining Cheng, Xianjun Chen, and Huiyan Qu

Application of Artificial Intelligence in the Academic Search Engine . . .   611
Guoqing Yue and Shaojie Peng

Artificial Intelligence and the Dilemma of Meaning of Life   617
Huaiqin Mu

Cultivating Creative Talents of Digital Media Art Under the
Background of Artificial Intelligence   622
Fei Li

Content, Operation, Laboratory: Application and Innovation
of Artificial Intelligence in Broadcast and TV Industry   629
Siwei Long and Dianyi Wu

Numerical Simulation Technology of Food Freezing and Refrigeration
Process Based on Supply Chain   635
Yihan Hou

Impact and Deconstruction of Artificial Intelligence
on Marriage Value   643
Zhen Zhang

Construction of Art Farmland and Beautiful Countryside Based
on Satellite Image Identification Technology   649
Hongpeng Yang

Drug Traceability System Based on RFID and Alliance Block Chain
Technology   655
Hongjin Li and Xiaohua Wang

The Application of Computer Aided Technology in Interior Design . . .   662
Siqi Lin

New Mode of Supply Chain Management and Financing Based
on Block-Chain 668
Jinyuan Zhang and Qun Cao

Intelligent Evaluation Method for the Credibility of Bank Digital
Transformation from the Perspective of Artificial Intelligence 676
Xie Chen and Junyi Zhang

Application and Development Prospect of Virtual Reality Technology
in Interior Soft Decoration Design Industry 684
Hongying Zhang and Yao Lu

Analysis and Construction of Visual Supply Chain for Internet Plus
Fresh Agricultural Products in the New Era 691
Ping Yang and Yanran Huang

Cloud Computing Aided Model Design of Urban and Rural Planning
and Design Based on OBE Results Oriented Theory 698
Peiyu You

Rural Meteorological Video Service Platform Based on Particle
Swarm Optimization Algorithm 703
Haimin Cheng

Optimization of Bank Credit Customer Financial Rating Model Based
on Artificial Intelligence Algorithm 710
Li Lin

General Freshness Recognition Method Based on Electronic Nose
and Improved Unsupervised Signature Projection Algorithm 715
Zhaojun Fan and Yongling Wang

Optimization Simulated Annealing Algorithm for High Strength
and Toughness Metal 720
Lijuan Zhu

RNN Neural Network for Recovery Characteristic System
of Resistant Polymer 725
Qi Wang

Real Time Thermal Error Compensation of CNC Machine Tools
Based on ACOBP Algorithm 730
Yan Huang

Brain Activity Recognition with Deep Convolutional
Neural Network 736
Zhengxing Yan

Machine Learning in Error Types of Machine Translation 743
Xueling Zhang

**Judgment Method of Landscape Design Rationality Based on Remote
Sensing Image Recognition** 748
Shihan Hu

**Environment Parameter Setting Control System for Sheep House
Based on Wireless Network** 753
Zhe Feng, Wen Zhong, and Rongxin Zang

Cyber Intelligence for Big Data

Impact of Big Data on Nursing Work and Application Prospects 761
Ke Bai

**Multi-dimensional Expansion of China's Economic Industry
Technology in the Age of Big Data** 770
Minglei Liu and Xiaoli Wang

**Modern Digital Technology Assisted Innovative Design of Chinese
Knot Button Modeling Art** 778
Yu Zhang

**Precise Design Research of Regional Cultural and Creative Products
Under Big Data** 784
Zhurong Wen

Expression of Internet Altruistic Behavior in Big Data Era 790
Jing Lin

**Logistics Supply Chain Management Under the Background
of Big Data** 796
Ru Zhao, Guoxin Gu, and Zhihui Yang

**Digital Protection and Development of Intangible Cultural Heritage
Gan Embroidery Based on Digital Technology** 803
Lanjian Zeng

**Construction of Digital Protection and Knowledge Integration
Platform for Intangible Cultural Heritage in the Context
of Big Data** 812
Ting Zhang and Chao Deng

**Safety and Emergency Management System for the Elderly Based
on Big Data** 820
Yongmei Tao

**Economic Laws and Regulations for the Development of Artificial
Intelligence Industry Based on Big Data** 826
Liuhong Chen

**Choice of Financing Mode for Serial Entrepreneurs in the Big
Data Era** 835
Yuyang Pan, Heyuzi Shi, and Guoying Niu

**Development and Innovation of Big Data Application in the Media
Industry from the Digital Perspective: 1950–2020** 841
Dianyi Wu, Siwei Long, Yan Zhou, and Sucheng Chen

**Tourism Development and Residents' Happiness Index Based on Big
Data Analysis** 848
Qi Zhou

**Precision Marketing Strategy of Insurance Market
from the Perspective of Big Data** 856
Ze Zhang and Qing Fan

Minority Patterns in Modern Interior Design Based on Big Data 866
Yuhan Zhang

**Qualitative and Quantitative Analysis of Financial Public Opinion
Risk Based on Big Data Analysis** 872
Xiuwen Wang and Zhen Wu

**Current Situation and Future Trends of Financial Management
Development in the Era of Big Data** 880
Zhou Yang

**Application of Transana Video Analysis Software in Teacher Case
Analysis Under the Background of Big Data** 887
Tongqing Yuan

**Enlightenment of Big Data Thinking on the Construction of Scientific
Research Performance Evaluation System for Humanities Teachers
in Local Universities** 895
Ermi Zhang and Wanbing Shi

Use Big Data to Analysis the Economic in China 903
Junming Chen

**Innovation of Modern Enterprise Logistics Management Model Based
on the Background of Big Data Era** 910
Jingde Weng

**On the Application of Digital Humanities in the Study of Classical
Literature** 916
Jie Zhao

**Analysis of the Integration of Big Data Technology
and Virtual Reality** 922
Xin Wang and Liang Wang

Precision Marketing Model Based on Big Data – Taking Xiaohongshu App as an Example 928
Daowen Ren and Xuejun Liu

Innovation of Machinery Manufacturing Model from the Perspective of Big Data 936
Shuai Tao

Internet Financial Innovation Under the Background of Big Data 943
Na Pu

Employment Environment for Overseas Students Based on Big Data 952
Jian Chen

Inclusive Finance for Intelligent Agriculture Development in Big Data 960
Jun Zhang

Convergent Operation of Traditional Media Under the Background of Big Data 967
Siwei Long and Dianyi Wu

Multi Maneuvering Target Tracking Based on Two Point Data Association Algorithm 973
Hong Wang, Cuijie Zhao, Nannan Zhang, Sheng Gao, and Qianqian Guo

Data Mining and Big Data Computing Platform for the Professional Identity 977
Jingjing Gao

Data Analysis for the Hogg and Max Weber Models 982
Lei Shen

Evaluation Index of Algorithm Performance in Building Energy Saving Optimization Design 986
Hai Zheng and Hongxia Yang

Author Index 991

Cyber Intelligence for Business and Management Innovations

Prediction of Total Investment in Fixed Assets Based on the SARIMA Model and X_12_ARIMA Seasonal Adjustment Method

Canyi Yang[✉]

School of Economics, Shanghai University, Shanghai, China

Abstract. Based on the monthly social fixed asset investment data of Shanghai from January 2003 to February 2020, this paper uses the SARIMA model and the X_12_ARIMA seasonal adjustment method to make predictions. The predicted value is compared with the actual value. Based on relative error, the predicted value based on the SARIMA model is more accurate and reasonable than the predicted value based on the X-12-ARIMA seasonal adjustment method. SARIMA (2, 1, 1) (0, 1, 1)12 can be used for future forecasts and can provide a basis for future fixed asset investment in Shanghai.

Keywords: SARIMA · X_12_ARIMA seasonal adjustment · Total investment in fixed assets · Time series forecast

1 Introduction

Since China's reform and opening up in 1978, the economy has achieved a "growth miracle" [1]. China's total investment in fixed assets has also continued to hit new highs. As my country's economy shifts from high-speed growth to high-quality development, my country's total social asset investment has become an important economic research indicator. Its prediction can provide a theoretical basis for further adjustment of the economic structure and regional distribution of productivity, and enhancement of economic strength [2]. It is of great significance to my country's socialist modernization.

2 Model Establishment and Indicator Description

2.1 Data Sources and Related Instruction

This article uses the monthly data of fixed asset investment in Shanghai from January 2003 to February 2020 as a sample. The total number of samples is 206. It should be noted that the statistical requirement of fixed asset investment data is to report the sum of January to February every year, because for the reasonable data forecast, the 1–2 sum data is averaged as the monthly fixed asset investment data for January and February.

© The Author(s), under exclusive license to Springer Nature Switzerland AG 2021
J. Abawajy et al. (Eds.): ATCI 2021, AISC 1398, pp. 3–12, 2021.
https://doi.org/10.1007/978-3-030-79200-8_1

Among them, the period from January 2003 to June 2019 (that is, the first 198 data) is selected as the training set, and the last 8 samples are used as the test set. The R language was used to establish the SARIMA model and the X-12-ARIMA seasonal adjustment method were used to predict the monthly data of fixed asset investment corresponding to the test set, compare the predicted value with the actual value, and compare the accuracy of the model through the relative error [3].

2.2   Judging Data Seasonality

The data is obviously unstable and may have a certain seasonality. Using a scatter plot, draw a scatter plot of Y_t and Y_{t-k} to see the linear relationship. The autocorrelation coefficient can indicate their difference. Whether there is a linear relationship between them, as shown in the figure below.

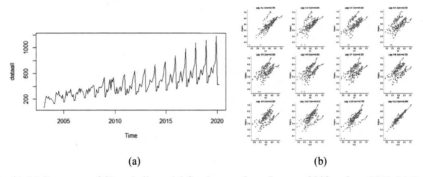

(a)   (b)

Fig. 1. (a) Data map of Shanghai's social fixed assets from January 2003 to June 2019 (b) Data autocorrelation graph

In Fig. 1(b), it can be clearly observed that the correlation coefficient between Y_t and Y_{t-12} is 0.96, and there is a good autocorrelation, indicating that this set of data has a seasonal trend, and the season length is 12 [4].

2.3   Model Establishment

2.3.1   SARIMA Model

The ARIMA model is better for short-term non-seasonal modeling and forecasting [5]. For seasonal time series, there are Seasonal ARIMA models based on the ARIMA model plus seasonal differences. The seasonal regression moving average model is called the SARIMA model. The general expression of the SARIMA model is:

$$\phi_p(L)\Phi_P(L^s)(1-L)^d(1-L^s)^D y_t = \theta_q(L)\Theta_Q(L^s)\mu_t \tag{1}$$

Where, d and D represent normal difference and seasonal difference, s represents the order of seasonal difference, q represents the moving average order, $\Phi_P(L^s)$ and $\Theta_Q(L^s)$ are the seasonal P-order autoregressive operator and seasonal Q-order moving average respectively, μ_t is a white noise process. The model uses lowercase letters to indicate the parameters of the non-seasonal model, and uppercase letters to indicate the parameters of the seasonal model [6, 7]. Because $\Phi_P(L^s)$ and $\Theta_Q(L^s)$ are multiplicative, the seasonal time series model is also multiplied as a product seasonal model. The short form of the model is:

$$SARIMA(p, d, q)(P, D, Q)_s \tag{2}$$

2.3.2 X_12_ARIMA Seasonal Adjustment Method

An economic time series is affected by many factors. These factors can usually be decomposed into trend, seasonal and random components. Seasonal adjustment is the process of correcting the influences implicit in the original data caused by seasonal factors. In terms of seasonal adjustment models, there are two classic time series models: Additive model and Multiplication model [8].

$$Y_t = T_t + S_t + R_t \tag{3}$$

$$Y_t = T_t \times S_t \times R_t \tag{4}$$

For a time series, which model is used for seasonal decomposition depends on the relationship between the components. Generally speaking, if the four components are independent of each other, the additive model is used, and the multiplicative model is used for mutual correlation; if the scale of the seasonal factors remains basically unchanged, and does not change with the increase or decrease of the original sequence level, the additive model is used; The scale changes in proportion to the original sequence level, using a multiplicative model.

In terms of seasonal adjustment methods, X-11 is an internationally accepted seasonal adjustment method. In the 1990s, the US Census Bureau developed the X-12-ARIMA program on the basis of X-11-ARIMA. The decomposition method is still the X-11 seasonal adjustment method. The forecast of the original sequence can be realized by adjusting the forecast of the decomposition factors of the season [9].

3 Model Prediction

3.1 Analysis and Prediction Using SARIMA Model

3.1.1 Seasonal and Normal Differences

First, perform seasonal difference on the data to eliminate the influence of season on data processing, determine the value of D, observe the data graph and autocorrelation function and partial autocorrelation function graph of the sample data after seasonal difference, and find the data according to unit root test It is not stable and does not meet the modeling requirements of the ARIMA model, so the data after the seasonal difference is then normalized. Here, the data is first-ordered and the data is stable [10] (Fig. 2).

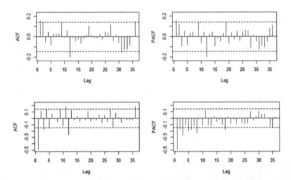

Fig. 2. Autocorrelation coefficients and partial autocorrelation coefficients after the first-order seasonal difference of the data and the first-order regular difference Correlation coefficient graph

3.1.2 Identification, Establishment and Testing of the SARIMA Model

To establish a proper SARIMA model, it is necessary to determine the values of the parameters P, Q, p, and q in the model after observing the ACF and PACF diagrams of the differential data. For the seasonal terms P and Q of the AR and MA models, we will see the difference in the lags of ACF and PACF. A prominent data can be seen at lag12 in the ACF, but not elsewhere, so here we make sure that Q is 1. There is also a prominent data at lag12 in the PACF diagram, but it is not very obvious, and there is attenuation in the periodic position in the PACF diagram, that is, it decreases at lags 12, 24, 36, so it conforms to the model The P value may be 0 and 1. For the normal AR and MA model parameters, it can be seen from the figure that ACF and PACF do not quickly approach zero after a certain order, and both are tailing. Choose p = 2 or 3 and q = 1.

For the established SARIMA(2,1,1)(1,1,1)$_{12}$, SARIMA(3,1,1)(1,1,1)$_{12}$, SARIMA(2,1,1)(0,1,1)$_{12}$. SARIMA(3,1,1)(0,1,1)$_{12}$ Four possible SRIMA model evaluations, through the output Akaike Information Criterion (AIC), Bayes Information Criterion (BIC) evaluation indicators, select Among them, the smallest AIC is used as the prediction model. AIC calculation criteria: where k is the number of model parameters and L is the likelihood function. The AICs of the four models are 1902.53, 1903.52, 1900.63, and 1901.62 respectively. Finally, SARIMA(2,1,1)(0,1,1)$_{12}$ is selected as the final prediction model.

Finally, the model is tested, mainly through residual test. The residuals meet normality, mainly because the residuals are concentrated in a certain value. If the value is very close to 0, it actually obeys a normal distribution with a mean of 0. That is, it is a white noise. The residuals satisfy the non-autocorrelation, mainly to no longer include the sequence generated by the AR or MA process in the residuals. As shown in the figure below, it can be judged that it is a white noise [11, 12] (Fig. 3).

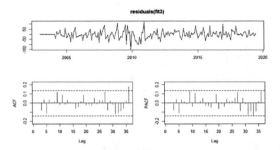

Fig. 3. SARIMA (2, 1, 1) (0, 1, 1)12 model residual test diagram

3.1.3 SARIMA Model Prediction

For the prediction of fixed asset investment in the next 8 months, the model forecast is more reasonable from the graph, and compared with the actual value, it shows that the model is very close to the true value and is more accurate. Although the forecast errors in January and February of 2020 are relatively large, it may be because the new crown pneumonia epidemic has had a certain impact on social fixed asset investment, resulting in low social fixed asset investment in early 2020 (Table 1).

Table 1. SARIMA (2,1,1)(0,1,1)12 model prediction results and true values and their relative errors

Time	Predicted value	True value	Relative error
2019.7	619.7649	616.2553	0.5695%
2019.8	602.5745	605.6287	−0.5043%
2019.9	709.2492	718.1374	−1.2377%
2019.10	757.7572	758.4522	−0.0916%
2019.11	781.9552	797.1309	−1.9038%
2019.12	1140.9846	1181.5827	−3.4359%
2020.1	489.2521	425.28603	15.0407%
2020.2	489.2568	425.28603	15.0418%

3.2 X_12_ARIMA Seasonal Adjustment Method Using Addition

3.2.1 Component Decomposition

In order to extract the trend component T_t, the seasonal component St, and the irregular component Rt in the quarterly data of Shanghai's fixed asset investment, we now use the additive X_12_ARIMA seasonal adjustment method through the seasonal expansion package to invest in fixed assets in Shanghai Province from January 2003 to June 2019. The monthly data for the current period is adjusted seasonally (Fig. 4).

8   C. Yang

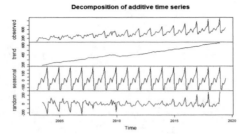

Fig. 4. Decomposition diagram of sample data using additive X_12_ARIMA seasonal adjustment method

3.2.2 Prediction of Composition

It can be seen from the decomposition diagram that the trend of each variable decomposed is clearer. Next, make predictions for each variable. First of all, for the seasonal component, it can be seen that the seasonal change trend is very stable, and through the seasonal data, it can be seen that the seasonal trend is the same every year. Therefore, this model assumes that the seasonal factor remains unchanged and directly intercepts the corresponding monthly seasonal component value.

For the prediction of irregular components, this article uses the ARIMA model to make predictions. After testing, the ARIMA (5,0,0) model is selected to obtain the predicted value of the irregular components.

For the prediction of trend components, the arima model and the least square method are used to predict. Observing the trend component image, it is found that the trend is similar to a linear form since 2013. For the two forecasting methods, it is found that the difference between the two is not very large, but the least squares method is more accurate. Therefore, the least squares method is selected for the prediction of the trend component, and January and February 2003 are used as independent variables and the value is 1, 2, and so on, the predicted result is: $y = 2.2046x + 428.1622$ (Fig. 5).

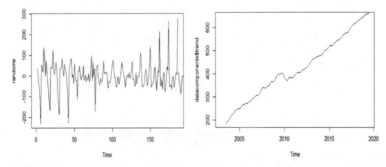

Fig. 5. ARIMA (5, 0, 0) model predicts irregular components; Trend component regression prediction chart

3.2.3 Data Prediction

According to the prediction results obtained by different prediction methods, the reduction of additive decomposition is performed, and the final prediction value is obtained according to the addition formula.

3.3 X_12_ARIMA Seasonal Adjustment Method Using Multiplication

The steps of the X_12_ARIMA seasonal adjustment method using multiplication are basically the same as the forecasting method used. The difference is that the final forecast is to follow the multiplication formula for data forecasting (Fig. 6).

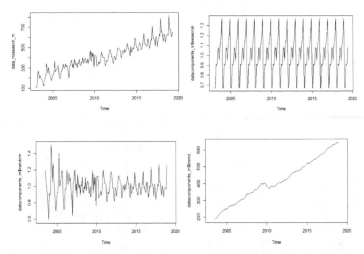

Fig. 6. Exploded diagram of X_12_ARIMA seasonal adjustment method using multiplication

3.4 Forecast Results and Comparison of X_12_ARIMA Seasonal Adjustment Method Using Addition and Multiplication

The seasonally adjusted predicted value and final predicted value of each component using addition and multiplication are as follows (Tables 2 and 3).

Table 2. Forecast results of X_12_ARIMA seasonal adjustment method using addition

Time	Seasonal component	Irregular component	Trend component	Sum forecast data
2019.7	16.6206	20.5398	648.1082	685.2686
2019.8	−29.3610	−22.9460	649.0981	596.7910
2019.9	35.7828	−52.9425	649.7262	632.5664
2019.10	65.4252	−87.9483	650.1135	627.5904
2019.11	98.0074	−54.3233	650.3478	694.0318
2019.12	190.4666	0.0278	650.4879	840.9823
2020.1	−133.6258	0.0279	650.5710	516.9730
2020.2	−135.6775	0.0278	650.6199	514.9702

Table 3. Forecast results of X_12_ARIMA seasonal adjustment method using

Time	Seasonal component	Irregular component	Trend component	Product prediction data
2019.7	1.097478	1.0753321	648.822041	765.7094872
2019.8	1.0599018	0.9506547	651.026641	655.974852
2019.9	1.0080785	1.0986979	653.231241	723.5017627
2019.10	0.955505	1.1461435	655.435841	717.7978378
2019.11	0.9155484	1.2059022	657.640441	726.0757085
2019.12	0.898055	1.3714917	659.845041	812.7146267
2020.1	0.9069334	0.6721116	662.049641	403.5592827
2020.2	0.9393033	0.6684297	664.254241	417.0574873

3.5 Automatic Prediction of X_12_ARIMA Seasonal Adjustment Method Based on R Language

There are many functions in the R language that can be used for automatic prediction. For the X_12_ARIMA method, the auto statement can be used to predict the data after the data is decomposed. In the process of automatic prediction, the computer will select the best fit method to predict the data.

4   Conclusion

As shown in Fig. 1, the SAIRMA model, X_12_ARIMA seasonal adjustment method of additive decomposition, multiplicative decomposition and automatic prediction method predicted values are compared with the true value, the smallest error is the SARIMA model, followed by the automatic prediction of X_12_ARIMA seasonal adjustment method. The large errors of the additive decomposition and multiplicative decomposition of the X_12_ARIMA seasonal adjustment method can also be related to the failure to find a suitable forecasting model, which does not mean that it is not suitable for forecasting the amount of fixed asset investment in Shanghai. Judging from the current results, the SAIMA model has the best prediction of monthly data on social fixed asset investment in Shanghai from 2003 to 2020 (Fig. 7).

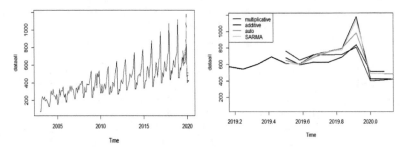

Fig. 7. Comparison of model forecast results

Investment in fixed assets is the direct driving force of economic growth. Reasonable investment scale and structure of fixed assets and good investment benefits play an important role in expanding domestic demand, adjusting economic structure, stimulating economic growth, alleviating employment pressure, and improving people's living standards (Table 4).

Table 4. Method forecast results relative error comparison table

Time/method of prediction	SARIMA prediction	X12 seasonal adjustment additive decomposition forecast	X12 seasonally adjusted multiplicative decomposition forecast	X12 seasonal adjustment method automatic prediction
2019.7	0.57%	11.20%	24.25%	5.57%
2019.8	−0.50%	−1.46%	8.31%	−0.66%
2019.9	−1.24%	−11.92%	0.75%	−4.30%
2019.10	−0.09%	−17.25%	−5.36%	−0.98%
2019.11	−1.90%	−12.93%	−8.91%	0.60%
2019.12	−3.44%	−28.83%	−31.22%	−16.44%
2020.1	15.04%	21.56%	−5.11%	15.46%
2020.2	15.04%	21.09%	−1.93%	14.89%

The results show that SARIMA $(2, 1, 1) (0, 1, 1)12$ can be used for future prediction and can provide a basis for future fixed asset investment in Shanghai.

References

1. Xu, X.: Accurately understand China's income, consumption and investment. Chin. Soc. Sci. (002), 4–24 (2013). (in Chinese)
2. Liang, Z.: Research on the relationship between fixed asset investment and economic growth in the whole society—an empirical analysis based on Zhejiang Province data. China's Collect. Econ. (3) (2017). (in Chinese)
3. Geng, J.: Research on CPI prediction based on X-12-ARIMA and SARIMA models and their combined models. Southwest Petroleum University. (in Chinese)
4. Oliphant, T.E.: Python for scientific computing. Comput. Sci. Eng. 9(3), 10–20 (2007). (in Chinese)
5. Huang, S., Li, W.: Research on the prediction of consumer price index based on GM (1,1), BP and ARIMA models. Econ. Perspect. (002), 258, 260 (2014). (in Chinese)
6. Li, H.: The application of ARIMA model in the forecast of my country's social fixed assets investment. Foreign Trade (07), 87–88 (2010). (in Chinese)
7. Long, H., Yan, G.: Research on GDP time series forecast based on the integrated model of SARIMA, GM (1,1) and BP neural network. Math. Stat. Manag. 32(5), 814–822 (2013)
8. Nielsen, S.F.: An introduction to analysis of financial data with R. J. Appl. Stat. 41(11–12), 2777–2778 (2014)
9. Shi, M.: The application of ARIMA model in Shanghai's social fixed assets investment forecast. Math. Stat. Manag. (2005). (in Chinese)
10. Survey and Statistics Department of the People's Bank of China, Time series X-12-ARIMA Seasonal Adjustment: Principles and Methods. China Finance Press (2006). (in Chinese)
11. Tsay, R.: Financial Time Series Analysis. Machinery Industry Press (2009)
12. Zhang, J.: Research on the forecast of my country's social fixed assets investment based on ARIMA model. Mod. Commer. (005), 77–78 (2019). (in Chinese)

Analysis of the Impact of Green Credit on the Operating Efficiency of Listed Commercial Banks Based on Python

Jiahan Song[✉]

School of Economics, Shanghai University, Shanghai, China

Abstract. At this stage, under the country's concept of advocating green development, a wave of green revolution with rich connotations has been set off. Among them, listed commercial banks are developing green credit business to achieve business transformation and innovation. This paper uses Python software to explore the impact of green credit on the profitability and asset quality of listed commercial banks by using a large amount of data from 2010 to 2018 through a random effect variable intercept panel model. The results show that green credit can improve the profitability of large state-owned commercial banks. Capability and asset quality have obvious positive effects.

Keywords: Green credit · Operational performance · Listed commercial banks · Python

1 Introduction

Since the reform and opening up, China's economic development has made world-renowned accomplishment, but environmental problems have become increasingly serious. Taking into account environmental protection while developing the economy has become a key challenge in the new era. How to reduce the random expansion of high-energy-consuming and high-polluting enterprises, and how to take into account the common development of economy and environment? The green credit (GC) policy answers these questions.

GC refers to banks and other financial institutions that provide welfare low-interest loans to companies engaged in environmental protection, resource recycling, green finance, etc. which are beneficial to promoting sustainable development (Macklon 2002) [1], and limited loans and formulate a relatively high interest rate credit policy to corporate projects that generate waste materials, cause pollution and waste (Baron 2010) [2]. The purpose of GC is to use financial means to establish entry thresholds with the environment as the core, reduce the inflow of funds to industries with high pollution, high pollution and overcapacity, and support and encourage environmental protection industries (Scholtens 2007; Weber 2018) [3, 4], which can realize the simultaneous progress of economy and environmental protection (Codsi 1988) [5]. In addition, GC policies can improve the operating performance of commercial banks (Richardson 2005;

J. Abawajy et al. (Eds.): ATCI 2021, AISC 1398, pp. 13–21, 2021.
https://doi.org/10.1007/978-3-030-79200-8_2

Platonova 2018; Lazzatoni 2017) [6–8] and help improve core competitiveness (Carmen 2019; Eshet 2017) [9, 10].

According to the Social Responsibility Report issued by the China Banking Association, in recent years, the GC balance of the banking industry has increased year by year. As of December 2018, the GC balance of 13 major banking institutions exceeded 28 trillion yuan, an increase of 14.24% over 2017. An annual increase of about 10% from 2010 to 2018 indicates that China has implemented GC guideline well. Thus, it has great importance and significance to research the impact of GC policy on the operating efficiency of commercial banks.

2   Data and Model

2.1   Variable Selection

2.1.1   Explained Variable

Return on total assets (ROA). The ROA is the net profit that the utilization of unit assets finally brings to the enterprise. The greater the return on total assets, the more benefit of listed commercial banks' unit assets; otherwise, the weaker.

Non-performing loan ratio (NPL). It is a key index in the risk control system of commercial banks. The NPL rate is a reflection of the quality of some commercial banks' loans. The higher the NPL rate, the greater the hazard of commercial banks recovering loans. The most first-hand display of commercial bank credit risk exposure is the dilation of non-performing loans and the rise of NPL ratios.

2.1.2   Explanatory Variables

Green credit scale (GC). The scale of GC is expressed by the bank each year. This paper selects the natural logarithm of GC as the core explanatory variable, which can not only directly reflect the degree of development of GC, but also avoid the influence of heteroscedasticity caused by data fluctuations in the empirical test.

2.1.3   Control Variable

This article selects the internal factors of the bank as control variables, including the size of assets (SIZE), capital adequacy ratio (CA), cost-to-income ratio (CI), and loan-to-deposit ratio (LD). The relatively perfect state of a commercial bank's steady operation is to maintain a moderate asset scale. Therefore, the asset scale (SIZE) is an effective guarantee for commercial banks to improve their profitability and asset quality. The capital adequacy ratio (CA) is the basic guarantee for the normal business and profitability of commercial banks, and directly affects the profitability and asset quality of commercial banks. The cost-to-income ratio (CI) can be a relatively intuitive way to see a bank's ability to use assets to achieve value-added, and refers to how much the bank pays to achieve established economic benefits. Obviously, the lower the CI ratio, the less cost that needs to be paid, and the more in line with the profitability characteristics of commercial banks. The loan-to-deposit ratio (LD) is a magic weapon for commercial banks to control asset quality. From the perspective of commercial bank risks, asset quality and loan-to-deposit ratio are positively correlated. The larger the loan scale and the smaller the deposit scale, it means the bank's asset quality It may be worse.

2.2 Data Source

China's GC policy was officially released in July 2007. Most banks began to reveal intact GC data since 2009. At the same time, some banks' financial statement data before 2010 were also lacking. Under the premise of ensuring the rigor and reliability of empirical analysis, this article selects 13 banks that have announced GC status during 2010–2018 for analysis.

2.3 Descriptive Statistical Analysis

The development of GC business is closely associated with the bank's operating performance, social responsibility, reputation, etc. Different banks have different motivations for GC. Large state-owned banks (LSB) may pay more attention to their contributions to community and their corporate duties, while small and medium-sized joint-stock commercial banks (SMCB) concern mote about the benefits. Therefore, this paper divides the banks into two parts: LSB and SMCB. The descriptive statistics of the variables are displayed in Table 1.

Table 1. Statistics summary

Category	Statistics	ROA	NPL	lnGC	lnSIZE	CA	CI	LD
All samples	Average	1.08	1.25	6.91	15.54	12.47	30.99	74.14
	Median	1.11	1.22	7.26	15.48	12.14	30.54	72.10
	Max	1.47	2.39	9.42	17.14	17.19	43.41	109.98
	Minimum	0.64	0.38	2.52	13.50	9.88	21.59	55.77
	Standard deviation	0.20	0.45	1.50	0.87	1.50	4.51	9.23
Large state-owned bank	Average	1.17	1.35	8.23	16.45	13.76	30.24	69.79
	Median	1.16	1.42	8.45	16.55	13.74	29.87	70.28
	Max	1.47	2.39	9.42	17.14	17.19	38.59	88.32
	Minimum	0.80	0.85	6.39	15.19	11.59	23.91	55.77
	Standard deviation	0.18	0.36	0.75	0.48	1.18	3.26	6.33
Small and medium joint-stock commercial banks	Average	1.03	1.18	6.09	14.98	11.67	31.46	76.86
	Median	1.02	1.14	5.97	14.99	11.50	31.12	73.10
	Max	1.46	2.14	9.04	15.72	15.68	43.41	109.98
	Minimum	0.64	0.38	2.52	13.50	9.88	21.59	61.95
	Standard deviation	0.20	0.49	1.23	0.51	1.05	5.10	9.73

Table 1 shows that for the overall sample, the average ROA is 1.081, the median is 1.110, and the median is relatively larger, indicating that the ROA of the sample is left-skewed. The standard deviation is 0.200, indicating that the ROA has less fluctuation. The average NPL ratio is 1.246 and the median is 1.220, indicating that the distribution of the NPL ratio is right-skewed, with a standard deviation of 0.445. Compared with the ROA, the difference in the NPL ratio between banks is greater. The average of the scale of lnGC is 6.913, and the standard error is 1.495, which is the relatively large standard deviation during all variables, showing that the scale of GC varies greatly during distinct banks.

Through contrast, we can see that the average value of lnGC of LSB is larger than that of SMCB, indicating that LSB are more active in developing GC business. For the two variables of lnSIZE and CA ratio, the average value of LSB is significantly larger than that of SMCB, indicating that the scale of GC may have a positive correlation with the scale of operations and CA ratio. In the two variables of cost-income ratio and LD, the average value of LSB is significantly lower than that of SMCB, indicating that the profitability and asset quality of LSB are higher than that of SMCB.

2.4 Model Setting and Selection

So as to study the impact of GC on the operating efficiency of listed commercial banks (Hypothesis I), the paper constructs the following model:

$$ROA_{i,t} = \partial + \beta_1 \ln GC_{i,t} + \beta_2 \ln SIZE_{i,t} + \beta_3 CAR_{i,t} + \beta_4 CB_{i,t} + \varepsilon_{i,t} \qquad (1)$$

Model (1) is a mixed regression model, which indicates that there is no difference in the data coefficients and intercept items of each bank every year, that is, the impact of various variables such as the scale of GC of each bank on the ROA is consistent.

$$ROA_{i,t} = \partial_i + \beta_1 \ln GC_{i,t} + \beta_2 \ln SIZE_{i,t} + \beta_3 CAR_{i,t} + \beta_4 CB_{i,t} + \varepsilon_{i,t} \qquad (2)$$

$$ROA_{i,t} = \gamma_t + \beta_1 \ln GC_{i,t} + \beta_2 \ln SIZE_{i,t} + \beta_3 CAR_{i,t} + \beta_4 CB_{i,t} + \varepsilon_{i,t} \qquad (3)$$

$$ROA_{i,t} = \partial_i + \gamma_t + \beta_1 \ln GC_{i,t} + \beta_2 \ln SIZE_{i,t} + \beta_3 CAR_{i,t} + \beta_4 CB_{i,t} + \varepsilon_{i,t} \qquad (4)$$

$\varepsilon_{i,t}$ is a random variable. If it is related to a variable in the model, it is a fixed-effect model. Model (2) is an individual fixed-effect model, which ∂_i means that it changes with individuals, but does not change with time. There are N different intercept terms for N sections; model (3) is a time-point fixed-effect model, which means it changes with time. It does not change with the individual, and there are T different intercept terms for T sections; model (4) is an individual time-point fixed-effect model, which means that the intercept terms vary with individuals and time. If $\varepsilon_{i,t}$ isn't related to the variables in the model, the above model is a random effects model.

In order to study the impact of GC on the NPL rate of commercial banks (Hypothesis II), the paper constructs the following model:

$$NPL_{i,t} = \partial + \beta_1 \ln GC_{i,t} + \beta_2 \ln SIZE_{i,t} + \beta_3 CAR_{i,t} + \beta_4 CB_{i,t} + \varepsilon_{i,t} \qquad (5)$$

$$NPL_{i,t} = \partial_i + \beta_1 \ln GC_{i,t} + \beta_2 \ln SIZE_{i,t} + \beta_3 CAR_{i,t} + \beta_4 CB_{i,t} + \varepsilon_{i,t} \qquad (6)$$

$$NPL_{i,t} = \gamma_t + \beta_1 \ln GC_{i,t} + \beta_2 \ln SIZE_{i,t} + \beta_3 CAR_{i,t} + \beta_4 CB_{i,t} + \varepsilon_{i,t} \qquad (7)$$

$$NPL_{i,t} = \partial_i + \gamma_t + \beta_1 \ln GC_{i,t} + \beta_2 \ln SIZE_{i,t} + \beta_3 CAR_{i,t} + \beta_4 CB_{i,t} + \varepsilon_{i,t} \qquad (8)$$

Models (5) are respectively mixed regression models; if $\varepsilon_{i,t}$ is related to variables, then models (6), (7), and (8) are fixed effects models, otherwise, they are random effects models.

2.5 Model Selection

The object of this article is 13 banks, the period contains 9 years, it is a typical panel data. For panel data, the model is selected in two steps: F test determines whether it's a mixed regression model or a fixed effect model; Hausman test determines whether it is a fixed effect or a random effect.

First, we use Python to establish a fixed-effects model regression, perform F test, and then establish a random-effects model regression, perform Hausman test. The results can be seen in Table 2 below.

Table 2. F test result

F test (Prob)	Cross-section F	Period F	Cross-section/Period F
Hypothesis I	0.0000	0.0000	0.0000
Hypothesis II	0.0000	0.0000	0.0000

It can be concluded from the Table 2 that the P value corresponding to the F test is 0.0000, so the null hypothesis of mixed regression effects is rejected, and an individual time-point fixed effect model should be established.

Table 3. H test result

H test result (Prob)	Cross-section random	Period random	Cross-section and period random
Hypothesis I	1.0000	1.0000	1.0000
Hypothesis II	1.0000	1.0000	1.0000

It can be concluded from Table 3 that after Hausman's test, both hypothesis I and hypothesis II accept the null hypothesis, that means, a random effects model is established.

To sum up the F test and Hausman test, for Hypothesis I and Hypothesis II, select model (4) and model (8) to build the model as follows:

$$ROA_{i,t} = \partial_i + \gamma_t + \beta_1 \ln GC_{i,t} + \beta_2 \ln SIZE_{i,t} + \beta_3 CAR_{i,t} + \beta_4 CB_{i,t} + \varepsilon_{i,t}$$

$$NPL_{i,t} = \partial_i + \gamma_t + \beta_1 \ln GC_{i,t} + \beta_2 \ln SIZE_{i,t} + \beta_3 CAR_{i,t} + \beta_4 CB_{i,t} + \varepsilon_{i,t}$$

Model Regression Results and Analysis

So as to study the impact of GC scale on bank profitability and asset quality, we conducted regression analysis on LSB, SMCB, and the total sample, respectively. All regressions are balanced panel data, and the regression results are displayed in Table 4.

Table 4. Regression results of the impact of GC scale on the ROA

Variable	State-owned bank (1)	Small and medium joint-stock banks (2)	Overall sample size (3)
C	2.752** (2.287)	2.701** (2.140)	1.472** (2.225)
lnGC	0.056** (2.265)	−0.027 (−1.305)	−0.007 (−0.418)
lnSIZE	−0.069 (−0.866)	−0.072 (−0.882)	−0.006 (−0.133)
CA	−0.034* (−1.757)	0.012 (0.718)	0.0149 (1.400)
CI	−0.015** (−2.230)	−0.018*** (−3.367)	−0.018*** (−3.367)

Note: *** means significant at the 1% level, ** means significant at the 5% level, * means significant at the 10% level

Table 4 shows the regression results of the impact of GC scale on profitability. The explained variable is the ROA.

It can be seen that in columns (2) and (3), the regression coefficients of the GC scale of SMCB and the overall sample are −0.027 and −0.007, and the impact on the ROA is not significant, indicating that banks are developing GC policy will not have a significant influence on profitability. However, the scale of GC and the ROA in column (1) show significantly positive correlation at the level of 5%, indicating that the development of this business will significantly increase the profitability of LSB. From other variables, asset size and CA ratio have no significant correlation with profitability; the CI has a significant negative correlation with commercial banks' profitability. The lower the CI, the less cost that needs to be paid. In line with the profitability characteristics of commercial banks.

It can be noticed from Table 5 that the scale of GC has a notable negative correlation with the NPL ratio at the level of 10%, indicating that GC business has indeed played

Table 5. Regression results of the impact of GC scale on NPL

Variable	State-owned bank (1)	Small and medium joint-stock banks (2)	Overall sample size (3)
C	2.103 (1.005)	0.741 (0.817)	−4.621*** (−4.430)
lnGC	−0.186* (−1.986)	0.013 (0.630)	−0.009 (−0.282)
lnSIZE	0.197 (1.438)	−0.051 (−0.754)	0.379*** (5.130)
CA	−0.069 (−1.300)	−0.021 (−0.876)	−0.09*** (−3.912)
CI	−0.022*** (−2.874)	0.018*** (4.532)	0.016*** (4.730)

Note: *** means significant at the 1% level, ** means significant at the 5% level, * means significant at the 10% level

a notable positive role in improving the asset quality of LSB. However, in general, the scale of GC has no obvious correlation with the NPL ratio of commercial banks, which means that GC business cannot improve the asset quality of commercial banks. From the regression results of other variables, the total bank assets and loan-to-deposit ratio have a notable positive effect on the NPL ratio, which means that the larger the asset size, the higher the NPL ratio and the worse the asset quality. The CA ratio has a notable negative correlation with the NPL ratio, indicating that the stronger the capital liquidity, the better the asset quality of commercial banks.

From the regression results, the scale of GC has a significant positive influence on the operating performance of LSB. This means that when LSB increase their investment in GC, their operating performance has improved. The reason for this phenomenon may be that LSB may pay more attention to their contributions to society and the industry and their corporate responsibilities. In recent years, the green environmental protection industry has developed vigorously. The public supports its development, and the industry has achieved good returns. Therefore, the bank's support for the green industry can improve its ability to resist credit risks and increase its possibility of obtaining potential benefits.

But on the whole, GC has no obvious impact on profitability, which means that GC business will not bring direct benefits to banks in the short term. This is the fundamental reason why banks lack the motivation to develop GC business. First, this is closely associated with the loan period of GC. GC is basically three to five years or longer. As a result, the liquidity of credit funds is relatively weakened, and the bank's capital utilization rate also declines. Realize short-term profit; secondly, GC incurs additional costs.

At the same time, on the whole, the scale of GC has not significantly improved the asset quality of commercial banks. The reason may be that my country has been developing GC business for a short period of time, and the scale of GC of listed banks is still relatively small, and there is not enough attention and scale effect has not yet been produced. However, the scale of GC has a notable negative impact on the NPL ratio of LSB, indicating that with the expansion of the scale of GC, the NPL ratio of LSB has declined, which will help improve asset quality. The reason may be that the implementation of GC by banks can pass the prior review of the corporate environment, thereby reducing the environmental risks faced by the banks and reducing the loss of non-performing loans; in addition, GC has enriched the variety of bank credit products and optimized the investment and investment of credit funds. The role of the structure of the loan industry in turn improved the bank's own asset quality to a certain extent.

3   Conclusions

First, as far as the profitability of banks is concerned, the implementation of the GC policy has no significant impact on the profitability of the overall commercial banks and SMCB, but it will improve the profitability of LSB. The reason may be that LSB may pay more attention to their contributions to society and the industry and their corporate responsibilities. Good environmental awareness helps to obtain more operating permits, thereby promoting revenue growth; at the same time, the reputation of green environmental protection can be improved The positive attention of the media will help to consolidate the cooperative relationship with external stakeholders.

Second, with regard to the asset quality of banks, the implementation of GC policies has no significant impact on the asset quality of overall commercial banks and small and medium-sized joint-stock banks, but the GC business of LSB has a significant effect on improving their asset quality. The reason is that, in accordance with the country's environmental and economic policies and industrial policies, commercial banks will restrict the loan lines of investment loans and working capital for new projects of polluting production and polluting companies, thereby effectively avoiding the risk of default by companies that violate environmental protection laws and regulations.. At the same time, GC has enriched the variety of bank credit products, optimized the investment of credit funds and the role of the loan industry structure, and thus improved the bank's own asset quality to a certain extent. However, the variety of GC products of listed commercial banks is small, and product innovation insufficient.

References

1. Macklon, N.S., Pieters, M.H.E.C., Hassan, M.A., Jeucken, P.H.M., Eijkemans, M.J.C., Fauser, B.C.J.M.: A prospective randomized comparison of sequential versus monoculture systems for in-vitro human blastocyst development. Hum. Reprod. **17**(10), 2700–2705 (2002)
2. Baron, D.P.: Private politics, corporate social responsibility, and integrated strategy. J. Econ. Manag. Strateg. **10** (2010)
3. Scholtens, B., Dam, L.: Banking on the equator. are banks that adopted the equator principles different from non-adopters? World Dev. **35**(8), 1307–1328 (2007)
4. Weber, O.: Corporate sustainability and financial performance of Chinese banks. Sustain. Account. Manag. Policy J. **8**(3), 358–385 (2017). https://doi.org/10.1108/SAMPJ-09-2016-0066

5. Codsi, G., Pearson, K.R.: GEMPACK: general-purpose software for applied general equilibrium and other economic modellers. Comput. Sci. Econ. Manag. 1(3), 189–207 (1988)
6. Richardson, B.J.: The equator principles: the voluntary approach to environmentally sustainable finance. Eur. Environ. Law Rev. 14(11), 280–290 (2005)
7. Platonova, E., Asutay, M., Dixon, R., Mohammad, S.: The impact of corporate social responsibility disclosure on financial performance: evidence from the GCC Islamic banking sector. J. Bus. Ethics 151(2), 451–471 (2016). https://doi.org/10.1007/s10551-016-3229-0
8. Lazzaroni, S., Ravelli, D., Protti, S., et al.: Photochemical synthesis: using light to build C-C bonds under mild conditions. C. R. Chim. 20(3), 261–271 (2017)
9. Fernández-Cuesta, C., Castro, P., Tascón, M., Castaño, F.: The effect of environmental performance on financial debt. European evidence. J. Cleaner Prod. 207, 379–390 (2019). https://doi.org/10.1016/j.jclepro.2018.09.239
10. Eshet, A.: Sustainable finance? The environmental impact of the 'equator principles' and the credit industry. Int. J. Innov. Sustain. Dev. 11(2–3), 106 (2017)

The Influence of Computer Network Technology on Trade Development

Ning Wang(⊠)

Department of Economic Management, Shandong Vocational
College of Science and Technology, Weifang, Shandong, China
albertwangning@sdvc.edu.cn

Abstract. With the continuous progress and development of the times and the continuous improvement of people's living standards, consumers' consumption concepts and behaviors have gradually changed. In various industries, the rapid development of network technology and computer technology, the gradual popularization of Internet technology, and the development of online shopping and e-commerce platform have also promoted the change of consumer behavior and consumption concept. Computer technology has a far-reaching impact on trade investigation. Under the background of rapid development of computer technology, consumer behavior has also undergone tremendous changes, which has promoted the change of the direction and mode of trade management. Consumers' consumption behavior is more independent, and their consumption concept pays more and more attention to service and quality. However, the traditional transaction management can no longer meet the consumer's consumption needs, and the consumer's consumption goals can not be understood in more detail, thus reducing the probability of consumer transactions. Therefore, in view of the change of consumer behavior, this paper discusses the influence of computer technology on trade development, and puts forward some problems, which need to be solved.

Keywords: Internet · Trade development status · Strategies and methods

Under the background of big data, with the rapid development of network technology and computer technology, Internet technology, trade development and trade marketing have become an inseparable whole [1]. The development of trade is inseparable from the statistics and analysis of information data, and e-commerce trade platform can also promote the progress and high-quality development of big data and Internet technology [2]. Therefore, the use of Internet and big data technology can effectively improve the efficiency and quality of trade marketing activities, make full use of information and data, intelligently analyze the needs of users, and can serve trade marketing activities. According to the different needs of different users, different trade methods and trade goods can be used to attract the attention of users, so as to increase the probability of consumption transactions and play a positive role in promoting precision trade activities. The application and development of Internet technology has created more economic

J. Abawajy et al. (Eds.): ATCI 2021, AISC 1398, pp. 22–27, 2021.
https://doi.org/10.1007/978-3-030-79200-8_3

benefits for enterprises. Only by analyzing the trade strategy under the background of Internet development and making full use of data and information resources of users, can the economic benefits of trade enterprises and trade market be really improved [3, 4].

1 Changes of Trade Market Environment under the Background of Internet Development

In recent years, the rapid development of economy has promoted the continuous improvement of people's living standards and quality [5]. Electronic products are gradually popularized to serve people's lives. Under the background of the development of Internet and big data, e-commerce industry is developing rapidly in China's trade and consumption market. The mobile trade and consumption platform enables people to buy products and all kinds of new products without leaving home Shopping style has two advantages of convenience and timeliness, which increases the satisfaction of shopping and consumption [6]. The trade market environment pays more and more attention to the consumption activities of mobile terminal, and China's trade and consumption market also shows a strong demand for mobile terminal shopping [7]. The demand of trade market environment promotes the improvement and development of trade activities. Internet technology also provides greater development opportunities for the development of trade market environment. The trade market environment of online shopping also leads to the direct fierce competition of e-commerce enterprises. Therefore, it is more necessary to analyze the development of trade under the background of Internet.

2 Application of Internet Technology in Trade Activities

In the trade marketing activities, the application of Internet technology effectively improves the success probability of consumption, especially promotes the progress and development of trade marketing activities on the mobile terminal line [8–10]. In the trade activities of mobile terminal, it is very important to analyze the user's positioning. Only by fully mastering the user's data and understanding the user's needs, can the trade marketing strategy play a directional role. For example, through the Internet e-commerce platform, you can directly obtain specific information such as the residence time of user pages and the jump time of commodity pages, and comprehensively analyze these data, reasonably apply these user data, and intelligently analyze the user's behavior, so as to realize the accurate positioning of users.

In the e-commerce trade and consumption trading activities, the design of trade marketing price and trade marketing advertisement play a decisive role in the trade activities. In the trade and consumption trading process of online platform, users can decide whether to consume or not through the comprehensive consideration of commodity information and commodity price on the merchant's advertisement page [11, 12]. Therefore, the design and trade of trade marketing price is very important the design of E-marketing advertising must meet the needs of users in order to promote the smooth progress of mobile terminal trade activities and make businesses get more economic benefits. In the design of trade commodity price and advertisement, through the use of user's commodity transaction amount and search browsing records and other specific data, we can complete the pricing of goods and the design of commodity advertisement. Through ganglia, yarn and other data management open source systems, we can directly call the data algorithm materials, providing a high-quality data environment and application environment. Internet technology is widely used in mobile trade Price and advertising design has been widely used.

For example, three squirrels used the Internet and new technologies to gain insight into the changes in the behavior of the new generation of consumers. Through accurate data analysis, dimension reduction, association analysis, classification and other big data technologies, they constructed accurate user portraits and predicted the needs of consumers. By analyzing the user's browsing discipline and the stay time in the advertising page, we can reduce the dimension of big data. Through the association analysis and classification of the information of the consumer's search records, we can accurately locate the user, and improve the final effect of trade marketing activities.

3   The Impact of Internet on the Development of Trade

In the context of the development of the Internet, the e-commerce industry is developing rapidly in China's trade and consumption market [13]. The e-commerce platform of online trade and consumption enables people to buy the products they need without leaving home. The new trade mode has the advantages of convenience and timeliness, which increases the satisfaction of shopping and consumption. The environment of trade and consumption market pays more and more attention to online consumption activities [14]. Compared with the traditional way of trade, online trade consumption is more convenient and fast, saving the time cost of people in the fast-paced society. Therefore, online trade consumption behavior of consumers is more common, online shopping has been closely related to people's life, online trade e-commerce platform is more diversified, and the content of trade goods is more abundant, which will have a great impact on consumption Consumers' consumption behavior has a certain impact, which increases consumers' desire to buy. The continuous development of the Internet and the rise of the e-commerce industry provide consumers with new ways of consumption and enrich the way of shopping.

With the continuous improvement of people's living standards, the development of the Internet makes people have new changes in the concept of trade and consumption. Traditional consumer care about commodity prices has also been adjusted, people pay more attention to the service behavior in the process of trade. Paying attention to the service concept in the process of trade consumption is also a new change in consumer behavior. In the process of consumption, we can get psychological satisfaction, and the proportion of spiritual consumption is gradually increasing. Therefore, businesses with better service concept are often favored by consumers, increasing the transaction rate of trade consumption.

In addition to new requirements for trade and consumption services, consumers have stricter standards for the brand and quality of trade goods. Especially under the background of big data and Internet popularization, more and more consumers pursue famous brand consumption. Brand effect has been integrated into people's consumption concept, and consumers can easily associate brand with product quality. At present, the trade consumption behavior of consumers is more and more inclined to brand consumption, and they also have their own unique consumption consciousness about the quality and cost performance between brands. The comparison between brand and brand quality can make consumers quickly determine the purchase goal. The diversification of brands also makes the trade consumption behavior of consumers more personalized (Table 1).

Table 1. Summary of the impact of internet on trade development

Category	Increase trade turnover rate	Improve the quality of trade	Promote the development of trade environment
Online consumption	98%	86%	91%
Trade service concept	78%	79%	86%
Trade brand effect	85%	91%	77%

4 Problems in the Application of Internet in Trade

The wide application of big data and Internet technology has great benefits to the e-commerce industry and the people's consumer market. Although the Internet facilitates people's life to a certain extent and promotes the transaction rate of trade and consumption, the data abuse degree of the e-commerce industry is too high, which also poses a great threat to the security of user data, and user information leakage incidents are frequent What's more, it shows that the current mobile trade marketing platform lacks security protection for users' data. Internet and big data technology can bring huge economic benefits, promote the positive development of the online consumer market, but also should strengthen the confidentiality of user data, pay attention to the security of information and data, in order to make the new trade marketing model develop more long-term. The leakage of personal information and privacy has always been a concern for consumers. Therefore, in the process of collecting and using user information and data, we should pay more attention to the legalization and standardization of the use of Internet technology.

In the current e-commerce platform, users' needs can be analyzed to a certain extent through the specific data of users' browsing records and search records. In order to pursue the success probability of trade and consumption transactions, e-commerce enterprises have excessive use of user data, which leads to excessive precision in commodity recommendation in trade marketing activities, abuse of user data platform, and even abuse of user data There will be the same situation of recommended goods, resulting in the reduction of users' consumption desire, and playing a negative role in trade marketing activities, which runs counter to the purpose of using Internet technology. It is not only difficult to improve the economic benefits of the application platform of e-commerce enterprises, but also impossible for the e-commerce platform without individuality and data abuse to have a foothold in the future online trading market.

5 Improve the Effectiveness of Internet Application in Trade Process

Under the background of Internet technology and big data, consumers are more active in online shopping activities. The main body of trade marketing activities is the masses, therefore, we should improve the participation of consumers in trade marketing activities. Compared with the traditional offline marketing mode, the current trade shopping with internet participation is more flexible. The diversified trade shopping mode requires that trade marketing strategies pay more attention to the dominant position of consumers. For example, in the trade marketing activities of the "double 11" large-scale e-commerce platform, consumers can actively participate in the activities organized by the platform and make use of diversified preferential ways, which highlights the dominant position of consumers in trade marketing activities.

With the continuous development of trade, consumption and shopping, the competition between e-commerce enterprises and platforms is more fierce, and the collection of user information is more difficult, which is very unfavorable to the trade marketing activities. The difficulty of user information collection seriously restricts the development of e-commerce platform. Therefore, it is necessary to establish the concept of Internet data sharing. Under the background of Internet, first of all, the security of user's data and information should be guaranteed. Each platform should actively implement the concept of data sharing, complete the circulation of data, and achieve a win-win situation between enterprises, which is more important to some extent, it increases the success rate of trade and consumption.

Users' data privacy and security have always been the focus of the society and the people. With the expansion of e-commerce industry, the amount of users' data is more and more huge, and the technical requirements for Internet data processing platform are also higher and higher. In the face of the development of trade and trading market, it is urgent to improve the efficiency of user data processing. We should develop advanced technology and intelligent data processing platform, only by constantly updating the trade marketing mode can the efficiency of data processing be improved. At the same time, we must strengthen the awareness of Internet data security, ensure that the user's data and privacy information are not leaked, and integrate the security technology and processing technology, so as to achieve the multiple goals of improving efficiency and ensuring security.

6 Conclusion

With the development of network technology and computer technology, Internet technology and big data are inseparable from trade activities. Trade marketing is facing development opportunities, but also bears huge pressure. Only by innovating Internet technology and trade marketing strategies under the background of big data can we actively promote the development of trade market and maximize the economic benefits brought by trade consumption Benefit.

References

1. Pan, J.: Research on the impact of Internet development on China's export trade. Natl. Trade Issues (12) (2018)
2. Zhao, J.: Research on precision marketing strategy of mobile online advertising in big data environment. Management Observation (2018)
3. Xu, T.: Research on precision marketing strategy under the background of big data mining. China Bus. Theory (8) (2019)
4. Xiao, Y., Xia, J.: The global rules game of digital trade and China's response. J. Beijing Univ. Technol. (Soc. Sci. Ed.) **21**(03), 49–64 (2021)
5. Zhang, W., Liang, Y.: Provincial differentiation and spatial effect of China's digital trade development level. Guizhou Soc. Sci. (12), 129–138 (2020)
6. Zang, W.: Research on foreign trade development of small and micro enterprises based on internet. Sci. Technol. Econ. Guide **28**(32), 240+197 (2020)
7. Zhang, B.: Research on the reform of applied talents training mode of logistics management specialty based on the integration of production and education. J. Beijing Inst. Printing **28**(06), 125–128 (2020)
8. Yang, H.: Research on the International Competitiveness and Influencing Factors of China's Digital Trade. Soochow University (2020)
9. Liu, J.: Research on the teaching reform of logistics course for improving innovation and entrepreneurship ability. Logist. Sci. Technol. **43**(05), 175–177 (2020)
10. Gu, A., Allen, W., Xu, G.: Research on China's international trade development under the internet background. Coop. Econ. Sci. Technol. (06), 108–109 (2020)
11. Bi, X., Hu, H., Xie, C.: Simulation of fractal segmentation of unsupervised texture remote sensing images. Comput. Simul. **37**(01), 22–26 (2020)
12. Kang, F., Cheng, Z., Bai, P.: Analysis of the Influence of internet plus on China's Foreign Trade. Mod. Enterp. (08), 99–100 (2019)
13. Wang, T.: The new trend and influence of software trade and China's response. Hubei Soc. Sci. (03), 67–72 (2019)
14. Zhang, J., Wang, Y., Zhang, W., Li, J., Liao, H.: Analysis of foreign trade development in the internet age. Public Investment Guide (05), 155–156 (2018)

The Application of Knowledge Map and the Construction of Enterprise Knowledge Map

Hua Xin[✉]

China Southern Grid Digital Grid Research Institute Co., Ltd.,
Guangzhou 510700, Guangdong, China

Abstract. With rapid development of the new generation of information technology, the exchanges and interactions between people, machines and machines, and people and machines become more and more frequent, and the information environment and data base of the development of artificial intelligence technology have undergone significant and profound changes. The rapid accumulation of massive data, the substantial improvement of computing power, the continuous optimization of algorithmic models and the rapid rise of industrial applications have comprehensively promoted the development of a new generation of artificial intelligence technology.

Keywords: AI · Knowledge map · Data · Informatization

1 Introduction

In the past, the construction of information system in enterprise was mainly promoted by integrated database and by the data warehousing system that are taken as effective data strategy of enterprises [1]. How to fully tap the potential knowledge and provide decision-making assistance for managers and improve the scientific decision-making level of enterprises is the key to the transformation of enterprises from traditional information system to a new generation of digital basic platform and Internet application. As a way of knowledge representation and management, knowledge graph is the foundation and bridge from enterprise informatization to intellectualization, and lays a solid foundation for intelligent interconnection of enterprise information system.

The technology in knowledge graph was originally introduced by Google as semantic enhancement for search engines [2]. With the continuous development of intelligent information services and applications, knowledge graph has been widely used in academia and industry, and plays an irreplaceable role in intelligent search, intelligent customer service, machine reasoning, recommendation system and other enterprise informationization applications. At present, domestic and foreign enterprises have started to build their own knowledge graph system, hoping to manage a large amount of data from companies and public sources through the knowledge graph system [3–5].

© The Author(s), under exclusive license to Springer Nature Switzerland AG 2021
J. Abawajy et al. (Eds.): ATCI 2021, AISC 1398, pp. 28–34, 2021.
https://doi.org/10.1007/978-3-030-79200-8_4

The life and entertainment knowledge graph built by the Meituan company that has accumulated 4 billion public evaluation data, 34.5 million global business data, 140 millions of pieces of data in restaurant dishes and 1 billion personalized labels. The BBC defines the ontology of knowledge in the sections of music, sports and entertainment, and transforms news information into machine-readable information sources for content management and automatic report generation. It can be seen that as a new method of knowledge representation and a new idea of knowledge management, knowledge graph has played an increasingly important role in more and more vertical fields in application, and is the cornerstone for enterprises to transfer from informatization to intelligence.

2   Knowledge Graph and Structure of the Enterprise Information System

We give our top priority on the IT architecture of modern enterprise informatization, which is based on the knowledge graph and deployed to the enterprise midplatform to provide a unified intelligent interface for enterprise AI-driven applications. The knowledge map serves as the package, that is, a collection of rules. The rule base could be edited and processed by a dedicated administration user interface. Besides, external resources are supported by the sensor after collection, and intelligent adapters are integral part of the component that can actively interact with the source during the reasoning process.

We adopted a specialized technical architecture that enabled most of the existing engineers with DBMS development experience to participate in the rule development of the knowledge graph. When we construct a knowledge map rule by adopting technology (which is rewritten as an interpretation), a query plan will be produced by us. Query plans can be optimized on criteria such as driving selection and projecting as close to the data source as possible. Finally, the query plan serves as an access plan, where the common rule nodes would be replaced by most appropriate conduction for the corresponding low-level operators (for example, selection, projection, interlinkage, aggregate, expression, etc.) [6]. We use a universal local cache in way of a wrapper to get access to nodes of plan and the factual location generated by each node can be stored.

At present, the system has been fully developed and has been used in many industrial applications [7]. It also allows partners to develop applications through extensible modules. Our company's internal system is also in detailed assessment on the related performance, according to the seven randomly generated data source of knowledge, from 10 to 1 million companies and five chart similar to the real world (density) and topology is similar to the actual environment, for each of the graphics is similar to the real world, we are totally real and totally real evaluation. The results demonstrates that good performance is generated by the engine in both batch processing and interactive applications on large knowledge graphs, which provides solution to the requirements on complex reasoning, and on turning the foundation of the core language into a practical executive strategy so as to achieves high performance and efficient memory utilization.

3 Application of Knowledge Graph

3.1 Problem Solving

Problem solving is a system that automatically offer solutions to questions. Human beings ask questions in the form of natural language. Question response system is capable of providing answers to a definite question through information searching on the Internet and knowledge map established by the collection of documents. Most question response systems are targeted at factual questions that can be given solutions by showing facts expressed [8]. For instance, the question "Where do I pay my electricity bill?" You can answer by looking for simple facts about "grid service stations." Knowledge maps are an important resource for question answering systems because the systems provide the knowledge to answer questions. Various open domain quality checking systems utilize a large number of facts by knowledge map that is available such as Freebase or DBpedia to conduct retrieval. Besides, a domain-targeted quality check system can create a domain-specific knowledge graph to retrieve answers. This project solves the question and answer problem by using the knowledge graph of the memory network framework mentioned above. This solution can expand the question based domain and answer system by offering new facts, which can be stored in memory without retraining, thus meeting the need of fast question solving.

3.2 Network Security

Cyber security is the technology used to predict, detect and protect systems, networks, computers, programs and technologies, and to protect enterprise information systems from digital attacks, sabotage or unauthorized access to data. Because of complexity, variety and dynamic nature of attacks, it is one of the most challenging tasks for enterprises to face. This project combines the network security knowledge base with deductive rules to discover the information entities related to network security. These extracted entities for other domain-specific classifiers can be utilized by us. The existing security network knowledge graph has been integrated into this work, including Common Vulnerabilities and Exposure (CVE), Vulnerability Database (NVD), Security Focus, etc. [9].

In addition, there is integration with Common Configuration Enumeration (CCE), Common Attack Pattern Enumeration and Classification (CAPEC), and other similar data sources. Through unified well-known network security knowledge base, knowledge map can be used to avoid external attacks, which is proved the effectiveness and efficiency in the external attack. Current or former employees, contractors or owned or used to grant way to the business partner based network of the organization, and somehow intentional beyond or abuse of rights to confidentiality, integrity, or availability, which negatively impact the information or information systems of organization. Given the contingency, insider attacks may pose a greater threat of influence than outsiders. Therefore, it is regarded as a more important network security issue [10, 11]. This project leverages a number of sources in data to finish the cyber security knowledge map that can take initial action and respond to different levels of network hazards (such as networking infrastructure, safety, online threats, and mission dependencies).

3.3 Intelligent Application

In addition, knowledge map serves as contributor to the development of many other intelligent systems [12]. In enterprise application, our partners use the enterprise knowledge graph for investment advising. They started by developing a knowledge map, which contains information about 40 million companies. By reference to details in the knowledge map, they can provide visualized information and analysis on enterprise investment for decisions in investment.

To sum up, a knowledge graph system of enterprise informatization is established to develop knowledge map based on IT system, which can meet the needs of various intelligent applications of enterprises.

4 Construction of Knowledge Map

Knowledge maps can be generally subdivided into two forms: ordinary knowledge graphs and enterprise graphs. The ordinary knowledge is geared to domain with eyes on Internet search, result recommendations, and additional scenarios in application, it is reflected in "breadth" of graph database of enterprises. The knowledge map is targeted at certain domains, and entities serves as professional terminology in the industry. Knowledge graphs provide high-precision searches of entities to assist complex analytical applications and decision support. The construction and generation of knowledge map include three steps: 1) extraction of knowledge, 2) knowledge fusion, and 3) processing of knowledge.

4.1 Knowledge Extraction

The major task in extracting knowledge is to extract the word information representing entities and attribute and conduct speech tagging. The learning is made by utilizing power dictionary matching to extract units from the data source. The first step is to extract the data source information, and the segmentation algorithm can be made by using the Hidden Markov Model (HMM).Digital dictionary is introduced to improve the reliability and precision of the extraction algorithm. The algorithm then matches a digital dictionary of generic digital dictionaries and corporate jargon, retrieves the matches, and extracts the words as entities for the knowledge map. In knowledge extraction, extracted information will be saved as a JSON intermediate file, and then the data will be mapped into triple information. In the application of this enterprise, it mainly involves the inspection information of connection state, operation and equipment, so the entity is divided into the following 8 categories: 1) Equipment terms, power equipment or components are represented as "E1";2) a system operator or a system management organization; 3) Nouns to embody manufacture, the manufacturer's name 4) The connection state between devices is expressed as "state; 5) The relations between verbs, conduction occurred during the inspection. 6) The verb that occurs in the process of production. 7) The results produced in the production process, such as natural calamity, switch fuses, equipment downtime, are displayed as "for instance since; 8) Mismatched expression can be ignored (Fig. 1).

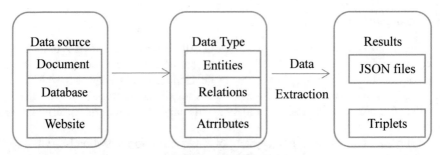

Fig. 1. Data extraction

4.2 Knowledge Integration

The major task of the integration can be made clear to identify corresponding relations between units, such as subordinated relationship, connection relationship and causality relationship. The relation types between entities are defined, and the relation extraction of knowledge integration is transformed into problem in classification. In this study, the relationship in electrical equipment and other entities is our concerns. As identification is conducted, a classification of entity pairs needs to be developed. Later on, the relationship between the entity pairs and the database needs to be matched. The relationship between the entities can then be identified. The specific implementation algorithm is as follows (Fig. 2):

Input: Entity E1,E2; Input Database D; Input vertex R
Output: The Relation between E1 and E2
(1) From the Database D , look for the Relations ;
(2) For each Table in D:
(3) A separated GPU unit and a distributed memory unit are assigned to R
(4) if (R contains E1 and E2)
(5) return R;
(6) end if
(7) end for

Fig. 2. Entity correspondence recognition algorithm

4.3 Developing Knowledge Map

In this study, equipment parameters, operating records, manufacturing power, and other equipment information are usually expressed in a single sentence. The construction method of knowledge graph is as follows:

1) Since the entity is limited to the electric industry that stipulates a clear standard in terminology, there is no problem in entity ambiguity so the disambiguation step is not chosen;

2) The number of units in a particular field is relatively small. To obtain more training samples, relevant parameter parsing is performed before relation extraction.
3) After extraction in relationship, relationships need to be filtered to avoid relationship redundancy, which will lead to repeated storage and affect search results;
4) To organize integration steps, merge and extract the triad information, and form the knowledge map of power equipment.

When the construction of the knowledge graph is completed, the search can be performed in the completed constructed knowledge graph providing that it exists in knowledge map. Once object is discovered, all related entities are compared. The search results can be utilized to determine whether there is a better result, and perform inference optimization. Entity, for example, "a", "b", "d" and "e" belongs to "E1", "c" is one factor of the "E2", "f" belong to "P". Relationships "belong" and "join" is geared to type "R1", while "operations" belong to "R2".As search object "g" is put into the knowledge map, no search results are displayed because there is no such entity in the knowledge graph.

Table 1. Experimental results

Task	Relational database		Knowledge graph		
	Time consumption	Number of records	Time consumption	Number of records	Performance improvement
Entity search	67	1	35	1	47.76%
Relation search	211	1	108	1	48.81%
Entities search	521	15	167	66	67.94%
Relations search	1021	17	214	17	79.04%
Relations between E1 and E2	351	1	132	1	62.39%

Analysis on experimental results: The experimental results are displayed in Table 1. It shows that through way of the graph, the query efficiency of entities and relationships has been greatly improved.

5 Conclusion

In the field of enterprise informatization, with the gradual deepening of the informatization level, a large number of enterprise data have been accumulated. An enterprise's knowledge of data can be an important aid in the effective use of smart applications, network security and smart customers. How to build the knowledge graph suitable for the enterprise information system, so as to provide services for the enterprise from the massive enterprise documents, pictures and other unstructured data, has a broad application prospect. From the perspective of enterprise informatization, this paper makes an in-depth to analyze architecture, application technology and implementation of enterprise knowledge graph.

References

1. Krathwohl, D.R.: A revision of bloom's taxonomy: an overview. Theory Pract. **41**(4), 212–218 (2002)
2. Feilmayr, C., Wöß, W.: An analysis of ontologies and their success factors for application to business. Data Knowl. Eng. **101**, 1–23 (2016)
3. Zhou, X.-G., Gong, R.-B., Shi, F.-G., Wang, Z.-F.: PetroKG: construction and application of knowledge graph in upstream area of PetroChina. J. Comput. Sci. Technol. **35**(2), 368–378 (2020). https://doi.org/10.1007/s11390-020-9966-7
4. Shan, S., Cao, B.: Follow a guide to solve urban problems: the creation and application of urban knowledge graph. IET Softw. **11**(3), 126–134 (2017)
5. Yoo, Y., Jeong, O.R.: Automating the expansion of a knowledge graph. Expert Syst. Appl. **141**, 112965 (2019)
6. Ehrlinger, L., Wöß, W.: Towards a definition of knowledge graphs. In: Proceedings of the 12th International Conference on Semantic Systems (SEMANTiCS 2016), Leipzig, Germany, 12–15 September 2016, pp. 1–4 (2016)
7. Nguyen, H.L., Jung, J.E.: Socioscope: a framework for understanding internet of social knowledge. Future Gener. Comput. Syst. **83**, 358–365 (2018)
8. Wilcke, X., Bloem, P., De. Boer, V.: The knowledge graph as the default data model for learning on heterogeneous knowledge. Data Sci. **1**(1–2), 39–57 (2017)
9. Färber, M., Bartscherer, F., Menne, C., Rettinger, A.: Linked data quality of dbpedia, freebase, opencyc, wikidata, and yago. Semant. Web **9**(1), 77–129 (2018)
10. Paulheim, H.: Knowledge graph refinement: a survey of approaches and evaluation methods. Semant. Web **8**(3), 489–508 (2017)
11. Pujara, J., Miao, H., Getoor, L., Cohen, W.: Knowledge graph identification. In: Alani, H., et al. (eds.) ISWC. LNCS, vol. 8218, pp. 542–557. Springer, Heidelberg (2013). https://doi.org/10.1007/978-3-642-41335-3_34
12. Ai, Q., Azizi, V., Chen, X., Zhang, Y.: Learning heterogeneous knowledge base embeddings for explainable recommendation. Algorithms **11**(9), 137 (2018)

Interaction Mechanism of Regional Tourism Industry and Information Industry

Xincheng Zhang$^{(\boxtimes)}$

School of Economics and Management, Northwest University, Xi'an, Shaanxi, China

Abstract. Based on the panel data of 11 cities in Shanxi Province from 2000 to 2016, the panel vector autoregressive model is used to further analyze the dynamic interaction, mechanism and region of the two major industries in the four regions of northern Shanxi, central Shanxi, southern Shanxi, and southeast Shanxi. Empirical analysis of differences. The results show that: From 2000 to 2016, the development level of tourism industry and information industry in 11 cities showed an overall upward trend year by year, but the development of information industry in most cities was obviously lagging behind; The results of the PVAR model show that the two major industries in each region are dependent on their own development. The phenomenon of inertia, but the duration and impact of inertia are weak; the dynamic response of the information industry to the tourism industry shows that areas with a relatively high level of information industry development are more active in responding to the impact of the tourism industry. The results of the dynamic response of the tourism industry to the information industry show that areas with low tourism industry respond more significantly to the impact of the information industry.

Keywords: Tourism industry · Information industry · Interaction mechanism

1 Introduction

In 2017, the National Tourism Administration issued the "Thirteenth Five-Year National Tourism Informatization Plan", which pointed out that "by 2020, the construction of "cloud, network, and terminal" tourism infrastructure will continue to improve, the application of information technology will continue to deepen, and tourism will be digitalized and informatized. Intelligentization has made significant progress ". Entering the new era, the development of my country's tourism industry is facing many challenges such as consumption upgrades, industrial upgrades, and the entire regional linkage upgrade. For this reason, the development of the tourism industry puts forward new requirements for the information industry, especially the use of information technology in promoting the transformation of consumption patterns, industrial integration and innovation, and regional sharing and co-construction to promote the transformation and upgrading of the tourism industry. Therefore, in-depth focus on the development of tourism informatization to promote the interactive development of the regional tourism industry and the

J. Abawajy et al. (Eds.): ATCI 2021, AISC 1398, pp. 35–41, 2021.
https://doi.org/10.1007/978-3-030-79200-8_5

information industry has become a practical problem that needs to be solved urgently in the development of the tourism industry.

The research on the development of the interaction between the tourism industry and the information industry is mainly carried out from the micro and macro perspectives, and mainly focuses on the research on the development of the tourism industry driven by the information industry. On the macro level, with the help of information technologies such as the Internet, the experience, visualization, and information level of tourist destinations can be enhanced, thereby enhancing the competitiveness of tourist destinations [1]. At the micro level, the research objects are the hotel industry [2], scenic spots [3], travel agencies [4], website construction [5], and tourists. It is believed that information technologies such as the Internet, e-commerce, and electronic social media have improved hotel management [6]. It has a significant driving effect in promoting the transformation of the travel agency industry, expanding the marketing and promotion capabilities of scenic spots, and promoting the upgrading of tourists' consumption behavior [7].

Based on this, this paper builds a panel vector autoregressive model (PVAR) from the two dimensions of time series and space. The northern, central, southern and southeastern parts of Shanxi Province are selected as the case sites. The dynamic relationship between the information industry, the mechanism of action, and the regional differences are systematically diagnosed and analyzed in order to provide a reference for promoting the coordinated development of the regional tourism industry and the information industry.

2   Research Methods

This article uses the comprehensive index method to measure the development level of the tourism industry and the information industry [8]. Using panel vector autoregressive model (PVAR), selecting panel data composed of 11 cities from 2000 to 2016, in-depth analysis of tourism industry, information industry and the relationship between the two. The vector autoregressive model (VAR) was first proposed by Sims. Its characteristic is to treat all variables as endogenous variables and measure the interaction between the variables. Holtz-Eakin further constructed a panel vector autoregressive model (PVAR) based on panel data measurement [9, 10]. This article uses this feature to analyze the interactive relationship between tourism and information industries in 11 cities in Shanxi Province. To this end, this paper constructs the following PVAR model:

$$U_{it} = \gamma_0 + \sum_{j=1}^{k} \gamma_j U_{it-j} + \alpha_i + \beta_t + \varepsilon_{it} \tag{1}$$

In the formula, $i = 1, 2, ..., N$, representing different cities; $t = 1, 2, ..., T$ representing the year; U_{it} represents a two-dimensional column consisting of tourism industry (U_{it}) and information industry (U_{2it}) Vector, γ_0 is the intercept term; k is the lag order, γ_j is the parameter matrix of U_{it} lag j order, α_i and β_t are the individual effect and time effect vectors, respectively representing the difference between cities and the impact of time changes on it, ε_{it} is a random disturbance term. In addition, logarithmic processing of U_{1it} and U_{2it} can not only maintain the authenticity of the data, but also avoid the problem of heteroscedasticity.

3   Evaluation System

The current measurement and evaluation index system for the development level of the tourism industry is relatively mature, while the information industry has yet to be clarified in terms of the composition of resource elements, organizational structure, and identification of industrial boundaries. This article combines the development reality of the information industry, the research results of existing scholars and the IDI Index (ICT Development Index) proposed by the International Telecommunication Union. Based on the "National Information Development Level Evaluation Index System", "Thirteenth Five-Year" National Informatization Plan, and "National Informatization Index Composition Program" are important supplements to the construction of the information industry indicator system, from technical foundation, application consumption, industrial scale, development knowledge support, etc. Comprehensively evaluate the development level of the information industry (Table 1).

Table 1. Evaluation system

Subsystem	First level indicator	Secondary indicators
Tourism industry	Industrial market	International tourism income
		Domestic tourism income
	Industry scale	Number of travel agencies
		Number of star-rated hotels
	Industrial effect	Receiving foreign tourists
		Receiving domestic tourists
	Knowledge support	Proportion of college graduates per thousand
		Number of students majoring in tourism
Information industry	Industrial market	Office and mobile telephone exchange capacity
		Number of Internet broadband access ports
	Industry scale	Fixed broadband penetration rate
		Number of sites
	Industrial effect	The added value of the information industry as a percentage of GDP
		Information Industry Practitioners
		Number of legal entities in the information industry
	Knowledge support	Number of R&D personnel
		Number of patents granted

4 Interaction Mechanism

4.1 Stationarity Test

Stationarity test can avoid the pseudo-regression problem in model analysis results. Use STATA 14.0 software to perform heterogeneous unit root and homogeneous unit root tests on lnU_{1it}, lnU_{2it} and the first-order difference sequence respectively. In the samples of the four major regions, lnU_{1it} and lnU_{2it} cannot reject the null hypothesis in most cases, the data is non-stationary. The first-order difference sequence accepts the alternative hypothesis, and the PVAR model can be constructed.

4.2 Selection of Lag Order

The determination of the optimal lag order can improve the accuracy of the estimation results of the PVAR model. The standard for the best lag order is, first of all, the lag order is determined by the minimum value of the AIC, BIC, and HQIC criteria. Secondly, the smaller the lag order, the better, to avoid the loss of freedom of the sample data if the lag order is too large. Finally, in order to facilitate the comparison of different regions, the lag orders of the four regions are unified. The best minimum order is also determined as the second order.

4.3 PVR Model Estimation

PVAR model estimation can determine the dynamic relationship between variables by calculating the influence of the lag value of each variable on each variable. Using the generalized moment estimation method (GMM), taking the tourism industry as the dependent variable, the first-order lagging tourism industry in the northern, central, southern, and southeast regions has an impact on itself: 0.651, 0.517, 0.783, 1.259, all of which have passed the P test. The development of the tourism industry in the four major regions has inertia characteristics. As the dependent variable, the information industry in the four major regions has a significant positive impact on its own development, indicating that it is affected by inertia. However, the influence of the tourism industry in the southeast region on the information industry is not significant. It comprehensively shows that, except for the central region, the interaction between the two major industries in other regions is not prominent. The second-order lag variables have no significant impact on the dependent variables, indicating that the inertial effect of the two industries in the four regions has a relatively weak duration and influence.

4.4 The Dynamic Response of Tourism Industry and Information Industry

From the perspective of the impulse response of the information industry to the tourism industry, in the face of the impact of the tourism industry, the information industry has obvious differences in various regions. Among them, the northern Shanxi, Central Shanxi, and southern Shanxi all showed a trend of "rising first and then descending" during the response period, while the southeast area of Shanxi showed the opposite trend of "falling first and then rising"; Both the southern Shanxi and Shanxi regions have a continuous positive impact, while the central and Southeastern Shanxi have shown two diametrically opposite phenomena: "turning from positive to negative" and "from negative to positive". From the perspective of direct effects, the immediate impact of each region is zero; from the perspective of cumulative effects, the northern and southern regions of Shanxi are always positive, while the central and southeastern regions of Shanxi have "turned from positive to negative" (Fig. 1 a-d). In areas with a relatively high level of information industry development, such as northern Shanxi, central Shanxi, and southern Shanxi, the response to the impact of the tourism industry has been more active. On the contrary, the southeastern part of Shanxi showed a negative inhibitory effect, and the two extreme phenomena had obvious characteristics of the "Matthew Effect". The reasons are as follows: firstly, in areas where the development of the information industry is relatively underdeveloped, insufficient investment in industrial construction and lack of competitive strength that can continuously enhance industrial development, leading to negative development of response effects. In this case, the only way to meet market demand is to undertake technology transfer from developed regions and technology emulation to meet market demand, causing the response effect to "turn from negative to positive" and fall into a "backward-catch-up" vicious development cycle model; second, considering that the information industry has Features such as frequent technical updates and strong timeliness. For this reason, the supply of information infrastructure invested in the early stage lags behind the current market demand. For example, although the development level of the information industry in Central Shanxi is higher than that of the other three regions, the previous investment in information infrastructure is surplus, and new technology iterations are urgent, resulting in resources Redundancy, and ultimately the intensification of the contradiction between supply and demand causes the response effect to "turn from positive to negative."

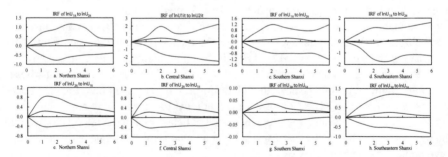

Fig. 1. Impulse response graph

There are also certain regional differences in the impact of the information industry on the tourism industry. The impulse response function in northern Shanxi showed a positive effect and reached its peak in the first period; in the Central Shanxi area, facing the impact of the information industry, the tourism industry showed a trend of "rising first, then falling and then rising", reaching its peak in the first period The tourism industry in the south of Shanxi has no obvious response; the tourism industry in the southeast of Shanxi reaches the maximum in the third phase; from the direct effect, the immediate impact of each region is 0; from the perspective of the cumulative effect, the southeast area is always positive, and the south Jin area has little impact. Although the response direction of the Central Shanxi area "turns from positive to negative", the overall cumulative response trend has changed slightly and is not significant (Fig. 1 e-f). It can be seen that the response effect caused by the impact of the information industry is closely related to the development of the regional tourism industry, and the cumulative response types formed by different tourism industry development levels are also diverse. The cumulative response to areas with a relatively high level of development in the tourism industry, such as Southern Shanxi and Central Shanxi, is not prominent. On the contrary, areas with a lower level of tourism industry development, such as northern Shanxi and Southeastern Shanxi, have more positive cumulative responses. The reason is that for the northern and southeastern regions of Shanxi, where the growth of the tourism industry is relatively low, the input of information resource elements accounts for a small share of the total investment in the tourism industry. Therefore, an appropriate increase in the proportion of information resource elements in the investment in the tourism industry can be significant Enhance the growth level of the tourism industry. As for the Central Shanxi and Southern Shanxi regions, the tourism industry has a relatively high level of development, coupled with relatively large investment in information resources, resulting in a cumulative response that is not prominent.

4.5 Analysis of Variance

Based on the variance decomposition of each region, it can be seen that the contribution of tourism industry to the development of the information industry in the northern, central, and southern Shanxi regions is greater than the contribution of the information industry to the tourism industry, while the southeast region is the opposite, indicating that tourism in general Industry has a greater impact on the information industry. This phenomenon is rooted in the fact that the development of the information industry in Shanxi Province lags behind the tourism industry as a whole. The reason is that the development of the information industry system in Shanxi Province is not yet mature, which has restrained the integration, synergy and integration of information technology and other industries to a certain extent; the second is that although the information industry investment has been soaring in recent years, The lack of core technology, low technology application rate, lack of high-end talents, and imperfect industrial policies still restrict the development of the information industry. These factors are not conducive to the sound and coordinated development of the two industries.

5   Conclusion

The development level of tourism industry and information industry in Shanxi Province has shown an upward trend year by year. In most cities, the development of the information industry is obviously lagging behind.

From the perspective of the impact of the information industry on the tourism industry, there are obvious differences in the four regions. From the perspective of the response direction, the central part of Shanxi presents alternating fluctuations of positive and negative effects, while the rest of the regions have positive effects; From the perspective of response speed, the response speed of central and northern Shanxi is the fastest, and the response speed of southeast Shanxi is the slowest; From the response trend, The range of fluctuations in central Shanxi is obvious; From the impact of the tourism industry on the information industry, there are also obvious differences between regions. From the perspective of the response direction, alternating fluctuations of positive and negative effects occurred in the central and southeastern parts of Shanxi, and the other regions showed positive effects. From the perspective of response speed, the response speed of central Shanxi was the fastest, and that of southeast Shanxi was the slowest.

Among the four major regions, the contribution of tourism to the development of the information industry in the northwest, central and southern areas of Shanxi is greater than that of the information industry to tourism, while the southeast region is the opposite.

Acknowledgments. This work was financially supported by National Social Science Foundation of China (19BGL141).

References

1. Almeida-Santana, A., David-Negre, T., Moreno-Gil, S.: New digital tourism ecosystem: understanding the relationship between information sources and sharing economy platforms. Int. J. Tourism Cities **6**(1), 1–7 (2020)
2. Novacka, L., Pícha, K., Navratil, J., et al.: Adopting environmentally friendly mechanisms in the hotel industry: a perspective of hotel managers in Central and Eastern European countries. Int. J. Contemp. Hosp. Manag. **31**(6), 2488–2508 (2019)
3. Bai, S., Han, F.: Tourist behavior recognition through scenic spot image retrieval based on image processing. Traitement du Signal **37**(4), 619–626 (2020)
4. Özturan, M., Mutlutürk, M., Çeken, B., Sarı, B.: Evaluating the information systems integration maturity level of travel agencies. Inf. Technol. Tourism **21**(2), 237–257 (2019). https://doi.org/10.1007/s40558-018-0138-3
5. Nguyen, N.-T., Tran, T.-T.: Optimizing mathematical parameters of Grey system theory: an empirical forecasting case of Vietnamese tourism. Neural Comput. Appl. **31**(2), 1075–1089 (2017). https://doi.org/10.1007/s00521-017-3058-9
6. Ivanova, M., Ivanov, S.: Tourism information technology. J. Tourism Futures **6**(2), 201–202 (2020)
7. Madhukar, D.V., Sharma, D.: The role of information technology applications in profitability: a study of the Indian travel and tourism industry. Worldwide Hosp. Tourism Themes **11**(4), 429–437 (2019)
8. Zhang, X.C., Liang, X.C., Song, X., Liu, J.S.: Spatial pattern of the mismatch degree and its causes of the high-quality development of tourism industry in the Yellow River basin. J. Arid Land Resour. Environ. **34**(12), 201–208 (2020). (in Chinese)
9. Fengju, X., Lina, M., Iqbal, N.: Interaction mechanism between sustainable innovation capability and capital stock: based on PVAR model. J. Intell. Fuzzy Syst. **38**(6), 1–17 (2020)
10. Bahmaei, A., Albaji, M., Naseri, A.A., et al.: Effect of irrigation type and interval on soil salinity in clay soils in Ahvaz. Arab. J. Geosci. **13**(21), 1–11 (2020)

Service Quality Evaluation of Office Property Management Based on SERVQUAL Model

Lin Zhu[(⊠)]

Business School, Shanghai Dianji University, Shanghai 201306, China

Abstract. On the basis of SERVQUAL model, combined with the requirements of domestic consumers for the quality of office property service, from the five dimensions of facility management, cleaning service, security service, maintenance service and window service, a set of property service quality evaluation index system including 23 three-level indexes is designed. In order to find out the key factors affecting the quality of property service, this paper uses AHP to analyze the weight of each index, and on this basis, makes an empirical analysis of X property in Shanghai in the form of questionnaire survey. The results show that the five most important factors affecting the service quality of office property are elevator operation, access management, reception attitude, maintenance quality and toilet cleaning.

Keywords: SERVQUAL model · Office property · Analytic hierarchy process · Service quality

1 Introduction

Property management originated in England in the 1860s. Modern property management came into being in the United States at the end of the 19th century. China's property management was born in Shenzhen in 1991. By the end of 2018, the national property management industry had a management area of 27.93 billion square meters, an operating income of 704.363 billion yuan, 127000 property service enterprises and 9.837 million employees [1]. By 2020, there are 24 listed companies in the property management industry [2], and the scale of China's property management industry is growing rapidly. The area under management of office property accounts for 9.1%. Although the proportion is not high, the service fee of office property is the highest compared with other types of business. The hardware of office building is important, but the soft power of office building property management is also important for the operation of office building. In fact, the success of commercial property largely depends on the later level of property management.

At the same time, with the continuous expansion of the scope of office property services and the increasing expectations of customers for the quality of office property, the problems of office property management are gradually emerging. Therefore, improving the level of office property management, paying attention to and analyzing the needs

J. Abawajy et al. (Eds.): ATCI 2021, AISC 1398, pp. 42–49, 2021.
https://doi.org/10.1007/978-3-030-79200-8_6

of customers, enhancing customers' loyalty and satisfaction to office property management services, and establishing a perfect office property management system will provide beneficial reference for improving the competitiveness of office property management enterprises and accelerating the transformation and upgrading of the industry to modern service industry.

2   Related theories of Office Property Management Service Quality

The research on service quality appeared in the early 1980s [3]. In 1982, Christian Gronroos first proposed that the level of service quality of consumers is based on perceived quality, which is formed by comparing service expectation with perceived quality. Gronroos further studies believe that service quality includes two dimensions: functional quality and technical quality. Functional quality is the way of service delivery to consumers, and technical quality refers to the objects that consumers get in the process of enjoying service. On this basis, a conceptual model of perceived service quality is constructed [4]. In 1985, Parasuraman, Zeithaml and Berry subdivided Gap 5 into four gaps (Gap 1–4) and built a service quality gap analysis model [5]. SQGAM was revised twice in 1988 [6] and 1991 [7], and finally formed SERVQUAL model including five dimensions of tangibility, reliability, responsiveness, trust and empathy [8]. The modified SERVQUAL model is widely used in various fields of service quality evaluation, and it is the most representative service quality evaluation method at present [9]. In China, the service quality of property management mainly focuses on the evaluation of customer satisfaction. Liu Chunyu (2014) divided the service items that affect the customer satisfaction of property management into 11 secondary indicators, such as municipal facilities [10]. Zhu Xueqin (2015) built an evaluation system of six indicators, including greening and cleaning, security services, according to the ACSI model [11]. At present, the research on the service quality of office property management is insufficient.

Based on the five dimensions of SERVQUAL model, combined with the property basic service items in the relevant property management service standards proposed by China Property Management Association, this paper constructs SERVQUAL evaluation model of office property management service quality, and analyzes the weight of office property management service quality factors by using analytic hierarchy process, the purpose is to find out the shortcomings of improving the service quality of property management, so as to provide better service for customers and enhance the competitiveness of enterprises.

3   Establishment of Service Quality Index System

3.1   Index Selection

According to the five dimensions and 23 specific evaluation index system scale proposed in SERVQUAL model [12], In this paper, based on the analysis and summary of the literature at home and abroad, through the discussion with experts, and consult property management personnel and third-party property evaluation institutions, combined with the characteristics and status of property management services in the domestic environment, based on the five dimensions of service quality as the basic structure, some indicators are modified, and the construction of 23 specific indicators of office property management service quality scale. The evaluation index system is shown in Fig. 1.

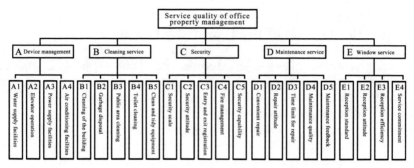

Fig. 1. Office property management service quality evaluation index

3.2   Weight Determination

It can be concluded from Fig. 1 that there are many evaluation indexes of office property management service quality. In order to obtain the relative importance (weight) of each index, the pairwise judgment matrix in AHP provides a practical and effective tool [13]. AHP decomposes the decision-making related elements into different levels such as objectives, criteria and schemes, and then solves them to determine the eigenvectors of the matrix. Then, the priority weight of each element in each level to a certain element in the upper level is obtained, and its consistency is tested. The specific operation steps are as follows:

(1) Analyze the relationship among the factors in the system and establish the hierarchical structure of the system. The first layer is the target layer, the second layer is the middle layer, and the third layer is the index layer. Among them, the target layer reflects the purpose of modeling, namely "office property management service quality evaluation"; the middle layer is the evaluation factor layer (see items A, B, C, D and E in Fig. 1 for details); the index layer is the specific 23 evaluation indexes (Fig. 1).

(2) The judgment matrix is constructed. For any two factors Ci and Cj, aij is used to express the influence of Ci on Cj, which is measured by the scale of 1–9. The ranks are scored according to: extremely important, strongly important, obviously important, relatively important, slightly important, and equally important, with 11, 9, 7, 5, 3 and 1 respectively. In particular, the elements on the diagonal of the square matrix are equally important relative to themselves. Finally, the scores of each index are summed up and standardized.

In order to get a reasonable weight, this paper uses AHP to determine the weight of all levels of indicators, and invites a number of experts and field managers to determine the judgment matrix through consultation. Among them, senior professors engaged in quality management research, middle and senior managers of property management enterprises, personnel of third-party evaluation institutions and owners' representatives are included, and the results are highly credible.

3.3 The Square Root Method is Used to Calculate the Eigenvector and Weight

The square root method is used to calculate the eigenvector and weight. Assuming that the comparative score of index i relative to j is aij, then the comparative score of j relative to i is aij = l/aij, so as to establish pairwise judgment matrix R1, R2, R3, R4, R5, R6.

The consistency ratio parameter CR = CI/RI is used to test the consistency and the rationality of the judgment matrix. The consistency index of judgment matrix C1 = (λmax − n)/(n − 1); RI is the average random consistency index given by T.L.Saaty. If CR < 0.1, it shows that there is good consistency, Wi can be used as the weight vector (Table 1).

Table 1. Average random consistency index

Matrix order n								
1	2	3	4	5	6	7	8	9
RI 0	0	0.58	0.90	1.12	1.24	1.32	1.41	1.45

$$R1 = \begin{bmatrix} 1 & 1/3 & 2 & 4 & 3 \\ 3 & 1 & 3 & 5 & 5 \\ 1/2 & 1/3 & 1 & 3 & 3 \\ 1/4 & 1/5 & 1/3 & 1 & 2 \\ 1/3 & 1/5 & 1/3 & 1/2 & 1 \end{bmatrix} \quad R2 = \begin{bmatrix} 1 & 1/4 & 1/2 & 2 \\ 4 & 1 & 2 & 5 \\ 2 & 1/2 & 1 & 3 \\ 1/2 & 1/5 & 1/3 & 1 \end{bmatrix}$$

$$R3 = \begin{bmatrix} 1 & 1/3 & 2 & 1/3 & 3 \\ 3 & 1 & 3 & 1/2 & 4 \\ 1/2 & 1/3 & 1 & 1/4 & 2 \\ 3 & 2 & 4 & 1 & 5 \\ 1/3 & 1/4 & 1/2 & 1/5 & 1 \end{bmatrix} \quad R4 = \begin{bmatrix} 1 & 1/3 & 1/5 & 3 & 4 \\ 3 & 1 & 1/3 & 3 & 5 \\ 5 & 3 & 1 & 5 & 7 \\ 1/3 & 1/3 & 1/5 & 1 & 2 \\ 1/4 & 1/5 & 1/7 & 1/2 & 1 \end{bmatrix}$$

$$R5 = \begin{bmatrix} 1 & 2 & 1/3 & 3 & 1/4 \\ 1/2 & 1 & 1/4 & 3 & 1/4 \\ 3 & 4 & 1 & 4 & 1/3 \\ 1/3 & 1/3 & 1/4 & 1 & 1/5 \\ 4 & 4 & 3 & 5 & 1 \end{bmatrix} \quad R6 = \begin{bmatrix} 1 & 1/3 & 2 & 1/2 \\ 3 & 1 & 4 & 2 \\ 2 & 1/4 & 1 & 1/3 \\ 1/2 & 1/2 & 1/3 & 1 \end{bmatrix}$$

The square root method is used to calculate the matrix eigenvector and the relative weight of each index factor. From the judgment matrix, CR < 0.1 are obtained, which pass the consistency test (Table 2).

Table 2. Weight scale scale of service quality

	W_i		W_i		W_i		W_i		W_i
A	0.234	B	0.457	C	0.168	D	0.078	E	0.063
A1	0.142	B1	0.161	C1	0.146	D1	0.134	E1	0.161
A2	0.507	B2	0.262	C2	0.240	D2	0.101	E2	0.466
A3	0.265	B3	0.098	C3	0.488	D3	0.263	E3	0.096
A4	0.086	B4	0.416	C4	0.079	D4	0.056	E4	0.277
λmax = 4.021 CR = 0.008		B5	0.062	C5	0.047	D5	0.446	λmax = 4.031 CR = 0.012	
		λmax = 5.608 CR = 0.015		λmax = 5.246 CR = 0.055		λmax = 5.297 CR = 0.066			

3.4   Evaluate the Service Quality of Office Property Management

According to the index system shown in Fig. 1 and the weight assignment of each index, the modified SERVQUAL model is used to calculate [14], where Wi is the weight of j attributes, Pi is the customer's perceived quality of i problem, Ei is the customer's expected quality of i problem, and the restaurant service quality is evaluated. The higher the score, the better the service quality and the higher the customer satisfaction. At the same time, we can also evaluate the other indicators of restaurant service one by one, so as to get the links that need to be improved in restaurant operation, so as to meet the needs of more consumers.

$$SQ = \sum_{i=1}^{n} u_i \sum_{j=1}^{m} w_{ij}(P_i - E_i) \tag{1}$$

4   Service Quality Satisfaction and Importance Evaluation

This paper selects x property in Shanghai as the research object. The questionnaire is mainly aimed at the clients of the property service. Face to face interviews and questionnaires are used. Starting from October 1, 2019, for a period of two months, 200 questionnaires are issued and 178 questionnaires are returned. After statistical analysis, there were 8 invalid questionnaires, and the effective rate accounted for 96% of the total.

According to the calculation and analysis of the importance indicators of each dimension, among the influencing factors of office property management service quality, the highest weight of cleaning service is 0.457, followed by 0.234 for equipment management, 0.168 for safety assurance, 0.078 for maintenance service and 0.063 for window service. Therefore, cleaning service is the key factor in the secondary indicators, which needs to be focused in the follow-up analysis. From the weight analysis of the three-level indicators, the five indicators of "elevator operation", "access management", "reception attitude", "maintenance quality" and "toilet cleaning" have the largest weight, indicating that these four indicators are the decisive factors affecting the service quality of chain restaurants. At the same time, it also shows that property management enterprises should focus on strengthening management in these five aspects, so as to improve the service quality of enterprises and meet the needs of customers.

According to Likert 5-point scale method, five different evaluation grades are set: very satisfied, satisfied, basically satisfied, dissatisfied and completely dissatisfied, and the corresponding scores are given: 100, 75, 50, 25, 0. According to formula (1), the scores of the first and second level indicators of the property are calculated, and the IPA quadrants of satisfaction values and weights of different dimensions are constructed.

Overall, the evaluation of each index is relatively high, which is at a high level. Among them, the owner's evaluation of "maintenance service" is relatively high, which is 89.75. The evaluation of "cleaning service" is relatively low. In the first quadrant, there are high importance and satisfaction indicators, including two dimensions of power supply facilities and building cleaning, which shows that these two dimensions have a greater weight in the service quality of office property management, and customers show strong satisfaction in these two dimensions. The second quadrant is the index of high importance and low satisfaction, which includes four dimensions: garbage disposal, access management, elevator operation and toilet cleaning. It shows that these dimensions are the dimensions that customers pay more attention to in the property management service, but the actual service quality does not meet the expectations. This will be the next dimension that needs to be improved in the property management service. There are many dimensions in the third quadrant, which indicates that the service quality of the property in these dimensions has been recognized by customers and needs to be maintained. The fourth quadrant is the index of low satisfaction and low importance, which shows that although these dimensions cause customer dissatisfaction, they have little impact on the overall service quality (Fig. 2).

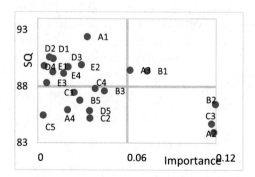

Fig. 2. IPA quadrant analysis of office property management service quality

5 Conclusions

First of all, we should pay attention to and improve the service quality with high importance and low satisfaction. In the secondary indicators, cleaning service has the highest weight in all indicators, which is the key factor, but the customer evaluation of cleaning service is relatively low, so improving cleaning service plays a vital role in improving customer satisfaction. Among the three-level indicators, the sanitary condition of toilets in cleaning service and the operation condition of elevators in facilities and equipment management are the key factors, but the service quality score is the lowest, which must be improved in time, and the operation condition of elevators also has a great impact on the evaluation level of facilities and equipment management. Customers report that the elevator often breaks down and needs to be repaired and checked in time. In addition, customers think that the waiting time of the elevator in rush hours is long. The property management should pay attention to the operation of the elevator. The maintenance of the elevator needs to be timely to provide convenience for customers, so that customers can feel the "people-oriented" service concept of the property management and improve customers satisfaction with property services.

Secondly, we should continue to maintain the service quality with high importance and high satisfaction and make continuous improvement. In all three levels of indicators, power supply facilities and building cleaning are the core elements of service quality, which must be continuously improved on the basis of maintaining the existing. We can't ignore the indicators that are not important and satisfied. If we keep low satisfaction for a long time, these indicators will have a great impact on the service quality of office property management.

Based on the SERVQUAL evaluation model and the characteristics of China's office property management service quality requirements, this study selects five dimensions of facility management, cleaning service, security, maintenance service and window service to establish a complete office property management service quality evaluation index system. The AHP is used to determine the weight of each index system in the evaluation system, and the SERVQUAL evaluation model is used to calculate the final evaluation results. At the same time, the evaluation index system has been applied to the empirical analysis of office property management service quality. Through the combination of qualitative analysis and quantitative analysis, this evaluation model is simple and easy to use, has strong practicability and operability, and provides reasonable and effective scientific basis for office property management enterprises to improve service quality and customer satisfaction.

References

1. China Property Management Association. 2019 property service enterprise development index evaluation report. Beijing (2018)
2. Li, W.: Exploration on the development trend of property management industry under the influence of epidemic situation – thinking from the perspective of property industry. Housing Real Estate (13), 57–60 (2020)
3. Grönroos, C.: An applied service marketing theory. Eur. J. Mark. **16**(7), 30–41 (1982). https://doi.org/10.1108/EUM0000000004859

4. Grönroos, C.: A service quality model and its marketing implications. Eur. J. Mark. **18**(4), 36–44 (1984). https://doi.org/10.1108/EUM0000000004784
5. Parasuraman, A., Zeithaml, V.A., Berry, L.L.: A conceptual model of service quality and its implications for future research. J. Mark. **49**(4), 41–50 (1985). https://doi.org/10.1177/002 224298504900403
6. Parasuraman, A., Zeithaml, V.A., Berry, L.L.: Servqual: a multiple-item scale for measuring consumer perceptions of service quality. J. Retail. **64**(1), 12–40 (1988)
7. Parasuraman, A., Berry, L.L., Zeithaml, V.A.: Refinement and reasesment of the SE-RVQUAL scale. J. Retail. **4**(8), 1463–1467 (1991)
8. Parasuraman, A., Zeithaml Valarie, A., Berry Leonard, L.: Alternative scales for measuring service quality: a comparative assessment based on psychometric and diagnostic criteria. J. Retail. **70**(3), 201–230 (1994)
9. Liu, J., Zhang, W.: Research review on perceived service quality. J. Dongguan Univ. Technol. **02**, 89–95 (2014)
10. Liu, C.: Preliminary study on customer satisfaction index system of property management. Sci. Technol. Innov. Guide **11**(20), 167 (2014)
11. Zhu, X., Wang, Y.: Customer satisfaction evaluation of property management: a case study of Zhengzhou Hengda Mingdu community. Manag. Eng. **20**(02), 11–14 (2015)
12. Yu, B., Du, G.: Research on fuzzy evaluation of online shopping express service quality based on SERVQUAL model. Ind. Eng. **02**, 127–133 (2013)
13. Xu, X., Liu, Y.: Customer satisfaction of the third-party logistics enterprise based on AHP: a case study. Int. J. Inf. Syst. Supply Chain Manag. **10**(1), 68–81 (2017). https://doi.org/10.4018/IJISSCM.2017010105
14. Taylor, S.A.: Measuring service quality: reexamination and extension. J. Mark **56**, 55–68 (1992)

Characteristics and Trends of Tourism Informatization Research - Based on Citespace Knowledge Map Analysis

Weitong Qiao[1], Nan Gao[1(✉)], and Xincheng Zhang[2]

[1] Faculty of Culture Tourism, Shanxi University of Finance and Economics, Taiyuan, Shaanxi, China
[2] School of Economics and Management, Northwest University, Xi'an, Shaanxi, China

Abstract. By using the data sources from CSSCI database, MS and PHD dissertation and conference papers during 1986 to 2017 and 'uses tourism' and 'Informatization' as research objects. With the help of Cite Space software from the age, periodicals, authors and author institutions Object, case, research methods and topics of bibliometric analysis. The results show: 1. The study of tourism informatization in our country began in 1986, and the volume of published papers increased rapidly after 2000. Although a core group of publications has been formed, but not yet have a stable core author group. The development of tourism information in the earlier regions of the universities and research institutes based; 2. The research objects of domestic tourism informatization tend to be diversified. In the selection of case studies, the research mainly focuses on domestic research and focuses on the developed areas in the east, lack of international research; 3. In the research methods on the theme of empirical research, new methods and methods need to be strengthened; 4. On the research theme, China's tourism informatization focuses on the topics of tourism informatization theory, travel website research and travel web information search.

Keywords: Tourism informatization · CiteSpace · The mapping knowledge domains

1 Introduction

In March 2017, the National Tourism Administration announced the 'Thirteenth Five-Year' National Tourism Informatization Plan, which aims to promote the widespread use of emerging information technologies represented by 'Internet plus, big data, and Internet of Things' in the tourism industry application, Increase the demand of tourists and the market for informatization, digitization, intelligence, etc., adjust the structure of the tourism industry, improve tourism products and services, meet the needs of tourists, and help the global tourism industry to flourish. For this reason, tourism informatization has become an important way to realize the transformation of my country's tourism industry from extensive to intensive, and it has become one of the hot issues studied by scholars at home and abroad.

© The Author(s), under exclusive license to Springer Nature Switzerland AG 2021
J. Abawajy et al. (Eds.): ATCI 2021, AISC 1398, pp. 50–56, 2021.
https://doi.org/10.1007/978-3-030-79200-8_7

Domestic exploration of tourism informatization began in the 1980s. Wang Yanfeng [1] proposed that tourism informatization research mainly includes three aspects: tourists, tourism resources, and organizations. Gong Yuming [2] pointed out that tourism informatization belongs to the category of socio-economic information research. These two propositions do not give a reasonable and scientific explanation for tourism informatization. Li Tianshun [3] and others defined various information that reflects changes in tourism activities as tourism informatization. Zhang [4] grouped the resource elements including scenic spots, travel agencies, hotels, transportation, tourists and so on in the process of tourism activities as tourism informationization. These two definitions fail to clarify the boundary between information and data and intelligence. Chen Zhihui [5] and others believe that tourism information is the reproduction of the law, nature, and characteristics of the development and change of tourism activities. Chunchang F [6] and others define tourism informatization as: tourism information is the description of the essence, characteristics and laws of the subject, object, and organizational media in tourism activities. Both of these definitions regard tourism informatization as a dynamic description of the development, changes, status, nature and laws of tourism activities.

In terms of the application of tourism information technology, Varjú, Viktor [7] and others build a tourism information development platform with the help of electronics, information, databases, and networks. Bastida U [8] pointed out that the tourism destination marketing system can significantly improve the level of tourism informatization. Qiu Xueqin [9] proposed countermeasures and suggestions for the construction of rural tourism informatization development. Pan B [10] and others have explored how to build a smart tourism framework.

The research on the development of tourism informatization from different perspectives not only reflects the urgency of improving the development level of tourism informatization, but also reflects the urgent need for in-depth research on domestic tourism informatization theory. It is particularly important to identify and excavate the frontier hotspots of tourism informatization development, and provide important reference for further analysis of the research goals, research objects, research hotspots and precise entry points of tourism informatization development. To this end, this paper selects the bibliometric method to objectively show the hotspot changes in the development of domestic tourism informatization on the basis of the existing domestic research review on tourism informatization, and records the development process of domestic tourism informatization research, in order to provide reference for the healthy development of tourism informatization in China.

2   Research Methods and Data Sources

In China National Knowledge Infrastructure (CNKI), a search for the title 'tourism' contains 'informatization' resulted in a total of 3279 journal documents, excluding newspapers, announcements and other documents unrelated to tourism informatization, and finally 899 documents were obtained. Among them, there are 298 master and doctoral papers, 412 CSSCI papers, and 189 conference papers. The retrieval time was from 1986 to 2017.

In order to further record the research trend of tourism informatization, this paper uses Citespace5.0.R1 to search related literatures to generate author institutions and co-occurrences of authors to conduct research on scholar groups.

3 Statistics and Analysis

3.1 Distribution of Domestic Tourism Information Research

Since 1986, the amount of travel information publications has slowly increased, and since 2000, the amount of publications has fluctuated upward. The number of articles published and the number of authors have increased in the same trend, but the number of authors is slightly higher than the number of documents. On the one hand, tourism informatization is an emerging tourism economy produced by the integration of the tourism industry and the information industry, which requires mutual cooperation in interdisciplinary research fields and a large number of cooperative publications. On the other hand, domestic tourism informatization research started late, the research force is scattered, and there are fewer prolific authors.

3.2 Distribution of Domestic Tourism Information Research Journals

Table 1. Distribution of domestic tourism information research source journals

Sorting	Periodical	Article volume
1	Tourism Tribune	32
2	Tourism Science	11
3	Business Management Journal	10
4	Economic Geography	7
5	Tourism Forum	7
6	Human Geography	6
7	Journal of Arid Land Resources and Environment	4
8	Scientia Geographica Sinica	3
9	Geography and Geo-Information Science	3
10	Journal of Intelligence	2

Table 1 shows the distribution of source journals of tourism informatization research literature. The top 10 journals published a total of 85 articles in the field of tourism informatization research, accounting for 20.63% of the total number of searched journals. The number one journal in the number of publications is 'Journal of Tourism', sponsored by the School of Tourism, Beijing Union University. The journal mainly focuses on the frontier hotspots of tourism academic research and the development of tourism industry practice. It has published a total of 32 articles in the field of tourism informationization research, and is the main front of tourism informationization research in China. The second place is 'Tourism Science', sponsored by the School of Tourism of Shanghai Normal University, which mainly explores the rules of tourism activities and reflects the latest developments in the tourism industry in a timely manner. Ranked third is 'Economic Management', sponsored by the Chinese Academy of Social Sciences, which mainly studies major management issues from the perspective of economics. It can be seen from Table 1 that the top 10 source journals are all professional journals in the field of tourism research. There are also a small number of economic management journals that publish articles on tourism informatization research, but the number is relatively small. It can be seen that with the continuous development of information technology, tourism informatization research, as a relatively new research field, has not yet fully entered the economic management journals, and can be expanded to more disciplines in the future.

3.3 An Analysis of the Frontier Hot Spots of Domestic Tourism Informatization Research

Highly-cited documents are generally classic documents in this academic research field, and the focus of this research field is discovered by studying highly cited documents. Among the 899 articles, 3 articles were cited more than 100 times. We analysed 12 articles that were cited more than 50 times. Through an in-depth analysis of these 12 documents, it is concluded that the highly cited documents mainly involve the following aspects: Li Yunpeng, Zhang Lingyun and others discussed the definition of tourism informatization, Basic theories such as classification and characteristics. Lu Zi, Xiong Jianping, Yu Xiaojuan, Wu Jinmei, Fu Yeqin, and other scholars conducted research on the development and application of information technology in the tourism industry.

According to the results of Citespace analysis, a certain team has been formed in domestic tourism informatization research. For example, the School of Tourism and Environment of Shaanxi Normal University has formed a research team centered on Ma Yaofeng, Li Junyi, and Yang Min. The Business School of Jishou University has formed a the research team centered on Wang Zhaofeng and Xie Juan. The School of Resources and Environmental Sciences of Hebei Normal University has formed a research team centered by Lu Zi and Li Yanli. Bohai University formed a research team with Zhang Yugai and Lan Guiqiu as the core. Team research is the main research force in the field of domestic tourism information. However, there are fewer connections between research teams in the map, indicating a lack of cooperation between team research.

According to the above analysis, there is no significant correlation between highly cited authors and highly published authors. Table 2 shows that Professor Li Yunpeng (School of Business Administration, Capital University of Economics and Business) has the highest citation frequency of 67 times, which has a great impact on other researchers in this field. Secondly, Xiong Jianping (Tourism Department of Hubei University Business School), Wang Zhaofeng (Business School of Jishou University), Li Junyi (School of Tourism and Environment of Shaanxi Normal University) and other people's citation frequency is relatively high, which has an important influence in the field of domestic

Table 2. Analysis of highly cited authors

No.	Author	Frequency	Years	Institution
1	LI Yunpeng	67	2014	Capital University of Economics and Business Department of Hospit
2	XIONG Jianping	64	2006	Tourism Management Department of Business School, Hubei University, Wuhan
3	WANG Zhaofeng	53	2012	School of Business, Jishou University
4	LI Junyi	48	2010	School of Tourism and Environment, Shaanxi Normal University, Xi'an
5	Du Peng	39	2013	Shenzhen university

tourism information research. In general, the frequency of mutual citation among domestic tourism informatization research scholars is high, and each scholar provides strong support for the system construction of tourism informatization theory from different disciplinary systems.

3.4 Trajectory and Trend Analysis of Domestic Tourism Informatization Research

According to the results of Citespace analysis, the first research hotspot of domestic tourism informatization research is 'informatization', which includes two different forms of 'tourism' and 'tourism informatization'. The frequency of three expressions is as high as 82 times. The second is smart tourism, which is different from the 'smart' redefined concepts such as tourism e-commerce and digital scenic spots. With the help of information technology such as Internet and cloud computing, in order to meet the new needs in the development of tourism industry, it is the advanced stage of the development of tourism informatization. In addition, the object of tourism information's 'e-commerce', 'tourism', 'tourism industry' and 'tourism enterprises' have a high frequency, indicating that it has attracted the attention of domestic tourism information scholars. With the rapid development of the domestic urban economy, the problems of unbalanced development between urban and rural areas and large income gap are highlighted, and the rural living environment is improved. The National Tourism Administration has identified 2006 as the 'Year of Rural Tourism in China'. At the same time, it has carried out agricultural tourism demonstration sites nationwide, and rural tourism has become one of the hotspots in tourism research. In general, with the continuous progress of social economy, single tourism has been difficult to meet the needs of tourists. Tourism informatization has evolved into an important form of integration of tourism industry and other industries. Networking, mobility and intelligence have become the future development direction of tourism informatization research.

4 Discussion and Conclusion

This paper uses the method of scientific knowledge mapping to conduct a scientific quantitative analysis of the literature on tourism informatization published in core journals of CSSCI from 1986 to 2017, and identifies the number of published papers and journal publishing of domestic tourism informatization, and draws the following conclusions and enlightenment:

(1) Since 1986, there has been a slow growth in the publication of tourism informationization. Although it has fluctuated since 2000, it has gradually accelerated in recent years. In terms of the distribution of research journals, professional journals in the field of tourism research and a few economic management journals published articles on tourism informationization, indicating that the future of tourism informationization can be expanded in a large area. At present, the study of tourism information has been widely concerned by domestic scholars. They are mainly distributed in universities and research institutes in Beijing, Kaifeng, Xi'an, Chengdu, Zhangjiajie, Jinzhou, Guilin and other provinces. In terms of the research sites, the domestic tourism informatization is mostly concentrated in the eastern economically developed regions, indicating the relationship between tourism informatization and the level of economic development and the spatial undertaking of information technology. In terms of research methods, empirical research methods have been paid attention to, and new methods and means need to be strengthened. In terms of research topics, there are various issues concerning tourism informatization in China, but they are mainly manifested in the theoretical research of tourism informatization, the research of tourism websites, and the search of tourism network information.

(2) Looking forward to the future, tourism informatization still needs to continue in-depth research: 1. The emerging concepts and technologies of tourism informatization come from different disciplines. Only by establishing a unified basic system of tourism informatization as soon as possible, can a variety of interdisciplinary disciplines be better integrated and further studied. 2. Research methods need innovation. The research on tourism informatization focuses more on technology introduction, less on innovation, application, feasibility and operability analysis, and lacks case studies. Moreover, case studies are mostly concentrated in areas with highly developed social economy, and the popularization and application of informatization is difficult to popularize. Therefore, the methodological system of tourism informatization can be constructed to gradually expand the promotion and development space of tourism informatization. 3. More emerging technologies are urgently needed to be applied to tourism. Due to the scalability, integration and integration of information technology, it will provide more and better popularization platform for the development of tourism in the future. 4. Tracking research on the integration of information technology and tourism. At present, neural network technology, virtual reality technology, big data analysis technology, cloud computing, Internet of Things will encounter some new problems in the process of integration with tourism, which will become the focus of future research.

Acknowledgments. This work was financially supported by National Social Science Fund Research on the Innovation of Rural Tourism Public Service Supply Mechanism under the Background of Rural Revitalization Strategy, NO. 19BGL141 fund.

References

1. Wang, Y.: Tourism information collection and tourism promotion propaganda. Nankai Econ. Stud. **04**(5), 5–56 (1986). (in Chinese)
2. Gong, Y.: The position and exploitation of tourism information in economic decision. Tourism Tribune **01**, 31–34 (1994). (in Chinese)
3. Li, T., Zhang, H.: Tourism Management, pp. 67–68. Shaanxi Normal University Press, Xi'an (1998). (in Chinese)
4. Zhang, R.Y.: Strategy of the information construction in modern tourism enterprise. Adv. Mater. Res. **201–203**, 759–762 (2011)
5. Cheng, Z., Chen, X.: Tourism Informatics. China Tourism Press, Beijing (2003). (in Chinese)
6. Fu, C.: The secondary development based on MapX—GIS of tourism design and implementation. Phys. Procedia **24**(1), 651–659 (2012)
7. Varjú, V., Suvák, A., Dombi, P.: Geographic information systems in the service of alternative tourism – methods with landscape evaluation and target group preference weighting. Int. J. Tour. Res. **16**(5), 496–512 (2013)
8. Bastida, U., Huan, T.C.: Performance evaluation of tourism websites' information quality of four global destination brands: Beijing, Hong Kong, Shanghai, and Taipei. J. Bus. Res. **67**(2), 167–170 (2014)
9. Qiu, X.: Research on rural tourism informatization in Yunnan Province. Inquiry Econ. Issues **08**, 157–161 (2009). (in Chinese)
10. Pan, B., Zheng, C., Song, F.: A comparison of the development of tourism information technologies between China and the United States. Inf. Technol. Tourism **21**(1), 1–6 (2018). https://doi.org/10.1007/s40558-018-0131-x

Policy Guarantee for the Development of FinTech Under the Flow of Information Technology in Guangdong-Hong Kong-Macao Greater Bay Area

Baoxin Chen, Kan Chen, and Xi Wang[✉]

School of Humanities and Social Sciences, Macao Polytechnic Institute, Macao 999078, China
xwang@ipm.edu.mo

Abstract. In the era of science and technology empowering the financial industry, to seize new opportunities in the construction of the Guangdong-Hong Kong-Macao Greater Bay Area, we must first vigorously develop financial technology, which is the core driving force for the development of the global digital economy. The key to the development of financial technology lies in the strong guarantee of policies. Policies can overcome certain lag in the law. Under the basic principles of the law, they can play a role in politics, management, finance and administration to achieve various goals. In other words, to develop financial technology in the Greater Bay Area, it is necessary to rely on policy guarantees to open up a convenient and orderly flow channel for talents, goods, funds, and information in the Greater Bay Area, and to promote regional integration in the Bay Area. Policy guarantees such as competition guarantee, platform construction guarantee, and risk prevention and response will realize the benign interaction among various production factors under the guarantee of financial technology policies in the Greater Bay Area.

Keywords: Guangdong-Hong Kong-Macao Greater Bay Area · Financial technology · FinTech · Policy · Production factors

1 Introduction

In 2020, the "Proposals for the 14th Five-Year Plan for National Economic and Social Development and Long-Term Goals for 2035" proposes to improve the level of financial technology and enhance financial inclusiveness. In 2019, the "Guangdong-Hong Kong-Macao Greater Bay Area Development Plan Outline" issued by the Central Committee of the Communist Party of China and the State Council emphasized the vigorous development of characteristic financial industries and described the Guangdong-Hong Kong-Macao Greater Bay Area (hereinafter referred to as the Greater Bay Area) to promote cross-border financial products with the help of science and technology. Create a blueprint for the Greater Bay Area Economic Community. According to the Hong Kong Institute of Asia Pacific Studies, Hong Kong willingness to Participate in the Economic

J. Abawajy et al. (Eds.): ATCI 2021, AISC 1398, pp. 57–63, 2021.
https://doi.org/10.1007/978-3-030-79200-8_8

Areas in the Greater Bay Area accounted for the largest proportion (27.7%, Fig. 1) [1]. It can take the lead in technology finance as a breakthrough point. Policy-oriented, optimize the allocation of resources such as talents, goods, capital and information [2], financial technology (Financial Technology, FinTech), provide innovative technologies and methods that are different from traditional ones in financial services [3]. This is an emerging industry that uses technology to improve financial activities [4]. That is, the financial industry uses big data, artificial intelligence, the Internet, blockchain, cloud platforms and other technological means to broaden financial services, such as auto-mated investment, online credit and online financial management, banking services and risk management [5]. It can also cover consumer healthcare, education, food, shopping, transportation, social networking, tourism and other scenarios. The development of Fin-Tech in the Greater Bay Area is a strategic plan and cannot be separated from the strong guarantee of policies. The FinTech industry is an emerging industry, and there are many obstacles to the development of unknown areas. In addition, the cross-border coordi-nated development of the Greater Bay Area under "One Country, Two Systems" faces restrictions from the system and borders. Therefore, this article focuses on the challenges faced by the development of FinTech in the Greater Bay Area. Design the corresponding policy guarantee model (Fig. 2).

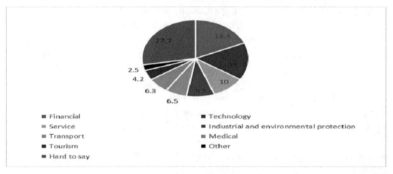

Fig. 1. The Percentage of Hong Kong willingness to participate in the Economic areas in the Greater Bay Area.

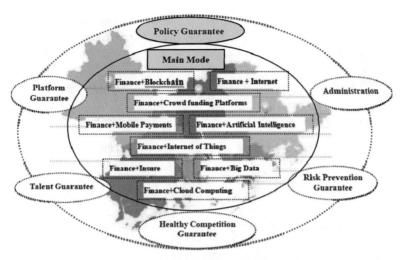

Fig. 2. Policy guarantee path for the development of finance technology

2 Challenges Facing FinTech Development in the Greater Bay Area

2.1 The Flow of FinTech Talents is Not Smooth

On the one hand, the circulation of talents mainly depends on job recruitment, project exchanges, academic exchanges, and joint training of talents. The Greater Bay Area has many well-known universities and scientific research institutions at home and abroad. It is the main source of FinTech talents, but universities and there is less cooperation between scientific research institutions, and there is very little cross-border cooperation. What follows is that good learning resources are concentrated in good universities and institutions, and there is a lack of resource circulation and sharing. On the other hand, the distribution of resources among cities in the Greater Bay Area is uneven. Different organizations and different positions often enjoy different treatment. Most talents are willing to stay on a good platform, compete for good resources and gradually develop themselves. Most talents are reluctant to flow to difficult and remote areas and the grassroots, making the employment of talents more solid. These two aspects make the circulation of talents fall into an endless loop.

2.2 Cross-Border Logistics Transportation is Not Smooth

There are convenient means of transportation in the Greater Bay Area, but the circulation of commodities on both sides of the strait often has to go through layers of complicated processes and long waiting times. Although GPS positioning and Internet of Things technologies are relatively mature, the cross-border real-time tracking of commodities is not possible, leading to potential safety hazards during commodity transportation, and additional freight costs for midway quarantine and transshipment, further reducing the willingness of cross-strait FinTech cooperation. In some high-tech fields, the smart chain of unmanned intelligent tractors, artificial intelligence product sorting, unmanned vehicles for product distribution, cargo transportation and customs clearance is still immature [6], under the impact of some major epidemics such as new crown pneumonia, Strict epidemic prevention and control measures, manpower shortage, increased logistics costs, and increased risk of collection, cross-border logistics are facing dual pressures from epidemic prevention and control and the global economic and trade situation.

2.3 The Cross-Border Use of Financial Funds is Not Smooth

From a macro perspective, the Greater Bay Area has three relatively independent separate tariff areas, two international free ports, and three currencies [7]. The financial development level, marketization degree, regulatory system and laws and regulations of Guangdong, Hong Kong and Macao are quite different. Moreover, the currency exchange rates and stock markets of Guangdong, Hong Kong and Macao are affected by the international market, which easily triggers financial market turbulence, which in turn has an impact on my country's economy. Specifically, first of all, cross-border funds inevitably have payment security issues, such as collection and refund, customs declaration mode, foreign exchange settlement and payment between different currencies, and third-party payment compliance issues. Second, scientific and technological enterprises need to invest a long period of time and large amounts of capital in research and development, and they are prone to face the dilemma of breaking the capital chain. In recent years, FinTech has developed technology credits (such as technology loans, high-tech loans, cloud tax loans, etc.), but there are still small coverage areas that still cannot meet the needs of technology companies.

2.4 The Communication of Information and Data is Not Smooth

There is an inevitable separation of information and data in Guangdong, Hong Kong and Macao. The first is that the development process in the technical field is different and it is difficult to synchronize. Furthermore, from the perspective of the country, funds, and information security, the interoperability of information and data is moving forward in the game of freedom and restriction. On the one hand, from a market perspective, data can only be valued when it is circulated and used. However, the risk of data leakage in multi-party sharing of information and data increases. On the other hand, restricting the exchange of information and data is not conducive to the true regional integration of the three places and sharing results. Finally, in terms of information supervision, there are many different regulatory bodies in Guangdong, Hong Kong, and Macao, resulting in obstacles to the exchange of information between individuals and enterprises under different regulatory models [8].

3   Policy Guarantee for the Development of FinTech in the Greater Bay Area

3.1   Administrative Management and Policy Guarantee

Administration is the guarantee for the development of FinTech in the Greater Bay Area. Mainly to realize its three functions, one of which is the overall planning function, to implement the "Guangdong-Hong Kong-Macao Greater Bay Area Development Plan Outline" and the spirit of the 14 five-year plans for national economic and social development, and through the administrative departments to jointly draft agreements and formulate policies to strengthen the district With international administrative cooperation and the advantages of the Greater Bay Area, the economic strength, scientific and technological strength, and overall national strength will be greatly improved. The second is to coordinate the function, balance the relationship between the government and the market, improve the risk mitigation capabilities of insurance and the financing risks of project cooperation, and cultivate a group of FinTech talents and talents in the Greater Bay Area who are responsible for the times. The third is the supervision function, which is suitable for the talent evaluation standards of the financial economy of the Greater Bay Area, fund security, anti-monopoly mechanism, and the direction of financial investment and industrial value.

3.2 Talent Guarantee for the Development of Dawan FinTech

Talent is the key to the development of Dawan FinTech. In terms of talent introduction, companies can work together to increase talent introduction. The government and policy makers should take the initiative to understand the needs of talent introduction, help solve the problems of talents' household registration, residence, spouse employment, childbirth and child education, retain talents, and develop talents Team. In terms of the talent training model, the FinTech model of school-enterprise alliance and industry-university-research integration can be implemented, focusing on the concepts of lifelong education and lifelong learning, and focusing on training young talent reserves. In terms of stimulating innovation and achievement transformation, scientific and technological achievements belong to institutions with a mixed-ownership nature. The sharing of achievements through various methods such as self-transformation, transfer, licensing, and equity participation is encouraged. In addition to traditional universities, support the development of high-level research institutions and establish innovative The quality, effectiveness, and contribution-oriented talent evaluation system makes full use of incentives such as training opportunities, wages, bonuses, promotion, leadership recognition, etc., to reward talents who have made positive contributions to the development of FinTech in the Greater Bay Area. The power of sustainable development. With regard to the orientation of the flow of talents, we will strengthen propaganda, create a good atmosphere for the whole society to care about and support talents, and guide the broad masses of talents to actively participate in the Development of FinTech.

3.3 Healthy Competition Guarantee for the Development of Insurance Technology in the Greater Bay Area

The benign competition to guide the development of FinTech in the Greater Bay Area is the secret to achieving the coordinated development of the Greater Bay Area. The market mechanism will inevitably have a competitive relationship. The future development of enterprises should shift from a sense of competition to a sense of win-win cooperation to prevent overcrowded markets in the same field from causing resource waste, vicious competition and other issues, and to maximize regional talent, materials, Technology and other factors of production allocation efficiency, the FinTech industry in the Greater Bay Area needs to have the spirit of regional common development, handle good competition and cooperation within and outside the region, and unify the industry rules for cross-border competition in the top-level design, and prohibit monopoly and trade Limit and protect consumer rights, such as payment and settlement methods, data monopoly, etc., and fully tap local advantages, learn from each other's strengths, and use the leader to drive regional development and strengthen teamwork.

3.4 Platform Guarantee for FinTech in the Greater Bay Area

Establishing an integrated FinTech platform in the Greater Bay Area is very effective. The function of the platform needs to realize several major functions. First, the FinTech government management service in the Greater Bay Area, simplifying the service process, is conducive to the implementation of policies, project management, and the realization of benefits and services for talents. Second, the exchange of FinTech talents in the Greater Bay Area is conducive to the cultivation and reserve of talents, the establishment of talent pools in different fields, the implementation of higher-level talent training, and the evaluation of talent demand forecasts, and recruiting talents in short supply according to demand. Third, FinTech supervision in the Greater Bay Area. The scope of supervision includes the supervision of personal credit and corporate credit, the supervision of intellectual property rights, and anti-monopoly supervision. The government, the people, industry associations and their own supervision are used to build a harmonious development of the FinTech industry in the Greater Bay Area. Attach importance to the power of the platform, open up online and offline, internal and external services, domestic and overseas services, and concentrate superior resources [9].

3.5 Risk Prevention of FinTech in the Greater Bay Area

Doing a good job in the risk prevention of FinTech in the Greater Bay Area is a condition for stable development. Any technology development and utilization is a double-edged sword [10]. Fintech brings us efficiency and hidden risks. We have to deal with the risks brought by advanced science and technology as well as the potential risks in the financial industry. The application of artificial intelligence multi-dimensional data analysis can evaluate the credit levels of different users, and at the same time, combined with the rational analysis ability of people, can greatly reduce the non-performing rate of financial institutions and optimize the flow of funds. But immature technology may also lead to systemic risks and affect financial stability. When encountering the threat of malicious code, we must strengthen supervision and improve the level of network security. When encountering legal risks, we must clarify the party responsible for the loss, and develop the FinTech insurance industry. When encountering the risk of personal privacy leakage, we need to balance and know The boundary between rights and personal privacy. When encountering data barriers between different companies and different credit reporting systems, strive to realize the sharing of government and market credit reporting data, and resolve the risk of mutual debt. When encountering intellectual property infringements, it is necessary to increase knowledge Strength of protection.

4 Conclusion

The development of FinTech in the Greater Bay Area is not a purely technical issue, nor is it a purely financial issue. Its policy guarantees must be based on the actual needs of the construction of the Greater Bay Area and the overall situation of the country's development, and make full use of the innovations brought by technology and technology to provide multiple entities for the "One Country, Two Systems" Finance goes hand in hand to escort. Realize the benign interaction between the various production factors under the guarantee of the FinTech policy of the Greater Bay Area; promote the Greater Bay Area to become a base for China's FinTech innovation and an international-level Greater Bay Area Economic Community demonstration zone, thereby contributing to the international cross-border market.

Acknowledgements. This research was supported by the Research Project Fund of Macao Polytechnic Institute (Approval No.: RP/ESCHS-03/2020).

References

1. Survey Findings on Public Opinion on the Development Plan of Guangdong-Hong Kong-Macao Greater Bay Area. http://www.hkiaps.cuhk.edu.hk/eng/press_release.asp?NewsCl ass1=-&Year=2018
2. Liu, J.: A study on the cooperative development path of technology finance in Guangdong-Hong Kong-Macao Greater Bay Area. South China Finan. (09), 57–65 (2020)
3. Van Loo, R.: Making innovation more competitive: the case of fintech. UCLA Law Rev. **65**(1), 232 (2018)
4. Schueffel, P.: Taming the beast: a scientific definition of fintech. J. Innov. Manag. **4**(4), 32–54 (2017). https://doi.org/10.24840/2183-0606_004.004_0004
5. Aldridge, I., Krawciw, S.: Real-Time Risk: What Investors Should Know About Fintech, High-Frequency Trading and Flash Crashes. Wiley, Hoboken (2017)
6. Li, N.: The construction and application of China-Laos-Thailand cross-border logistics intelligent chain based on "blockchain + artificial intelligence" technology. Foreign Econ. Trade Pract. (01), 89–92 (2021)
7. Huang, Z., Liao, D.: Study on the construction of a new mechanism for the coordinated development of the Guangdong-Hong Kong-Macao Greater Bay Area. South China J. (11), 18–20+62 (2020)
8. Tang, Z.: Constructing open banks: strategic measures for financial innovation and development in the Guangdong-Hong Kong-Macao Greater Bay Area. South China Finan. (05), 73–81 (2019)
9. Ma, Y.: Technology promotes the upgrading of banking platform services. China Finan. (08), 64–66 (2020)
10. Guo, W., Wang, W.: On spatial spillover effects of financial agglomeration on science and technology innovation and heterogeneity in Guangdong-Hong Kong-Macao Greater Bay Area. J. Guangdong Univ. Finan. Econ. **33**(02), 12–21 (2018)

Training Mode of Applied Talents in Tourism Management Specialty Under Artificial Intelligence

Yan Li[✉]

Guangzhou University of Science and Technology, Guangzhou 510440, Guangdong, China

Abstract. In recent years, with the rapid development of AI, deep learning, mobile Internet and other technologies, products related to AI have been applied to many places, such as our work, life, study and so on, so that our work, study, life is more convenient and convenient. Whether colleges and universities can catch the fast train of AI era, improve their talent training mode, improve the teaching quality, enhance the overall competitiveness of the school, and transport more outstanding talents for the society is an important problem. Therefore, through practical experiments on two professional classes in a university in our city, we can know that AI technology can help students to improve their academic performance to a certain extent.

Keywords: AI · College education · Talent training · Social experiment

1 Introduction

In recent years, with the economic and social transformation and the continuous improvement of residents' living standards, tourism has become a part of people's life, and mass tourism has a blowout development. On the one hand, the number of domestic and outbound tourists and consumption are increasing year by year. However, compared with the development of tourism practice, the theoretical research of tourism education is slightly lagging behind. On the other hand, the supply side reform of higher education, the transformation of local undergraduate colleges and universities to application-oriented, the contradiction between supply and demand of tourism talents has become increasingly prominent, the enrollment scale of tourism undergraduate education has declined, the number and structure of teachers have changed slowly, the professional characteristics are not bright enough, and various problems such as "double low" of students' recognition and employment rate in the industry have become increasingly prominent [1].

From the origin of medieval universities, to the classical universities and then to the modern universities, the scale, form and structure of universities are constantly developing and changing, and their functions are increasingly enriched in this process. The increasingly rich functions of knowledge imparting, personnel training, scientific research, cultural heritage, social service and so on make the concept and mission of

J. Abawajy et al. (Eds.): ATCI 2021, AISC 1398, pp. 64–71, 2021.
https://doi.org/10.1007/978-3-030-79200-8_9

the university constantly renewed. As the first important meaning of running a university, personnel training is the backbone of leading the development of higher education and runs through the whole process of Higher Education development. Its reform and improvement is related to the overall level of Application-oriented Undergraduate construction. The vigorous development of tourism has put forward greater demand and better requirements for talents, and the contradiction between supply and demand of high-quality tourism talents has become increasingly prominent [2]. In China, the task of training tourism talents is mainly undertaken by tourism colleges and universities. Therefore, it is urgent for tourism colleges and universities to alleviate the problems of quantity, quality and structure of tourism talents and explore the direction and Countermeasures of training tourism talents. At present, the employment of college students has gradually become the focus of common concern of the society, and the cultivation of tourism professionals has attracted more and more scholars' attention. By reviewing the development process of tourism higher education in China, combing the problems, and carrying out practical exploration, the tourism academic circles have investigated and studied tourism higher education in developed countries in Europe and America, and gained a lot of valuable experience [3].

This study mainly adopts the methods of literature research, case analysis and multidisciplinary comprehensive research. Through the preliminary preparation and on-the-spot investigation, a large number of information retrieval and collation are carried out from searching literature on the Internet, consulting materials in the library and looking at relevant books. The main purpose is to search and collect the current situation of private colleges and universities, the origin and related technologies of AI, the existing application of intelligent teaching, the innovation and reform of talent training mode, and other domestic and foreign literature Analysis and interpretation. This paper analyzes some successful cases of applying AI to teaching at home and abroad, and then makes a detailed analysis of how to apply AI technology to talent training mode, which will promote the innovation of talent training mode.

2 Relevant Concepts

2.1 AI Concept

AI is man-made. Man-made is something that man makes through thinking and processing. But man made things can be smarter and more intelligent than man. The translation of intelligence is intelligence, which includes self-awareness, thinking, learning, thinking and so on. The reason why people can rule the earth is that people have high intelligence and ability, and can use tools, create tools, etc., so people are intelligent [4]. But we have a very limited understanding of what intelligence is made of and what the limits of intelligence are. So what is AI?

Generally speaking, AI is a subject that studies the simulation of human intelligent behavior by computer, such as the study of human thinking process, how to learn, how to reason, how to make decisions, etc., mainly including how to realize the intelligence of calculation, how to make computers similar to human brain intelligence, and how to use computers to complete deeper application. It can be seen that AI includes many subjects, such as computer science, psychology, linguistics, neurology, sociology, etc. [5].

2.2 Applied Talents of Tourism Major

The characteristics of tourism industry determine the characteristics of tourism talents and the law of talent cultivation. On the one hand, tourism application-oriented talents have the characteristics of multi-level knowledge structure, diversified ability level and multi-dimensional quality literacy. On the other hand, due to the particularity of tourism products, tourism service objects, tourism industry development and tourism service industry attributes, the cultivation of tourism application-oriented talents should follow the laws of comprehensiveness, practicability, adaptability and internationalization [6]. Based on the above understanding, this paper believes that tourism application-oriented talents are "professional" + "practical", that is, the application-oriented talents who use knowledge, specifically refers to the compound talents who master the theoretical basis of tourism specialty, have the practical ability of the industry, and are competent for the work of tourism enterprises [7].

2.3 Talent Training Mode

Talent training mode refers to the theoretical model and operation mode of talent training process, which is designed under the guidance of specific education concept of training theme and certain guarantee training system to achieve specific talent training objectives. It is composed of several elements and has systematic characteristics, such as purposiveness, intermediary, openness, diversity and imitatability [8].

It can be seen that talent training mode is a concept with profound meaning and broad content, and all the elements related to talent training should be included in it. This study believes that the concept of talent training mode mainly involves three aspects: first, the concept of talent training. The concept of talent cultivation is an important basis for promoting talent cultivation and an important support for guiding talent cultivation practice. It should exist as an abstract element in talent cultivation mode [9].

Second, the carrier of talent training. The realization of talent training objectives must rely on relevant carriers. Curriculum, teaching and teachers, as the three essential elements in the process of talent training, are the important carriers for the realization of talent training objectives in application-oriented universities. Therefore, curriculum and curriculum content, teaching methods and evaluation system, teachers' ability and level and other relevant elements should be included. Third, the environment of talent training. In addition to the concept of guiding the practice of talent cultivation and the carrier of talent cultivation, the educational environment and cultural construction also play an invisible role in the process of talent cultivation, which should be a key in the talent cultivation mode.

To sum up, this study believes that the talent training mode refers to a set of talent training paradigm formed by integrating all the abstract and specific elements involved in the teaching activities and environment with teachers and students as the main body. The goal concept of talent cultivation, the course content selected to achieve the goal, the teaching principles followed by teaching activities, the teaching methods adopted to improve the efficiency of talent cultivation, the evaluation system formulated to evaluate the quality of talent cultivation, the educational environment and culture created by the school, and all the educational resources and services provided [10].

2.4 Basic Algorithm

(1) Mean square error

$$RMSE = \frac{\sqrt{\sum_{(u,i) \in T} (r_{ui} - r'_{ui})^2}}{|T|} \tag{1}$$

(2) Mean absolute error

$$MAE = \frac{\sum_{(u,i) \in T} |r_{ui} - r'_{ui}|}{T} \tag{2}$$

(3) Accuracy

$$Precision = \frac{\sum_{u \in U} |R(u) \cap T(u)|}{\sum_{u \in U} |R(u)|} \tag{3}$$

3 Practical Research on the Talent Training Mode of Tourism Management Major from the Perspective of AI

From some cases of the application of AI technology to the talent training mode, we can find that the application of AI technology to the talent training mode will create a new talent training mode, improve the teaching quality and the school running level.

3.1 Subjects

According to the requirements, a total of 39 students, 23 males and 16 females, were selected from class a of Tourism Management Major in Tourism College of a university in our city. The average score of tourism related courses is 67, and 12 people fail. According to the teacher's reflection, the students' enthusiasm in class is not high. According to the statistics, there are an average of 3 students late or leave early in each class; there are 43 students in class B, with an average score of 71 points in the practical class and 11 students failing. According to the teacher's reflection, the students' enthusiasm in class is not high, with an average of 5 students late or leave early in each class. Two classes were tested under the background of "AI+".

3.2 Experimental Steps

The process of improving and strengthening the learning effect of students is carried out in different steps at different times between students and teachers. The main means is through online implementation, through the setting of teacher controlled question answering discussion area; according to the learning situation of individual students, tutoring, refreshing and storing the interaction information of students and teachers and students' learning information, and storing the data in the database as a specific file.

Class A is set as the control class, and class B is set as the experimental class. Through a three-month simulation test of the two classes, this paper analyzes the effect of AI technology on students' performance and learning attitude.

4 Statistical Analysis of Data

4.1 Students' Understanding of AI in Teaching

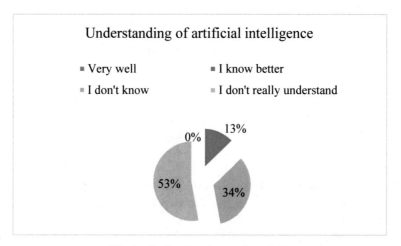

Fig. 1. Students' understanding of AI

As shown in Fig. 1, the students in class a of the experimental group don't know much about the application of AI in teaching. Only 13% of the students in this class know much about it, and 34% of the students don't know much about it. The remaining 54% of the students don't know much about the application of AI in teaching. The students in this class know little about AI and are more unfamiliar with the education platform. On the one hand, teachers need to do a simple training for students when using the platform for blended teaching, introduce the relevant concept knowledge, and improve their understanding of this kind of learning, which is conducive to the smooth implementation of blended teaching. On the other hand, the experimental group a students know less about the application of AI in teaching, which can be more conducive to the comparative analysis of learning situation and attitude towards this kind of education mode after education. Just entered the university gate, for all things full of curiosity students, this is a new teaching method. Therefore, the new teaching is easier to attract students' attention and stimulate students' interest in learning, which is its advantage.

4.2 Students' Experience Based on AI

(See Table 1).

Table 1. The completion of each process of class A and class B

	Class A	Class B
Participation in answering questions in class	82%	64%
Number of people completed the operation	93%	74%
Discuss interaction	62%	12%
Review after class	74%	13%

According to Fig. 2, classroom data is recorded through real-time follow-up, Statistics show that the experimental group a class in the application of AI in teaching, the proportion of classroom participation and interaction between teachers is 82%, the number of homework is 93%, the number of discussion and interaction between students in class is 62%, the proportion of autonomous after class pre review is 74%; the control group B class in the traditional education mode, the proportion of classroom participation and interaction between teachers is 64%, the proportion of homework is 93% 74% of the students completed their homework, only 12% of the students discussed and interacted with each other in class, and 13% of the students independently reviewed after class in their spare time. Through the analysis of this data statistics, it can be concluded that under the new mode of education and the traditional mode of education, the two classes of students in the classroom and the teacher interaction enthusiasm gap is not big, but the autonomous learning enthusiasm difference is obvious, under the new mode of education, the number of students in the classroom autonomous interaction is significantly more than that in the traditional education, and after class autonomous learning situation of class a students is more than that of class B students extremely, the application of AI in teaching, on the basis of the traditional education mode, can make students have the power of thinking and solving in and out of the classroom, which is the lack of traditional education mode.

4.3 Assessment Result Analysis

According to the relevant data in Fig. 2 and Table 2, it can be analyzed that under the traditional mode of education, the efficiency of knowledge absorption in students' courses is not high, and after three months of courses, the scores are basically not improved, indicating that most students do not seriously complete the classroom practice tasks; in the application of AI in teaching education mode, students can improve their

Table 2. Average scores of students in two classes in different periods

	Theory performance of class A students	Theory special results of class B students	Practice performance of class A students	Practice performance of class B students
Check out	73	72	75	73
Day 5	77	77	82	84
Day 15	88	79	83	85
Day 25	95	84	93	87
Final Assessment	97	86	92	89

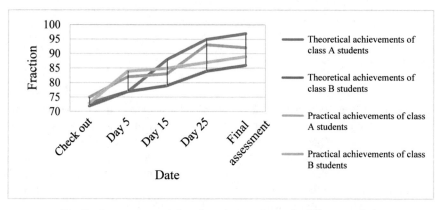

Fig. 2. Average scores of students in two classes in different periods

scores by 20% in three months compared with the assessment results It shows that we have actively completed the corresponding learning tasks in the course teaching. Under the monitoring of the data platform, it is impossible to know whether the students are forced to attend class seriously or to improve their interest, but it is an objective fact that the final score is improved.

5 Conclusion

With the development of AI technology, the application of AI in teaching is also increasing, which has brought changes to education and teaching. It is changing the traditional teaching methods. Based on the research on the reform of Undergraduate Tourism Management Talents Training Mode from the perspective of AI, this paper first analyzes the AI technology and the environment of private colleges, and discusses the application status of AI technology in teaching, and puts forward that AI can change teaching tools, teaching and learning methods, teaching evaluation and talents training mode. It is clear that private colleges and universities should reform the personnel training mode and improve the teaching quality through four aspects of AI technology to change the teaching tools, teaching and learning methods, teaching evaluation and management innovation, so as to improve the level of running a school and shape the excellent brand strategy, and put forward feasible methods and safeguard measures for the implementation of the strategy. Through the practical experiment, we can see that the application of AI in teaching can improve the academic performance to a certain extent, and achieve the effect of assisting students to learn knowledge.

References

1. Lu, H., Li, Y., Chen, M., et al.: Brain intelligence: go beyond artificial intelligence. Mob. Netw Appl. **23**(7553), 368–375 (2017)
2. Ghahramani, Z.: Probabilistic machine learning and artificial intelligence. Nature **521**(7553), 452–453 (2015)
3. Moravčík, M., Schmid, M., Burch, N., et al.: DeepStack: expert-level artificial intelligence in heads-up no-limit poker. Science **356**(6337), 508–513 (2017). https://doi.org/10.1126/science.aam6960
4. Raza, M., Khosravi, A.: A review on artificial intelligence based load demand forecasting techniques for smart grid and buildings. Renew. Sustain. Energy Rev. **50**(Oct), 1352–1372 (2015). https://doi.org/10.1016/j.rser.2015.04.065
5. Crawford, E.D., Batuello, J.T., Snow, P., et al.: The use of artificial intelligence technology to predict lymph node spread in men with clinically localized prostate carcinoma. Cancer **88**(9), 2105–2109 (2015)
6. Parkes, D.C., Wellman, M.P.: Economic reasoning and artificial intelligence. Science **349**(6245), 267–270 (2015)
7. Bundy, A.: Preparing for the future of Artificial Intelligence. AI & Soc. **32**(2), 285–287 (2016). https://doi.org/10.1007/s00146-016-0685-0
8. Davis, E., Marcus, G.: Commonsense reasoning and commonsense knowledge in artificial intelligence. Commun. ACM **58**(9), 92–103 (2015)
9. Rigas, E.S., Ramchurn, S.D., Bassiliades, N.: Managing electric vehicles in the smart grid using artificial intelligence: a survey. IEEE Trans. Intell. Transp. Syst. **16**(4), 1619–1635 (2015)
10. Zang, Y., Zhang, F., Di, C.A., et al.: Advances of flexible pressure sensors toward artificial intelligence and health care applications. Mater. Horiz. **2**(2), 25–59 (2015)

Risk Analysis and Development Research of Cross-Border E-commerce Payment Under the Background of Internet Finance

Rui Lin[✉]

School of Economics, Shanghai University, Shanghai, China

Abstract. Under the background of Internet finance, electronic payment has become the core payment method of China's cross-border e-commerce transaction system. However, there are serious problems in the third-party payment in China's Internet finance industry, which are manifested in the circulation and monitoring of funds, imperfect credit information system, personal information leakage and so on. To solve these problems, the government should gradually improve the construction of relevant legal systems, fill the blank areas of Internet financial supervision and eliminate the gray areas of supervision. At the same time, the platform is required to control the scale of personal information disclosure and create a benign consumption and investment environment for consumers and investors. Through the above methods, the credit risk, technical risk and liquidity risk in cross-border payment can be reduced, making positive contributions to the long-term development of cross-border e-commerce electronic payment in China.

Keywords: Internet finance · Cross-border electronic payment · Third-party payment platform · Payment risk

1 Research on the Background of Cross-Border E-commerce Payment in China

Cross-border payment generally refers to the transnational and cross-regional transfer of funds in international investment and international trade by two or more countries or regions through certain settlement instruments, payment systems and financial instruments. In recent years, with the digitalization of modern payment means and the emergence of the third-party cross-border electronic commerce payment platform, payment in cross-border electronic commerce has developed rapidly. According to the Research Report on China's Third Party Cross-border Payment Industry in 2020 released by iResearch, the scale of China's cross-border e-commerce industry reached 5.5 trillion yuan in 2019. The report predicts that by 2022, the online payment scale of China's third-party cross-border payment C-end market is expected to increase by 11.1%. It is estimated that online payment will surpass offline payment for the first time by 2023. The main settlement methods of domestic cross-border payment and remittance are foreign exchange purchase including third-party transactions and cross-border income settlement including third-party collection and settlement. This mode mainly includes the traditional commercial bank transfer mode and two Internet payment modes involving third-party payment institutions [1].

1.1 Traditional Payment Cross-Border E-commerce Platform

Traditional cross-border electronic payment platforms mainly include bank wire transfer, professional remittance companies (such as Western Union Remittance and Express) and common international credit cards (such as Visa and American Express, etc.) [2], and their payment methods are mainly online orders and offline payments. In the payment process, consumers need to purchase foreign exchange through offline banking counters or online banking according to the requirements of overseas merchants, select credit cards with corresponding transaction currencies, fill out remittance application forms, and remit the order amount to designated accounts [3]. This payment institution has the characteristics of direct contact with banks, immediate direct payment and high security. However, they have a single mode and lack good security in transaction fraud, return and exchange, etc.

1.2 Third-Party Payment Cross-Border E-commerce Platform

The existing third-party cross-border payment platforms mainly include PayPal, Alipay and China UnionPay [4]. Compared with traditional payment methods, the third-party cross-border electronic commerce payment platform has many business entities, diverse service features and weak substitutability. Mutual benefit accounts for a large proportion of group relationships, which indicates that the payment platform and other partners are benefit-driven, rather than interdependent. Behind the development is the internal driving force of demand. In the case that both parties generally lack mutual understanding, mutual credit and legal protection, as an "intermediate platform" for the transition of capital payment, it ensures the safety of capital flow and transaction process between buyers and sellers and banks. Under the background of higher requirements for the security, confidentiality and transaction speed of transaction data, the advantages and risks of third-party payment have also become more prominent. According to the analysis data, as of the first half of 2018, the two giants of third-party online payment accounted for a total of 35.2%, and the third-party mobile payment accounted for a total of 93.1%. With the accelerated reshuffle of the industry, the proportion of third-party mobile payment will still rise further.

2 Internet Finance Promotes Cross-Border Electronic Payment

Internet finance is a new business model developed on the basis of traditional finance in recent two years. Financial institutions combine big data and information communication technology to realize financial functions such as financial payment, investment and financing and information intermediary. Internet finance is not a simple combination of the Internet and the financial industry, but a new mode and new business naturally produced to meet the new demands after being familiar with and accepted by users (especially the acceptance of e-commerce) at the level of realizing network technologies such as security and mobility. It is a new field combining traditional financial industry with Internet technology [5].

In China's financial system, payment service plays an extremely important role. The deep integration of Internet finance can just drive the sustainable development of payment Internetization. In the field of cross-border payment, the role of the third-party payment platform is becoming more and more prominent, the market development is becoming more and more stable and perfect, and it gradually shows its advantages in the payment market, becoming the main payment channel in cross-border electronic commerce. Compared with traditional finance, both sides of the third-party payment platform can use the platform to screen information, match demand and price products, and finally complete the transaction between the two sides without any traditional intermediary, which reduces the transaction cost to a certain extent. Similarly, relying on the development of big data financial technology, the development of payment business is gradually scaled and convenient, the efficiency of transaction process is improved, and the constraints of transaction time and space on the development of payment business are also weakening. Because third-party payment can integrate the needs of several industries into a service, third-party payment also plays a boosting role in supporting the development of various industries and integrating industry resources, and may also have an impact on the real economy. For example, by integrating the back-office resources of banks and the huge advantages of customers, we can promote cooperation with public utilities. Therefore, the third-party payment provides convenient financial services and great payment convenience. However, while cross-border electronic payment brings convenience, it also brings new cross-border transaction risks and security problems of domestic and foreign capital transfer.

3   Existing Problems and Risk Analysis of Cross-Border E-commerce Payment

Payment service is the most basic link in the financial industry, and the traditional transaction mode is restricted by time and space, which can no longer meet the huge transaction demand. The emergence of third-party payment institutions such as Alipay has opened up new funding channels for China's banking system. At the same time, due to the lack of comprehensive supervision of cross-border payment, this provides a hotbed for criminal activities caused by cross-border transactions, such as cross-border money laundering, information theft and illegal transactions [6]. Although the development of Internet finance based on big data technology provides technical guarantee for the convenience of third-party transactions and the improvement of credit reporting mechanism, the corresponding multi-party information is still opaque, personal privacy is still lacking, and the supervision of data circulation is imperfect. These risks will not only affect the operation of the third-party payment platform, but also have a negative impact on the sustainable development of the Internet finance industry [7].

3.1 Credit System Defects and Payment Credit Risk

At present, cross-border electronic commerce mostly uses the third-party payment platform and adopts the delayed net settlement method. The funds of the users of the third-party payment platform will stay in the third-party payment institutions for a period of time, resulting in deposited funds. With the continuous expansion of funds in transit, the credit risk index of the third-party payment platform itself also increases. The third-party payment platform provides guarantee for both parties in online transactions, while the third-party payment institution itself lacks guarantee. If the deposited funds lack effective liquidity management, it may lead to payment risks. Deposited funds are delayed in storage mode and transaction time, lack of supervision, huge amount and credit risk of unauthorized call within enterprises, which will not only affect customers' own interests, but also pose a challenge to financial supervision.

3.2 Regulatory Risks of Big Data and User Privacy

With the accumulation of user information and transaction data, the third-party payment platform will generate a lot of real data in the process of participating in transaction payment. While using big data to provide basic financial services, it also faces management risks of information services.

First of all, as a kind of valuable resource, data itself may lead to malicious attacks and information theft by competitors or hackers at home and abroad, thus increasing the difficulty and potential risks for organizations to manage payment data. As far as the platform itself is concerned, the system security management is relatively weak, which easily leads to the acquisition and tampering of the transaction information itself and the transaction data in the process of transaction transmission, and there are risks such as user funds fraud or account theft.

Second, the existing data supervision is not in place, and there are problems such as standard generalization and implementation vacancy, which can easily lead to user privacy leakage. At the legal level, the existing laws and regulations on Internet finance still have many defects, such as the lack of laws and regulations on the safety of consumers' personal information, the lack of effective supervision on the platform's access to and use of users' sensitive information, and the lack of reasonable norms on the operation of payment platforms.

3.3 Risk of Cross-Border Money Laundering and Illegal Flow of Funds

At the legal level, the current Anti-Money Laundering Law of the People's Republic of China and other laws and regulations need to be further improved, in which the illegal content of cross-border electronic payment is not clearly defined, and there is a corresponding gray transaction risk. Similarly, third-party payment institutions have not formulated standardized standards and specific regulations on their delivery contents and methods. The boundary between illegal money laundering and normal and legal cross-border capital flows is unclear, difficult to distinguish and costly. At present, there is still a lack of corresponding technical means to distinguish. For example, the same cross-border transaction subject can be registered as a third-party payment institution in

China, and can also register multiple identities overseas or even in different countries. It can bypass domestic foreign exchange control restrictions through proprietary trading, and carry out cross-border capital transfer and money laundering activities. With the popularization and high frequency of the use of third-party payment, many third-party payment institutions have begun to get involved in transnational payment business. The frequent renewal and competition of market commercial entities will, to a certain extent, create opportunities for criminals to liquidate other countries' funds in the form of third-party payment, which will flow into the country's capital market and disrupt the market transaction order [8].

3.4 Legal Risks of Cross-Border Electronic Payment

In the process of cross-border transactions, the ownership of the transaction subject is different from that of legal tender, the legal system on which the risk control is based conflicts, and the laws and regulations of sovereign countries and related international electronic payment are also different. In the transaction process, there is no strict requirement and uniform definition for which country or region the two parties should adopt the electronic payment supervision law, and this direct difference will bring potential risks of trade handling problems. With the acceptance of "The Belt and Road" Initiative by more and more countries, there are new problems in cross-border payment in China, such as the scope of foreign exchange management in transactions, and the formulation of laws to define the tax zone of cross-border transactions.

4 Strategic Analysis and Development Research of Cross-Border Electronic Payment Industry

China's Internet finance industry has been officially developed for eight years. Although specific businesses such as third-party payment, big data finance and online crowdfunding have shown a good situation, it still faces serious problems such as the circulation and monitoring of third-party payment funds, imperfect credit reporting system and personal information disclosure. In order to make the Internet financial industry develop better, it is necessary to gradually improve the construction of relevant legal systems, fill the blank areas of Internet financial supervision, eliminate the gray areas of supervision, standardize the platform, control the scale of personal information disclosure, create a benign consumption and investment environment for consumers and investors, and reduce and solve many risks or problems in cross-border payment, such as credit risk, technical risk, liquidity risk, privacy protection and regulatory difficulties.

4.1 Strengthen the Supervision of Cross-Border Electronic Commerce Payment

The development of cross-border e-commerce business involves traditional international trade and e-commerce transaction mode, which includes cash flow, logistics, customs inspection, foreign exchange management, third-party payment and other fields. Due to the numerous businesses and complex processes, in the process of formulating corresponding laws and regulations and improving policies, the tax authorities, the State

Administration of Foreign Exchange, the central bank and other government regulatory agencies should establish a joint working mechanism, establish an appropriate supervision and service system, optimize the supervision and service work, and comprehensively and standardly supervise the cross-border payment platform. In view of the authenticity and legitimacy of the transaction in the operation process, we should communicate with relevant government departments and legislative bodies in time to ensure that the transaction payment is protected by law. Because funds are transferred through the third-party payment platform, many existing supervision systems put the supervision responsibility of capital circulation on the institutions themselves, but the roles of market subjects and supervision subjects are different, and there is self-conflict to a certain extent, which integrates the dual functions of power exercise and supervision, which easily leads to unfair supervision problems. Therefore, it is necessary to clarify the legal status of multinational third-party payment institutions, rationally allocate the supervision responsibilities of payment institutions, and ensure that the roles of institutions are clear and the powers and responsibilities are unified, so as to improve and standardize the system.

4.2 Improve the Security of Third-Party Payment Information in the Era of Big Data

In the era of big data, we should improve the information security mechanism of third-party payment. First, improve the network information protection system. In today's economic environment, data is both a resource tool and an infrastructure. As the fundamental guarantee in the era of big data, the government should further solve the problems of data ownership, data access and processing, data transaction and transmission, and severely punish customer information disclosure, transfer, sale and data privacy. Second, from the ideological point of view, payment enterprises must establish the enterprise values of information security first, and clearly assume the responsibility of properly protecting customer information and maintaining platform security. In the operation and maintenance of the platform system, enterprises should formulate management measures in strict accordance with the standards of financial institutions, and establish security procedures, including risk monitoring, early warning analysis, system backup and recovery emergency measures, so as to protect customer identity and transaction data [9]. Third, regulators should also use big data technology to conduct real-time regulatory analysis on these enterprises, control the boundaries of trading networks, and find potential problems in a timely and effective manner.

4.3 Improve the Credit Information Work and Establish an Online and Offline Credit Information System

The establishment of credit system can effectively improve the credit awareness of the public and enterprises, thus strengthening the construction of internet credit system and related market environment. The key to the establishment of credit system is to improve the handling mechanism of dishonesty, increase the cost and punishment of default, increase the dimension of credit evaluation, and establish the stable standard of public credit system, so as to maintain the security of internet financial ecosystem and transactions [10]. In the industry, we should give full play to the central bank's leading position in the credit field, improve the industry information sharing mechanism, and clarify the scope and content of information disclosure.

4.4   Improve the Access System for Third-Party Cross-Border Payment Services

The regulatory authorities may, in accordance with the Measures for the Administration of Payment Services of Non-financial Institutions, establish a systematic access system, standardize the qualification management of foreign exchange business conducted by third-party payment institutions, and improve the access and exit management system of cross-border business. As the existing payment institutions have complicated business models and unclear rights and responsibilities, they should also standardize the corresponding cross-border business operations, bring third-party payment institutions into the scope of foreign exchange management, and stipulate the responsibilities and obligations of third-party payment platforms, including the standardization of business operations, the management of customer information, the security of funds, and the timely disclosure of information.

4.5   Attach Importance to and Strengthen Anti-money Laundering Work

The trend of capital flow in cross-border transactions reflects the way and purpose of transactions to a certain extent. The Internet platform should cooperate with relevant institutions to improve the supervision mechanism of RMB cross-border capital flows, give early warning of abnormal RMB capital flows and take further control measures. Cross-border payment institutions should pay attention to multi-level and multi-channel anti-money laundering to ensure that they will not become tools for criminals to carry out illegal business such as smuggling and money laundering [11].

References

1. Wang, Q., Yang, D.: A preliminary study on the main payment methods of cross-border e-commerce in China. Mark. Modernization **12**, 90–91 (2017)
2. Xue, C., Li, S., Cao, W., Cao, H.: Construction of cross-border payment e-commerce ecosystem. J. Field Actions **19**, 143–150 (2019)
3. Zhou, L., Yu, P.: The status quo, risks and regulatory countermeasures of cross-border e-commerce payment. Shanghai Finan. **5**, 73–78 (2016)
4. Lang, L., Li, Z.: Research on the development status and payment problems of cross-border e-commerce in China. Coop. Econ. Sci. Technol. **2**, 124–127 (2019)
5. Wang, S.: Development status and trend of internet finance. Bus. News **16**, 91–92 (2020)
6. Weng, D.: Research on the development strategy of third-party cross-border payment institutions. Asia-Pac. Econ. **6**, 39–47 (2018)
7. Yang, B.: The status quo of internet finance in China. Time Finan. **10**, 44–45 (2019)
8. Zhang, C.: The impact of internet finance on traditional financial system and its regulatory countermeasures: a case study of third-party payment. New Finan. **10**, 43–48 (2015)
9. Zhu, Q., Zhu, H.: A SWOT analysis of cross-border e-commerce development under the background of internet – a case study of Chongqing. J. Sichuan Univ. Sci. Technol. (Soc. Sci. Ed.) **31**(5), 73–82 (2016)

10. Zhao, Z., Lin, B., Bai, Y.: Research on internet finance support in China (Hangzhou) cross-border e-commerce comprehensive test area. Zhejiang Finan. **3**, 10–16 (2016)
11. Wang, J.: Strategies for cross-border e-commerce talent cultivation under the background of "Internet Plus." Knowl. Econ. **1**, 172 (2017)

The Construction and Application of Financial Sharing Service Center of Enterprise Group

Zijian Han(✉)

School of Accounting, Shandong University of Finance and Economics, Jinan 250220, Shandong, China

Abstract. Financial Sharing Service Center is a new financial management mode that uses big data, cloud computing, artificial intelligence, mobile internet and other new generation information technology to reengineer and standardize the financial business of enterprise groups, improve management efficiency, reduce operating costs, enhance management and control, and realize transparent supervision. The construction of Financial Sharing Service Center adopts a set of unified information system and workflow processing platform, unifies accounting standards, business standards and information standards, integrates the design of core system with business system, so as to realize financial digital transformation, break information island, improve service value and support management decision.

Keywords: Financial sharing · System construction · Application research

The Financial Sharing Service Center aims at strengthening the centralized financial management and control and realizing the creation of financial value. By using the new generation of information technology such as big data, cloud computing, artificial intelligence and mobile Internet, the group enterprises are scattered in various regional operation units, and the process reengineering and standardization of the financial business with high repeatability, heavy workload, easy standardization and standardization are carried out, and the group enterprises are concentrated in the regional operation units It is a new financial management mode that a new independent business unit is unified to form scale effect, so as to improve management efficiency, reduce operating costs, enhance management and control, and realize transparent supervision.

1 The Strategic Significance of Building Financial Sharing Service Center to Strengthen Financial Control of Enterprise Groups

(1) The construction of Financial Sharing Service Center is conducive to strengthening financial control and promoting financial transformation. The application of financial sharing service is the premise and foundation of management accounting. It creates favorable conditions for the application of management accounting from the aspects of organizational structure, personnel reserve, data base, system architecture, etc. Through the construction of Financial Sharing Service Center, optimizing process, unifying data, improving data quality and integrity, it has laid a solid foundation for "running to management accounting".

J. Abawajy et al. (Eds.): ATCI 2021, AISC 1398, pp. 80–86, 2021.
https://doi.org/10.1007/978-3-030-79200-8_11

(2) The construction of Financial Sharing Service Center is conducive to improving the quality and efficiency of financial services and reducing operating costs. The Financial Sharing Service Center of enterprise group provides financial basic services in a centralized way. On the basis of shared financial software of enterprise group ownership, it realizes unified accounting subject, unified information system, unified financial approval process, unified financial accounting standard and unified data standard. Realize the standardization and process operation of financial accounting, realize the sharing of resources in tax, capital and financial data, improve the work efficiency and quality of financial department, and reduce the operation cost.

(3) The construction of Financial Sharing Service Center is conducive to improving the quality of financial information and ensuring the authenticity and effectiveness of data. On the basis of realizing the reimbursement business, the budget management, cost management, fund management, tax management, debt management, performance analysis and other management are included in the business scope of the Financial Sharing Service Center, forming the closure of the whole business management chain, so as to improve the quality of information and ensure the authenticity and effectiveness of financial data.

2 Principles and Key Points of Financial Sharing Service Center Construction

(1) Principles of Financial Sharing Service Center construction

1) Overall planning, top-level design principles. Bring the construction of Financial Sharing Service Center into the strategic planning of enterprise group, and plan the construction plan of Financial Sharing Service Center in the enterprise development strategy, business operation, financial management and other large systems.

2) According to the enterprise conditions, the principle of step-by-step implementation. Fully consider the enterprise group size, main business, accounting basis, business management and information work level and other factors, scientifically design the positioning and construction mode of Financial Sharing Service Center. Adhere to the strategy of "first pilot, then promotion, first solidification, then optimization", and actively and steadily promote the construction of Financial Sharing Service Center.

3) Integration of industry and finance and technical support. Adhere to the business needs as the leading, the new generation of information technology as the support, and process reengineering as the basis to achieve a high degree of integration of business and finance, give play to the value of data assets, promote the level of financial informatization, and promote the digital transformation of enterprises.

4) The principle of value creation and financial transformation. Through the construction of Financial Sharing Service Center, promote the transformation of financial accounting to management accounting, make the financial personnel

liberated from accounting work more engaged in decision-making support type of financial management, make the financial sharing service center gradually change from cost center to profit center, and realize the creation of financial value.

5) The construction principle of "platform unification, system unification, standard unification, implementation unification and data unification". In order to "help enterprise groups strengthen management and control, standardize management" as the construction goal, according to the overall planning, distributed implementation of the construction strategy, the construction of Enterprise Group Financial Sharing Service Center.

(2) Key points of Financial Sharing Service Center construction

1) Adopt a set of unified information system and workflow processing platform, establish multiple shared service centers according to regions or ownership enterprises, and each center provides accounting services for corresponding regions or ownership enterprises; the group company level mainly plays the leading role of guidance, coordination and balance.

2) Standardized construction, unified accounting standards, business standards, information standards. The most standardized, low value-added, non-core business, large business volume, high repeatability, automation and easy to effect business in accounting are included in the business scope of Financial Sharing Service Center by stages and batches.

3) Centralized deployment, unified information platform. Effectively solving the problem of information island can realize standardization and normalization in business, data, interface, analysis, user operation and many other aspects, provide the most basic support for the application of big data, and reduce the total cost of its construction and operation.

3 Core Systems and System Integration of Financial Shared Service Center Construction

As an innovative management mode, the essence of financial sharing service is the change of enterprise operation management and control mode driven by information technology. It needs safe, efficient, stable and advanced financial sharing system as support. The design of financial sharing core system and its integration with business system are directly related to the operation efficiency and effect of Financial Sharing Service Center.

(1) The specific core functions of financial sharing core system include the following contents:

1) Online reimbursement platform. It is a platform to collect and integrate the basic information of enterprise group, realize the business process tracking through the interconnection with the business system, and reduce the financial basic

workload such as the collection and sorting of original documents through the full reimbursement.

2) Business operation platform. On the basis of combing and integrating the business and financial systems, a platform supporting the integration of financial and business systems is established to realize the organic integration of financial process and business process.

3) Operation management platform. It is a platform for design, operation, evaluation and improvement of products or services of Financial Sharing Service Center.

4) Operation support platform. To manage the basic information for the business development and operation of Financial Sharing Service Center, and realize the integration of type definition and business customization.

5) Capital settlement platform. It provides a unified platform for fund settlement business under different financial sharing service center modes.

6) Electronic image system. In the financial sharing mode, the original document is replaced by the document image, and the document image is transferred along with the flow of electronic documents, which can be viewed by the approval leaders and financial sharing service personnel.

7) Invoice management system. Through the system integration docking financial system, contract management, tax information system, or through OCR (optical character recognition) and other intelligent acquisition mode, automatically obtain the tax system invoice information and push sales invoice information.

8) Mobile application system. Through the intelligent mobile terminal, it provides convenient mobile application services for staff reimbursement, leadership approval, image acquisition, reimbursement query, etc.

9) Financial analysis system. Using financial big data, improve the automation level of financial processing in the aspects of statement issuance, bank reconciliation, analysis report formation, financial analysis, financial early warning and prediction.

(2) Financial sharing system integration. The core system integrates with ERP system, capital settlement system, golden tax system, business travel system, contract system and electronic filing system through system integration.

1) ERP system. After the financial sharing service platform processes the accounting business, it automatically pushes the voucher to the ERP accounting system, and the ERP system pushes the system number of the accounting voucher back to the financial sharing service platform, establishes the association of the accounting voucher between the two systems, and forms the integrated process of industry and finance.

2) Fund settlement system. Payment instructions are integrated into the fund settlement core system through the finance enterprise direct link, and the payment status is written back to the financial sharing system; collection document information is integrated from the fund settlement core system to the sharing system through the finance enterprise direct link, and the collection claim information

is written back to the fund settlement core system; fund plan information is integrated into the financial sharing system through the finance enterprise direct link.

3) Golden tax system. Automatic synchronization of VAT input tax invoice information, authentication of VAT input tax invoice information, automatic integration of invoice information, integration of tax declaration information, etc.

4) Business travel system. In the financial sharing service platform, single sign on business travel system of enterprise group can realize the functions of train ticket, air ticket, hotel, taxi reservation, etc. the formed expenses are automatically summarized and apportioned on a monthly basis to form reimbursement forms, which are directly pushed to the financial sharing service platform for approval and payment, greatly reducing the processing capacity of travel documents and improving the processing efficiency.

5) Contract system. When auditing the contract payment, the contract management system pushes the basic information and accessories such as contract number, contract name, contract amount, paid amount, payment stage and contract supplier to the financial sharing service platform. After the approval of payment is completed, the financial sharing service platform automatically writes the payment information back to the contract management system to form the actual situation of contract performance and payment progress. It is a closed-loop system with time-dependent control.

6) Electronic accounting file system. Through the electronic management and utilization of the whole process of accounting archives, it provides information support for daily accounting archives collection, sorting, query, borrowing and destruction, and realizes the information management of the whole life cycle of accounting archives.

4 Management and Control Mode and Expected Effect of Financial Shared Service Center

(1) Financial Sharing Service Center Management and control mode. According to the principle of separation of management and operation and consistency of responsibilities and rights, after the Financial Sharing Service Center is established, the "Trinity" management and control mode of group headquarters, ownership enterprises and financial sharing service center can be established.

1) Financial management department of enterprise group: focus on top-level financial strategic management, be responsible for the formulation of the company's financial strategic planning and objectives, and be responsible for guiding the construction of ownership enterprises and Financial Sharing Service Center.

2) Financial Sharing Service Center: it focuses on standardized services and business support, and is responsible for providing standard, efficient and standardized accounting and statement issuing services for service enterprises in accordance with the unified accounting system and standards of enterprise groups and the service agreement signed with enterprises.

3) Ownership member units: focus on the front-end financial management of business, implement the business performance assigned by the enterprise group, and be responsible for the financial budget, prediction, analysis and risk control of the enterprise.

(2) Expected effect of Financial Sharing Service Center

1) Optimize and reengineer business process to improve risk control ability. By providing standardized financial services, the Financial Sharing Service Center has established a complete accounting process specification, formed a standardized, process oriented and systematic work process, and achieved the four unification of business process, accounting method, data caliber and business operation, so as to greatly improve the quality of financial work and business flow efficiency, as well as the timeliness and efficiency of financial business processing. The timeliness of business process was effectively improved.

2) Break the information island, realize interconnection and enhance service value. With the help of the Financial Sharing Service Center, the timeliness and accuracy of data collection and data mining are realized, the data standards are unified and the caliber is consistent, the "information island" no longer exists, and the "data fight" phenomenon is completely solved, so as to gradually build a "big data leading, big data driven" command decision-making and management model, and provide a large number of standards and data for big data analysis standardize and unify the same source data, reconstruct the enterprise wisdom with data, and move towards data sharing.

3) Management supports decision making and realizes financial digital transformation. The operation of Financial Sharing Service Center makes the basic accounting work concentrated in the hands of some accounting personnel, liberates a large number of financial personnel from the tedious accounting work, releases energy, promotes the transformation of financial team to business finance and management accounting, and focuses on budget management, fund management, performance management, tax planning, decision analysis and other financial management work, so as to make decisions for the operation. We need to provide stronger financial support.

5 Financial Sharing Service Center Construction Risk and Countermeasures

(1) Risk of change: Financial Sharing Service is a new financial management and control mode, which is bound to cause great adjustment to the existing concepts, systems, mechanisms and the responsibilities and rights of relevant organizations. Response: the main leaders of the enterprise group personally take the lead, set up the financial sharing service leading group, clarify the objectives and strategies, make overall planning and promote the implementation of financial sharing service, change the concept from top to bottom, unify the thinking; do a good job in reform training, popularize the concept, knowledge and value of sharing service; continue to

communicate, promote the company wide publicity, correct guidance and eliminate the problems Misunderstanding.

(2) Platform selection risk: the stability, scalability and ease of use of the core platform are the key to the smooth implementation of financial sharing services.

Response: under the premise of the existing ERP system of the enterprise group, it is necessary to fully refer to the success and failure cases in the industry, and select the core system of financial sharing service with high performance, excellent integration, easy expansion and flexible configuration, so as to effectively avoid the risks caused by the increase of user scale, business adjustment and ERP upgrade.

(3) System integration risk: the financial sharing service technology platform involves many supporting systems and back-end business systems, and the integration technology is complex. Whether the systems can be effectively integrated will have a great impact on the implementation effect of sharing.

Response: Based on the principle of standardizing business operation and reducing operation complexity, reasonably design the integration scheme, and do a good job in requirement analysis and architecture design. On the one hand, it combs and controls the personalized business and requirements of enterprises, on the other hand, it strengthens the support for differentiated business through the scalability of system architecture.

References

1. Wang, X.: Financial Sharing in Digital Transformation. Electronic Industry Press (2018)
2. Construction and operation innovation of Financial Sharing Center in management accounting practice of large coal enterprise group. http://czt.shandong.gov.cn/art/2019/3/21/art_17107_4939392.html. Accessed 21 Mar 2019
3. Chen, H., Guo, Y.: Financial digital infrastructure enabling enterprise transformation. Monthly Finan. Account. (13), 15–21 (2020)
4. Wang, N.: Analysis of data analysis under financial sharing mode. Mod. Commer. (36) (2018)
5. Xie, Y.: Research on enterprise accounting electronic archives management under financial sharing mode. Zhejiang Arch. (3), 62–63 (2019)

Analysis on Employment Situation of Applied Universities in Transformation and Development

Peng Li[1,2], Hongjia Yu[1,2], Haitao Chang[1,2], and You Tang[1,2(✉)]

[1] Electrical and Information Engineering College, Jilin Agricultural Science and Technology University, Jilin 132101, Jilin, China
[2] Smart Agricultural Engineering Research Center of Jilin Province, Jilin 132101, Jilin, China

Abstract. At present, the employment comprehensive index is an important parameter to measure the quality of running a university and cultivating talents, and it is the quantitative performance of the comprehensive quality of colleges and universities. With the development of society and the improvement of national comprehensive strength, the industries are booming, the talents that the society can accommodate and need are increasing year by year, and the employment breadth and refinement degree have changed significantly. This paper focuses on the analysis of the employment data of our university over the years, how to adapt to the needs of social development in the transformation and development of applied universities, puts forward the employment work model of accurate positioning, hierarchical promotion, ideological guidance and overall planning. Adhering to the guidance of teaching reform to meet the needs of the society and taking firm ideals and beliefs as the cornerstone, the quality of employment in the college has been achieved good results.

Keywords: Employment rate · Work innovation

1 Introduction

Comprehensive employment rate is one of the important indexes to measure the overall quality of colleges and universities [1]. In recent years, with China's comprehensive strength and economic situation has undergone profound changes [2]. On the one hand, the breadth of employment continues to widen, social demand surges bring new economic growth points, and new industries and new occupations emerge; on the other hand, employment units have more accurate positioning of talent demand [3]. The comprehensive literacy of talents is higher and higher, the purpose of school recruitment is stronger, and the degree of adaptation between talents and posts is unprecedented.

The main purpose of this study is to break the inherent employment work model, to combine employment work with ideological guidance, student management and student cultivation closely, to highlight the characteristics of schools and talents, to adjust the work model and direction, to guide employment work through the principles of accurate positioning, grading and overall planning, and to adapt to the needs of enterprises according to the students' comprehensive quality and ability.

© The Author(s), under exclusive license to Springer Nature Switzerland AG 2021
J. Abawajy et al. (Eds.): ATCI 2021, AISC 1398, pp. 87–92, 2021.
https://doi.org/10.1007/978-3-030-79200-8_12

2 Main Factors Affecting the Employment Rate of Colleges and Universities

2.1 The Increase of College Enrollment Leads to the Decrease of the Gold Content

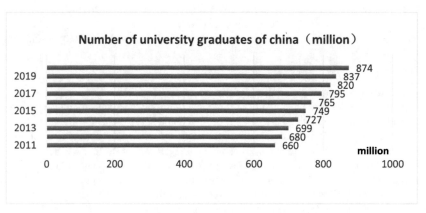

Fig. 1. Statistics of college graduates in China

Referring to the data of China Intelligence Network (Fig. 1). By 2020, the number of college graduates has reached 8.74 million, and 110 million undergraduate graduates have been transferred to various industries in the past two decades.

2.2 The Leading Role of University Education is Challenged by the Development of Internet

Students can study independently on the network, and the network has gradually become an important position knowledge expansion and self-improvement [4]. The advantages of hardware teaching environment in colleges and universities are challenged by simulation software and VR technology.

2.3 The Speed of Knowledge Renewal in Colleges and Universities Lags Behind the Development of Industry

In recent years, China's scientific and technological development has made brilliant achievements, China began to lead the international trend of science and technology [5]. It is difficult to avoid the cyclical lag of curriculum knowledge and technology renewal in ordinary colleges and universities, and the gap between the knowledge and skills acquired by students and the of work needs [6].

2.4 The Motive Force for Colleges and Universities to Seek Employment Growth Point is Not Strong

Compared with the former college students, the shortage of graduates has gradually changed into oversupply, and ordinary colleges and universities have gradually increased their attention to employment work, but the current job demand of applied college graduates can still basically cover the graduates of that year.

2.5 Higher-Educated People are More Attractive to Businesses than Students

For enterprises, the core competitiveness of enterprises mainly comes from the of science and technology innovation, the main recruitment target of core R & D talents comes from key domestic colleges and universities [7]. Most college students' professional knowledge reserve and comprehensive learning accomplishment cannot meet the professional requirements of R & D innovation.

2.6 The Orientation of Employment of College Students is Not Accurate

Under the new situation, the employment orientation of ordinary college students is not accurate, the willingness to seek employment is often higher than the actual situation of market supply and demand, and the position and content are difficult to achieve self-expectation. Popular education leads most college students to concentrate on manufacturing, production and management. Failure to recognize the current situation will cause a large psychological gap, resulting in passive employment.

3 Analysis of Employment Data of Jilin Agricultural Science and Technology Institute

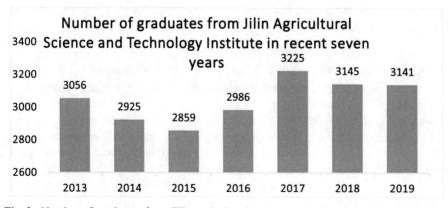

Fig. 2. Number of graduates from Jilin agricultural science and technology institute in recent seven years

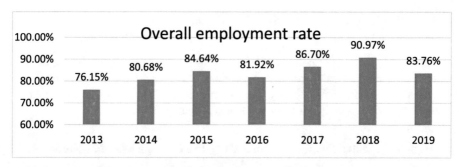

Fig. 3. Overall employment rate of Jilin agricultural science and technology institute

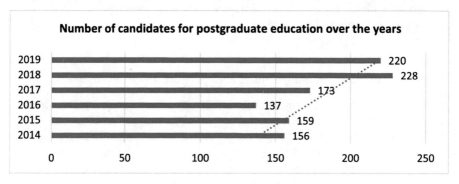

Fig. 4. Number of candidates for postgraduate education over the years

According to Fig. 2, 3 and 4, First of all, the number of graduates of our school continues to grow as a whole; grade 09 to grade 11, the number of graduates is decreasing due to the decrease in school enrollment, from grade 11 to grade 15, Steady growth of undergraduate graduates. Our school's overall employment rate has risen steadily in the past six years, Over 80 per cent for grade 10 to 15, Breaks 90 per in grade 14.Reflecting the beneficial combination of theory and practice of applied universities.

4   Exploration of the Work Model Which is Favorable to the Comprehensive Quality Improvement of Employment

Through the analysis of the employment rate of the school and college in the past years, combined with the experience summed up in the actual work, we put forward the employment work model of accurate positioning, hierarchical promotion, ideological guidance and overall planning, and put forward the educational goal of accurate teaching, based on demand and Bacon's soul, and strive for the participation of all staff and joint efforts to ensure it (Fig. 5).

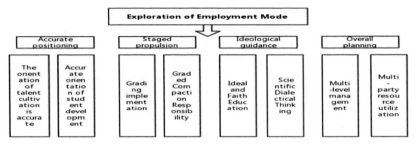

Fig. 5. Exploration of employment mode

4.1 Accurate Positioning

The orientation of schools and students should be accurate [8]. Only by having a clear orientation on the school running concept and training goal can we grasp the employment work and student management work macroscopically and as a whole.

Compared with research universities, the students' practical operation ability and theoretical application ability of applied universities have advantages [9]. Compared with higher vocational colleges, the students of applied universities have a solid theoretical foundation and can explain the mechanism and causes in essence.

4.2 Graded Advance

College should make clear the leadership responsibility, take the college leader as the first responsible person to [10], make the college form the work mode of leading, working group cooperation and full participation of the college, so that the employment work can be promoted at different levels and carried out in an orderly manner.

4.3 Ideological Guidance

In the daily management, taking the great rejuvenation of the Chinese nation as the program, the students are guided by ideology and politics, so that the students are determined to devote themselves to the construction of the motherland and produce the endogenous motive force of active employment and high quality employment.

4.4 Comprehensive Plan

Employment work is the end point and key point of the two levels of work in the school and college [11] and it is the concentrated embodiment of the comprehensive strength of the college; the employment work is meticulous and complex, which requires a large number of staff members to work together and cooperate together.

5 Summary and Outlook

In this paper, we analyze the employment data of schools and colleges over the years, put forward the main factors restricting the promotion of employment rate, combined with the actual situation, summed up the important factors conducive to the improvement of employment quality. Finally, the innovative working mode of applied university employment is put forward to solve the problem.

Acknowledgements. This work was supported by Cross-discipline Cultivation of Digital Agriculture Project of Jilin Agricultural Science and Technology University (JLASTU Contract Number [2020] No. XSNY019).

References

1. Cao, Q., Zhong, M.: A study on construction of employment service systems for university students in new era. In: Conference Proceedings of the 8th International Symposium on Project Management, China (ISPM2020) (2020)
2. Li, J., Zhang, R.: An analysis of the employment difficulties of college students based on the theory of labor market segmentation. In: Proceedings of 4th International Conference on Modern Management, Education Technology and Social Science (MMETSS 2019). Advances in Social Science, Education and Humanities Research, vol. 351 (2019)
3. Wang, C., Shen, Z.: Information asymmetry in college students' employment market. In: Proceedings of 2019 3rd International Conference on Economics, Management Engineering and Education Technology (ICEMEET 2019) (2019)
4. Zhang, L.: Study on employment intention and influencing factors of Xinjiang graduates studying outside Xinjiang. In: Proceedings of 2019 3rd International Conference on Education Technology and Economic Management (ICETEM 2019) (2019)
5. Mo, C.: Analysis on the employment of college graduates under the new economic normal. In: Proceedings of 1st International Conference on Business, Economics, Management Science (BEMS 2019) (2019)
6. Yue, C., Bai, Y.: An empirical study on the employment status of college graduates in 2017. J. East China Normal Univ. (Educ. Sci. Ed.) (2018)
7. Wang, P., Chang, L.: Research on graduate employment—tracking survey of 2005–2009 graduates based on human resource management major of Zhejiang sci-tech university. In: Proceedings of 2019 International Conference on Management, Education Technology and Economics (ICMETE 2019) (2019)
8. Li, W.: Influence of father capital on the university students' employment quality. In: Proceedings of 2018 5th ICMIBI International Conference on Training, Education, and Management (ICMIBI-TEM 2018) (2018)
9. Jing, L.: Research of precision employment service of Chinese college graduates in the big data era. In: Proceedings of 2019 International Conference on Management, Finance and Social Sciences Research (MFSSR 2019) (2019)
10. Li, J.: A study on problems in employment service system for university students and relevant countermeasures. Academic Forum (2017)
11. Li, Y., Yin, F.: Research on gender difference in college students' cognition on their employment ability investigation based on national college graduates. In: Proceedings of 2018 International Conference on Economy, Management and Entrepreneurship (ICOEME 2018). Advances in Economics, Business and Management Research, vol. 70 (2018)

Corpus-Based Comparative Study on Chinese and English Subjectivity

Weiyan Ma[⊠]

Department of Literature, Bohai University, Jinzhou, Liaoning, China
maweiyan@qymail.bhu.edu.cn

Abstract. Subjectivity is widely considered as an important element in modern syntactic analysis, especially for medieval Chinese which focuses on spoken language. This study aims to investigate the use of subjectivity in one of Chinese classics A Dream of Red Mansions written by Chinese writer (CX), translated by British sinologists Hawkes & Minford and Chinese translator Yang Xianyi with British translator Dai Naidie (HD) through a corpus-based study. The study examines the typical Chinese subjective adverb "fan" in CX and its equivalent "instead" in HD through comparative way to uncover their difference, frequency and function. A total of 1039 word tokens obtained from this Chinese-English Parallel Corpus would be analyzed The findings showed that there was significant difference in the use of "fan" and "instead". Meanwhile, this study suggested that Chinese subjective adverbs not belonging to the periphery outside the core clause can express subjective meanings, which is different from the subjective using of English subjective adverbs. Therefore, translators used different expressions to translate these subjective forms.

Keywords: Corpus-based study · Subjectivity · Chinese and English parallel corpus · Statistical analysis

1 Introduction

Subjectivity is becoming more and more important in semantic and pragmatic analysis of languages. Benveniste (1971) believed subjectivity should be determined by the linguistic status of "person", meanwhile, "language is marked so deeply by the expression of subjectivity that one might ask if it could still function and be called language if it were constructed otherwise" [1]. We can see the important role of subjectivity in language expression and analysis. Traugott (1989) claimed that in the process of semantic change, "meanings tend to become increasingly situated in the speaker's subjective belief, state or attitude toward the proposition" [2]. Nowadays, linguists have given much focus on the the speakers' point of view. Subjectivity is an ambient synchronic state coding the speaker's beliefs and stance towards what is said [3, 4]. It can be seen that subjectivity exits with languages. However, knowing the scopes of subjectivity is still a problem to be solved in language research especially comparative study of different languages.

J. Abawajy et al. (Eds.): ATCI 2021, AISC 1398, pp. 93–98, 2021.
https://doi.org/10.1007/978-3-030-79200-8_13

1.1 Problem and Research Question

In language it's difficult to classify the meanings of the subjective forms. For Chinese subjective expression "fan", experts proved it has strong subjectivity when it's used as adverbs. The adverb "fan" usually means counter-expectation of speaker. It seems it has only one meaning and one sentence position which is in front of verbs. "Instead" can be the equivalent form of "fan", then how much frequency the two forms will match with each other is what we want to explore. Also in what circumstances will the translators use "instead" to translate Chinese "fan"? Will the translation forms express subjective meanings as the original word? For the collected data, we use statistical instrument JASP to make a comparison of the two subjective expressions for the parallel corpus [5]. Through analyzing of difference and frequency, we deeply explored the functions of Chinese and English subjective adverbs. We hope to inspire the later Chinese language and English language specification research, and provide a reference for historical study of the two languages.

1.2 Review of Literature

Yan & Xiao (2016) has compared the four similar subjective adverbs including "fan", "dao", "fandao", "fan'er", which showed that "fan" is usually used in traditional Chinese to express counter-expectation [6]. From the frequencies of the four expressions in modern Chinese and in spoken circumstances, "fan" is usually replaced by "fan'er" or "fandao". Zou (2010) tried to analyze the different contexts of the adverb "fan'er", which would be helpful for foreign students to know in what circumstance the subjective adverb "fan'er" can be used [7]. This gives enlightenment to analyze grammar when considering the elements of context. Besides, comparative study has been used in the analysis of two Chinese subjective adverbs "shenzhi" and "fan'er" [8].

The adverbial expression "instead", which has flexible positions, can be used as the equivalent of Chinese "fan". However, "fan" and "instead" have some differences, which make the two cannot be translate to each other all the time. Loghmani, Ghonsooly & Ghazanfari (2020) proved that the adverb "instead of" can be used as subjective expression, which here conveys counter-expectation [9]. Some linguists believe that some pragmatic markers of different languages have subjective and inter-subjective functions at the left and right periphery [10–12], and Chinese "fan" can be used with denial forms "bushuo" or "bushi" together either to express unexpected views or to emphasize the focus.

This corpus-based study was intended to explore the different use for Chinese and English subjective expressions. Meanwhile, the study would examine whether there will be difference between Chinese and English subjective expressions. We assume that subjective expressions are usually unique in a language, thus different languages can use different ways to express their subjective meanings.

2   Research Methods

2.1   The Corpus and Data

In order to provide fresh insights of the study of medieval Chinese language and find out the features used on subjective expressions between Chinese and English, corpus-based research has been adopted for this study.

The corpus we used is Chinese-English Parallel Corpus of A Dream of Red Mansions, which was written by Chinese writer CaoXueqin (CX), translated by British sinologists Hawkes & Minford and Chinese translator Yang Xianyi & British translator Dai Naidie (HD). We search the Chinese subjective adverb "fan" from CX and its nearest English equivalent subjective adverb "instead" from HD in the parallel corpus. A total of 1039 word token (includes 547 tokens of "fan" and 492 tokens of "instead") obtained from this Chinese-English Parallel Corpus will be analyzed in this study.

2.2   Data Collection and Analysis

We use "fan" and "instead" as searching words in the parallel corpus. For "fan", we only choose the adverbial usage and removed the adjective and other usages manually. This Chinese novel in the parallel corpus was translated into two versions. We analyze both of the two versions. All the adverbial usage of "fan" we kept in Chinese has subjective meaning. What we want to collect is how much percentage it will keep when translated into English. And according to its translation, subjective meanings were divided into 5 groups, which are unexpected meaning, emphatic meaning, concessive meaning, denial meaning, and the meaning which combined two or three of the former meanings. Then the other usages including not translating or objective translating would be put into non-subjective meaning group. Then we use the opposite way to collect data. We would like to see when translators would choose the subjective word "instead" to translate Chinese expressions. The same five groups of "fan" were used. And we get the distribution of "fan" and "instead" from the corpus. See the following table (Table 1).

Table 1. Distribution of "fan" and "instead"

Adverbs	Subjective in translation					Non-sub
	Unexpected	Emphatic	Concessive	Denial	Combined	Other
Fan	51	80	52	7	22	335
Instead	106	56	6	112	30	183

2.3   Results

We use the statistical tool JASP 0.9.2.0 to do chi-squared test for the data collected. Altogether there were 547 tokens of "fan" and 492 tokens of "instead" analyzed in this study. For the 547 tokens of "fan", 212 tokens have been translated into English with subjective meanings, and 335 tokens haven't been translated into English with subjective meanings. For the 492 tokens of "instead", 309 tokens were used to translate Chinese subjective meanings, and 183 tokens haven't been used to translate Chinese subjective expressions. For the unexpected meaning "instead" is almost twice more than "fan", and for the denial meaning, "instead" is sixteen times more than "fan". While, for the concessive and emphatic meanings, "fan" has been used more than "instead". The nearest one in subjective expression is the combined forms, which is 22 tokens of "fan" and 30 tokens of "instead". The contingency table of distribution shows that chi-squared test can be used to determine whether there' difference in the using of the two subjective expressions in Chinese and English. Contingency table of "fan" and "instead" can be seen in the following table (Table 2).

Table 2. Contingency table of "fan" and "instead"

Expression		Subjective	Non-sub	Total
Fan	Count	212.0	335.0	547.0
	Expected count	274.3	272.7	547.0
	% within row	38.8%	61.2%	100.0%
Instead	Count	309.0	183.0	492.0
	Expected count	246.7	245.3	492.0
	% within row	62.8%	37.2%	100.0%
Total	Count	521.0	518.0	1039.0
	Expected count	521.0	518.0	1039.0
	% within row	50.1%	49.9%	100.0%

A chi-squared test was analyzed to determine if there was statistical difference in the translation of the two subjective expressions between Chinese and English. In the contingency table, we can see that the least expected count of the two "fan" and the two "instead" is 245.3, which is > 5, and the total tokens we used is $N = 1039 > 40$, thus we can just choose the result of chi-squared test. From Table 3, we can see that $X^2 = 59.92$, $p < .001$. It was indicated that there's significant difference between the subjective and non-subjective using of "fan" and "instead".

Table 3. Chi-squared test

	Value	df	p
X^2	59.92	1	$< .001$
N	1039		

In this chi-squared test, p < .001 which is smaller than .05, thus the result shows clearly that the difference is significantly different for the translation of Chinese subjective expression "fan" and English subjective expression "instead".

3 Discussion

We use the parallel corpus to get the basic data of subjective expressions in Chinese and English, and analyze whether there is significant difference for them. Then we try to find out the differences in translation and the specific meanings of subjective expressions.

The adverbial usage of "fan" in Chinese is typical subjective expression, while translators only make 38.8% of "fan" be translated with subjective meanings, which includes unexpected, emphatic, concessive, negative and the combined ones with unexpected plus emphatic, negative plus unexpected and so on. However, 61.2% of "fan" haven't been translated with subjective meaning, most of them were removed in English expressions. For example, "fan" in this Chinese sentence "dayu linde shuiji shide, ta fan gaosu bieren 'xiayule, kuai biyu qu ba'" in CX was translated into "he got drenched himself in the rain and advised someone else to take shelter" in HD. In Chinese sentence, "fan" is used to express the behavior of Jia Baoyu is out of people's expectation, while in the translation, there's no subjective meaning. We traced reasons for this phenomenon, and we found that on one hand subjective "fan" has no literal meaning, it only has emotional meaning. That means if we remove "fan" from the sentence, it will not change the literal meaning of it. On the other hand, English adverbs used not in front of sentences will be hard to be seen as subjective expressions. This also proved Traugott's analysis for adverbs used as VAdv has less subjectivity than IPAdv or DM in English [13].

The English subjective "instead" can be used at the beginning, in the middle or at the end of a sentence. No matter in what position, it can be used to translate the subjective expressions of Chinese. For example, "instead" in this sentence "why bring this filthy creature here instead to pollute this domain of immaculate maidens?" is adopted to translate "fan". In the sentence "now instead of showing yourselves properly grateful, you try to lord it over me", "instead" is used at the beginning to mean "not..., on the contrary..." Subjective "instead" can also be found in the following sentence translating Chinese subjective adverb "dao", "I told you to go and hurry them with the water, but you said you were busy and made us go instead." The flexible position of "instead" can also prove the different expressions in different languages.

In the study, we found that both Chinese and English subjective expressions have different ways to translate in different contexts. It's difficult to find complete equivalents for these expressions, which shows different languages choose various ways to express emotional meanings. Meanwhile, this research helps people to analyze the scopes of subjective meanings through a quantitative method.

4 Conclusion

To sum up, the results of this study revealed that Chinese subjective adverbs tend to be omitted when translated into English, while English subjective words tend to have subjective meanings when used to translate Chinese. It suggested that subjective expressions have strong feelings in a language may be difficult to be translated faithfully to another language, which will make the emotional meaning different with the original meaning, which should be given much attention in future.

Yet it can be seen from the study that the data collected was focused on two Chinese and English subjective expressions, which is small-sampled. In the future study, larger-sampled research for other typical subjective expressions from different languages and covering different periods would be made.

Acknowledgments. This work was supported by Liaoning Social Science Planning Foundation Project (L19BYY010).

References

1. Benveniste, E.: Subjectivity in language. In: Meek, E. (ed.) Problems in General Linguistics, pp. 223–230. University of Miami Press (1971)
2. Traugott, E.C.: On the rise of epistemic meanings in English: an example of subjectification in semantic change. Language **65**, 31–55 (1989)
3. Lyons, J.: Deixis and subjectivity: loquor, ergo sum? In: Jarvella, R.J., Klein, W. (eds.) Speech, Place, and Action: Studies in Deixis and Related Topics, pp. 101–124. Wiley, New York (1982)
4. Traugott, E.C.: Revisiting subjectification and intersubjectification. In: Davidse, K., Vandelanotte, L., Cuyckens, H. (eds.) Subjectification, Intersubjectification and Grammaticalization, pp. 29–71. De Gruyter Mouton, Berlin (2010)
5. Ren, L., Sun, H., Yang, J.: Chinese-English parallel corpus of a dream of red mansions (2010). http://corpus.usx.edu.cn/. Accessed 15 Dec 2020
6. Yan, M., Xiao, X.Q.: A contrastive study on four synonyms: "fan", "dao", "fandao", and "fan'er." Hanyu Xuebao **4**, 50–58 (2016). (in Chinese)
7. Zhecheng, Z.: On Fan'er. Studies in Language and Linguistics **4**, 59–61 (2010)
8. Yulin, Y.: Counter-expectation, additive relation and the types of pragmatic scale: the comparative analyses of the semantic function of "shenzhi" and "fan'er." Contemp. Linguist. **2**, 109–121 (2008)
9. Zahra, L., Behzad, G., Mohammad, G.: Engagement in doctoral dissertation discussion sections written by English native speakers. J. Engl. Acad. Purp. **45**, 1–20 (2020)
10. Chen, J.: (Inter) subjectification at the left and right periphery: deriving Chinese pragmatic marker bushi from the negative copula. Lang. Sci. **66**, 83–102 (2018)
11. Shansheng, T.: "Bushuo" used as an adverb and a conjunction. TCSOL Stud. **3**, 13–18 (2019). (in Chinese)
12. Traugott, E.C.: Investigating, "periphery" from a functionalist perspective. Linguist. Vanguard **1**, 119–130 (2015)
13. Traugott, E.C.: The rhetoric of counter-expectation in semantic change: a study in subjectification. In: Blank, A., Koch, P. (eds.) Historical Semantics and Cognition, pp. 177–196. Mouton de Gruyter, Berlin (1999)

Analysis on the Development of Regional Innovation Ability Driven by Internet Under the Background of "Internet Plus"

Rongting Long[✉]

School of Economics, Shanghai University, Shanghai, China

Abstract. This paper makes a preliminary analysis on the Internet-driven regional innovation development in China under the background of China's "internet plus" from the background introduction, theoretical analysis and current situation analysis, and finds that the development of Internet can affect regional innovation capability through cost effect, competition effect and spillover effect. Through the analysis of China's reality, it is found that China's Internet is developing rapidly, and under the planning of "internet plus", China uses the Internet as a favorable medium and tool to achieve regional innovation-driven development and promote regional industrial transformation and upgrading. However, there is spatial heterogeneity between the development level of the Internet and the distribution of innovative regional capabilities, and there is a correlation between them. The coordination mechanism shows that the development of Internet not only has local positive effects, but also has positive spillover effects. Based on the above analysis and conclusion, this paper draws the corresponding enlightenment.

Keywords: Internet plus · Regional innovation ability · Regional development difference · New engine

1 Introduction

In the trend of the world's "fourth industrial revolution", the development of the Internet has played an important role, and its characteristics such as connectivity and immediacy have greatly improved the speed of information flow, making the contact and communication between the main bodies in various regions more frequent. As an efficient information transmission carrier and communication medium, the convenience and advantages brought by the Internet are enormous. Therefore, related issues in this field have been widely concerned by the practical and academic circles. Nowadays, the Internet has been closely integrated with many fields of modern economy and social activities, and has a strategic and important impact on the innovation-driven development of China's economy [20]. The State Council of China has clearly stated that it is necessary to give full play to the innovative driving role of the Internet, so as to make "internet plus" an important driving force for social and economic innovation and development. Undoubtedly, the Internet will play an increasingly important role in the construction of an innovative

J. Abawajy et al. (Eds.): ATCI 2021, AISC 1398, pp. 99–105, 2021.
https://doi.org/10.1007/978-3-030-79200-8_14

country and the evolution and development of regional innovation systems. According to relevant data, the number of Chinese netizens increased from 137 million in 2006 to 854 million in 2019, and the Internet penetration rate soared from 10.43% in 2006 to 62.2% in 2019. It can be seen that China's Internet is developing very rapidly, which has paved the technical way for promoting regional innovation and shortening the innovation gap between regions [17]. Then, what is worth paying attention to is, what is the impact of the rapid development of the Internet on the innovation of regional innovation subjects in China? What effects may exist? Is there a spatial correlation between such influences? The answers to these questions help us to understand more clearly the mechanism of Internet in promoting regional innovation and its spatial spillover.

2   Theoretical Analysis

In the past literature, the related issues of regional innovation were mostly analyzed from the aspects of human capital, government support, financial development and international technology spillover [3, 13, 16], but there are few direct studies on the Internet and regional innovation, and more literatures focus on the relationship between the Internet and economic growth [5, 14, 15], there are also literatures that discuss the relationship between the Internet and enterprise innovation from a theoretical perspective [1, 9]. The above research has laid a theoretical and empirical foundation for revealing the role of the Internet, but the research on the influence of the Internet on regional innovation and its spillover is still relatively lacking [18], so it is necessary to analyze and discuss it theoretically and empirically.

2.1   Cost Effect

The Internet has the characteristic of "being able to spread information at a very low cost", which can almost lower the distance cost that innovation subjects in different regions need to pay for innovation activities to zero. And the immediacy of using this type of technical medium to obtain information can greatly improve the efficiency of all aspects [2]. This can not only greatly reduce the R&D innovation cost of innovative subjects, and increase the innovation output under the constraint of existing resources, but also contribute to the exchange and development of innovative technologies in different regions, reduce the cost of cooperative R&D process of innovative subjects in different regions, improve the comprehensive performance of collaborative innovation and promote the evolution and development of the whole innovation system [4]. Moreover, with the increase of innovation output, it can expand the market scale of its innovative products, drive further innovation in the production field and stimulate the consumption demand of innovative products. When the market scale increases to a certain extent, it can also produce economies of scale and specialized division of labor [6, 7], which will further reduce the production cost to obtain more excess profits, encourage them to expand innovation fields, and then improve the overall innovation ability of the innovation subject, and realize the regional innovation-driven development to a higher level.

2.2 Competitive Effect

With the development of the Internet, consumers have diversified channels for obtaining information. Consumers can quickly and easily obtain market information and market prices of products by using the Internet, and can purchase goods across regions, which impacts local protectionism in some regions to a certain extent. The more information consumers get, the weaker enterprises are in market transactions, which will reduce the excess profits of enterprises, make the competition among enterprises in the market more intense, and force enterprises to innovate in technology, management and service [8, 10]. At the same time, the lowering of China's market entry barriers, the development of financial platforms and the government's policy guidance also make more enterprises participate in competition, enhance market vitality and intensify market competition. In this process, there will be the effect of escaping from competition among enterprises, which will become the internal driving force to stimulate enterprise innovation, so as to obtain "irreplaceable" in the market and stabilize their market share position. In this process, market competition has effectively promoted technological innovation, which is conducive to the rational and optimal allocation of social resources. And the Internet has played an important role of adding fuel to the flames.

2.3 Spillover Effect

For scientific research institutions, the development of Internet can facilitate the communication and cooperation between scientific research institutions and personnel, strengthen the spillover effect of innovation activities, and increase the generation of innovative thinking and the organic integration of technology in this process, thus enhancing the regional collaborative innovation capability and the overall system innovation level [11, 19]. As far as enterprises are concerned, the wide application of the Internet promotes the contact and cooperation between enterprises in different regions, increases the opportunities and frequency for enterprises to exchange and learn different technologies, and helps to enhance the strength and scope of technology spillover between high-tech enterprises and from high-tech enterprises to low-tech enterprises, thus stimulating the production technology level of industries and industries and reducing the production costs in the industry. In addition, innovation participants can quickly and effectively acquire new knowledge and technology and realize the accumulation of human capital, which not only promotes the full exchange and diffusion of knowledge and technology among innovation participants, but also accelerates the process of upgrading human capital structure with the rapid accumulation of knowledge stock in innovation system [12], which will further strengthen the overflow of knowledge and technology, form a benign cycle promotion process, and finally accelerate the improvement of the overall innovation ability of innovation system, while the Internet has always played a role in this process.

3 Analysis of Existing Circumstance

After the above preliminary theoretical analysis of the impact of Internet development on regional subject innovation, this paper refocuses the perspective on China's region, trying to link China's Internet development with regional innovation-driven development, and analyze the current situation.

When Premier Li Keqiang attended the first world internet conference in November 2014, he pointed out that the Internet is a new tool for mass entrepreneurship and innovation. Among them, "mass entrepreneurship and innovation" is called the "new engine" for economic upgrading. At the Third Session of the 12th National People's Congress in 2015, Premier Li Keqiang put forward the "internet plus" action plan for the first time in his government work report, promoting the integration of internet technology with traditional industries and modern industries, promoting the healthy development of e-commerce, industrial Internet and Internet finance. "internet plus" represents a new economic development path. It relies on Internet information technology to realize the organic combination of Internet and traditional industries, and completes economic transformation and upgrading by optimizing production factors, updating service system and reconstructing business model. The purpose of the "internet plus" plan is to give full play to the advantages of the Internet, apply this advantage to traditional industries, promote economic productivity through industrial upgrading, and search for growth points under the new economic growth model, thus activating the vitality of the market economy and accelerate the transformation of economic development models.

According to the 45th Statistical Report on Internet Development in China, under the cooperation of technology and policy, China's "internet plus" accelerates the integration with industry, and the digital economy has become the new engine of China's regional development. It is manifested in its scale rising year by year. According to relevant reports, by the end of 2019, its scale has reached 31.3 trillion yuan, ranking first in the world. It can be seen that the new economic industries spawned by the Internet are extremely developing and constantly add vitality to China's economy. With the continuous innovation of Internet cutting-edge technology and the rapid transformation and upgrading of regional industries, the consumption potential has been further explored, which drives the continuous follow-up of the production field and makes more consumption demands satisfied. By March 2020, the number of online shopping users in China reached 701 million, and the transaction scale reached 1.063 billion yuan. Online consumption played a positive role in expanding domestic demand. At the same time, the development of Internet speeds up the flow of information and provides a good platform for technical exchange, study and cooperation. According to the relevant data of China Statistical Yearbook and China Information Yearbook, the average number of Internet pages in China increased from 274 million in 2007 to 9.606 billion in 2019. As far as regional distribution is concerned, there is spatial heterogeneity in China's Internet development level. The Internet development level in eastern coastal areas is generally higher than that in the central and western inland regions. The Internet development tends to gather in eastern coastal areas, and the eastern coastal areas tend to have more obvious spatial proximity and spatial interaction in Internet development level. From the industrial structure point of view, the intensive degree of high-tech industries in the eastern coastal areas is obviously higher than that in the central and western inland regions, which is closely related to the improvement of infrastructure such as Internet in the eastern coastal areas. In addition, according to the Evaluation Report of China's Regional Innovation Capability in 2020, Guangdong, Beijing, Jiangsu, Shanghai and

Zhejiang rank in the top five, and these provinces and cities are located in the eastern coast of China. It can be seen that there is a close relationship between the degree of Internet development in China and the level of regional innovation capability. High-tech talents are also inseparable from the development of the Internet. From previous theoretical research and empirical research, it can be seen that regions with developed information often attract more talents, which is also a major reason for the abundant human capital and vigorous innovation in the eastern coastal areas. Generally speaking, in the past 20 years, the gap of innovation ability in the eastern, central and western regions of China is almost in a solid state. The gap between the north and the south shows the evolution law of expansion-stability-expansion-stability, and the overall situation is expanding. This not only involves the existence of regional infrastructure level, economic development level and openness, but also reflects the regional gap in technology and hardware facilities. For example, from the long-term perspective of 20 years, especially in the past five years, the innovation ability of Beijing-Tianjin-Hebei region has basically not changed; In recent years, the Yangtze River Delta region has developed steadily, the gap of innovation ability in the region is steadily narrowing, and the degree of regional integration is gradually increasing; Three provinces in Northeast China have become the region with the most serious industrial isomorphism. It can be seen that there is still regional imbalance in China's regional synergy, and the problems are more prominent. The authorities should further promote the implementation and development of the "internet plus" plan, so that the region can develop new growth vitality of generate.

At present, under the background of innovation driving development, promoting industrial transformation and upgrading, shortening regional development gap and enhancing regional collaborative innovation capability, the proposal of "internet plus" conforms to the requirements of productivity development and social upgrade. The interconnectivity of the Internet makes the resources such as technology, information and capital flow more openly, and drives China's regional innovation and development in the main ways of low cost, competition promotion and technology spillovers. The new information technology revolution, represented by the Internet, big data and 5th generation mobile networks, is breaking the geographical limitations of regional innovation. Data has become a new factor of production, and innovation models, characteristics of innovation activities, innovation inputs and outputs are changing. In turn, this will also affect the formulation of government policies. In the next few decades, China's regional innovation and development will promote and influence each other with the development of Internet.

4 Conclusion and Enlightenment

In this paper, the background introduction, theoretical analysis and current situation analysis are carried out to make a preliminary analysis of the Internet-driven regional innovation and development under the background of China's "internet plus". The main conclusions are as follows:

First, China's Internet has developed rapidly, and under the plan of "internet plus", China has used the Internet as a favorable medium and tool to achieve regional innovation-driven development, promote regional industrial transformation and upgrading, reduce production costs and improve technical efficiency.

Second, the spatial distribution of Internet level in China shows obvious correlation and difference. The eastern coastal areas are Internet developed areas, while the central and western regions are relatively Internet underdeveloped areas. Correspondingly, the improvement of internet technology drives the development of regional innovation, and the level of innovation in the eastern coastal areas is higher than that in the central and western inland regions, with high innovation vitality and intensive technical links. Therefore, the development of the Internet is significantly related to the regional innovation ability, and influences the regional innovation through a series of effects.

Third, there are some related mechanisms and coordinated division of labor in China's regional innovation activities. Internet development has positive promotion effect and positive spatial spillover effect on regional innovation ability, which shows that China's Internet development can not only promote the innovation ability of the region, but also help to strengthen the spatial externalities of R&D innovation activities, thereby enhancing the level of regional collaborative innovation and narrowing the differences in innovation development between regions.

Based on the analysis and conclusions, this paper has a preliminary understanding of the current situation of China's Internet development. Simply put, while the Internet technology is advancing rapidly, there are also some problems and deficiencies in certain regions and fields. In order to exert its positive effects, the enlightenment can be listed as follows:

First, China should accelerate the construction of relevant infrastructure, especially in information technology fields, fill the shortcomings of information infrastructure in areas where Internet development lags behind, and fill the inter-regional "digital gap". Seize the opportunity of the "internet plus" era, vigorously develop high-tech information technology, and promote the deep cooperation between the Internet and traditional industries and regional innovation systems.

Second, government departments should strengthen the top-level design and policy planning of "internet plus", build an efficient collaborative innovation platform based on "internet plus", reshape the regional collaborative innovation model, pay attention to the training of high-tech talents, and increase relevant policy guidance.

Third, strengthen the construction of regional innovation system, eliminate local protectionism, expand regional openness, strengthen the systematic integration and optimal allocation of innovation resources in the process of R&D innovation, break the institutional barriers of regional collaborative innovation cooperation, and improve regional collaborative innovation ability.

References

1. Audretsch, D.B., Heger, D., Veith, T.: Infrastructure and entrepreneurship. Small Bus. Econ. **44**(2), 219–230 (2015)
2. Audretsch, D.B., Feldman, M.P.: R&D spillovers and the geography of innovation and production. Am. Econ. Rev. **86**(3), 630–640 (1996)
3. Junhong, B., Fuxin, J.: Collaborative innovation, spatial correlation and regional innovation performance. Econ. Res. **7**, 174–187 (2015). (in Chinese)
4. Cui, T., Ye, H.J., Teo, H., Li, J.Z.: information technology and open innovation. Inf. Manag. **52**(3), 348–358 (2015)

5. Czernich, N., Falck, T., Woessmann, L.: Broadband infrastructure and economic growth. Econ. J. **121**(552), 505–532 (2011)
6. Elhorst, J.P.: Spatial Econometrics: From Cross-sectional Data to Spatial Panels. Springer, Berlin (2014)
7. Fischer, M.M., Varga, A.: Spatial knowledge spillovers and university research: evidence from Austria. Ann. Reg. Sci. **37**(2), 303–322 (2003)
8. Freund, C., Weinhold, D.: The effect of the internet on international trade. J. Int. Econ. **62**(1), 171–189 (2004)
9. Guire, T.M., Manyika, J., Chui, M.: Why big data is the new competitive advantage. Ivey Bus. J. **7–8**, 1–13 (2012)
10. Glavas, C., Mathews, S.: How international entrepreneurship characteristics influence internet capabilities for the international business processes of the firm. Int. Bus. Rev. **23**(1), 228–245 (2014)
11. Xianfeng, H., Wenfei, S., Boxin, L.: can the internet become a new driving force for China's regional innovation efficiency? China Ind. Econ. **7**, 119–136 (2019). (in Chinese)
12. Hansen, R., Siew, K.S.: Hummel's digital transformation toward omnichannel retailing: key lessons learned. MIS Q. Executive **14**(2), 51–66 (2015)
13. Hsu, P., Tian, X., Xu, Y.: Financial development and innovation: cross-country evidence. J. Financ. Econ. **112**(1), 16–135 (2014)
14. Lin, F.: Estimating the effect of the internet on international trade. J. Int. Trade Econ. Dev. **62**(3), 171–189 (2014)
15. Mallick, H.: Role of technological infrastructures in exports: evidence from a cross-country analysis. Int. Rev. Appl. Econ. **28**(5), 669–694 (2014)
16. Seyoum, M., Renshui, W., Yang, L.: Technology spillovers from Chinese outward direct investment: the case of Ethiopia. China Econ. Rev. **33**, 35–49 (2015)
17. Bingzhan, S.: Internet and international trade: empirical analysis based on bilateral two-way website link data. Econ. Res. **5**, 172–187 (2016). (in Chinese)
18. Sydney, T.: Transnational politics: contention and institutions in international politics. Annu. Rev. Polit. Sci. **4**(1), 1–20 (2001)
19. Mingfeng, W., Juan, Q.: Analysis of the Inter-provincial differences and convergence of the growth of internet users in China. Geogr. Sci. **31**(1), 42–48 (2011). (in Chinese)
20. Zhen, Z.: "Internet plus" cross-border management: the perspective of creative destruction. China Ind. Econ. **10**, 146–160 (2015). (in Chinese)

Influence Mechanism of Mobile Social Network Users' Product Recommendation Information on Consumers' Intention to Participate Sharing Economy

Ying Dong[1] and Li Dong[2(✉)]

[1] School of Management, Changchun Institute of Technology, Changchun, Jilin, China
[2] School of Management, China Women's University, Beijing, China

Abstract. Under the Internet environment, mobile social network has become an effective platform to post product recommendation information. This paper aims to analyze how mobile social network users' product recommendation information influence consumers' intention to participate sharing economy and find its working mechanism. Based on the theory of planned behavior and source credibility theory, from the source level, it is put forward that the second-order variable source credibility, which is explained by intimacy, expertise, trustworthiness, attractiveness and past behavior, influences consumers' sharing economy participation intention via influencing consumer attitudes, subjective norm and perceive behavior control. At the same time, perceive behavior control and subjective norm influence attitude. This paper also put forward the management implication, hoping to provide certain reference for firms to carry out precise target marketing strategy.

Keywords: Mobile social network · Product recommendation information · Participation intention · Sharing economy

1 Introduction

Sharing economy, also known as cooperative economy or collaborative consumption, is an economic model in which idle resources such as space, skills and goods are shared for monetary or non-monetary benefits [1]. Under the guidance of green and sustainable development policies, China entered the fourth consumption stage, forming a transition from improving resource productivity to controlling the total amount of material consumption, thus emerge sharing travel and sharing housing. So far, sharing economy has developed rapidly in China. Sharing economy firms hope to promote consumers' participation in sharing economy and enhance the engagement of collaborative consumption through effective product recommendation information dissemination of mobile social network users, so as to realize the strategy of sustainable economic development.

Under the Internet environment, mobile social network has become a new element of the marketing mix. Recommendations of enterprise products (and services) published by mobile social network users are the main factors influencing consumers' purchase decisions. Research shows that product recommendation information published by network users can directly affect users' attitudes towards products and purchasing behaviors [2], and is the core element for increased sales revenue [3] and enhance business value [4]. However, current research mainly focus their attention on studying consumer's reaction to new product, seldom studies how consumers behave facing sharing economy circumstance. Sharing economy firms lack theoretical guidance in formulating and carrying out the recommendation strategy through mobile social network, which does not achieve the desired effect and result in ineffective marketing.

Therefore, it is urgent to study the influence mechanism of product recommendation information of mobile social network users on consumers' willingness to participate in sharing economy and find out its internal rules. Based on the theory of planned behavior and source credibility theory, this paper will put forward an influence mechanism model and the corresponding marketing strategy, which will has important theoretical and practical significance for firms to formulate scientific and accurate marketing strategies.

2 Theoretical Bases

2.1 Theory of Planned Behavior

The theory of planned behavior (TPB) explains the general decision-making process of individual behavior from the perspective of information processing on the basis of expected value theory. According to TPB, all factors that may affect behavior indirectly will influence behavior performance through behavior intention, which is influenced by three related factors, namely attitude, subjective norm and perceived behavior control [5].

Generally speaking, the more positive an individual's attitude towards a certain behavior is, the stronger his behavioral intention will be. The more positive the subjective norm for a certain behavior is, the stronger the individual's behavioral intention will be. The more positive the attitude and subjective norm are and the stronger the perceived behavior control is, the stronger the individual's behavioral intention will be. TPB can effectively reflect a person's possibility to perform a certain behavior or willingness to engage in this behavior.

2.2 Source Credibility Theory

Source credibility theory explains how the persuasiveness of information is affected by the perceived credibility of the source, information sender. Credibility is a multidimensional concept that affects consumers' attitudes, intentions and behaviors. Ohanian proposed a source credibility model [6], pointing out that three factors lead to source credibility: Perceived Expertise, Trustworthiness, and Attractiveness, which has been widely accepted and used in related fields.

3   Influencing Mechanism Analysis

3.1   Influencing Mechanism of TPB Factors

(1) Attitude and Consumers' Willingness to Participate in the Sharing Economy
   The study of Ajzen proved that there is a significant positive correlation between an individual's attitude towards a specific behavior and its intention to perform this behavior [5]. The study of Chen verified Ajzen's findings [7]. Therefore, the more mobile social networks users agree with the product recommendation information posted by their contacts, the more willingly they will be to participate in the sharing economy and the more inclined they are to buy the products recommended.

(2) Subjective Norm and Consumers' Willingness to Participate in the Sharing Economy
   The approving attitude of important others to purchase behavior would greatly enhance consumers' purchase intention [7]. The subjective norm will affect consumers' response intention to product recommendation information posted via mobile social networks: that is, the stronger the influence of subjective norm, the greater the possibility of consumers' willingness to participate in the sharing economy, and the greater their purchase intention to the recommended products.

(3) Subjective Norm and Attitudes
   Subjective norm affect consumers' attitude towards organic food [8] and consumers' brand attitudes [9]. Therefore, the more the important others agree with the product recommendation information, the more positive the consumer's attitude will be. Subjective norm will affect the consumer's attitude towards the product recommendation information.

(4) Perceived Behavioral Control and Consumers' Willingness and Attitude to Participate in the Sharing Economy
   Perceived behavioral control, i.e., the degree of difficulty perceived by an individual to perform a certain behavior, has an important influence on his behavioral intention. In 1991, Ajzen found that the stronger an individual's perceptual behavior control, the stronger his/her intention to behave [5]. Therefore, the perceived behavior control of mobile social network users affects their response intention to information. In other words, when users believe that they have high perceived behavior control ability, that they have time, and that they are skilled in using mobile social network, they will be more willing to participate in the sharing economy. Moreover, the more willingly they are to form a more favorable attitude towards the product recommendation information posted.

3.2   Influence Mechanism of Source Credibility Factors

Source credibility persuades the receiver through the internalization process. The higher the credibility of the source, the more trust the audience has on the information provided, and the easier it is to form a positive attitude, leading to acceptance and persuasion. On the basis of Ohanian's model, this research divides source credibility into five dimensions: intimacy, expertise, trustworthiness, attractiveness and past behavior. The five dimensions of credibility affect the consumer's perceived trust degree to the information poster.

(1) Intimacy

Strong tie relationships in interpersonal relationships significantly positively affected the effect of word-of-mouth communication [10]. The research on the forwarding of mobile coupons found that recommendations from friends are easier to be accepted by consumers than direct marketing from merchants [11]. Users are more willing to trust product recommendation information posted by close friends and form a positive attitude.

In addition, Asch. B. E. (1956) revealed the social conformity brought about by oral information transmission and pointed out the normative influence of information [12]. According to the research on contextual word-of-mouth in virtual communities, individuals are more inclined to get social support and emotional consolation from others with strong ties, and are more willingly to identify the views of the strong tie group and want to belong to it, thus forming the normative influence of the group with strong ties. Therefore, consumers feel the pressure of the social environment and adopt the consistent behavior under the influence of subjective norm.

(2) Expertise

The product information transmitted by highly specialized information sources has a more significant effect. Consumers are more willing to trust online product review information published by those with certain product experience, and believe that their reviews are more valuable [13]. In addition, when information sources with expert characteristics provide reliable information, they will exert strong influence on other members of the group, thus generating social normative influence. Expertise positively affects the subjective norm of brand switching of negative information receiver [9].

(3) Trustworthiness

Trust means that behaviors and results can be relied on, which contributes to the formation of positive attitudes towards product recommendation information by consumers and important others. Trust affects online shopping intention by influencing consumers' attitudes towards online shopping [12]. It is found that trust affects the intention of knowledge sharing behavior by influencing the subjective norm of knowledge sharing behavior [14].

(4) Attractiveness

Consumers are more likely to trust product advertisements endorsed by attractive celebrities, so they are more likely to form positive attitudes towards products and purchase intentions [15]. If the consumer appreciates the taste of product recommendation information posters, and considers their shopping style to be attractive, users will form a positive attitude towards the product recommendation information and those who have an important influence on them will also support their comments.

(5) Past Behavior

Individual's past behavioral experience plays an important role in predicting their behavioral intentions and behaviors. Matook S. et al. pointed out in their research that users' previous interaction experiences with contacts on online social networking sites significantly influence users' trust attitude towards product recommendation information, and then influence users' behavioral intention [16]. If the product recommendation information posted in the past is satisfactory and pleasant, then consumers will form a positive attitude towards the current information, and others who are important to consumers will also approve consumers' behavior intention.

Based on the theory and current researches, this paper believes that the second-order variable source credibility, which is explained by five dimensions of intimacy, expertise, trustworthiness, attractiveness and past behavior, is correlated with consumers' attitude towards product recommendation information and subjective norm. When consumers believe that those who post or disseminate product recommendation information have a high degree of trustworthiness, they will form a positive and favorable attitude towards the information, and at the same time, affect the role of subjective norm.

3.3 The Theoretical Model

Based on the comprehensive application of TPB and source credibility model, this paper analyzes the influence mechanism of mobile social users' product recommendation information on consumers' participation intention of sharing economy from the perspective of information source. The theoretical model is shown in Fig. 1.

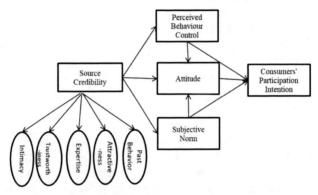

Fig. 1. The Research Model: The Influence Mechanism of Mobile Social Network Users' Product Recommendation Information on Consumers' Intention to Participate Sharing Economy

4   Management Implication

4.1   Improve Consumers' Perception of Source Credibility

(1) When designing marketing strategies, firms must release consumer-oriented product recommendation information through the analysis of consumers' preferences and with a responsible attitude. Firms must pay attention not to distribute excessive incentive to make consumers' excessive dissemination. Avoid information bombardment and discard fatigue marketing. Only by choosing a scientific and responsible information communication strategy from the perspective of consumers can firms win consumers' emotional trust and thus gain their support.

(2) Formulate strategies to improve consumers' perception of the expertise and attractiveness of product recommendation information disseminators. In terms of expertise improvement, firms can select opinion leaders from existing loyal consumers, provide them with new product experience, and train them, so that they have professional technical knowledge, and their consumption taste can be more appreciated. In terms of attraction enhancement, firms should pay attention to timeliness and novelty of information transmission. Firms not only need to pay attention to the "quantity", but the "quality" of product recommendation information transmission. Attract users' attention through innovative and unique marketing approaches, and win consumers' recognition and preference for products through professional and high-quality information content.

(3) Firms should pay attention to the role of intimate relationship and formulate strategies to guide consumers to recommend products through mobile social networks. Firms can give specific policies or rights to relevant consumers, allowing them to use Internet channels to carry out certain communication activities in their "social circles". For example, firms can set up a website navigation of "recommend the products you are satisfied with to your close friends" to guide consumers' recommendation behavior, carry out precise marketing, and promote existing and potential consumers to buy the recommended products, forming a recommendation multiplication effect.

At the same time, firms can establish close enterprise personnel - consumer relationship with existing and potential customers to strengthen consumers' emotional trust on the enterprise, to provide product experience and deliver product value, to form consumers' identity to firms and their products, take the initiative to disseminate the products information for the enterprise.

4.2 Improve Consumers' Perceived Ease-of-Use to Mobile Social Network Apps

Firms should actively cooperate with mobile social network App platforms, pay attention to the simplicity and convenience of information interface design, so that consumers can acquire the perception of convenience and ease of use, and will not ignore product recommendation information due to time resources. For example, information text design should be simple and clear, easy to read; pictures should be directly deliver product image and value. Set up a simple navigation button, users can forward or share useful and interesting product recommendation information through one click, so as to provide a convenient channel for the wider dissemination of product information.

5   Conclusions

Combining the theory of planned behavior with source credibility theory; this paper put forward an influence mechanism model, which explains how product recommendation information posted by mobile social network users influence on consumers' participation in the sharing economy. In this model, the second-order variable source credibility has been expanded to include two more variable: intimacy and past behavior. Source credibility influences consumers' sharing economy participation intention via attitudes, perceived behavior control and subjective norm. This paper also put forward the management implication, hoping to provide certain reference for firms to carry out precise marketing under the Internet environment.

Acknowledgements. This work was supported by Social Science Foundation Project of Jilin Province: Research on Promotion Strategy of Consumers' Participation Stickiness under Sharing Economy Model (2019B64), and Social Science Project of "the 13th Five-Year Plan" sponsored by the Education Department of Jilin Province: A Study of the Influencing factors on Mobile Social Network User's Response Intention towards Product Recommendation Information Posted by the Contacts (JJKH20181319SK).

References

1. Felson, M., Spaeth, J.L., Structure, C., Consumption, C.: A routine activity approach. Am. Behav. Sci. **21**, 614–624 (1978)
2. Zuo W., Wang, X., Fan, F.: The relationship between online word-of-mouth and purchase intention based on social capital in the context of social e-commerce. Nankai Manag. Rev. **17**(4), 140–150, 160. (2014). (in Chinese)
3. Gong, S., Liu, X., Zhoa, P.: How online consumer reviews affect product sales? – an empirical study based on online book reviews. China Soft Sci. **6**, 171–183 (2013). (in Chinese)
4. Gallaugher, J., Ransbotham, S.: Social media and customer dialog management at starbucks. MIS Q. Exec. **9**(4), 197–212 (2010)
5. Ajzen, I.: The theory of planned behavior. Res. Nurs. Health **14**(2), 137–144 (1991)
6. Ohanian, R.: Construction and validation of a scale to measure celebrity endorsers' perceived expertise, trustworthiness, and attractiveness. J. Advert. **19**(3), 39–52 (1990)
7. Chen, M.F.: Consumer attitudes and purchase intentions in relation to organic foods in Taiwan: moderating effects of food-related personality traits. Food Qual. Prefer. **18**(7), 1008–1021 (2007)
8. Tarkiainen, A., Sundqvist, S.: Subjective norm, attitudes and intentions of Finnish consumers in buying organic food. Br. Food J. **107**(11), 808–822 (2005)
9. Guoqing, G., Zhongke, Z., Kai, C., et al.: The impact of word-of-mouth communication on consumers' intention to switch brands: the mediating effect of subjective norm. Manage. Rev. **22**(12), 62–69 (2010). (in Chinese)
10. Wangenheim, F.V., Bayón, T.: The effect of word of mouth on services switching: measurement and moderating variables. Eur. J. Mark. **38**(9), 1173–1185 (2004)
11. Xuefeng, Z., Qing, T., Fen, L.: Mobile coupon forwarding model based on electronic word-of-mouth marketing. J. Manag. **10**(11), 1657–1662 (2013). (in Chinese)
12. Asch, S.E.: Studies of independence and conformity: I. A minority of one against a unanimous majority. Psychol. Monogr. **70**(9), 1–70 (1956)
13. Cheung, M., Luo, C., Sia, C., et al.: Credibility of electronic word-of-mouth: informational and normative determinants of on-line consumer recommendations. Int. J. Electron. Commer. **13**(4), 9–38 (2009)
14. Yonghui, G.: Design chain knowledge continuous sharing model based on the theory of planned behavior. Sci. Res. **26**(s1), 159–165 (2008). (in Chinese)
15. Xia-qi, d., Hua-ming, W., Mou-Chao, M.: The influence of celebrity recommenders' moral reputation on celebrity advertising effect. J. Psychol. **37**(3), 382–389 (2005). (in Chinese)
16. Matook, S., Brown, S.A., Rolf, J.: Forming an intention to act on recommendations given via online social networks. Eur. J. Inf. Syst. **24**(1), 76–92 (2015)

Optimizing System Management Based on Trust of Domain

Xiaozhu Wang$^{(\boxtimes)}$ and Ping Song

Department of Information Technology and Business Management, Dalian Neusoft University of Information, Dalian, Liaoning, China
wangxiaozhu@neusoft.edu.cn

Abstract. This paper first introduces the conventional way of APG40/43 account management, points out the shortcomings of the conventional way, puts forward the centralized account management, and gives the configuration details.

Keywords: Trust · Local group · Global group

1 Account Management Issues

The input and output system of Ericsson AXE10/AXE810 switch mostly adopts the IOG20C, APG40 or APG43. The IOG20C does not support the centralized management of users.

Windows NT operating system, Windows 2000 and windows 2003 are products of Microsoft Corporation, and they are compatible with each other. APG40 uses Windows NT server operating system and APG43 use Windows Server 2003, each group of APG40 constitutes a domain, a domain is usually composed of multiple domain controllers. A computer with Windows 2000 server or windows 2003 server installed with an active directory can also act as a domain controller. In a complex network, there may be multiple domains. In order to enable users in one domain to access resources in other domains, we can establish a trust relationship between domains. Trust relationship is a bridge between domains. When a domain establishes a trust relationship with other domains, the two domains can not only manage each other as needed, but also allocate device resources such as files and printers across the network. According to the principle of domain trust, this paper proposes an optimized configuration scheme of APG40 account management.

1.1 APG40 Account Management

In the APG40 system, the administrator can still manage the account data in a way similar to the method of IOG20C, as shown in Fig. 1. In this scheme, all maintenance accounts are defined on APG40 respectively. The administrator of APG40 adds these accounts to the local group of APG40 to authorize the corresponding maintenance rights of the account. This scheme has many disadvantages. First of all, whether the account

J. Abawajy et al. (Eds.): ATCI 2021, AISC 1398, pp. 113–119, 2021.
https://doi.org/10.1007/978-3-030-79200-8_16

configuration process is cumbersome. If a maintenance personnel is added, the APG40 administrator should log in to all APG40 to add the account, and then add it to the corresponding local group, so as to authorize the corresponding operation authority of the user. Secondly, the password Whether the maintenance is troublesome, when the maintenance personnel modify the login password, they should log in to all APG40 for operation.

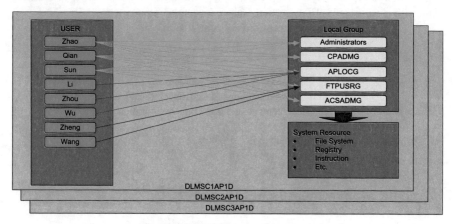

Fig. 1. APG40 account management

1.2   Centralized Account Management Scheme

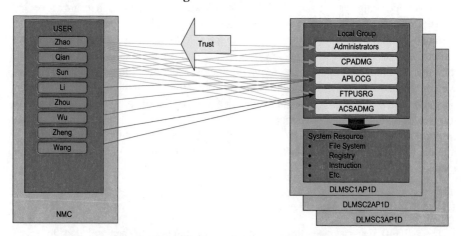

Fig. 2. Centralized account management scheme

In this scheme, we set up a domain NMC for managing users. All users are defined in this domain, and all APG40 establish a trust relationship with this domain. Then the administrator of APG40 adds the account in NMC to the local group of APG40.

In this way, the maintainer can log in to all APG40 for operation and maintenance by using the account in NMC. In this scheme, the password management of the maintainer is simplified. When the maintainer changes the password, he only needs to log in to the NMC domain to change it. The operation of changing the password is effective for all APG40 that trust the NMC domain. In this scheme, the allocation of permissions is for each maintenance personnel, so it is still troublesome to add an account. When the APG40 administrator needs to add a management account, he needs to log in to all APG40 and add the new users in NMC to the local group of APG40. Therefore, it is necessary to further simplify this scheme.

1.3 Improved Authority Distribution

Many of the maintenance personnel have the same authority to operate the equipment. As can be seen from Fig. 2, the maintenance permissions of accounts Zhao, Qian and sun on APG40 are the same, the permissions of Li, Zhou and Wu are the same, and the permissions of Zheng and Wang are the same. Therefore, we can group the above maintenance accounts, divide the maintenance personnel with the same permissions into a group, and then assign permissions to groups in APG40, instead of assigning permissions to everyone. In this way, the account configuration is nearly simplified. Figure 3 is an improved account management method. The administrator of APG40 can establish three global groups lnwg, dlwg and boss on NMC domain according to the actual maintenance, and then add Zhao, Qian and sun to the global group lnwg, Li, Zhou and Wu to the global group dlwg, and Zheng and Wang to the global group boss. In APG40, add the three global groups in NMC to the corresponding local groups in APG40, then the account in NMC has the permissions of the corresponding global groups in APG40. The schemes represented by Fig. 2 and Fig. 3 are actually equivalent, but the scheme represented by Fig. 3 is more optimized and easier to configure.

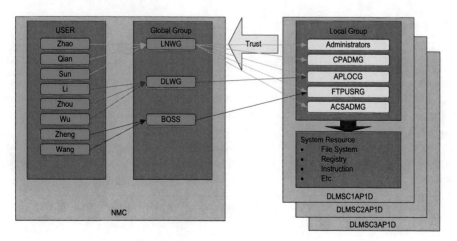

Fig. 3. Improved authority distribution

2 Data Configuration

The domain controller used for account management can use the operating system running Windows NT server or the computer with Windows 2000/2003 server installed with active directory. Therefore, we can use a set of existing APG40 to manage the APG40 maintenance account of the whole network. We can also create a new domain or use other existing domain controllers in the network. The following is the process of data configuration. Here, the domain controller of the management account uses Windows 2000 server. The configuration mode of other versions is not different from this example.

2.1 Configuration of Domain Controller

Install the windows 2000 server operating system on a computer, and install the active directory to upgrade it to a domain controller. If you want to improve the reliability, you can install multiple domain controllers.

2.2 Configuration of APG Network

There may be multiple network terminals in the network management. In order to accurately locate each other between APG40 and the domain controller of the management account, the administrator of APG40 should configure DNS and wins server address for APG40. The specific configuration method is as follows:

Network neighborhood > Properties > TCP/IP

When modifying the APG40 network configuration, the system must be rebooted before the new configuration can take effect. Therefore, we suggest that the standby side be modified first, and then the execution side be modified to reduce the impact on system maintenance.

2.3 Building Trust

Netdom is a command-line tool that is built into Windows Server 2008 and Windows Server 2008 R2. It is available if you have the Active Directory Domain Services (AD DS) server role installed. It is also available if you install the Active Directory Domain Services Tools that are part of the Remote Server Administration Tools (RSAT). For more information, see How to Administer Microsoft Windows Client and Server Computers Locally and Remotely.

To use netdom, you must run the netdom command from an elevated command prompt. To open an elevated command prompt, click Start, right-click Command Prompt, and then click Run as administrator.

```
netdom trust /d:NMC DLMSC1AP1D /add /Ud:NMC\admin /Pd:* /Uo:DLMSC1AP1D\admin /Po:*
```

2.4 Assign User Rights

The permissions of a maintenance account are determined by the local group of APG40 to which it belongs. Therefore, adding an account to the local group of APG40 can complete the assignment of the account permissions. The administrator of APG40 can assign permissions to a single account or a group. This account can be a local account or group, an account or a global group in a domain trusted by APG40.

In the following example, Liu is an account defined in NMC, and all APG40 in the network trust the NMC domain. Then the administrator of APG40 can join the user Liu in NMC to the local group of APG40.

```
NET LOCALGROUP FTPUSRG NMC\Liu /add
NET LOCALGROUP APLOCG NMC\Liu /add
NET LOCALGROUP CPSADMG NMC\Liu /add
NET LOCALGROUP CPUSRG NMC\Liu /add
```

In the following example, dlwg is a global group defined in NMC, and APG40 in the network all trust NMC domain, so the administrator of APG40 can join the global group lnwg in NMC to the local group of APG40.

```
NET LOCALGROUP CPSADMG NMC\DLWG /add
```

3 Using Centralized Accounts for Maintenance Operations

After the account is defined in the NMC domain, and the administrator of APG40 assigns the corresponding permissions to the account, the maintenance personnel can make the centralized account on NMC for APG40 maintenance.

3.1 Image Interface Using APG40

When logging in to APG40, we select the domain controller to log in to from the domain drop-down list. To use a centralized account, we can choose to log in to the domain NMC.

3.2 Log in to APG40 Using Telnet

When using telnet to log in to APG40, the system can input not only the account name and password, but also the login domain. Here, we can enter the domain NMC to indicate the authentication to the domain NMC.

```
C: > telnet dlmsc11ap2a
Pragma Systems Inc.
Welcome to Pragma TelnetServer 2000 for Windows NT
(C) Copyright 1994-2000 Pragma Systems, Inc.
Login name: wangliguo the user is defined in NMC
password: *******
Windows NT domain: NMC indicates to verify with NMC
User Account verification is being performed. Please Wait.
Client reported TERM=ansi
Microsoft(R) Windows NT(TM)
(C) Copyright 1985-1996 Microsoft Corp.
C:\>
```

3.3   Use Port 5xxx to Access CP Directly Through APG40

When accessing CP directly, we can use port 5000. When the system prompts domain, you can enter the domain name of the user.

```
Pragma Systems Inc.
Welcome to Pragma TelnetServer 2000 for Windows NT
The user is defined in NMC
PASSWORD: *******
Domain: NMC indicates to verify with NMC
WO DLMSC11R10/JB/0/0/09/05 AD-602 TIME 061011 1608 PAGE 1
```

3.4   Access to FTP Function of APG40

When using the FTP function of APG40, we need to add the domain of the account before the login user name, and separate the domain and account name with a slash.

```
C: \ > FTP dlmsc11ap2a
Connected to dlmsc1ap2a.
220 dlmsc11ap2a Microsoft FTP Service (Version 4.0).
User (dlmsc11ap2a:(none)): NMC\wangliguo
331 Password required for NMC\wangliguo.
Password:******
230 User NMC\wangliguo logged in.
ftp> bye
```

4   Concluding Remarks

This paper introduces the principle of APG40 account centralized configuration and the specific data configuration method in detail. I hope this article is helpful for your maintenance work.

References

1. Yuan, C., Lin, Y., Bo, F.Z., Feng, Y.: Identity management architecture: paradigms and models. Appl. Mech. Mater. (40), 56–76 (2011)
2. Rissanen, T.: Electronic identity in Finland: ID cards vs. bank IDs. Iden. Inf. Soc. (1), 345–354 (2010)
3. Manders-Huits, N.: Practical versus moral identities in identity management. Ethics Inf. Technol. (1), 213–221 (2010)
4. Zhu, D., Li, X.-B., Wu, S.: Identity disclosure protection: a data reconstruction approach for privacy-preserving data mining. Decis. Supp. Syst. (1), 112 (2009)
5. Birch, D.G.W.: Psychic ID: a blueprint for a modern national identity scheme. Ident. Inf. Soc. (1), 79–84 (2008)
6. Sarma, A., Matos, A., Girao, J., Aguiar, R.L.: Virtual identity framework for telecom infrastructures. Wirel. Pers. Commun. (4), 46–54 (2008)
7. Goto, A., Bhargav-Spantzel, A., Camenisch, J., Gross, T., Sommer, D.: User centricity: a taxonomy and open issues. J. Comput. Secur. (5), 97 (2007)
8. Machanavajjhala, A., Kifer, D., Gehrke, J., Venkitasubramaniam, M.: L-diversity: privacy beyond k-anonymity. ACM Trans. Knowl. Discov. Data (TKDD) (1), 43–58 (2007)
9. Mandelbaum, Y., Fisher, K., Walker, D., Fernandez, M., Gleyzer, A.: PADS/ML: a functional data description language. ACM SIGPLAN Notices (1), 213–214 (2007)
10. Miklau, G., Suciu, D.: A formal analysis of information disclosure in data exchange. J. Comput. Syst. Sci. (3), 119–121 (2006)

Service Clustering Based on GSDMM Topic Model

Bing Wang[✉]

School of Information Science and Technology, Qingdao University of Science and Technology, Qingdao, Shandong, China

Abstract. Service clustering is an important part of realizing accurate service discovery and efficient service management. Research on how to improve the accuracy and precision of clustering has become the main task in the field of service clustering. The clustering method proposed in this paper first preprocesses the service description text, extracts feature words that have positive significance for category division, and reduces data sparsity; further, uses the GSDMM topic model to generate the corresponding topic vector; finally, through AGNES algorithm performs the final clustering effect analysis on the vector. Experiments have proved that the GSDMM topic model method proposed in this paper is effective for service clustering.

Keywords: Service clustering · Topic model · Clustering algorithm · GSDMM · AGNES

1 Introduction

With the rapid development of SOA applications, a large number of Web services on the Internet have emerged, showing the characteristics of diversity and complexity. Therefore, finding suitable services from a large number of similar services and performing effective management and organization have become an urgent problem to be solved. In order to speed up the accuracy and speed of service search, service clustering is usually used. Dividing the services in the datasets into multiple sub-categories with similar functions, narrowing the comparison scope of service selection, and realizing fast search of services on the basis of service clustering can effectively improve the ability of Web service discovery. To a certain extent, it is the premise and foundation of service search, service execution, and service composition, and is of great significance for subsequent service composition and service recommendation. Therefore, improving the effect of service clustering is a hot issue in the field of service computing.

The clustering process is the process of classifying services according to certain rules and requirements, and dividing a collection containing a large number of services into different classes or clusters. In this process, there is no prior knowledge about category information, only the similarity between services is used as the standard for classification of categories. Finally, the similarity of data objects in the same cluster is as large as possible, and the similarity of objects in different clusters is as small as possible.

© The Author(s), under exclusive license to Springer Nature Switzerland AG 2021
J. Abawajy et al. (Eds.): ATCI 2021, AISC 1398, pp. 120–127, 2021.
https://doi.org/10.1007/978-3-030-79200-8_17

The data in the clustering process is unlabeled, and there is no prior knowledge to divide it. Clustering is an effective way to perform unsupervised learning partition. It is hoped to discover the potential information in unlabeled data, that is, the data of the same type should be gathered together as much as possible, and different data should be separated as much as possible.

2 Related Work

The topic model is an effective way of clustering. It is an important issue to explore the construction of effective topic clustering through a large amount of unlabeled data. The topic model is a statistical model that clusters the implicit semantic structure of the data set in an unsupervised way. Many researchers have conducted extensive research on topic models. The implicit Dirichlet distribution LDA [1, 2] is the most common topic model, but it does not perform well in the field of short text due to the sparseness of the data. The researcher proposes that the Dirichlet hybrid GSDMM topic model based on the Gibbs sampling algorithm shows good performance in short text clustering applications [3]. Some researchers consider word embedding for clustering research, modeling topic relevance in a continuous word embedding space, and replacing words in documents with meaningful word embedding [4, 5]. Zou proposed DeepWSC, a heuristic-based Web service clustering framework, which combines the deep semantic features extracted from the service description by the improved recursive convolutional neural network with the service composability features obtained by the symbolic graph convolutional network from the service call relationship to generate integrated implicit features for Web service clustering to accurately perform service clustering jointly [6]. Existing clustering algorithms are divided into two categories: clustering research based on functional attributes and clustering research based on non-functional attributes. Kotekar uses K-Means clustering algorithm and biologically inspired CSO clustering algorithm to improve the performance related to time and accuracy in the Web service discovery process [7]. In addition, some scholars use the related structure of services to cluster, hoping to propose better service clustering methods by using structural indicators. Pan proposed a service clustering method based on structural similarity and genetic algorithm clustering algorithm, and applied the algorithm to quantify the structural similarity between each pair of Mashup services [8]. Zhou et al. proposed a community discovery algorithm to organize services into clusters, grouped services into different clusters based on the service clustering method based on structural measurement, and constructed a joint network of atomic services based on service calculations. The similarity measurement method for atomic service sharing similarity measurement [9, 10]. However, the structure-based calculation methods proposed above all have the disadvantage of high space complexity.

2.1 Topic Model

The topic model is a generative probability model for discovering abstract graphs. It is a text mining tool used to extract hidden topics from a large number of documents. Widely used in the field of machine learning and natural language processing, the basic idea is that text is a mixture of one or more topics, and the topic is a probability distribution of feature words. The topic distribution of the text is the mapping of the text vector space and the topic of the inferred text can be calculated by the frequency of feature words. The specific method of calculating the similarity between texts is to calculate the corresponding topic probability distribution. Selecting a suitable topic model for different data sets is the key to achieving a good clustering effect.

2.2 Clustering Algorithm

Common clustering methods include partitioning method, hierarchical method. In addition, there are relatively novel clustering based on neural network and clustering based on graph theory. These clustering algorithms have their own advantages and disadvantages.

The partition-based clustering algorithm, which is usually called "heuristic algorithm", first determines the number of clusters and the initial cluster center point, iteratively resets the cluster from the data point to the center point, and divides the categories according to a certain threshold.

Based on the hierarchical clustering algorithm, the text is clustered layer by layer by using a nested type of tree data structure. It can be roughly divided into a bottom-up aggregation method and a top-down split method. The initial center point is located at the top or bottom of the cluster tree.

3 Research Contents

3.1 Preprocess

Service description is a short text describing the service, but because the text contains a lot of useless information, the information that can be extracted is limited. In order to improve the accuracy and accuracy of text clustering, preprocessing is required, including the following steps:

(1) Feature extraction. Extract all the characteristics describing the Web service from the service description.
(2) Participle. Split the compound words in the extracted text.
(3) Remove stop words. Remove stop words, meaningless symbols, punctuation in the web service description text to avoid affecting clustering accuracy and reduce data sparsity.
(4) Lemmatization. Restore various forms of words to stem prototypes.

3.2 Topic Vector

Definition 1: Define all services in the data set as $D = \{s_1, s_2, s_3, \cdots, s_m\}$, extract each service description text after all services are preprocessed, and construct a service description corpus $SiS = \{s_{i1}, s_{i2}, s_{i3}, \cdots, s_{im}\}$. $SiD = \{d_{i1}, d_{i2}, d_{i3}, \cdots, d_{im}\}$ is the vector set of $SiS = \{s_{i1}, s_{i2}, s_{i3}, \cdots, s_{im}\}$, where d_{im} is the subject vector corresponding to s_{im}.

This paper chooses GSDMM short text model to generate topic vectors. The GSDMM topic model is an unsupervised classification model for modeling and dividing text topic information. It is generated according to a hybrid polynomial model, in which the topic and the document are in a one-to-one correspondence.

The GSDMM model can essentially be regarded as a search heuristic algorithm for text clustering. The process of solving the model is divided into two steps: firstly, the document is generated based on the Dirichlet Mixture Model (DMM), and then the Gibbs Sampling algorithm is used to approximate the model, which has a good application effect in the field of short text clustering.

Use K to indicate the number of topics, and D to indicate the number of documents in the corpus, $p(d)$ represents the probability of each document, $p(d|t = k)$ represents the probability of each document belonging to topic k, and $p(t = k)$ represents the probability of each topic k. For each topic t, when the document d is generated, a category k is first determined according to $p(t = k)$, and then a document is determined according to the probability $p(d|t = k)$ of each document in the category. The DMM sampling process is shown in Fig. 1.

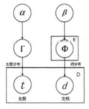

Fig. 1. DMM sampling process

The probability of each document appearing is calculated by formula (1):

$$p(d) = \sum_{k=1}^{K} p(d|t = k)p(t = k) \tag{1}$$

When the topic to which a document belongs is known, all words w in the document are independently and identically distributed. The probability of document d is obtained from topic k and calculated by formula (2).

$$p(d|t = k) = \prod_{w \in d} p(w|t = k) \tag{2}$$

In order to obtain the topic distribution of the words in the document, assuming that the topic is a polynomial distribution on the word, Φ indicates that the probability that the word belongs to the k topic is calculated by formula (3).

$$p(w|t = k) = p(w|t = k, \Phi) = \phi_{k,w} \tag{3}$$

In the above formula, Φ is the word distribution matrix, and $\phi_{k,w}$ is the word distribution of topic k, the sum of the topic distribution of all words in a document is 1, which is:

$$\sum_{w=1}^{n} \phi_{k,w} = 1 \qquad (4)$$

(n represents the word vector dimension of document d).

The Dirichlet distribution formula is (5):

$$p(\Gamma|\vec{\alpha}) = Dir(\vec{\theta}|\vec{\alpha}) \qquad (5)$$

In formula 6, Γ is the topic distribution matrix, θ_k represents the topic distribution, in a document:

$$\sum_{k=1}^{k} \theta_k = 1 \qquad (6)$$

Definition 2: $Sil = \{l_{i1}, l_{i2}, l_{i3}, \cdots, l_{im}\}$ is the subject vector set generated by $SiS = \{s_{i1}, s_{i2}, s_{i3}, \cdots, s_{im}\}$. The generation process is based on the Dirichlet Mixture Model (DMM) to generate the document, and then the Gibbs sampling algorithm is used to approximate the model through continuous sampling of a word on different topics to obtain the topic distribution matrix θ of the word, and then obtain the document. The topic distribution is θ_k, and finally a topic vector $Sil = \{l_{i1}, l_{i2}, l_{i3}, \cdots, l_{im}\}$ based on the GSDMM topic model is generated.

The GSDMM method uses the Gibbs Sampling algorithm to approximate the model, and continuously samples a word on different topics, and finally obtains the topic distribution matrix of the word, thereby obtaining the topic distribution and topic word distribution of the document.

The experimental results show that the GSDMM clustering model can get a good clustering effect, and multiple iterations make the clustering develop in a wonderful direction.

3.3 Clustering Algorithm

This paper uses the AGNES algorithm as the final clustering algorithm. AGNES is a bottom-up aggregation strategy hierarchical clustering algorithm. First, each service in the data set is regarded as an initial cluster, and each iteration takes the closest two clusters are merged and iterated many times until the preset number of clusters is reached.

The advantage of the AGNES algorithm is that the logical structure is simple, easy to understand, and does not rely on the choice of initial values.

When calculating the distance between clusters, the bottom-up principle is adopted. First, each cluster is taken as a sample set, and the gradual merging between clusters is realized by calculating the minimum distance between clusters. Given the distance between the clusters c_i and c_j can be calculated by the formula (7):

$$d_{\min}(c_i, c_j) = \min_{x \in c_i, y \in c_j} dist(x, y) \qquad (7)$$

4   Research Results

4.1   Datasets

ProgrammableWeb is a popular online community of Web services generated by users, where users can publish, use, and evaluate Web APIs. The data in this article is service data crawled from the ProgrammableWeb.com website. Each Web API contains information such as service name, service tag, and service description (Fig. 2).

Fig. 2.   ProgrammableWeb service example

4.2   Evaluation Criteria

This paper uses a combination of multiple clustering indicators to evaluate the clustering results, using Calinski-Harabaz index, Silhouette coefficient, and Davies-Bouldin index to evaluate the clustering effect and verify the clustering effect.

The Calinski-Harabaz index measures the tightness within a class by calculating the sum of the squares of the distance between each point in the class and the center of the class. The larger the class, the tighter the class itself, and the more scattered between the classes, which is the better clustering result. Silhouette profile coefficient, which measures the degree of similarity between a node and its genus cluster compared to other clusters. The value ranges from -1 to 1. The larger the value indicates that the node is more suitable for its genus cluster and not adjacent clusters match. The Davies-Bouldin Index is calculated by calculating the sum of the average distance of any two categories and dividing by the distance between two cluster centers to find the maximum value. The smaller the DB means the smaller the intra-class distance and the larger the inter-class distance.

4.3   Experimental Results

(1)   This paper verifies the clustering effect of different topic models under the KMeans algorithm. Experiments show that the GSDMM topic model has the best clustering effect.

(2)   The service vector generated by GSDMM model is used to represent the vector data set, and multiple clustering algorithms are verified. The experimental results prove that AGNES has better performance in the data set in this article (Figs. 3 and 4).

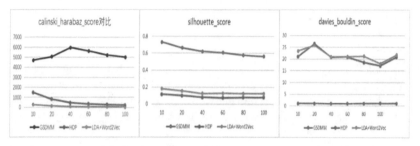

Fig. 3. Comparison of different theme models

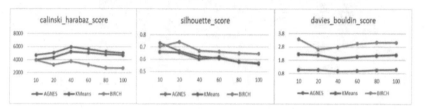

Fig. 4. Comparison of different clustering algorithm

5   Conclusions

In order to improve the effect of service clustering effectively, this paper proposes a service clustering method based on the GSDMM topic model. The topic vector is generated from the GSDMM topic model, and the AGNES clustering algorithm is used for the final service clustering.

We will further optimize the GSDMM topic model in future work, accelerate the AGNES clustering process, and make the clustering effect more reasonable. In addition, in the follow-up work, the clustering evaluation indicators will be further optimized, and more diversified indicators will be used to quantify the clustering effect.

References

1. Jelodar, H., et al.: Latent Dirichlet Allocation (LDA) and topic modeling: models, applications, a survey. Multimed. Tools Appl. **78**(11), 15169–15211 (2019). https://doi.org/10.1007/s11042-018-6894-4
2. Pavlinek, M., Podgorelec, V.: Text classification method based on self-training and LDA topic models. Expert Syst. Appl. **80**, 83–93 (2017)
3. Agarwal, N., et al.: Evaluation of Web service clustering using Dirichlet multinomial mixture model based approach for dimensionality reduction in service representation. Inf. Process. Manag. **57**(4), 102238 (2020)
4. Liang, W., et al.: GLTM: a global and local word embedding-based topic model for short texts. IEEE Access **6**, 43612–43621 (2018)
5. Dieng, A.B., Ruiz, F.J.R., Blei, D.M.: The dynamic embedded topic model. ArXiv Preprint arXiv:1907.05545 (2019)

6. Zou, G., et al.: DeepWSC: Clustering web services via integrating service composability into deep semantic features. IEEE Trans. Serv. Comput. (2020)
7. Kotekar, S., Kamath, S.S.: Enhancing service discovery using cat swarm optimisation based web service clustering. Perspect. Sci. **8**, 715–717 (2016)
8. Pan, W., Chai, C.: Structure-aware mashup service clustering for cloud-based internet of things using genetic algorithm based clustering algorithm. Futur. Gener. Comput. Syst. **87**, 267–277 (2018)
9. Zhou, S., Wang, Y.: Clustering services based on community detection in service networks. Math. Prob. Eng. **2019**, 1–11 (2019)
10. Wu, J., et al.: Clustering web services to facilitate service discovery. Knowl. Inf. Syst. **38**(10), 207–229 (2014)

Opportunities and Challenges of Hotel Marketing Under the Background of Big Data

Jingfeng Jiang and Ziwei Yu[(⊠)]

Department of International Hospitality, Fuzhou Melbourne Polytechnic, Fuzhou, Fujian, China

Abstract. The data entered each sector of activity and became a major factor of production. The significance and application of big data herald a new wave of productivity growth and consumer demand. In the context of Big Data, hotel marketing pays greater attention to the use of extensive data and applies it to the formulation of the marketing strategy. With the development of mobile Internet and the continuous improvement of communicative smart device, software, functions, the behavior of netizens has changed from the traditional PC-side to the "PC-side + mobile-side" with emphasis, showing a trend of cross-screen interaction. Hotels can take full advantage of the extensive data to seek out target populations, build product associations, anticipate market trends and find partners. In addition, hotels should also recognize the shortcomings of big data applications and overcome the difficulties of these applications. To the extent a hotel can effectively apply it to big data. It can use to do good work in hotel sales.

Keywords: Big data · Hotel marketing

1 Introduction

Big Data is an information technology development concept. This means that computers will collect large volumes of diversified data. In general, people use smart tools to analyse and process this data and then draw useful information from it. Marketing under the background of big data refers to the marketing methods applied to the Internet advertising industry by marketing departments based on a large amount of data from multiple platforms and relying on big data technology. The key is to advertise on the network to the right person at the right time, through the right carrier and in the right manner.

2 Characteristics of Hotel Marketing Under the Background of Big Data

Big data is a kind of data information. Under modern technological conditions, not only can you see data from different industries and dimensions, but also traffic and cross-correlation among data, so as to analyze the correlation of various behaviors behind the

J. Abawajy et al. (Eds.): ATCI 2021, AISC 1398, pp. 128–133, 2021.
https://doi.org/10.1007/978-3-030-79200-8_18

data. The characteristics of the Big Data era are the completeness, richness, depth and assemblage of data [1].

Overall, applying large data has the following advantages: 1) the ability to collect data on request. In the past, when people were evaluating market data, because it was difficult to collect more accurate data, information, people can only rely on experience to judge some information, such as people can only judge that the current market may need a product or not need a product [2]. Such a judgment method is too subjective. In the background of large data, hotels can make questionnaires, collect data as needed, or use the consumer behavior association analysis of other industry-related consumer goods, and then use the data information to know if a product is needed in the market. (2) The establishment of a set of mechanisms for monitoring and assessing data is easy. At the moment, the computer can complement the data collection. In order to facilitate the collection of data by computer, people can classify the data and then monitor it in a classified way. The data will be quantitative, qualitative, and then, based on changes in the data, understand the information reflected by the current data. (3) Visual reflection of the information, the present computer can display the collected data in a graphical manner. People can intuitively understand the dynamic changes in data information through graphics, and analyse the information they need

In the era of big data, hotel marketing can label consumers differently through multiple dimensions such as age, interests, circle hobbies, and then directly reach users [3]. Although there are many successful cases, it becomes increasingly difficult to achieve the expected results. This is because of the drawbacks of big data marketing. (1) Data sources are prone to problems, and the value of a single data is low. a hotel, a single data information from a single large data information may have a lower value. This is because hotels are collecting huge amounts of information, so the information they get may or may not be in effect. If people can't relate data to data. Analyzing information for your needs doesn't really add much value to this definite data. (2) The data amount information is complex, and the data obtained by the hotel include different types of data information such as video, audio, text, pictures and so on. Hotels will use intelligent software to automatically collect information, classify it, and then manage it. The hotel has superior talent and technical requirements. Many hotels are not configured in place, but the opposite is true. (3) Accurately locating target users mean abandoning non-target users. Big data can really give hotels precise targeting of target users, but it reduces the possibility of non-target users buying their own products, which make hotels tend to ignore user experience more or less. For example, hotel insight into customers through big data to do some activities only for this part of the people, but the opposite is true. Ignoring the other part of users will lead to non-obvious consequences.

The pros and cons of big data marketing exist at the same time. On the positive side, analysis of user behavior and characteristics, accurate matching of marketing materials and pushing information on users can better guide products to their best, improve users' experience, on the negative side, too much reliance on data analysis can also lead to wrong decisions, which will affect some users' consumption.

3 Marketing Opportunities in the Context of Big Data

3.1 Mining Target Population from Data

In the past, people used to base their experience on identifying the target population, so the method of identifying the target population of a product was difficult to quantify. In the era of big data, hotels can use information technology to collect product-related target population data, and then target the target population to develop products [4].For example, in the past, people could only use the experience to analyze a product market that might be a promising product market, but could not say exactly what the future is. In the context of large data, people can understand the monthly and annual sales of this product [5]. In addition, it can refine the classification of product sales data, and analyze the consumption of different types of consumers, such as age, gender, income, etc. With these data, hotels can not only know if a product meets the needs of a certain group, but also refine the target group according to the data.

Once the hotel understands the target population, it can upgrade the product to make it more suitable for the current population based on business need [6]. For instance, after understanding the consumer needs of a particular group of people, hotels can tailor a set of products to suit their consumption. In order to optimize the product service work of the target population.

3.2 Mining Information Associations from Data

In the context of large data, hotels can analyze the association between data and data [7]. Take, for example, the association between Hotel analysis of customer data for young consumers. Through analysis, the hotel found that most young consumers would choose to re-consume product B after selecting product A. So now the hotel can analyze the market prospects of product B according to the consumption of product A, or analyze the consumption prospects of product A by collecting data of product B.

After the hotel establishes a data association, the enterprise establishes a cross-group of clients and provides them with products that make such products acceptable to everyone in the cross-group and meet the needs of cross-group consumers. Hotels can collect such data to research and develop products according to their own development needs, so that products can meet different groups of consumer need, so as to improve the economic efficiency of the hotel [8].

3.3 Mining Market Trends from Data

From large data, by analogical reasoning and summary, it is possible to derive some data trends from which future market trends can be inferred [9]. For example, hotels can collect data about product demand, production and sales to ascertain whether a product is oversaturated in the market or has broad prospects. After understanding the persuasion of the market, the hotel can develop products to meet the needs of the market, or innovate products to let a brand new product occupy the market. Hotels can only take the lead in the market and improve the one-touch power if they understand the trend of the market and research and develop products and marketing products according to the needs of the market.

3.4 Mining Partners from Data

From large data, hotels can discover partners to work with [10]. For example, a hotel can analyze the products it needs to produce according to its development goals, and if it wants to optimize the production of products, it needs to optimize the industry chain. In the context of large data, hotels can collect a variety of data, which contains data information from the same industry and different industries. When a hotel establishes upstream and downstream industry chain, hotels can use large data to understand upstream and downstream hotel operations and development, and make a comprehensive evaluation of upstream and downstream hotels. Hotels can select the best hotel to be their dominant partner and establish partnerships to achieve a win-win goal. In the market economy environment, under the background of more and more intense one-touch between hotels and hotels, only by optimizing the industrial chain and increasing economic profits with partners in the industrial chain, can a hotel reduce the cost of operation, improve economic benefits and thus improve the competitiveness of the hotel.

4   Marketing Challenges in the Context of Big Data

4.1   Challenges in Data Sources

Currently, people will use various channels to collect large data. It is important to see that there is a quality risk to the statistical quality of large data. For example, former hotels used to send out web-based questionnaires to teenagers to get data, while using this channel to collect data often failed to get high-quality data. When the accuracy of large data is not high, it is difficult for a hotel to derive valuable information from the data.

Hotels need to optimize their data information in the following ways: First, use diverse channels to collect and compare data. For example, use the methods of issuing network and entity questionnaires to collect information, and then compare the information; use data from different regions to find errors in the data, and then correct the data. Second, use intelligent software to correct information, such as hotels that can optimize the application of smart software so that the software can automatically analyze errors in the data and correct the data. Third, fine data collection, as far as possible to collect the latest data information. Such a method can be used to improve the quality of data collection as possible.

4.2   Challenges in Data Analysis

When collecting large data information, hotels need to develop data collection schemes based on data survey needs to obtain the data information they need. At this time, artificial analysis and judgment are of great significance. For example, marketers need to develop a set of data survey information questionnaire to get the information they need by optimizing the content of data information options. At this point, the quality of the questionnaire affects the results of the data survey. After obtaining a large amount of data to information, the staff need to dig the data information, dig out the required information, and make a judgment on it. If there are errors in data analysis when you analyze data, you will be unable to get the data base you need from the data information.

It must be pointed out that in the context of large data, hotels cannot use computers to replace manual work. Hotels need to strengthen the literacy of their marketers so that they can effectively use larger amounts of data as a basis for judgment. In the work, the hotel can use big data as the work of optimizing marketing, but cannot completely rely on big data to carry out marketing work instead of manual analysis. Therefore, the hotel should improve the literacy of the marketing personnel, so that the marketing personnel can effectively use the big data information, master the method of collecting and analyzing data under the big data background, and effectively use this information work.

4.3 Challenges Posed by Data Management

In the context of large data, data management can have an impact on Hotel Operations. For example, if the hotel does not do a good job in safety management, leading to the theft of data information, it may leak the business secrets of the hotel, thus bringing business risks. If the hotel's data is stolen, there may be personal privacy of the customer stolen, so the hotel will face some legal risks. For example, in 2018, 240 million users of China Hotels were stolen, which caused great repercussions. Users generally believe that the reason why the information was stolen in China Liquor Cafe is that they did not do a good job in safety management. Management capability of an enterprise which can not do well in safety management is questionable. The event affected the marketing work of the hotel.

Under the background of more and more developed information technology, security issues have become a common concern. When a hotel should carry out marketing management with big data, it must do a good job of security protection to avoid the business risks brought by data stealing. In order to optimize the management of data, the hotel must set up a security protection system so that the network of the hotel can not be invaded illegally, can be monitored and managed immediately, and have a reasonable data backup scheme. Hotels should pay attention to the management of data information to avoid business risks.

5   Conclusion

Under the background of big data, big data have become an important work for hotel marketing. On the one hand, hotels can discover marketing data based on big data, which can become an important basis for hotel marketing strategy. If a hotel can effectively apply big data, it can optimize its strategy to win the first chance in the process of market operation, thereby improving the competitiveness of the hotel. On the other hand, there are restrictions on the application of big data. Hotels must face challenges, overcome difficulties, eliminate the shortage of big data information, and utilize the marketing advantages brought by big data information. How to effectively use big data to carry out marketing work, give full play to the advantages brought by big data information, overcome the shortcomings of its application, and enable marketers to use these data information to optimize marketing programs is a problem that hotels need to seriously think about and face.

Acknowledgment. This work was supported by 2020 Educational research projects of Fujian Provincial Department of Education (Grant No: JAS20785).

References

1. Liu, J., Li, J., Li, W., Wu, J.: Rethinking big data: a review on the data quality and usage issues. ISPRS J. Photogramm. Remote. Sens. **115**, 134–142 (2016)
2. Liu, Y., Teichert, T., Rossi, M., Li, H., Hu, F.: Big data for big insights: Investigating language-specific drivers of hotel satisfaction with 412,784 user-generated reviews. Tour. Manag. **59**, 554–563 (2017)
3. Zhao, Y., Xu, X., Wang, M.: Predicting overall customer satisfaction: big data evidence from hotel online textual reviews. Int. J. Hosp. Manag. **76**, 111–121 (2019)
4. Talón-Ballestero, P., González-Serrano, L., Soguero-Ruiz, C., Muñoz-Romero, S., Rojo-Álvarez, J.L.: Using big data from customer relationship management information systems to determine the client profile in the hotel sector. Tour. Manag. **68**, 187–197 (2018)
5. Zhang, Y., Shu, S., Ji, Z., Wang, Y.: A study of the commercial application of big data of the international hotel group in China: based on the case study of marriott international. In: 2015 IEEE First International Conference on Big Data Computing Service and Applications, pp. 412–417. IEEE. (2015)
6. Ban, H.J., Kim, H.S.: Semantic network analysis of hotel package through the big data. Culinary Sci. Hosp. Res. **25**(2), 110–119 (2019)
7. González-Serrano, L., Talón-Ballestero, P., Muñoz-Romero, S., Soguero-Ruiz, C., Rojo-Álvarez, J.L.: Entropic statistical description of big data quality in hotel customer relationship management. Entropy **21**(4), 419 (2019)
8. Parvez, S.J., Moyeenudin, H.M., Arun, S., Anandan, R., Janahan, S.K.: Digital marketing in hotel industry. Int. J. Eng. Technol. **7**(2.21), 288–290 (2018)
9. Pan, B., Yang, Y.: Forecasting destination weekly hotel occupancy with big data. J. Travel Res. **56**(7), 957–970 (2017)
10. Leung, R.: Smart hospitality: Taiwan hotel stakeholder perspectives. Tourism Review. **74**(1), 50–62 (2019)

The Impact of Technological Innovation on Regional Economic Development Under the Background of Internet

Xia Zhong[✉]

School of Economics, Sichuan University, Chengdu, Sichuan, China
zx610200@xueshumail.cn

Abstract. The use of computers in people's work and life has become more and more extensive. While its use has improved the people's living standards, it has also promoted the development of various industries. Technological progress has laid the foundation for the development of Internet technology. With the maturity of network technology, the regional economy has also developed very well. China has maintained a relatively high economic growth rate since the reform and opening up, and internal technological innovation is the basic guarantee for the healthy development of the regional economy. This shows that traditional economic growth methods can no longer meet the economic needs of modern society, and sustainable economic advancement with technology as the core is the main way to continue economic development. The positive promotion effect of technological innovation is very significant in regional development and has profound practical significance.

Keywords: Internet · Technological · Innovation · Region · Economy

1 Introduction

Internet technology has outstanding advantages and can have an important impact on regional economic development factors, development structure, new technology derivation, and economic growth patterns in practical applications [1]. In order to effectively use the advantages of Internet technology to promote the healthy and long-term development of the regional economy, it is necessary to maintain the harmonious progress between Internet technology and the regional economy, increase the usage of Internet technology in the process of regional economic development, and promote technological innovation, expand the future development space of the regional economy.

2 The Influence of Internet Technology on Regional Economic Development

2.1 Internet Technology Affects Regional Economic Development Factors

The integration of Internet technology in various development activities of the regional economy will directly affect the elements of economic development. In the actual development process of traditional regional economy, it pays more attention to the rational

J. Abawajy et al. (Eds.): ATCI 2021, AISC 1398, pp. 134–140, 2021.
https://doi.org/10.1007/978-3-030-79200-8_19

distribution of elements such as talent, capital, and technology in the regional economy. After the usage of Internet technology, talent, capital, and technology can interact with each other in a timely and rapid manner. It has severely impacted the traditional regional economic development model, and promoted the rapid circulation of resources within and between different regions [1]. At the meantime, the usage of Internet technology will profoundly affect the rationality and overall efficiency of regional economic resource allocation, and will directly affect the overall quality of regional economic development. Internet technology effectively guarantees the integrated development of the regional economy, efficient allocation of resources in all aspects, and deep implantation of multiple economic resources to ensure that data and information can better meet the requirements of social and economic development to promote the healthy and sustainable social economy Development, as shown in Fig. 1. The organic application of Internet technology can effectively extend the boundaries of the world of commodities, promote the expansion of market space, and promote the rapid development of regional economy [2].

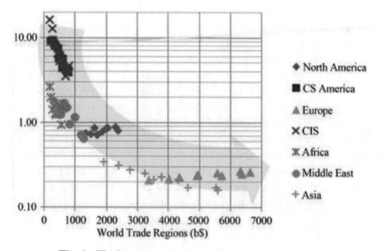

Fig. 1. The Internet promotes regional economic growth

2.2 Internet Technology Has Changed the Structure of Regional Economic Development

The social development can accelerate the optimization and improvement of the industrial structure of the regional economy at the technical level, promote the speed of transformation of the traditional industrial structure, and promote the development of agricultural production in the direction of industrialization, and promote industrial upgrading. Traditional production methods have been unable to keep up with the rapid development of the modern economy, and the usage of Internet technology can promote the transformation and upgrading of traditional production technologies. Information processing of all aspects of production processes promotes more precise production control and improves the rationalization of production processes [3]. In order to ensure that the technology applied in regional economic development meets the objective requirements, and comprehensively promote the technicalization and informatization of China's regional development and economic activities. For example, the usage of Internet technology in the management industry can promote information and data in the management process to effectively improve the interactive ability and aggregation ability, improve the management service ability and level, and on the basis of all aspects of data provided by the usage of Internet technology, more accurate and fast carry out a series of management service activities; the use of Internet technology English in agricultural production activities can promote more intensive and modern production, make up for the shortcomings and shortcomings of traditional agricultural production methods, guide farmers' production activities more scientifically, and gradually improve agricultural production efficiency and overall output [4]. Utilizing the advantages of various aspects of Internet technology can dig deeper into the factors that are conducive to regional economic development.

2.3 Internet Technology Promotes the Birth of High-Tech

Under the usage of Internet technology, a variety of high and new technologies can be produced more efficiently. With the support of related high and new technologies, the transformation of traditional economy will be accelerated, and the comprehensive integration of traditional industries and modern information technology and electronic technology will be promoted. The productivity has achieved leapfrog development, and on this basis, the industrial structure has been fully optimized and the industrial quality has been improved [5]. Under the usage of high-tech, it can effectively enhance the added value of labor services and products, and at the same time innovate more subsidiary products, prompting traditional industries to further innovate production methods and expand marketing channels [5]. Under various high-tech applications, the production cycle of products will be effectively shortened, thereby enhancing the competitiveness of the product market, which has an important impact on promoting regional economic development.

3 The Impact of Technological Innovation on Regional Economic Development

3.1 Technological Innovation Investment and Regional Economic Progress

Technological innovation investment is essentially a cost increase process, but the cost occupancy in this process is obviously within a reasonable "threshold" range, which is a prerequisite for the application and development of technological achievements, which is essential for the realization of technological innovation and regional economic development [6]. Although increased investment will increase operating costs in a short period of time, from the perspective of long-term benefits, it can have a positive effect on promoting economic and even social development, as shown in Fig. 2. From the perspective of measures, it is possible to make linkages between and to promote technological innovation on this basis. Scientific research activities and technical activities (equipment research and development, production technology research, etc.) in sub-fields are regarded as the main direction of technological innovation, and in accordance with the social The actual needs of development refine adaptive technical means. In addition, the current capital investment management system can also be improved, focusing on increasing investment in technological innovation. Government departments should also implement the function of organizing economic construction, change the fragmented application of funds, and activate production factors and development potentials in different regions in a point-by-point manner [6]. Even rural areas have good development prospects in modern society, especially some undeveloped rural areas have good development space. Using technology to drive regional development can also create a new form of society. Therefore, the process of connecting the industrial chain and the innovation chain can achieve sufficient resources, meet the reasonable needs of social development in the social ascent cycle, form regional scale development benefits, and deeply tap the development potential of some regions.

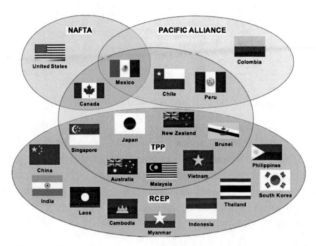

Fig. 2. China's participation in regional economic integration

3.2 Technological Innovation Output and Regional Economic Development

The investment of resources in technological innovation will be transformed into specific results under certain conditions, and these innovation results will become an important factor to promote the economic development of the region, as shown in Fig. 3. In other words, technological innovation output can be regarded as a manifestation of technological capabilities and a reference basis for measuring innovation achievements. Take intellectual property as an example. The intellectual property management system is one of the core contents in the development. It will also focus on protecting some patented contents. With the support of international property rights rules, it will continue to increase its efforts and build local intellectual property rights. The property rights protection system was established, and control plans were resolutely adopted for some behaviors that disrupt market development and violate the laws of market economic development [7].

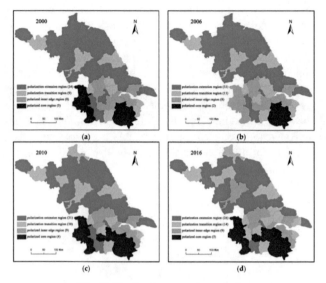

Fig. 3. Regional economic evolution

In general, the management of innovation results should be orderly and scientific, and the innovation platform should also formulate corresponding measures to reasonably evaluate the results, so that the value of the product and its market value should be equal. In the case of changes in the regional economic development environment in the future, reasonably respond to and control market risks, and focus research on the output of technological innovation results [7].

3.3 Featured Operations at the Regional Level

From the perspective of the region itself, it is necessary to establish a technical cooperation development model between different regions, so that a coordinated and coordinated development model between various regions should be formed [8]. Different regions in China have differences in technological innovation capabilities. Therefore, after the establishment of a cooperation mechanism, some regions with a low level of technological development can achieve high-quality innovation through access to resources, and obtain cutting-edge knowledge and policies. Therefore, a corresponding technology research and development part should be built within the region to participate in the monitoring of policy formulation and development and operation, and strive to bring out the advantages of different regions, on the one hand, make full use of resources, on the other hand, give play to the active mediation of the market effect [8]. Specifically, while improving their own development level, each province should also pay attention to the spatial diffusion of development, and rationally adjust its development strategy based on the geographical environment of its own region and the technical level of other regions, and integrate the industries with regional characteristics [9]. Coordinated and innovative operations in other regions will play a positive role in promoting regional economic development through spatial spillover effects. In future research, we can also analyze the relationship between technological innovation and operation development from the perspective of demand, and analyze the relationship between technological innovation and regional operations from the perspective of supply relations, especially in the dynamic changes of the system and the environment [9].

3.4 Coordinated Development at the National Level

The development of modern social technology has made technological innovation capability a key factor for a region to gain competitive advantage. Therefore, from a national perspective, in order to meet the development requirements of the regional economy, it is necessary to adhere to the innovative ideas based on the coordinated regional development model [10]. Correctly recognize the development differences of different regions, coordinate economic, cultural, and institutional aspects, promote the exchange of technological innovation resources, allow regions to learn from each other's strengths, and further utilize the advantages of natural resources and scientific and technological knowledge. In this way, different regions can build innovative cooperation mechanisms and formulate targeted development policies according to different situations. Specifically, the eastern part of China has the most outstanding technological innovation capabilities, followed by the central part, and the western part last. The reason is that the eastern region has obvious advantages in investment funds for innovation and environmental advantages. China needs to adopt corresponding support policies in response to the development trend of the central and western regions, and to stably play the regulatory role of the market mechanism [10]. Under the socialist market economy model, the allocation method of scientific and technological resources will inevitably undergo a fundamental adjustment. Technological innovation in the region will inevitably involve the exchange of funds and resources, increase market-level research and development, and ensure that resources are in a reasonable supply and demand system. The reasonable distribution of the downstream has become an inevitable demand for regional economic development.

4 Conclusion

In the process of technological innovation and economic development, it is necessary to accurately locate the relationship between development factors and regional economy, aiming to achieve steady development in the period of social transformation through institutional construction and system construction. China's economy is transitioning from a stage of rapid growth to a stage of stable and high-quality development, and the original theoretical system of economic development can no longer meet the requirements of contemporary development. Establishing a sound economic development system to clarify the connotation of regional development and constructing development standards in line with national conditions can effectively solve economic contradictions, drive the coordination of various elements to achieve quality and efficiency changes, and make innovation truly become the driving force for guiding development.

References

1. Lu, M.F., Xu, Y.Y.: Internet + strategic research on the optimization of regional economic structure. Mod. Econ. Inf. **6**(02), 86–89 (2017). (in Chinese)
2. Wang, W., Zhang, W.J.: China's regional economic development and research in the new era. Chin. Foreign Entrepr. **26**, 113–116 (2017). (in Chinese)
3. Ge, Q.S., Xi, J.C.: Some thoughts on China's regional tourism development strategy under the new normal. Adv. Geogr. Sci. **12**, 97–100 (2018). (in Chinese)
4. Xing, W.Y.: Research on the promotion of "Internet+" to local economic development under the new normal. Bus. News **15**, 211–214 (2018). (in Chinese)
5. Li, X.Q.: Research on the impact of technological innovation on regional economic development under the background of the Internet. Chin. Foreign Entrepr. **10**, 63–65 (2018). (in Chinese)
6. Chen, X.T.: The relationship between Internet resources and regional economic development. Account. Study **8**(07), 44–46 (2018). (in Chinese)
7. Jing, K.K.: Analysis of the causes and countermeasures of the imbalance of China's regional economic development. Mod. Econ. Inf. **6**(08), 67–70 (2015). (in Chinese)
8. Li, X.W.: Research on the Impact of technological innovation on regional economic development under the internet background. Chin. J. Econ. **12**, 53–55 (2018). (in Chinese)
9. Zhang, Y.L.: Relying on the internet to drive Shanxi's economic development. J. Econ. **6**(05), 76–79 (2018). (in Chinese)
10. Jiang, W.: Internet + e-commerce: promoting the coordinated development of China's regional economy. Mod. Econ. Inf. **11**, 17–20 (2019). (in Chinese)

Digital Inclusive Finance and Innovation of Small and Micro Enterprises Under Mobile Internet

Qun Cao[✉] and Jinyuan Zhang

Macau University of Science and Technology, Macau, China

Abstract. Small and micro enterprises are an important part of China's economy. At present, there are more than 12 million small and micro enterprises in China, which play a prominent role in stabilizing growth, expanding employment, prospering economy, meeting public demand and promoting innovation. Based on the research background, this paper deeply analyzes the financial services of small and micro enterprises at home and abroad. This paper discusses the problems existing in the important financial services. Taking the small and micro enterprises in R region and relying on R's marketing project "small and micro enterprises in villages" as an example, this paper studies the financial services of small and micro enterprises, learn from the advanced experience of small and micro enterprises in the same industry, and explore new paths and new models for the development of financial services of small and micro enterprises.

Keywords: Small and micro enterprises · Financial services · Digital inclusive · Practice survey

1 Introduction

Internet finance is a new model of emerging Internet technologies based on big data and cloud computing in the financial field. Internet finance has brought a lot of changes to the business model, to the market environment, the main demand and service mode has brought rapid changes, which are benefited from the integration of Internet and finance.

In 2014, the development had a great impact on China's commercial banks and their related businesses, which not only brought a certain impact, but also led to the transformation, upgrading and innovative development of commercial banks. At the beginning of 2014, General Secretary Xi made important instructions on China's network security and informatization. At the meeting, he stressed that we should strive to build China into a network power, which has also become the development goal of China's network construction. In addition, Premier Li Keqiang said in the government work report that to build an important network, we need to improve its regulatory mechanism on the basis of realizing the healthy operation of Internet finance [1]. It can be seen that China's Internet development and security construction has become an urgent task of the current socialist construction. At the same time, Internet finance also arises at the historic moment and has a rapid development. According to the development trend, China's Internet finance

J. Abawajy et al. (Eds.): ATCI 2021, AISC 1398, pp. 141–150, 2021.
https://doi.org/10.1007/978-3-030-79200-8_20

development mode will gradually step onto the historical stage in the future. According to the current development situation, small and micro enterprises have made great contributions to the development of China's national economy, provided important support for China's economic growth, and are also moving towards a deeper and broader direction. According to some data, by the end of 2013, the number of registered users of small and micro enterprises in China had exceeded 56 million, including individual businesses, and occupied the first place of enterprise registration with a 94% share. It played an immeasurable role in both the economic contribution rate and the social contribution rate (including employment opportunities, product development and creation, etc.), which shows the rapid development trend. The economic contribution rate is still increasing, and the financing demand is also increasing, but the financing difficulty is also increasing accordingly. The development of digital Inclusive Finance in China is still facing many uncertainties [2–4].

The research of this paper is carried out under this background. Although Internet finance has the advantages, high efficiency and more controllable risk, it brings a turning point to the development bottleneck of small and micro enterprises in China. However, the existing several Internet financial financing modes still need the support of national policies and the supervision of regulatory departments. Under the existing several Internet financial modes, we should develop strengths and avoid weaknesses, and integrate a new financing mode suitable for China's national conditions [5–8].

2   Related Concepts

2.1   Digital Inclusive Finance

Generally speaking, digital Inclusive Finance is the result of the superposition of three meanings. When you take it apart, it is the three terms of digital, inclusive and finance. First of all, digital Inclusive Finance is still finance in essence, which is not separated from the framework of finance. On the contrary, finance is just the starting point of digital Inclusive Finance. Secondly, finance with the modifier of "inclusive" highlights the equality of finance. Inclusive Finance is committed to working for the welfare of the majority of vulnerable groups through financial services, so that the poor can get out of their plight, improve their living standards and quality, and the small, medium and micro enterprises with financing difficulties can get funds to promote social equity. Finally, the addition of the word "digital" brings new features to Inclusive Finance. Such as big data, blockchain and other technologies can not only promote the Internet to better serve Inclusive Finance, but also make inclusive finance more efficient to serve the corresponding groups and improve the level of Inclusive Finance. The emergence of digital Inclusive Finance is the inevitable result of the development of the times, which breaks the time and space constraints of Inclusive Finance [9, 10].

The conference proposed eight high-level principles and outlined the framework of digital Inclusive Finance from eight different perspectives. Principle 1 discusses the key points of using digital technology to better serve and Inclusive Finance; principle 2 explains how to balance the risk control in the process of innovation and development of digital Inclusive Finance; principle 3 creates a regulatory and legal framework for digital Inclusive Finance and stipulates market access standards; Principle 4 is committed to advocating countries to build a basic ecosystem of digital Inclusive Finance Principle 5 designs the content related to consumer protection of digital financial services and establishes a consumer complaint mechanism; Principle 6 defines the education of digital technology and financial knowledge that consumers need due to the continuous popularization of digital Inclusive Finance; principle 7 constructs the identification system of customers' identity, which is committed to protecting customers' privacy and security; principle 8 defines the system of customer identification The main purpose is to monitor the implementation of the principles and supervise the development of the whole digital inclusive financial system.

2.2 Small and Micro Enterprises

Due to the active entrepreneurship of the public, many individual businesses and enterprises have appeared in the public eye. Professor Lang Xianping defined these as small and micro businesses. In our country, there was no concept of enterprises before. With the increasing number of enterprises, the government departments in our country divide the enterprises into small and medium-sized enterprises by subdividing them. Banks and other financial institutions are divided according to the amount of loans, such as micro enterprises with loans less than 1 million, small enterprises with loans less than 5 million, and large and medium-sized enterprises with loans more than 5 million. Therefore, there are differences between large and medium-sized enterprises and enterprises in many aspects, such as the production and operation status, financial status and industrial structure of enterprises.

2.3 Financing Mode of Small and Micro Enterprises

Endogenous financing comes from the internal of the enterprise. The profits obtained by the enterprise through production and operation activities are put into the reproduction of the enterprise. In the future scale expansion or investment, the enterprise can meet the future production and operation, investment activities and operation needs. Endogenous financing has two characteristics: the use of funds by enterprises is not affected by the outside world, and can be controlled by themselves; the second is that the funds are raised within the enterprise, and there is no interest and other financial costs, so they are more flexible and free. The exogenous investment of enterprises includes two aspects. The first is bank lending and issuing bonds, which become direct financing. This way of obtaining funds has a big defect, that is, it improves the debt ratio of enterprises. Well run enterprises can increase their profits by manipulating financial leverage. Well run enterprises have to pay fixed financial expenses, but their income is not enough to pay interest and other expenses, so their financial risks will increase, and they will fall into the dilemma of operation Dead end, serious will lead to enterprise bankruptcy. As a kind of exogenous financing, direct financing is also known as the financing in the primary market. It is a process in which enterprises obtain funds by issuing securities.

3 Small and Micro Enterprise Financial Project Practice Research

3.1 Practice Background

The purpose of the scheme is not only to implement the relevant policies of the government on vigorously supporting the development of enterprises, practice the measures of Inclusive Finance of R bank, provide better financial services for enterprises, and alleviate the financing bottleneck of enterprises; but also to deal with the unoptimistic situation of small and micro credit business of R bank and solve the continuous fermentation of "assets" We need to innovate effective business model.

3.2 Project Promotion Plan

The starting and ending time of the whole project is from April to August in 2019, in which the end of April to the beginning of May is the preliminary preparation stage, the middle of May to the end of July is the key marketing stage, August is the community development stage, and the end of August is the summary stage.

3.3 Questionnaire Survey Statistics

Through the questionnaire collection and collation, we can clearly understand the financing situation of enterprises. Through the investigation of the financing demand of enterprises, the development of the company, the financing channels of enterprises, and other financial services needs of enterprises, we can analyze them.

First, the development of enterprises. In terms of the time of establishment, about 18% of the enterprises have been established for more than 10 years. The proportion of the enterprises established for 1–5 years is the highest, reaching more than 41%, followed by the enterprises established for 6–10 years, accounting for about 32%. In terms of the number of employees, 71% of enterprises have less than 50 employees, and 97% have less than 100 employees. In terms of main products, there are labor-intensive industries with low technology content, such as home appliance manufacturing, packaging and printing, gas cookers, electronic products, etc. The products of small and micro enterprises are mainly sold at home, accounting for about 10% of the total exports, and only 1% of the total exports. According to the annual sales revenue, most of the annual sales revenue is less than 20 million yuan, and only about 3% of the annual sales revenue is more than 50 million yuan. Generally speaking, the establishment time is generally short, the number of people is not too large, the technical content of the main products is not high, the main products are domestic sales, and the annual sales revenue is not high, indicating that the overall strength of small and micro enterprises is not strong.

Second, the demand for credit funds of small and micro enterprises. From the analysis of the credit fund gap of small and micro enterprises, 15% of them have capital, accounting for 31%, of which only half of the capital demand is the demand for working capital; from the perspective of guarantee conditions, 63% of small and micro enterprises have housing, plant, mechanical equipment and other guarantee conditions; from the perspective of financing methods, nearly 40% of small and micro enterprises obtain funds through short-term loans, and nearly 20% of small and micro enterprises obtain funds through medium and long-term loans or loans Bank acceptance to obtain funds. Through the analysis, it is found that there is a capital gap in small and micro enterprises, and their capital demand is mostly short-term liquidity demand. At the same time, 63% of small and micro enterprises have guarantee conditions, which has a certain degree of security.

Third, the financing channels of small and micro enterprises. According to the financing situation of small and micro enterprises, only 17% of them have applied for loans from banks, and only 12% of them have loan balance in banks. Through the analysis, it can be found that although 31% of small and micro enterprises have a capital gap, the number of small and micro enterprises that actively apply for bank loans and have bank loan balance is far lower than the number of small and micro enterprises that have demand. This shows that the development potential of credit business of small and micro enterprises is still large, and there is a certain space to tap the potential. The number of small and micro enterprises that can solve the problem of financing demand through bank loans is lower than 40% of small and micro enterprises with capital gap.

Fourth, small and micro enterprises need other financial services. According to the questionnaire, except for loans, the demand for other financial services is relatively small. About 75% of small and micro enterprises have no demand for other financial services, while only 25% of small and micro enterprises have certain demand for discount, online banking, POS, financial management, collection and payment, SMS, receipt box and other financial services. This shows that small and micro enterprises are in the initial stage of development, the demand is relatively single, and the most important demand is capital demand.

4 Statistical Analysis of Practice Data

4.1 Sample Analysis

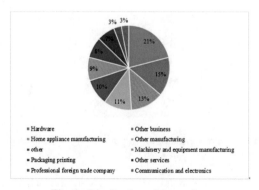

Fig. 1. Distribution of enterprises

The total industrial area of the community is more than 2200 mu, including more than 1250 companies including household appliances, printing and packaging plants, chemical coating plants, catering services, etc. the specific distribution and proportion of enterprises are shown in Fig. 1.

4.2 Project Completion Statistics

In the process of this activity, the small and micro account managers visited 700 enter-prises in total, 55 of which had financing needs, with the amount of 62.11 million yuan, of which 21 had been put in, with the loan amount of 15.55 million yuan; 6 of which were under review, with the loan amount of 6.65 million yuan; 6 of which were in data collection, with the loan amount of 4.8 million yuan; 25 of which had new intention, with the loan amount of 36.65 million yuan. Based on the initial goal, the completion of the project is shown in Table 1 and Fig. 2.

Table 1. Proportion of project development completed

Project	Target	Number of extensions	Percentage completion
Number of targeted customer visits	500	700	140%
New Small Loan Delivery	10000	1358	11%
New micro-loan delivery	1500	197	13%
Number of new small businesses	50	11	14%
New micro-loan households	80	10	12.5%

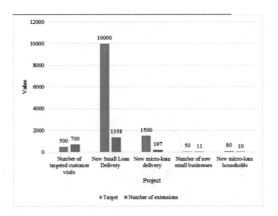

Fig. 2. Proportion of project development completed

In the process of this activity, in addition to expanding loan business, the sub branch also specially strengthened cross marketing of other businesses, completing 82.8% of the total task, as shown in Table 2.

Table 2. Proportion of achievement of linkage indicators

Project	Linkage indicators	Number of targets	Execution	Percentage completion
Increase in linkage indicators	Basic account	250	26	82.8%
	Cyber bank		40	
	Public SMS		35	
	Payment		2	
	POS		4	
	Discount		38	
	Manage money matters		17	
	Back to Single Box		22	
	Credit card		23	
	Total		207	

4.3 Suggestions on Financial Services Under the Background of Digital Inclusive Finance for Small and Micro Enterprises

(1) Giving full play to the functions of the government

International practice has proved that commercial operation is the successful road and development trend of microfinance, and the government should use policy mechanism to play the role of guidance and supervision to ensure a good market environment. Government policy regulation can be divided into direct and indirect, the main tools are administration, finance, supervision and other ways to guide the allocation of financial resources to small and micro enterprises, which has the characteristics of short-term and quick effect.

(2) Reducing government's improper intervention

Through the commercial loan interest rate, we can cover the operating costs and business risks of financial institutions, accelerate the credit business, and support the healthy and sustainable development of small and micro enterprises. On the one hand, restricting the pricing of credit funds by local governments will reduce the enthusiasm of bank loans and the total amount of financing of small and micro enterprises; on the other hand, it will produce rent-seeking space and adverse selection, increasing the credit risk of banks. If the local government forces banks to issue loans in the form of credit, it will directly increase the credit risk of banks, and at the same time, it may lead to improper financial behavior. In a word, the direct administrative means of both price type and quantity type will bring about the loss of the efficiency of financial resource allocation, which is contrary to the policy intention. Therefore, local governments should try their best to reduce direct administrative intervention and avoid "doing bad things with good intentions".

(3) Make good use of financial policy guidance

Fiscal policy is an important tool for the government to play a regulatory role. The government should increase its support for the financial services of small and micro enterprises, use the financial government to guide all kinds of guarantee funds and risk compensation funds to provide a good guarantee environment for small and micro enterprises, and give preferential treatment to small and micro enterprises in terms of loan discount, reward and tax preference, so as to carry out two-way regulation from two aspects of capital supply and capital use. Through the leverage of fiscal policy, the financing cost of small and micro enterprises can be reduced, and the capital use efficiency of small and micro enterprises can be improved. At the same time, with the help of fund leverage, the government can set up special funds for loan turnover of small and micro enterprises to alleviate the problem of capital turnover of small and micro enterprises. The core problem is the proportion of risk-taking between banks and government, and how to find the balance point of risk sharing between the government and banks. The government can try to use tax preferential policies to influence banks to expand the credit scale of small and micro enterprises, support more small and micro enterprises, and provide positive incentives to small and micro enterprises and financial institutions, so as to build a win-win development model.

(4) Establish a hierarchical supervision system

In order to obtain scale efficiency and achieve greater business results, financial institutions have the internal driving force of scale. However, after its scale has grown to a certain extent, the willingness to carry out financial services for small and micro enterprises has begun to decline, especially when there are many management levels and long chains, and the system and mechanism also begin to not adapt to the characteristics of small and micro enterprises' microfinance and rapid demand. On the contrary, small and medium-sized financial institutions have local advantages and mechanism advantages, which are more conducive to the financial services of small and micro enterprises. Therefore, as an economically active Shunde region, first of all, from the perspective of financial development strategy and regulatory philosophy, we must clarify and adhere to the differentiation strategy, and adhere to serving the local economy and small and micro enterprises. Secondly, we should appropriately develop small and medium-sized financial institutions represented by local banks to promote the standardization of private capital lending. Thirdly, we should give appropriate preferential policies in terms of tax burden and supervision, encourage different financial institutions to carry out differentiated financial services in their respective fields, and carry out diversified financial services for small and micro enterprises.

(5) Build financing service platform for small and micro enterprises

An important reason for the financing difficulties of small and micro enterprises is the information asymmetry between the two sides and the lack of communication platform between the two sides. It is of great significance to build a financing service platform for small and micro enterprises, which enables the two sides of funds to communicate with each other and promote information matching and capital matching. It should be led by the government, supported by matching banks, and participated by guarantee institutions to establish a financing service platform for small and micro enterprises, establish an information base for both the supply and demand sides of funds, and effectively match them, so as to realize the direct transfer of funds from the supply side to the demand side, reduce intermediate links, and promote the normalization of private capital. In particular, through the establishment of the platform, the fund information can be made public, and the channels of supply and demand can be widened, the efficiency can be improved, and the operation can be standardized. At the same time, we should steadily promote relevant innovations in local financial markets, lead and drive all kinds of industrial investment funds and venture capital funds to participate in financial services for small and micro enterprises through government mechanisms, and open up new ways of financing financial services for small and micro enterprises.

(6) Establish the concept of "centralization, intensification and cluster" to support small and micro enterprises

Enterprises in industrial clusters generally have the following characteristics: closer geographical location, greater industrial relevance, more familiar with enterprise managers and closer cooperation. From the practical point of view, we should explore new models such as batch credit, cluster financing and overall wind management based on the above characteristics and from the perspective of overall service. One is to organically connect government professional institutions, industry organizations and business circle markets, establish an overall service and linkage service mode, and build a multi-dimensional industrial supporting system. Second, it is necessary to strengthen the segmentation research of small and micro enterprises, and carry out group marketing and development through the government, large supply chain enterprises, chamber of commerce, etc. it is necessary to carry out batch credit and scale development for small and micro enterprise customer groups with similar characteristics, demand and risk control by means of batch credit scheme, adaptive product portfolio, overall marketing and centralized risk control. Third, innovate the supply chain products centered on core enterprises, and actively explore "industry association + guarantee institution + banking institution + small and micro enterprises".

5 Conclusion

The Internet Digital inclusive financial financing mode is a transformation and upgrading of traditional credit intermediary in the network ecology. In the era of big data, according to the core competitiveness of data mining, the credit resource allocation of small and micro enterprises is adjusted and re planned, so as to solve the financing problem of small and micro enterprises, break through the shackles of cost shackles, and realize the financing of small and micro enterprises Innovation and other issues bring strong impetus. Based on the actual follow-up analysis of the practice projects in R area, this paper puts forward various suggestions and countermeasures for policies, platforms and enterprises after data statistical analysis.

References

1. Li, J., Wu, Y., Xiao, J.J.: The impact of digital finance on household consumption: evidence from China. Econ. Model. **86**(3), 317–326 (2020)

2. Borisov, O.I., Karamova, O.V., Naidenova, E.M., et al.: Development of digital competencies among students of higher educational institutions. J. Phys: Conf. Ser. **1691**(1), 12–16 (2020)
3. Arner, D.W., Buckley, R.P., Zetzsche, D.A., Veidt, R.: Sustainability, fintech and financial inclusion. Eur. Bus. Organ. Law Rev. **21**(1), 7–35 (2020). https://doi.org/10.1007/s40804-020-00183-y
4. Wu, H., Patel, C.: Adoption of Anglo-American models of corporate governance and financial reporting in China. Stud. Manag. Finan. Account. **29**(8pt2), 1–255 (2015)
5. Mortimer-Schutts, I.: Traditional retail at the crossroads of inclusive payment services: how serving the needs of the traditional retail sector can accelerate financial inclusion—examples from Indonesia. J. Payments Strategy Syst. **9**(4), 246–255 (2016)
6. Ren, B., Li, L., Zhao, H., et al.: The financial exclusion in the development of digital finance — a study based on survey data in the Jingjinji rural area. Singap. Econ. Rev. **63**(1), 12–16 (2018)
7. Ren, B., Li, L., et al.: The financial exclusion in the development of digital finance - a study based on survey data in the Jingjinji rural area. Singap. Econ. Rev. **63**(1), 65–82 (2018)
8. Goanta, M.A.: Digital development of products with NX9 for academical areas. In: IOP Conference Series: Materials Science and Engineering, vol. 95, no. 4, pp. 120–122 (2015)
9. Michaels, L., Homer, M.: Regulation and supervision in a digital and inclusive world. In: Handbook of Blockchain, Digital Finance, and Inclusion, vol. 1, 9, no. 13, pp. 329–346 (2018)
10. Hedrick-Wong, Y.: Inclusive growth as democratizing productivity - sciencedirect. In: Handbook of Blockchain Digital Finance & Inclusion, vol. 3, no. 2, pp. 463–470 (2018)

Artificial Intelligence Technology in Enterprise Economic Management

Tingting Li[✉]

Department of Economics and Management, North China Electric Power University, Baoding, Hebei, China

Abstract. The application and research of artificial intelligence has been extensively explored abroad. This research mainly discusses the research and application of artificial intelligence technology in enterprise economic management. In the project approval stage, a large amount of information must be prepared, AI-related technologies are used, targeted investment research plans are formed for different industries, and research reports are issued quickly, so as to achieve good results in improving work efficiency and saving costs. Through scientific mathematics theory and efficient AI model to get the best risk assessment results. After the project risk assessment is passed and the funds are invested, it enters the later stage of project investment management, using AI technology, dynamically collecting and analyzing data, timely warning, and timely stop loss. Regarding the key development needs of future digital transformation, 52% of A company financial sharing center managers believe that AI intelligent economic management is a major development direction for future digital transformation. This research is helpful to promote the popularization of artificial intelligence technology in enterprise economic management.

Keywords: Artificial intelligence technology · Enterprise economic management · Risk assessment · Dynamic data collection

1 Introduction

Most of the domestic applications stay in the era of weak artificial intelligence, only replacing simple mechanical repetitive processes with artificial intelligence, and most of them are used to improve efficiency. Of course, this is also limited by artificial intelligence software and hardware. It usually consists of corporate finance, stock trading, industry research, wealth management, risk control and other departments.

With the development of the economy, the number of many small and medium-sized enterprises has shown an upward trend. Therefore, the problems are also found in these many small and medium-sized enterprises, of course, including the companies studied in this article [1, 2]. The problems we found can be summarized as the cost issues required for big data processing, big data technology, big data system support, etc. [3, 4]. Under the background that the country attaches great importance to network security, although AI security companies have the ability to provide overall security software for

J. Abawajy et al. (Eds.): ATCI 2021, AISC 1398, pp. 151–157, 2021.
https://doi.org/10.1007/978-3-030-79200-8_21

the industrial Internet, they have cloud security technologies and products. However, with the increasingly fierce market competition, AI security has also been troubled by business volume that cannot meet the company's development plan, increased peer competition, and service convergence [5, 6]. This article uses a combination of theory and practice, supported by marketing related theories, to elaborate on the current situation of AI security companies' operations, and at the same time through the current network security market conditions, AI's internal advantages and disadvantages of the system Analysis and put forward the existing problems of AI security [7, 8]. Comprehensively analyze the applicable business development strategy, and give corresponding strategic countermeasures in market promotion, strategic innovation, and enhancement of the core competitiveness of the company, in the hope of helping AI companies form the company's characteristic strategy in the cyber security industry circle [9, 10].

The cross-regional development has brought huge profits to the enterprise, but also brought many management problems. Based on the company's financial management, with the establishment of corporate branches and subsidiaries, the development of various new businesses has brought about a rapid increase in the amount of financial business processing. The business processing personnel at the grass-roots level are faced with massive repetitive labor, low added value and error-prone. The company's geographical differences bring more inconvenience in communication. Various factors make the financial processing of multinational companies extremely complicated, and also bring risks to corporate financial management.

2 Enterprise Management

2.1 Artificial Intelligence

Big data and artificial intelligence technology have brought new growth points to the times, and the talent recruitment system continues to make breakthroughs in the direction of digital informatization. The optimization of organizational structure and the reasonable deployment of personnel mark the way for enterprises to upgrade and transform. In this process, if the internal human resource management of the enterprise cannot keep up with the pace of development in a timely manner, so as to meet the new strategic goals of the enterprise, then it will be difficult for us to reap the benefits and convenience brought by technological development. In the actual situation, it is still trapped in the problems of low screening efficiency, poor matching of job positions, and superficial technology that exist in traditional recruitment work.

When using smart IoT technology for data collection, we often monitor the changes in the data.

$$\delta = e^{-\frac{1}{2\sigma}} \left(\tau_{(t)} - T \right)^2 \tag{1}$$

In the time T, the change value of the signal with time is discretely distributed. In the process of intelligent decision-making, the specific output value of the data is often detected.

$$\mu = \frac{y_{(t)} - y_{(min)}}{y_{(max)} - y_{(min)}} \tag{2}$$

The change of the coefficient ratio μ reflects the difference in the measured value.

$$P_{(n)} = 0.55 \cos \frac{2\pi n}{N} + 0.45 \cos \frac{4\pi n}{N} \tag{3}$$

Among them, $P_{(n)}$ represents the value of technology application in the process of economic management. Artificial intelligence technology provides a new calculation and problem solving model.

2.2 Corporate Economic Management

Under the premise of the advent of the era of artificial intelligence scenario applications, it is of practical significance to explore particularly important asset management companies in the financial field. This article's research on A company may be of reference significance to Chinese asset management companies. So this is a forward-looking study in order to plan ahead for the future. Because the company has a large amount of transaction data, artificial intelligence deep learning algorithms based on these data can make financial business better implementation, simulated transactions, data analysis, not only first-class performance, greatly reduced error rate, but also 24 h of work. Taking stock analysis, one of the main businesses of asset management companies, for example, the brains of human stock analysts can remember the trading rules and price trends of hundreds of stocks at most, and can at most according to the thousands of factors in the stock market that affect the stock price, work out a simplified trading strategy.

3 Enterprise Economic Management Experiment

(1) Efficient and intelligent investment research and strategy formulation plan. In the project approval stage, a large amount of materials must be prepared, including business due diligence and financial due diligence. The entire process takes a long time and generates a large amount of travel costs, such as accommodation, air tickets, and travel subsidies. Use AI-related technologies to form targeted investment research plans for different industries, intelligently collect and organize data, achieve differentiated and targeted due diligence, and quickly issue research reports, so as to improve work efficiency and save costs good results.

(2) Improve the dimension of risk assessment. Use AI technology to reorganize and supplement risk assessment factors, and obtain the best risk assessment results through scientific mathematical theories and efficient AI models.

(3) Effective project risk assessment. After the project risk assessment is passed and the funds are invested, it enters the later stage of project investment management, but the market environment is unpredictable. Using AI technology, dynamic data collection and analysis, timely warning, timely stop loss. Table 1 shows the comparison of the average working time of the account check change process.

Table 1. Comparison of the average working time of the account check change process

Transition before optimization	Average working time/S	Transition process after optimization	Average working time/S
TO	240	T0	15
T1	1 440	T1	15
T2	120	T2	15
T3	1 800	T3	10

4   Intelligent Enterprise Economic Management

4.1   Indicators Before and After Tax Management Business Process Optimization

Table 2 shows the comparison of indicators before and after tax management business process optimization. It is not difficult to see from Table 2 that in the tax management optimization process, compared with the manual operation of the financial sharing center of A company, the introduction of the processing technology of robot process automation based on the AI platform not only shortens the transition process of the entire process, but also time efficiency. An increase of 97.6%. It only used less than 3% of the original time to complete the entire business process, which greatly reduced the time waste in the tax management process, improved work efficiency, and reduced the time cost of this link.

Table 2. Comparison of indicators before and after tax management business process optimization

Optimization process	Value before process optimization	Value after process optimization	Optimize efficiency
Average process time	11820	272	97.6%

4.2   AI Professionals

Due to the lack of strategic height of financial artificial intelligence technology of A company, combined with the talent demand of financial artificial intelligence technology transformation of financial sharing center management personnel of A enterprise, 33% of financial sharing center managers believe that enterprises urgently need financial artificial intelligence technology strategy leaders, more than half of managers believe that companies need strategic leaders to implement financial artificial intelligence technology from a strategic height, leading the strategic transformation of artificial intelligence technology in corporate financial sharing centers. In addition, the current personnel in the financial sharing center of A company are basically financial personnel related to financial work. Professionals with in-depth research in the professional field of financial artificial intelligence technology are still relatively scarce, such as artificial intelligence technology project managers, robots and intelligence. This has a certain relationship with the lack of strategic height of financial artificial intelligence technology of A company, lack of professional leadership at the strategic level, fail to deploy enough financial artificial intelligence technology professionals according to development needs, and the development of financial sharing center artificial intelligence technology is hindered to a certain extent. The artificial intelligence technology project manager is familiar with artificial intelligence technology and business, and can implement the latest artificial intelligence technology management methods in the enterprise according to the strategic planning and development of the enterprise, promote the transformation of artificial intelligence technology in the financial sharing center, and improve the effect of increasing efficiency and reducing costs. Robots and intelligent engineers can ensure the normal operation of the financial robots in the financial sharing center, and rationally optimize the financial robots according to the business and development characteristics of the enterprise to make it more suitable for the development needs of the A enterprise. The demand for AI professionals is shown in Fig. 1.

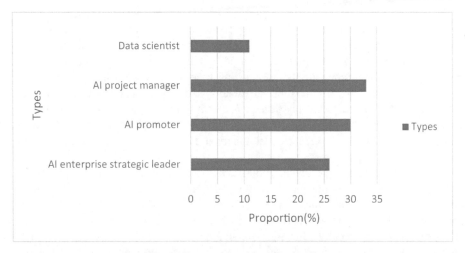

Fig. 1. Demand for AI professionals

Regarding the key development needs of digital transformation in the future, 52% of A company financial sharing center managers believe that AI intelligent economic management is a major development direction for digital transformation in the future. A company can learn from the more mature shared center operating experience of internationally well-known companies. From the current relatively single financial shared service, it has been optimized and developed into a multi-functional shared service center covering finance, procurement, sales, salary, administration, human resources, etc., which is also in line with A. Future development needs and development plans of the corporate financial sharing center. Figure 2 shows the key demand analysis of digital transformation.

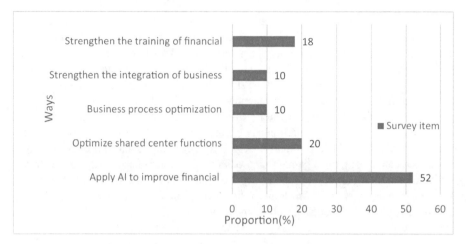

Fig. 2. Analysis of key needs for digital transformation

5   Conclusion

This research mainly discusses the research and application of artificial intelligence technology in enterprise economic management. In the project approval stage, a large amount of materials must be prepared, including business due diligence and financial due diligence. The entire process takes a long time and generates a large amount of travel costs, such as accommodation, air tickets, and travel subsidies. Use AI-related technologies to form targeted investment research plans for different industries, and quickly issue research reports, so as to achieve good results in improving work efficiency and saving costs. Through scientific mathematical theories and efficient AI models, the best Risk assessment results.

Under the premise of the advent of the era of artificial intelligence scenario applications, it is of practical significance to explore particularly important asset management companies in the financial field. After the project risk assessment is passed and the funds are invested, it enters the later stage of project investment management, using AI technology, dynamically collecting and analyzing data, timely warning, and timely stop loss. This research is helpful to promote the popularization of artificial intelligence technology in enterprise economic management.

References

1. Thrall, J.H., Li, X., Li, Q., et al.: Artificial Intelligence and Machine Learning in Radiology: Opportunities, Challenges, Pitfalls, and Criteria for Success. J. Am. Coll. Radiol. **15**(3), 504–508 (2018)
2. Miao, Z.: Investigation on human rights ethics in artificial intelligence researches with library literature analysis method. Electron. Libr. **37**(5), 914–926 (2019)
3. Hansen, E.B., Iftikhar, N., Bgh, S.: Concept of easy-to-use versatile artificial intelligence in industrial small & medium-sized enterprises. Procedia Manuf. **51**(7), 1146–1152 (2020)

4. Kim, K.Y., Jung, J.H., Yoon, Y.A., et al.: Designing a performance certification test for automatic detection equipment based on artificial intelligence technology. J. Appl. Reliab. **20**(1), 43–51 (2020)

5. Jiang, H., Cui, C., Chen, Y., et al.: The application of artificial intelligence to drug sensitivity prediction. Chin. J. **65**(32), 3551–3561 (2020)

6. Feshchur, R.V., Tymoshchuk, M.R., Kopytko, S.B.: Management of social and economic stakeholders engagement in business environment. Finan. Credit Activity Prob. Theory Pract. **1**(24), 469–477 (2018)

7. Wagner-Tsukamoto, S.A.: Economics, management ethics, and business history: adam smith - then and now. Interdisc. J. Econ. Bus. Law **7**(1), 120–154 (2018)

8. Portna, O.V.: Economic business partnerships within Industry 4.0: new technologies in management. Montenegrin J. Econ. **17**(1), 151–163 (2021)

9. Annisa, H.Z., Utami, F.L., Ali, A.J.: Women's economic empowerment through home business financial management in North Kembangan. ICCD **2**(1), 264–268 (2019)

10. Mazandarani, M., Ma, A.R.: Political connections and financial performance of companies accepted in tehran stock exchange. Int. J. Bus. Econ. Manag. **6**(1), 81–84 (2018)

Face Detection of Innovation Base Based on Faster RCNN

Haixing Guan, Hongliang Li, Rongqiang Li, and Mingyang Qi[✉]

Jilin Agricultural Science and Technology University, No. 77, Hanlin Street, Changyi District, Jilin, China

Abstract. In order to improve the efficiency of teachers' attendance and save teachers' energy and time, this paper proposes a face detection algorithm based on improved Faster-RCNN, in order to overcome the problems of gradient disappearance and gradient explosion caused by too deep network depth. In this paper, ResNet50 is used to replace VGG16 backbone feature extraction network, and soft non-maximum suppression method is used to improve the recognition rate of overlapping faces based on the principle of Faster-RCNN algorithm, ResNet50 residual network model and RPN network structure principle are described in detail. Training and testing on Wider Face data set, carrying out a variety of network comparison experiments, provide a strong basis for the experimental results. At last, the experimental data show that ResNet50 feature extraction network is relatively ideal and the detection accuracy on the test set reaches 85.3%, which is 3.9% higher than that of VGG16 on average. The detection time of a single picture is 0.34 s, which meets the real-time monitoring requirements and provides a new idea for the face attendance system.

Keywords: Face detection · Faster-RCNN · Soft non-maximal suppression · Residual network

1 Introduction

Face detection system is an important part of computer vision system, which has a wide range of applications, such as security check, target tracking and so on, which brings a lot of convenience to life. The facial supervision system with the theme of innovation base can supervise students' learning, save teachers' time and energy, save a lot of manpower and material resources for the construction of school innovation base which greatly improves students' consciousness and learning efficiency.

In recent years, the network based on computer vision face recognition algorithm is constantly upgrading. The mainstream detection network is mainly divided into two stages: yolo [1], ssd [2] and Faster-RCNN [3]. Faster-RCNN network is improved on the basis of target detection framework RCNN [4], which greatly improves the overall computing ability and training performance and saves a lot of time. At the same time, face detection methods in deep learning have achieved good results in recent years, for example, the Basic Region Suggestion Network (RPN) and RCNN proposed by Chen

© The Author(s), under exclusive license to Springer Nature Switzerland AG 2021
J. Abawajy et al. (Eds.): ATCI 2021, AISC 1398, pp. 158–165, 2021.
https://doi.org/10.1007/978-3-030-79200-8_22

[5] et al. The multi-scale Faster-RCNN method proposed by Le [6] et al. fuses the feature images of different layers to realize multi-scale face detection. Dai [7] et al. put forward an improved R-FCN method, the core idea of which is to extract feature images according to the position-sensitive characteristics of faces to improve the performance of face detection. However, in normal life, there are various situations such as overlapping and small size of face information, and the effect of R-FCN convolution network in processing deep semantic information is not ideal.

Based on the above description, the Faster-RCNN network adopted in this paper can solve the shortcomings of the above network. Generally, the feature extraction network of Faster-RCNN uses VGG16 [8]. Because VGG16 network is prone to gradient disappearance and gradient explosion, it leads to the loss of training and information. The requirement of face detection accuracy is relatively high. Therefore, in this paper, Faster-RCNN uses residual network (ResNet50) instead of traditional VGG16 network, and uses deeper ResNet50 network [9] to deal with some details more carefully. Residual network can also prevent the loss of original picture information and keep more original information to improve the picture information of trunk extraction network, which can achieve more accurate prediction results.

2 Faster-RCNN Model Structure

2.1 Basic Principles of Faster-RCNN

The main feature extraction network of Faster-RCNN can use various networks to extract information, while the traditional extraction network is VGG16 network, and the network used in this paper is ResNet50 extraction network. The function of using the deeper residual network ResNet50 is to convolute and compress the input picture to obtain the characteristic picture, which provides the foundation for the following network.

To generate prediction frames, ResNet50 can be used to generate feature pictures, and then 9 prior frames will be generated in each feature picture grid, and the prior frames will be convolved to generate prediction frames. Then study how the prediction frame predicts whether the inside of each prediction frame on each grid point on the common feature layer contains human face information.

To obtain Proposal suggestion box is to adjust these prediction boxes by convolution on the basis of prediction boxes and judge whether the prediction boxes contain targets. If there are targets, a rough box will be generated in the nearby position.

In the process of decoding the Proposal suggestion box, although the prior box can represent the position information and size information of a certain box, it is not accurate enough, so it needs to be adjusted. The prediction results of RPN network are transformed into suggestion boxes and then slice is used for segmentation to prevent the suggestion boxes from exceeding the edge of the image, and some suggestion boxes with the highest scores are taken out. These suggestion boxes with high scores are the final output results.

The whole implementation flow chart of Faster-RCNN is shown in the following Fig. 1:

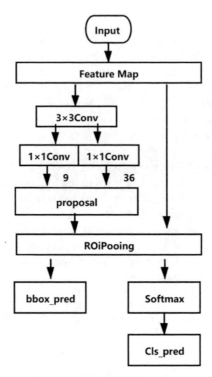

Fig. 1. Faster-RCNN flow chart

2.2 RPN Principle

RPN [10], whose full name is Region Proposal Network, is mainly used to extract candidate frames. The principle of RPN is that after a feature graph is processed by sliding window, an n × n sliding window appears on the shared feature layer Feature Map, which can be selected as n = 3 according to picture analysis, that is, a 3 × 3 sliding window. After moving, M × N feature graphs are generated, each feature graph is a fully connected feature with a length of 256-dimensional vector and then two branched fully connected layers are generated after the features in its dimension. The main function of reg-layer is to generate proposal coordinates x, y and width and height w, h corresponding to the central anchor point of proposal for prediction; The main function of cls-layer is to judge whether the proposal is a foreground or a background.

The output results are divided into two scores and four coordinates, in which the two scores represent the probability of predicting foreground and background respectively, the four coordinates are offset coordinates relative to the original image and then combined with the preset anchor, the candidate frame is obtained after post-processing.

RPN structure is shown in Fig. 2:

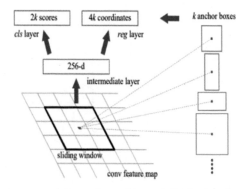

Fig. 2. RPN network structure

For the boundary box regression, this paper adopts the following four coordinate parameterization formulas:

$$Tx = (X - Xa)/Wa$$
$$Ty = (Y - Ya)/Ha$$
$$Tw = \log(W/Wa)$$
$$Th = \log(H/Ha)$$
$$Tx* = (X * -Xa)/Wa$$
$$Ty* = (Y * -Ya)/Ha$$
$$Tw* = \log(W * /Wa)$$
$$Th* = \log(H * /Ha)$$

$$(1)$$

In formula (1), x, y, w and h represent the center coordinates of the box and its width and height, variables x, Xa and X* are prediction box, anchor box and truth box respectively.

2.3 ResNet50 (Residual Network)

In the general deep-level network training, the deeper the network is, the more difficult it is to train, and gradient disappearance and gradient explosion are easy to occur. That is to say, when the gradient disappears, the hidden layer closer to the output layer is relatively normal in weight updating while the hidden layer closer to the input layer is slow and suspended in weight updating due to its gradient disappearance, resulting in training loss and information loss. Gradient explosion is similar to gradient disappearance. The difference is that gradient explosion occurs when the weight is too large and gradient disappearance occurs when the weight is too small. The essence of both is the same, which is caused by too many layers of training network.

In order to solve this problem, the residual network is introduced. The structure of the residual network is shown in Fig. 3.

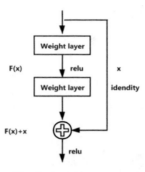

Fig. 3. Residual network

2.4 ReLU Activation Function

Commonly used activation functions are Sigmoid [11], Tanh, ReLU [12] and so on. Neurons using ReLU activation function only need to add, multiply and compare which is more efficient in calculation.Sigmoid-type activation function will lead to a non-sparse neural network, while ReLU has good sparsity, and about 50% of neurons will be activated. In the aspect of optimization, because both ends of Sigmoid type function are saturated, ReLU function is left saturated function and its derivative is 1 when x > 0, which alleviates the gradient disappearance of neural network to a certain extent and accelerates the convergence speed of gradient descent. However, Tanh activation function will cause gradient disappearance due to saturation, so this paper finally decided to adopt ReLU activation function.

Equation (2) of ReLU activation function is as follows:

$$f(x) = \max(0, x) = \begin{cases} x, x > 0 \\ 0, x \leq 0 \end{cases} \tag{2}$$

In which: x represents input and f(x) represents output.

2.5 Soft Non-maximum Suppression (Soft-NMS)

Because NMS algorithm will delete candidate frames larger than the specified threshold (a certain preset confidence level) for overlapping candidate frames, if the overlap degree of faces in the image is high, the score of face detection frame in the occluded part will be set to 0, which will lead to the failure of face detection and reduce the map value of the algorithm, as shown in Fig. 5 (Fig. 4):

In view of the existing problems of NMS, this paper adopts a new Soft-NMS algorithm. For the case of overlapping detection frames, an attenuation function is set instead of directly setting its score to 0. That is to say, if a detection frame has some overlapping parts with m, it will have a very low score; However, if the detection frame does not overlap with M, its original detection score will not be affected.

The NMS score reset function formula (3) is as follows:

$$S_i = \begin{cases} S_i, \text{iou}(M, b_i) < N_t \\ 0, \text{iou}(M, b_i) \geq N_t \end{cases} \tag{3}$$

Fig. 4. Schematic diagram of overlapping face missing detection

The formula (4) of Soft-NMS fractional attenuation function is as follows:

$$S_i = \begin{cases} S_i, \text{iou}(M,b_i) < N_t \\ S_i(1 - \text{iou}(M,b_i)), \text{iou}(M,b_i) \geq N_t \end{cases} \tag{4}$$

3 Experiment and Experimental Analysis

3.1 Experimental Environment and Experimental Details.

Server operating system is Ubuntu18.04 LTS, loading software environment is cuda10.0 + cudnn7.42 + anaconda + tensorflow-GPU = 1.2.1 + keras-GPU = 2.1.6 + python = 3.6.2 + screen and GPU is NVIDIA Tesla k80. Epoch is 100 and batch_size is 32 in the experiment. Wider Face face data set is used for training. The data set contains 16,106 pictures, of which 806 are used as test sets, and all the images are collected in natural scenes, which can meet the training requirements of this experiment.

In the training stage, the data set is trained on the model for 100 generations, and the initial learning rate is set to 0.0001. After training to the 20th generation, the learning rate is reduced to 0.00001. In RPN network, the number of Anchor is 9, which includes three basic dimensions with aspect ratio of 1: 1, 1: 2, 2: 1, 64 × 64, 128 × 128 and 256 × 256. In the testing stage, for each test image, RPN network will generate 128 prior frames and then get the suggestion frame after being adjusted by soft non-maximum suppression. In this paper, the threshold of Soft-NMS is set to 0.3.

3.2 Experimental Analysis

In order to test the recognition effect of each network, different comparative experiments were carried out to test the effectiveness of the network. The experimental results are shown in Table 1:

It can be seen from the above table that the accuracy of ResNet50 model is improved by 3.9% on average compared with VGG16 model and the effect of Soft-NMS method is better than NMS method with the accuracy increased by 1.8% on average. Although NMS algorithm and VGG16 can improve the speed of image detection, this paper finally

Table 1. Experimental results of accuracy comparison of different strategies

Model	Single picture detection time (s)	Accuracy (%)
VGG16 + NMS	0.23	79.6
VGG16 + Soft-NMS	0.25	81.3
ResNet50 + NMS	0.31	83.4
ResNet50 + Soft-NMS	0.34	85.3

chose Faster RCNN + ResNet50 + Soft-NMS network model compared with its lost accuracy.

In the training stage, this paper tries to carry out iterative training with different learning rates, which are 0.001, 0.0001 and 0.00001 respectively. The results show that when the learning rate is 0.001, the model iteration is faster and the training speed is faster, but the final detection effect is not ideal. When the basic learning rate is 0.0001 and 0.00001, the iteration time of the model is delayed, but the final detection effect is ideal. The detection effect is shown in Fig. 5:

Fig. 5. Schematic diagram of detection effect

4 Results and Discussion

In this paper, based on the Faster-RCNN network model, an improved Faster-RCNN model is proposed. The residual network ResNet50 is used as the backbone network to extract image features, which effectively avoids the problem of image over-fitting, and at the same time makes the model more suitable for face image detection of different scales. In addition, the soft non-maximum suppression method is used to solve the problem of overlapping and missing detection of face targets. Experiments show that the improved Faster-RCNN network model has a good effect on face detection, and the average detection accuracy on Wider Face dataset is 85.3%. Due to the limitation of experimental conditions, there may be some errors in the experimental data. The detection accuracy of this paper needs to be improved, and the future work will further optimize the network structure to better extract image face features and improve the accuracy of face detection.

Acknowledgements. Fund Project: Supported by Jilin Agricultural Science and Technology College Students' Science and Technology Innovation and Entrepreneurship Training Program (No.202011439004).

References

1. Redmon, J., Divvala, S., Girshick R., et al.: You only look once: unified, real-time object detection. In: Proceedings of the IEEE Conference on Computer Vision and Pattern Recognition, pp. 779–788 (2016)
2. Liu, W., et al.: SSD: single shot multibox detector. In: Leibe, B., Matas, J., Sebe, N., Welling, M. (eds.) Computer Vision – ECCV 2016: 14th European Conference, Amsterdam, The Netherlands, October 11–14, 2016, Proceedings, Part I, pp. 21–37. Springer International Publishing, Cham (2016). https://doi.org/10.1007/978-3-319-46448-0_2
3. Ren, S., He, K., Girshick, R., Sun, J.: Faster R-CNN: towards real-time object detection with region proposal networks. In: Advances in Neural Information Processing Systems, pp. 91–99 (2015)
4. Girshick, R., Donahue J., Darrell, T., Malik, J.: Rich feature hierarchies for accurate object detection and semantic segmentation. In: Proceedings of the IEEE Conference on Computer Vision and Pattern Recognition, pp. 580–587 (2014)
5. Chen, D., Hua, G., Wen, F., et al.: Supervised transformer network for efficient face detection. In: Proceedings of the IEEE Conference on Computer Vision and Pattern Recognition, pp. 122–138 (2016)
6. Le, T.H.N., Zhang, Y., Zhu, C., et al.: Multiple scale faster-RCNN approach to driver's cellphone usage and hands on steering wheel detection. In: Proceedings of the IEEE Conference on Computer Vision and Pattern Recognition, Workshops, pp. 46–53 (2016)
7. Dai, H., Mao, Y.: An improved face detection algorithm based on R-FCN model. Comput. Modern. **276**(8), 16–19+24 (2018)
8. Simonyan, K., Zisserman, A.: Very deep convolutional networks for large-scale image recognition. In: ICLR (2014)
9. Theckedath, D., Sedamkar, R.R.: Detecting affect states using VGG16, ResNet50 and SE-ResNet50 networks. SN Comput. Sci. **1**(2), 18–37 (2020). https://doi.org/10.1007/s42979-020-0114-9
10. . Uijlings, J.R.R., Sande, K.E.A., Gevers, T., Smeulders, A.W.M.: Selective search for object recognition. Int. J. Comput. Vis. (2) (2013)
11. Langer, S.: Approximating smooth functions by deep neural networks with sigmoid activation function. J. Multivariate Anal. (2020, prepublish)
12. Liang, X., Xu, J.: Biased ReLU neural networks. **423**, 71–79 (2021)

Application of Computer Information Technology and Liang Ping's New Year's Paintings in the Field of Modern Design

Yao Lu[✉]

Chongqing College of Architecture and Technology, Chongqing, China

Abstract. Chinese folk New Year pictures have diverse themes, rich content and bright colors, and are deeply loved by people. Chinese folk New Year pictures use contrast methods such as light and shade, cold and warm, and the pictures are vivid and painted in primary colors. The content revolves around praying for blessings and warding off evil spirits, expressing people's emotions and wishes. Liang Pingming's New Year paintings have extremely high aesthetic and artistic value, and are of great significance to the study of local customs and lifestyles in Liangping. This article introduces the characteristics of Chinese New Year pictures and Liang Ping New Year pictures, and then introduces the application of computer information technology in the design of New Year pictures.

Keywords: Computer · Information technology · Liang ping new year pictures · New year pictures · Design

1 Introduction

New Year pictures created a precedent in Chinese folk art. People like to post New Year pictures during the Spring Festival to pray for happiness and peace. Folk New Year pictures have the meaning of eliminating disasters in people's minds. The forms are very rich and have high artistic value. Folk New Year pictures were originally carved and printed on wood plates. With the changes of the times, folk New Year pictures developed into color printing, which is convenient and beautiful to use. They are often used as decorations to create a festive atmosphere [1]. The forms of folk New Year pictures are relatively diverse, and the "door god paintings" that have been handed down to this day are also a kind of folk New Year pictures. Legend has it that during the reign of Emperor Taizong Li Shimin of the Tang Dynasty, the palace was haunted. In order to make the emperor feel at ease, generals Qin Shubao and Yu Chigong volunteered to stand guard to ensure safety. Two generals, one holding a mace and the other holding a whip, guarded the palace gate day and night to brace for Li Shimin. Later, Li Shimin felt that they had worked too hard, so he asked the artist to paint their mighty image on the palace gate. Later, the portraits of Qin Shubao and Yu Chigong spread to the people and became the "door gods" of thousands of households. In addition, people in ancient times were very afraid of thunder and lightning and beasts, and they liked to use New Year

J. Abawajy et al. (Eds.): ATCI 2021, AISC 1398, pp. 166–172, 2021.
https://doi.org/10.1007/978-3-030-79200-8_23

pictures with the symbolic meaning of exorcising evil to eliminate fear in their hearts [1]. Therefore, the "door god" appeared earlier as a form of New Year pictures. With the changes and development of society, people look forward to peace and happiness. Therefore, folk New Year pictures have gradually added styles that symbolize good luck and success, such as multi-children and longevity, dolls, and women [1].

2   The Color Characteristics of Chinese Folk New Year Pictures

Chinese folk New Year pictures have a strong artistic flavor and rich connotation of color expression. The use of colors can highlight the unique artistic sense of New Year pictures, which is favored by more and more people [2]. Our in-depth study of the color characteristics of folk New Year pictures can inherit and absorb the essence of the culture, which has very important reference value and significance for innovation in modern painting.

2.1   Symbolic

Color is an important part of folk New Year pictures. There are qualitative differences in various colors, and the cultural connotations expressed are also different. Therefore, folk New Year pictures have symbolic characteristics and have a certain inheritance effect on colors.. New Year pictures draw on the Chinese culture's understanding of the meaning of color to use colors. Red is usually used to express joy and auspiciousness. Yellow, green, and purple are used after red. The auspicious meaning is second only to red. The color in folk New Year pictures is a kind of inheritance of symbolism. The use of color can make people feel the meaning and certain emotions [3]. The color can express the character of the character. For example, green symbolizes tenacity and indomitability; red symbolizes integrity and majesty; gold and silver Most of them mean ghosts and gods; yellow means tyranny and so on. The expression of color can more directly affect people's subjective consciousness.

2.2   Popularity

The production of folk New Year pictures needs to consider the aesthetic habits of the public in different periods. Folk New Year pictures should make full use of colors in the production to show the richness of New Year pictures. In the drawing of folk New Year pictures, the popular color formulas are summarized. These color-matching formulas are the experience continuously summarized in the production of folk New Year pictures, and the working people's deep understanding of color based on aesthetics [4].

2.3   Performance

The painting production of folk New Year pictures is eclectic in the use of colors, and is not restricted by the color of traditional Chinese paintings. It can richly express the connotation of New Year pictures and is deeply loved by people. The colors of folk New Year paintings are mostly three primary colors [4]. In order to highlight expressiveness, color contrast is often used to add visual impact. Generally, there are color contrasts in terms of coldness and warmth, brightness, and purity. For example, the combination of red and green colors gives people a feeling of beaming, which enhances the tension and expressiveness of folk New Year pictures as a whole, and reflects people's enthusiasm for life [5].

3   The Unique Color System of Liang Pingmingjian's New Year Paintings and Its Application in Design Works

The use of colors in folk New Year pictures has high artistic expressiveness. After years of inheritance and development, it has been recognized and loved by the public. Folk New Year pictures can attract people's attention through the use of colors. Excellent New Year pictures can not only enhance the artistic atmosphere, but also promote the innovative development of contemporary art design.

3.1   The Unique Color System of New Year Pictures

The most emotional resonance of Chinese folk New Year pictures is the use of colors. The colors used in Chinese folk New Year paintings are different from those in Western paintings. Western paintings often use conditional colors, while Chinese folk New Year paintings do not use intermediate colors and often use the unique style of contrasting primary colors, which has a visual impact, as shown in Fig. 1. Folk New Year pictures have always been subjective, strong, and distinctive, which are directly related to the use of their color system [5]. Visually, they can feel the resonance of colors and are

Fig. 1. Unique color system of New Year pictures

deeply loved by the Chinese people. In the color system of folk New Year pictures, the design is often centered on the "year". People are used to sticking New Year pictures on the Spring Festival to get good blessings. Therefore, the main color of folk New Year pictures is red, so that people can feel the New Year pictures bring. The warmth and joy [6].

3.2 Application of Colors in New Year Pictures in Design Works

Folk New Year pictures can be passed down to this day and are loved by people mainly because they have unique color styles, and their bold colors and creative shapes are unforgettable. Folk New Year pictures have profound historical charm and cultural connotation, and have high artistic value and inheritance value [6]. The colors of folk New Year pictures are often used in decoration production by later generations, not only highlighting Chinese culture, but also enhancing the artistic value of decorations. The color system of New Year pictures brings a bright and lively feeling. In contemporary graphic design, the use of colors in folk New Year pictures is often used for reference to improve the appreciation and practicality of the works.

4   The Characteristics of Computer New Year Picture Image Design

4.1   Speed up the Drawing Speed and Quality

The traditional New Year picture image design is mainly designed by designers using their own hands to draw in their minds according to the needs described by the customer, and then draw the imagined New Year pictures and images on paper with a brush. This process requires a lot of time to design. In addition, energy, if the design of the image New Year pictures does not meet people's needs, a large amount of work needs to be modified. In addition, use computer to design New Year pictures, according to the actual needs of customers, make overall plans for New Year pictures, and then design New Year pictures images that meet customer needs and various styles through color matching, which injects new vitality into the design of New Year pictures and brings people [7].

4.2   High Precision and Small Error

Computer as a tool for processing information, although the structure is relatively complex, it is very simple to use as long as you master the method of use. With the development of various drawing software, many software are very convenient to use [8]. The computer is very fast and accurate when processing data, and can accurately draw the size of the New Year picture with high precision and small error. In addition, with the continuous upgrading of computer drawing software, the image New Year pictures drawn by the computer is very realistic and vivid, and can present a good three-dimensional effect.

5   The Application of Computer Technology in the Image Design of New Year Pictures

5.1   Improved Processing Accuracy of New Year Pictures

The authenticity of the picture quality is improved. The picture quality processing in the traditional New Year picture production often suffers from distortion, and it looks like it is not smooth [8]. The application of computer technology has improved the processing precision of the picture quality of the New Year pictures, and the authenticity of the picture quality has been improved. Each object has a more independent activity in the illusory world [9]. The comprehensive application of computers, kinematics, electron optics and other disciplines promotes the visual effects of the face-to-face design to be more realistic, and viewers can see more realistic and perfect Picture quality, as shown in Fig. 2.

Fig. 2. Computer software used in the design of New Year pictures

5.2   Enrich the Design Content of New Year Pictures

In the design of New Year pictures, the use of computer drawing software can achieve the purpose of selling goods. In the design of New Year pictures, information can be effectively disseminated through New Year pictures animation [9]. The competition in the New Year picture market is becoming more and more fierce. If people want to be concerned in the fierce market competition, it is the most feasible way to promote the New Year picture animation. The use of multimedia to spread the New Year picture animation designed by computer can arouse people [10]. The demand for commodities promotes user consumption and constitutes consumption. In the New Year picture animation, all the benefits and advantages of the product are shown. New Year picture animation can not only sell New Year pictures, but also sell a certain industry or a certain service [10].

5.3   Design to Improve the Image Quality of New Year Pictures

The processing of New Year pictures using computers has also been widely used in our country. With the continuous improvement of people's quality of life, people are paying more and more attention to spiritual enjoyment. What is satisfied in modern times is people's sensory enjoyment. Making full use of the computer New Year picture image design can realize the right virtual design. The image of New Year pictures designed by computer has strong expressive ability and can make the real construction more convenient [10]. During the virtual design, you can design in the virtual reality performance scene to enhance the overall design effect, as shown in Fig. 3.

Fig. 3. Computer software design improves the image quality of New Year pictures

6   Conclusion

Due to the continuous development of computer technology, computer New Year picture image design has been widely used in New Year pictures. Using computers to draw various renderings is not only simple and convenient, data processing is very accurate, and it has a good performance effect. The colors of Liang Ping's New Year's paintings are symbolic and popular, and the use of different colors expresses the different emotional desires of the people.

Acknowledgement. Humanities and Social Sciences Research Project of Chongqing Education Commission (No. 20SKGH358); Project name: Research on the Application of Digital Edition in Liang Ping New Year Pictures.

References

1. Zhang, S.: The application of digitization in the protection and dissemination of intangible cultural heritage of traditional handicrafts. Art Design **5**(05), 212–215 (2019). (in Chinese)
2. Wang, Y.: Looking at traditional Chinese color views from Yangjiabu's woodcut New Year pictures. Tradit. Innov. **12**, 85–88 (2012). (in Chinese)
3. Men, D.L., Tang, L.: The worship of red in traditional Chinese color research. Southern J. **11**(10), 45–47 (2010). (in Chinese)
4. Jin, N., Cheng, J.: An analysis of the value of traditional folk new year pictures from a cultural perspective. J. Heihe Univ. **9**(09), 191–193 (2019). (in Chinese)
5. He, D.: Analysis of the performance characteristics and application of traditional Chinese new year pictures. Fine Arts Educ. Res. **10**(07), 35–37 (2019). (in Chinese)
6. Chen, H.: The "color" and "color" of Chinese folk woodblock new year pictures. Grand View Fine Arts **4**(07), 87–91 (2019). (in Chinese)
7. Wang, H.: Research and engineering application of computer animation technology. Inf. Comput. (Theoret. Ed.) **15**(03), 165–167 (2019). (in Chinese)
8. Zeng, L.K.: The application of computer virtual simulation technology in three-dimensional animation production. Mod. Inf. Technol. **3**(08), 92–93 (2019)
9. Zhang, S.: Application of computer technology in 3D animation production. Education **8**(02), 85–87 (2019). (in Chinese)
10. Zheng, R.Q.: Research on the application of computer information technology in the production of new year pictures. New Educ. **7**(02), 44–46 (2019). (in Chinese)

Exploration of Innovative Development of Logistics Management in the Internet Age

Jingde Weng[⌧]

Yang-En University, Quanzhou, Fujian, China

Abstract. In the current society with the rapid development of computer technology, the speed, type, and quantity of information dissemination in people's lives have reached unprecedented heights. The Internet has gradually integrated into people's lives and played an important role. Due to the great tolerance of Internet technology Due to its nature, it has been widely used in various industries. Taking logistics management as an example, the application of Internet technology can significantly improve work efficiency. This not only proposes a new development direction for enterprise logistics management, but also promotes a new era. The transformation and development of enterprise logistics management

Keywords: Internet · Innovation · Logistics · Management

1 Introduction

The current era of the Internet, the development of logistics management enterprises must keep pace with the times and keep up with the progress of the times. If modern logistics management companies want to stand out in the fiercely competitive market and create more social and economic benefits, they must continuously improve their core competitiveness and influence, and firmly grasp the challenges faced by logistics management companies in the era of big data. It is the only opportunity to take effective solutions in time, innovate and improve the logistics management model, actively introduce and apply advanced Internet technologies, and cultivate professional Internet logistics management talents [1].

2 The Impact of the "Internet+" Environment on the Logistics Industry

According to data released by the National Bureau of Statistics of China, the ratio of total social logistics costs to GDP has been declining in recent years, and the growth rate of social demand for the logistics industry has slowed down. However, the ratio of total social logistics costs to GDP is still about 5 percentage points higher than the global average, and even about 8 percentage points higher than economically developed countries such as the United States, Japan, and Germany [1]. Therefore, the logistics industry urgently needs to accelerate its transformation and upgrading.

© The Author(s), under exclusive license to Springer Nature Switzerland AG 2021
J. Abawajy et al. (Eds.): ATCI 2021, AISC 1398, pp. 173–179, 2021.
https://doi.org/10.1007/978-3-030-79200-8_24

In such an environment of changing demand, the most intuitive impact of "Internet+" on small and medium logistics companies is to accelerate the transformation and upgrading of the logistics industry. In such an environment of changing demand, the most intuitive impact of "Internet+" on small and medium logistics companies is to accelerate the transformation and upgrading of the logistics industry. At present, new technological breakthroughs are being produced in the fields of transportation, information, and modern logistics; intelligent transportation, intelligent warehousing, Internet of Things, and next-generation information technology will be applied in the logistics field; big data, intelligent logistics, cloud computing, etc. Deeply integrate into the logistics industry [2]. Therefore, the logistics industry will face a major transformation and upgrading.

3   The Advantages of the Internet in Logistics Management

3.1   Fast Information Connection

In the enterprise logistics management work, the most obvious advantage of the application of Internet technology is the rapid connection between information, to avoid management problems caused by information asymmetry. In the early stage of the development of logistics management, lack of perfect technical means and more reliance on manual work, resulting in low efficiency of logistics management and very slow information connection. After the application of Internet technology to carry out logistics management work, it not only meets the needs of enterprise logistics management system for modernization transformation, but also can monitor the actual operation process of logistics in real time [2]. The increase in information connection speed effectively improves the efficiency of enterprise logistics management. Regarding the loopholes in logistics transportation and management work, managers can use Internet analysis technology to find out in time and arrange for relevant technical personnel to repair them, so as to realize the risk control of physical management to ensure the smooth development of all work of the enterprise.

3.2   Increase Stickiness with Customers

In logistics management work, the work is mainly for customers, and the stickiness between customers and enterprises has naturally become the standard for measuring logistics management work. In traditional logistics management work, due to the lack of contact with customers and Information is asymmetry, resulting in low stickiness between the company and its customers and difficult for further development. After the application of Internet technology, companies and customers can communicate more closely, thereby speeding up the efficiency of logistics management, while increasing the stickiness with customers, understanding customer needs and developing targeted products [3]. Combining the obtained data as a basis can effectively improve the company's logistics management capabilities and provide customers with better marketing services.

4   Problems in Logistics Management in the Internet Era

4.1   Data Security Faces Threats

Today, with the highly developed information technology, Internet technology is gradually being proposed and applied to all occupations, and then comes the problem of data security. Although in the logistics management of enterprises, the application of Internet technology provides a great convenient, but at the same time, the issue of data security has become a major issue that has to be considered during its development [3]. In the Internet operating system, information damage, loss, distortion and other issues are the most common, so there are higher requirements for environmental safety. However, it is difficult to prevent information leakage in a fabricated operation environment, which has also caused great obstacles to logistics management.

4.2   The Inventory Management Model is Relatively Backward

The effectiveness of the inventory management model is very important to the logistics management of enterprises. However, as far as the current development background of logistics management is concerned, the application of Internet technology is still at the preliminary stage, so there is still a lack of a more complete inventory management model. The efficiency and quality of management work cannot be effectively improved because of the lack of advanced inventory management models. In addition, business managers do not have a comprehensive grasp of Internet technology, which also causes difficulties in obtaining inventory management information [4]. Since there is no timely feedback of inventory management information to customers, which arouses customer disgust, this problem is very serious. Therefore, attention should be paid to the improvement of inventory management models.

4.3   Distribution Management Issues

In the current enterprise logistics management work, the distribution management system lacks the support of advanced technology, and the distribution work has not achieved the desired effect. The lack of a complete distribution system is the key factor that causes the low distribution rate, and ultimately the actual needs of the customers are not met. The longer the customer waits for the logistics, the worse the experience is. The trust in the service quality of the enterprise decreases, and it is easier to choose other enterprises for cooperation [4]. It should also be noted that the logistics management of enterprises should pay attention to the improvement of logistics distribution status and distribution information.

4.4   Lack of Professionals

Since the logistics management industry has only become popular in recent years, the current society does not have enough talents in the logistics management field to meet market demand [5]. The application time of Internet technology is shorter, and there are currently few professionals with strong technical capabilities, and companies lack training in related technologies, which also hinders the development of corporate logistics management. Any industry or enterprise needs talents if it wants to develop. If you don't want the logistics management development of the enterprise to be restricted in the future, then it is essential to establish a highly professional logistics management team [5].

5   Innovation and Improvement of Logistics Management Services

For all things to develop and progress continuously, innovation and reform must be carried out. No matter what field or industry is inseparable from innovation, only innovation can better meet the needs of human society. The same is true for the logistics industry, it is necessary to innovate and change the management method reasonably in order to develop better [6]. The following are specific ways of innovation and improvement in the logistics industry.

5.1   Technicalization of Logistics Management

Nowadays, all work in all occupations requires high-efficiency and high-quality completion. In the information age, the management of the logistics industry should be scientifically rationalized. Scientific management and technology are inseparable. Logistics management not only requires advanced storage and transportation tools and a large enough storage space, but also requires a scientific and reasonable combination with network technology and communication technology to ensure that its transportation efficiency is greatly improved. Network technology and communication technology are very important to the logistics industry. It can closely link the various departments of the logistics industry, can communicate better, and it is very convenient to release and obtain information [6]. All aspects of logistics management can also be it is well connected to ensure work efficiency and quality, as shown in Fig. 1. For example, barcode technology and automatic sorting technology have brought great convenience to logistics management services. In addition, through multimedia technology, consumers can know the transportation status of the goods at any time, and know when the goods are expected to arrive, so that consumers can feel more at ease [6]. The Chinese government should

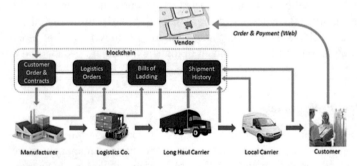

Fig. 1. Information technology management and transportation links

support logistics companies in this regard, help build larger and more advanced logistics venues, add advanced equipment, and establish an automated and intelligent logistics management service model with Chinese characteristics.

5.2 Informationization of Logistics Management Services

Traditional paper documents are filled out manually, which are prone to errors and are kept in large quantities, which are prone to confusion and loss, which will cause some problems of loss of goods. However, the informatization of logistics management can prevent such problems. Hand-filled documents can be turned into electronic documents. Logistics information can be released in a timely manner without filling errors. The filling content is also very clear, and the internal management of the enterprise will be orderly [7]. The informatization of logistics management can keep up with the pace of the times and adapt to the development needs of the society. During the management process, all information is safe and reliable, and the processing and sharing of various data is convenient and quick, as shown in Fig. 2. Through the rapid and sensitive communication of information technology, the goal of zero inventory can be achieved, and the operation cycle can be reduced [7].

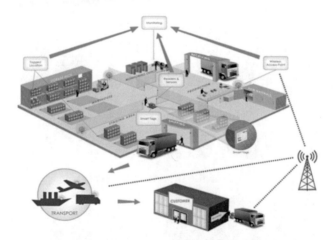

Fig. 2. Informationization of logistics management services

5.3 People-Oriented Logistics Management Services

The logistics industry with service as its purpose should be more humane and people-oriented. This concept is very applicable in any industry. Enterprise staff must put the customer first, think from the perspective of the customer, sort the goods according to the needs of consumers, and then package and transport them to ensure the quality of the goods [8]. This one-person-oriented service approach can better meet the needs of consumers, thereby improving efficiency and work efficiency. At the same time, corresponding programs should be formulated for special customers to maximize customer needs. Under the influence of e-commerce, the logistics chain can have different functions, whether it is the supply of raw materials or consumers, relying on the logistics chain can better carry out management work.

5.4 Integration of Logistics Management Services

For each link of logistics management, the staff should regard the logistics work as a whole and realize the integration of logistics management services, as shown in Fig. 3. Each production enterprise establishes close contact with consumers through logistics enterprises and sales departments, which may benefit and bring convenience to consumers [9]. Therefore, logistics enterprises must strengthen the staff's own quality and skills, introduce advanced equipment, and proceed from all aspects from all angles to achieve the best results in logistics management services.

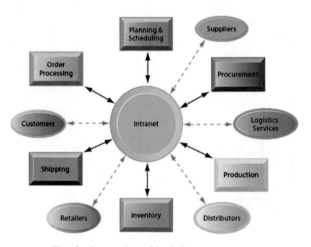

Fig. 3. Integration of logistics management

5.5 Internationalization of Logistics Management

Although China has established logistics and trade relations with more than 200 countries, there is no domestic logistics company that can provide global delivery services. In this era of economic globalization, communication between countries is close, and logistics management must Realization of globalization can meet the needs of all consumers. To make logistics management international, it is necessary to globalize technology, implement globalization, and service globalization [10]. Therefore, logistics enterprises should first improve the service level of their staff, and then establish a logistics management system adapted to international needs.

6 Conclusion

Nowadays, the development of the logistics industry has become the lifeblood of China's economic operation. Its innovative development must rely on the Internet in order to achieve efficient, orderly and error-free management and transportation, and better meet the needs of humanity and society. The logistics industry must rely on innovation to strengthen its management work, improve its work efficiency and operational capabilities, so that China's logistics industry can achieve sustainable development, stand firm in the domestic and foreign markets, and promote China's economic development.

References

1. Su, Z., Zhao, C.: Talking about the construction of enterprise logistics management system under the Internet background. Shangxun **9**(04), 112–115 (2020). (in Chinese)
2. Shao, Q.: Analysis of the optimization of the Internet-based enterprise logistics management system. Knowl. Econ. **11**(01), 87–90 (2020). (in Chinese)
3. Jiang, J.L.: Analysis of enterprise logistics management under the internet background. Int. Public Relat. **8**(11), 52–54 (2019). (in Chinese)
4. Wen, C.: A glimpse into the future development of WeChat business based on the current development of WeChat business. Shop. Mall Modern. **8**(03), 17–20 (2015). (in Chinese)
5. Zhou, R.: Talking about the current situation and development trend of my country's C2C e-commerce market. Mod. Econ. **6**(03), 72–75 (2011). (in Chinese)
6. Li, H.X.: Research on the development path selection of e-commerce industry. Shandong Soc. Sci. **9**, 125–129 (2015). (in Chinese)
7. Chen, Y.Q.: An analysis of the development law and path of E-commerce in the internet era. Hebei Acad. J. **12**(05), 219–222 (2016). (in Chinese)
8. Chang, Z.: Analysis of logistics management strategies for small and medium-sized enterprises based on Internet thinking. Enterp. Technol. Dev. **10**(05), 292–294 (2018). (in Chinese)
9. Li, X.Q.: Analysis and thinking on the current situation of logistics management of small and medium-sized enterprises in my country. Taxation **21**, 60–63 (2017). (in Chinese)
10. Miao, S., Huang, J.X.: "Internet + agricultural industry chain" model construction and path innovation. Logist. Technol. **11**(05), 11–14 (2018). (in Chinese)

Optimizing Short Video Platform Based on User Portrait Theory——Take KuaiShou APP as an Example

Han Li[(✉)]

College of Fine Arts, Henan University, Kaifeng 475001, Henan, China

Abstract. Since 2016, the network short video industry has progressed rapidly, and short video has become a hot network culture form. Its novel communication mode and unique perspective not only expand the depth and breadth of cultural production, dissemination and exchange, but also shapes a brand-new mass culture form. With the advent of the 5G era, the network technology has become more developed, the short video self-media industry has continued to reform, the Internet platform has also ushered in the "e-commerce era". In the era of big data, the continuous and accurate improvement of User portrait technology has optimized the operation mode of short video self-media platforms, and has also enriched the diversification and more possibilities of the platform. This article is based on the theory of User portrait, combined with Kuaishou APP as an example, to explore the optimization of its industry content.

Keywords: User portrait · Short video · Anchor · E-commerce

User portrait is a tagged user model abstracted based on the user's individual attributes, living habits, consumption behavior and other information. It can accurately locate the needs of Internet users. Through data calculation, the platform can provide users with faster and more accurate Service [1]. In the 5G era with developed network technology, mobile intelligent terminals are constantly being updated, and the short video self-media platform is becoming more and more popular. The public can use mobile phones to do self-media without leaving home, and such a convenient way also allow short video users to present a blowout phenomenon. Combined with the ever-increasing aesthetics and needs of the public, the content quality of short videos is getting higher and higher. Since its development in 2013, Kuaishou APP has been successfully listed in 2020, which has established its important position in the field of short video self-media platform. And it's inseparable from the tremendous changes in Internet technology. Based on the technology and theory of User portrait, this paper attempts to analyze and study the route of APP optimization in KuaiShou and predict its future trend.

1 User Portrait Technology in the Era of Big Data

The author of the User portrait is Alan Cooper, and the results of Alan Cooper's research have a decisive influence in the field of interaction design. He said: "The essence of

© The Author(s), under exclusive license to Springer Nature Switzerland AG 2021
J. Abawajy et al. (Eds.): ATCI 2021, AISC 1398, pp. 180–185, 2021.
https://doi.org/10.1007/978-3-030-79200-8_25

User portraitis the representative image of marketing target" [2]. That is, the virtual image of an individual user is the concept of the User portrait itself, a target user model constructed from a large amount of real data. User portrait, also known as user role, is a way of individually designing accurately for users based on their needs and goals. Nowadays, User portrait technology is widely used in various fields, and the user roles generated by the virtual image of real users is closely related to commercial activities. As an increasingly convenient data analysis tool, User portrait can extract all kinds of information from users in an all-round way, and also carefully understand and track the changes of users' needs, analyze how users' needs change.Eventually, more precise marketing can be achieved. The characteristics of the big data era provide more suffi-cient conditions for User portrait to mine and discover user preferences, characteristics, needs and other comprehensive information.In the era of e-commerce, embedding User portrait technology into major e-commerce platforms can capture user information more accurately and recommend the commodity information they need. For example, when we search for the name of a product with Taobao, many similar products will appear when we open the homepage of Taobao APP again, which is the most commonly used part of User portrait technology. Major e-commerce platforms such as JD, Alibaba, Suning, and Xiaohongshu all have mature User portrait technology.

2   Technical Outline and Case Analysis of User Portrait of KuaiShou APP

2.1   KuaiShou APP User Portrait is Accurately Outlined

The User portrait can capture the multi-dimensional user label system, such as the user's gender, hobbies and interests, location, online time, etc. Then establish a multi-dimensional label for the user's behavior. Kuaishou APP initially used short video traffic as its business model. On the one hand, users can shoot short videos to upload, get flow pushes, and other users can see the video comments and forwards, thus building their own short video social circle; On the other hand, if users are not confident enough about themselves, they can only repost and comment on other users' videos, and also develop their own online social circle. The User portrait technology can obtain the user's video features and audience through the short video content shot by the user, and accurately push it to the appropriate community. In addition, for tourists who don't shoot short videos but just watch other users' videos, they can get the classification of users' preferences through their comments, collections, likes and forwards, quickly and accurately push bloggers who meet users' preferences to the user's homepage for viewing.

According to the report "Douyin & Kuaishou: Where are the users and what are their characteristics" released by "Penguin Think Tank" in November 2018, it is pointed out that the number of users with bachelor degree or above in KuaiShou is 31.9%, and the Douyin is 41.9%, and users with high school education or below in KuaiShou is 14% more than Douyin, KuaiShou accounts for 46.3%, and Douyin is 32.2%, Overall, Douyin's users have a higher education level than Kuaishou. In terms of city-level dis-tribution, Kuaishou's user rate in third- and fourth-tier cities has reached 61.2%, while

Douyin's user rate is 54.7% [3]. Kuaishou has a higher utilization rate in fourth-tier and lower cities. Douyin is more commonly used in Tik 1 and Tier 2 cities than Kuaishou. Kuaishou APP's short videos are mainly classified into lifestyle, food, and folklore, and most of the users' categories are from rural areas, or from second, third, and fourth-tier cities and villages in the Northeast. Such precise user positioning has brought distinctive tags to Kuaishou, such as "vulgar video". The vocabulary that often appears in the video includes "vicissitudes", "wandering", and other words, highlighting the difficulty of drifting in the city and the difficulty of survival. In this way, it arouses the resonance of the majority of users, thereby increasing the usage [4].

2.2 Analysis of the User Portrait Approach of Kuaishou APP

In terms of gender, the proportion of male users in Kuaishou APP is 58%, and the proportion of female users is 42%. From the perspective of age level, according to the information of Kuaishou short video bloggers, the main age level of Kuaishou users is between 16 and 45 years old. The object of Kuaishou APP User portrait is mainly divided into two parts, one is more well-known video bloggers, and the other is ordinary tourist users. From the current business model of Kuaishou, the roles of the two types of users for the Kuaishou APP are complementary.

Kuaishou APP mainly relies on data to make User portrait for users, including user comments, reposts, likes, favorites, online duration, online hours, etc. Then based on these data to determine the age class and professional hobbies of users. With the reform of e-commerce, the data of User portraitalso includes browsing product records, purchase records, shopping cart records, and viewing anchor-carrying market records. Through the above data, shopping groups can be more accurately divided and users' consumption needs and preferences can be found out (Fig. 1).

Fig. 1. User portrait

3 Interpretation of the Optimization Value of the KuaiShou APP Platform Under the Theory of User Portrait

3.1 The Labels of the Two Categories of Users Attract Each Other to Enhance the Commercial Value of the Platform

The first type of user takes Xin Youzhi, who currently has the largest number of fans on the Kuaishou app, as an example, He is 30 years old, born in Heilongjiang, educated in high school, and employed as a businessman, character characteristics for inspiration, enthusiasm, hard work, etc. He labels himself as "Son of a Farmer". Xin's original short video content on the KuaiShou APP ranged from inspirational stories to rural life, and gained a group of like-interested fans. After the live broadcast function was started in KuaiShou, Xin began to reward the major anchor live broadcast rooms to rapidly increase the number of fans. After the fans were attracted to Xin Youzhi's own live broadcast room, they began to promote their business content. However, at that time, they did not have a platform for selling goods by live broadcast, but only recommended their contact information. Later, KuaiShou ushered in the era of e-commerce. At this time, Xin's fans were firmly ranked among the top in KuaiShou. When selling goods, he gave full play to his character label "Son of a Farmer", as well as inspirational feelings, which are same with the KuaiShou most of the tourist users, that is, the User portrait of the second category of user attributes. The platform will recommend him to more users, give him more clout, so accurate User portrait to Xin has brought great success and profit.

The propaganda characteristics of short videos are evident in the official media, marketing ID and broadcasters with a certain fan base. Because of their attention, they have mastered the main publicity resources, and then become the first choice for major brands and product sales. For ordinary tourist users, KuaiShou is accurately pushed to the favorite blogger or marketing anchor of each tourist user through data calculation. Before the early era of e-commerce, KuaiShou was the leader of short video platform, relying on the tags of these users to connect the relationship between tourists and bloggers closely. Only when more users interact with each other can the platform have more clout. One feature of KuaiShou, which is different from other short video platforms, is "lower threshold". The "lower threshold" mentioned here is not a derogatory term, but that users can show themselves as bloggers in KuaiShou without fancy video editing ability and high-standard aesthetic ability. Su Hua, CEO of KuaiShou, once said, "One of KuaiShou's values is universal. There is no higher or lower life, and everyone deserves to be recorded and remembered." It is precisely because of such inclusiveness that every ordinary people's cognition and understanding of life are respected, and every ordinary people are given the right to express and show their own life. Accurate user positioning has brought a huge number of users to KuaiShou, which is extremely beneficial to the future transformation into an e-commerce platform.

3.2 The Value of Improving the Sense of Experience of User Service

In the environment of information big data, all kinds of information surround users and form an information ocean, this information is nerve lack of junk information. Users want to obtain valuable information need to distinguish between useless spam, which brings discomfort to the user service experience. However, it is the ubiquitous spam that makes it difficult for users to build trust in some platforms. KuaiShou APP's intelligent User portrait service breaks the problem of geographical classification which is difficult to solve in the traditional short video platform information service mode, and accurately provides intelligent information services for users. KuaiShou divided life, food and other columns on the platform very early. Although many platforms already have this point. At that time, KuaiShou completed this classification very efficiently, so that users can quickly find the works they are interested in according to their own needs. KuaiShou reduced the embedding of advertising spam. Of course, for a period of time, KuaiShou also fell into the stage of chaotic management. Many minors used their parents' mobile phone bank cards to blindly reward and support their favorite bloggers. To solve this problem, KuaiShou immediately introduced the protection mode for minors, so that minors were not allowed to reward anchors casually, thus avoiding many unnecessary losses.

When users are shopping in e-commerce, through the user's online time, as well as the purchase record, KuaiShou will quickly capture the user's preferences, and recommended to the user similar attributes of the live rooms and merchandises. This is not blindly recommended like some irresponsible platforms. Some platform anchors can get more clout as long as they buy promotional services. KuaiShou will strictly control the reputation and service quality of anchors, and only recommend anchors whose clout and fan base are quite different from the anchors that users originally bought. Because the prices of goods sold by anchors with a similar fan base will also be similar, which fully takes into account the economic strength of users. This recommendation is inseparable from the accurate calculation of User portrait. Compared with the traditional information service mode, the User portrait information service is more accurate, efficient and proactive, and provides users with information services actively [5] (Fig. 2).

Fig. 2. User portrait

Short videos inject new vitality into pluralistic society. Mr, Fei Xiaotong said in the opening of Earthbound China: "From the grassroots level, Chinese society is earthbound." The integration of urban and rural areas makes some short video bloggers emphasize the excitement all the time. The video on KuiaShou also reflects this characteristic. KuaiShou APP provides us with a multi-cultural display space, increases the way for us to experience social diversity, and at the same time injects new vitality into the diversified development of our society.

In the year after the epidemic, for the rapid recovery of the economy, the government also strongly supported the live commerce. The e-commerce era of the Internet ushered in the spring after the epidemic. Under the correct guidance of the government, the live commerce by e-commerce became more and more standardized. As a short video platform with the most users in rural areas and second, third and fourth tier cities, KuaiShou has also brought new vitality to the sales of agricultural and sideline products. Man is the yardstick of all things. The current era is not only the era of "things", but also the era of "people". Human beings continue to develop comprehensively and freely, pay more attention to the emotional experience of "people" and pay more attention to the needs of people. As a short video platform, several anchors with a large fan base, led by Xin Youzhi, have set several record highs for influencer marketing to e-commerce throughout the country since 2019. During the Singles' Day holiday in 2019, the total amount of goods brought by Xin Youzhi team reached about 3 billion, and the commercial value of Kuaishou platform has also been rapidly improved. As a short video e-commerce platform, driven by the increasingly accurate and mature User portrait base, as long as it provides more detailed services to users, provides a better shopping environment and continuously optimizes the platform, it is believed that it can create higher commercial value.

References

1. Shen, J.: Application of user portraits in internet finance. Mod. Bus. (33) (2017)
2. Zhang, X., Shen, W., Du, C.: Research on Bayesian networks in the construction of user profiles. Mobile Commun. (12) (2016)
3. http://www.doc88.com/p%2D66216953365177.html
4. Zhao, X., Lu, H.: Research on modern information service model based on information ecological theory (06) (2015)
5. Chen, D., Luo, Y., Wu, Z.: Research on the personalized service of university libraries based on big data mining and user profiles. Libr. Res. Work (04) (2019)

Fault Analysis and Diagnosis Method for Intelligent Maintenance of OPGW Optical Cable in Power Systems

Jing Song[1(⊠)], Ke Sun[1], Di Wu[1], Fei Cheng[1], and Zan Xu[2]

[1] Zhejiang Huayun Power Engineering Design Consulting Co., Ltd., Hangzhou, Zhejiang, China
[2] Nokia Communication System Technology (Beijing) Co., Ltd., Hangzhou, Zhejiang, China

Abstract. The development and improvement of China's power communication system promote the smooth operation of the State Grid, which not only speeds up the operation of the power grid but also improves the service quality of the power communication department. OPGW optical cable is an important part of the power communication system, its common faults are relatively more, which will directly affect the safety performance of power grid operation. Therefore, the staff should take effective operation and maintenance methods to ensure that the OPGW optical cable is in a normal operation state. This paper briefly describes the common faults of OPGW optical cable, analyzes the critical point, management mode, specific operation and maintenance contents and methods of OPGW optical cable operation and maintenance.

Keywords: OPGW optical cable · Fault detection · Fault diagnosis · System maintenance

1 Introduction

Optical Fiber Composite Overhead Ground Wire (OPGW) optical cable is a kind of composite optical fiber composite overhead ground wire, and also an important part of power communication system, which has important influence on the efficiency of power grid operation. At present, the power communication system in China is stable, but there are still many common faults, which will directly threaten the security and stability of the power communication system. Therefore, it is of paramount importance to investigate the common faults and operation and maintenance methods of OPGW optical cable, solve the fault problems in time and ensure the reliability of power communication system operation [1–3].

Optical fiber makes use of the difference of refractive index between core and cladding materials to transmit light energy in optical fiber, which has become a major revolution in the history of communication. Optical fiber and cable are light in weight and small in volume, which have been used in the power system to transmit dispatching telephone, telecontrol signal, relay protection, television image and other information between the substation and central dispatching office. To improve the stability and reliability of optical fiber and cable, foreign countries have developed a composite structure

J. Abawajy et al. (Eds.): ATCI 2021, AISC 1398, pp. 186–191, 2021.
https://doi.org/10.1007/978-3-030-79200-8_26

of optical fiber and cable, phase conductor of the transmission line, overhead ground wire and power cable [4–6]. The OPGW optical cable is more reliable, stable and firm because it is wrapped by the metal conductor. Because the overhead ground wire and optical cable are combined into one, compared with other optical cables, the construction period is shortened and the construction cost is saved. Also, if the OPGW made of aluminum-clad steel wire or aluminum alloy wire is used, it is equivalent to erecting a good conductor overhead ground wire, which can reduce the secondary arc current of the transmission line, reduce the power frequency overvoltage, and improve the interference and dangerous influence of power line on the communication line. Due to the characteristics of anti-electromagnetic interference and lightweight, the optical fiber can be installed on the top of the transmission line tower without considering the optimal hanging position and electromagnetic corrosion. Therefore, OPGW has high reliability, superior mechanical properties and low cost. This technique is particularly suitable and economical for new laying or replacement of existing ground wires [7–10].

OPGW optical cable is mainly used on 500 kV, 220 kV and 110 kV voltage grade lines, and is mostly used in new lines due to the influence of power failure and safety factors. The applicable characteristics of OPGW are summarized as follows and illustrated in Fig. 1:

(1) The high voltage line with a voltage of over 110 kV has a large span (generally above 250m);
(2) It is easy to maintain, easy to solve for the problem of line crossing, and its mechanical characteristics can meet the requirements of large line crossing;
(3) The outer layer of OPGW is metal armor, which does not affect high voltage electrical corrosion and degradation;
(4) OPGW must be cut off during construction, and the loss of power cut is large, so OPGW should be used in the new high voltage line above 110 kV;

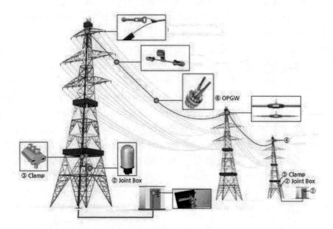

Fig. 1. Illustration of application of OPGW system

(5) In the performance index of OPGW, the larger the short-circuit current, the more need to installed with a good conductor, the tensile strength is reduced correspondingly. In the case of certain tensile strength, to increase the short-circuit current capacity, only increase the metal cross-section area, which leads to the increase of cable diameter and cable weight, so the safety problem is raised for the strength of the line tower.

The rest of the paper is organized as follows: Sect. 2 discusses the typical faults of OPGW and the maintenance of the OPGW system; Sect. 3 presents the fault location methods adopted in power systems. Finally, the conclusive remarks are given in Sect. 4.

2 Typical OPGW Faults and Maintenance Method

In the power communication system, the operation environment of OPGW cable is relatively bad, so there are many problems such as icing and galloping, which not only threaten the operation safety but also have an important impact on the power communication system. Icing refers to being placed in the external environment.

The OPGW cable in the pipeline is damaged by ice, which leads to the fracture of the cable and increases the difficulty of repair work. The optical cable fault caused by icing is also reflected in the poor quality of optical signal transmission, which will affect the power supply. Besides, the OPGW cable fault caused by lightning is also more frequent, especially in areas with more lightning. Under the action of lightning, the outer monofilament of the OPGW cable is easy to be damaged by melting, and under the synergistic effect of the current effect, the problem of strand breakage will be directly caused. At the same time, due to the influence of external factors, OPGW cable failure is more, such as theft, typhoon, etc., can cause OPGW cable failure.

The criticality and management mode of OPGW optical cable operation and maintenance for OPGW optical cable, there is a critical point between its operation management and maintenance. To better carry out the operation and maintenance work, the staff must grasp the dividing point of the two. At present, OPGW.

The implementation of optical cable in the power communication system is as follows: through the terminal junction box and non-metallic connection, directly into the communication room. Based on this, the cut-off point of optical cable operation and maintenance is set at the junction box of the frame terminal, and the operation and maintenance work at both ends of the cut-off point is respectively in the charge of different departments. On the one hand, the OPGW optical cable from the demarcation point to the station end is operated and maintained by the power communication department; on the other hand, the OPGW optical cable from the demarcation point to the transmission line is operated and maintained by the transmission line department.

In the power communication system, OPGW optical cable and high voltage transmission line are erected on the same tower, so they are endowed with more functions, such as transmission communication signal, control signal and relay protection signal. Therefore, the operation and maintenance management of OPGW optical cable is relatively complex. The departments of the transmission line, communication, automation, and dispatching and relay protection generally require effective cooperation to manage and maintain the OPGW system. At present, all departments of power enterprises have clear provisions on the operation and maintenance of OPGW cable. For the OPGW cable operation management mode of power communication system, the transmission line specialty and power communication specialty should be combined to maintain the OPGW cable.

3 Fault Location Solutions of OPGW Systems

(1) Single-phase to ground fault location technology in power distribution networks
In the fault location of the transmission line, the application of single-phase grounding fault location technology in the distribution network is very important. This technology location system is composed of traveling wave line selection equipment, feeder terminal equipment and fault location center. In addition, each piece of equipment realizes wireless network communication through OPGW cable, which provides favorable basic conditions for fault location behavior. The implementation of single-phase grounding fault location technology in the distribution network is to use the bus residual voltage to judge the possibility of line fault, and the corresponding threshold value should be set, so that when the line fault occurs, the zero-sequence current can be prohibited from entering the 10 kV feeder side to achieve the purpose of maintaining the fault. Besides, a large number of practical studies show that the fault location process of single-phase to ground fault location technology in the distribution network is to detect the transient traveling wave, and then determine the traveling wave arrival time and fault location. This fault location technology can find the fault problems in line operation in time. Thus, it should be able to strengthen the application of this technology by relevant technical personnel, to improve the stability of line operation and achieve the purpose of safe operation.

(2) Network traveling wave positioning technology
The application of network traveling wave positioning technology requires a high number of traveling wave positioning units. Therefore, in this context, the relevant technical personnel should reasonably set the number of units according to the fault detection conditions in the application process of this technology, to improve the reliability of the fault positioning system. Also, to improve the accuracy of fault location, priority points should be set according to the principle that the greater the fixed-point degree, the greater the attenuation of traveling wave signal amplitude, to achieve the goal of network traveling wave location. In addition, the network traveling wave location can determine the actual operation status of transmission lines according to the fault location of two location units on the endpoint of the largest nonsingular set element of the transmission system topology, and timely monitor the existing fault problems, and then effectively solve the fault problems, to achieve stable operation status of transmission lines. In the actual operation of the industrial site, there will be the problem of circuit breaker tripping. The network traveling wave positioning technology can be used to determine the fault point, and the corresponding fault maintenance measures can be implemented with the help of the network traveling wave status information.

(3) Communication technology of online inspection and monitoring system
The communication technology of online inspection and monitoring system is based on tower terminal, background monitoring host and optical fiber communication system. To achieve the purpose of fault location, the relevant technical personnel are required to place corresponding tower terminal on the transmission tower, collect line operation status information through the operation of equipment, and based on OPGW, the collected data will be transferred to the background monitoring host, that is to provide the corresponding monitoring data for the relevant staff, to prompt them to find the fault problems existing in the line operation in time through the integration of data, and carry out effective treatment. Also, the optical fiber communication system in the on-line inspection and monitoring system realizes EPON Technology Based on OPGW, which makes the tower terminal realize real-time monitoring behavior in the actual operation process, and then improves the stability of line operation, and achieves the goal of fault location.

4   Conclusions and Discussions

The combination of OPGW cable and ground wire, and the same tower with the transmission line, is conducive to reduce the economic cost, rational use of resources, at the same time, give full play to the positive role of OPGW cable in the operation of the power communication system. In the process of the construction of a power communication system, due to the special performance of OPGW cable, its common faults are relatively more, which will directly reduce the operation quality of the power system. Therefore, in practical work, the relevant Personnel should seriously study the operation and maintenance methods of OPGW cable, to effectively solve the OPGW cable fault and give full play to the positive role of OPGW cable in the power communication system.

OPGW optical cable is a combination of optical cable and ground wire. By erecting with transmission line on the same tower, it can greatly save the resources and construction cost of the line corridor, so it plays an increasingly important role in the construction of communication channels in the power system. At the same time, due to the particularity of multi-purpose OPGW cable, it also brings cross-professional technical barriers to its operational management and maintenance. Therefore, in the operation and maintenance process of OPGW optical cable, we must strengthen the system optical fiber test and optical cable line inspection work, through the development of a reasonable and timely maintenance scheme and emergency repair scheme to solve and eliminate the OPGW optical cable fault.

Acknowledgments. This work was supported by the Science and Technology Development Project of Zhejiang Huayun Power Engineering Design Consulting Co., Ltd.

References

1. IEEE Approved Standard for Testing and Performance of Hardware for Optical Groundwire (OPGW), in IEEE P1591.1/D3, November 2011, pp.1–55, 29 February 2012

2. Madge, R.C., Barrett, S., Grad, H.: Performance of optical ground wires during fault current tests. IEEE Trans. Power Delivery **4**(3), 1552–1559 (1989). https://doi.org/10.1109/61.32642
3. Lu, L., Sun, X., Bu, X., Li, B.: Study on passive, wide area and multi-state parameter monitoring and diagnosis for power transmission lines. In: 2018 International Conference on Power System Technology (POWERCON) 2018, Guangzhou, China, pp. 3903–3907 (2018). https://doi.org/10.1109/POWERCON.2018.8601868
4. McMaster, R.: A tutorial on optical ground wire ratings analysis for protection engineers. In: 2019 72nd Conference for Protective Relay Engineers (CPRE), College Station 2019, TX, USA, pp. 1–9 (2019). https://doi.org/10.1109/CPRE.2019.8765885
5. Jie, L., Gang, L., Xi, C.: Heat properties algorithm research and software development of OPGW cable. In: 2009 WRI World Congress on Software Engineering 2009, Xiamen, China, pp. 334–339 (2009). https://doi.org/10.1109/WCSE.2009.272
6. Gonzalez, L., Dmitriev, V.: Electrothermal analysis of modified OPGW cables using Multiphysics model. In: 2015 SBMO/IEEE MTT-S International Microwave and Optoelectronics Conference (IMOC) 2015, Porto de Galinhas, Brazil, pp. 1–5 (2015). https://doi.org/10.1109/IMOC.2015.7369208
7. Alamo, C.G., Pardinas, J.A.: Georeferential video for located access road and stringing tension equipment places to install 1366 km OPGW cables in existing transmission lines. In: 2006 IEEE/PES Transmission & Distribution Conference and Exposition: Latin America 2006, Caracas, Venezuela, pp. 1–4 (2006). https://doi.org/10.1109/TDCLA.2006.311555
8. Hooper, D., Bailey, T.: Aerial Surveys Calculate Vegetation Growth, Transmission & Distribution, October 2004
9. People's Republic of China Electric Power Industry Standard DL/T832-2003. Optical Fiber Composite Overhead Ground Wire (1999)
10. Jakl, F., Zunec, M., Ticar, I., et al.: Synopsis distribution of current density in layers of overhead bare conductors. In: Power Systems and Communications Infrastructures for the Future, Beijing, September 2002

Impact of Margin Trading on the Liquidity of China's Stock Market: Based on VAR Model

Yiting Shen[✉]

School of Economics, Shanghai University, Shanghai, China
cindysyt@shu.edu.cn

Abstract. Since China's margin lending and short selling system was formally implemented in 2010, scholars have conducted many researches and discussions on whether it has played an expected role in the securities market. This paper studies the impact of margin trading on the liquidity of China's A-share market under the background of the leaping development of margin trading in China's stock market. A conceptual model is established through theoretical analysis. The mechanism of the effect of margin trading on stock market liquidity through the volatility channel is proposed, and the research hypothesis is established accordingly. On the basis of data collection and variable setting, the VAR model and Granger causality test are used to conduct an empirical test on the actual effect of margin trading on stock market liquidity. The test results show that margin trading can enhance the liquidity of the stock market through the volatility channel.

Keywords: Stock market · Securities margin trading · Liquidity · Big data analysis

1 Introduction

Margin and short selling, also known as securities credit trading, refers to the way that investors provide collateral to securities companies with margin and short selling qualifications, borrow money to buy securities or borrow securities and sell them. Liquidity is the soul of the market. With the main purpose of "promoting liquidity and reducing volatility", the China Securities Regulatory Commission (CSRC) simultaneously launched margin trading and short selling business in the Shanghai and Shenzhen stock exchanges on March 31, 2010. As of November 7, 2019, the Shanghai and Shenzhen Stock Exchanges have expanded the capacity of the underlying stocks in margin trading for six times, increasing the capacity of the underlying stocks from the initial 90 to 1600. At the same time, the balance of margin trading and short selling has reached a scale of trillions. As the rapid development of margin trading, has it actually increased market liquidity?

There are different opinions in academic researches on the impact of margin lending and short selling on stock market liquidity. Some scholars believe that margin lending and short selling have a positive effect on the promotion of stock market liquidity. Frino et al. [1] took the bid-ask spread of the market as a liquidity indicator and pointed out that

J. Abawajy et al. (Eds.): ATCI 2021, AISC 1398, pp. 192–199, 2021.
https://doi.org/10.1007/978-3-030-79200-8_27

the liquidity level of the market with less restrictions on short selling was significantly higher than that with strict restrictions on short selling. Senchack et al. [2] took NYSE and ASE listed companies as samples, selected data from 1980 to 1986 to build a panel data model, and found that short selling trading significantly enhanced the liquidity of underlying stocks. Kong Xiangyu et al. [3] used the Wilcoxon rank sum test to investigate the influence of the adjustment of the underlying stocks on liquidity. The results showed that margin and short selling, market trading amount and market return fluctuations constituted the determinants affecting the liquidity of underlying stocks. Zhang Bo et al. [4] used VAR model and Granger causality test method to prove that short selling capital and short selling can improve the liquidity of stock market through the trading channel. Lin Xiangyou [5] studied the difference of ETF fund market liquidity before and after stocks entered the margin. The conclusion is margin and short selling improved the liquidity of ETF fund market. Gao [6] believes that margin trading and short selling system can reduce the cost of related transactions and thus improve market liquidity. Huang [7] studied the Shanghai and Shenzhen A-share markets by using the differential model and the fixed-effect regression model. She found that due to the existence of short selling restrictions, financing transactions played an obvious role in improving the stock market liquidity compared with short selling transactions.

Some researchers believe that the positive effect of margin trading on promoting stock market liquidity is not obvious. Miller [8] thinks that banning short selling will cause pessimistic and informed traders with bad interest rates to leave the market, resulting in insufficient stock supply and resulting in a decline in liquidity. Cai and Xia [9] analyzed the high-frequency data of China's Hong Kong main board market and found that noisy traders became more cautious and even withdrew from the market because they were worried that the introduction of margin short selling mechanism would increase the risk of loss, thus reducing the market activity and liquidity. Some scholars believe that the government imposes too many restrictions on short selling, resulting in too small a trading scale and insignificant impact on market liquidity. Yu [10] thinks in long-term, financing transaction has a greater impact on the stock market than short selling transaction, which will reduce the liquidity of the stock market.

On the basis of existing research, this paper selects Shanghai Securities A-share index and Shenzhen Securities A-share index related to the margin trading market data from 2010. 3. 31 to 2020. 12. 31, and use VAR models to further analysis of the impact of margin trading on the liquidity of China's stock market. On the one hand, this paper enriches the empirical evidence about the impact of margin trading on market liquidity, and on the other hand, it provides feasible ideas for regulators to develop and improve the stock market and improve the quality of the market.

2   Methodology and Data

2.1   Variable Description

Market liquidity was selected as the explained variable in this paper. Liquidity generally refers to the rapid entry of a large number of transactions at a relatively low trading cost, while producing a relatively small shadow on the trading price range. Different scholars describe liquidity indicators in different dimensions. Some scholars describe liquidity through trading volume and take turnover rate as the liquidity index. However, this index only considers the liquidity of assets, without the price fluctuations of the securities themselves. Liquidity is the transfer between securities assets and cash assets. The ability of exchange includes both the ability of securities assets to be converted into cash assets (liquidity) and the ability of cash assets to be converted into securities assets(the ability to change coupons). When considering the dimension of liquidity index, this paper thinks that the stock liquidity should be described from three aspects: trading volume, price and time. The calculation formula of relative price difference is:

$$LIQ = \left| \frac{(InP_t - InP_{t-1})}{TM/TMV} \right| \tag{1}$$

P_t and P_{t-1} respectively represent the closing price of the CSI 300 Index on the day and the previous day, while TM and TMV respectively represent the daily transaction amount and the daily closing market value in circulation. LIQ reflects the influence of unit stock trading on the price and combines with the connotation of liquidity. LIQ attribute can be determined as a reverse indicator: The larger the value is, the lower the market liquidity is.

According to the requirements of theoretical analysis and modeling, financing transaction, short selling transaction and market volatility were selected as explanatory variables. The list of relevant variables is shown in Table 1.

Table 1. Variable list

Variables	Proxy variable	Computational formula	Attribute
Financing transaction	Daily financing balance MP		Positive indicators
Short selling transaction	Daily short selling balance SS		Positive indicators
Market volatility	Volatility Vol	$VOL_t = \frac{P_t^h - P_t^l}{(P_t^h + P_t^l)/2} * 100$	Positive indicators

a P_t^h, P_t^l respectively denotes the highest and lowest price of CSI 300 Index on t day.

2.2   Data Description

The data collection of market volatility and liquidity is based on the CSI 300 Index, which mainly takes into account that the constituent stocks of the index cover both the Shanghai and Shenzhen markets. As for the proxy variable of margin trading, the daily margin trading balance of Shanghai and Shenzhen stock markets is selected. In this paper, 2431 data of from March 31, 2010 to March 31, 2020 were selected as the sample interval of the study.

In order to prevent the occurrence of "false regression", before co-integration analysis of time series data using VAR model, unit root test should be carried out on each variable to check whether the variable is stable. In this paper, ADF unit root test was adopted, which was judged by various forms of test and information criteria. The test results are shown in Table 2. In the test form, C represents a constant term, T represents a time trend term, and K represents the lag order. At the significance level of 1%, MP and SS all accept the original hypothesis with a single root, and the dMP and dSS variables after the first-order difference are stable.

Table 2. Panel unit root test results

Variables	Inspection form (C,T,K)	ADF test value	Prob	1% critical value	The inspection results
MP	(0,0,1)	−2.3924	0.0162	−2.7653	Stable
SS	(0,0,1)	−3.3789	0.0185	−3.6752	Stable
VOL	(C,0,0)	−5.7892	0.0194	−3.1264	Stable
LIQ	(C,T,6)	−7.2067	0.0000	−3.6789	Stable

2.3 Model Specification

There are many ways to analyze the interaction between variables, but there are many and complex factors affecting the liquidity of the securities market. Margin and short selling is only one aspect of the factors. Therefore, VAR model is adopted in this paper. By establishing multiple equations, endogenous variables are used to regress the lag values of all endogenous variables in the model. Then the dynamic relationship among endogenous variables is estimated.

Based on the high order lag regression, the composite criterion is used to test and determine the optimal lag order as the first order. dMP, dSS, VOL and LIQ were set as endogenous variables to construct the VAR model. The matrix form of the modeling results was shown as follows:

$$y_t = \sum_{i=1}^{P} A_i y_{t-1} + Bx_t + \varepsilon_t \qquad (2)$$

3 Empirical Analysis

3.1 Granger Causality Test

In order to further analyze the relationship between financing transactions, short selling transactions, market volatility and market liquidity, Granger causality test was conducted on these variables. Lag order 1 was selected as the lag length of each variable. The test results are shown in Table 3.

dMP of LIQ exist significant negative impact, while the dSS of LIQ doesn't exist significant negative impact. Granger causality test also shows that dMP can Granger cause LIQ, while dSS can not. Based on the reverse of LIQ index properties, financing deals have significantly increased the function of market liquidity, and securities trading has not significantly enhance the role of market liquidity. This suggests that margin trading has positive influence on the overall market liquidity. It is worth noting that the dMP and LIQ exists a two-way Granger causality in the VAR model, which supports the momentum effect based on the expectation of margin trading and liquidity in the market is the positive interaction between the point of view.

Table 3. Granger causality test

Hypothesis testing	Null hypothesis	F value	P value
Population effect Granger Causality test	LIQ does not cause Granger dMP	12.8711	0.0006
	dMP does not cause Granger LIQ	110.7890	0.0015
	LIQ does not cause Granger dSS	3.6E−05	0.9965
	dSS does not cause Granger LIQ	0.13600	0.7260
"Volatility Channel" Granger Causality test	VOL does not cause Granger dMP	8.7651	0.0039
	dMP does not cause Granger VOL	12.3760	0.0004
	VOL does not cause Granger dSS	5.6E−05	0.9950
	dSS does not cause Granger VOL	5.0976	0.0378
	LIQ does not cause Granger VOL	3,2765	0.0776
	VOL does not cause Granger LIQ	12.789	0.007

In the transmission mechanism test results, both dMP and dSS have a significant negative impact on VOL, and constitute the Granger cause of the latter, while VOL has a significant positive impact on LIQ, which indicates that margin trading can inhibit the volatility of the stock market and improve the level of market liquidity.

3.2 Impulse Response Analysis

When giving 1 unit positive shoot on the dMP, LIQ always presents a negative response. In the first two periods, the negative reaction is almost the same. In the third period, the negative reaction reaches the maximum and then gradually decreases, which indicates that the financing transaction will significantly and continuously increase the market mobility. Consistent with the market liquidity shock caused by dMP, LIQ always presents a negative response to the dSS shock. This negative shock presents a fast attenuation trend on the whole, which indicates that short selling transactions can also improve market liquidity, but the continuity is weak.

The impact of VOL on dMP from the beginning to the end showed a negative reaction. The negative reaction reached the maximum before the second phase and then gradually decreased, indicating that financing transactions significantly and continuously reduced market volatility. VOL also always presents a negative response to dSS shock, and this negative shock rapidly decreases after reaching the maximum in the second phase, indicating that short selling trading can also reduce market volatility, but the continuity is weak. In addition, the response of LIQ to VOL always presents a positive response, which indicates that reducing market volatility continuously improves market liquidity.

3.3 Variance Decomposition

The variance decomposition of market liquidity (LIQ) is carried out, and the results are shown in Table 4.

Table 4. Market liquidity (LIQ) Variance decomposition

Time limit	dMP	dSS	VOL	LIQ
1	2.227855	6.182332	19.73012	67.93404
2	4.186113	6.035761	21.12189	64.60845
3	6.256322	6.378125	21.08045	62.06612
4	7.896534	6.436783	20.99677	60.56734
5	8.567231	6.472956	20.90961	59.68230
6	8.610345	6.481234	20.84620	58.75342
7	8.875345	6.480145	20.80523	58.56297
8	9.030784	6.477290	20.77045	58.47612
9	9.115627	6.475623	20.74914	58.39761
10	9.181302	6.472905	20.73499	58.27893

As can be seen from Table 4, the variance contribution rate of LIQ itself reached 67.93% in the first phase and then gradually decreased, indicating that a considerable part of the market liquidity variance was explained by other variables. The contribution rate of VOL to LIQ square difference is the largest and reaches the maximum value in the second period (21.12%), indicating that the variance of LIQ is explained by itself, and VOL has the largest contribution rate, indicating that market volatility has a significant impact on market liquidity, which is consistent with the results of VAR model and Granger causality test. The contribution rates of DMP to the variance of LIQ increased progressively. The above variance decomposition results show that financing transactions, short selling transactions and the investor sentiment changes and market volatility caused by them all have an impact on market liquidity, among which market volatility has the most prominent impact on market liquidity.

4 Conclusion

Margin lending, based on expected momentum effect, positively influences market liquidity through the volatility channel. In the meantime, the expected reversal effect is not significant in this process. Investigate its reason, the author proposes the biggest difference between short selling and traditional stock trading is the short selling mechanism. When the stock price is overvalued, the short selling trader will sell the stock by virtue of his own judgment on the market, and induce a new round of short selling behavior on the basis of promoting the fall of the stock price, thus forming the expected momentum effect and strengthening the market liquidity.

This paper establishes a conceptual model on the basis of literature review, and puts forward the hypothesis that margin financing and short selling act on market liquidity based on volatility channel. The empirical test using A-share data shows that financing transactions have a significant effect on improving market liquidity, while short selling transactions have a positive effect on market liquidity, but it is not significant. It shows that margin lending and short selling system has indeed played the institutional reform goal of improving market liquidity. However, at the same time, short selling is difficult to adapt to the change of short selling demand and increase effectively, which greatly reduces the policy effect of establishing a two-way trading mechanism through short selling and margin financing system. And this also leads to the insignificant effect of short selling on improving market liquidity.

Margin short selling and short selling are indispensable parts of the stock market, which is the only way for the market to mature. The results of this paper add to the empirical evidence that margin short selling and margin short selling can improve market liquidity. Margin and short selling business of stock exchanges is beneficial to increase trading activity and improve stock market liquidity. However, at present, the 1600 stocks allowed for margin trading and short selling only account for 30% of the total number of A-share stocks, and margin trading and short selling in China account for about 10% of the turnover in the A-share market, which is relatively low compared with 15% ~ 20% in western developed markets. Therefore, we should continue to open the scope of underlying stocks in margin trading and short selling, expand the scale of margin trading and short selling, and increase the supply and demand of stocks, so as to further enhance market liquidity.

References

1. Frino, A., Lecce, S., Lepone, A.: Short-sales constraints and market quality: evidence from the 2008 short-sales bans. Int. Rev. Financ. Anal. **20**(4), 225–236 (2011)
2. Senchack, A.J., Starks, L.T.: Short-sale restrictions and market reaction to short-interest announcements. J. Financ. Quant. Anal. **28**(2), 177–194 (1993)
3. Kong, X., Bi, X., Zhang, S.: The Impact of margin and short selling on liquidity: an empirical study based on the trading data of China's stock market. Bus. Econ. **06**, 165–170 (2014). (in Chinese)
4. Zhang, B., Yanping, L., Yang, A.: Empirical study on the impact of margin and short selling system on stock market liquidity. J. Xi'an Univ. Technol. **33**(01), 119–126 (2017). (in Chinese)
5. Lin, X.: The impact of margin trading on the quality of ETF fund market: a study based on the double difference model. Rev. Investment Stud. **33**(06), 138–148 (2014). (in Chinese)
6. Hardouvelis, G.A.: Margin requirements, volatility, and the transitory component of stock prices. Am. Econ. Rev. **80**(4), 736–762 (1990)

7. Huang, W.: Study on the impact of margin trading on stock market liquidity: based on the empirical analysis of Shanghai and Shenzhen a-share markets. China Prices **10**, 34–37 (2018). (in Chinese)
8. Miller, E.M.: Risk, uncertainty and divergence of opinion. J. Financ. **4**, 1151–1168 (1997)
9. Jinghan, C., Le, X.: Short selling system, liquidity and information asymmetry: the case of the Hong Kong market. J. Manage. Sci. **02**, 71–85 (2011). (in Chinese)
10. Xiaojian, Y.: The Impact of margin trading on liquidity and volatility of Chinese stock market: taking shanghai stock market as an example. J. South China Univ. Technol. (Soc. Sci. Ed.) **14**(02), 1–7 (2012). (in Chinese)

Computer Information Processing Technology in Financial Accounting Information System

Ran Lv[✉]

Audit Division, Shandong University of Arts, Jinan 250014, Shandong, China

Abstract. Under the traditional accounting process, the information provided by the accountant still cannot meet the needs of corporate management. This research mainly discusses the application of computer information processing technology in financial accounting information system. Based on the business data and analysis data, the pre-defined data warehouse model can use data conversion service tools such as SQL Server to extract, organize and load data from the business database to the data warehouse. The system will generate accounting vouchers according to the established rules. Accounting personnel only need to maintain the corresponding classification and summary mechanism, which realizes the accounting information systematization of all tasks from data collection to data output, greatly liberating the basic workload of accounting personnel. Authorization approval is carried out by means of digital signatures, realizing a networked and digital financial process. After the implementation of network finance, business processing speed has been accelerated by 27.5%, and the accuracy of production and procurement plans has been increased from 30% to more than 90%. This research will help advance the process of enterprises to realize computerized accounting.

Keywords: Computer information processing technology · Financial accounting information system · Financial process · Data warehouse

1 Introduction

In the procedural approval process, there are long approval procedures and low approval efficiency. There are ambiguities in the business handover nodes, and the problem of lagging in the handover of front and back links sometimes occurs. There is a lack of necessary feedback links when faced with problems that are not regulated. At present, the company as a whole lacks a prevention and feedback mechanism for possible problems, which results in no corresponding system guarantee as a solution when problems occur.

There are deficiencies in organizational settings and related department management, and similar situations cannot be avoided [1, 2]. Generated from the concept of entrusted responsibility under the background of large-scale production of natural resource economy and socialization and from the elements: the market is relatively developed, and the market plays an important role in the allocation of resources. The useful view of decision-making under the environment has played an important role in the function of accounting

J. Abawajy et al. (Eds.): ATCI 2021, AISC 1398, pp. 200–205, 2021.
https://doi.org/10.1007/978-3-030-79200-8_28

[3, 4]. In the network environment and network economy, knowledge, information, and intelligence have become the most important resources. The new requirements of economic globalization for accounting functions make accounting goals must adapt to the new economic environment [5, 6]. The accounting information system process in the network environment adopts event-driven processing, and the needs of accounting information users can be fed back to the business event process in time [7, 8]. When users of accounting information cannot obtain relevant information, their information is reflected in the model library and business processes in a timely manner to guide the collection of accounting information [9, 10].

In order to solve the current shortcomings of the company in the management of accounting information, improve the degree of standardized management of accounting information, and help the company to complete the standardization of internal accounting information in a targeted manner, the company has issued relevant information to the accounting personnel within the company's financial department through actual investigations. Questionnaire on the status quo of standardization management of company accounting information. By investigating the degree of satisfaction of current accounting information by in-service accountants, it is possible to accurately understand and analyze the current company's accounting information management problems based on the rich practical work experience of front-line business personnel, and more targeted analysis of internal reasons and Formulate.

2   Financial Accounting Information System

2.1   Financial Accounting

There are many problems in accounting practice, especially in the consistency of accounting information. For example, in the existing design of accounting subjects, especially when the company's project cost accounting problems are involved, the classification of accounting subjects is not specific and cannot accurately reflect the project cost. At the same time, the original document management standard is not perfect and the accounting vouchers fail to reflect all accounting information. Project cost accounting issues. The design of accounting books has the problem that the corresponding relationship between local currency and foreign currency is not close, and it is prone to problems such as foreign currency nuclear deviation, confusion of local currency and foreign currency correspondence. Third, the actual implementation effect in the entire three-level accounting process in the accounting process is not satisfactory. The process design is only a general regulation, and does not play its due role in the actual financial work of the department. At the same time, the accounting process only makes general provisions for the transmission of information, and there is no unified process management mechanism to solve the poor communication between departments caused by the intersection of departments.

2.2 Computer Information Processing Technology

In the network environment, accounting information changes greatly from form to content. According to the survey content, the judgment matrix Y can be described:

$$Y = \begin{pmatrix} F/F & F/C & F/B \\ C/F & C/C & C/B \\ L/F & L/C & L/L \end{pmatrix} \quad (1)$$

Among them: F is a financial perspective, C is a customer perspective, B is an internal process, and L is a learning and growth perspective. Assign points to each individual's options for comparing and judging the importance of the four aspects of the balanced score card, and the average value is substituted into the matrix Y to obtain a comparison matrix.

$$Y_i = \begin{pmatrix} 1 & a_{12} & a_{13} \\ a_{21} & 1 & a_{23} \\ a_{31} & a_{32} & 1 \end{pmatrix} \quad (2)$$

The eigenvector corresponding to the maximum eigenvalue h of the above matrix is found to be (a_1, a_2, a_3). Take the customer perspective as an example: In the customer perspective, customer satisfaction, customer, and waiting time have been investigated in two pairs, forming the following judgment matrix:

$$y_i = \begin{pmatrix} 1 & b_{34} & b_{35} \\ b_{43} & 1 & b_{45} \\ b_{53} & b_{54} & 1 \end{pmatrix} \quad (3)$$

The eigenvector corresponding to the maximum eigenvalue of the above matrix is (b_3, b_4, b_5).

3 Financial Accounting Information System Experiment

The company's main accounting data from 2018 to 2020 is shown in Table 1.

Table 1. The company's main accounting data from 2018 to 2020.

Project	2018	2019	2020
Operating income (ten thousand yuan)	17219. 64	16170.82	15415.98
Operating cost (ten thousand yuan)	7991.60	7398.48	7381.18
Sales expenses (ten thousand yuan)	1599.56	2035.60	1504.84
Management expenses (ten thousand yuan)	6387.42	6288.65	6395.55

4 Financial Accounting Information System Analysis

4.1 Asset-Liability Data Analysis

Figure 1 shows the results of asset-liability data analysis. According to the important data and related indicators of the company's balance sheet from 2018 to 2020, it can be seen that the total assets of the company have shown a downward trend in the past five years. Assets decreased from 706,203,500 yuan in 2018 to 683,563,700 yuan in 2020, a decrease of 3.21%. The current assets of enterprises have shown an upward trend. Compared with 2018, the current assets of enterprises have increased by 5.23%.

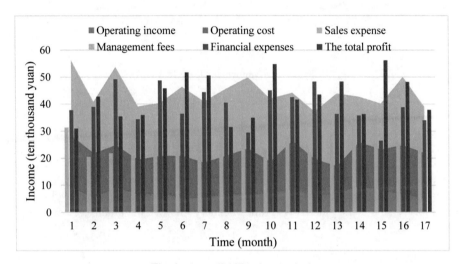

Fig. 1. Asset-liability data analysis

4.2 Implementation Effect of the Scheme

The implementation effect of the program is shown in Fig. 2. The company's financial operation performance shows that the company's financial stability and efficient operation. The main product, household air-conditioning, sales revenue increased by 18.9% in 2019, and the market share reached about 68%; in 2020, sales revenue increased by 25.7%, and the market share has reached 75%.

5 Conclusion

Financial accounting is composed of basic elements such as data, carrier and transmission. These elements have different contents and manifestations, different functions and quality characteristics, different development methods and management methods under different environments and conditions. The company has not established an effective internal accounting information standardization system.

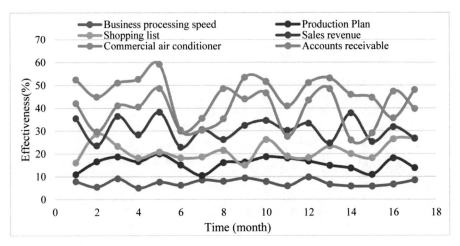

Fig. 2. Plan implementation effect

From the beginning of the generation of accounting information to the final submission process, in addition to the corresponding accounting practices that stipulate general principles, no relevant organization is responsible for irregularities in the processing of accounting information, and there are problems that cannot be dealt with in time. The phenomenon of unsolved problems with accumulation. With the intelligent network as the operating platform and a brand-new model of full resource sharing, the development and improvement of this model will fundamentally improve the quality of accounting information.

References

1. Zhou, J.Y., Fan, J.X., Lin, C.K., et al.: A Cost-efficient approach to storing users' data for online social networks. J. Comput. Sci. Technol. **34**(1), 234–252 (2019)
2. Motohashi, Y., Sakagami, H., Kamei, S.: Information processing device, information processing system, information processing method and computer-readable medium. J. Am. Soc. Inform. Sci. Technol. **55**(4), 333–347 (2018)
3. Pei, S., Guan, J., Zhou, S.: Fusion analysis of resting-state networks and its application to alzheimer's disease. Tsinghua Sci. Technol. **24**(04), 88–99 (2019)
4. Alkaam, N.O.: Image compression by wavelet packets. Oriental J. Comput. Sci. Technol. **11**(1), 24–28 (2018)
5. Wang, Q., Zhang, S., Wang, S.L., et al.: Comment on "Controlled mutual quantum entity authentication with an untrusted third party." Quantum Inf. Process. **19**(4), 1–6 (2020)
6. Mbindzoukou, P.M., Arsène Roland, M., Naccache, P.D., et al.: A stochastic model for simple document processing. Int. J. Inf. Technol. Comput. Sci. **11**(7), 43–53 (2019).
7. Susanto, A.M.: The role of computer assisted audit technique. Int. J. Sci. Technol. Res. **7**(11), 99–102 (2018)
8. Guo, J.M., Seshathiri, S.: Watermarking in dot-diffusion halftones using adaptive class-matrix and error diffusion. ECTI Trans. Comput. Inf. Technol. (ECTI-CIT) **13**(1), 1–8 (2019)

9. Massaro, A., Vitti, V., Galiano, A., et al.: Business intelligence improved by data mining algorithms and big data systems: an overview of different tools applied in industrial research. Comput. Sci. Inf. Technol. **7**(1), 1–21 (2019)
10. Saw, T.: Feature selection to classify healthcare data using wrapper method with PSO search. Int. J. Inf. Technol. Comput. Sci. **11**(9), 31–37 (2019)

The Improvement Path of Police's Scientific and Technological Literacy Based on Multi-objective Optimization Algorithm

Yujie Wu[✉] and Lan Sun

School of Public Security Law, People' Police University of China, Hebei, China

Abstract. At this stage, the national public security is implementing the big data strategy, and actively building intelligent public security and data policing. The role of police big data in combating crimes, maintaining social stability, and promoting public security is becoming increasingly prominent. However, due to the imperfect connotation of police data literacy and the lack of professional and technical personnel, the in-depth mining and application of public security big data is seriously restricted. Through a detailed interpretation of the concept and connotation of police's data literacy, this paper puts forward an education model of police's data literacy from four aspects of data awareness, data knowledge, data skills and data application. Combined with the construction of police vocational education system, through reasonable top-level design, it carries out comprehensive development from awareness, education, training and application.

Keywords: Big data · Data literacy · Data mining · Education model

1 Introduction

In the Internet era, with the rapid development of new media, it has gradually become the main channel of communication between the police and the public. At the same time, it is also an important channel to help form a good public security image and publicize public security ideology and culture, which can promote the harmonious development of society, It is necessary for the police to improve their media literacy so as to better cope with the development of the new era. In real life, many optimization problems have multiple objectives and must be optimized at the same time. The traditional technology can not solve the problem of multi-objective optimization. Meta heuristic algorithm provides a new idea for solving multi-objective problems. Particle swarm optimization is a representative algorithm of meta heuristic algorithm. The idea of particle swarm optimization comes from the predatory behavior of birds in nature [1].

With big data becoming the growth point of new police force, the data literacy of police is becoming more and more important to promote the implementation of big data strategy. It is an urgent problem to carry out data literacy education for public security police and guide them to actively analyze data and deeply mine data in actual combat.

J. Abawajy et al. (Eds.): ATCI 2021, AISC 1398, pp. 206–210, 2021.
https://doi.org/10.1007/978-3-030-79200-8_29

2   Fundamental Theory

2.1   Multi Objective Optimization Concept

Consider a minimization problem. The general form of multi-objective optimization problems (MOPs) with K objective functions is as follows.

$$\min F(x) = [F_1(x), F_2(x), \cdots, F_k(x)] \tag{1}$$

According to the duality principle, any objective function can be transformed from the minimum form to the maximum form, and vice versa, as shown below.

$$\max F_i(x) = \min(-F_i(x)) \tag{2}$$

$$\min F_i(x) = \max(-F_i(x)) \tag{3}$$

2.2   The Concept of Multi Objective Evolutionary Algorithm

Since the pioneering work of computer simulation evolution in 1954, with the contribution of various researchers in the development of new computer simulation that combines evolutionary theory with computational methods, a new field of evolutionary computing has emerged. In evolutionary computing, the algorithm is based on the population. The population goes through the process of iterating to guide it to achieve the expected goal. These processes can be inspired by concepts different from mathematics or computer fields, such as biological evolution mechanism or social behavior. In the computing technology of evolutionary computing, EAS adopts the evolution mechanism inspired by biological principles. EAS includes some well-known technologies, such as, Genetic algorithm, evolutionary programming, evolutionary strategy and genetic programming all adopt evolutionary mechanism, but their implementation is different [2].

The main disadvantage of using traditional algorithms and other mathematical programming techniques to solve the optimal solution is that they are mostly designed for specific problems. In one run, they can only find one optimal solution at most, and need to run many times to find the approximate Pareto front. EAS can overcome this disadvantage.

The main idea of EA is to model the basic mechanism of evolution and optimize it by using the concept of evolution. The five main mechanisms of EA are reproduction, natural selection, survival of the fittest, crossover and mutation. In EA, candidate solutions as individuals are encoded as genes in chromosomes, and a group of candidate solutions are called population. In a series of iterations, individuals will be evaluated to determine their fitness values. According to their fitness values, those considered to be more suitable are selected by the selection operator, because they have a higher probability of producing "more suitable" individuals (offspring). Therefore, two individuals randomly selected are called parents. Next, the parents are crossed, The mutation operation is occasionally carried out to produce new individuals or offspring, which is suitable for all selected individuals. Figure 1 illustrates the main steps of the single generation evolutionary algorithm.

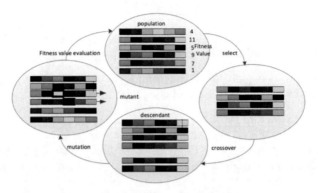

Fig. 1. Diagram of evolutionary algorithm

Since 1980s, all kinds of MOEAs have been developed. Vector Evaluated Genetic Algorithm (VEGA) has become the pioneering work of MOEA. In each generation, the whole population is divided into subpopulations of equal size. The number of subpopulations depends on the number of objective functions of mops, and these subpopulations are combined, The crossover operator and mutation operator are applied to the transformed population to obtain a new population. The advantage of Vega is that it is easy to implement, but the disadvantage is that it tends to generate good solutions for one goal rather than for all goals, because the selection operator tends to select a subpopulation with more suitable values than others.

3   Effective Measures to Improve Police's Media Literacy in the Internet Era

3.1   Changing the Concept of Media

The police in the new era is a kind of service-oriented public security, and the relevant wechat and microblog channels of the police need to actively analyze the public needs, provide the public with the information they need, and promote the effective interaction between the public and the police. In addition, the police also need to strengthen the management of network public opinion, solve all kinds of network events as soon as possible, and reduce the influence of the network. Police should form the concept of spokesperson in the network, pay attention to the online communication with Internet users, and become the network masses. For the problems raised by the social masses using the network platform, we should take targeted measures to solve them as soon as possible. Even for the things that can not be done for the time being, we also need to explain the reasons for the masses with a sincere attitude and obtain the recognition of the masses [3].

Traditional media, such as television broadcasting, periodicals and magazines, can comprehensively control and supervise their communication process, and at the same time, they can repeatedly check the relevant information content, so as to enrich the forms of information communication. As long as people have the corresponding media operation technology, they can become information communicators. The particularity of the police profession requires them to have good judgment ability in front of the media information, to treat all kinds of media information calmly, and to form a more mature discrimination ability. There is a close relationship between rational judgment ability and mature values. Therefore, it is necessary to further improve the comprehensive quality of police and form a correct media concept.

3.2 Improve Media Professionalism

If the police want to effectively deal with the network public opinion in the new era, they need to constantly improve their professional quality. First, the police need to strengthen the study of relevant legal knowledge, so as to be able to respond effectively to various questions of Internet users. Second, to cultivate the police's good ability of agenda setting and crisis management, which is also a necessary ability in the context of the Internet. They should exercise their ability of controlling netizens' psychology, setting topics, and early warning of Internet public opinion, so that they can seize the opportunity and have the initiative of discourse after the occurrence of public opinion [4]. Third, improve their own network language application level. It can flexibly use all kinds of network languages to communicate equally with netizens, achieve humor and wisdom, and form a good communication environment. Fourth, improve their ability to distinguish the network public opinion. It can quickly control the public sentiment in the mass information, pay attention to opinion leaders, focus on all kinds of public opinion hot spots and focus content, and strengthen the management of some information that can trigger network public opinion. Fifthly, effective communication with the media can treat the media well, make rational use of the media, and promote the comprehensive improvement of their media response ability. The public sentiment tends to be weak and trust the media reports too much, and the truth of the event is not the focus of public consideration. Under the background of the expanding influence of network communication, it is easy to damage the image of the whole police group. Even if the media published a correction report later, it is still unable to recover the losses that have occurred. In view of this phenomenon, the public security police need to actively learn advanced Internet technology, and be able to reasonably publicize through the network.

4 Conclusion

They gradually rely on big data policing, and even doubt the original hard-working spirit of the police and the thinking mode of self handling cases. They are not willing to stick to and carry forward the fine tradition. Although the improvement of police data literacy promotes the efficiency of the public security department, it has a negative impact on the traditional fine quality and work attitude. In the short term, it promotes the improvement of public security business efficiency, In the long run, the decline of the original work style and the loss of sense of responsibility will also affect the upgrading of the overall work efficiency. Therefore, we want to promote the development of big data policing by improving the police's data literacy. We also require the public security organs to continue to carry forward the traditional fine style while strengthening the police's data literacy. While ensuring the technical progress, the ideological construction will not retreat, so as to promote the healthy development of public security big data.

Acknowledgements. Research plan of public security theory and soft science: Research on the cultivation mechanism of police science and technology literacy from the perspective of invigorating the police with science and technology (Subject number: 2020LLYJJCDX026).

References

1. Tingyu, Z.: Analysis of police media literacy elements and ability construction from the perspective of police related network public opinion guidance. J. Yunnan Pol. Coll. **2**, 76–79 (2017)
2. Hua, J.: The necessity of improving public security police's media literacy in the Internet era. J. Liaoning Publ. Secur. Judicial Manage. Cadre Coll. **3**, 86–88 (2015)
3. Wei, L., Yin, W.: Research on media literacy of newly recruited police in public security organs: a sample of 450 new police. J. Hubei Pol. Coll. **28** (2015)
4. Chen, R.: Challenges faced by public security organs in combating cyber crime and countermeasures. Publ. Secur. Res. (2016)

Design and Application of Tourism Marketing Platform Based on Cloud Computing

Yuhua Jiang(✉), Haizhi Yu, and Jing Wang

Hulunbeir College, Hulunbeir, Inner Mongolia, China

Abstract. "Smart tourism" is the main task and hot research issue of the informatization construction of China's tourism industry. Its "wisdom" is reflected in the "wisdom of tourism service", "wisdom of tourism management" and "wisdom of tourism marketing". Tourism service recommendation is a popular marketing strategy in today's tourism industry, which has been widely studied. To some extent, it is the embodiment of wisdom. Nowadays, tourism data is becoming more and more huge, and the storage and processing of these data has become a difficult problem. At the same time, for information retrieval, search engines can not meet the diversified and personalized needs of users, and can not effectively solve the problem of information load. As another way to solve the information load, recommendation engine can not only find suitable or potential information for users through recommendation technology, but also bring better experience for users. But facing the challenge of big data, we need a new platform and appropriate and effective algorithm to deal with it.

Keywords: Smart tourism · Cloud computing · Tourism service recommendation

1 Introduction

In recent years, the development of cloud computing is hot. The concept of cloud computing has a long history. As early as 1959, there was a paper describing it, but it has been in the academic palace. It was not until Google company in the United States published three papers on MapReduce, GFS and BigTable that this concept was paid attention to by the world again. Since then, cloud computing has become hot. The reason why cloud computing can be a blockbuster is inseparable from the background of the times [1]. With the development of the Internet, the data in the Internet has become more and more huge. Storing and processing these data has brought new challenges in technology. Because of its own characteristics, cloud computing is destined to become the mainstream of future development, Cloud computing has strong distributed parallel computing ability and stable large-scale storage capacity, which can solve the problem of massive data computing and storage. There are more and more people studying cloud computing. In addition to business organizations, cloud computing is also a hot issue in the research of universities and scientific research institutions.

J. Abawajy et al. (Eds.): ATCI 2021, AISC 1398, pp. 211–215, 2021.
https://doi.org/10.1007/978-3-030-79200-8_30

2 The Basic Theory of Tourism Service Recommendation

2.1 Personalized Recommendation System

At present, with the rapid development of the Internet, the access to data and information is explosive growth. Most of the web pages in the website are dynamic and the frame structure is complex, and the data and information are scattered in the Internet. Information load is one of the problems existing in the Internet. One way to solve the information load is to retrieve information through search engines. However, search engines have low degree of automation and need to select keywords. Users can only query passively, and the number of query results is so large that they need to be screened, which may eventually take a lot of time to find the appropriate information. In addition, if users can't come up with keywords or appropriate keywords for the information they are looking for, then the information can't be found, which is time-consuming and ineffective. Personalized recommendation engine is another way to solve the information load, it makes up for some problems in the retrieval. First of all, from the aspect of information acquisition, from passive to active, users can directly receive the interest knowledge pushed, saving the search time. In addition, personalized recommendation is also helpful to the discovery of potential interest [2].

2.2 Social Network Analysis

As early as the 1930s, social network theory began to rise. Its initial development was based on the graph theory. Later, it continued to deepen its development, absorbed the knowledge of many disciplines, and finally gradually formed a set of systematic theories, methods and technologies. Now it has become an important paradigm of social structure research. Like the Internet, social network is an invisible network, which describes the relationship between social actors. Social network is usually composed of points and lines. Points represent social actors and lines represent relationship of social actors. A social network is a collection of relationships. Actors in social networks can be any entity object, usually the relationship between people, the relationship between countries, the relationship between cities and so on.

3 Framework Design of Tourism service Recommendation System in Cloud Computing Environment

3.1 Data Analysis Module

This paper studies the time distribution characteristics of inbound tourist flow, mainly discusses the changes of the number of tourists with months, the number of tourists and the length of stay, and the number of tourists and the number of accompanying people, and analyzes the seasonal intensity index and the characteristics of tourists [3].

The calculation formula of seasonal intensity index is as follows (1):

$$R = \sqrt{\sum_{i=1}^{12} (x_i - 8.33)^2 / 12} \qquad (1)$$

The closer the R value is to zero, the more uniform the time distribution of tourism demand is; the larger the R value is, the greater the time change is.

3.2 Analysis of Tourism Flow Based on Social Network

In the whole tourism flow network, the scale of tourism network refers to the number of tourism network nodes (tourist attractions or scenic spots). The number of all possible relationships in the directed network is formula 2. K represents the number of tourism nodes.

$$k \times (k - 1) \tag{2}$$

The number of possible relationships in undirected network graph is formula 3, and K represents the number of tourism nodes.

$$[k * (k - 1)]/2 \tag{3}$$

The value is the ratio of the actual total number of connections to the theoretical total connection coefficient, which reflects the closeness of the connections between all nodes. The tourism network density formula 4 is given.

$$D = (2 \sum_{i=1}^{k} d_i(n_i)/(k * (k - 1)) \tag{4}$$

k is the number of nodes, D is the density of tourism network, and the value is 0 to 1. 0 means that there is no connection between nodes, and 1 means that all scenic spots are closely related in ideal state.

4 Cloud Computing Platform Hadoop

Hadoop is an excellent cloud computing platform, developed by Apache Software Foundation, and widely used by commercial institutions and scientific research organizations. It is one of the best platforms to deal with big data problems. Its MapReduce and HDFS are the main parts of Hadoop. MapReduce can use distributed computing to process massive data quickly, and HDFS is stable and safe, and can store large amounts of data [4]. Hadoop is a top-level project, and there are many subprojects below, such as hve and HBase, which are super capable of storing and operating massive data, and scoop, which is a relational data ETL tool for data exchange between relational databases (my SQL, Oracle) and HDFS. Next, we will give a detailed explanation of the two core HDFS and MapReduce under Hadoop, as well as swoop, which is a relational data ETL tool.

4.1 Distributed File System HDFS

HDFS (Hadoop distributed file system) is a widely used distributed file system. It is safe, stable and has high fault tolerance performance, so it is a sharp tool to solve the problem of mass storage. Another reason to choose HDFS is low cost, because it can be deployed on cheap hard disk. HDFS supports high throughput and is suitable for massive data applications.

The design of HDFS architecture has four characteristics: separation of metadata and data; master/slave architecture; write once and read many times; mobile computing is more cost-effective than mobile data 21. First of all, separation of metadata and data. For traditional system files, metadata and data are independent, metadata records the attributes of the file body, and data is the content of the file. The design of HDFS is different from common sense, but it is not unreasonable. This is a smart design. For files with a large amount of data, it does not store them in a hard disk like a common storage system. Its storage method is to divide large files into several small files and store them in different nodes. If metadata and data are integrated, it is not easy to manage them. If part of the data is damaged, it will be difficult to manage them, Metadata will also be damaged, and the fault tolerance of data will be very poor. Then, the master/slave architecture and HDFS design adopt the master-slave mode. There are only two types of nodes in the whole cluster, one is namenode, the other is datanode, and the rest of the machines are datanodes. Generally, namenode does not store data, It stores metadata and the distributed information of each small data file. The real storage of data is datanode. Big data files are stored in HDFS, which are divided into several small data first. These small data are stored in several different nodes. HDFS has fault tolerance function, so every file in HDFS has backup, and the number of backup can be set, In this way, if one node in the cluster is damaged, the backup data in other nodes will become a substitute. In addition, one write and multiple read is HDFS's design for access. At the same time, only one writer can operate on the same data. This design avoids the complicated concurrency problem. HDFS also supports high throughput data access. Finally, mobile computing is more cost-effective than mobile data. With the increasing amount of data, the movement of a large amount of data will cause high IO overhead, which will lead to higher requirements for devices. If the amount of data movement has been increasing, it will bring heavy economic costs, so mobile computing becomes more cost-effective. HDFS supports high throughput and allows computing in each node, which is very reasonable.

4.2 Summary of this Chapter

This chapter describes the two core parts of Hadoop, distributed file system (HDFS) and parallel programming model (MapReduce), explains the four features of HDFS architecture design and the she instructions under HDFS, and analyzes the model, programming and operation of MapReduce in detail. Finally, a case of SCOP, a relational data eil tool, is demonstrated.

5   Conclusion

It is an important and significant work to solve the big data problem by writing data mining algorithm in the cloud environment. In theory, we can study the algorithm model of data mining in the cloud platform. In practice, it is an attempt to solve practical application problems.

Acknowledgements. (1) Inner Mongolia Autonomous Region Project of the 13th five year plan of Education Science in 2019: Research on the Practical Teaching of Hotel Management Specialty and the Cultivation of Entrepreneurship and Innovation Ability in Colleges Based on VBSE [NGJGH2019030];

(2) "The Second National History and Culture Research Special Project" of HulunBuir College:

Research on the Protection of HulunBuir Folk Cultural Heritage and Tourism Development Path from the perspective of cultural and tourism integration [2020MZZC02].

(3) Transverse Project: Research on the Design and Application of Marketing Platform on Travel Agency.

References

1. Yang, L., Wu, N.: Discussion on the natural adaptability of tourism industry and e-commerce. J. Chongqing Bus. Univ. (Soc. Sci. Ed.) **20**(06), 97–100 (2003)
2. Guoxia, W., Heping, L.: Summary of personalized recommendation system. Comput. Eng. Appl. **07**, 66–76 (2012)
3. Xia, P.: Research on Collaborative Filtering Algorithm in Personalized Recommendation Technology. Ocean University of China (2011)
4. Chen, Y.: Research and Application of Two Personalized Recommendation Algorithms. Jilin University (2007)

Analysis on Application Framework Construction of Transportation and Tourism Integration Under Big Data Platform

Jing Wang[✉], Yuhua Jiang, and Haizhi Yu

Hulunbeir College, Hulunbeir, Inner Mongolia Autonomous Region, China

Abstract. With China entering a well-off society in an all-round way, the national living standard has been continuously improved. As an important comprehensive industry, tourism has gradually become an important component of the national economy. With the rapid development of tourism, the number of tourists and the corresponding traffic demand are also increasing. Tourism traffic has become a more and more important special component in transportation, with distinct periodicity and seasonality. As an important carrier of tourism transportation, tourism transportation network not only undertakes the daily traffic demand, but also needs to face a series of traffic congestion and environmental problems caused by the surge of tourism transportation. In order to achieve the sustainable development of tourism transportation network as the goal, this paper comprehensively considers the concepts of innovation, coordination, green, open and co prosperity of tourism transportation resources, provides new ideas for the coordinated development of tourism and transportation, and establishes a more systematic planning method for the sustainable development of tourism transportation network.

Keywords: Road network optimization · Resource sustainability · Road network balance · Big data

1 Introduction

In the new stage of entering a well-off society in an all-round way, expanding domestic demand and promoting consumption have become the long-term goal of national economic development under the guidance of the Party Central Committee and the State Council. Tourism is an important comprehensive industry and an important part of the tertiary industry, which can meet people's growing material and cultural needs. It is known as "sunrise industry". In 2017, six ministries and commissions jointly issued several opinions on promoting the integrated development of transportation and tourism, which put forward the new development concept of innovation, coordination, green, open and sharing [1].

Tourism is an important industry which relies on the construction of tourism resources and facilities to attract tourists to visit, shopping, accommodation and other activities for the purpose of entertainment. As one of the fastest growing emerging industries in the world, tourism has gradually become an important component of China's national

© The Author(s), under exclusive license to Springer Nature Switzerland AG 2021
J. Abawajy et al. (Eds.): ATCI 2021, AISC 1398, pp. 216–220, 2021.
https://doi.org/10.1007/978-3-030-79200-8_31

economy, which has an important impact on economic development, traditional customs, land development and transportation demand. Its main business includes tourism, transportation and accommodation. However, for the tourism resources with short travel cycle, such as half day tour or one-day tour, the surrounding accommodation industry is not developed. Therefore, tourism industry and transportation industry are important basic pillar industries for tourism development. In addition, the three major elements of tourism are tourism resources, tourism facilities and tourism services, which are characterized by distinct volatility, seasonality and driving. Their existence has greatly promoted the development of transportation, commerce and catering industry, and promoted the progress of social economy and culture.

2 Database Design

For the information system, the database design is the key, it is the core of the system, just like the foundation of the house, so the quality of the database structure will have a great effect on the function realization and the corresponding performance of the whole system. In order to design the database, we need to make a comprehensive and detailed analysis of the business needs that the system needs to meet, the data information involved, such as the relevance between different departments involved in the system, the attributes of related equipment, employee information, etc. Through the scientific analysis of the database structure, it can better meet the needs of users with diversified functions, which not only helps to improve the storage efficiency of data, but also makes the data more coherent and complete [2].

At the beginning of database development, the conceptual model should be designed first. Through this model, we can show the correlation between different entities in the real environment. At the same time, we can make clear the storage, extraction and modification requirements of specific entity information. These conceptual models can be represented by means of abstract diagrams. In essence, these conceptual models use computer language to model the real world abstractly. They need to be intuitive and can show the related concepts in the information environment.

Big data is mainly divided into main data, business structured data, unstructured data and massive historical quasi real-time data. Main data storage mainly stores users, roles, systems, dictionary management and other data that are not frequently changed and relatively fixed. Business structured data mainly stores work order, business flow, log, business operation form and business related metadata. Unstructured data storage adopts unified service management, such as documents, videos, pictures, and related drawings.

The research content of this unit is the construction of system development environment and the design and implementation of the system. This paper first describes the steps of building the development environment, and then analyzes the system database design strategy, which involves the structure of data flow and data reading.

In the study of the influencing factors of tourism transportation service level, the utility theory is the most likely level of tourism transportation service level under the comprehensive effect of various factors. If there are tourism traffic service level M and tourism traffic service level n for the roads around the tourism node, then the final service level is evaluated as m level by considering various factors:

$$U_{km} < U_{kn} \tag{1}$$

Utility function consists of two parts: deterministic utility term and random term (or error term):

$$U_{km} = V + \varepsilon_{km}, m \in M \tag{2}$$

The attribute characteristics of tourism node K include seven indicators of tourism resources and transportation resources:

$$V_{km} = C_{km} + a_m^T X_k \tag{3}$$

The sustainable development optimization model theory of tourism transportation network. Firstly, a multi-objective optimization model is established based on the social economic benefits, the balanced benefits of tourism resources and the green ecological benefits.

3   Road Network Planning Method Based on Tourism Traffic Matching Curve

In civil engineering, it is generally believed that the soil skeleton is formed by solid particles. The size and percentage of particles will affect the physical properties of the structure, so the particle gradation is of great significance to the stability of the subgrade [3]. In engineering, the phenomenon of discontinuous gradation is often eliminated by adjusting the particle gradation of soil, so as to achieve good gradation and achieve the bearing capacity required by safe use. Generally, the distribution and gradation of soil particles are accurately judged by drawing the cumulative curve of soil particle size. The cumulative curve of soil particle size is represented by the abscissa of soil particle size, and the ordinate represents the percentage of soil particle mass less than a certain particle size in the total, which is drawn according to the screening test results. According to the curve, the non-uniformity coefficient and curvature coefficient can be calculated to realize the comprehensive evaluation of particle gradation.

At present, the existing models often obtain the reasonable scale of road network through multi-objective optimization, which is difficult to provide guidance for the over-all balance of road network. According to the corresponding relationship, it is the common goal of engineering application in the two disciplines to seek a relatively reasonable mix proportion. Through the introduction of the concept of particle gradation, the paper will establish relevant indicators to provide an intuitive and effective planning scheme for the sustainable development of tourism transportation network.

3.1   Basic Steps of Curve Establishment

The upper layer establishes a multi-objective optimization model and uses genetic algorithm to solve the problem; the lower layer establishes relevant indicators on the basis of the upper layer optimization to plan the sustainable development of tourism transportation network [4].

The main establishment process of tourism traffic matching curve is as follows:

(1) According to the characteristics of the planning area, the standard tourism traffic matching curve is drawn for the multi-objective optimization results;
(2) Calculating the reasonable index of tourism traffic matching curve;
(3) Determine the reasonable tourism traffic matching curve;
(4) Based on the node importance degree, the current road network index of tourism traffic is calculated;
(5) The results were compared with the reasonable indicators to determine the current resource allocation of the road network;
(6) Through dynamic adjustment, the final optimization scheme is determined.

4   Basic Parameters of Tourism Traffic Matching Curve

In order to master the particle size and distribution uniformity of soil particles, the particle composition and gradation of soil are deeply analyzed by the particle size composition analysis method in geotechnical test in civil engineering, so as to determine the classification and properties of engineering soil and ensure the good exertion of soil skeleton function. The composition of soil particle size is generally expressed by the cumulative curve of soil particle size, and the judgment indexes are nonuniformity coefficient and curvature coefficient respectively. The judgment criteria are obtained by combining practical engineering experience with experimental method. This paper will establish the basic parameters of the tourism traffic network based on two indicators - the uneven coefficient of the scale of the tourism traffic network and the curvature coefficient of the scale of the tourism traffic network.

The scale non-uniformity coefficient of tourism transportation network is derived from the non-uniformity coefficient in the particle size distribution curve, which reflects the uniformity of soil particle distribution. It is one of the criteria to determine the particle size distribution of soil. It is usually used to analyze the possibility of adverse geological phenomena, such as the appearance of sand pipe. The non-uniformity coefficient is usually calculated according to the ratio of the limited particle size to the effective particle size. In practical engineering, it is reflected as the ratio of the corresponding particle size when the cumulative weight of soil particles reaches 60% and 10% of the total.

5   Conclusion

In this paper, the sustainable planning method of road network based on tourism traffic matching curve is proposed. Firstly, the influencing factors of tourism traffic and the analysis methods considering heterogeneity at home and abroad, as well as the research status of tourism traffic optimization model and planning method based on sustainable are summarized. Then it analyzes the heterogeneity sources of influencing factors of tourism transportation service level from two aspects of tourism resources and transportation resources, and establishes a multi factor analysis model of tourism transportation service level considering heterogeneity.

Acknowledgements. (1) Inner Mongolia Autonomous Region Project of the 13th five year plan of Education Science in 2019: Research on the Construction of Teaching Quality Evaluation System of School-Enterprise Collaborative Education Based on Model SERVQUAL [NGJGH2019003].

(2) Transverse Project: Study on the Individual Spell Group Model of Travel Agency.

(3) Local Key Projects of Hulunbuir College: Research on the Construction and Realization of the Matching Pattern of Hulunbuir TourismCommunity Development Based on the Theory of Cultural Adaptation [2019FDZD06].

References

1. Chen, Y.: Coordinated Development and Optimization Path of Tourism Economy and Transportation Coupling. Chang'an University (2019)
2. Zhang, N.: Prediction of Highway Tourism Traffic Demand Based on Network. Southwest Jiaotong University, Chengdu (2003)
3. Wang, J., Wang, W., Du, H.: Tourism traffic volume prediction based on four stage method. Highway **1**(01), 167–170 (2011)
4. Guo, W.: Co Evolution and Difference Analysis of Inbound Tourism Flow and Air Transport Network. Shanghai Normal University (2014)

Data Mining Method of Intelligent Civil Aviation Cabin Sales

Feng Guo[⊠]

Sanya Aviation and Tourism College, Sanya 572000, Hainan, China

Abstract. With the continuous development of air transport industry, the competition among airlines and between airlines and other modes of transportation is becoming increasingly fierce. Especially, private airlines have weak market competition advantages, and private airlines are an important part of China's civil aviation transportation industry. Therefore, it is of great significance to study the development and operation of private airlines, It is not only conducive to the further development of private airlines, but also conducive to the transformation and adjustment of China's economic structure. Social network based on Internet technology has become an important platform for civil aviation cabin marketing, which has been widely concerned by industry and academia. On the other hand, the Internet makes information and data more transparent. Based on the mining and analysis of a large number of data, it can provide support for civil aviation cabin marketing decision-making.

Keywords: Civil aviation company · Marketing strategy · Development strategy · Data mining

1 Introduction

At present, China's civil aviation transportation industry has been developing rapidly, and has gradually become one of the main ways of China's foreign exchanges and economic and trade exchanges. Especially under the background of China's economic restructuring and the gradual development of the tertiary industry as China's pillar industry, the development speed and economic scale of China's national shipping industry are gradually approaching the international level [1]. With the continuous improvement of China's residents' living standards, more and more passengers are pursuing comfort and time efficiency of transportation means. The proportion of residents who choose air travel in long-distance travel is greatly increased. This demand for air transportation greatly promotes the vigorous development of China's civil aviation transportation industry. made outstanding achievements, which not only brings great convenience to the life of Chinese residents, but also greatly promotes the improvement of China's economic level, especially since the reform and opening up, China's civil aviation transportation industry has made remarkable achievements in terms of total passenger and cargo transportation, number of routes and development scale of airlines.

© The Author(s), under exclusive license to Springer Nature Switzerland AG 2021
J. Abawajy et al. (Eds.): ATCI 2021, AISC 1398, pp. 221–225, 2021.
https://doi.org/10.1007/978-3-030-79200-8_32

2 Data Mining Technology

Data mining technology is actually the result of the development of information technology, is the result of people's long-term research and development of database technology. From the beginning, all kinds of commercial data were only stored in the computer database, developed to query and access the business data in the database, and then developed to the real-time traversal of the database. Data mining is an advanced stage of database technology application. It is a revolutionary change. It can not only query and traverse the past data, but also find the potential relationship between the data, so as to increase the depth of information application. With the development of mass data storage, powerful multiprocessor computers and data mining algorithms, three basic technologies have been developed, Data mining technology has been widely valued in commercial applications.

2.1 Decision Tree

The learning of decision tree is a process of using information gain theory in information theory to find the field with the largest amount of information in the data set, establish a node of the decision tree, and then set up the branches of the tree according to the different values of the field, and repeat the process of building the lower level nodes and branches of the tree in each branch subset [2].

The regression equation can be determined by least square estimation.

The overall model of binary linear regression was as follows:

$$Y_i = \beta_0 + \beta_1 X_{1i} + \beta_2 X_{2i} + u_i \tag{1}$$

The sample regression model was as follows:

$$Y_i = \widehat{\beta}_0 + \widehat{\beta}_1 X_{1i} + \widehat{\beta}_2 X_{2i} + e_i \tag{2}$$

The linear regression equation is as follows:

$$Y_i = \widehat{\beta}_0 + \widehat{\beta}_1 X_{1i} + \widehat{\beta}_2 X_{2i} \tag{3}$$

According to the regression, the sum of squares of the deviation between the estimated value Y' and the actual value y reaches the minimum:

$$\sum (Y_i - Y')^2 = \sum e_i^2 \tag{4}$$

2.2 Market Segmentation

Market segmentation is a kind of means for enterprises to carry out refined marketing. It is usually based on the relevant needs of customers to further classify customers with different needs, so as to design relevant products or adopt different sales strategies for different customers with different needs, so as to meet the needs of different types of customers. The customer demand in each market segment is different. When the enterprise subdivides the market, it is necessary to determine the type of the segmentation first, and at the same time, it should further do a good job in customer demand survey, on this basis, design the relevant products that can better meet the needs of customers, so as to achieve the purpose of precision marketing. Generally, the main factors affecting market segmentation are population, geographical environment, psychological role and customer behavior. Therefore, enterprises should rely on these four dimensions in order to achieve the purpose of accurate market segmentation.

3   Main Methods of Precision Marketing

The methods and tools of precision marketing are very rich, and they are still developing. In this paper, these methods are classified into the following three categories.

3.1   Marketing Method Based on Database

The establishment of a potential consumer database with a certain scale and relatively complete relevant information is an important foundation for precision marketing. It is a long-term and arduous work to establish a potential consumer database, which requires continuous accumulation and continuous efforts of enterprises, If the enterprise has not established its own independent consumer database, it can use the consumer database of other organizations (such as postal database, social security database, and other intermediary agencies' database) to screen the information of potential consumers meeting the needs of the enterprise to carry out its own precision marketing activities.

3.2   Internet Based Approach

Precision marketing based on database is a kind of very good method, but this kind of method also has its limitations: ① it is difficult for an enterprise to directly have a database of potential consumers reaching a certain scale at the beginning of the enterprise; ② the construction of the database usually takes a relatively long time; ③ the database needs to be updated timely, Otherwise, it is easy to produce a lot of junk information (invalid information due to the change of customer conditions). Therefore, many enterprises are difficult to use the precise marketing method based on customer database at the beginning [3].

Precision marketing based on the Internet is to identify the consumer psychology and behavior characteristics of Internet users, and relevant enterprises carry out targeted precision marketing activities according to the significant characteristics of these Internet users.

At present, precision marketing methods based on the Internet mainly include:

(1) The vast majority of Internet portal websites have opened many different channels. For example, sina.com has opened channels for sports, automobile, real estate, digital, tourism, health and so on. Enterprises can choose channels that match their own product characteristics to launch advertisements. Internet users entering a certain

channel are generally interested in relevant content, At this time, the enterprise's advertising shows accuracy. Of course, this precision is relative to the advertising in the mass media.

(2) Keyword search advertising. Baidu, Google, Yahoo and other major search sites currently provide keyword search advertising services. When they want to buy a certain kind of product or service, many consumers will search the relevant information through the search website. The product information of enterprises can appear in front of the consumers in need through the search website, and the pertinence and accuracy are very strong, If you want to buy a digital camera, you are likely to search the website for the relevant information of digital camera. At this time, if the product information of Sony digital camera appears in front of you, you can easily understand it. This realizes the accuracy of marketing.

(3) Blog. Blog is a new application of Internet Web2.0 technology. Netizens can not only browse the information on the Internet, but also send the information they want to spread (such as tourism information) to the Internet, so that other netizens can browse and discuss with each other. In this way, people who are interested in this field tend to gather together in the blog area of the Internet.

For example: Blogcn blog website is divided into music blog, sports blog, tourism blog, etc. in this way, music lovers gather together, sports fans gather together, and pet owners gather together On the basis of blog CN, users can actively cooperate with each other to form a strong common marketing platform, Nike has launched a blog called "art of speed", which invited 15 video workers to express their interpretation of speed, and invited influential blog authors to participate in the discussion, It has attracted the continuous participation of many target consumers [4]. The whole activity is oriented to the sports blog group and has strong pertinence.

(4) E-mail advertising. Users will be required to leave a lot of information when they apply for email mailbox on the EMLI website. When enterprises carry out marketing activities, they can cooperate with the email website, select users in their user groups that meet the product characteristics of the enterprise according to the user information, send the relevant advertisements of the enterprise to these users, and leave various contact information of the enterprise, In order to further communicate with these email users, as long as the selected user group features are highly consistent with the characteristics of enterprise products, the enterprise's marketing activities will be very accurate.

4   Conclusion

Airlines have great deficiencies in marketing strategy and marketing network. At present, the development of advanced technologies such as Internet technology and big data technology poses more severe challenges to the further development of airlines. This requires that airlines should further explore and develop their e-commerce model, in order to promote the development of airlines with the help of the rapid development of Internet technology. The establishment of e-commerce model can not only expand the customer service of Shanghai Jixiang airlines, but also effectively reduce the operating costs of airlines.

References

1. Xu, H.: On the System and Theory of Precision Marketing. China Post, No. 914 (2006)
2. Wang, W.: Case Study: Research on Aviation Marketing Management in Shenzhen. South China University of technology (2016)
3. Zhang, L.: Research on Internet marketing in education and training industry. Master's thesis. two thousand and eleven
4. Wang, F., Qu, C.: Precision Marketing: A New Trend of Marketing Communication. Private economy news, 19 September 2005

BP Neural Network Based on Genetic Algorithm in Economic and Industrial Structure Analysis and Prediction

Yajun Yang[(⊠)]

School of Management, Henan University of Science and Technology,
Luoyang 462000, China

Abstract. The analysis and prediction of economic industrial structure can grasp the future development characteristics and trend of industrial structure, and then help to optimize and upgrade the industrial structure. This paper analyzes the basic principles of BP neural network and genetic algorithm, and further analyzes the prediction of economic industrial structure by BP neural network based on genetic algorithm.

Keywords: BP neural network · Industrial structure · Genetic algorithm

1 Basic Principle Analysis of BP Neural Network

In this paper, was proposed an error feedback of the concept of BP neural network, i.e., multilayer feedforward neural network multiplication training algorithm. Current neural network error diagnosis is widely applied. The British Petroleum Corporation neural network feeds back an error to the multiplication algorithm that reflects input output relationships and nonlinear structures. This is easy to educate because it is easy to operate. Image processing signal processing and mode recognition [1].

BP neural network model is each layer is connected with each other. It is a feedforward neural network model composed of non-phenomenon units. Its model learning consists of back propagation and forward propagation. After passing through the input layer, the forward propagation information is processed through the hidden layer and then transmitted to the output layer. If the error between the output value and the real value of the output layer is large, the weights of the neurons in each layer are modified by back propagation according to the original path, and then the error values are reduced and corrected repeatedly, until the error value meets the standard [2, 3].

2 Analysis of Basic Principles of Genetic Algorithm

Genetic algorithm was proposed in the 1970s. Its principle is mainly designed according to the evolution law of nature. It simulates a computational model designed by Darwin's concept of genetics and related natural selection. Therefore, it is called genetic algorithm.

© The Author(s), under exclusive license to Springer Nature Switzerland AG 2021
J. Abawajy et al. (Eds.): ATCI 2021, AISC 1398, pp. 226–230, 2021.
https://doi.org/10.1007/978-3-030-79200-8_33

Genetic algorithm is widely used in signal processing, artificial life, prediction and combinatorial optimization. The algorithm calculates the fitness value by objective function, and the calculation process does not require other derivation and additional information. BP neural network based and finally obtain the optimal threshold and weight value. According to the prediction results of previous data, the optimization effect is roughly shown in Table 1.

Table 1. Comparison of error results between genetic algorithm optimized bp neural network model and basic neural network model

Projected year	Prediction error (%) of genetic algorithm optimization BP neural network model	(%) of Prediction error BP neural network model
First year	1.49	26.63
Second year	1.71	31.73
Third year	1.77	43.82
Fourth year	2.06	45.93
Fifth year	3.15	31.59

3 Genetic Algorithm Based BP Neural Network Analysis and Prediction of Economic Industrial Structure

3.1 Genetic Algorithm Based BP Neural Network Prediction Model

The analysis and prediction of economic industrial structure is usually an important index of economic development. When the per capita GDP is at different levels, the proportion of each industry can be analyzed. The per capita GDP can reflect the social comprehensive productivity, income level, total production volume and the development level of the national economy. Therefore, this paper takes the per capita GDP as the input object of the BP neural network prediction model, takes the economic industrial structure as the model output object, and uses the relevant data as the training sample to train the BP neural network [2, 3].

3.1.1 Training Sample Data Processing

The data of 31 provinces in the last six years of the statistical yearbook are selected as the relevant data for model training. The total number of samples is 186. The per capita GDP is converted into comparable price according to the price index in the statistical yearbook. Unified processing of per capita GDP and industrial ratio data, data range within the range of 0.1–0.9.

3.1.2 BP Neural Network Creation

The characteristic of single hidden layer BP neural network is that its nonlinear mapping ability is strong. If the number of neurons in the hidden layer of BP neural network is not limited, it can approximate any function under any precision condition. This paper uses a simulation software with neural network toolbox, which can construct BP neural network model or calculate the model. The software is used to design a single hidden layer neural network, and its programming process is relatively simple. Therefore, the model is created by using the software, the per capita GDP data is used as input layer, the industrial specific gravity of per capita GDP is used as input layer neuron.

3.1.3 BP Neural Network Training

A training method for BP neural network is to iterate the input sample data, then find a suitable weight matrix, and finally minimize the total error of the network. According to the corresponding algorithm theory, in the process of training the sample data, there will be a certain error between the actual output and the expected output. At the same time, the threshold and weight of the neural network are adjusted and modified to provide new parameters for the next signal input. After the training of neural network, the training error curve and training result can be observed and compared, and then the training effect can be analyzed. If the error of neural network training is small, the training accuracy is higher. When the training result of neural network is very close to the sample value, the training effect is good.

3.1.4 BP Neural Network Testing

In the actual process of neural network testing, we should select the appropriate test samples, and then input the test sample data to simulate the neural network. There will be a certain error between the simulation results and the real values. The size of the error is analyzed and evaluated.

3.1.5 Forecasting

By using the trained neural network to predict the industrial proportion of different per capita GDP levels, taking the per capita GDP of each province and the industrial proportion of the primary industry as an example, the per capita GDP is taken as the input layer neuron of the model. The training error can be obtained by training the BP neural network. Through the analysis, we can see that after 200 iterations of training, the error value begins to slide rapidly, and then continues the iterative training. After 6000 iterations, the final value of the BP neural network training error reaches 0.0038. By analyzing the training results, it can be seen that the training error of BP neural network can be controlled in a very small interval, and the output value of its neural network is basically consistent with the real value of sample data training. Therefore, it can be judged that the model can accurately fit the changes of sample data.

To further analyze and evaluate the training effect of BP neural network, we should select certain test samples to test the trained BP neural network. The relevant data of three provinces in a certain year can be selected as the sample of this test. The BP neural network model is tested. In addition, when selecting test samples, we should pay attention to transforming per capita GDP into comparable prices. In the process of simulation and prediction of BP neural networks, we can select sample data within a suitable range. After prediction, the proportion of primary industry can be obtained, and then the industrial structure can be analyzed.

According to the above forecast and analysis method, the proportion of three industries can be simulated and analyzed, and the change value of industrial structure can be obtained under different per capita GDP level, and the change of its proportion can be analyzed. The future economic industrial structure can be predicted.

3.2 Analysis of Forecast Results

The analysis of the forecast results shows that with the rising per capita GDP in China, the economic and industrial structure will show certain characteristics in the future. When the per capita GDP reaches a certain standard, the proportion of the primary industry will begin to decline continuously, and will be basically stable within a certain range when it falls to a certain standard; the proportion of the secondary industry will rise slowly, but the development will be more stable.

Conclusion: to sum up, the prediction accuracy of the BP neural network prediction model based on genetic algorithm has been improved, which has played a great advantage in the process of forecasting the economic industrial structure. From the analysis of this paper, we can see that the BP neural network based on genetic algorithm includes: prediction model construction, data processing, neural network creation, neural network training, neural network testing, prediction.

4 Conclusion

Based on the study of BP neural network and genetic algorithm, a multi-objective GA-BP algorithm is established by fusing the feedforward neural network with excellent approximation, classification and self-learning functions and the genetic algorithm based on natural selection. The genetic algorithm is used to adjust the topology and parameters of the network adaptively and dynamically in the global solution space, so as to obtain the optimal design of the network The plan. In addition, the Pareto optimal solution set obtained by multi-objective genetic algorithm provides great convenience for design decision. The GA-BP algorithm is applied to the multi-objective dynamic job shop scheduling problem. The experimental results show that the GA-BP algorithm is effective and feasible.

Although the algorithm proposed in this paper has the above excellent characteristics compared with the traditional evolutionary algorithm, there are still many unsatisfactory places. For example, when the problem is quite complex, the calculation workload is large and the calculation time is long. Further research work can be carried out in the future.

References

1. Sheng, W., Zhao, H., Sun, Y.: Sales prediction model of bp neural network optimization based on improved genetic algorithm computer systems applications **28**(12), 200–204 (2019)
2. Yao, J., Huang, Y.: A neural network optimization calculation method based on improved genetic algorithm electronic technology and software engineering (08), 138–140 (2019)
3. Zhou, Z., Cao, C.: Neural Network and Its Application. Tsinghua University Press (2004)

Implementation of ID3 Algorithm in Asset Inventory Model

Liuqing Ye[✉], Canhui Zhang, Dan Zhao, Longmin Bu, and Huiting Xu

State Grid Hunan Electric Power Company Limited, Power Supply Service Center (Metrology Center), Hunan Province Key Laboratory of Intelligent Electrical Measurement and Application Technology, Changsha 410004, China
{zhangch10,xuht2}@hn.sgcc.com.cn

Abstract. This paper aims to solve the problem of optimal portfolio of venture capital investment by intelligent algorithm. Using the actual data, this paper constructs the optimal portfolio models based on different risk measures and termination conditions: unit risk return maximization model, VaR risk control model, risk and return preference model. Genetic algorithm and simulated annealing algorithm are used to optimize the models.

Keywords: Mathematical finance · Optimal portfolio · Genetic algorithm · Simulated annealing algorithm

1 Introduction

In the classical Markowitz mean variance model, the risk is defined as the variance of the total return of the portfolio. This risk measurement method has its irrationality, because it regards both upward and downward volatility as risk. In fact, when the portfolio yield is large, investors generally do not care about the size of variance, so we can consider unilateral volatility, that is, the lower half variance as a risk measure. Value at risk (VaR) is a commonly used risk measure in practice. Var refers to the worst expected loss of an asset or portfolio at a certain confidence level in a certain holding period in the future under market fluctuation. It is a quantile in essence. More details on the application of value at risk.

A more general optimization model is utility optimization model, which considers the different preferences of investors for risk and return, and aims to maximize the final utility of investors.

2 Introduction of Intelligent Algorithm

Genetic algorithm (GA) is an optimization method, which simulates the natural selection of Darwinian biological evolution theory and the biological evolution mechanism of genetic mechanism. From the given original solution group, it continuously evolves to produce new solutions until it converges to an optimal solution of the problem [1]. The

J. Abawajy et al. (Eds.): ATCI 2021, AISC 1398, pp. 231–235, 2021.
https://doi.org/10.1007/978-3-030-79200-8_34

profit maximization model of unit reinsurance can be written as the following quadratic programming problem:

$$\max f(\omega) = \omega^T \mu / \sqrt{\text{var}(\omega^T X)} \tag{1}$$

It is the risk control function based on VaR risk. The optimal portfolio model based on VaR risk control function is obtained:

$$\min VaR = -E(r_p) + \sigma_p \varphi^{-1} \tag{2}$$

$$E(X_p) = \mu^T \omega \geq r_p \tag{3}$$

The steps of genetic algorithm are as follows: initialization and selection of a population, that is, the set of hypothetical solutions of the problem. Generally, the set of individuals is generated by random method, and the optimal solution of the problem is obtained by the evolution of these initial hypothetical solutions. Choose the next generation of individuals according to the survival of the fittest principle. When the fitness of the optimal individual reaches the given threshold, or the fitness of the optimal individual and the fitness of the population no longer rise, the iterative process of the algorithm converges and the algorithm ends. Otherwise, the new generation population obtained through selection, crossover and mutation will replace the previous generation population, and return to the second step, that is, the selection operation, and continue to cycle. Figure 1 flow chart of genetic algorithm.

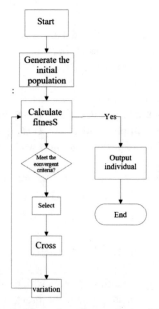

Fig. 1. Flow chart of genetic algorithm

3 Check Out and Set Up Relevant Forms

First, before the asset inventory, the accounting personnel should comprehensively sort out the outstanding and completed business items, record all the completed economic business into the relevant account books, and settle the increase and decrease amount and balance, so as to meet the requirements of account certificate, account and account table, so as to provide accurate and reliable basis for the inventory work. The second is to set up the inventory list of fixed assets according to the contents and requirements of the inventory. The main contents include: inventory time, asset category, storage location, number, name, unit of measurement, quantity, unit price, amount and remarks, which are used as the written document to record the physical inventory results and the original evidence to reflect the actual assets, To provide guarantee for the account verification of fixed assets, the "report form of inventory profit and loss of fixed assets" is set up [2]. The main contents include number, category, name, book quantity and amount, actual quantity and amount, profit and loss quantity and amount, remarks and other items, which are used as the basis for analyzing the causes of differences, finding out the economic responsibility and adjusting the account book records.

4 The Main Methods are as Follows

It mainly introduces the functions.

The inventory of fixed assets shall adopt the method of on-the-spot inventory, and the physical storage personnel shall go to the site to carry out the inventory one by one according to the classification, so as to fill in the "inventory list of fixed assets" while cleaning up, and strictly prevent the occurrence of missing points, key points, missing records, wrong records and deliberate cheating. After the completion of the inventory, the reference personnel shall sign or seal the form in triplicate, one for the village accountant, one for the material object keeper, and one for the person in charge of the inventory team, which is used for bookkeeping, reference and responsibility. The inventory list of fixed assets shall be checked with the relevant assets in the book, and the report form of inventory gain and inventory loss of fixed assets shall be filled in to determine the amount of inventory gain or inventory loss of fixed assets, The members of the inventory team shall sign or seal the form in triplicate, one for the village accountant, one for the material object keeper, and one for the person in charge of the inventory team, which shall be used for account adjustment and reference. 6. The village collective shall publicize the inventory of assets and accept the supervision of the masses. The investigation team is responsible for timely review and feedback of the doubts and objections raised by the villagers, so as to ensure the villagers' right to know, decision-making and supervision in the operation and management of collective assets, and prevent the occurrence of violations of law and discipline.

5 Results Processing

5.1 Key Points of Inventory

The inventory team is responsible for sorting out the inventory gains and losses of fixed assets found in the inventory one by one. Through investigation and research, the paper analyzes whether the causes of the differences are caused by natural causes or by dereliction of duty of the custodians, corruption, theft or natural disasters, accounting errors or lax warehousing procedures, and finds out the causes, Distinguish the responsibility, determine the nature according to the relevant provisions, put forward the handling opinions or suggestions, and submit them to the villagers' meeting or the villagers' representative meeting for discussion and voting. After the approval of more than two thirds of the villagers, the handling opinions, together with the fixed assets inventory profit and loss report form and the written information for villagers' opinions, shall be submitted to the township (town) government or the competent business department for examination and approval, and then the relevant profit and loss matters shall be adjusted handle. According to the relevant provisions of the financial and accounting system and the "fixed assets inventory profit and loss report form" and the township (town) audit opinion, the accounting adjustment and treatment of the inventory profit and loss assets are carried out [3]. Inventory surplus assets. If it is really impossible to find out the cause, the inventory surplus fixed assets shall be charged to the "fixed assets and other income" account according to the market price of similar assets, and the assets that have been purchased and constructed but have not yet been carried forward shall be charged to the "fixed assets" account according to the final price of completion acceptance and delivery for use, and carried forward to the construction in progress account; if the donation has been accepted but the account has not been recorded.

5.2   Check the Assets of Inventory Loss

The damage caused by natural disasters is recorded in the accounts of "fixed assets", "fixed assets liquidation", "receivables", "internal transactions", "other income" or other expenses "; the inventory loss caused by the dereliction of duty of the custodians is recorded in the accounts of" internal transactions of other expenses "and" fixed assets"; the inventory loss caused by corruption and theft cases, It should be handled together with relevant departments or units. If the original items cannot be forced back, it should be recorded in other expenses, "fixed assets", "internal transactions," receivables "and other accounts to check the implementation of the asset management system [4]. According to the problems and deficiencies existing in the asset inventory, the inventory team put forward rectification opinions and measures, further improved the relevant management system, continuously improved the quality of accounting, improved the level of operation and management, fundamentally reversed the undemocratic procedures of asset purchase and disposal, lax entry and exit procedures, and unclear economic responsibilities, The bad situation of ineffective regulatory measures constantly tap the potential of assets to better serve the development of rural economy.

6   Conclusion

In this paper, we first accept and explain the selected risk asset model, and then propose an artificial intelligence algorithm with optimized risk model. Finally, we select the return rate of stocks (risk assets) as the analysis object, and construct the optimal portfolio model based on different risk measures and constraints. When it is extremely difficult to find the analytical solution directly, we can find the optimal portfolio model, The intelligent algorithm is used to optimize the model and a satisfactory solution is obtained.

Acknowledgements. Deepen the research and application of HPLC application technology for intelligent perception (5216A020005K).

References

1. Huang, G.: How to strengthen the inventory management of fixed assets of rural commercial banks. China Rural Credit Cooperation Daily, 08 September 2020 (2007)
2. Haixia, G.: On the inventory of the company's fixed assets. Inner Mongolia Sci. Technol. Econ. **13**, 54–55 (2020)
3. Chi, N.: The actual operation of fixed assets inventory based on information platform. In: China Medical Equipment Association, China Medical Equipment Magazine, China Medical Equipment Conference and 2019 Medical Equipment Exhibition, vol. 3 (2019)
4. Yuanjun, Z.: Research on the application of new fixed assets inventory technology in power enterprises. Enterp. Manage. **S2**, 92–93 (2018)

Computerized Accounting System of E-commerce Platform

Yuehui Hu[✉]

Yunnan Vocational College of Finance and Economics Accounting College,
Kunming 650222, Yunnan, China

Abstract. In 1979, China first referred to the application of computer in accounting as "accounting by electronic data processing". After learning from foreign references, the concept of "electronic data processing accounting" appeared. Relationship between E-business and accounting computerization becomes more and more prominent. Based on the in-depth analysis of e-commerce and accounting computerization, this paper systematically discusses how the computerized accounting information system with management department and financial department as the center can effectively adapt to e-commerce, so as to guide the development of enterprises.

Keywords: Accounting computerization · E-commerce · Platform

1 Introduction

E-commerce, as a very sensitive network topic, always appears frequently in all kinds of T media [1]. In essence, e-commerce is not the only favorite in the commercial field, but also the epitome of e-commerce in various fields, including internal units. Especially in today's steady development of accounting by electronic data processing, the relationship between e-commerce and accounting computerization is becoming more and more prominent. Through the in-depth analysis of the related problems in the field of e-commerce and accounting computerization, this paper systematically discusses how the computerized accounting information system centered on the financial department of the management department should marry the "e-commerce" problem, so as to provide guidance for the development of enterprises.

2 Design of E-commerce Platform System

At present, the author does not agree with the e-commerce and its classification in many media [2]. Nowadays, many media divide e-commerce into three types: BtoB (business and business), BtoC (business and customer), CtoC (customer and customer), or from the transaction form of e-commerce, it can be divided into group price and sale along. Obviously, this is e-commerce from the business area, that is, sales. At present, the manifestation of e-commerce is mainly concentrated in this field. Therefore, in

J. Abawajy et al. (Eds.): ATCI 2021, AISC 1398, pp. 236–240, 2021.
https://doi.org/10.1007/978-3-030-79200-8_35

many impression concepts, e-commerce in the commercial field is regarded as the real e-commerce. The e-commerce that the author wants to talk about here is a kind of e-commerce based on the commercial field. At the same time, it is a real e-commerce facing enterprises. It not only aims at the commercial field, but also aims at the e-commerce among the internal links or subsystems of the unit's computerized accounting information system.

The current e-commerce with its "fierce momentum, rapid development, promising prospects" has begun to be approved by all walks of life, recognized and widely used in all walks of life. For enterprises, e-commerce mainly shows the following unique advantages:

It improves the operation rate of funds and shortens the operation cycle. E-commerce, with its fast, accurate, convenient and flexible characteristics of Internet, which is the basis of its survival, can make the funds arrive in the account in the shortest time after the transaction, and there is no need to over consider the previous funds in transit. The high-speed operation of capital and the shortening of its cycle show the relativity and flexibility of time. It not only provides a strong financial guarantee for the follow-up production of enterprises, but also injects vigor and vitality into the whole enterprise and business field, thus accelerating the capital flow in the market, It provides a favorable technical guarantee for the benefits of computerized accounting information system of enterprises in the information age.

The cost of production and sales is reduced. In the modern market economy society, reducing the production and sales costs of products has become a common problem. According to the survey, statistics and verification of a Japanese financial institution, the emergence of e-commerce has reduced the production and sales costs of products by 2%–36%. For enterprises, the emergence of e-commerce not only accelerates the purchase and transportation of raw materials, but also reflects the transfer of products and semi-finished products. E-commerce will transfer semi-finished materials to the next sub-system module in the form of data, participate in the operation of the module, and store the data through the enterprise's computerized accounting information system for the enterprise's overall data management and operation.

Some intangible transactions and electronic internationalization are the process of e-commerce, and both sides understand each other through network transactions. Remind each other of the result of the transaction by clicking the mouse, according to the international commercial standards and Internet data transmission protocol, but they have no idea whether the other party is in the next room or on the other side of the earth. Similarly, in the process of e-commerce transactions, both sides of the transaction are in accordance with the agreed rules and procedures. Each program section confirms the legality and feasibility of the other party through e-commerce data. Therefore, the whole e-commerce transaction process and transaction procedure are intangible, international and electronic. The "invisibility" here is a kind of image, only relative to human senses.

3 Research on Internal Control Content of Computer Information System

3.1 General Control of Information System

According to the third international audit standard of the U. S. patent Accountants Association, the internal control of sea is divided into generality and internal control. Control mechanism. The first problem of the application is the electronic data processing behavior of all types. 3) general control of special accounts transactions is the control of the computerized accounting system. Applied control - trading and trading. Unlike this classification, this paper examines the concrete contents of the sea internal control at the start [3].

3.2 Technical Connotation of Computerized Accounting System Under Internet Environment

Under the Internet environment, computerized accounting system is very different from the traditional network financial system. It integrates LAN and WAN The third rate technique is a technology that shares system components and tertiary services (like client servers, application servers, database servers, shown Fig. 1 of the client system/server, and the application server is an important technology. By combining logic, we resolve the relationship between the application and the client and database servers, and greatly improve the network's efficiency and security capabilities.

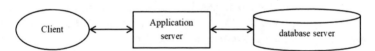

Fig. 1. Three layer C/S structure

Softmax layer is usually placed in the last layer of the network structure, that is, after all nodes in the last layer are fully connected. The input of softmax layer is T * 1-dimensional vector (where T * 1 vector is the output of full connection layer and t is the total number of training data set categories). The probability of each type is calculated by formula (1).

$$S_j = \frac{e^{a_j}}{\sum_{k=1}^{T} e^{a_k}} \tag{1}$$

Where, S_j is the probability corresponding to the j-Class sample after the softmax formula, with the range of [0, 1]; a_j represents the value of class j output by the node of the upper layer after inputting the full connection layer.

The softmax formula is used to calculate the probability of each sample class in the sample, and then the softmax loss function is used to calculate the sample loss value. The mathematical expression is as follows: formula (2).

$$L = -\sum_{j=1}^{T} y_i \log S_j \tag{2}$$

CS system, we can know that under the Internet environment, the data of the computerized accounting system flows through the client, flows to the database server through the application server, and is sent to the application server after being temporarily stored by the database server, These data are processed by the application server and then returned to the database server for classification and storage. The data is uploaded to the Internet by legal means on the premise of ensuring its security and integrity, and then transmitted to relevant customers (including business departments or branches of enterprises, information users, government functional departments, etc.) In these open and confidential information systems, to ensure the integrity and security of the whole accounting computer system, in addition to the specific features of SEEA internal control, under the above circumstances, with special conditions, in the accounting system and network environment, the new internal control computer is digitized become [4].

Of course, for the internal control of computerized accounting system, whether in the general environment or in the Internet environment, in addition to the above control measures, we must promulgate and improve various corresponding laws, regulations, policies and systems as soon as possible, Efforts should be made to improve the professional quality of accounting staff (including accounting knowledge, computer knowledge, etc.) and strengthen the research and implementation of computerized audit, especially to deal with the attacks of hackers who are frightening in the Internet environment [5, 6]. In addition to adopting the above control measures and countermeasures, we should also give full play to the strength of the whole society to prevent them from legal, moral and technical aspects, So as to further strengthen and improve the internal control of enterprise computerized accounting system.

4 Conclusion

This paper mainly studies the concept, function and characteristics of internal control of accounting system, and the analysis and research details of general control and application control of computer information system. This paper analyzes the process and characteristics of the computer accounting system and internal control under the network environment, helps the internal control of the computer accounting system under the network environment, improves the risk awareness of the enterprise in the construction of the accounting computerization system, seriously formulates the internal control system and practically implements the corresponding control measures, so as to ensure the responsible quality to the maximum extent.

References

1. Yang, B.: Accounting Information System, pp. 17–98. Higher Education Press (2001)
2. Zhao, X.: New field of accounting development in the information age. Acc. Monthly (2), 19–21 (1999)
3. Xia, S.: Enlightenment of the historical development of internal control on the construction of internal control system in Chinese enterprises. Bus. Acc. (9), 37–39 (2004)
4. Wang, X.: How to establish and improve the internal control under the environment of accounting computerization system. Friends Acc. (8), 87–88 (2004)

5. Yang, L.: Challenges and changes of enterprise financial management in the context of big data. Mod. Mark. (11), 133–134 (2017)
6. Cui, H., Li, Q.: Challenges and changes of enterprise financial management in the context of big data. Manage. Observ. (30), 166–167 (2017)

Big Data in the Network Live Platform of Enterprise Brand Communication

Ou Yangli and Yue Wang[✉]

Jingchu University of Technology, 33th Xiangshan Avenue, Jingmen,
Hubei, People's Republic of China

Abstract. The rapid rise of mobile network video live broadcasting platform brings challenges and opportunities to enterprise brand communication. This paper expounds the advantages of network live broadcast platform in information dissemination, such as high digitization, interactivity and personalization, which has become the mainstream trend of brand communication. It brings a brand-new communication revolution, meets the requirements of brand communication precision and efficiency, personalized service and push algorithm, and has better communication effect. The live broadcast industry chain has ushered in a huge growth space, bringing innovation in media channels, profound changes in information structure and audience groups. Enterprise brand communication should be integrated with the network live broadcast platform to maintain its advantages in the market competition.

Keywords: Network broadcast platform · Brand communication · Integration development · Situation

1 Research Background and Interpretation of Basic Concepts

1.1 Research Background

The rapid rise of mobile network video live platform and the rapid development of live network with goods have brought about rapid changes in the whole Internet business. With the rapid changes of domestic and international situation, global economic growth is facing more and more challenges. Domestic enterprises are faced with various competitive pressures and the pain of transformation and upgrading. As a strategic tool for enterprise development, entrepreneurs have to think deeply about the way of brand communication and the improvement of related effects. In the era of traditional media, enterprise brand advertising information is transmitted to the general public consumers through traditional media. Ordinary consumers generally receive information passively, even if they have multiple choices, they still receive information passively. Traditional consumers are passive and powerless, unable to participate in the communication process and have no idea of their own, and media channels do not support consumers' self-development. In the past few decades, entrepreneurs have carried out one-way communication, communication and marketing separation by means of commercial advertising, corporate public relations and promotion. These characteristics have become the brand of the era of traditional media and brand communication.

J. Abawajy et al. (Eds.): ATCI 2021, AISC 1398, pp. 241–245, 2021.
https://doi.org/10.1007/978-3-030-79200-8_36

1.2 Basic Concepts

For "brand", the American Marketing Association defines a brand as a name, term, mark, symbol or design, or a combination of them. Its purpose is to identify a producer or a producer's products or services, and to distinguish them from competitors' products and services [1]. David ogway, a representative of the theory of brand image, believes that brand is a complex symbol, which is an invisible combination of brand attributes, name, package, price, history and reputation. In a word, brand is an important guarantee for an enterprise to compete, establish brand loyalty and tree image.

In the process of communication, another brand is created. In the era of consumer led communication, the object of brand communication is gradually generalized, including suppliers, government units, the public, peer competitors and enterprise internal personnel. In brand communication, enterprises can find opportunities to communicate with consumers anytime and anywhere, and quickly develop personalized and approachable content for information dissemination. The channels of information media are greatly enriched, and the dependence of media is reduced, and the communication characteristics of communication content itself become the most important. Communicators and receivers, producers and consumers, online and offline gradually integrated into a body. So far, enterprise brand communication has undergone earth shaking changes.

2 Comparative Advantages of Live Network Platform in Enterprise Brand Communication

2.1 The Brand Communication of Network Live Broadcasting Platform is Precise, and the Fans Group is Transformed Into Purchasing Power Customers

In the process of brand communication, business owners invest a lot of brand building funds on media, hoping to occupy the commanding height and radiate a large number of mass groups. Its communication effect is generally fed back through marketing research methods. It is difficult to accurately and data the influence between enterprise brand and consumers, let alone track and contact. At the same time, traditional media time, layout, program content and other aspects affect the content and effect of brand communication. On the live network platform, digitization and interactivity become typical features. The reading environment further presents the characteristics of fragmentation and subculture, and the market is changing rapidly. In the new era, the environment needs media that can accurately disseminate content, accurately base, accurately spread time, and accurately segment the market.

2.2 The Live Network Platform Interacts with Fans in a Two-Way Manner, With High Stickiness and Highly Personalized Consumer Groups

Compared with traditional brand communication methods, online live broadcast platform is easier to establish stickiness with fans, create demand and experience in real-time interaction, form communication topics conveniently, and create word-of-mouth communication effect. In the process of communication, the media is passive and passive, and in the process of communication, the media is passive and powerless. In this basic situation, corporate brand communicators only set issues according to their own judgment, and consumers passively receive the information. The marketing planning idea behind this is 4P theory, which always takes profit sales as the first starting point rather than the consumer led mode. For a long time, the brand communication of enterprises has been far away from the real needs of consumers and the real feedback.

2.3 Fuzzy Comprehensive Evaluation Method and Situation Value Calculation

The membership degree of each feature on each fuzzy subset is calculated. After M membership functions are obtained, for each data $x = (x_1, x_2, \ldots x_n)$ According to the membership relation, the fuzzy relation matrix R is (1).

$$R = \begin{pmatrix} r_{11} & r_{12} & \cdots & r_{1m} \\ r_{21} & r_{22} & \cdots & r_{2m} \\ \cdots & \cdots & \cdots & \cdots \\ r_{n1} & r_{n2} & \cdots & r_{nm} \end{pmatrix} \tag{1}$$

The fuzzy relation operation is performed on the fuzzy relation matrix R. There are two methods of fuzzy relation operation, i.e. taking big and small. The largest method is to select the item with the largest membership degree of each feature as the output, and the smaller method is the opposite. Since the maximum method can retain the information of each feature to the greatest extent, the fuzzy operation is adopted to output the maximum value of each line in R, and the maximum operation result is obtained by transposition, as shown in formula (2).

$$FR = (fr_1, fr_2, \ldots fr_n) \tag{2}$$

3 The Influence of Network Live Broadcast Platform in Enterprise Brand Communication

3.1 The Rise of Live Network Platform Brings the Innovation of Brand Media

In the traditional media brand communication, business owners put a lot of brand building funds on the media, hoping to occupy the high point and radiate a large number of public groups. The communication effect is generally fed back through marketing research methods. It is difficult to accurately and data the influence between enterprise brand and consumers, let alone track and contact. At the same time, traditional media time, layout, program content and other aspects affect the content and effect of brand communication [2]. On the live network platform, digitization and interactivity become typical features. The reading environment further presents the characteristics of fragmentation and subculture, and the market is changing rapidly. The environment of the new era needs the media that can accurately spread the content, the precise region, the accurate communication time and the accurate market segmentation. The online live broadcast platform P account fan group is also the target audience of enterprise brand communication. Brand communication activities have relatively accurate controllability, and the communication effect can be detected immediately. Enterprises can communicate equally with the target consumers in the soft intelligence short video and live broadcast room by using the technology innovation of network live broadcast platform.

3.2   Fans Two-Way High Stickiness Link, Consumer Groups Highly Personalized

Compared with traditional brand communication methods, online live broadcasting platform is easier to establish stickiness with fans, create demands and experience in real-time interaction, form broadcast topics conveniently, and create word-of-mouth broadcast effect. In the process of communication, the media is passive and passive, and in the process of communication, the media is passive and powerless. In this basic situation, corporate brand communicators only set issues according to their own judgment, and consumers passively receive the information. The marketing planning idea behind this is 4P theory, which always takes profit sales as the first starting point rather than the consumer led mode. For a long time, the brand communication of enterprises has been far away from the real needs of consumers and the real feedback [3, 4]. In the era of mobile network video live broadcast, enterprise brand communication needs to be transformed and upgraded, and truly move to the communication mode of audience demand orientation. The market is further subdivided, consumer preferences are further classified, and communication is faster and more interactive. According to the scientific and technological algorithm, the network live broadcast platform pushes homogeneous consumers to a live broadcast space, providing space-time conditions for free communication. Under certain conditions, the communication effect of brand communicators with opinion leader attribute or Internet celebrity entertainment attribute shows geometric progression growth. Highly personalized opinion leaders bring the brand communication dozens of times the influence of traditional advertising.

4   Conclusion

Mobile network video broadcast platform brings revolutionary progress in digital technology, reducing the communication barriers between enterprises and consumers. Love Kwai, tiktok and other live software are being used to make more money and spend more time on the consumption and social interaction. Under the background of mobile network video live broadcast, the real life is further digitized and virtualized, and the digital content in the network live broadcast platform is more and more deduced into the trend of real life. Brand and consumer contact, communication, trading and other ways are changing with each passing day. These changes should arouse the attention of business owners, transform and upgrade to the road of digital economy. In other words, corporate brand communication should be integrated with the live network platform to maintain its advantages in the market competition.

References

1. Kotler, P.: Marketing Management. 9th edn. Shanghai People's Publishing House (2002). Translated by Mei Ruhe, Mei Qinghao and Zhang Heng

2. The 45th statistical report on China's Internet development. https://baike.baidu.com/item/
3. Media, society and the world: social theory and digital media practice. Fudan University Press (2014)
4. Tang, L.: The trend of brand communication in the new media environment. Chin. Youth Soc. Sci. (2015)

Machine Learning Differences in Machine Translation of Urban Publicity Texts

Zhang Hong[⊠]

Nanchang Institute of Technology, Nanchang, Jiangxi, China

Abstract. Through the parallel evaluation of domestic commonly used machine translation platforms, this paper studies the shortcomings of the existing machine translation platforms in the translation from Chinese to English. Through the four machine translation platforms Sogou, Youdao, Google and Baidu, the selected case texts are translated into Chinese into English, and compared with the manual translation, the typical machine translation problems that are easy to appear in the Chinese English translation of domestic machine translation platforms are analyzed. Through the parallel comparison between machine translation and manual translation of a Chinese English speech translation example on α hina Daily website, it is found that there are some common problems in the translation of Chinese into English in the four machine translation platforms tested, such as translation errors at the word and sentence level, omission of translation, and case and case format errors. These translation problems need to be solved by the improvement of machine translation platform and the translator's post editing.

Keywords: Machine translation · Chinese English conversion · Speech · Translation Comparison · Translation differences · Omission

1 Introduction

The ideal goal of machine translation is to realize the natural language pair conversion without human participation with the help of computer technology, and output the translation with the translation quality and language expression level required by the end user. Since the beginning of the 21st century, although machine translation "is moving towards the direction of practicality and commercialization", and the theory and Application Research of machine translation have made vigorous development, it is undeniable that, due to "the complexity of translation itself", "the complexity of natural language" and "the limitations of machine itself", machine translation still has deficiencies in the translation of words and sentences, In order to improve the quality of translation, post-processing links such as machine translation error recognition and post editing are often needed. In order to clarify the shortcomings of the current machine translation platform, many scholars have carried out evaluation and comparative research on machine translation [1]. For example, he Wenzhao and others investigated the effect and problems of machine translation in the translation of English relative clauses by using Google

J. Abawajy et al. (Eds.): ATCI 2021, AISC 1398, pp. 246–250, 2021.
https://doi.org/10.1007/978-3-030-79200-8_37

machine translation as the tool, questionnaire survey on machine translation acceptability and sample comparison of machine translation cases, It is found that when there are attributive relative clauses, core words in prepositional phrases and complicated logical relations, multiple items in parallel, and omission of relative pronouns, the translation of English Chinese machine translation is prone to translation problems; klubicka et al. Take translation accuracy and translation errors as the measurement index, A comparative study of traditional machine translation and machine translation based on neural network and deep learning is carried out.

2   Statistics of Error Types in Case Text Machine Translation

This study makes a sentence by sentence parallel comparison between the Chinese daily manual translation of the case text and the machine translation of Sogou, Youdao, Google and Baidu. The original Chinese case text "deepening the exchange and mutual learning of civilizations and building a community of shared destiny in Asia" has 3608 Chinese characters; the translation was obtained by four machine translation platforms on October 7, 2019. For the classification of machine translation errors found in the process of comparison, this paper refers to the two-level classification system and error category induction framework adopted by Luo Jimei et al. 0 and Li Mei et al. 01. At the same time, combined with the actual case text comparison, the specific classification of the involved error types is added and deleted. In order to clarify the classification of machine translation error types and the statistics of error types obtained by horizontal comparison of translation of various machine translation platforms, this study marks all types of machine translation secondary error types as error types according to the statistical order in the table.

3   Machine Translation Vocabulary Translation Errors

The current machine translation platform is equipped with cloud data dictionary and text corpus with large amount of data and constantly updated, which makes the machine translation platform have a high level of ability of word capture, recognition, meaning discrimination and transformation. However, due to the high degree of mechanization or imperfection of machine translation platform, the multi-dimensional personality of different languages and the semantic and pragmatic supersignature of languages, there are still some errors or deficiencies in the process of word translation in machine translation, which need to be improved systematically or rely on the translator's post editing [2]. This paper summarizes and analyzes the typical vocabulary translation errors found in the process of this machine translation investigation.

4   Difference Calculation

Due to the difficulty of the optimal matching block and the sample block, there will be certain differences. With the continuous image restoration, these differences will be constantly enlarged, so that there are bad phenomena such as block effect in the repaired image. In this paper, the gradient value of pixels is used to construct the texture difference measurement model, and the confidence level is updated by the texture difference measurement model to optimize the repair difference generated in the repair process.

Let T_{pi}. Is the gradient value corresponding to the center pixel of the sample block, T_{qi} is the gradient value of the optimal matching block at the corresponding pixel points, and the total number of known pixels in the sample block is M. Then the texture difference measurement model can be expressed as:

$$G = \frac{\sum_{i=1}^{M} |T_{pi} - T_{qi}|}{M} \tag{1}$$

When G is smaller, it means that the texture difference between the sample block and the optimal matching block is smaller, and the corresponding pixel confidence is greater. The confidence update is expressed as follows:

$$G\,(q) = C(p')\exp(-C) \tag{2}$$

4.1 Terminology and Quasi Term Machine Translation Errors

In the current machine translation platform based on big data corpus and deep learning mechanism, the corpus size, expansion speed and self-renewal ability of big data corpus are important indicators affecting translation function. The lack of corpus or functional defects in the above corpora will lead to the decline of translation quality or significant translation errors. The resulting machine translation errors mainly include terminological and quasi terminological errors, part of speech or tense errors and so on [3]. These errors will cause the translation to deviate from the original meaning and reference.

4.2 The Difference of Meaning Choice of Common Polysemous Words

Catford pointed out that "the meaning of a word is actually a hierarchical semantic structure or network, which is composed of multiple senses of a word", which makes the understanding of lexical meaning in actual discourse or conversation refer to the sentence in which the word is located and the context related sentence. Machine translation is based on lexicon, parallel corpus and mathematical statistical model. Therefore, it has strong ability to identify and transform the meanings of words with few branching meanings or high frequency applications, while it has a weak ability to identify and transform multi branched, implicit or low-frequency applications of words. In addition, the weak context discrimination function of machine translation also restricts the accuracy of meaning selection of common words. For example, in example 3, the hidden subject of the original sentence is "Chinese civilization", in which "accumulation" actually means "representative and symbol", rather than "superposition number" in physical sense. Among the four machine translation platforms tested, only Youdao has translated the correct meaning of the original word in the context, while the other three machine translation platforms rigidly translated the original word into "accumulate", which is a mistranslation, failing to correctly translate the node meanings of the words in the context of the text, resulting in the translation being inconsistent with the original in terms of semantics and reference.

4.3   Differences in Machine Translation of Idioms and Idioms

Idioms and idioms are a kind of habitual expressions with fixed structure and complete meaning, and often have cultural load. The semantics of idioms and idioms includes three basic levels: literal meaning, extended meaning and cultural connotation. Considering that even manual translators are faced with great difficulties in dealing with the transformation and transmission of language and cultural connotations due to language and cultural heterogeneity, the statistical standard for machine translation errors of idioms and idioms in this study is: as long as the extended meaning of the original words or idioms is translated, it is regarded as correct translation. According to this standard, a considerable number of idiom or idiom translation errors still exist in the four machine translation platforms tested [4]. when translating the idiom "from generation to generation", the four machine translation platforms tested are all mechanically translating word for word, resulting in the translation meaning remaining at the surface literal meaning, but not translating the deep meaning and implied meaning of the original expression, which is easy to cause readers to understand the semantic content of the original text in confusion, It also results in the difference of semantic information integrity and expression function between the original text and the translated version.

5   Conclusion

In this case text machine translation investigation, we found that Sogou, Youdao, Google and Baidu four machine translation platforms tested have common vocabulary and sentence translation problems in Chinese English translation. The types of machine translation errors at the lexical level mainly focus on the translation errors of terminologies and quasi terminologies, the wrong choice of meanings of common polysemous words, improper mechanical selection of words or repeated words, translation errors of idioms and idioms, etc.; the types of machine translation errors at the sentence level mainly focus on the structural and logical confusion of short and long sentences, semantic translation errors in sentence context, and subject reduction errors of non subject sentences. In addition, the four machine translation platforms tested also have serious problems of missing translation and case format. For the tested machine translation platform, the degree of omission is not only related to the nature of the corpus and the performance of the platform itself, but also related to the unit input length when the translator inputs the original text. When inputting the original text, the translator can input the original text to the machine translation platform with a small unit quantity, which can reduce the rate of missing translation, but will increase the labor workload. However, the problem of capitalization is most prominent in Youdao translation, which increases the workload of cleaning the translated text. Generally speaking, in this case translation practice and investigation, Sogou translation.

Acknowledgements. Study on the C-E Translation of City Publicity Materials from the Perspective of Linguistic Adaptation Theory (Social Science "13th Five-Year Plan" Project of Jiangxi Province (2016)).

Project Number: 16YY16.

References

1. Wang, X., Yang, Y.: 60 years of research on machine translation of national meat; analysis based on core foreign language journals. J. Hunan Univ. Soc. Sci. Ed. vol. **33**(4), 90 = 96 (2019)
2. Zhiwei, F.: Parallel development of machine translation and artificial intelligence. Foreign Lang. **41**(6), 35–48 (2018)
3. Mei, L., Ximing, Z.: Classification and statistical analysis of English Chinese machine translation errors [J]. J. Shanghai Univ. Technol. (Soc. Sci. Ed.) **35**(3), 201–207 (2013)
4. Wenzhao, H., Defeng, L.: Machine translation of English relative clauses: a case study of Google machine translation. China Sci. Technol. Transl. **32**(3), 30–34 (2018)

Cyber Intelligence for Network and Cloud Technologies

Analysis on the Construction of Cultivation Mechanism of the College Counselors' Scientific Research Ability with the Computer Technique Support in the Internet Age

Chunsheng Deng[1], Yuebing Huang[2], and Jing Yu[3](✉)

[1] School of Computer Science of Jiangxi University of Traditional Chinese Medicine, Nanchang, Jiangxi, China
[2] School of Economics and Management of Jiangxi University of Traditional Chinese Medicine, Nanchang, Jiangxi, China
[3] School of Humanities of Jiangxi University of Traditional Chinese Medicine, Nanchang, Jiangxi, China
20000425@jxutcm.edu.cn

Abstract. With the development of the Internet age, the improvement of scientific research ability within college counselors becomes the basic content of the construction of college counselors and It is also an inevitable choice to promote the professionalization and specialization of counselors' development. Based on the scientific definition of the counselor's scientific research ability, this paper analyses the constituent elements of counselors' ability. As a result, a mechanism for cultivating the scientific research ability of college counselors is constructed, which is based on the guarantee mechanism, focused on the education mechanism, promoted by the computer technology support mechanism, supported by the discipline platform construction mechanism, and tested by the interaction of theory and practice. And this mechanism aims at providing theoretical reference for strengthening the cultivation of scientific research ability of college counselors in the Internet age.

Keywords: Internet age · College Counselors · Scientific research ability · Cultivation mechanism

In order to strengthen the construction of college counselor troops, a series of documents and measures like "Occupational Ability Standard for College Counselors (Provisional)" has been introduced by the Ministry of Education. Especially the initiative that proposed to implement the spirit of the National College Ideological and Political Work Conference, which raised that colleges should take the strengthening of moral education and cultivating people as their basic requirements, colleges and universities are attaching increasing importance to the scientific research ability cultivation of college counselors, and scholars are paying more and more attention to the scientific research problems of counselors. In recent years, domestic scholars' research on the scientific research ability of college counselors has mainly focused on the definition, current situation, influencing

J. Abawajy et al. (Eds.): ATCI 2021, AISC 1398, pp. 253–259, 2021.
https://doi.org/10.1007/978-3-030-79200-8_38

factors, promoting approaches, and cultivating strategies of the scientific research ability of counselors. When it mentioned about the factors that affect the improvement of scientific research ability of college counselors, some scholars believed that the influencing factors were college counselors have low scientific research awareness, low enthusiasm, lack the support of academic background, lack the foundation of scientific research knowledge, lack the proper conditions of scientific research, lack the professional training and guidance, and they worked with heavy pressure that did not have sufficient time for scientific research [1]; and when it mentioned the cultivating mechanism of scientific research ability of college counselors, some scholars have proposed a "the Three-sphere Integrated" mechanism including the guarantee mechanism, the promotion mechanism and the strengthening mechanism for building a team of research counselors; and other scholars have proposed strengthening the scientific research assessment and evaluation mechanism of counselors. Then, aiming at the current situation of the college counselors' overall scientific research ability that needs to be improved, this article constructs the mechanism for cultivating the scientific research ability of college counselors that are based on the guarantee mechanism, focused on the education mechanism, promoted by technical support mechanism, supported by discipline platform construction mechanism, and tested by the interaction of theory and practice, which aims at providing the theoretical reference for strengthening the cultivation of scientific research ability of college counselors.

1   Definition of Scientific Research Ability of College Counselors

At present, there is no unified discussion and accepted definition on the concept of college counselors' scientific research ability in domestic academic circles, but a consensus has been formed on the following elements [2]: firstly, it is the practical need of college counselors; secondly, it has a certain theoretical foundation of the related disciplines; thirdly, it applies the working experience to theory research by the use of scientific methods; and finally, it could help to apply the research results to daily work and achieve practical results. For example, Sha Jun [3] believes that college counselors' scientific research ability is that of counselors who summarize the scientific working rules and methods according to their own work experience, and apply them to guide their work with the theories they concluded. According to Deng Deqiu [4], the scientific research ability of counselors refers to the ability of counselors to carry out academic, applied and investigative research related to their works on the basis of mastering certain theoretical knowledge according to practical work needs.

In conclusion, this paper holds that the scientific research ability of college counselors refers to the skills and level of counselors in developing academic and applied research by utilizing the theoretical knowledge of relevant disciplines according to the practical needs of ideological and political education and management of college students.

2   Analysis on Constituent Elements of Scientific Research Ability of College Counselors

The scientific research ability of college counselors is composed of four parts: the basis of theoretical knowledge, the research ability, the scientific research awareness and the scientific research spirit [5].

The basis of theoretical knowledge includes Socialist theories with Chinese characteristics, Management, Psychology, Sociology, Statistics and other related disciplines, as well as the methods and theories of scientific research. Among them, research methods exert a pivotal part in clarifying research ideas, accomplishing research content and achieving research objectives. Research ability mainly includes the ability of literature research, the ability of investigation, the ability of data analysis and processing, the ability to ask, analyze and solve problems, as well as the ability of innovative thinking and writing. Thereto, the ability to ask questions involves the topic selection of scientific research, and the valuable questions play a decisive role in the innovative value and significance of scientific research. Scientific research awareness refers to the awareness that college counselors who are adept at discovering problems actively in their work, and adapting scientific research approaches to solve problems and summarize rules and methods. It is one of the important prerequisites for counselors to achieve innovative research results, which mainly includes the awareness of innovation and collaboration [6]. And scientific research spirit refers to the relatively stable will and quality formed by researchers in their scientific research work, which is the internal guarantee condition for the formation of college counselors' scientific research ability. However, the achievement of an important scientific research result requires not only a long and arduous process but also a lot of painstaking efforts by researchers. Therefore, college counselors should have the indomitable will and the spirit of seeking truth from facts.

3   The Construction of the Cultivation Mechanism of Scientific Research Ability of College Counselors

The cultivation mechanism of college counselors' scientific research ability includes the guarantee mechanism, the education mechanism, the technical support mechanism, the discipline platform construction mechanism, the interaction of theory and practice mechanism, etc.

3.1   The Guarantee Mechanism

The guarantee mechanism of college counselors' scientific research ability cultivation consists of four parts, which includes establishing organizations, concentrating on institutional construction, guaranteeing enough funds, and ensuring enough time.

3.1.1   Establishing Organizations

To ensure the effectiveness of the cultivation of counselors' scientific research ability and establish the platform for cultivating counselors, colleges and universities should set up the Ideological and Political Educational Research Association of College Students, which is managed by functional departments such as the Students' Affairs Office or the Publicity Department. At the same time, colleges and universities should focus on establishing vigorous counselors research teams with several research directions, and having one or two people in charge of various research directions. Then, the counselors could join relevant research directions respectively according to their research interests and research bases.

3.1.2   Concentrating on Institutional Improvement

In order to regularly and orderly promote the cultivation of counselors' scientific research ability, a set of scientific and reasonable cultivation systems should be established. They include the management system such as counselors' participation in cultivating, academic meetings, and daily communication activities of scientific research, the assessment and evaluation system, the incentive system, etc. [7]. For example, the counselors are scientifically evaluated and assessed according to their performance in training, project application and paper publishing. Meanwhile, the evaluation and assessment results of counselors are taken as the basis for selecting scientific research backbone, so as to raise the counselors' enthusiasm and initiative to participate in scientific research, which aims at improving the cultivating effect.

3.1.3   Guaranteeing Enough Funds

Adequate funds are the foundation and important assurance for developing the scientific research ability of counselors. And after getting positive support from the school leaders, it is important to strengthen the communication and coordination with the functional departments such as the Scientific Research Department, the Personnel Department and the Discipline Construction Department of college, and implement the investment of funds such as the school-level project approval fund, the training fee, academic exchange fee, expert lecture fee, academic paper publication fee and monograph publication fee of the university. Most importantly, it must make an overall plan of fund utilization systematically and purposefully.

3.1.4   Ensuring Enough Time

If work tasks are too heavy at their routine duties, it will make the counselors impossible to guarantee their time and energy of participating in scientific research, which will seriously affect the improvement of their scientific research ability. Therefore, it is necessary to coordinate with each faculty for reducing the counselors' work tasks appropriately, especially the non-professional work tasks, in order to reduce their workload and ensure that they have enough time to participate in training, academic communication and promote scientific research practice.

3.2   The Educational Mechanism

The educational mechanism of scientific research ability cultivation for college counselors consists of three parts, such as designing cultivating objectives, implementing training programs and promoting scientific research practice. This mechanism is the core of the cultivation of scientific research ability of college counselors that determines the effect of the cultivation.

3.2.1 Designing Cultivating Objectives

This part mainly includes the design of training objectives in the aspects of counselors' theoretical knowledge foundation, research ability, scientific research awareness, and scientific research spirit, among which the core content is the cultivation of research ability. And the whole training work is mainly promoted with the requirements of improving counselors' research ability. Meanwhile, the training objectives should be designed in stages, with three or five years for a period, to promote counselors' scientific research ability orderly according to the overall situation of counselors' scientific research foundation, ability and level.

3.2.2 Implementing Training Programs

This part mainly focuses on improving the research ability of counselors and strengthening their study on theoretical knowledge foundation. Firstly, colleges and universities should employ high-level experts to promote various levels and types of humanities and social sciences subject application lectures, academic paper writing training, research methods, data statistics and analysis training, and knowledge training about the theory of Socialism with Chinese characteristics, Management, Sociology, Psychology, other related disciplines as well. Secondly, arranging the backbone of scientific research to participate in all kinds of training classes such as Ideological and Political Education and Management Scientific Research training classes, scientific research training classes, academic communication, and academic visits [8]. Thirdly, holding counselors' salons should regularly strengthen scientific research and academic communication between counselors, in which they can share their successful experiences and practices.

3.2.3 Promoting Scientific Research Practices

The scientific research ability of counselors is formed in the practice of scientific research. By means of setting up university-level project approval and encouraging counselors to apply for projects at provincial level or above, the counselors are encouraged to undertake certain scientific research tasks [9]. So, they would voluntarily devote part of their time and energy to scientific research, and constantly improve their scientific research awareness and level. In the meantime, it will not only urge counselors to complete scientific research tasks on schedule for strengthening their practical ability and spirit of scientific research, but also constantly improve their subject knowledge and research ideas of the counselors according to their quarterly, mid-term and final reports on the project.

3.3 The Technical Support Mechanism

The technical support mechanism for the cultivation of college counselors' scientific research ability refers to the mechanism of inviting experts and scholars to guide and assist college counselors' scientific research [10]. This mechanism plays a great role in accelerating the improvement of college counselors' scientific research ability. Employed experts and scholars, with high attainments and achievements in the field of ideological and political education and management, to train counselors' scientific research ability with high enthusiasm as the guiding teacher. Firstly, experts and scholars should guide counselors to apply for and complete scientific research projects at the provincial and ministerial level or above, including the topic selection, declaration modification and the process of project completion. Secondly, experts and scholars should guide counselors to write high-level academic papers and work. Thirdly, experts and scholars should guide counselors to apply for various levels of scientific research achievement awards.

3.4 The Discipline Platform Construction Mechanism

The discipline platform construction mechanism of college counselors' scientific research ability cultivation mainly includes the application and construction mechanism of the key discipline of college students' ideological and political education and relevant research bases. After accumulation of training, scientific research achievements of counselors will constantly emerge, so, colleges and universities should fully combine and utilize the power of teachers of the Ideological and Political course. And then, the academic departments and the School of Marxism should jointly declare the key disciplines and related research bases of ideological and political education for college students at all levels, build a high-level subject platform, and deepen the self-enhancement function. In this way, the counselors' scientific research ability and level can be continuously developed.

3.5 The Interaction of Theory and Practice Mechanism

The Ministry of Education requires colleges and universities to strengthen the construction of counselors' teams and further enhance their scientific research ability, which aims at applying the scientific research results of counselors to ideological and political education as well as management of college students, so as to achieve the fundamental task of strengthening moral education and cultivating people. Therefore, the scientific research ability of the counselors is not only reflected in the application of subjects and the publishing of papers, but also reflected in the application of scientific research achievements, that is, whether the results would be effective in practice. Therefore, colleges and universities should establish a test mechanism of the interaction of theory and practice in the cultivation of counselors' scientific research ability. It can be shown primarily in two aspects. On the one hand, it emphasizes the practical value of scientific research achievements in the application of projects, which requires counselors to solve the practical problems existing in students' ideological and political education and management; on the other hand, at the conclusion of the project, it should pay attention to the practical application of scientific research results, whether it has been applied in accordance with the requirements of the project application and solve the actual problems.

4   Conclusion

Scientific research is an activity with strong practicality, and the ability of scientific research can only be continuously improved in scientific research practice. And it is a long process to improve the scientific research ability of college counselors, and it is a complex project because the cultivation of scientific research ability is affected by many factors. As a result, colleges and universities should clarify the relationship between the various elements of the counselors' scientific research ability cultivation by constructing and implementing a series of measures, such as the guarantee mechanism, the education mechanism, the technical support mechanism, the discipline platform construction and interaction mechanism, and exerting the integrated effect of the above mechanisms. Only in this way, can the scientific research ability of counselors be improved and developed continuously.

Acknowledgments. This work was financially supported by the Project of the Humanistic and Social Science "Research on Team Building Mechanism of College Counselors' Scientific Research Based on the Concept of Shared Collaboration" in Jiangxi Province (No. JY19127) fund.

References

1. Sun, Y.: Analysis on the status quo of scientific research ability of local college counselors, J. Party Build. Ideological Educ. Sch. 04, 69–70+73 (2017). (in Chinese)
2. Yaojie, Y.: Research summary on the scientific research ability of college counselors. J. J. Guangxi Youth Leaders Coll. **26**, 58–61 (2016). (in Chinese)
3. Jun, S.: Status quo and countermeasures of the scientific research ability of college counselors. J. Ideological Theor. Educ. **05**, 90–93 (2013). (in Chinese)
4. Deqiu, D.: Research on the Improvement of college counselors' scientific research ability. J. Henan Sci. Technol. **11**, 280–281 (2014). (in Chinese)
5. Bai, L.: Working mechanism of the formation of college counselors' scientific research ability. J. Ideological Theor. Educ. **10**, 84–87 (2013). (in Chinese)
6. Lu, F.: Thinking on the cultivation of scientific research ability of college counselors in the new period. J. Ability Wisdom **34**, 119 (2014). (in Chinese)
7. Hao, L.: Research on the function and training mechanism of university counselor's scientific research ability. J. Manage. Observer. **07**, 124–125 (2019). (in Chinese)
8. Hu, B.: On the Cultivation of scientific research ability of college counselors, J. J. Changchun Univ. Technol. (High. Educ. Study Ed.), 32, 76–76+83 (2011). (in Chinese)
9. Zhanren, W., Weiwei, C.: Strategies for improving the scientific research ability of university counselors. J. Heilongjiang Res. High. Educ. **03**, 81–84 (2016). (in Chinese)
10. Wei, C.: Research on cultivation of scientific research ability of college counselors based on career development. J. Ind. Sci. Tribune. **14**, 186–188 (2015). (in Chinese)

Optimal Strategy of Front-End Development Technology of Computer Website

Rong Wu[✉]

Guangdong Polytechnic of Science and Trade, Guangzhou 510430, Guangdong, China

Abstract. With the advent of the era of knowledge economy, the explosive growth of information technology, timely and accurate information has become an important support for scientific decision-making and business success. At the same time, people pay more attention to the learning and accumulation of knowledge. Knowledge has become a valuable resource that individuals and organizations compete to possess. Due to the shortcomings of outdated design concepts, delayed information release, and low credibility, traditional websites cannot truly meet people's requirements for the Internet in the era of knowledge economy. The progress of the times calls for a brand-new, efficient and intelligent communication platform and knowledge management platform. In addition, with the attention of major Internet companies to front-end development, the number of front-end developers has gradually increased, which has caused many problems such as front-end performance optimization and coding standards. The front-end modular technology has successfully solved the above-mentioned problems with its advanced ideas and good design patterns, and has been favored by more and more people. This article is mainly based on the development of computer front-end website, focusing on the research and analysis of the application of HTML5, CSS3 and responsive design technology and other web technologies in it. By understanding the specific needs of users and responsive design ideas, analyze the root cause of the problem, and seek a solution strategy. According to the specific needs of ordinary websites, this paper draws up the application strategy of responsive design technology, designs and implements an actual responsive video website.

Keywords: Computer network · Website front-end · Front-end development technology · Optimization strategy

1 Introduction

The web front-end development is developed in the production of web pages, and its name has obvious characteristics of the times. Web page creation is a product of the Web 1.0 era, the website is displayed as static content, and the behavior of users who use the website is mainly the browser [1]. After 2005, the Internet entered the Web 2.0 era. Ajax brought non-refreshing data interaction and reduced the number of web page jumps. Web front-end applications will strive to reduce web page jumps, instead of data display, it is through front-end technology to complete more functions on a single page. From the perspective of basic technical support or the necessity of cross-platform compatibility,

© The Author(s), under exclusive license to Springer Nature Switzerland AG 2021
J. Abawajy et al. (Eds.): ATCI 2021, AISC 1398, pp. 260–267, 2021.
https://doi.org/10.1007/978-3-030-79200-8_39

the front end of the website will bring more prosperity to the future [2]. More web front-end leading products are on the stage, and there will be many web front-end engineers on the stage, which is expected to produce a lot of creativity and innovation. The main content of the front end of the website includes page structure, content and style, which is the page display of the website. Therefore, the design of the front-end part should be used to guide the needs of customers [3].

The spiritual essence of front-end modular development is to divide the front-end of the website into multiple modules according to the layout and functions, and each module completes specific sub-functions [4]. All modules are constructed in a specific way to form a whole and have the function of completing the entire system. The biggest advantage of front-end modularity and non-development creativity is that in the development process, the degree of interference between programmers is minimized and multiple people are jointly developed [5]. Through the research and realization of this topic, it can provide more reference for community-based and knowledge-based website development and the solution of frequently occurring problems in the front end of the website [6]. Due to the rapid development of Front-end technology and the complexity of network applications, network applications are compatible with multiple terminals and are developing towards high performance. In order to quickly repeat products according to changing user requirements, improve user experience, and enhance corporate competitiveness, network application development technology takes into account performance requirements and requires more effective development, which facilitates rapid iteration and simple maintenance management [7].

Although there are many excellent front-end frameworks for computer websites that can be used by developers at home and abroad, most of the excellent front-end frameworks for computer websites in China provide services for the internal products of enterprises and meet the individual requirements of enterprises. They are not suitable for several general application system development of [8]. Most front-end development basically uses external open source frameworks, but most front-end frameworks are not accurate in data processing. When there is a lot of data work and data work is frequent, DOM work is very much. At the same time, in the management information system, if you want to deal with similar business logic processes, you need to build a lot of repetitive codes, so the code repetition rate will increase [9, 10].

2 Algorithm Establishment

2.1 Basic Particle Swarm Algorithm

Particle Swarm Optimization (PSO) is a combination-based metadata algorithm that relies on the cooperation and information sharing between all individuals in the group in the process of finding the best solution. In order to further improve the optimization ability of the algorithm, Shi proposed in 1998 to use the inertia weighted value to control the speed change of the particles. The PSO that introduces the inertia weighted value is called the standard PSO optimization algorithm. During the repetition, the velocity and particle position are determined by the following formula:

$$v_{ij}(k+1) = \omega v_{ij}(k) + c_1 r_1 \big(pbest_{i,j}(k) - x_{i,j}(k)\big) + c_2 r_2 \big(pbest_j(k) - x_{i,j}(k)\big) \quad (1)$$

$$x_{ij}(k+1) = x_{i,j} + v_{ij}(k+1) \quad i = 1, 2, ..., N \quad j = 1, 2, ..., n \tag{2}$$

The basic particle swarm algorithm process is as follows:

(1) Initialize a population, and initialize the speed and position of each particle in the population;
(2) Evaluate the fitness value of particles;
(3) Initialize the individual optimal pbest to the current position of the particle, and initialize the global optimal gbest to the position of the best particle in the current population;
(4) For each particle, update its speed and position according to formulas (1) and (2) and re-evaluate the fitness value of the particle;
(5) Update pbest and gbest;
(6) If the algorithm reaches the maximum number of iterations, the algorithm ends. Otherwise, go to (4) to continue execution.

2.2 Discrete Particle Swarm Algorithm

The particle swarm algorithm was originally designed for continuous problems. The vector count in continuous space cannot reflect the particle path in the discrete state. Therefore, in order to overcome the combinatorial optimization problem, a discrete particle swarm algorithm is needed. In this algorithm, each particle uses a binary value of 0 or 1 to indicate its position in outer space, and the probability that the particle speed is defined as the particle state is 1. In other words, the particle position in the BPSO model is used to determine the relationship between "yes" and "no", and the speed and position are updated as follows:

$$v_{ij}(k+1) = sig\left(v_{ij}(k)\right) = \frac{1}{1 + e^{v_{ij}(k)}} \tag{3}$$

$$x_{ij}(k+1) = \begin{cases} 1 & if \quad r_{ij} < sig\left(v_{ij}(k+1)\right) \\ 0 & otherwise \end{cases} \tag{4}$$

In this paper, in order to solve the multi-constrained combinatorial optimization problem, an improved exchange particle clustering algorithm is used. This method increases the convergence speed of the algorithm and the optimization ability of the algorithm when dealing with the combinatorial optimization problem. In this method, the speed and position update formulas of the basic PSO are improved to formulas (5) and (6):

$$v_{id}^{k+1} = w \otimes v_{id}^{k} \oplus c_1 \otimes \left(p_{id}^{k} \ominus x_{id}^{k}\right) \oplus c_2 \otimes \left(p_{gd}^{k} \ominus x_{id}^{k}\right) \tag{5}$$

$$x_{id}^{k+1} = x_{id}^{k} \oplus v_{id}^{k} \tag{6}$$

(1) Position subtraction operation $\ominus$

The subtraction of two positions indicates the speed. When x minus y, replace y directly with x, which is expressed as $x \rightarrow y$.

(2) Speed plus operation

The velocity is added to get the first new velocity, assuming $v = a \rightarrow b, w = x \rightarrow y$, then:

$$v \oplus w = (a \rightarrow b) \oplus (x \rightarrow y) = \begin{cases} a \rightarrow b, & \text{if } a \neq x \\ a \rightarrow y, & \text{if } b = x \end{cases} \tag{7}$$

(3) Position acceleration operation

$$a \oplus v = a \oplus (x \rightarrow y) = \begin{cases} a, & \text{if } a \neq x \\ a, & \text{if } b = x \end{cases} \tag{8}$$

(4) Coefficient multiplying speed operation

Both d and coefficient c are random numbers between (0,1).

$$c \oplus v = c \oplus (x \rightarrow y) = \begin{cases} x \rightarrow y, & \text{if } d > c \\ x \rightarrow rand, & \text{if } d \leq c \end{cases} \tag{9}$$

After a series of derivations, the following formula is finally obtained:

$$x_{id}^{k+1} = \begin{cases} P_{id}^k & \text{if } \phi_1 > rand\,(0, 1) \\ P_{gd}^k & \text{if } \phi_1 \leq rand\,(0, 1), \text{ if } \phi_2 > rand\,(0, 1) \\ rand & \text{if } \phi_1 \leq rand\,(0, 1), \text{ if } \phi_3 > rand\,(0, 1) \end{cases} \tag{10}$$

According to the idea of learning from individual optimal particles and global optimal particles, formula (10) is updated as follows:

$$x_{id}^{k+1} = \begin{cases} P_{id}^k & \text{if } P_{id}^k! = x_{id}^k \\ P_{gd}^k & \text{if } \phi_1 > rand\,(0, 1) \\ rand & else \end{cases} \tag{11}$$

3 Modeling Method

3.1 Linear Two-Degree-of-Freedom Manipulation Model

In the process of website front-end optimization, it is necessary to design and rehearse the detailed process of experimental operation in advance to ensure the success of the experiment. However, the prerequisite for solving these problems is to design a complete model. After obtaining the low-frequency sub-image I_0' of I_{0LL}' in the above, next, use the SIFT feature extraction method to extract the feature points of I_{0LL}', which mainly includes the following steps:

Scale space extreme value detection:

$$L(x, y, \sigma) = G(x, y, \sigma) * I_{0LL}'(x, y) \qquad (12)$$

Among them, * is a convolution operator. The larger the value, the more it can describe the overall content of the detection process, and the smaller the value, the more it can describe the details of the detection process in detail. The expression of Gaussian function $G(x, y, \sigma)$ is shown in formula (13):

$$G(x,y,\sigma) = \frac{1}{2\pi\sigma^2} e^{-(x^2+y^2)/2\sigma^2} \qquad (13)$$

Then, the Gaussian difference function is convolved with the image I_{0LL}' to get the scale difference function++.

$$D(x, y, \sigma) = (G(x, y, k\sigma) - G(x, y, \sigma)) * I_{0LL}'(x, y) = L(x, y, k\sigma) - L(x, y, \sigma) \qquad (14)$$

$$D(x, y, \sigma) = D(x, y, \sigma) + \frac{\partial D^T}{\partial x} x + \frac{1}{2} x^T \frac{\partial^2 D^T}{\partial x^2} x \qquad (15)$$

$$x = \frac{\partial^2 D^{-1}}{\partial x^2} \frac{\partial D}{\partial x} \qquad (16)$$

After completing the above steps, construct a linear combination of classifiers:

$$f(x) = \sum_{m=1}^{M} a_m T_m(x) \qquad (17)$$

The final data cloud processor obtained from the scientific research is:

$$T(x) = \text{sign}(f(x)) = \text{sing}\left[\sum_{m=1}^{M} a_m T_m(x)\right] \qquad (18)$$

Direction distribution. For the feature point (x, y), the scale image with the scale I_{0LL}' of the feature point is:

$$L(x, y) = G(x, y, \sigma) * I_{0LL}'(x, y) \qquad (19)$$

The gradient magnitude m(x, y) and direction ++ can be approximated by the difference operation optimized near the front end of the website, as shown in formulas (20) and (21):

$$m(x, y) = \sqrt{(L(x + 1, y) - L(x - 1, y))^2 + (L(x, y + 1) - L(x, y - 1))^2} \qquad (20)$$

$$\theta(x, y) = \tan^{-1}((L(x, y + 1) - L(x, y - 1))/(L(x + 1, y) - L(x - 1, y))) \qquad (21)$$

4 Evaluation Results and Research

We deployed websites of different natures in three ways: no cache, simple cache, and memory optimization model (Method 1, Method 2, Method 3), and randomly and continuously inserted new data into the website, while recording the insertion the corresponding time when it arrives in the array (in order to realize that the data generation time can be directly seen on the web page, which is convenient for experimental recording), and the experimental data is collected. The interval between each visit is 10 min.

CPU: The average usage rate of the server-side CPU when visiting the website.

Memory: The average usage rate of server-side memory when visiting the website.

Response time: from the start of the visit to the rendering of the web page (that is, the data has been returned).

Data update: After the data is inserted, how long will the data appear on the page (Table 1 and Figs. 1, 2).

Table 1. Access to experimental data records (CPU and memory)

Visits	Do not use cache		Use simple cache		Use memory optimization model	
	CPU	RAM	CPU	RAM	CPU	RAM
1	66%	41%	46%	63%	50%	66%
2	67%	48%	42%	66%	56%	69%
3	75%	45%	38%	69%	53%	72%
4	74%	49%	50%	72%	55%	70%
5	80%	51%	45%	70%	59%	63%
6	69%	57%	42%	65%	61%	65%

Method 1: When not using any cache, the server's average CPU reaches 72%, which has exceeded the CPU risk value of the WEB server. It means that when the cache is not used, the generated resources need to be recalculated for each request, and Database communication causes excessive CPU load. At this time, the memory usage is low, the memory usage of the WEB server is not enough, and the idle value is too high. Due to the need to recalculate each time, the response time has reached an average of 3 s. These are not ideal.

Method 2: The simplest caching method is adopted, namely the principle of caching and invalidation in 5 min. The CPU cache has dropped significantly, the utilization of the memory has also risen obviously, and the website's response time is also very good, but the data takes an average of 5.7 min to update once, which makes the website's data update speed too slow.

Method 3: After adopting a self-built caching system, although the CPU usage is slightly higher than the simple caching mode, it is significantly lower than 70% of the dangerous

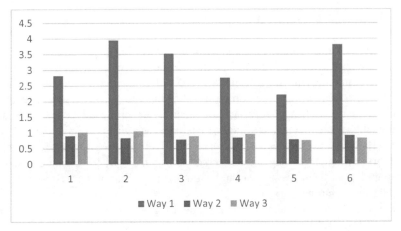

Fig. 1. Response time (unit: s)

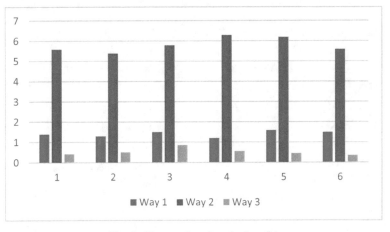

Fig. 2. Data update time (unit: min)

line, the memory usage is also more appropriate, and the website response time is also very reasonable. The most important thing is Realize the timely update of data, and control the data update speed at about 1 min.

Method 3 has improved in terms of CPU, memory usage, response time, and data timeliness. Under limited conditions, it can greatly improve the performance of the front-end of the website.

5   Conclusion

The current society is the information age where PC clients are the mainstream of work and the mobile Internet is the mainstream of life. Web front-end development technology is the basic technical requirement of this era. As we all know, the production of website pages cannot be separated from the corresponding Web front-end development technology. With the continuous development and maturity of technology, users' requirements for website experience are also increasing. This also puts forward higher requirements for website web front-end developers, who need to continue to lay a solid foundation, clarify their ideas, and meet various requirements for website construction and user experience while improving the quality of web front-end development. In summary, based on the business of the website production platform, the development, design and research of the web front end. From the perspective of future development, relevant manpower should pay attention to the development of technical capabilities and demand levels, and control the power. In order to better control the serious problems posed by technical forces and development space, the quality improvement of website design has been a technical realization. To achieve more web front-end development technologies and adapt to future development trends, it is necessary to use web front-end development technologies to solve more website production problems and fully optimize them.

References

1. Prvan, M., Oegovi, J.: Methods in teaching computer networks: a literature review. ACM Trans. Comput. Educ. **20**(3), 1–35 (2020)
2. Akinola, A.A., Adekoya, A.A., Kuye, A.O., et al.: Quantitative evaluation of cyber-attacks on a hypothetical school computer network. J. Inf. Secur. **10**(3), 103–116 (2019)
3. Kim, M., Park, K.T., Kim, J.: A study on the analysis and classification of cyber threats according to the characteristics of computer network of national·public organizations. J. Inf. Secur. **20**(4), 197–208 (2020)
4. Pratama, Z.A., Suharnawi, S.: Perancangan website badan narkotika nasional provinsi jawa tengah. JOINS (J. Inf. Syst.) **4**(1), 31–40 (2019)
5. Ricardez, G.A.G.: Key technologies for collaborative robots in front-end retail automation. J. Robot. Soc. Jpn. **37**(10), 934–937 (2019)
6. Zubaedah, R., Putra, N.P.: Website pengarsipan dokumen dan surat pada jurusan sistem informasi universitas musamus. Mustek Anim Ha **8**(1), 41–48 (2019)
7. Kim, J.H., Jin, H.W.: Virtio front-end network driver for RTEMS operating system. IEEE Embed. Syst. Lett. **12**(3), 91–94 (2020)
8. John, P.K., Dann, S., Gregor, S., et al.: Designing a visual tool for teaching and learning front-end innovation. Telev. New Media **10**(9), 16–26 (2020)
9. Elmotasem, H., Awad, G.E.: A stepwise optimization strategy to formulate in situ gelling formulations comprising fluconazole-hydroxypropyl-beta-cyclodextrin complex loaded niosomal vesicles and Eudragit nanoparticles for enhanced antifungal activity and prolonged ocular delivery. Asian Journal of Pharmaceutical Sciences **15**(5), 617–636 (2020)
10. Montoya, O.D., Gil-González, W., Grisales-Noreña, L.F., Ramírez Vanegas, C.A., Molina-Cabrera, A.: Hybrid optimization strategy for optimal location and sizing of DG in distribution networks. Tecnura, **24**(66), 47–61 (2020)

SAR Image Change Detection Based on Complex Neural Network

Xia Wen[✉], Huiyong Jiang, Yanghong Mao, and Hongqin Wang

Software Engineering Institute of Guangzhou, Guangzhou 510990, Guangdong, China
wenxia12_15@sina.com

Abstract. With the deepening of the world exploration, more and more observation projects are needed, and the requirements of observation are also higher and higher. The development of remote sensing image change detection requires higher accuracy. Synthetic aperture radar (SAR) has a special advantage in the field of image change detection because it is easy to obtain images and is not affected by atmospheric and light conditions. Therefore, in recent decades, SAR images have been successfully applied to environmental monitoring, urban research and disaster assessment. In recent years, neural network deep learning has received great attention and rapid development. The existing deep learning is basically based on the operation of real number. We need to consider how to connect the original images captured by multiple satellites to the network. Most of the existing methods use the direction of image color channel to splice images, but the color channels are not related therefore, this method will hide the association between multiple images. In this paper, a SAR image change detection method based on complex neural network is proposed. The complex neural network is used in SAR image change detection. Through the deep learning of complex neural network, the correlation between images is obtained, which effectively improves the effect of SAR image change detection. The principle of this method is that the original two groups of data in the detection problem are naturally combined into complex numbers, and then the complex convolution and full connection are used to process the data in the network, which effectively improves the accuracy of change detection.

Keywords: Complex neural network · Change detection · SAR image · Deep learning

1 Introduction

Change detection in remote sensing monitoring has always been an important research technology in the development of science and technology in our country. The technology compares the images taken at the same place at different times by analyzing, so as to get the landmark changes in the detection area [1]. This technology is widely used in natural landform detection and analysis, urban land cover change, environmental monitoring and urban expansion assessment. Therefore, change detection in remote sensing images has been paid more and more attention [2]. Because the use of synthetic aperture radar

J. Abawajy et al. (Eds.): ATCI 2021, AISC 1398, pp. 268–276, 2021.
https://doi.org/10.1007/978-3-030-79200-8_40

(SAR) technology has nothing to do with sunlight and a series of weather factors, SAR image technology is the most important image change detection technology in China [3]. Although SAR image has the advantage of tolerant detection conditions, it also has speckle noise, which affects the accuracy of image change detection. Tolerant detection conditions will make the detection work very easy, but the lack of detection accuracy makes the significance of detection work lower, so we should have both tolerant detection conditions and a certain level of detection accuracy [4].

With the layer by layer unsupervised pre training method proposed by relevant scholars, deep learning has been paid more and more attention. With the use of convolutional neural network, SAR image has achieved a breakthrough in improving the accuracy, and from then on, deep learning broke out completely. Due to the automatic feature extraction ability of deep learning, good robustness to noise and imitating the working structure of human brain, it has achieved breakthrough application results in more and more scenes, such as image processing, speech recognition and so on [5]. In recent years, more and more algorithms based on deep learning have appeared in the research of SAR image change detection. However, the current deep network is basically based on real number, and although complex neural network was proposed earlier, due to the lack of systematic research and the obscure association between complex number and real data, people seldom study complex number network after the breakthrough success of real number network [6]. However, a paper published in 2020 proposed the systematic formulaic derivation of the basic components of complex network. Experiments show that complex network has the same expressive ability as real network in real data sets, and has achieved the effect of surpassing real network in special fields (such as music prediction and voice analysis data contain amplitude and frequency, which can be expressed by complex numbers). The original data in the field of change detection contains two parts (two images to be compared), which can naturally form a group of complex data and train a complex network to complete the task of change detection [7].

Among the existing change detection technologies, the operation based change detection methods mainly include the difference method, ratio method (R) and log ratio method (LR) with a single point as the processing object, the mean ratio method (MR) with a local area as the processing object, some methods with other elements as the processing object and related combination and fusion methods, Including change vector analysis (CVA), spectral angle mapper (SAM), combined difference image (CDI), wavelet fusion (WF), etc. [8, 9]. As an unsupervised change detection method, fuzzy C-means (FCM) clustering method is widely used in image clustering. This algorithm mainly uses fuzzy membership matrix to represent the probability of each sample point belonging to a cluster. The greater the similarity between objects, the higher the possibility of classification. FCM clustering method is easy to operate, but because it ignores the spatial characteristics of pixels, the algorithm is easily affected by speckle noise, texture and abnormal points. Although the traditional clustering and segmentation algorithms have achieved good results in change detection, with the expansion of data acquisition channels and application scope, some existing clustering and segmentation algorithms can not meet the higher accuracy and more flexible application requirements. In this paper, based on the optimization and improvement of the algorithm and combined with complex neural network deep learning processing image [10].

2 Algorithm Establishment

2.1 Complex Convolution Algorithm

For a complex convolution kernel $w = a + ib$ and a two-dimensional complex feature $h = x + iy$, where a, b, x, y is a real valued matrix.

Because convolution operation is essentially composed of multiplication and addition, it meets the allocation rate. Then we can use the following matrix operation to represent the complex convolution operation as described above:

$$\begin{bmatrix} R(w * h) \\ V(w * h) \end{bmatrix} = \begin{bmatrix} a & -b \\ b & a \end{bmatrix} * \begin{bmatrix} x \\ y \end{bmatrix} \tag{1}$$

The operation of the above matrix form can be further transformed into a more intuitive equation form, as follows:

$$w * h = (a * x - b * y) + i(b * x + a * y) \tag{2}$$

2.1.1 Complex Fully Connected Layer

For a complex weight $w = a + ib$ and a one-dimensional complex feature $h = x + iy$, the process of complex full join can be expressed as follows:

$$w \cdot h = (a \cdot x - b \cdot y) + i(b \cdot x + a \cdot y) \tag{3}$$

You can see that complex full join is also easy to implement with real full join.

2.1.2 Multiple Initialization Method

The variance of complex network weight initialization is derived. We can express the complex weight in polar coordinate form of complex number

$$w = |w|e^{i\theta} = R\{w\} + iV\{w\} \tag{4}$$

Where θ and w represent the amplitude and phase parameters respectively.

Variance is the difference between the square of the magnitude and the expected Square:

$$\mathrm{Var}(w) = E[ww^*] - (E[w])^2 = E\left[|w|^2\right] - (E[w])^2 \tag{5}$$

When w symmetry distribution is around 0, it will be simplified to $E\left[|w|^2\right]$. We don't know the value of $E\left[|w|^2\right]$, but we know the relevant Var(w). Because the amplitude $|w|$ of complex normal distribution satisfies Rayleigh distribution, its value is:

$$\mathrm{Var}(|w|) = E\left[|w||w|^*\right] - (E[|w|])^2 = E\left[|w|^2\right] - (E[|w|])^2 \tag{6}$$

From the above formula, we can get the following results

$$\text{Var}(|w|) = \text{Var}(w) - (E[|w|])^2 \tag{7}$$

$$\text{Var}(w) = \text{Var}(|w|) + (E[|w|])^2 \tag{8}$$

Now we can calculate the mean value and variance of the weight amplitude through a parameter of Rayleigh distribution

$$E[|w|] = \sigma\sqrt{\frac{\pi}{2}}, \ \text{Var}(|w|) = \frac{4-\pi}{2}\sigma^2 \tag{9}$$

The variance of weights can be expressed by the above calculated values

$$\text{Var}(w) = \frac{4-\pi}{2}\sigma^2 + \left(\sigma\sqrt{\frac{\pi}{2}}\right)^2 = 2\sigma^2 \tag{10}$$

2.2 Fuzzy C-Means Clustering Method

Set $X = \{x_1, x_2, \ldots, x_i\} \subset R^1$, the characteristic vector is $X^k = (x_{k1}, x_{k2}, \ldots, x_{kl})^T \in R^1$, the test data is divided into c classes, corresponding to c clustering centers c, each sample j belongs to a certain class of i membership is u_{ij}, FCM objective function (11) and its constraint conditions (12) are defined as:

$$j = \sum_{i=1}^{c}\sum_{j=1}^{n} u_{ij}^m d_{ij}^2 \tag{11}$$

$$\sum_{i=1}^{c} u_{ij} = 1, j = 1, 2, \ldots, n \tag{12}$$

M is the membership factor, and $d_{ij} = ||x_i - c_i||$ is the distance from sample x_j to the center c_i of cluster i.

Taking Eq. (12) as the constraint condition, the following results are obtained:

$$J = \sum_{i=1}^{c}\sum_{j=1}^{n} u_{ij}^m ||x_i - c_i||^2 + \lambda_j \sum_{i=1}^{c} u_{ij} \tag{13}$$

The above formula calculates the derivatives of u_{ij} and c_i respectively, and the iterative updating formula of u_{ij} and the iterative updating formula of cluster center c_i can be deduced as follows:

$$u_{ij} = \frac{1}{\sum_{k=1}^{c}\left(\frac{||x_j - c_i||}{||x_j - c_k||}\right)^{\left(\frac{2}{m-1}\right)}} \tag{14}$$

$$c_i = \frac{\sum_{j=1}^{n}\left(x_j u_j^{mi}\right)}{\sum_{j=1}^{n} u_{ij}^m} \tag{15}$$

2.3 Algorithm Optimization

Aiming at the problem of SAR image change detection, this paper proposes a change detection method based on complex neural network. This method makes use of the characteristics of the original data of the change detection problem, and naturally transforms the splicing of the original data into the construction problem of complex number, so that the change detection can be completed by using the complex number network, and the network can learn the association between the data more easily. Through experiments, we find that complex network can achieve better results than real network. Moreover, the idea and framework of this method have good expansibility, and can be applied to other data source change detection tasks.

3 Modeling Method

The basic concept of automatic coding neural network is to approach the input layer and output layer as much as possible, and to propagate the algorithm by automatically adjusting network parameters in reverse direction. The network structure of automatic encoder is mainly composed of input layer, hidden layer and output layer. The hidden layer is called feature extraction layer. The forward conduction of the automatic encoder is divided into encoding and decoding. Encoding is the process of obtaining the excitation value of each neuron in the hidden layer through the input layer, and decoding is the process of obtaining each neuron in the output layer through the input layer.

The modeling steps are as follows.

Let $x \in R^{n_0}$ (n_0 is the number of neurons in the input layer) be mapped to the hidden layer by the function $f(\cdot)$ to obtain the excitation value $h \in R^{n_1}$ (n_1 is the number of neurons in the hidden layer)

$$h = f(W_1 + B_1) = \frac{1}{1 + \exp[-(W_1 + B_1)]} \tag{16}$$

W_1 is the weight matrix from input layer to hidden layer, and B_1 is the matrix of offset term from input layer to hidden layer.

The hidden layer h is mapped to the output layer to get the value $\tilde{x}$ of each neuron

$$\tilde{x} = f(W_2 h + B_2) \tag{17}$$

Adjust the parameters W_1, B_1, W_2, B_2 and calculate the average activity of the hidden neurons j in the automatic coding neural network.

$$\hat{\rho}_j = \frac{1}{m} \sum_{k=1}^{m} \left[h_j(k) \right] \tag{18}$$

Add a sparsity constraint to Eq. (18) so that $\hat{\rho} = \rho$, where ρ is the sparsity parameter, usually a value close to 0, and m is the number of sample data. In order to make the sparsity constraint effective, an additional penalty factor is added to the optimization objective function

$$KL\left(\rho || \hat{\rho}_j\right) = \rho \log \frac{\rho}{\hat{\rho}_j} + (1 - \rho) \log \frac{1 - \rho}{1 - \hat{\rho}_j} \tag{19}$$

The cost function of the neural network is as follows

$$J(W, B) = \frac{1}{m} \sum_{k=1}^{m} ||h(k) - x(k)||^2 + \lambda \left(||W_1||_F^2 + ||W_2||_F^2 \right) + \beta \sum_{j=1}^{n_1} kL(\rho\hat{\rho}_j)$$

(20)

Where $kL(\rho||\hat{\rho}_j)$ is the sparse penalty term; $||W_1||_F^2 + ||W_2||_F^2$ is the regular term; λ and β are the weight coefficients of the regular term and the sparse penalty term respectively.

4   Evaluation Results and Research

We can see that the results based on deep learning method are better than traditional methods. Through the sample selection strategy, we can make the samples with high reliability detected by the classical methods be used to train the network. After learning the characteristics of the changed and unchanged samples, the deep network can achieve better results in the parts of the remaining traditional methods with poor detection effect relying on its strong robustness and generalization ability.

In this paper, mean filter, Lee filter, frost filter, gamma filter, sigma filter, enhanced Lee filter and enhanced frost filter are used to suppress the speckle noise of an area data set, and the normalized index (nm) and peak signal noise (PSN) are used PSNR, enl and EPI were used to evaluate the filtering results. Because the two images in the same group of data are obtained by the same satellite and the same sensor, the filtering experiment unifies the previous one phase image as the research object. The results of image filtering quality evaluation are shown in Table 1.

Table 1. Quality evaluation of filtering results of a certain data set

Filtering method	NM	PSNR	ENL	EPI
Original image	1.0000	1.0000	1.1910	1.0000
Mean filter	1.0061	22.9685	1.1165	0.4074
Median filter	1.0241	23.6802	1.3254	0.4321
Lee filter	0.7961	18.4712	0.8841	0.9654
Frost filter	0.7649	15.6311	0.9051	0.6745
Gamma filter	0.7484	15.8165	0.9345	0.6414
Sigma filter	0.8582	20.2941	0.8546	0.9548
En-Lee filter	0.7554	15.9654	0.9154	0.6415
En-Frost filter	0.7652	15.6324	0.9045	0.6754

274   X. Wen et al.

It can be seen from Table 1 that the normalized mean index (nm) and peak signal-to-noise ratio (PSNR) values of meanfilter, mediafilter, leefilter and sigmafilter are relatively large, which indicates that the four filtering methods are better in terms of nm and PSNR evaluation; the value of equivalent number of views (enl) is relatively large except sigmafilter. This shows that the image effect of other filtering methods is better than that of sigmafilter; the filtering methods with larger EPI value are leefilter and sigmafilter, and the larger the EPI value is, the better the image filtering effect is, so the filtering effect of leefilter and sigmafilter is better than that of other filtering methods. Through the quality evaluation and comparison of the above filtering methods, we can see that the filtering image obtained by leefilter is better than other methods.

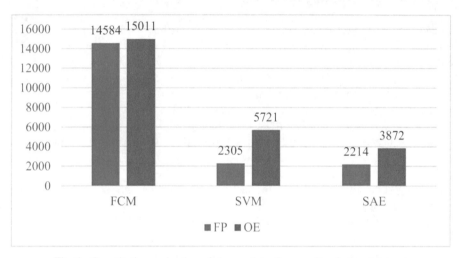

Fig. 1. Quantitative evaluation of change detection results of a local dataset

As can be seen from Fig. 1, for the data of the detection area, the FP and OE of the change detection map obtained by FCM are 14584 and 15011, which are larger than those obtained by other methods, and the change detection map contains more noise. It can be seen that the experimental effect of FCM is the worst. The FP and OE of the change detection map obtained by SVM are 2305 and 5721 respectively, and the FP of the change detection map obtained by SAE is 2214, the OE is 3872, which shows that the effect of SAE method is the best compared with other methods.

It can be seen from Fig. 2 that the PCC and kappa coefficients of the change detection map obtained by SAE method are slightly higher than those of the other two methods. Combined with FP, OE and data in Fig. 1, for the data set of this detection area, the accuracy of SAE method for change detection is better than that of traditional FCM and SVM methods.

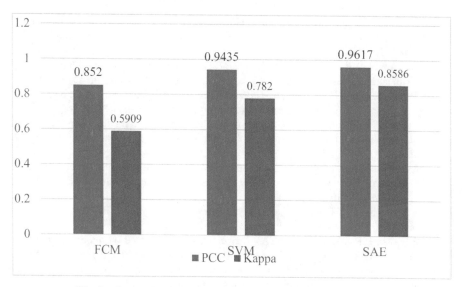

Fig. 2. Comparison of PCC and kappa values of different methods

5 Conclusion

In this paper, the principle of automatic encoder, one of the deep learning architecture, is summarized. On this basis, the implementation process of automatic encoder is introduced in detail. Based on the survey area data set, the application of automatic encoder method proposed in this paper in change detection is studied. The effectiveness of this method is proved by experiments, at the same time, the experimental results are compared with the SAR image change detection methods based on FCM, SVM and kappa. In order to reduce the influence of speckle noise on the accuracy of SAR image change detection, this paper applies SAE technology to SAR image change detection based on complex neural network. A SAR image change detection method based on automatic encoder is proposed. Firstly, the logarithmic ratio method is used to obtain the difference image, then SAE is used to extract the deep features of the image, and FCM is used to cluster the extracted features, and median filter is used to process the clustering results to obtain the change detection image. Compared with FCM, SVM and SAE methods, the proposed algorithm has higher robustness to speckle noise of SAR image and can effectively improve the change detection accuracy.

Acknowledgements. Research on complex neural network modulation(ky202003).

References

1. Zhukovyts'Kyy, I.V, Pakhomova, V.M, Ostapets D.O., et al.: Detection of attacks on a computer network based on the use of neural networks complex. Sci. Trans. Prog. Bull. Dnipropetrovsk Nat. Univ. Railway Transp. **5**(89), 68–79 (2020)

2. Ou, X., Yan, P., He, W., et al.: Adaptive GMM and BP neural network hybrid method for moving objects detection in complex scenes. Int. J. Pattern Recogn. Artif. Intell. **33**(2), 1950004.1–1950004.16 (2019)
3. Kim, K., Ahn, J., Kim, S., et al.: Adaptive neural network controller design for a blended-wing UAV with complex damage. J. Korean Soc. Aeronaut. Space Sci. **46**(2), 141–149 (2018)
4. Wu, P., Liu, J., Shen, F.: A deep one-class neural network for anomalous event detection in complex scenes. IEEE Trans. Neural Networks and Learn. Syst. **31**(7), 2609–2622 (2020)
5. Alcantarilla, P.F., Stent, S., Ros, G., Arroyo, R., Gherardi, R.: Street-view change detection with deconvolutional networks. Auton. Robot. **42**(7), 1301–1322 (2018). https://doi.org/10.1007/s10514-018-9734-5
6. Le Hégarat-Mascle, S., Ottlé, C., Guérin, C.: Land cover change detection at coarse spatial scales based on iterative estimation and previous state information. Remote Sens. Environ. **95**(4), 464–479 (2018)
7. Khalile, L., Rhinane, H., Kaoukaya, A., et al.: Forest cover monitoring and change detection in nfifikh forest (Morocco). J. Geogr. Inf. Syst. **10**(2), 219–233 (2018)
8. Li, M.Q, Xu, L.P., Xu, N., et al.: SAR image segmentation based on improved grey wolf optimization algorithm and fuzzy C-means. Math. Prob. Eng. **2018**(PT.10), 1–11 (2018)
9. Singh, P., Shree, R.: A new homomorphic and method noise thresholding based despeckling of SAR image using anisotropic diffusion. J. King Saud Univ. -Comput. Inf. Sci. **32**(1), 137–148 (2020)
10. Jaegul, C., Shixia, L.: Visual analytics for explainable deep learning. IEEE Comput. Graph. Appl. **38**(4), 84–92 (2018)

Time Series Forecasting Model Method Based on Neural Network

Peng Hua$^{(\boxtimes)}$

School of Economics, Shanghai University, Baoshan District, Shanghai, China

Abstract. Forecast is the basis of decision-making. One of the existing forecasting methods is to build a time series model based on historical data, and then predict the future based on the forecast model. This is a time series forecast. Neural network is a hot research topic in the field of artificial intelligence research in recent years, among which multilayer forward neural network has been widely used. We have analyzed and studied the structure and algorithm of neural network, researched and put forward the normalization method and related problems of establishing time series prediction neural network model, and also proposed a comprehensive prediction method to avoid the system error of neural network prediction model and improve the prediction accuracy. Experiments have proved the superiority of neural network time series and the superiority of the method of predicting time series by joint neural network. Finally, this article analyzes the problems of the neural network-based time series prediction model method, and at the same time conducts research on reinforced learning, and proposes that the problem of time series real-time modeling can be regarded as a delayed reinforced learning problem, so it can be used to solve reinforced learning the method solves this real-time modeling and forecasting problem. As with enhanced learning, the time series real-time modeling and prediction problem based on neural network also involves time credit assignment and structural credit assignment.

Keywords: Neural networks · Time series · Artificial intelligence · Predictive models

1 Introduction

With the development of science and technology and the continuous progress of society, the amount of social information continues to increase, and rapid changes have occurred in the corporate world and other fields. As a result, people's demand for forecasts continues to increase in order to make decisions as quickly as possible. For this reason, time series forecasting has been established, but traditional time series forecasting models cannot handle semantic values and small-scale data sets [1]. In order to solve this situation, the academic circle proposed the concept of time series forecasting model. Due to the excellent performance of the model in processing semantic values and small-scale data sets, more and more scholars have begun to engage in research in this area [2]. However, the existing time series forecasting models still need to be improved in terms of forecasting accuracy and scope of application. The main reason is that the data format

© The Author(s), under exclusive license to Springer Nature Switzerland AG 2021
J. Abawajy et al. (Eds.): ATCI 2021, AISC 1398, pp. 277–284, 2021.
https://doi.org/10.1007/978-3-030-79200-8_41

and generality are not considered enough, and the statistical information in the data is not fully excavated. For example, multi-attribute time series, insufficient consideration of external time series, ignoring trend factors in time series, and adopting a more subjective form to divide the universe. In addition, uncertain data, which is becoming more and more popular, cannot be processed by existing models [3].

A time series is a series formed by arranging the values of a certain statistical indicator of a certain phenomenon in chronological order at different times. Time series can reflect the development laws of social economic phenomena and natural phenomena, and through further research on the laws, the past and current values can be used to predict the future state of the research object [4]. The traditional time series forecasting model is based on the assumption that there is a clear and accurate functional relationship between the future value, the past value and the current value, and has achieved high accuracy in the time series forecasting. However, in the face of time series composed of semantic values or when the size of the time series data set is limited, the performance of the traditional time series model is not satisfactory [5]. In order to deal with the time series composed of semantic value data and to deal with the small number of samples in the time series, the concept of time series is proposed, and the corresponding time series forecasting model is constructed on this basis. Because this type of model has semantic value processing capabilities and can be applied to small-scale time series, it has received more and more attention from scholars [6].

With the progress of science and technology and society, more and more data are stored in the database in different forms for different applications. Many of these data are stored in the form of intervals rather than a definite value, which leads to the possibility of data at a time being classified into one or more categories at the same time in the process of fuzzification, resulting in uncertainty [7]. Unfortunately, the current time series forecasting models are mainly aimed at time series composed of certain semantic values, and there is no way to deal with this type of data. More unfortunately, in order to preserve the amount of information in the data as much as possible, and with the increase of storage space, more and more data sets begin to adopt this interval storage method, and they are often represented by a statistical method before [8]. For example, the data collected by the sensor is polluted by the measurement accuracy and the transmission process, and the value collected at the terminal is often not true. At this time, it is often a better choice to use the probability interval to indicate the value [9]. How to model the time series expressed in interval form, and on this basis, propose a method of fuzzy time series rule extraction and denervation which is suitable for improving the use range of time series models. Making full use of the existing advantages of time series in dealing with semantic value time series and small-scale time series to realize the prediction of interval series has strengthened the connection between theoretical research and reality, and highlighted its practical value [10].

2 Algorithms and Methods

2.1 Fitness Function Algorithm

The threshold of the weight of the neural network is optimized to affect the error of the neural network. Take the reciprocal of the error square as the fitness function. Then, when the sum of squares of the error is the smallest, the adaptation will be the highest. The Fitness function is expressed as follows:

$$fitness = \frac{1}{E}$$ (1)

Where E is the output error of the neural network:

$$E = \frac{1}{N} \sum_{k=1}^{N} (y_k - \hat{y})^2$$ (2)

After calculating the applicability based on the adaptation function, the population evolution operation is performed. In the population evolution process, all species in the group are arranged according to their adaptability and the possibility of individuals being selected, and the order is determined. The specific procedures are as follows:

1. Sort the individuals in descending order according to their fitness;
2. According to actual problems, according to the results of the above plan, compile a probability distribution table and assign probability values to individuals. The formula is:

$$q = r(1 - p_{max})^{N(k-1)}$$ (3)

$$r = \frac{p_{max}}{1 - (1 - p_{max})} N(k)$$ (4)

3. Then use the wheel selection method for chromosome selection.

2.2 Combination Forecast

In predictive experiments, different prediction models can be used to predict certain prediction problems. Any single prediction method is used to test the authenticity of dynamic changes. In order to make better use of the information of each model, different models can be used. The comprehensive method is to obtain a comprehensive forecasting model, which is better than any single forecasting model. The traditional method of predicting time series exponential smoothing has the advantages of simple, stable and reliable algorithm; the prediction method of neural network is well prepared for any complicated functions, is sensitive to signal changes, and is suitable for medium and short-term prediction. These methods are applicable to different forecast time ranges and show different characteristics. Using neural network technology to combine the prediction results of the two methods can illustrate the benefits of each method, obtain better prediction results, and improve the accuracy of prediction.

2.3 Partial Training

Training imbalance often occurs in multi-class training. That is to say, some types of samples are trained faster and mature soon, while some types of samples are slower and require a lot of training. In this case, if you continue to train, the mature class will have overfitting; if you stop early, the class that is not fully trained will not have the opportunity to mature. There are two reasons for the imbalance in training: one is the randomness of the weight initialization, the initialized weight may be inclined to the training samples of certain classes; the second is the number of training samples of each class is different, and the number of samples is large. The class gets more training opportunities, so it is easy to train mature, and the class with a small number of samples has relatively fewer training opportunities, and it is not easy to train mature. In response to this phenomenon, the error of only the part of the class that is behind training can be used in each training cycle. The specific method for training is: at the end of each training cycle, the various errors are sorted, and the first k classes with the smallest error are selected. In the next training cycle, the sample errors of these k classes are not Train again. This method stops training for fast-training classes, effectively avoiding the imbalance. In the experiment, it is found that when the k value is large, the training of each class is more balanced, but the training speed is slower; when the k value is small, the training the speed is faster, but the balance of network training decreases.

3 Model Establishment

3.1 Linear Stationary Model

In short, for time series $\{r_t\}$, there are two equivalent time series representations. One is based on the meaning of the past time series, and the other is based on the white noise of the time series. We believe that the white time series are generated from the linear prediction function. Forecast previous time series values.

One is the p-order autoregressive model AR (p), which represents the relationship between the dependent variable r_t and the independent variables $r_{t-1}, \ldots, r_{t-p}$ namely:

$$r_t = \sum_{i=1}^{p} a_i r_{t-i} + a_t \tag{5}$$

Specifically, given the first p observations, we have:

$$r_t = \phi_0 + \phi_1 r_{t-1} + \cdots + \phi_p r_{t-p} + a_t \tag{6}$$

$$t = p+1, p+2, \ldots, T \tag{7}$$

The corresponding residual is:

$$\hat{a}_t = r_t - \hat{r}_t \tag{8}$$

Call $\{\hat{a}_t\}$ the residual sequence, and get,

$$\hat{\sigma}_a^2 = \frac{\sum_{t=p+1}^{T} \hat{a}_t^2}{T - 2p - 1} \tag{9}$$

For multi-step forward prediction: Generally, we have:

$$r_{h+1} = \phi_0 + \phi_1 r_{h+t-1} + \cdots + \phi_p r_{h+t-p} + a_{h+t} \tag{10}$$

The forward l-step prediction obtained based on the minimization of the mean square loss function is the r_{h+1} conditional expectation under the given $\{r_{h-i}\}_{i=0}^{\infty}$ condition, which can be obtained by the following formula:

$$\hat{r}_h(l) = \phi_0 + \sum_{i=1}^{p} \phi_i \hat{r}_h(l-i) \tag{11}$$

The second is the q-order sliding mean model abbreviated as MA (q), which represents the relationship between the dependent variable r_t and white noise $a_{t-1}, a_{t-2} \ldots, a_{t-q}$, namely:

$$r_t = \sum_{i=0}^{q} \beta_i a_{t-i} \tag{12}$$

Let's look at the following, let the prediction origin be h. For the one-step forward prediction of the MA (1) process, the model is:

$$\hat{r}_{h+1} = c_0 + a_{h+1} + \theta_1 a_h \tag{13}$$

Taking conditional expectations, we have:

$$\hat{r}_h(1) = E\big(r_{h+1}|r_h, r_{h+1}, \ldots \big) = c_0 - \theta_1 a_h \tag{14}$$

$$e_h(1) = r_{h+1} - \hat{r}_h(1) = a_{h+1} \tag{15}$$

4 Results

4.1 Experimental Investigation Results

From the data analysis in Fig. 1, it can be seen that there are still major problems in current combination forecasting, and there are many combination forecasting scholars who are not satisfied with the combination forecasting. Only a few scholars are satisfied with the current combination forecast. Most of them think that the current combination forecast is good or not bad, but relatively satisfactory or average. From the data in the figure, it can be seen that the research on neural network-based time series forecasting model methods cannot allow the majority of scholars to realize that combined forecasting is an advantage, so we must make a little in this respect and strengthen the promotion of neural network time series forecasting model methods. So that scholars can better understand this information, so that they can develop better.

From the data analysis in Fig. 2, it can be seen that there are some specific problems in local training. The existence of these problems affects people's experience and makes these people have more opinions on local training. If local training is to better enable people to have a better user experience, these problems need to be overcome, so that it can bring these people a good user experience and use environment, so that people can

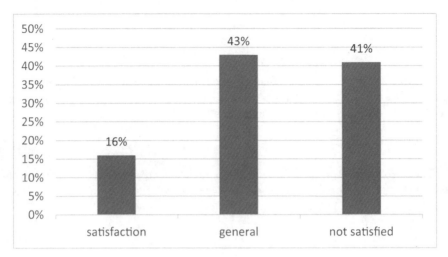

Fig. 1. Are you satisfied with the combination forecast

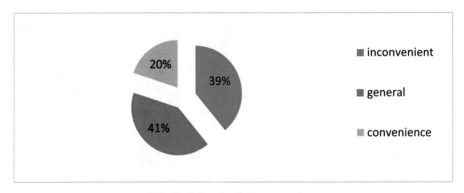

Fig. 2. Is local training convenient

Table 1. Men and women's understanding of this method

Gender	Understanding		
	Understand	Do not understand	Unknown
Male	23	14	13
Female	9	13	28

better learn and improve their own various items. Ability, so that neural network time series forecasting model methods can also develop better.

We randomly checked 100 people, with a male to female ratio of 1:1, and conducted a survey on the understanding of neural network time series forecasting model methods as follows:

According to the questionnaire survey data in Table 1, there are gender differences in the degree of understanding of neural network time series forecasting model methods between men and women. Boys are better than girls on neural network time series forecasting model methods. It is said to be relatively familiar, and girls know very little about this aspect. Therefore, gender will have a very different degree of cognition for neural network time series forecasting model methods, which is also an important factor we have learned based on the results of this survey. Therefore, the knowledge dissemination of neural network time series forecasting model methods should be strengthened, so that the development of neural network-based time series forecasting model methods will be of great help, so that people can understand them and support their development.

5 Conclusions

Using the observed time series data to predict the future changes of the research object, people make reasonable adjustments in advance. Therefore, it is very important to predict the time series wisely and effectively. Previously, most forecasting methods were based on general statistical methods. However, due to the complex nonlinear relationship in time series forecasting, traditional forecasting methods often did not work. In addition, traditional forecasting methods usually use sampling to process large-scale data sets. This processing will result in loss of information and increase in forecast errors, and the use of all data will take a long time. With the development of cloud computing, mobile Internet and Internet, the amount of data is increasing rapidly, and how to predict quickly and accurately is difficult to solve. The rapid development of artificial neural networks in recent years has provided people with new methods. Neural network has become the most widely used model of artificial neural network and its network. Among various neural networks, neural network has become the most widely used neural network model due to its advantages. Scientists at home and abroad have already had many successful time series examples, but neural networks have some birth defects. Therefore, methods to improve the accuracy of neural networks have become an important content of research.

References

1. Tavanaei, A., Maida, A.S.: Training a hidden Markov model with a Bayesian spiking neural network. J. Sig. Proc. Syst. **90**(2), 211–220 (2016). https://doi.org/10.1007/s11265-016-1153-2
2. Chunlei, Z., Kazuhito, K., Hansen, J.H.L.: Text-independent speaker verification based on triplet convolutional neural network embeddings. IEEE/ACM Trans. Audio, Speech, Lang. Proc. **26**(9), 1633–1644 (2018)
3. Karim, F., Majumdar, S., Darabi, H., et al.: LSTM fully convolutional networks for time series classification. IEEE Access **6**(99), 1662–1669 (2018)
4. Fazliana, R.N., Mahmod, O., Rajalingam, S., et al.: Forecasting Crude Palm Oil Prices using Fuzzy Rule-Based Time Series Method. IEEE Access, 1–1 (2018)

5. Rufin, P., Müller, D., Schwieder, M., et al.: Landsat time series reveal simultaneous expansion and intensification of irrigated dry season cropping in Southeastern Turkey. J. Land Use Sci. (3), 1–17 (2021)
6. Muhammad, M., Wahyuningsih, S., Siringoringo, M.: Peramalan nilai tukar petani subsektor peternakan menggunakan fuzzy time series lee. Jambura J. Math. **3**(1), 1–15 (2021)
7. Thrall, J.H., Li, X., Li, Q., et al.: Artificial intelligence and machine learning in radiology: opportunities, challenges, pitfalls, and criteria for success. J. Am. Coll. Radiol. **15**(3), 504–508 (2018)
8. Lee, J.N., Cho, H.C.: Development of artificial intelligence system for dangerous object recognition in X-ray baggage images. Trans. Korean Inst. Electr. Eng. **69**(7), 1067–1072 (2020)
9. Gore, R.W.: Re: prediction model to estimate presence of coronary artery disease: retrospective pooled analysis of existing cohorts. BMJ **5**(3), 368–375 (2020)
10. Wu, Y., Yu, L., Song, G., Xu, L.: Feasibility study of fuel consumption prediction model by integrating vehicle-specific power and controller area network bus technology. Transp. Res. Rec. **2341**(1), 66–75 (2018)

Improve the Application of EMD and SVM in the Recognition of Emotional EEG Signals

Zhiyi Li[✉]

Institute of Information and Communication, North University of China,
Taiyuan 030000, Shanxi, China
18050041218@st.nuc.edu.cn

Abstract. Aiming at the non-linear and non-stationary characteristics of emotional EEG signals, an improved Empirical Mode Decomposition (EMD) method based on correlation coefficient is proposed for feature extraction. In the process of empirical mode decomposition, the spectral components of each intrinsic mode function are analyzed, the original EEG signal is reconstructed and filtered, and calculates the correlation coefficient between every intrinsic mode function and the original signal, and effective IMF components are screened. Finally, emotion recognition is realized by support vector machine (SVM). The experimental consequences exhibit that compared with the traditional empirical steady-state decomposition, the algorithm greatly improves the accuracy of emotion recognition, which proves the actual value of the algorithm in the field of emotion recognition.

Keywords: Emotion EEG · Correlation coefficient · Empirical mode decomposition · Support vector machine

1 Introduction

A great deal of subjective cognitive experience is called by emotion, which is a person's attitude experience to objective things and corresponding behavioral responses. It has a strong practical application meaning in daily life. Therefore, emotion recognition plays a crucial role in human-computer interaction. In medical treatment, doctors can guide patients according to their emotional state. In teaching, teachers can choose appropriate teaching methods according to the emotional state of students to improve their learning efficiency.

Since human beings can control their own behavior, it is often difficult to ensure the authenticity and effectiveness of emotion recognition by using non-physiological signals such as facial expressions, actions, and voice tones, and cannot accurately reflect human emotion changes. The rich information contained in physiological signals can intuitively reflect the changes in human emotions, and has the advantage of not being easily disguised [1]. Among physiological signals such as ocular electricity, myoelectricity, and brain electricity, EEG signals have the advantages of accurate recognition effect and simple operation in emotion recognition, and they have been widely used

J. Abawajy et al. (Eds.): ATCI 2021, AISC 1398, pp. 285–293, 2021.
https://doi.org/10.1007/978-3-030-79200-8_42

[2]. Relevant studies have shown that the forehead of the brain is related to complex cognitive behaviors and decision-making, and can intuitively reflect human emotional information [4].

In 1998, Huang et al. improved EMD on the basis of Hilbert. Any signal is decomposed into a few intrinsic mode functions (IMF) and a residual by EMD algorithm. Compared with wavelet transform, EMD decomposition does not need to formulate basis functions, but is based on the signal itself, which has high efficiency. It plays a very good role in processing non-linear and unstable EEG signals. However, in the process of selecting IMF to reconstruct EEG signals, there are problems of introducing useless information or missing important information. In 1964, V.N. Vapnik et al. proposed Support Vector Machine (SVM). SVM has been applied in aspects such as portrait recognition and text classification. SVM uses computational empirical risk, and introduces regularization items in the solution process to optimize structural risk. SVM solves the non-linear classification problem well through the kernel method.

Aiming at the problems of traditional EMD, this paper proposes a combination of conditional empirical mode decomposition of correlation coefficients and support vector machines to recognize emotion signals. Use the EEG signals stimulated by different movie clips, use EMD to decompose and filter, get the reconstructed EEG signal, use the correlation coefficient for IMF screening to obtain the effective IMF component, and further extract the fluctuation index and average energy of the IMF as features and send it to the support vector Machine for emotion recognition. Experiments with multiple sets of data show that this method can greatly improve the accuracy of emotion recognition.

2   Experimental Data

This article uses the SEED data set released by Shanghai Jiaotong University. The SEED data set records the brain electrical signals of subjects watching movie clips. Watching different movie clips to arouse different types of emotions, including positive, negative, and neutral emotions. A 62-channel ESI NeuroScan system was used to collect the EEG data of 15 Chinese subjects (7 males, 8 females, average age 23.27, standard deviation 2.37), and the sampling frequency was 100 Hz (Fig. 1).

Fig. 1. EEG signal channel distribution diagram

In the SEED data set, each subject needs to conduct an experiment every about a week, and each experiment needs to watch 15 movie clips. A total of three experiments are performed, for a total of 45 trials. Each movie fragment is about four minutes long. Due to the large number of data points, this article selects the middle 1024 points of data for processing. The acquired EEG signal is preprocessed, and the data is down-sampled to 200 Hz to avoid the interference of ocular electricity and EMG noise, and a band pass filter of 0–75 Hz is used. Through the Fp1 and Fp2 channels to obtain data, Fp1 and Fp2 are located in the forehead of the brain, which can reflect human emotion changes [3].

3 Processing EEG Signal and EEG Feature Extraction and Classification

3.1 Principle of Conditional EMD

EMD is an adaptive nonlinear signal processing method. Because this method performs signal decomposition based on its own time scale characteristics, there is no need to set the basis function ahead of schedule. The key of EMD is decomposing any complex signal into a few IMF and a residual component. Each IMF component obtained after decomposition reflects the local characteristics of the original signal at different time scales, which is beneficial to highlight the local characteristics of EEG signals The detailed process of EMD is as follows, assuming that the original signal is $y(t)$:

Step 1: Search for all the minimum as well as maximum points of the original signal $y(t)$.
Step 2: Using cubic spline function to construct the maximum envelope of the signal $n_1(t)$ by using all the maximum points. Using cubic spline function to construct the minimum envelope of the signal $n_2(t)$ by using all the minimum points.
Step 3: Calculate the average value $c(t)$ of the upper and lower envelopes.

$$c(t) = \frac{1}{2}[n_1(t) + n_2(t)] \tag{1}$$

Step 4: Subtract the average value of the upper and lower envelopes from the original signal to obtain a new signal $f(t)$.

$$f(t) = y(t) - c(t) \tag{2}$$

Step 5: If the new signal sequence meets the requirements of the IMF, it will be the first IMF component. If it does't meet the requirements, repeat steps (1) (2) (3) until the two conditions of the IMF component are met, and get The first IMF component is denoted as $s_1(t)$. Since the mean value of the upper and lower envelopes cannot be zero in actual situations, Huang believes that when the Cauchy screening stop criterion is met, the cycle can be stopped.

$$SD = \frac{\sum_{t=0}^{T} |s_{k-1}(t) - s_k(t)|^2}{\sum_{t=0}^{T} s_{k-1}^2(t)} \tag{3}$$

Generally, when the value of SD is between 0.2 and 0.3, the screening can be stopped.

Step 6: The IMF component $y(t)$ is subtracted from the original signal $s_1(t)$ to acquire the residual component $r(t)$.

Step 7: Regard the residual component $r(t)$ as a new original signal, repeat the above process n times in sequence until the residual function $r_n(t)$ obtained is a monotonic function or there is only one pole. The EMD of the original signal can be decomposed into:

$$y(t) = \sum_{i=1}^{n} s_i(t) + r_n(t) \tag{4}$$

The IMF components of each order obtained by the EMD decomposition are arranged in order of frequency from high to low, and the key features of the original signal are concentrated in the first few IMF components. For the purpose of overcome the problem of missing effective information or doping noise during traditional EMD decomposition. This paper introduces correlation coefficients and selects effective IMF components.

The correlation coefficient can be defined as

$$r = \frac{Cov(X, Y)}{\sigma_X \sigma_Y}$$

$$= \frac{\sum_{i=1}^{n} (x_i - \overline{x})(y_i - \overline{y})}{\sqrt{\sum_{i=1}^{n} (x_i - \overline{x})^2 (y_i - \overline{y})^2}} \tag{5}$$

The closer the absolute value of the correlation coefficient is to 1, the stronger the correlation is. Calculate the correlation coefficient between every intrinsic mode function and the original signal, and use the absolute value of the correlation coefficient as the condition for screening the intrinsic mode function, and extract the IMF component containing the main brain telecommunication signal. The conditions to satisfy the screening IMF are:

$$|r| \geq \lambda \tag{6}$$

In the formula, if the constant λ is satisfied $0 \leq \lambda \leq 1$, it will be set as the screening threshold.

3.2 Correlation Screening of EMD Decomposition of EEG Signal and IMF

Studies have shown that the main reaction of the EEG signal of the movie clip leading to emotional changes is on three frequency bands. When positive emotions are generated, the power of the wave becomes larger [4]. Most of the waves located on the frontal lobe appear in a relaxed state, which can reflect the different valence of emotions [5, 6]. Wave is closely related to positive ideology, concentrated in the frontal area, and can reflect the valence of emotions.

Because the brain electrical signal is very small, it is easy to be interfered by other signals, such as myoelectric signal and ocular electrical signal. We need to filter the noisy

EEG signal first. In order to remove noise, first the original signal was decomposed by EMD to obtain the IMF component as shown in the figure. Then perform FFT on the IMF component to obtain each spectrogram. It can be seen from the figure that the higher the order of the IMF component, the lower the frequency. The IMF1 component is mainly noise above 30 Hz, so we remove the IMF1 component and reconstruct the signal, and the reconstructed signal is smoother than the original signal. Avoid the interference of other physiological signals on the EEG signal (Figs. 2 and 3).

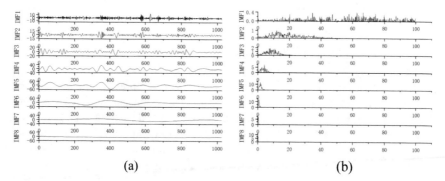

(a)   (b)

Fig. 2. IMF component time domain diagram (a) and IMF component spectrogram (b)

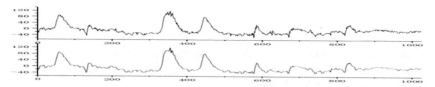

Fig. 3. Signal comparison before and after filtering

In order to obtain effective IMF classification, calculate the correlation coefficient between the original EEG signal and each IMF component, and the effective IMF components are screened. The distribution of the absolute value of the correlation coefficient of each IMF component is shown in the figure (Fig. 4).

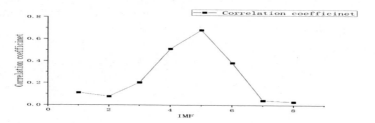

Fig. 4. Correlation coefficient distribution diagram of each level of IMF

According to the above figure and analysis of the results of many experiments, the threshold is set to 0.2. The main IMF components (IMF3-IMF6) are screened out.

3.3 Emotional EEG Signal Feature Extraction

Extract the feature components of the IMF components after sorting, and use the combination of the fluctuation index and the average energy as the feature vector of the IMF component classification and send them to the vector machine for training and classification.

The volatility index is used to indicate the degree of change in brain electrical signals. When mood fluctuations occur, the fluctuations of the EEG signal will fluctuate greatly. Define the volatility index as

$$F_{ij} = \frac{1}{n} \sum_{t=1}^{n} |x_i(t) - x_j(t)|, \tag{7}$$

In the formula, x_i represents the i-th IMF component, x_j represents the j-th IMF component, n represents the number of IMF component data points, as well as F_{ij} represents the fluctuation index between the i-th IMF component and the j-th IMF component.

The activity level of the cerebral cortex reflects the amplitude of brain electrical signals, which in turn reflects changes in energy. Therefore, the average energy of each IMF component is taken as the characteristic value. Define the average energy as

$$E_i = \frac{1}{n} \sum_{t=1}^{n} |x_i(t)|^2 \tag{8}$$

Where E_i is the average energy of the i-th IMF component, x_i is the i-th IMF component, as well as n is the number of IMF component data points.

In summary, each group of EEG signal data consists of 4 average energy characteristics and 3 fluctuation index characteristics.

3.4 Libsvm Classification Algorithm

SVM was first proposed by Cortes and Vapnik in 1995. As a second-class classifier, it finds a hyperplane and uses two types of data as far as possible from the hyperplane to achieve the purpose of accurately classifying new data. The SVM model is defined as the maximum interval linear classifier in the feature space, the interval maximization is used as its learning strategy, and finally it can be converted to a convex quadratic programming problem.

The basic steps are as follows:

Step 1: Convert the original problem to a convex optimization problem

$$\min_{w,b,\xi} \frac{1}{2}\|w\|^2 + C\sum_{i=1}^{N}\xi_i \tag{9}$$

$$s.t. \quad y_i(w.x_i + b) \geq 1 - \xi_i, \quad i = 1, 2, \ldots, N \tag{10}$$

$$\xi_i \geq 0, \quad i = 1, 2, \ldots, N \tag{11}$$

Step 2: Construct a Lagrangian function, dualize the original problem and choose appropriate kernel function $K(x_i, y_j)$ and penalty parameters C, then the original problem can be converted to

$$\min_{\alpha} \frac{1}{2}\sum_{i=1}^{N}\sum_{j=1}^{N}\alpha_i\alpha_j y_i y_j K(x_i, x_j) - \sum_{i=1}^{N}\alpha_i \tag{12}$$

$$s.t. \quad \sum_{i=1}^{N}\alpha_i y_i = 0 \tag{13}$$

$$0 \leq \alpha_i \leq C, \quad i = 1, 2, \ldots N \tag{14}$$

The Gaussian kernel function is selected as the kernel function, and its expression is $K(x_i, x_j) = \exp\left(-\frac{\|x_i - x_j\|}{2\sigma^2}\right)$, σ as the control parameter. By changing the control parameters σ, the Gaussian kernel function has high flexibility. Then an optimal solution α^* is obtained through the SMO algorithm.

Step 3: Select a component $0 < \alpha_j^* < C$ in the vector α^*, and then calculate b

$$b^* = y_j - \sum_{i=1}^{N}\alpha_i^* y_i K(x_i, x_j) \tag{15}$$

Step 4: Obtain the classification decision function

$$f(x) = siin\left(\sum_{i=1}^{N}\alpha_i^* y_i K(x, x_i) + b^*\right) \tag{16}$$

This paper uses cross-checking to search for the Optimal parameter of parameter penalty C and kernel parameters, and uses the optimal parameters to train the entire data set.

3.5 Algorithm Flow

The specific process of the algorithm in this paper is shown in the figure (Fig. 5).

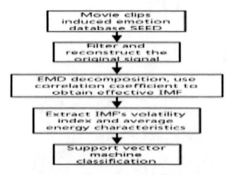

Fig. 5. Algorithm flow chart

4 Experimental Results and Analysis

This paper extracts the features of 45 groups of EEG signals collected, and sends the features to the support vector machine for training and testing. Use the first 30 data sets as the training set and the last 15 data sets as the prediction set to observe the classification accuracy. As a comparison, the traditional EMD method is selected for feature extraction and classification. The classification results are exhibited in the table.

As shown in the table, using the traditional EMD method, the classification accuracy of SVM is 84.38%. Using empirical modal decomposition combined with correlation coefficients, the accuracy rate can reach 93.75%. It shows that compared with the traditional EMD method, the improved empirical mode decomposition method used in this paper has a higher recognition rate for feature extraction, and can well identify the positive, negative and neutral emotions, which has practical application value (Table 1).

Table 1. The recognition accuracy of traditional EMD and conditional EMD

Instructions	Accuracy/(%)
EMD	84.38
Condition EMD	93.75

5 Conclusion

As one of the important components of human-computer interaction, emotion recognition has good application value and development prospects in many aspects in real life. In this paper, an EMD method combined with correlation coefficients is used to realize the feature extraction of emotional EEG signals, and a high accuracy rate is obtained in emotion recognition. First, EMD is used to decompose and reconstruct the original EEG signal to achieve the purpose of filtering noise. Subsequently the correlation coefficients between each IMF component and the original signal were calculated and screen the effective IMF. And according to the characteristics of the EEG signal, the average energy and fluctuation index are extracted as the characteristic quantities. Finally, the SVM algorithm is used to realize emotion recognition. The experimental results show that compared with the traditional EMD feature extraction method, the empirical mode decomposition method combined with correlation coefficients used in this paper has a higher correct rate of emotion signal recognition. Therefore, the empirical mode decomposition method combined with the correlation coefficient has high practical application value in the field of emotion recognition. At the same time, this method can also be extended to other fields to provide a reference for the classification and recognition of other physiological signals.

References

1. Shahiri, M., Ghadrir, R., Hosseinnia, S.H., Momani, S.: Chaotic fractional-order coullet system: synchronization and control approch. Commun Nonlinear Sci Number Simul 15, 665–674 (2010)
2. Alarcao, S.M., Fonseca, M.J.: Emotions recognition using EEG signals: a survey. IEEE Trans Afffect Comput (2017). https://doi.org/10.1109/TAFFC.2017.2714671
3. Chanel, G., Kierkels, J.J.M, Soleymani, M., et al.: Short-term emotion assessment in a recall paradigm. Int J Hum.– Comput. Stud. 67, 607–627 (2009)
4. Sammler, D., Grigutsch, M., Fritz, T., et al.: Music and emotion: electrophysiological correlates of the processing of pleasant and unpleasant music. Psychophysiology 44, 293–304 (2007)
5. Davidson, R.J.: What does the prefrontal cortex do in afffect: perspectives on frontal EEG asymmetry research. Biol Psychol. 67, 219–234 (2004)
6. Yuvaraj, R., Murugappan, M., Ibrahim, N.M., et al.: Emotion classifification in Parkinson's disease by higher-order spectra and power spectrum features using EEG signals: a comparative study. J Integr Neurosci 13, 89–120 (2014)

College Students' Internet Altruistic Behavior from the Perspective of Big Data

Jing Lin[(✉)]

School of Management, Wuhan Donghu University, Wuhan 430000, Hubei, China
elle_lynn@sina.com

Abstract. Objective: As the current focus of high attention, the research on college students' internet altruistic behavior from the perspective of big data was carried out in this paper. Methods: A questionnaire survey based on 203 respondents was used and the results were analyzed by statistical software. Results: It was shown that the average value of the respondents' internet altruistic comments in the past year was 3.48. The main reason prevented them from posting altruistic comments was not knowing the whole story, and the desire to express their own thoughts was the primary reason that motivated them posting altruistic comments. Conclusion: Focusing on the college students' internet altruistic behavior in big data era can not only enrich the research content of internet altruistic behavior from the perspective of psychology, but also put forward educational strategies for the guidance of college students' internet altruistic behavior from the perspective of ideological and political education.

Keywords: College students · Internet altruistic behavior · Perspective of big data · Moral identity education

1 College Students' Internet Altruistic Behavior

The concept of altruistic behavior was put forward in 1851 by French sociologist Comete to describe individuals' selfless and voluntary behavior [1]. Internet altruistic behavior referred to a voluntary behavior occurring in internet environment that benefit others without obvious selfish motives [2]. The openness, anonymity and time-space cross of the network had reduced the helpers' cost of time consumption and ensured the timely feedback of individual behaviors, which had promoted altruistic behavior [3]. The factors affecting internet altruistic behavior mainly included the helper [4], the environment [5] and the recipient [6] through sorting out the previous research literature. In addition, scholars had also carried out related studies on the internet altruistic behavior of college students. It was shown that there were significant differences in college students' internet altruistic behavior in terms of gender and origin, while there were no significant differences in whether they were the only child, grade or online age [7]. Family factors had a significant impact on adolescents' online interpersonal relationship and Internet satisfaction [8]. Moral identity could significantly predict the college students' internet altruistic behavior [9].

J. Abawajy et al. (Eds.): ATCI 2021, AISC 1398, pp. 294–299, 2021.
https://doi.org/10.1007/978-3-030-79200-8_43

2   Expression of College Students' Internet Altruistic Behavior in Big Data Era

Questionnaires had been distributed to 210 college students in order to investigate the college students' internet altruistic behavior and 203 valid questionnaires had been collected with effective recovery rate of 96.7%. In addition, only 'making good comments or replies on the Internet' was selected as the representative of internet altruistic behavior due to the broad concept and scope of it in this survey.

2.1   Frequency of Internet Altruistic Behavior

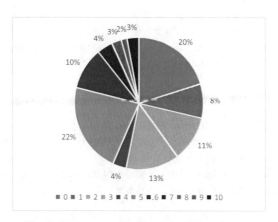

Fig. 1. Frequency of internet altruistic comments

In the past year, the frequency of altruistic comments by respondents was shown in Fig. 1, where 0 represented never comment and 10 represented every comment. 56.7% of the respondents had published altruistic comments with a frequency between 0 and 4, and only 21.2% of them had a frequency greater than (including) 6.

On the other hand, the news topics that the respondents showed more altruistic behaviors mainly focused on leisure, entertainment and people's livelihood (food, clothing, housing and transportation) in the past year, and the altruistic comments on national security (politics, military) and economy were less.

2.2   Platform Selected by Internet Altruistic Behavior

There were three types of online platforms that provide users to obtain and reply to information. One was for acquaintances represented by Moments and Qzone. Most of the information seen on these platforms was published by friends. The second was for strangers represented by Weibo and official accounts. Most of the information seen on these platforms was posted by strangers. The third type was the mixed one represented by Tik Tok, where you can see information posted by both friends and strangers. According to the survey, 67.49% of respondents said that their internet altruistic comments were

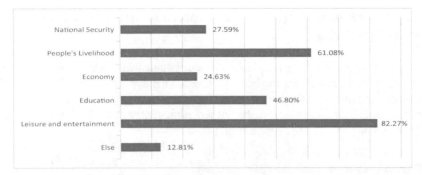

Fig. 2. News topics that the respondents showed altruistic comments

published on the acquaintances platforms in the past year, 15.27% on the strangers platforms. Therefore, people were more willing to help acquaintances than strangers, which might be influenced by interpersonal distance.

2.3   Expression of College Students' Internet Altruistic Behavior

Internet altruistic behavior could be divided into four types: network support, network guidance, network sharing and network reminder. The type distribution of respondents' altruistic comments in the past year was shown in Fig. 3. As can be seen from the curve, the frequency of the four kinds in descending order was network support, network reminder, network sharing and network guidance.

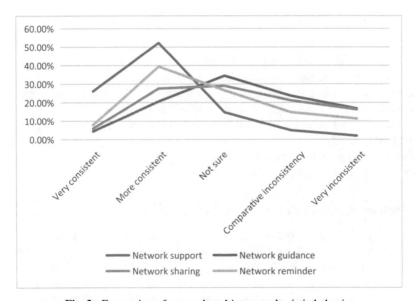

Fig. 3. Expression of respondents' internet altruistic behavior

2.4   Reasons for Hindering Internet Altruistic Behavior

The average value of the respondents' internet altruistic comments in the past year was 3.48, far lower than the median value. The main reasons that hinder the respondents from putting forward altruistic comments were shown in Fig. 4.

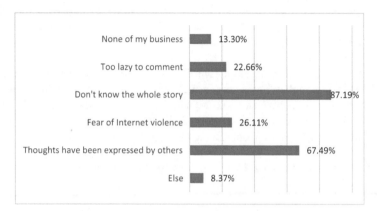

Fig. 4. Reasons for hindering internet altruistic behavior

2.5   Reasons for Promoting Internet Altruistic Behavior

Although the frequency of respondents' internet altruistic behavior in the past year was not high, it could be found that the following reasons can promote the respondents' internet altruistic behavior to some extent, as shown in Fig. 5.

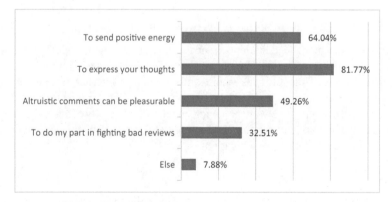

Fig. 5. Reasons for promoting internet altruistic behavior

3   Influence of Big Data on Internet Altruistic Behavior

With the low threshold of information platform and the gradual popularization of infor-mation means, the rapid growth of public opinion data made it more and more difficult to manage. Firstly, huge amounts of data had challenged new data processing techniques. Most of the public opinion data on the Internet are disordered, worthless, or even false, which had put forward higher requirements for new technological means. Secondly, the development of big data technology provided a convenient and fast channel for netizens to express their personal opinions and emotions. Due to the limited identification ability, netizens might always choose and accept the same views with their own opinions, and even copy and spread without thinking. Thirdly, in big data era, network information had been no longer monopolized by a certain person or department, the public had gained more discourse power. In this case, we must take effective measures to actively guide the internet altruistic behavior of netizens, especially college students, so as to maximize the important role of big data in optimizing the network platform.

4   Guiding Strategies of College Students' Internet Altruistic Behavior from the Perspective of Big Data

4.1   Strengthening Internet Supervision

In the big data era, in the face of massive data resources, people's discrimination ability was very limited, which made false information waste the network resources and even posed a threat to people's property safety, which resulting in the decline of network interpersonal trust. College students, in particular, were not discriminating enough in the face of online fraud and were more likely to be cheated, thus hindering the generation of internet altruistic behaviors. Therefore, it is necessary to perfect the network norms, strengthen the network supervision, severely crack down on the network cheating and enhance the network interpersonal trust so as to create a good and clean network space for the occurrence of internet altruistic behavior.

4.2   Strengthening College Students' Network Moral Education

The moral system of college students is not stable, as well as the ability of thinking independently, which made them easy to be affected by bad values. Previous studies had shown that moral self-identity was significantly related to internet altruistic behavior. So it's necessary to cultivate the students' network moral accomplishment, help them establish correct concept of network behavior, enhance the capacity of analysis, judgment and evaluation of network information, and guide students to establish good relationships with others in the network communication.

4.3   Giving Full Play to Family Function

Family education is a lifelong education which affects the formation of altruistic behav-ior gradually. Harmonious family atmosphere, moral education and parents' altruistic behavior can be conducive to the formation of children's altruistic awareness and per-formance of altruistic behavior. Parents should pay more attention to communicate with their children in daily life and cultivate individual empathy level.

5   Conclusions

As a popular medium of information dissemination, the Internet had been widely and deeply integrated into people's life, and college students had become the main body of cyberspace. It had been found that college students' internet altruistic behavior was common by scholars' research, but the frequency was not high, so it is necessary to give appropriate guidance to college students' internet altruistic behavior in the era of big data.

Acknowledgements. This work was supported by the grants from Youth Foundation Project of Wuhan Donghu University (project number: 2020dhsk003).

References

1. De Waal, F.B.: Putting the altruism back into altruism: the evolution of empathy. Annu. Rev. Psychol. **59**(1), 279–300 (2008)
2. Qinghong, P., Fumin, F.: College Students' Internet Altruistic Behavior and Its Enlightenment to College Moral Education. Lead. J. Ideological Theoret. Educ. **12**, 51–53 (2005). (in Chinese)
3. Qinsheng, L.: Internet altruistic behavior of adolescents from the perspective of positive psychology. Theory Pract. Educ. **37**(23), 26–28 (2017). (in Chinese)
4. Xianliang, Z., Wei, Z.: Relationship between empathy, self-efficacy and internet altruistic behavior. Chin. J. Clin. Psychol. **23**(2), 358–361 (2015). (in Chinese)
5. Guitao, L.: A review of internet altruistic behavior of college students. Academy **10**, 1–2 (2016). (in Chinese)
6. Xianliang, Z., Yaqin, W.: Relationship between internet altruistic behavior and subjective well-being among adolescents: a mediated moderation model. J. Psychol. Sci. **1**, 70–75 (2017). (in Chinese)
7. Qing, Y., Jieru, L.: Characteristics of Adolescents' family function and Its Influence on Internet behavior. Chin. J. Health Psychol. **24**(6), 850–855 (2016). (in Chinese)
8. Jiang, H., Lin, L., Ma, S.: The influence of family function on the network altruism of college students: the intermediary role of gratitude. Journal of Guizhou Normal University(Natural Sciences), **34**(3), 106–110 (2016). (in Chinese)
9. Qian, Z.: College students' moral identity and internet altruism:the role of online morality and gender. Chin. J. Clin. Psychol. **26**(6), 182–185 (2018). (in Chinese)

Monitoring and Prediction Mechanism of Artificial Intelligence Technology in Online Language Violence

Zhihui Yu[1] and Weimin Wang[2(✉)]

[1] College of Marxism, Shanghai University of Political
Science and Law, Shanghai 201701, China
[2] Dean and Professor of School of Government at Shanghai University of Political Science and
Law, Shanghai 201701, China

Abstract. Online language violence is a new product of the Internet's information exchange function. It refers to the influence of violent language, including the most serious judgmental language, profanity and defamatory language on the Internet, gossip, human flesh search research, etc. People who are experiencing online language violence will be mentally hurt. At the same time, as artificial intelligence technology continues to be adopted and researched by everyone, artificial intelligence technology is inevitably selected for the research on the monitoring and prediction mechanism of online language violence. This paper mainly adopts the literature reference research method, the analytic hierarchy process and the investigation and research method, uses artificial intelligence technology to study the monitoring and prediction mechanism of online language violence, establishes a mathematical model, uses the analytic hierarchy process to solve it, and monitors online language violence Evaluate the research status of the prediction mechanism, and use historical data to revise the model to improve the accuracy of the monitoring of online language violence and the evaluation of the prediction mechanism research status. The experimental results show that the analytic hierarchy process has increased the research efficiency of the monitoring and prediction mechanism of online language violence by 23%, and reduced the rate of false positives and false positives. Finally, by comparing the social harm analysis of online language violence and the analysis of the evaluation index system of online language violence, it illustrates the influence of artificial intelligence technology on the research of the monitoring and prediction mechanism of online language violence.

Keywords: Artificial intelligence technology · Cyber language violence ·
Literature reference research method · Analytic hierarchy process

1 Introduction

1.1 Background and Significance

In the 21st century, with the rapid development of the Internet and computers, the way people express their opinions is more convenient than in the past. The Internet provides

J. Abawajy et al. (Eds.): ATCI 2021, AISC 1398, pp. 300–307, 2021.
https://doi.org/10.1007/978-3-030-79200-8_44

people with countless information and greatly expands the space for people to express their opinions [1]. We see many different comments on news comments, Weibo messages, video barrage and game messages. It is precisely because of the openness, virtuality and concealment of the Internet that we often see a lot of violent language on the Internet [2, 3]. This violent Internet language infringed and severely damaged the spirit and psychology of other people, but most Internet platforms were not very effective in their management [4, 5]. The strategy they adopted is mainly to block a few common words related to cyber violence. However, the language of cyber violence is still everywhere. Therefore, it is necessary to find a new method to find the language of cyber violence [6, 7].

1.2 Related Work

Nabi R M provides a method that can evaluate relevant and participatory innovations in a complex and relevant environment to solve essential problems [8]. Based on the principle of common value creation, he proposed a research framework. This framework illustrates the research process of artificial intelligence technology in the monitoring and prediction mechanism of online language violence. In this process, the relevant parties integrate their resources and capabilities to develop the innovative analytic hierarchy process [9, 10]. In order to evaluate this analysis framework, Nabi R M collected multiple data in the research. This case also represents the significance of the research and system realization of the monitoring and prediction mechanism of artificial intelligence technology in online language violence [11, 12]. Because of the relatively few data variables in his research, the results are not convincing.

1.3 Main Content

The innovation of this article lies in the literature reference research method, analytic hierarchy process and investigation research method. Based on the research on the monitoring and prediction mechanism of online language violence under the background of artificial intelligence, the research on the monitoring and prediction mechanism of artificial intelligence technology in online language violence is analyzed through the analytic hierarchy process. Establish a calculation method combining the analytic hierarchy process and the literature research method to provide guidance for the research on the monitoring and prediction mechanism of cyber language violence in the background of artificial intelligence.

2 Methods of Artificial Intelligence Technology in the Monitoring and Prediction Mechanism of Online Language Violence

2.1 Literature Research Method

Literature research method is a way to understand scientific literature information through scientific research. In the process of collating theories and cases about the monitoring and prediction mechanism of online language violence, the author needs a lot of data and useful literature. Consult the electronic resources and paper materials of the school library, purchase relevant books, and use the Internet to search for relevant materials, and extensively collect reports, policies and statistics on the monitoring and prediction mechanism of online language violence to ensure a comprehensive and correct understanding.

In the process of research and writing, according to the research topics identified, with the help of Wanfang and CNKI China databases, such as Vipu and SDOS full-text database, many books, master and doctoral papers, journals and other materials have been consulted, and the domestic a large number of foreign theoretical results on the monitoring and prediction mechanism of online language violence have played a role in theoretical support for the writing of this paper and provided a reference for in-depth research on this topic.

2.2 AHP

AHP is a weight loss method that combines quality and quantity tests. This method is based on the principle of understanding things from simple to complex, and divides complex issues into several orderly and reasonable levels. By analyzing each relatively simple level layer by layer, the relative decision goals of each index are obtained. Through mathematical methods, the relative importance of each index is converted into weight values, and different decision-making schemes are sorted to find the best decision-making scheme.

Normalize the required feature range to classify the significance of each evaluation factor, that is, the weight value. The test shall use the following formula:

$$Q_A = \gamma_{\max} A \tag{1}$$

The random sequence ratio of the judgment matrix is represented by EI; the general sequence index of the estimation matrix is represented by M.

$$EI = (\gamma_{\max} - M)/(M - 1) \tag{2}$$

I represents the average random consistency index in the standard table.

The reason for adopting the analytic hierarchy process is that the analytic hierarchy process treats the research object as a system, and makes decisions according to the way of decomposition, comparison, judgment, and comprehensive thinking. It has become an important tool for system analysis developed after mechanism analysis and statistical analysis. The idea of the system is not to cut off the influence of each factor on the result, and the weight setting of each layer in the analytic hierarchy process will eventually directly or indirectly affect the result, and the degree of influence of each factor on the result is quantitative and very clear.

2.3 Investigation and Research Method

Investigation and research methods refer to methods that directly obtain relevant materials through the understanding and understanding of objective conditions and analysis of these substances. Through a large number of online materials, practical cases, etc., this article fully investigates the current situation and reasons of the monitoring and prediction mechanism of online language violence in the era of traditional models and artificial intelligence, analyzes, synthesizes, compares and summarizes the large amount of data collected in the survey. Therefore, we have a comprehensive and profound understanding of how to better construct the model of the monitoring and prediction mechanism of online language violence in the era of artificial intelligence.

3 Monitoring and Prediction Mechanism of Artificial Intelligence Technology in Online Language Violence

3.1 Experimental Design of the Monitoring and Prediction Mechanism of Online Language Violence

The Internet provides convenient technology and open space for personal language communication. People can use the Internet to communicate in language at any time, any place, and any time to achieve the purpose of language communication. The simplicity of this communication makes people in the virtual network society continue to gather together and regard the network society as an important living space. As the online community becomes a huge interactive space, its unprecedented openness and freedom fully realize people's freedom of speech, but this has also become the basis of online language violence. In this experiment, we use the word vector calculation method and the similarity based on the corpus to create a dictionary of online language violence words. When using this tool to train the word vectors in words, the CBOW model is not used, but the Skip-Gram model.

3.2 Experimental Data Collection of Monitoring and Prediction Mechanism of Online Language Violence

After completing the calculation of the word vector, the distance between the word vectors is used as the basis for calculating the semantic similarity of the two words. The result of the experiment is the cosine distance between words. The larger the cosine distance, the smaller the spatial distance, and the higher the similarity of words. Compare the language model optimization rules with the rules before the language model optimization. The experiment uses a rule-based language detection language to test a fixed set of 400 tests. The experiment also selected 3 common violent words as the introductory words, and calculated the 3 furthest words from the violent word "SB". The details are shown in Table 1:

It can be seen from Table 1 that compared with the rules before the optimization of the language model, the accuracy of the optimized rules of the language model has increased, while the recall rate is basically unchanged, and the experiment has achieved the expected effect. In order to be able to see the relationship between the data, the data analysis of Table 1 is carried out, and the analysis results are shown in Fig. 1.

It can be seen from Fig. 1 that the recall rate of the experiment is not very good. The reason is that many violent languages are derived from violent words. Another reason

Table 1. Experimental result data table of test set

Candidate	Word distance	Number of words	Accuracy (P)	Recall rate (R)
Grass mud horse	0.733772	5	37.3%	37.2%
Mentally retarded	0.698580	8	42.8%	37.5%
Stupid	0.696195	24	67.9%	37.9%

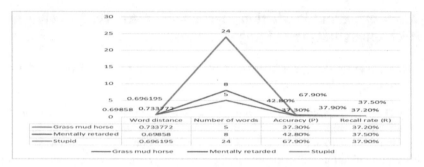

Fig. 1. Data analysis diagram of test set experimental results

is that the training data is not large. Since one of the purposes of this experiment is to find the location of online vocabulary or online violent phrases, it is effective to use semantic-based research methods.

4   Monitoring and Prediction Mechanism of Artificial Intelligence Technology in Online Language Violence

4.1   Social Harm of Online Language Violence

Nowadays, the age of idol worshipers is gradually decreasing, which is called younger age. Young people do not have enough social experience to distinguish and judge social phenomena and information discussed on the Internet. In this case, capitalists will strengthen and control the development trend of online language violence. When it comes to Internet mobs, they use some language that they feel bad about in an uncontrolled Internet environment. They only think that such bad language will not cause much harm. They think that these bad words are just an inconspicuous episode in life. After that, they continued to move forward without scruples, continue to engage in online language violence, and establish online language violence groups from time to time.

In recent years, there have been reports that more and more people are addicted to online virtual games or online novels, and some people are even unable to extricate themselves. This shows that people have been in the virtual world for a long time. It is easy to bring it into a virtual life that is not helpful to real life. The fun and enemies of the virtual environment not only lack chivalry in real life, but also become one of the root causes of social instability. Internet users who like to use online language violence have the same mentality, that is, they cannot fully understand things, and they are emotional when evaluating and making decisions every day. The most common online language violence is the phenomenon of "whole Internet black" by some celebrities. In such a phenomenon, most Internet users will use violent language when following the trend, and even they are not surprised by the Internet environment. Once they become commonplace in daily life, it is difficult to decline from the moral level.

4.2 Evaluation Index System of Online Language Violence

Current systems for monitoring and analyzing verbal violence basically use subjective judgments based on statistical indicators or manually to evaluate verbal violence on the Internet. However, there is no complete evaluation index system yet. Scholars at home and abroad have begun to study indicators for evaluating online language violence. Many conceptual indicators have become more and more difficult to evaluate, and it is impossible to measure the weight of each indicator. Given the best solution, it is difficult to obtain a true quantitative evaluation.

In order to quantify it, this paper studies the preprocessing technology used to improve the violent language network on the Internet. Generally speaking, the violent language networks that are harassed on the Internet are mainly website navigation information, actual violent networks, content-related violent language networks, content-independent violent language networks, and data-based violent language networks. The method described in this article is mainly aimed at the use of online language violence. The text obtains relevant information from websites on the Internet, with the purpose of understanding the importance of symbols in violence language in web page cleaning. Therefore, after creating the blocked DOM tree, only two <div> <div> tags are considered, and the statistical results are shown in Table 2:

Table 2. Data table of Chinese punctuation statistics on Internet language violence web pages

Chinese punctuation	Period	Comma	Exclamation mark	Question mark
Page appearances	4135	8372	513	1134
Number of occurrences of text	3587	7153	194	366
Percentage in the body	86.7%	85.4%	37.8%	32.2%

It can be seen from Table 2 that various Chinese punctuation marks appear more frequently in web pages than various Chinese punctuation marks appear in online language violence, and the period accounted for 86.7%, which is the highest proportion among the four types of symbols. However, whether it is webpage or online language violence, the most frequent occurrence is the comma. In order to be able to see the relationship between the experimental class and the control class, data analysis was carried out on Table 2. The analysis results are shown in Fig. 2:

It can be seen from Fig. 2 that periods and commas appear the most frequently in webpages and online language violence, which shows that the preprocessing technology

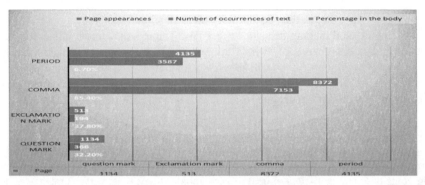

Fig. 2. Data analysis of Chinese punctuation statistics on Internet language violence web pages

of purification of online language violence web pages can accurately capture online violence language, and then take precautions Processing technology is used to predict and supervise online violent language.

5   Conclusions

Although this article has achieved certain research results through literature reference research method, analytic hierarchy process and investigation research method, there are still many shortcomings. The research on the monitoring and prediction mechanism of artificial intelligence technology in online language violence still has much content worthy of in-depth study. There still many steps in the research process that cannot be covered due to space and personal ability. In addition, the actual application effect of the improved algorithm can only reach the level of comparison with the traditional model from the level of theory and simulation.

Acknowledgements. This work was supported by Youth Fund for Humanities and Social Sciences Research of Chinese Ministry of Education under Grant No. 18YJC710092 and General Project of Philosophy and Social Science Planning in Shanghai under Grant No. 2017BKS011.

References

1. Manfeng, C., Nan, X., Lei, L., et al.: Research on the monitoring and coping mechanism of seismic network public opinion. Plateau Earthquake **030**(002), 53–57 (2018)
2. Luyao, Y.: Research on the realistic obstacles and innovation mechanism of chinese enterprises' direct investment in developed countries. Henan Soc. Sci. **025**(006), 81–87 (2017)
3. Mackinnon, L.: Artificial stupidity and the end of men. Third Text **31**(5–6), 603–617 (2017)
4. Winters, N., Eynon. R., Geniets, A., et al.: Can we avoid digital structural violence in future learning systems? Lear. Media Technol. **45**(3), 1–14 (2019)
5. Dong, S., Fangxu, X.U., Tao, S., et al.: Research on the Status Quo and Supervision Mechanism of Food Safety in China. Asian Agricultural Research, **10**(02), 36–37+42 (2018)
6. Hui, T., Lihua, X.: Research on the cohesive mechanism of undergraduate and graduate education for the vocational teacher in China. Contemp. Vocat. Educ. **000**(003), 83–88 (2019)

7. Martinez, V.R, Somandepalli, K., Singla, K., et al.: Violence Rating Prediction from Movie Scripts// Association for the Advancement of Artificial Intelligence (www.aaai.org), 671–678 (2019)
8. Nabi, R.M., Ab, S.M., Haron, H.: Artificial intelligence techniques and external factors used in crime forecasting in violence and property: a review. J. Comput. Sci. **16**(2), 167–182 (2020)
9. Dong, S., Fangxu, X.U., Tao, S., et al.: Research on the status quo and supervision mechanism of food safety in China. Asian Agric. Res. Eng. Ed. 000(002), P.32–33 (2018)
10. Wang, C.: Research on the performance and evaluation mechanism of department offices in private colleges. J. Jiamusi Educ. Coll. 000(008):453, 455 (2017)
11. Kong, W., Guo, S.: Research on the function and mechanism of social organizations participating in medical disputes. Chin. Health Serv. Manage. **035**(003), 182–184,189 (2018)
12. Wu, M.: The administration of management in chinese universities and the thinking of de-administration—research on the mechanism of university governance abroad. Adv. Soc. Sci. **06**(11), 1341–1346 (2017)

Medium Access Control Protocol Based on Continuous Listening and Hierarchical Scheduling

Tongfei Shang[1(✉)], Liqiong Yang[2], and Jing Liu[1]

[1] College of Information and Communication, National University of Defense Technology, Xi'an, Shaanxi, China
[2] Provincial Military Region, Xi'an, Shaanxi, China

Abstract. Aiming at the problem that excessive energy consumption of nodes in each layer protocol design of wireless sensor network will affect the transmission speed of data packets, this paper proposes a medium access control protocol based on continuous listening and hierarchical scheduling, and the upper network is obtained through the interlayer interface Topological structure and routing information to assist in obtaining the channel contention scheme of this layer, and to obtain and maintain a unified hierarchical time schedule based on the clock synchronization mechanism. The data sent and received in the regular active phase can be transmitted to the upper level step by step without stagnation, and when the network traffic exceeds the threshold, the continuous listening mechanism is turned on to offload the sending and receiving of data packets and reduce the probability of receiving conflicts. Simulation analysis shows that this method can dynamically adjust the local time schedule to achieve the goal of reducing energy consumption as much as possible while ensuring real-time transmission.

Keywords: Continued listening · Hierarchical scheduling · Medium access control protocol

1 Introduction

In order to effectively reduce the energy consumption of a node, a common method is to put the node into a dormant state as much as possible. When the node needs to send, forward or receive data packets, it will wake up and return to the full-speed working state. The node cannot perform any data receiving and sending work in the dormant state, so it brings a contradiction problem, that is, to make the data transmission in the network as fast as possible, the delay is as small as possible, and the real-time performance is as high as possible. Wake-up state, and to make the life of the network as long as possible, it is necessary to put the node into a sleep state as much as possible. Moreover, even if the nodes are awake, that is, data packets can be sent and received at any time, if the channel allocation mechanism is unreasonable, it will still affect the transmission performance of the network [1–3].

© The Author(s), under exclusive license to Springer Nature Switzerland AG 2021
J. Abawajy et al. (Eds.): ATCI 2021, AISC 1398, pp. 308–314, 2021.
https://doi.org/10.1007/978-3-030-79200-8_45

2 CLHS-MAC Protocol Framework

In most wireless sensor networks, the analog data collected by most nodes in the network cannot be directly sent to the concentrator, because the radio transmission distance of each node is limited, and the temperature and humidity will also be received., Electromagnetic interference and other environmental factors. Therefore, the data in the network needs to be relayed through other nodes to forward the data layer by layer, and finally send the data to the concentrator, and the concentrator will upload the data to the monitor or the Internet or the cloud [4]. This layer-by-layer forwarding method is the multi-hop method commonly used in wireless sensor networks. At the network layer, many routing schemes will choose to send the collected values of many nodes to the cluster head node, and the cluster head node will perform a series of processing such as data fusion and then send the data to its superior node. This process is similar to a tree structure. The nodes sending the collected values can be regarded as child nodes, and the first-level cluster head nodes are their parent nodes, but at the same time these parent nodes (or some other nodes at the same level)) Is also the child node of the upper cluster head, and so on, forming a tree structure [5, 6].

The overall structure of the protocol is shown in the Fig. 1 below. In this structure, the depth of the tree is 3, where the highest level of aggregation node (or concentrator) is at the highest level, and the layer depth is set to n, and the remaining nodes are deep (from top to Below) are n + 1, n + 2, n + 3 in sequence. The data is forwarded layer by layer along the established route indicated in the figure to the sink node. The tree is constructed and maintained by the routing protocol in the network layer, so the relevant details of the tree are beyond the scope of this article.

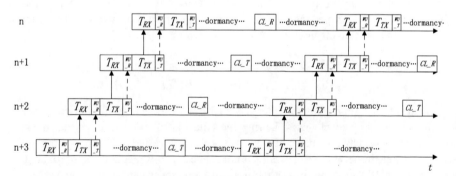

Fig. 1. CLHS-MAC protocol framework

In CLHS-MAC, when the concentrator is positioned at a depth of 0, the node with a depth of n will be forwarded (Excursion) by an amount nE based on the time schedule of the concentrator, and the size of E is adjacent The calculation formula for the relative drift of two-level nodes is:

$$E = T_{RX} + T_{WU} \tag{1}$$

In the formula: TWU represents the duration of a wake-up time slot.

When network traffic is large, this mechanism can save a lot of energy for nodes. Since the length of the time slot is fixed, the time length expression of the frame is:

$$
\begin{aligned}
T_{frame} &= T_{wakeup} + T_{shutdown} \\
&= T_{RX} + T_{TX} + T_{WU\_R} + T_{WU\_T} + T_{CL} + T_{sleep} \\
&= 2\left(T_{RX} + T_{WU\_R}\right) + T_{CL} + T_{sleep}
\end{aligned}
\tag{2}
$$

The Tsleep in the above formula is divided into two sections, which are located on both sides of the CL time slot. For the calculation needs in the following text, they are respectively recorded as Tsleep1 and Tsleep2. From the above formula, the division can be continued:

$$
\begin{aligned}
T_{frame} &= 2\left(T_{RX} + T_{WU\_R}\right) + T_{CL} + T_{sleep1} + T_{sleep2} \\
&= 2\left(T_{RX} + T_{WU\_R}\right) + T_{CL} + T_{TX} + T_{WU\_T} + T_{sleep2} \\
&= 3\left(T_{RX} + T_{WU\_R}\right) + T_{CL} + T_{sleep2}
\end{aligned}
\tag{3}
$$

The time synchronization mechanism is not service-oriented in the media access control layer. It is also applied to other protocol layers. Secondly, the typical length of Tframe is usually at the level of 10ms, while the synchronization accuracy of the existing clock synchronization mechanism can reach less than 8 μs.

3 Dynamic Selection of Continued Listening Stage

The CLHS-MAC protocol's continued listening duration TCL is dynamically adjusted. The design of each layer protocol of the WSN network is to reduce energy consumption as the main goal. Therefore, energy consumption is the main consideration. The amount of energy consumption change:

$$
\begin{aligned}
\Delta E &= P\left(T_{CL\_Success}\right) \cdot E_1 - P\left(T_{CL\_Fail}\right) \cdot E_2 \\
&= P\left(T_{CL\_Success}\right) \cdot E_1 - \left(1 - P\left(T_{CL\_Success}\right)\right) \cdot E_2
\end{aligned}
\tag{4}
$$

In Eq. (4), E1 represents the energy saved due to the successful adoption of the continuous listening mechanism. During the data transmission time period, because fewer nodes have data transmission requests, the energy wasted on receiving conflicts is reduced. When the network When traffic is not congested, this reduction is approximately equal to the product of one-fifth of the transmission power consumption generated by the average value of cluster members in the network during the TTX time period and the ratio of the length of the two periods, so:

$$
\begin{aligned}
E_1 &= \frac{1}{5} \cdot T_{TX} \cdot P_{TX} \cdot \bar{x} \cdot \frac{T_{CL}}{T_{left}} - P_{RX} \cdot T_{CL} \\
&= \left(\frac{T_{TX} P_{TX} \cdot \bar{x}}{5 T_{left}} - P_{RX}\right) \cdot T_{CL}
\end{aligned}
\tag{5}
$$

E2 represents the energy wasted due to the continuous listening mechanism. This part of the energy consumption is mainly because the data transmission request from the child node is not received during the continuous listening period, so this part of the energy is wasted [7, 8]. The expression of E2 is as follows:

$$E_2 = T_{CL} \cdot P_{RX} \tag{6}$$

P(TCL_Success) and P(TCL_Fail) correspond to the probability of occurrence of these two events respectively. Suppose the node traffic is β, $g(\beta)$ which is a function of the traffic. The expression for the probability that the node can receive the request from the child node during the continuous listening period is:

$$P(T_{CL\_Success}) = \frac{T_{CL}}{T_{frame}} \cdot g(\beta) \tag{7}$$

Therefore, formula (4) can be further expressed as:

$$\Delta E = \frac{T_{CL}}{T_{frame}} \cdot g(\beta) \cdot \left(\frac{T_{TX} P_{TX} \cdot \bar{x}}{5 T_{left}} - P_{RX} \right) \cdot T_{CL} - \left(1 - \frac{T_{CL}}{T_{frame}} \cdot g(\beta) \right) T_{CL} \cdot P_{RX} \tag{8}$$

T_{left} represents the remaining time left for continued listening in this frame.

Obviously, ΔE is a function of T_{CL} opening downward, and the extreme point of ΔE is:

$$\frac{d}{dT_{CL}} \Delta E = \frac{2g(\beta) T_{TX} P_{TX} \bar{x}}{5 T_{left} T_{frame}} T_{CL} - P_{RX} \tag{9}$$

Therefore, the extreme points of T_{CL} are:

$$T_{CL} = \frac{5 T_{left} T_{frame} P_{RX}}{2g(\beta) T_{TX} P_{TX} \bar{x}} \tag{10}$$

To reduce the energy consumption after the use of continuous listening, there are:

$$\Delta E > 0 \Rightarrow T_{CL} > \frac{5 T_{left} T_{frame} P_{RX}}{g(\beta) T_{TX} P_{TX} \bar{x}} \tag{11}$$

In summary, the dynamic expression of the piecewise function T_{CL} is:

$$T_{CL} = \begin{cases} 0 & , \text{ if } \dfrac{5 T_{left} T_{frame} P_{RX}}{g(\beta) T_{TX} P_{TX} \bar{x}} \leq 0 \\[3ex] \dfrac{5 T_{left} T_{frame} P_{RX}}{2g(\beta) T_{TX} P_{TX} \bar{x}} & , \text{ if } \dfrac{5 T_{left} T_{frame} P_{RX}}{2g(\beta) T_{TX} P_{TX} \bar{x}} > 0 \\[3ex] T_{left} & , \text{ if } \dfrac{5 T_{left} T_{frame} P_{RX}}{2g(\beta) T_{TX} P_{TX} \bar{x}} > T_{left} \end{cases} \tag{12}$$

Construct a random network topology. There are 50 nodes in the network randomly distributed in a square area of 800 m*800 m. According to the actual situation of the substation, the concentrator (node 0) is placed at one end of the entire area. The node obtains the tree link structure as shown in the following Fig. 2 according to the routing

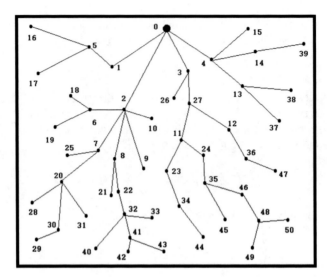

Fig. 2. Network topology structure diagram

algorithm of the network layer. After the node collects the data, it will follow the line identification The path is reported and forwarded layer by layer, and finally, all the messages collected and forwarded by all nodes are collected at the concentrator 0 [9, 10]. It is assumed that the storage capacity and energy of all nodes in the network are infinite, so there is no data overflow and topology changes.

After the topological structure is constructed, the next step is to analyze the different experimental results produced by different protocols in the same network, and compare and analyze them.

4   Simulation Analysis

Assuming that the probability of each node generating data that needs to be sent at each moment is 20% ~ 30%, in simulation experiment 2, the change of network traffic starts from 0.25 pkts/s, gradually increases to 5 pkts/s and then ends. The time for all data packets to be sent is within one frame after the start of the time. The probability is that the number of experiments is 10 in total, and the experimental results are compared with the average of the 10 experiments.

The change of the average energy consumption on nodes under the five MAC protocols as the traffic increases. Observed from the general trend in the Fig. 3, the energy consumption of all nodes under the control of the protocol increases with the increase of traffic in the network, but the correspondence between these two variables under each protocol is very different. When the traffic in the network is low, the average power consumption of nodes under the five protocols is roughly the same, but there are some subtle differences. The reason for the difference is that T-MAC reduces the idle listening time due to its own duty cycle adjustment mechanism. The duty cycle of nodes under T-MAC can be reduced to as low as 4%, although CLHS-MAC can also Adaptive adjustment can

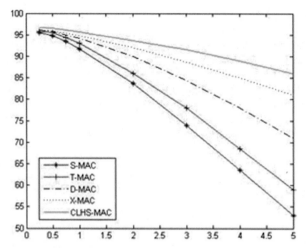

Fig. 3. The relationship between the success rate of data packet reception and the change of traffic

even make the duty cycle of the node lower than 4%, but the duty cycle is relatively large in the initial adjustment stage, which affects the overall energy consumption. Among the five protocols, D-MAC has the highest node energy consumption under small traffic.

5   Conclusion

CLHS-MAC should flexibly change the duty cycle of the time schedule through self-adaptation in a wide range, to achieve a low flow rate and a low duty cycle, and save energy through sleep mode. Large flow is a high duty cycle, and the sub-nodes compete for channels by dispersing Time to reduce channel competition, thereby avoiding energy waste caused by receiving conflicts, thereby saving node energy consumption. At the same time, through the hierarchical scheduling method, the data packets transmitted in the regular time period avoid the sleep delay and improve the efficiency of data transmission on the channel.

References

1. Berkowitz, B.: Technology catches up to runners. Washington Post, 4–20(1) (2001)
2. Mainwaring, A., Szewczyk, R., et al.: Wireless sensor networks for habitat monitoring. Proceedings of ACM WSNA 2002 (2012)
3. Szewczyk, R., Osterweil, E., Polastre, J., et al.: Application driven systems research: habitat monitoring with sensor networks. Commun. ACM Spec. Issue Sens. Netw. **47**(6), 34–40 (2014)
4. Basha, E.A., Ravela, S., Rus, D.: Model-based monitoring for early warning flood detection. In: Proceedings of ACM SenSys 2008, pp. 295–308 (2008)
5. Werner-allen, G., Lorincz, K., Johnson, J., et al.: Fidelity and yield in a volcano monitoring sensor network. In: Proceedings of the 7th USENIX Symposium on Operating Systems Design and Implementation, pp. 381–396 (2006)

6. Werner-allen, G., Lorincz, K., Ruiz, M., et al.: Deploying a wireless sensor network on an active volcano. Internet Comput. **10**(2), 18–25 (2016)
7. He, T., Krishnamurthy, S., Stankovic, J.A., et al.: VigilNet: an integrated sensor network system for energy-efficient surveillance. ACM Trans. Sens. Netw. **2**(1), 1–38 (2016)
8. Ledeczi, A., Nadas, A., Volgyesi, P., et al.: Countersniper system for urban warfare. ACM Trans. Sens. Netw. **1**(2), 153–177 (2015)
9. Hart, J.K., Martinez, K.: Environmental sensor networks: a revolution in the earth system science. Earth Sci. Rev. **78**(3–4), 177–191 (2006)
10. Delin, K.A.: The sensor web: a macro-instrument for coordinated sensing. Sensors **2**(7), 270–285 (2012)

Analysis of Network Public Opinion Based on BiLSTM and Self-attention Fusion Mechanism

Jianming Sun and Yibo Sun[✉]

School of Computer and Information Engineering, Harbin University of Commerce, Harbin, Heilongjiang, China

Abstract. With the advent of the digital era, it is of great significance to control the emotional tendency of network to maintain social stability. LSTM network can only one-way through the information, but the information cannot be encoded from the back to the front. For the analysis of network public opinion, the sequence value will also have a certain influence on the emotional tendency of the whole sentence. Therefore, this paper uses bidirectional memory network to extract the semantic features of microblog text. Because the general attention mechanism needs more parameter dependence, and ignores the dependence within the text. For increasing the accuracy of emotional tendency control, the self-attention mechanism model is used. Compared with LSTM, BiLSTM and BiLSTM attention model, the network public opinion analysis model based on BiLSTM and self attention mechanism can achieve good results.

Keywords: Bidirectional long short-term memory · Self attention mechanism · Internet public opinion · Emotional tendency

1 Introduction

Many netizens can not only obtain relevant information in real time on the Internet, but also interact with each other and express their views on social focus issues and emergencies, thus generating a lot of public opinion information. Public opinion is mainly based on text information. The analysis of public opinion is to analyze the tendency of netizens to express their opinions around public opinion events [1]. The analysis of emotional tendency of network public opinion has important research value and far-reaching significance.

2 Related Research

Internet public opinion analysis is mainly aimed at specific events or specific people to analyze the emotional tendency of netizens' opinions or opinions [2]. The development of deep learning and hardware computing power, the sentiment analysis method which based on deep learning has gradually become the mainstream sentiment analysis

J. Abawajy et al. (Eds.): ATCI 2021, AISC 1398, pp. 315–321, 2021.
https://doi.org/10.1007/978-3-030-79200-8_46

method. A sentiment analysis method of deep learning does not need a lot of manual annotation, and has strong semantic expression ability. Word vector embedding technology largely determines the quality of neural network model in deep learning. Such as one hot representation, word2vec word vector representation and negative sampling, more and more scholars have applied deep learning to the task of sentiment orientation analysis. In 1990, the first mock exam was proposed by Jeffrey L Elman, which is a recurrent neural network (recurrent neural networks, RNN) model, which is more useful in dealing with time-series data [3]. Because LSTM can't encode from back to front, and the post value of the sequence will also have a certain impact on the judgment of emotional tendency, Wu Peng proposed a bi-directional short-term memory (BiLSTM) micro blog emotional analysis model by combining word vector with emotional semantics, which can effectively identify three kinds of emotions [4]. Compared with SVM and LSTM, it has better effect of emotion recognition. Although the above methods have a good effect on general sentiment classification, the traditional deep learning model is trained by assigning the same weight to all features, so it cannot pay attention to the features that the sentiment classifications of sparse articles.

In 2014, the Google mind team first proposed the attention mechanism [5]. It can focus on the local and important areas of the image, and extract the key features of the image which distinguish it from other images, which can improve the accuracy of image recognition. In 2016, bahdanau applied attention mechanism to the field of natural language processing (NLP) for the accuracy of translation was significantly improved compared with the traditional network model [6].By learning the semantic features from the positive and negative directions of sentences, we can capture the useful semantic information more efficiently and apply it to the analysis of emotional tendency of short articles. Based on this, this paper proposes an analysis model of online public opinion sentiment tendency combination of BiLSTM and self-attention mechanism, which is dedicated to providing a more accurate method model for the analysis of online public opinion sentiment tendency.

3   Related Models

3.1   LSTM

Compared with RNN, LSTM could effectively resolve the problem on gradient disappearance and gradient explosion and avoid long-term dependence due to the threshold mechanism and the existence of cells. The conformation of LSTM is indicated in Fig. 1.

The LSTM consists of four neural network layers and three gate units. The three gate units are forgetting-gate, memory-gate, output-gate. The overall computation process are as follows:

The forgetting-gate is calculated to choose the forgotten information. By inputting hidden layer state h to the forgetting gate h_{t-1}. Input word X_t for the time of current. The value f of forgetting-gate is gained f_t. Calculation formula is shown in formula (1). σ is the activation function. In this paper, we use sigmoid activation function, which can map a real value to (0, 1) interval. W_f is a weight matrix of a output of update, b_f is a weight matrix of a output of update which is the offset of updating.

Compute memory-gate to select information which will be remembered. By inputting a hidden layer state h_{t-1} to memory-gate and input word X_t of current time. Get a value of memory-gate i_t and a temporary cell state $\widetilde{C}_t$ where W_i, W_c are two weight matrices of updated output, b_i, b_c are the weight matrix of update output. Tanh is a hyperbolic tangent function, which can compress the value between - 1 and 1. The tanh function is expressed as follows. Sigmoid-layer calls input-gate to determine which data needs to update. And there will be a new candidate value vector is created through the tanh-layer and appended to the current state. These calculation formulas are shown in formula (2), (3), (4).

The forgetting-gate and memory gate are combined to calculate the cell state. By forgetting the value f_t of the gate and last cell state C_{t-1} multiplication, memory gate value i_t and temporary cell state $\widetilde{C}_t$ multiplication, and the addition of the two are a cell state C_t. Calculation formula is shown in formula (5).

Determines a output of output-gate and the current hidden layer state. The output-gate value O_t and a hidden layer state h_t are gained by inputting hidden-layer state h_{t-1}, the current time input X_t and the cell state C_t. Among them, W_o is a weight matrix of update output and b_o is offset value of update output. The calculation formula is shown in (6) (7).

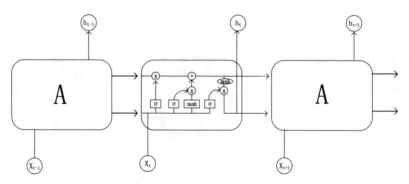

Fig. 1. LSTM conformation diagram

$$f_t = \sigma(W_f \bullet [h_{t-1}, X_t] + b_f) \tag{1}$$

$$i_t = \sigma(W_i \bullet [h_{t-1}, X_t] + b_i) \tag{2}$$

$$\widetilde{C}_t = \tanh(W_c \bullet [h_{t-1}, X_t] + b_c) \tag{3}$$

$$\tanh(x) = \frac{e^x - e^{-x}}{e^x + e^{-x}} \tag{4}$$

$$C_t = f_t * C_{t-1} + i_t * \widetilde{C}_t \tag{5}$$

$$O_t = \sigma(W_o \bullet [h_{t-1}, X_t] + b_o) \tag{6}$$

$$h_t = O_t * \tanh(C_t) \tag{7}$$

3.2 BiLSTM

Since LSTM cannot encode information from back to front, not only the pre-order value will affect the final polarity judgment, the post-order value will also affect the general text. Therefore, this paper uses BiLSTM to extract semantic features of microblog text. The status of each time in BiLSTM is shown below, and outputs are determined by the LSTM in both directions at same time.

3.3 Attention Mechanism

Attention mechanism was proposed by Google team [5]. Because the general semantics of a sentence can be expressed by one or several words, we can simulate the characteristics of human brain through attention mechanism, browse all the contents quickly and focus on the important contents to obtain the key information for related tasks. Attention mechanism is essentially composed of many query and key value pairs. When the query is an element in a given target, by computing correlation or similarity between the query and each key. The weight of the value corresponding to each key could be gained. Finally, the final attention value is obtained by weighted sum of value.

In the first stage, the correlation between a query and a key is calculated to get the corresponding weight. This paper adopts the method of vector dot product. In the second stage, the SoftMax function is introduced to convert the scores of the first stage. We can normalize a score to a distribution, then the sum of the weights of all elements is 1. The internal mechanism of SoftMax can make the weight of important elements more prominent. The result is a_i is the weight coefficient of value, and then weighted sum to get the final attention value.

3.4 Self-attention Mechanism

Self-attention is a attention model when $Q = K = V$. Self-attention mechanism is fundamentally different from traditional attention mechanism. The result of traditional attention is the dependency between each word in the source side and each word in the target side, but it ignores the dependency between words in the source side or target side sentences. But self-attention is different. This mechanism can not only get dependency between the source and target, but also get dependency between the source and target.

Self-attention can also capture certain syntactic or semantic features between words in the same sentence. Therefore, after applying self-attention, it is easier to capture a remote interdependence features in target sentence.

In addition, a self-attention only needs to train its own information to update parameters, so self-attention can help increase the parallelism of computing as well.

4   BiLSTM Affective Tendency Analysis Model Based on Self-attention Mechanism

In order to improve the accuracy of sentiment analysis, a BiLSTM sentiment trend analysis model on attention is proposed. Input layer.

Transforming the text into a mathematical structure that can be processed by the model is the premise of completing the corresponding text task. The most common method is to use distributed word vector to represent a word. Word2vec model is an open source tool of Google in 2013 [7]. By training every sentence in the data set, and sliding a fixed size window on it, we try to predict the center word of the window, so as to give other words. Using the loss function and other optimization programs, the model can generate a unique word vector for each word.

① Word embedding layer

Word embedding layer, there are two ways to deal with word embedding: static embedding and dynamic embedding. Static embedding means that word vectors are trained in advance; dynamic embedding means that they are trained together with neural networks. This paper uses static embedding Taking the word as the smallest unit, the text features of the sentence where the word is located are obtained through BiLSTM.

② BiLSTM. layer

Using BiLSTM. to calculate the embedded word vector, we can get a higher level sentence vector. Finally, the output of BiLSTM. layer is: assuming that the forward output after LSTM is $\overrightarrow{h_t}$, the reverse output is $\overleftarrow{h_t}$, then the t-th vector after BiLSTM is $h_t = [\overrightarrow{h_t} \oplus \overleftarrow{h_t}]$.

③ Self - attention layer

Q, K, V all come from the same input, namely Q = K = V, the model becomes self-attention model. d_k is the adjustment factor, which is generally the dimension of the input vectors.

④ Output layer

It mainly calculates the polarity of emotions. In the calculation process, the steepest descent algorithm is used as optimization function. Use backpropagation mechanism to optimize the parameters of this model.

5   Experiment

5.1   Experimental Data

On October 13, a graduate student in the College of chemical engineering, Dalian University of Technology who hanged himself in the laboratory in the early morning, which caused the regret and lament of the broad masses of people in the society. At the same time, many netizens expressed their views on the education system, and generated a large amount of public opinion information on the Internet. Based on this, this paper collects the relevant micro blog comments after the event broke out, cleans them, and then manually annotates them to form 3491 positive texts and 2678 negative texts. The proportion of training set and test set is 8:2.

5.2 Evaluation Criteria

The evaluation criteria of this experiment are commonly used indicators, including accuracy (Acc), accuracy (Pre), recall (Rec) and F1 score (F1). The accuracy rate evaluation model's ability to correctly classify emotional tendencies; the accuracy rate can evaluate the accuracy rate; the recall rate evaluation data set detects the proportion of all correct samples; the F1 index is a comprehensive evaluation index, which is the harmony of the pre and Rec values.

5.3 Comparative Experiment

To verify the superiority of the proposed model, BiLSTM self-attention model is compared with the LSTM model, BiLSTM model, BiLSTM attention model and BiLSTM self-attention model under the same conditions. The results of comparative experiments are shown in Table 1.

Table 1. Data of comparative experiment

Model	Acc	Pre	Rec	F1
LSTM	76.56%	77.78%	75.39%	76.56%
BiLSTM	78.91%	77.05%	78.34%	77.68%
BiLSTM-Attention	80.82%	81.16%	82.35%	81.75%
BiLSTM-Self-Attention	**83.59%**	**81.67%**	**83.05%**	**82.35%**

5.4 Analysis of Comparative Experiment Results

① The comparative experiments between LSTM and BiLSTM show that BiLSTM is usually slightly higher than LSTM in accuracy, precision, recall and F1 value because it can recognize semantic information in two directions. In the contrast experiment, the accuracy of BiLSTM is 2.35% higher than that of LSTM.

② Through the comparison experiment of BiLSTM and BiLSTM attention, it is shown that adding attention mechanism can effectively help the model to accurately identify the key information in the text, thereby increasing accuracy of this model.

③ By comparing the results of BiLSTM attention and BiLSTM self-attention, attention mechanism, self-attention mechanism can capture long-distance interdependence features in sentences and is more accurate than attention mechanism. In the contrast experiment, compared with BiLSTM, BiLSTM self-attention improves the accuracy by 2.77%, accuracy by 0.51%, recall by 1.7% and F1 by 0.6%.

6   Conclusion

This paper proposes a bi-directional long short-term network sentiment tendency analysis model based on self-attention mechanism. The features of the text are obtained by BiLSTM, and the features of the sentence and the whole sentence are weighted by self-attention mechanism. Through analysis to the self-hanging incident in the graduate Laboratory of Dalian University of technology, which caused an uproar on microblog, LSTM is compared with other methods. The results show that BiLSTM self-attention model is effective in the analysis of online public opinion sentiment tendency.

In this paper, we don't label emotion multi classification when we analyze emotion tendency. In the future research, researchers will pay more attention to the multi classification of emotional tendencies, and further study in the larger data and more complex network public opinion environment.

References

1. Chen, J., Liu, Y.P., Deng, S.L.: Research on the influencing factors of the spreading effect of rumor-refuting information on government microblog. Inf. Sci. **36** (317 01), 91–95 (2018)
2. Yu-Liang, Z.H.: Network opinion risk evaluation index system based on the cycle of emergency. Inf. Sci. **07**, 73–76 (2012)
3. Elman, J.L.: Finding structure in time. Cogn. Sci. **14**(2), 179–211 (1990)
4. Wu, P., Ying, Y., Shen, S.: Research on the classification of Netizens' negative sentiment based on the two-way long short-term memory model. J. Inf. **37**(08), 845–853 (2018)
5. Vaswani, A., et al.: Attention is all you need. Adv. Neural Inf. Proc. Syst. **22**(7), 139–147 (2017)
6. Bahdanau, D., Cho, K., Bengio, Y.: Neural machine translation by jointly learning to align and translate. Comput. Sci. **5**(6), 69–78 (2014)
7. Mikolov, T., Chen, K., Corrado, G., et al.: Efficient estimation of word representations in vector space. Comput. Sci. **21**(9), 226–234 (2013)

Prediction and Analysis of Saturated Electricity Consumption Based on Logistic - BP Neural Network

Xiang Cui[✉], Zhenghao Jia, Ping Xue, Qidan Xu, Shuang Li, and Liankai Zhou

Economic Technology Research Institute of State Grid Xinjiang Electric Power Co., Ltd., Urumqi, Xinjiang, China

Abstract. In this paper, the grey relational degree method is used to analyze the various macroeconomic and social factors that affect electricity consumption. The results show that social labor productivity and resident population have the greatest impact on electricity consumption. Logistic - BP neural network is used to predict saturated electricity consumption of Beijing, and the traditional method of using influencing factors to predict is abandoned. The prediction results show that the growth rate of electricity consumption begins to be less than 2% in 2023, and the growth rate is decreasing year by year, which can be considered that the power consumption of Beijing will enter the low-speed saturation stage in 2023. The corresponding development suggestions are put forward at the same time.

Keywords: Logistic equation · BP neural network · Saturated electricity consumption · Prediction model

1 Introduction

As a basic industry of the national economy, the electric power industry is an important guarantee to improve people's living standards and can provide strong support for the long-term development of the economy. In order to ensure the balance of power supply and demand and the normal development of society and economy, it is necessary to make scientific analysis and prediction of regional electricity consumption. The prediction of electricity consumption includes traditional prediction methods and modern prediction methods. The traditional prediction methods mainly include regression analysis, time series analysis, growth rate method, etc. [1, 3]. Modern prediction methods mainly include: grey prediction method, Logistic prediction method, neural network prediction method, etc. [4, 9]. The prediction of saturated electricity consumption is a medium and long term prediction, which is affected by various factors. The above single prediction method is difficult to obtain relatively satisfactory prediction results. In this paper, Logistic - BP neural network is used to predict saturated electricity consumption of Beijing, and the traditional method of using influencing factors to predict is abandoned. On the one hand, Logistic equation can fully reflect the long-term development law of electricity consumption; on the other hand, it can achieve better prediction accuracy by using the powerful learning ability of neural network.

J. Abawajy et al. (Eds.): ATCI 2021, AISC 1398, pp. 322–327, 2021.
https://doi.org/10.1007/978-3-030-79200-8_47

2 Analysis of Influencing Factors of Saturated Electricity Consumption Based on Grey Relational Degree

According to the development law of general things and the experience of developed countries, the saturated electricity consumption of cities generally appears in three stages, the first is the slow development stage, the second is the high-speed growth stage, and the last is the low-speed saturation stage. Only when the growth rate of electricity consumption is less than 2%, the city's electricity demand can be judged to have reached the saturation stage [10] (Table 1).

Table 1. Grey correlation analysis results

Influencing factor	Grey correlation
Gross regional product	0.73
Residents' consumption level	0.82
Social labor productivity	0.91
Permanent resident population	0.90
Per capita disposable income	0.82
General public budget revenue	0.63
Investment in fixed assets of the whole society	0.79
Total value of imports and exports of goods	0.66
Gross industrial output value above designated size	0.75
Total retail sales of consumer goods	0.75

Since saturated electricity consumption is formed in a long time, it is mainly affected by macroeconomic and social development. This paper collected the macro economic and social development data, then the grey relational degree method is used to analyze each factor to determine the factor with higher degree of synchronous change with electricity consumption. Social labor productivity and resident population have the greatest influence on electricity consumption through grey correlation analysis.

After analyzing the changes of social labor productivity and permanent resident population, it is found that there is a difference in the growth of the two main influencing factors, which makes it difficult to judge the future growth of electricity consumption according to social labor productivity and permanent resident population. In the future, the growth rate may be less than 2%, entering the low-speed saturation stage.

3 Saturated Electricity Consumption Prediction Model

Belgian mathematician Pierre Francois Verhulst developed the Logistic equation, a mathematical description of the S-shaped curve, which fully represents the three stages of electricity consumption. Therefore, Logistic prediction method can be applied to electricity consumption prediction.

The Logistic equation is defined as follows:

$$x_t = \frac{1}{c + ae^{bt}} \qquad (1)$$

Wherein, t usually represents the time variable and a, b, c are the parameter of the model. The parameters can be estimated by Yule algorithm.

The key of Logistic prediction method is how to estimate the parameters of the equation. Because there are many estimation methods, the accuracy of the prediction results is different. In order to improve the prediction accuracy of Logistic equation, BP neural network is introduced in this paper. BP neural network is a multi-layer feedforward neural network with strong learning ability. It can realize nonlinear mapping from input to output, and the error can be as small as possible through continuous training. In this paper, the influence factors are not used to predict the electricity consumption and the interference of other factors is excluded. Instead, the Yule algorithm is used to calculate the parameters of Logistic, so that the Logistic model can be used to predict the saturated electricity consumption. Then, the BP neural network was used to take the logistic prediction result as the input and the electricity consumption as the output, and the prediction accuracy was improved through the strong learning ability of the neural network.

4   Prediction Result Analysis and Development Suggestion

4.1   Prediction Results of Saturated Electricity Consumption

Beijing is the political center, cultural center, and scientific and technological innovation center of China. By the end of 2019, the electricity consumption of the whole society in Beijing was 116639.64 million kWh, an increase of 191.64% compared with 2001, with an average annual growth rate of 10.65%. With the adjustment of economic structure and upgrading of industrial structure, the proportion of electricity consumption in the primary industry decreases year by year, so does the proportion in the secondary industry, while the proportion of electricity consumption in the tertiary industry increases year by year. With the continuous improvement of residents' living standards, the proportion of urban and rural residents' electricity consumption also increases year by year.

In this paper, the data of electricity consumption in Beijing from 2001 to 2019 are analyzed, and the prediction model proposed in this paper is used to predict the saturated electricity consumption in Beijing. The sample size of BP neural network test is 15, the hidden middle layer is 6, and the learning rate is 0.035. Meanwhile, the grey prediction method and logistic prediction method are used to make the prediction, and the prediction accuracy of different methods is judged according to the average absolute percentage error. The prediction results are shown in the following table.

The average absolute percentage errors of the three methods are 4.16%, 6.24% and 1.59%, respectively. The results show that the error of the proposed method is smaller. Therefore, this method is used to predict the electricity consumption of Beijing in 2020–2030 and calculate the growth rate. It is found that the growth rate of electricity consumption in 2023 starts to be less than 2%, and the growth rate is decreasing year by year. It can be considered that the electricity consumption of Beijing will enter the low-speed saturation stage in 2023 (Tables 2 and 3).

Table 2. Prediction results from 2001 to 2019

Year	Electricity consumption	Grey prediction	Logistic prediction	Logistics - bp neural network prediction
2001	39994	39994	36200	40405
2002	43996	48090	39670	44001
2003	46761	50925	43347	47807
2004	51318	53927	47221	51807
2005	57054	57105	51276	55977
2006	61157	60472	55493	60293
2007	66701	64037	59850	64722
2008	68972	67811	64318	69230
2009	73915	71809	68868	73779
2010	80990	76042	73467	78331
2011	82171	80525	78082	82848
2012	87428	85271	82678	87295
2013	91311	90298	87221	91637
2014	93705	95621	91679	95843
2015	95272	101258	96023	99888
2016	102027	107228	100225	103750
2017	106689	113549	104263	107413
2018	114238	120243	108118	110864
2019	116640	127331	111775	114096

4.2 Development Suggestions

According to the prediction results, the growth rate of Beijing's electricity consumption will gradually decrease, and may even show negative growth. Therefore, the construction of power grid should be kept moderately ahead of time in order to meet the social demand for electricity. The scale of the current power grid should be maintained and the coordinated development between regional economy and power grid should be fully considered. Power grid construction is mainly to optimize the grid structure, update and transform backward lines, improve the reliability of the power grid, and then timely meet local electricity demand, strengthen the analysis of power grid operation, and scientifically and rationally arrange the operation mode of the power grid in summer. Power grid construction should be carried out reasonably through the above methods.

In the absence of rapid growth in electricity consumption, the energy structure should be optimized. On the one hand, the backward capacity of thermal power should be eliminated continuously, the minimum power of the power plant should be reduced and the start-up time should be shortened. On the other hand, we will vigorously develop photovoltaic power generation and wind power generation, make full use of green energy in

Table 3. Prediction results for 2020–2030

Year	Electricity consumption	Growth rate
2020	117108	2.64%
2021	119897	2.38%
2022	122469	2.15%
2023	124831	1.93%
2024	126991	1.73%
2025	128959	1.55%
2026	130747	1.39%
2027	132366	1.24%
2028	133828	1.10%
2029	135146	0.98%
2030	136331	0.88%

areas rich in power resources, and stabilize social production and living electricity consumption. In addition, biofuel ethanol, biodiesel and biomass direct combustion power generation can be developed according to local conditions, and systems for collecting, transporting, storing and transforming biomass resources can be improved, so as to explore a new mode of multi-energy complementary development.

5   Conclusion

This paper analyzes the current situation of electricity consumption in Beijing, and the results show that the average annual growth of electricity consumption in Beijing is 10.65%. Then the grey relational degree method is used to analyze the various macroeconomic and social factors that influence the electricity consumption. The results show that the social labor productivity and the permanent population have the greatest influence on the electricity consumption. Then, in order to improve the prediction accuracy of Logistic equation, this paper proposes Logistic - BP neural network prediction method for the prediction of urban saturated electricity consumption, and the prediction accuracy is significantly improved. It is predicted that the growth rate of electricity consumption will start to be less than 2% in 2023, and the growth rate is decreasing year by year. The reasonable construction of the power grid, reducing the ineffective investment, optimizing the energy structure and studying the development of the multi-energy complementary mode are put forward at last.

References

1. Chaoyong, J., Yurong, P., Fuquan, X.: Prediction of annual electricity consumption in Guangzhou based on ARIMA model. J. Bengbu Univ. **8**(05), 72–75 (2019)

2. Yurong, P., Chaoyong, J.: Prediction model of urban residents' electricity consumption in Anhui province based on principal component regression. J. Foshan Univ. Sci. Technol. (Nat. Sci. Edn.) **37**(02), 1–6 (2019)
3. Haoran, J., XiaoCong, L., Yang, L.: Medium and long term electricity forecasting of provincial power companies based on multivariable grey forecasting model. Power Demand Side Manag. **21**(01), 37–41 (2019)
4. Zezhong, Y., Shoujing, Y., Bin, Y.: Power consumption prediction based on grey relational analysis and BP neural network. Value Eng. **37**(35), 30–33 (2018)
5. Chen, Z., Yang, D.: Predicting viscosities of heavy oils and solvent–heavy oil mixtures using artificial neural networks. J. Energy Resourc. Technol. **143**(11) (2021)
6. Mitchell, S.M., Cero, I., Littlefield, A.K., Brown, S.L.: Using categorical data analyses in suicide research: considering clinical utility and practicality. Suicide Life-Threatening Behav. **51**(1) (2021)
7. Wang, M., Wu, X.: Online car-hailing service quality evaluation based on BP neural network. IOP Conf. Ser.: Earth Environ. Sci. **638**(1) (2021)
8. Xu, S., Alturki, R., Rehman, A.U., Tariq, M.U.: BP neural network combination prediction for big data enterprise energy management system. Mob. Netw. Appl. (2021, prepublish)
9. Tang, Q., Liu, X., Ji, X., Li, J., Chen, Y., Lu, B.: Using seabed acoustic imagery to characterize and classify seabed sediment types in the pockmark area of the North Yellow Sea, China. Appl. Acoust. **174** (2021)
10. Xin, X., Yuhui, Z., Ning, Z.: Review on analysis technology and application of urban power saturation consumption. Power Autom. Equip. **06**, 146–152 (2014). (in Chinese)

The Sustainable Development of the Internet Economy Under the Background of Big Data

Da Kuang[1] and Mo Kuang[2(✉)]

[1] School of Marxism, Huazhong University of Science and Technology, Wuhan, Hubei, China
[2] South China Business College, Guangdong University of Foreign Studies, Guangzhou, Guangdong, China
kuangmo86@yeah.net

Abstract. Big data has brought unlimited possibilities to the development of the Internet economy. Big data can realize rapid and accurate analysis of data. By analyzing and processing data, it can formulate more reasonable marketing strategies and reduce transaction costs. Consumers can achieve fast transactions and fast consumption. The good application of big data in the development of the Internet economy has had a relatively large impact on traditional marketing models, and has also promoted the transformation and integration of marketing, making more companies pay more attention to Internet marketing, optimizing and improving big data in the Internet application strategies in economic development, thereby promoting the rapid development of the Internet economy. This article mainly analyzes the sustainable development of the Internet economy in the context of big data.

Keywords: Big data · Internet · Economy · Sustainable development

1 Introduction

In the context of big data, modern technologies such as information, network, and digitization have shown a leapfrog development trend. Various industries are also seeking their own development paths under this background. With the advent of the era of big data, the development of human civilization has also ushered in huge development opportunities and challenges. From the perspective of big data itself, in the process of continuous development and enrichment, it has become a major engine to promote the development of the Internet economy. Its large-capacity data storage and application functions have also brought people new business development opportunities [1]. In modern society, only by learning to use big data can people accurately segment the basic information of the market and customers, dig in-depth what they need, and seize new opportunities to promote their own good development.

2 Characteristics of Big Data

2.1 Huge Amount of Data

The amount of data is extremely large, which is one of the basic characteristics of big data. In the era of rapid development of Internet information technology, science and

J. Abawajy et al. (Eds.): ATCI 2021, AISC 1398, pp. 328–333, 2021.
https://doi.org/10.1007/978-3-030-79200-8_48

technology have led people into the data age. The amount of data generated by big data is very huge, and it continues to grow. People experience the changes brought to life by the data age, and more and more people are aware of the convenience brought by big data functions [1]. The scale and number continue to expand, as shown in Fig. 1. In the past, the amount of data was generally calculated in MB or GB, but now the arrival of the era of big data has caused the amount of data to increase dramatically, requiring TB, PB, EB, or even ZB for calculation.

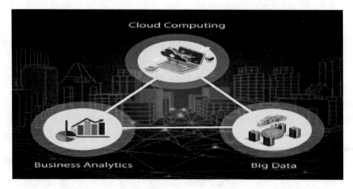

Fig. 1. Internet economic model under big data

2.2 Rich Types of Data

Data in traditional forms can generally be reflected in the form of text. In big data, the types of data are not only in text form. New types of unstructured data such as pictures, videos, graphics, and sounds are gradually derived. More formats and standards [2]. At this stage, the traditional text data format is still used, which can no longer meet the impact of Internet information technology. Therefore, the types of data need to be expanded, so that the data can be transformed from an abstract perspective and reflected in a more specific form of expression, so as to meet the basic requirements of data representation [2].

2.3 The Data Speed is Very Fast

The fast data speed is not only reflected in the fast data acquisition speed, but in the fast data processing speed. In the Internet, user data can be transmitted to the data center at a very fast speed, and the data center can quickly transmit the data to the upstream economy. The current market economy requires that upstream economies can quickly process data to ensure the timeliness of data, which is a factor in ensuring the Internet economy development [3]. Big data requires data processing speed to be associated with economic development. Once the data processing speed is slow and the data processing is not timely, the data will lose its effect.

2.4 The Complexity of the Data is High

The huge amount of data and the continuous enrichment of types have gradually increased the complexity of the data. The data source channel in the traditional form is relatively single, the data complexity is not high, and it is easier to process. In the era of big data, the data source channels are very wide, with relatively strong diversity, which makes the complexity of the current stage continue to increase, and the difficulty of data processing is also higher [3]. A more comprehensive analysis of the data is needed to ensure the rationality of the data analysis, so as to make the data useful.

3   The Characteristics of the Internet Economy Development

3.1 The Efficiency of Completing the Transaction is Relatively High

The transaction speed of the Internet economy is relatively fast, and some economic transactions can be completed in the form of virtual network, which greatly improves transaction efficiency and enhances the conversion efficiency of the Internet economy. For example, at the current stage of e-commerce, people only need to select products in the e-commerce platform, and then can quickly complete the transaction. Different from traditional trading methods, people no longer need to trade in person to complete value conversion [4]. The transaction model under the Internet background can achieve three-party transactions, even multi-party transactions, and a single user can complete transactions with multiple users, which greatly improves transaction efficiency under the Internet background.

3.2 The Cost of Completing the Transaction is Relatively Low

E-commerce is an important achievement of the development of the Internet economy, and the e-commerce platform is the main trading platform for e-commerce. Consumers' consumption patterns in traditional forms mostly need to go to physical stores for consumption. Consumers need to sort out more time to find their favorite products. If they want to buy products from other regions, they also need to spend a certain amount of transportation expenses [5]. This causes consumers to invest more time and material costs. Economic cost. At the same time, merchants need to rent or purchase physical stores, design and arrange stores, and recruit service personnel, which will spend a lot of financial, manpower and material resources. This part of the cost is invisibly added to the product, which increases the transaction cost. In the development of the Internet economy, consumers can directly find their favorite products through the e-commerce platform in e-commerce, which not only realizes shopping around, but also realizes transparent transactions. Consumers can complete consumption anytime and anywhere, saving a lot of time and transaction costs [5]. At the same time, sellers save a lot of operating costs and reduce the transaction cost of goods.

3.3   The Risk of Completing the Transaction is Relatively High

Although the Internet economy has improved transaction efficiency and transaction costs, due to its virtual nature, the risk of completing transactions is relatively high. In the traditional transaction mode, people need to conduct transactions face-to-face. Buyers will pay the fees after inspecting the goods, which reduces the possibility of transaction risks [6]. Both parties to a transaction do not need to conduct face-to-face transactions. They only need to complete a simple check of the items on the platform, and then the fees can be paid. This requires sellers to have high market credit [6]. At the same time, the interoperability of the Internet increases the possibility of users facing intrusion by criminals, and it is easy for users to face relatively large economic and property losses.

4   The Role of the Internet Economy in the Sustainable Development of the Big Data

4.1   The Big Data Application Reduces the Cost in the Development of the Internet Economy

Compared with the traditional form of economic development, the rapid development of the Internet economy has caused tremendous changes in the form of transactions. Buyers and sellers have transformed from the traditional physical store form to the current online store form, which is the current online e-commerce platform [7]. The Internet economy is a huge impact on the traditional real economy, but it has a very large promotion effect on the overall social and economic development. Through the online shop, the speed of buying and selling is accelerated, manpower and material resources are greatly saved, and transaction costs are reduced.

4.2   The Big Data Application Improves Transaction Efficiency

There is a very obvious obstacle, that is do not know each other, which to a certain extent makes buyers have a great deal of authenticity and reliability of products [8]. There are certain obstacles to normal transactions due to mutual mistrust. Through the flexible use of big data, detailed quality analysis of the items that buyers often purchase can be carried out, so as to Increase the probability of successful product transactions.

4.3   The Big Data Broadens Application in the Trading Space

The traditional form of transaction makes us be restricted by the place of life when we buy goods, especially in some places where the economic development is relatively backward, many products cannot be purchased, or we need to go far to buy [8]. It brings great inconvenience to people's normal life. Big data and the Internet economy has just solved this problem. Sellers can use Internet e-commerce platforms to promote their products to all parts of the country. Buyers can also purchase the products they need according to their actual needs. Compared with transaction forms, buyers and sellers have more choices. This is also the role of big data in the Internet economy [9]. The expansion of transaction space has given more impetus to the rapid development.

4.4 Further Promoted the Reorganization of Business

Compared with the traditional business model, Internet economic trade can effectively reduce the time for customers to purchase goods, thereby greatly improving the convenience of various transaction tasks in consumers' daily lives. As consumers' lives continue to improve, China's online economy continues to grow. Internet marketing is a business activity in itself, which contains a lot of information about big data [9]. In other words, consumers can rely on the big data model to carry out various business activities on the Internet to meet many of the psychological needs of consumers. According to comprehensive computer programs, relevant suppliers can use computers or other multimedia devices to create various e-commerce platforms to fully display their products. The expansion of Internet trade has broadened the sales channels of various products, and has also greatly affected the development of China's online economy [10]. As a result, consumers can purchase the products they need overseas without going abroad, and extensive online transactions make it easier for buyers and sellers to conduct daily economic transactions.

4.5 The Application of Big Data Improves the Trading Experience

In the past transaction development, when companies were carrying out product publicity work, they usually carried out untargeted publicity [5]. This form of publicity work did not play a practical role in publicity, and it would cost a lot of money. Manpower, material resources and financial resources cannot stimulate customers' desire to buy after seeing the product, which creates a bad buying and publicity experience for both parties. But through the big data platform, companies can make accurate positioning based on the personal interests of consumers before publicity, and can accurately send some product information according to the actual needs and preferences of customers [10]. This kind of publicity can make advertising content Practical publicity is in place. In addition, because the content of the advertisement is very suitable for the actual needs of users, it can also win the favor of users to a certain extent, can stimulate buyers' desire to buy, and make the transaction between the two parties more smooth (Fig. 2).

Fig. 2. Internet trading system under big data

5   Conclusion

Big data has penetrated into people's lives. In short, the emergence of big data has brought more vitality and vitality to the development of the Internet economy, and has also promoted the development of the Internet economy to a certain extent. In terms of long-term development, integrating big data into the development of the Internet economy is a necessary trend of the times and will also become one of the important driving forces for future economic development.

References

1. Li, Y.Z., Chen, Y., Hu, Y.R.: The application of big data in the development of the Internet economy. Comput. CD Softw. Appl. **8**(08), 87–89 (2014). (in Chinese)
2. Wei, Q.Q.: Application analysis of big data credit investigation in internet finance. Financ. Econ. **11**(08), 11–13 (2015). (in Chinese)
3. Hu, M.T.: The role of big data in the development of the Internet economy. Chinese Market **21**, 192–194 (2019). (in Chinese)
4. Yang, J.J.: Research on the application of big data in the development of the Internet economy. Natl. Circulation Econ. **11**(31), 9–10 (2018). (in Chinese)
5. Li, J.Z.: The application of big data in the development of the Internet economy. Technol. Market **23**(11), 230–234 (2016). (in Chinese)
6. Zhang, K.M.: Discussion on the application of big data in the development of the Internet economy. Managers **10**(18), 272–275 (2017). (in Chinese)
7. Zhao, H.T.: Application and research of computer big data in Internet learning. Modern Econ. Inf. **13**(03), 400–402 (2019). (in Chinese)
8. Li, Z.Y.: Talk about the application of big data in the field of economic management. Commun. World **12**, 120–123 (2019). (in Chinese)
9. Yan, S.H.: Analysis of the interactive efficiency between the development of Internet technology and the current economic development in China. Sci. Technol. Econ. Tribune **9**(21), 216–219 (2018). (in Chinese)
10. Wang, G.Q.: Analysis of the interactive effectiveness of Internet technology and current Chinese economic development. China Bus. Rev. **9**(32), 22–24 (2017). (in Chinese)

Modeling and Simulation of Gansu Province Emergency Logistics System Based on Petri Net

Xueqin Zhang[✉]

Lanzhou Institute of Technology, Lanzhou 730050, Gansu, China

Abstract. With the rapid development of the logistics industry, people have gradually realized the importance of the logistics industry in the national economy. This paper uses the relevant theories of the logistics industry and productivity measurement methods, according to the actual situation in Gansu Province, to study the changing trends of total factor productivity, capital input, and labor input in the logistics industry in Gansu Province, and analyze the development status of the logistics industry in Gansu Province, and analyze separately from transportation, logistics, wholesale and retail, third-party logistics and logistics informatization. This article mainly introduces the modeling and simulation of Gansu Province emergency logistics system based on petri net. This paper uses petri net-based modeling and simulation of Gansu emergency logistics system, and uses petri net for emergency logistics planning and design, and rationally analyzes the feasibility of Gansu emergency logistics system modeling. Discuss and analyze the evolution of logistics concept and the development process of logistics warehousing system. The gate describes the relevant concepts, types and functions of the emergency logistics system in Gansu Province; secondly, it introduces the main facilities and equipment of logistics; discusses the role of logistics in modern logistics. The experimental results of this paper show that the modeling and simulation of the emergency logistics system in Gansu Province based on petri nets increases the response rate of emergency incidents by 26%. The limitations of the design and research for the modeling of emergency logistics system in Gansu Province provide for the application of petri nets. The methods and approaches of a good emergency logistics system are analyzed, discussed and summarized, thereby enriching the academic research results.

Keywords: Petri nets · Emergency events · Emergency logistics · System modeling

1 Introduction

With the rise of information technology and network platforms, in various industries such as military operations, transportation, machinery production, and rapid enterprise information processing, some "complex system models" are often revealed [1, 2]. For example, automated production line systems, aerospace technology, network platform trading systems, large-scale company information management systems, airport airline flight systems, etc. The characteristics of these system models have many similarities,

© The Author(s), under exclusive license to Springer Nature Switzerland AG 2021
J. Abawajy et al. (Eds.): ATCI 2021, AISC 1398, pp. 334–342, 2021.
https://doi.org/10.1007/978-3-030-79200-8_49

that is, discrete events, rather than continuous variables, drive system operations in the platform [3, 4]. It can also be understood that discrete events have a critical impact during the operation of the system, and what it obeys is not ordinary laws of physics, but man-made rules. These rules are usually very complicated, not easy to express, and the process is difficult to simplify [5, 6].

In recent years, due to the promotion of modern logistics industry in Gansu Province, professional logistics companies have undergone transformation, and they have transformed into logistics companies required by the modern logistics industry in terms of corporate management models. Now they have become the main component of the development of logistics industry in Gansu Province [7, 8]. Liang H believes that the logistics system is a typical discrete system. First of all, the rules followed by the logistics warehousing system are human rules, not physical laws. Secondly, the critically influential processes in all operational processes of the logistics warehousing system are not continuous variables, such as inbound operations, outbound operations, and other operations process points are all discrete events [9]. Kumar N thinks that the system can cause new discrete events according to the normal operating mode, which will change the state of the system in the next operation. It is inferred that the economic growth of the logistics industry in Gansu Province is capital-driven, the technology is not yet obvious, and the quality of economic growth is relatively stable [10]. However, there are errors in their experimental process, which leads to inaccurate results.

The innovation of this paper is to propose a petri net-based modeling and simulation of Gansu Province's emergency logistics system. Research on the design of emergency logistics system in Gansu Province, and analyze the effective countermeasures of emergency logistics algorithms. In-depth analysis of the results of emergency logistics in Gansu Province, put forward strategic measures and suggestions suitable for the growth of Gansu Province's logistics industry. Using petri net's non-parametric index method and model, it analyzes the impact of total factor productivity of Gansu's emergency logistics industry on economic growth, and compares and analyzes the contribution of technological progress and technical efficiency to economic growth. The aim is to find a new path suitable for the development of the current indoor path planning system in the complex environment through this research [11].

2 Emergency Incident Analysis Under Petri Network

2.1 Emergency Planning Analysis

The rapid development of transportation infrastructure in Gansu Province has provided a solid foundation and guarantee for the development of the logistics industry. The construction of logistics and transportation infrastructure such as highways, railways, and aviation has developed rapidly, and a relatively modern comprehensive transportation system has been initially formed. The emergency logistics system of Gansu Province through petri nets, for emergency incidents, transportation of emergency materials when emergencies occur, especially large natural disasters such as earthquakes and snow disasters, which often cause roads and bridges in the affected areas to be interrupted, so how to transport emergency supplies to the disaster area is a big problem. When necessary, even measures such as helicopter airdrop of materials and armed police personnel reaching the disaster area on foot with heavy load are used to ensure emergency response.

The emergency logistics center is divided into administrative regions or geographical locations. Its main responsibility is to collect emergency materials provided by the material supply entities within the region, and at the same time distribute the functions of the center, according to the needs assessment or the collected emergency material demand information Emergency materials perform operations such as picking, sorting, packaging, processing, distribution, and shipping. The materials needed in the disaster area are directly transported to the demand center or material transfer station, and the relief materials that are not needed in the disaster area are classified, packaged, and stored. The supplies provided by the main body of supply have clear classifications, labels, and complete packaging. Materials that meet the requirements of direct supply can be directly transported to the end demand center of the supply network without passing through the emergency logistics center to speed up the supply of materials. The petri net is a simple and reliable machine learning method, which optimizes the petri net and then uses it for disaster prediction, which effectively improves the accuracy of the prediction. Appropriate application of this function can design an effective algorithm in the modeling of emergency logistics system, which has the following form.

$$X_t = \sum_{j=0}^{q} \theta_i \varepsilon_{i-j} \tag{1}$$

The calculation value of the algorithm of emergency logistics system modeling is as follows;

$$3I_{01} = I_{A1} + I_{B1} + I_{C1} = 3U_{0j}\varpi C_{01} \tag{2}$$

The Petri net test should use the following formula:

$$dN_1 = -\mu N d\theta \tag{3}$$

2.2 Characteristics of Petri Nets

The calculation of the total factor productivity of the logistics industry in Gansu Province provides an objective and accurate reference for improving the total factor productivity of the logistics industry in our province. Before using Petri Net to model the logistics system, we must first combine the basic terminology of Peri Net to define the resources, warehouses and changes in the logistics system. For the logistics system, the goods, logistics staff, equipment and tools used in the operation of the system that enter or will enter the system can be regarded as resources; the location or state of these resources is the warehouse; changes in the location or state of resources caused by the use, consumption, and transfer of resources in the process of system operation are changes. The Petri net modeling method is used to construct the job flow model. It can combine graphical analysis and mathematical analysis, and has great advantages compared with other DEDS models.

The Petri net graphical interface has simple and intuitive characteristics, and has a strong mathematical foundation and rich analysis methods. But for more complex systems, the complexity of the Petri net model will increase with the increase of the complexity of the system, resulting in the phenomenon of state combination explosion. Objectivity: It can accurately describe the sequence relationship and concurrency relationship between events. Moreover, this relationship between events exists objectively. Compatibility: The network system can be used in various fields with different interpretations, and it is a bridge of communication between various fields; asynchronous concurrency: it is suitable for describing asynchronous concurrency systems. Such as the unified description of different types of information, the concurrency relationship between different subsystem events, the conflict between the local state and the global goal, etc.

3 Petri Net's Emergency System Algorithm

3.1 Emergency Logistics in Gansu Province

The rapid development of Gansu's emergency logistics industry must improve the development level of Gansu's logistics industry, improve the supply environment of logistics facilities, and promote the development of Gansu's logistics industry by optimizing the allocation of resources and improving the investment environment, and thus promote the economic development of Gansu Province. Through the establishment of Petri nets, information collection risk emergency logistics project is launched, and the collection, acquisition, transportation, and distribution of rescue materials and vehicles should be collected and released through multi-directional and multi-angle technical means to collect and release logistics-related information.

When an emergency occurs, the means of transportation is the carrier of emergency rescue materials and personnel transportation, including roads, railways, aviation and many other means of transportation. In the daily management stage, no means of transportation are reserved. After entering the start-up phase, the emergency logistics executive agency will recruit railway, aviation, and transportation management departments to organize their own or temporary expropriation of transportation means to perform the task of transporting rescue materials. The traffic layout at the location of an emergency is directly related to whether emergency materials can be delivered to the disaster area in time. The emergency materials needed in the disaster area are often huge, and the joint transportation of railways and roads is needed to meet the requirements. However, if there is no railway extending to the disaster area, the transportation speed and quantity of emergency supplies will be greatly reduced.

3.2 Response Methods for Emergencies

The emergency logistics system modeling and simulation of Gansu Province using petri nets. In the face of large-scale emergencies, especially natural disasters and public health incidents, there is no emergency logistics stage division, and there is no corresponding emergency logistics system group and In the design plan, all the people will go to battle, use administrative means to collect various resources from the whole society to organize the supply of rescue materials, and use the collected resources for a long time to ensure the supply of emergency materials. Establish a complete emergency logistics transportation system: the availability of transportation means, the rationality of transportation layout, the degree of damage to transportation channels, the rationality of transportation means arrangements, and the rationality of the construction of emergency logistics centers. Emergency logistics information risk: information collection risk, information transmission risk, information sharing risk.

The unimpeded transportation of transportation routes is the key to whether emergency supplies can be delivered to the disaster area in time. When a disaster occurs, the transportation lines are often severely damaged, making it difficult to reach the transportation vehicles. At this time, it is necessary to use helicopters to urgently transport the wounded and materials, which greatly increases the difficulty of disaster relief. The rationality of the arrangement of transportation means is restricted by the characteristics of emergencies and the environment. The arrangement of emergency logistics transportation means is different from ordinary logistics projects. The primary goal is how to deliver emergency supplies in time. For example, in the case of road interruption, helicopters need to be used to transport emergency supplies; according to the road conditions on the spot, the choice of transport model: emergency dispatch of trains to the disaster area, etc. Therefore, whether the arrangement of transportation means is reasonable is related to the efficiency and safety of emergency logistics operations. The specific results are shown in Table 1.

Table 1. The responsibility of information and warehouse department

	Information department	Warehouse department
Main duty	Information management and data maintenance	Goods circulation operations
Specific responsibilities	Production and review of documents	Item warehousing, item inventory control
	Receipt and transmission of system information	Storage location management and shelving
	System maintenance of spare parts/warehousing information	Tally and inventory work: order picking and picking

4   Modeling and Analysis of Petri Offline System

4.1   Emergency Logistics System Model

The development of modern logistics in Gansu Province is still in its infancy. The logistics facilities are relatively outdated and backward. Compared with developed cities and countries, the gap is obvious. The modeling method of Gansu Province's emergency logistics system establishes the required model when discussing the problem of discrete event dynamic system. The system model expresses the relationship between the variables of the running state and the constituent parameters of each link in the running time. The model is expressed by various methods such as computer programs, different forms of curves, mathematical equations, and graph tables. According to the established model, analyze the working status of the system, as well as the relationship between the status, system construction and various parameters, and optimize and control the model. The working conditions of discrete event dynamic systems cannot be expressed by difference equations or differential equations, which are different characteristics from CVDS. The specific results are shown in Fig. 1. The error forecast shows fluctuating changes, and the occurrence of specific emergency events maintains stable fluctuations with the number of event simulations.

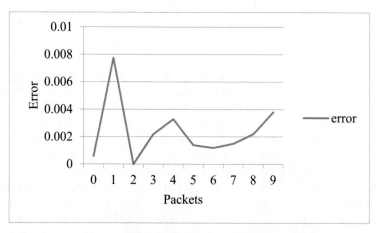

Fig. 1. Mental model work

4.2   Emergency Logistics Modeling under Petri Network

The focus of the emergency logistics model is based on the roots of operations research, intelligent labor, stochastic process, cybernetics and other fields. Through the establishment of model analysis and discussion of discrete event dynamic system, calculation performance analysis. A series of analysis of the mutual influence of activities in a discrete event dynamic system from the logic stage. Analyze the flow of the discrete event dynamic system from the algebraic stage. Forms of establishing emergency logistics system models include: communication sequential processes, finite recursive processes, minimax algebra, etc.; from the perspective of computing performance stage characteristics, the resolution forms for discussing the process characteristics of discrete event dynamic systems include: perturbation research, queuing network, generalized Marl Cove process and so on. When analyzing the system model, the two elements of state and event are usually used to describe the discrete event dynamic system. Events often occur instantaneously and lead to the transition of state. This discrete mechanism is determined by factors outside the system. The work space does not use other topological structures, and some discrete symbols are usually used to express the analysis of the discrete event dynamic system in the statistical phase, focusing on the optimization and research of the random quality of the discrete event dynamic system. The specific results are shown in Table 2.

Table 2. Universal in warehousing system

Serial number	Operation	Serial number	Operation
p01	Receiving operation	P05	Order processing
p02	Store job	P06	Picking
p03	Replenishment operations	p07	Shipment operation
p04	Inventory work	p08	Assembly work

Explore the source of Gansu's economic growth, and put forward targeted and operable policy recommendations in practice. Often use perturbation research method, simulation research method and random research method, as well as approximate research method, average research method and operation research method to analyze queuing network, generalized Markov process and other random dynamic systems in the algebraic stage. The study of structure focuses on the flow of time. The mathematical tool of maximum algebra is used to discuss the operation process and algebraic characteristics of the discrete event dynamic system. For the study of the discrete event dynamic system model at the logic stage, the focus is on the logic stage. Discuss the interconnection between the state and event symbols in the discrete event system. In the logical phase, the system states and events are mainly analyzed according to the sequence of logical time, which has nothing to do with the concept of physical time. Some tools have very strong logical function characterization capabilities, such as finite automata, Peti nets, etc., for studying and describing these elements and logical sequences. The specific results are shown in Fig. 2. By using the evaluation model, the comprehensive risk evaluation value of the emergency material transportation can be calculated to be 0.679, which corresponds to the normal range of the risk area in the sample data.

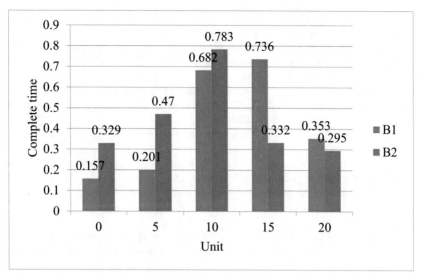

Fig. 2. Questionnaire

5   Conclusions

This paper models and simulates the emergency logistics system of Gansu Province based on petri nets. According to the relevant data of the China Statistical Yearbook and Gansu Development Yearbook, the parameters needed for empirical analysis were determined, and a total factor productivity measurement model for the emergency logistics industry was constructed. The design of emergency logistics system not only requires extensive theoretical knowledge, but also a solid theoretical foundation and competence. There are a lot of in-depth content for the modeling and simulation of the emergency logistics system in Gansu Province for petri nets. There are still many steps to study petri net analysis because of space and personal ability, etc., which are not covered. In addition, the practical application effect of the relevant experiments of the emergency logistics system can only be compared with the traditional model from the level of theory and simulation.

Acknowledgements. Project Name: Modeling and Simulation of Emergency Logistics System in Gansu Province Based on Petri Net (Item No. 2020A-148).

References

1. Li, S., Huang, J.X., Tohti, T.: Fake plate vehicle auditing based on composite constraints in internet of things environment. IOP Conf. Ser.: Mater. Sci. Eng. **322**(5), 204–205 (2018)
2. Liang, H., Liu, Y., Wan, L., Sheng, G., Jiang, X.: Penetrating power characteristics of half-wavelength AC transmission in point-to-grid system. J. Mod. Power Syst. Clean Energy **7**(1), 123–130 (2018). https://doi.org/10.1007/s40565-018-0401-9

3. Wang, X., Han, S., Yang, L., et al.: Parallel internet of vehicles: ACP-based system architecture and behavioral modeling. IEEE Internet Things J. **7**(5), 3735–3746 (2020)
4. Lee, D.G.: A multi-level behavior network-based dangerous situation recognition method in cloud computing environments. J. Supercomput. **73**(7), 3291–3306 (2017). https://doi.org/10.1007/s11227-017-1982-1
5. Jing, M., Jie, Y., Shou-yi, L., Lu, W.: Application of fuzzy analytic hierarchy process in the risk assessment of dangerous small-sized reservoirs. Int. J. Mach. Learn. Cybern. **9**(1), 113–123 (2015). https://doi.org/10.1007/s13042-015-0363-4
6. Chen, H., Feng, S., Pei, X., et al.: Dangerous driving behavior recognition and prevention using an autoregressive time-series model. Tsinghua Sci. Technol. **22**(006), 682–690 (2017)
7. Ruddy, J., Meere, R., O'Donnell, T.: Low Frequency AC transmission for offshore wind power: a review. Renew. Sustain. Energy Rev. 14–15 (2016)
8. Bhattacharyya, B., Raj, S.: Swarm intelligence based algorithms for reactive power planning with flexible AC transmission system devices. Int. J. Electr. Power Energy Syst. **78**, 158–164 (2016)
9. Scott, J.K., Laird, C.D., Liu, J., et al.: Global solution strategies for the network-constrained unit commitment problem with AC transmission constraints. IEEE Trans. Power Syst. 1 (2018)
10. Liang, H., Liu, Y., Wan, L., et al.: Penetrating power characteristics of half-wavelength AC transmission in point-to-grid system. J. Mod. Power Syst. Clean Energy 277–299 (2019)
11. Kumar, N., Rodrigues, J.J.P.C., Chilamkurti, N.: Bayesian coalition game as-a-service for content distribution in internet of vehicles. IEEE Internet Things J. **1**(6), 544–555 (2017)

Overview of Cyber Threat Intelligence Description

Liang Guo[1], Senhao Wen[2(✉)], Dewei Wang[1], Shanbiao Wang[1], Qianxun Wang[1], and Hualin Liu[1]

[1] Beijing Data Star Observatory Technology Co., Ltd., Beijing 102300, China
[2] National Computer Network Emergency Response Technical Team/Coordination Center of China (CNCERT/CC), Beijing, China

Abstract. The description of cyber threat information is the basis of threat intelligence sharing. Focusing on the description of threat information, this paper analyzes the threat intelligence standards at home and abroad, summarizes the description and analysis methods of threat intelligence and their advantages and disadvantages. Finally, the direction and development trend of cyber threat intelligence research are prospected and summarized.

Keywords: Cyber Threat Information (CTI) · Cyber Threat Intelligence Sharing (CTIS) · Threat intelligence description · CTI analysis · Network security

1 Introduction

With the rapid development of information technology, the scale of cyberspace has expanded. Cyber attacks show a trend of distribution, scale, and complexity [1]. It is necessary for cyber security to transform from passive event-driven to proactive intelligence-driven [2]. Cyber threat intelligence (CTI) comes into being. According to the definition proposed by Gartner, threat intelligence is evidence-based knowledge including context, mechanisms, indicators, implications and responsive advice, about an existing or emerging menace or hazard to assets that can be used to inform decisions regarding the subject's response to that menace or hazard [3]. A typical CTI system is shown in Fig. 1.

Threat intelligence sharing is an important part in the cyber threat intelligence system, while intelligence description is the basis of threat intelligence sharing.

Threat intelligence description refers to the activity of analyzing, selecting, and recording the data of threat information according to the needs of relevant organizations to form a unified threat intelligence. Normative and standardized threat intelligence is conducive to the establishment of threat information sharing and data transmission

J. Abawajy et al. (Eds.): ATCI 2021, AISC 1398, pp. 343–350, 2021.
https://doi.org/10.1007/978-3-030-79200-8_50

flexibly and dynamically, improving efficiency in handling and responding to security incidents and reducing time delay, and facilitating automated response strategies [4]. Threat intelligence analysis mainly revolves around data association and visual display, and in-depth mining of important information in threat intelligence.

In view of the positive effects of cyber threat information description on CTI management, this paper focuses on threat information description and analyzes the threat intelligence standards at home and abroad, combs the description and analysis methods of CTI, and elaborates their respective characteristics. Finally, the direction and development trend of CTI research are prospected and summarized.

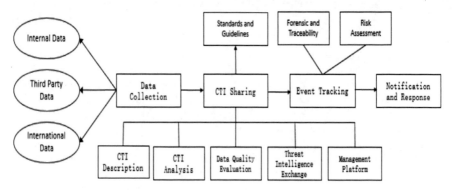

Fig. 1. Cyber threat intelligence system

2 Research on Threat Intelligence Standards

Threat information standards are the basis of CTI technology. They are used to standardize threat intelligence data, improve the efficiency and accuracy of threat intelligence exchange, reduce misunderstandings in communication, and automatically process threat intelligence.

Currently, influential frameworks and standards include NIST SP 800-150, STIX, TAXII, CybOX, MAEC, etc. China also issued a format specification for cyber security threat information in 2018.

The "Guide to Cyber Threat Information Sharing" (NIST SP 800-150) issued by NIST is an expansion of "NIST SP800-61: Computer Security Incident Handling Guide", providing guidelines to assist organizations in establishing, participating and maintaining cyber threat information sharing and coordination [5]. It clarifies the importance of trust and specific data processing precautions.

Structured Threat Information eXpression (STIX) is issued by MITRE and DHS [6]. The purpose is to regulate cyber threat information including threat intelligence collection, characteristics and communication. It expresses the various characteristics of threat intelligence through the object and description relationship, supports the process and automation of cyber threat management in a structured way, and promotes the sharing of threat intelligence.

Trusted Automated eXchange of Indicator Information (TAXII) is a set of services and information exchange mechanisms that enables the sharing of cyber threat intelligence across product lines, services, and organizational boundaries [7]. It is based on HTTPS to exchange threat intelligence information. It is originally designed to support the exchange of threat intelligence described by STIX, but they are two independent standards. TAXII can also be used for data sharing in other formats [8].

Cyber observable expression (CybOX) specifies a common structured schematic mechanism for cyber observables, the intent is to enable the potential for detailed automatable sharing, mapping, detection and analysis heuristics. It has been integrated into STIX 2.0.

Open indicator of compromise (OpenIOC) is a specification for sharing threat intelligence released by MANDIANT. It is an open source and flexible framework that can identify threats, attackers' methods or other threat indicators [9].

Malware attribute Enumeration and Characterization (MAEC) is a structured language for encoding and sharing high-fidelity information about malware [10]. MAEC dictionary defines malware elements from three different levels, including low-level action level, intermediate behavior level, and high-level mechanism level. MAEC aims to eliminate the ambiguity and inaccuracy that currently exists in malware descriptions, and enable correlation, integration, and automation.

China publishes the national standard "Information Security Technology- Cyber Security Threat Information Format" in 2018. It means China has gone further in terms of laws and regulations in cyber security, and conformed to the current development and trends of threat intelligence [11].

At present, the mainstream threat intelligence standards have strong expressive ability and comprehensive coverage, but are difficult to apply in practice with the limitation of complex content and low formal level [12]. Currently, there is no authoritative universal threat intelligence standard. The following table compares the mainstream standards in terms of application scenarios, applicability, advantages and disadvantages (see Table 1).

It can be seen that various standards can solve the problem of cyber threats to a certain extent, but they have some emphasis. After the standards are implemented, they need to be continuously improved and perfected according to the application situation.

Table 1. Comparison of threat information standards

Standards (Country)	Introduction	Application scenario	Applicability	Advantages	Disadvantages
NIST SP 800-150 (USA)	Provide guidance for information sharing planning, implementation, and maintenance. Improve efficiency of incident response	Establish, participate in and maintain information sharing relationships. Coordinate with external agencies	Suspicious commands, malicious files, emails and malware etc.	A complete guide for cyber threat information sharing	Low readability
STIX (USA)	Quickly describe the relevance and coverage of events, and structural cyber threat information	Collaborative threat analysis, automated threat exchange, automated detection and response, and more [13]	Threat factors, security activities, intelligence sharing	Flexibility, scalability, automation and interpretability	Low transmission efficiency, limited interoperability, and not fully mature
TAXII (USA)	Organizations share threat intelligence by defining APIs corresponding to the common sharing model	For producers and consumers of security intelligence and threat management agencies	Types of services, types of information and information exchange	High-efficiency of threat intelligence transmission. As a supplement to STIX	Cyber attack attribution is difficult
CybOX (USA)	Used to encode and communicate network observables [14]	Threat assessment, log management, malware characterization, indicator sharing and incident response	Attacks to file, processor and memory etc.	Interoperability and overall high situational awareness ability	No details and patterns describing complex attacks
OpenIOC (USA)	Rapid sharing of different types of threat intelligence in a machine-readable format	Capture incident response information for multiple threats	Abnormal behavior of the host or network	Open and flexible. New attribute descriptions can be added as needed	Developed from the field of digital forensics, only a few attacks can be applied
MAEC (USA)	Malware description. Utilizes element dictionaries, output formats and vocabulary syntax	Improve communication about malware. Allow for the faster development of countermeasures	Malware	Fine information granularity. Can be used as a supplement of STIX	Only focus on the attributes and behaviors of malware

(*continued*)

Table 1. (*continued*)

Standards (Country)	Introduction	Application scenario	Applicability	Advantages	Disadvantages
Cyber Security Threat Intelligence Format (PRC)	Using eight components and three domains to establish a CTI expression model	Generation, sharing and use of threat intelligence. Construction and operation of a CTI sharing platform	Automatically share the latest threat samples, events, detection and protection rules	Automated interaction. Rapid establishment of situational awareness mechanism	Need to be further promoted

3 Research on Threat Intelligence Description Methods

3.1 Method Based on Meta-model

The more common way to describe threat intelligence is to build a meta-model. By defining threat specifications, describe the elements that make up the model and the relationship between the elements. Most standards are based on meta-models, and STIX is the representative one. STIX is based on XML language and consists of content and tags. As follows [13]:

<indicator: Description> Content </indicator: Description>

Gong 666–682 believes that STIX and TAXII are two viable cutting-edge interoperability standards for cyber threat intelligence sharing, but there are barriers when adopting and using them [14]. Specific barriers for adoption emerging from the interviews include: initial setup and learning curve; organizational compatibility and maturity; understanding of cyber threat vocabulary; and lack of conformity in notating data. At the same time, recommendations for future research are proposed.

The meta-model-based method can obtain a wide range of threat information, but it is not enough for attack attribution. It cannot cover the description of anti-traceability technologies such as anonymous networks, botnets, and dark webs. Thus unable to provide in-depth support for automated attack attribution [15].

3.2 Graph-Based Methods

The field of security analysis has begun to use portrait analysis method for cyber attack recognition. At present, research and application in this field are still in the preliminary stage, and the research content is scattered and lacks systemicity [16].

Yang et al. 136–143 designed an intelligence data expression specification that satisfies the portrait analysis of attacks, using threat elements in threat intelligence as the portrait skeleton and threat attributes as portrait elements [17]. Associate relevant elements and attributes stored in the threat intelligence database to form a richer and more complete attack feature, and establish a threat intelligence portrait.

The graph-based method can intuitively show the logic of threat intelligence and provide data for establishing a visual analysis method. However, due to the use of graph storage, it consumes more storage space than other forms.

3.3 Method Based on Expert Knowledge

Literature believes that human perception provides an additional source of information for detected events, which may be an information source for events that cannot be detected by automated programs [18]. On the text-based event information report, the author has established a method to systematically report perceived abnormalities or events in a structured manner.

Expert-based methods require high professionalism and are difficult to use on a large scale and automatically. In addition, forming a complete and standardized structured expression is a problem that needs to be solved continuously.

3.4 Semantic-Based Approach

Huang applied the semantic web technology to the description of CTI, adding semantic information to scattered, messy and huge threat data, which let the computers can understand the data semantic.

The semantic-based threat information description method needs to abstract threat data and give specific semantics, which is relatively weak in automation and promotion. But in terms of the description of various threats and the mining of threat intelligence, the efficiency and accuracy are higher.

4   Research Progress of Threat Intelligence Analysis

The analysis of threat intelligence is based on correlation analysis. At present, more popular method is around the knowledge graph [19].

The knowledge graph can visually display the relationship between the development process of threats and the attack structure, extract high-value threat characteristics from massive information, correlate multi-dimensional intelligence, and dig out the hidden connections between multi-dimensional clues, thereby making a clear description of the system's overall security situation.

Wang proposed a supervised deep learning model that automatically extracts entities and entity relationships, and visualizes the knowledge graph [20]. Thus, integrate a large amount of isolated intelligence, actively adjust defense strategies, and predict attacks that have not occurred.

The current efficiency of using knowledge graphs is still not high, and there are problems such as slow data generation.

5   Outlook on Future Research

CTI research has been one of the important directions of cyberspace security research. However, due to the complexity of the attack and the inconsistent description format of the CTI, they cannot be read mutually; the content framework is incompatible and cannot be understood by each other. This shows that there is still a long way to go in the process of systemizing CTI. Table 2 summarizes some of the problems in CTI research and possible solutions.

Table 2. Problems and methods faced by CTI research

Problem	Method
Threat information description theory	Complete description system and process
Threat intelligence system construction	Combination of personnel management and technical management
Data quality assessment	Data description and value measurement
Threat intelligence sharing	Decentralized technology
Information management platform	Situational awareness method
Threat intelligence correlation	Knowledge graph algorithm
Intelligence data set	Establish a public data set that can be used as a research benchmark
Automated threat intelligence discovery	Use machine learning to monitor in real time and automatically discover threats
Standards and information industrialization	Combining different industries to make the standards practicable

6 Summary

Threat intelligence plays an important role in defending against cyber threats. But the TI description limits the sharing process. The TI description is too complex, which is not conducive to rapid popularization and application. It will also increase the load of TI analyzing and processing. TI description should be able to correctly express the features of threat factors, threat activities, security incidents etc., provide a consistent and unambiguous semantic framework and map rules to achieve information understanding and communication among multiple intelligence subjects.

Acknowledgements. Foundation Item: The National Key R&D Program China (2018YFB0804701).

References

1. Li, W.J., Jin, Q.Q., Guo, L.: Research on security situation awareness and intrusion intention recognition based on threat intelligence sharing. Jisuanji Yu Xiandaihua **01**(03), 65–70 (2017)
2. Liska, A.: Building an intelligence-led security program, p. 24 (2016)

3. Shi, Z.X., Ma, Y.R., Zhang, Y., et al.: Overview of threat intelligence standards. J. Inf. Secur. **5**(07), 560–569 (2019)

4. Liu, Z.Y., Li, Q.Y., Gong, Y.: Information sharing model of network security threat based on large-scale private network. Comput. Eng. Appl. **32**, 18–20 (2018)

5. Johnson, C., Badger, M., Waltermire, D.: Guide to cyber threat information sharing. https://csrc.nist.gov/publications/detail/sp/800-150/final

6. Brown, S., Gommers, J., Serrano, O.: From cyber security information sharing to threat management. In: Proceedings of the 2nd ACM Work-shop on WISCS 2015, pp. 43–49. ACM, New York (2015)

7. Gong, N.: Barriers to adopting interoperability standards for cyber threat intelligence sharing: an exploratory study. In: Arai, K., Kapoor, S., Bhatia, R. (eds.) Intelligent Computing. SAI 2018. Advances in Intelligent Systems and Computing, vol. 857, pp. 667–668. Springer, Cham (2018). https://doi.org/10.1007/978-3-030-01177-2_49

8. OASIS CTI TC. Cyber Observable eXpression (CybOX™) Archive, pp. 563–566. https://cyboxproject.github.io/

9. Burger, E.W., Goodman, M.D., Kampanakis, P., et al.: Taxonomy model for cyber threat intelligence information exchange technologies. In: Proceedings of the 2014 ACM Workshop on Information Sharing & Collaborative Security, pp. 51–60. ACM (2014)

10. Li, H., Du, Z.Y., Zhang, L.: Analysis and application of the network security threat information standard. J. Inf. Eng. Univ. **19**(06), 762–768 (2018)

11. Information security technology—Cyber security threat information format. GB/T 36643-2018 (2018)

12. Chen, J.F., Fan, H.B.: Ontological threat intelligence sharing in cyberspace security. Commun. Technol. **51**(01), 171–177 (2018)

13. The Mitre Corp. STIX-structured threat information expression. https://stix.mitre.org/. 765-768

14. Cyber Observable eXpression. http://cybox.mitre.org/language/version2.0/

15. Yang, Z.M., Li, Q., Liu, J.R.: Research of threat intelligence sharing and using for cyber attack attribution. J. Inf. Secur. Res. **1**(01), 31–36 (2015)

16. Yang, P.A., Liu, B.X., Du, X.Y.: Portrait analysis of threat intelligence for attack recognition. Comput. Eng. **46**(1), 136–143 (2020)

17. Vielberth, M.: Human-as-a-security-sensor for harvesting threat intelligence. Cybersecurity **2**(1), 23 (2019)

18. Huang, W.: Application of semantic web technology in network threat intelligence analysis system. North China University of Technology (2017)

19. Qin, Y., Shen, G.W., Zhao, W.B., et al.: A network security entity recognition method based on feature template and CNN-BiLSTM-CRF. Front. Inf. Technol. Electron. Eng. **20**(6), 872–884 (2019)

20. Wang, T.: Research and implementation of threat intelligence knowledge map technology. Electronic Science Research Institute of China Electronic Technology Group Corporation (2019)

Snooker Match Outcome Prediction Using ANN with Inception Structure

Shanglin Li, Bo Li, Haiyang Lu, and Juan Xiao[✉]

Department of School of Software and Information Engineering, Xiangnan University, Chenzhou, Hunan, China

Abstract. Machine learning is widely used in the prediction of the outcomes of sports matches such as football, soccer, basketball, etc. However, there are very few studies on the prediction of snooker. Although neural networks have achieved excellent performance in this field, To achieve higher accuracy prediction, a novel artificial neural network (ANN) was proposed. First, the features for the input of ANN was extracted from the player and match datasets. Then, we proposed an ANN with inception structure, which can combine and train these features more effectively. Finally, we evaluate the predictability by comparing our model with other widely used prediction models. The results show that our model has a better performance to the rest of models.

Keywords: Artificial neural network · Snooker · Prediction · Inception model

1 Introduction

Snooker is a cue sport, which is said to have been devised in 1875 by army officers in India experimented with variations on billiards. The game quickly became a mainstream game in the United Kingdom, and has enjoyed much success since the late 1970s, with most of the ranking tournaments being televised. A match of snooker game generally consists of a predetermined number of frames and the player who wins the most frames wins the match. After each game played, The World Professional Billiards and Snooker Association (WPBSA) will record a large amount of statistical data describing the performance of each player who played in the match. The statistics can be used to characterize the performance of each player over time, or give reliable predictions on the outcome of upcoming matches.

In a snooker game, players strike the white ball (or "cue ball") to pot or pocket the remaining balls in the correct sequence, accumulating points for each pot. Every shot has a probability of failure. As a result, the snooker game is full of suspense, make prediction for the outcome is a big challenge. Past relevant academic studies on the prediction of sports games can be divided into the following categories:

Rule based system. The simplest model is the rule based system, such as Football Result Expert System (FRES, used widely by media and book makers). The problem with FRES is that it uses a rule based approach or an if-else statement based approach that considers and rejects various factors one by one [1].

© The Author(s), under exclusive license to Springer Nature Switzerland AG 2021
J. Abawajy et al. (Eds.): ATCI 2021, AISC 1398, pp. 351–359, 2021.
https://doi.org/10.1007/978-3-030-79200-8_51

Rating model. Another prediction system is called ELO rating system. The Elo rating system is a method for calculating the relative skill levels of players in zero-sum games such as chess, but can also be used for sport games prediction [2, 3].

Network-based models. Network-based models is another possible way to build a prediction system. E. Ben-Naim [4] presents an statistical analysis to quantify the predictability in five major sports in the United States and England: soccer, baseball, hockey, basketball, football, it can be used to estimate the upset frequency from standings data. Vaz [5] model the NBA as a complex network and develop metrics that predict the behaviour of NBA teams.

Machine learning and data mining methods. Machine learning methods are the most widely used methods for sport game prediction areas. In terms of artificial intelligence methods, there's no essential difference between snooker and other sports. The prediction mission can be done with a classifier which is trained using enough history data. Such kinds of classifier include SVM, random forest, bayesNet, Markov methods, Logistic regression, and artificial neural network (ANN). Goddard [6, 7] proposed ordered probit regression models for football match prediction. Constantinou [8, 9] proposed probabilistic graphical models to predicts football match outcomes. One of such model, which is named Dolores [10], is a mixture of two methods: dynamic ratings and Hybrid Bayesian Networks. Sujatha [1] used ANN to predict the football match outcome. Different prediction systems use their own feature selection and computation. The common complexity is that how to choose the proper features extracting from various of data. Vaidya [11] discussed the features used in football match winner prediction, and proposed the most significant features and the way of computing new parameters from these features.

The approach described in this paper takes advantage of the fact that neural networks are good at recognizing underlying relationships in a set of data. Experiments have proved that simply increasing the depth of the neural network cannot improve its performance. In order to obtain higher prediction accuracy than traditional neural networks, we designed a neural network with inception module [12], which means it can be used to map these patterns to outputs. Meanwhile, we designed corresponding features to make the neural network work more effective.

2   Data and Feature Engineering

All the data we used comes from WPBSA. The data consist of a training dataset which incorporates 10,000 match instances from different professional snooker tournaments in the past ten seasons (2011–2021) held by WPBSA. In addition, we build a datasets for major professional tournaments, and a dataset for personal information of players. Based on these dataset, we can extract some important feature for machine leaning in the next step.

2.1   Match Data

WPBSA holds dozens of tournaments each year (or season). The venue, the total number of games, and the amount of prize money vary greatly in different tournaments. These factors will be considered as important features for our machine learning method. Table 1 shows some typical tournaments and their features, which include:

Date. The season (year) when the tournaments is held. In our dataset the range is 2011–2021.

Prize money. High prize money is the biggest motivation to attract top professional players. For example, The World Snooker Championship has total prize money in 2020 of £2,395,000, including £500,000 for the winner. Almost all the top-level players participated in this event. On the contrary, the India open has only £50,000 for the winner, Many non-professional players participated in this tournament. Our dataset choose the 20 highest-level events each season to ensure the quality of training data.

Location. The location of a tournament is a feature to define whether the match is going to be played at home or away for a player.

Rank. Most of the tournaments are ranked matches, which means the final result of the match will affect the player's world ranking. A small part of the matches is invitational, take Master for example, only the world top-16 players in the current season will be invited to participate the match.

Table 1. The important features for some tournaments

Event name	Date	Prize money for winner	Location	Rank
World championship	2020	£500,000	UK	Y
Shanghai masters	2019	£250,000	China	N
Scottish open	2020	£70,000	UK (Scottish)	Y

2.2 Player Data

We collected personal data and performance data from world top-100 snooker players. These characteristics are obtained through statistical analysis of the historical big data of players, which are important indicators of a player's overall competitive level. Table 2 gives an example of world top-5 snooker players. These features include:

Rank. The current ranking of the world, which is an important feature of the overall level of a player.

Birthday (Age). The average duration of a frame is about 15 min. Usually an average of 9 frames are played in a match (best of 13). As a result, snooker match consume a lot of mental and physical energy. The influence of age on competitive status is also very important.

Nationality. This feature and the venue feature mentioned in 2.1 together determine whether a match is home or away.

100 break ratio. Probability of scoring more than 100 in a single stroke.

Win rate of short matches. Most professional matches require a player to win four frames (best of 7) or five frames (best of 9). Short matches is more likely to increase the uncertainty of the game outcome.

Win rate of longer matches. Some professional matches require a player to win 10 frames (best of 21) or even more frames. For example, World Championship uses longer matches – ranging from best of 19 in the qualifiers and the first round up to best of 35 for the final frame. Longer matches can minimize the uncertainty of the game as much as possible, however, put forward higher requirements on the players' physical fitness.

Clean sheet rate. Clean sheet means a player win a match without losing a frame. The rate denotes a player's ability of keep competing at the highest level.

Turnaround rate. Turnaround means when a player falls behind, it finally win back the game. Turnaround rate denotes competitive spirit and stable mentality of a player. Mark Selby, for example, is called "king of turnaround", because of his high turnaround rate of 18.75%.

Table 2. The important features for some players*

Player	Birthday	Rank	Nationality	100 break ratio (%)	Win rate of short matches	Win rate of longer matches	Clean sheet rate (%)	Turnaround rate (%)
Mark Selby	1983	1	UK (England)	6.98	0.70	0.63	45.88	18.75
Mark J Williams	1975	2	UK (Welsh)	4.11	0.70	0.69	33.20	12.4
Ronnie O'Sullivan	1975	3	UK (England)	9.03	0.78	0.74	44.02	17.04
John Higgins	1975	4	UK (Scottish)	6.24	0.72	0.72	40.96	14.25
Judd Trump	1989	5	UK (England)	8.92	0.69	0.64	57.14	17.26

* The statistics are from 2019–2020 season.

2.3 Feature Normalization

The features mentioned above have different value ranges. In order to speeds up learning and leads to faster convergence, feature normalization is an essential step in data pre-processing. We use linearly scale normalization as follow:

$$f* = \frac{f}{f_{max} - f_{min}} \tag{1}$$

Where f is the value of a feature, fmax is the maximum value of the feature, fmin is the minimum value of the feature. Equation 1 make sure all the value of feature is in range [0,1]. It's significantly helpful as it transposes the input variables into the data range that the sigmoid activation functions lie in.

2.4   Feature Vector

Although it is difficult to take into account all features that influence the results of the matches, we choose 20 features mentioned above for each training sample by combining the history data with match feature and the player feature. The final feature vector we used as in Fig. 1, 8 features for player A, 8 features for player B, and 4 features for match information.

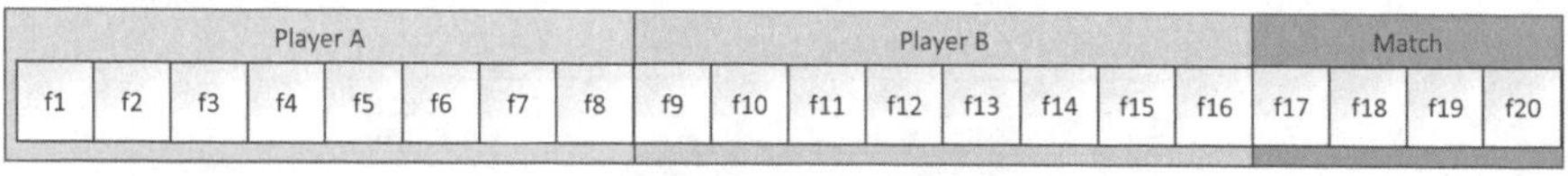

Player A								Player B								Match			
f1	f2	f3	f4	f5	f6	f7	f8	f9	f10	f11	f12	f13	f14	f15	f16	f17	f18	f19	f20

Fig. 1. Feature vector.

3   Prediction Model

Although neural networks have achieved excellent performance in the field of sport outcome prediction, to achieve higher accuracy prediction, we tried to find a better model with our features.

3.1   Network Structure

The traditional ANN solutions patch all the features together as input vector, just like Fig. 1 shows. The player feature denotes the personal ability for a players, while the match feature denote the level of a tournament. We believe that these two types of features should be optimized separately in the first few layers. Inspired by inception model [12], we designed a model which has an inception model, which has 2 input branches for our features. In addition, we use batch normalization after each hidden layer to makes the optimization landscape significantly smoother [13]. The structure of the network is showed in Fig. 2, with a total number of trained parameters of 1046.

3.2   Loss Function

The categorical cross entropy is used to assess the errors between the probability distribution and the ground true distribution obtained from the current training:

$$Loss = - \sum_{i=1}^{output-size} y_i \cdot \log \hat{y}_i \qquad (2)$$

Where y_i is the output of the ANN, and $\hat{y}_i$ is the ground true label for the train sample. Giving the loss function, the AdaGrad [14] stochastic optimization algorithm is used to update the parameters in the ANN.

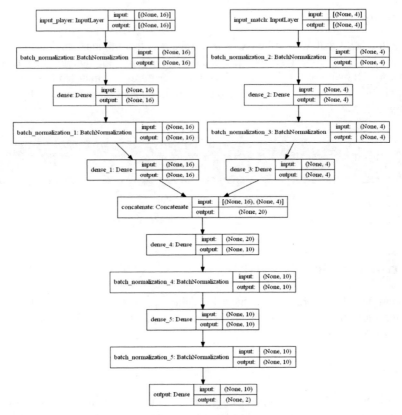

Fig. 2. The ANN structure.

3.3 Training Parameters

The dataset is split into 2 part: 80% for training and 20% for validation. The batch size is set to 128. Moreover, we also use drop out to prevent overfitting. The total optimization iteration (epochs) is set by 100, however, early stopping technology is used to terminate the training when necessary. Finally, we change learning rate as below:

$$\eta_t = \eta_{\min} + \frac{1}{2}(\eta_{\max} - \eta_{\min})(1 + \cos(\frac{T_{cur}}{T_{\max}})\pi) \tag{3}$$

Where $\eta_{\max}$ is the initial learning rate, T_{cur} is the learning rate for current epoch, $\eta_{\min}$ is set by 0.

The loss function of training for 50 iteration is show in Fig. 3. The green line shows the train loss, whine the black line shows the validation loss. Notice that the optimization is early stopped before epoch 50.

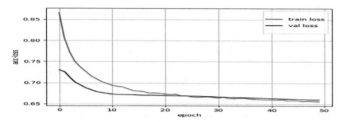

Fig. 3. The loss function of training.

4 Predictability

We use 33 matches of the 2020 World Grand Prix, which are not in our training dataset, as a test dataset for our prediction model. The location for this tournament was in UK, and it provided £100,000 for the winner. We choose some widely used prediction model, including logistic regression [6], support vector machine (SVM), deep forest [15], simple ANN [1], to compare with our model. Each model was trained 10 times with cross-validation, and the best performed model among them is kept. The details of comparison results for these models show in Table 3.

All the models get a accuracy over 66% on the test data. It is worth noting that the accuracy on the test data is higher than on the validation data. That's because compare with the World Grand Prix, some small rank tournaments in our training dataset have a higher shock rate—which we will discuss in the next section.

Compared with other machine learning methods, logistic regression is the worst model of all, while ANN has achieved significant advantages. Inception ANN has the highest accuracy of all the models, for the inception structure bring additional performance improvements to the ANN.

5 Discussion

Is accuracy over 70% good enough for the snooker prediction? Especially, the state-of-the-art model in football match prediction has achieve a accuracy over 80% [11]. How to evaluate the prediction accuracy of a model in snooker prediction?

Unlike other sport game, snooker is "almost a game of probability", a failure of any single shot may lead to the failure of a frame. The most hardest prediction for the prediction in snooker is "Shock". The biggest shock of 2020 World Grand Prix is listed in Table 4. No one believes that Zhao Xingtong, No. 68 in the world, will defeat John Higgins, No.4 in the world. In some small tournaments, the shock rate is even more higher. It's not strange that none of the model can predict these shocks. This can also be explained by the distribution of the history data—The probability of shock is too small to infect the prediction of a classifier.

So far, it's still a big challenge to predict the shock for current prediction model. One possible solution is performing data mining on the dataset to dig deeper connections between features. This will be the next goal we need to achieve.

Table 3. The comparison with other models

Prediction model	Logistic regression[a] [6]	SVM[b]	Deep forest[c] [15]	Simple ANN[d] [1]	Inception ANN (our)
Prediction accuracy in test data	0.66	0.67	0.69	0.71	0.78
Prediction accuracy in validation data	0.65	0.66	0.65	0.66	0.66
Trained parameters	–	–	–	1282	1046
Max optimization iterations	–	–	–	100	100

[a] Logistic Regression model uses stochastic average gradient descent.
[b] SVM model uses radial basis function as kernel.
[c] Deep forest model uses an implement of gcForest, with window size of 10.
[d] ANN model has 2 hidden layers, with a total trained parameters number of 1282.

Table 4. The shock of 2020 World Grand Prix

Player A*	Player B*	Win frames of player A	Win frames of player B
Robert Milkins (32)	Neil Robertson (10)	4	2
Hossein Vafaei (39)	Ding Junhui (8)	4	1
Martin Gould (27)	Mark Allen (6)	4	3
Zhao Xintong (68)	John Higgins (4)	4	3
Jack Lisowski (16)	Mark Selby (1)	6	4

* the number after player name means the current world rank of the player.

Acknowledgments. This work was supported in part by the Natural Science Foundation of Hunan Province for Youths under Grant 2019JJ50564, and in part by Scientific Research Project of Education Department of Hunan Province for Youths under Grant 20B536, and in part by Chenzhou Science and Technology Bureau Project under Grant ZDYF2020163, all support is gratefully acknowledged.

References

1. Sujatha, K., Godhavari, T., Bhavani, N.P.: Football match statistics prediction using artificial neural networks. Int. J. Math. Comput. Methods **3**, 1–8 (2018)
2. Leitner, C., Zeileis, A., Hornik, K.: Forecasting sports tournaments by ratings of (prob) abilities: a comparison for the EURO 2008. Int. J. Forecast. **26**(3), 471–481 (2010)

3. Hvattum, L.M., Arntzen, H.: Using ELO ratings for match result prediction in association football. Int. J. Forecast. **26**(3), 460–470 (2010)
4. Ben-Naim, E., Vazquez, F., Redner, S.: Parity and predictability of competitions. J. Quant. Anal. Sports **2**(4) (2006)
5. Vaz de Melo, P.O., Almeida, V.A., Loureiro, A.A.: Can complex network metrics predict the behavior of nba teams? In: Proceedings of the 14th ACM SIGKDD International Conference on Knowledge Discovery and Data Mining (2008)
6. Goddard, J.: Regression models for forecasting goals and match results in association football. Int. J. Forecast. **21**(2), 331–340 (2005)
7. Goddard, J., Asimakopoulos, I.: Forecasting football results and the efficiency of fixed-odds betting. J. Forecast. **23**(1), 51–66 (2004)
8. Constantinou, A.C., Fenton, N.E., Neil, M.: Profiting from an inefficient association football gambling market: prediction, risk and uncertainty using bayesian networks. Knowl.-Based Syst. **50**, 60–86 (2013)
9. Constantinaou, C., Fenton, N., Neil, M.: A Bayesian network model for forecasting association football match outcomes. Knowl.-Based Syst. **36**, 322 (2012)
10. Constantinou, A.C.: Dolores: a model that predicts football match outcomes from all over the world. Mach. Learn. **108**(1), 49–75 (2018). https://doi.org/10.1007/s10994-018-5703-7
11. Vaidya, S., Sanghavi, H., Gevaria, K.: Football match winner prediction. Int. J. Comput. Appl. **154**(3), 31–33 (2016)
12. Szegedy, C., et al.: Going deeper with convolutions. In: Proceedings of the IEEE Conference on Computer Vision and Pattern Recognition (2015)
13. Qian, Y., Zhang, H., Furukawa, Y.: Roof-GAN: learning to generate roof geometry and relations for residential houses. arXiv preprint arXiv:2012.09340 (2020)
14. Duchi, J., Hazan, E., Singer, Y.: Adaptive subgradient methods for online learning and stochastic optimization. J. Mach. Learn. Res. **12**(61), 2121–2159 (2011)
15. Zhou, Z.H., Feng, J.: Deep forest: towards an alternative to deep neural networks. In: Twenty-Sixth International Joint Conference on Artificial Intelligence (2017)

Construction of Secure and Stable Communities in Higher Vocational Colleges in the "Internet+" Era

Yunshan Liu[✉]

School of Marxism, Wuxi Vocational Institute of Commerce, Wuxi, Jiangsu, China
lubianinfo@eixueshu.cn

Abstract. In the "Internet+" era, there are problems in the work of security and stability of higher vocational colleges. In order to improve the current situation, it is necessary to start from three aspects: the supervision mechanism of late-return inspection, sanitation inspection and high-power electrical appliance inspection, the timeliness of class feedback and the closeness of parental contact. It is essential to intensify the work of communities' security and stability in vocational colleges, and promote the security and stability.

Keywords: "Internet+ era" · Higher vocational colleges · Security and stability

1 Introduction

The "Internet+" era is an important new trend of social development in the world today. Its essence is to make full use of the optimal combination and integration of Internet technologies in the allocation of social resources, integrate the latest achievements of the Internet in various economic and social fields, improve the productivity and creativity of the society, so as to form a new form of Internet economic development with a wider scope and higher efficiency. For more than ten years, many industries at home and abroad have deeply integrated into the "Internet+" field, such as "Internet+" entrepreneurial innovation, industrial integration, services that benefit citizens, governance system modernization, key technology research and development and infrastructure construction.

2 The Meaning and Characteristics of the Concept of "Internet+" Era

The meaning of "+" in "Internet+" era refers to cross-industry integration, which is opening, reform and reshaping. In the Internet age, innovation can only be achieved if people dare to cross-border; development can only be achieved if people dare to integrate. As far as the world is concerned, to realize the sustainable development of the economic society, innovation is crucial. It is necessary to make full use of Internet thinking to

© The Author(s), under exclusive license to Springer Nature Switzerland AG 2021
J. Abawajy et al. (Eds.): ATCI 2021, AISC 1398, pp. 360–366, 2021.
https://doi.org/10.1007/978-3-030-79200-8_52

achieve change and self-innovation, and pay highly attention to and grasp the innovation drive.

An important feature of the "Internet+" era reshaping. In the Internet age, the human community has undergone a major structural remodeling. The original economic, social, geographical, cultural, and information structures have been broken, and the original power, voice, and rules have been changed. Therefore, social governance mechanisms in the "Internet+ era" are emerging in endlessly, continuously affecting all walks of life, including the security and stability of higher vocational colleges.

3 The Status Quo of the Security and Stability Work in Higher Vocational Colleges in the "Internet+" Era

3.1 The Status Quo of the Security and Stability Work in Higher Vocational Colleges

The security and stability work of higher vocational colleges in the "Internet+" era is mainly to test the theoretical study of students and the application of theoretical knowledge in practice. The security and stability work of higher vocational colleges in the "Internet+" era is to insist on understanding, caring, respecting, and cherishing 'vocational college students are the new generation who were born in the "Internet+" era. They have a wide range of getting information. In the "Internet+" era in China, it is more difficult to do security and stability work for college students than in the past [1]. Under such a specific background, it is necessary to take into account the overall status quo of society and the particularities of students. At the same time, it is necessary to carry out a variety of related activities according to students' new interests, enrich their lives, and actively carry out various security and stability activities, so that students not only increase their knowledge in various aspects, but also exercise their abilities, and enhance their awareness of security and stability.

In the specific security and stability work in the "Internet+" era, higher vocational colleges should take the promotion of students' all-round development as the starting point and end point, fully respect the individual differences of students in practice, and respect the creation of students. Based on basic rights, colleges should evaluate each student objectively and scientifically, and treat each student fairly. In the era of "Internet+", colleges and universities must also keep up with the pace of the times, strive to achieve the innovation of security and stability work, give students room for development, and fully mobilize students' enthusiasm, initiative and creativity in innovation and entrepreneurship to establish the long-term effective security and stability working mechanism that promotes the comprehensive development of students, allows them to make better use of their advantages, better correct their shortcomings, and achieve sustainable development [2].

3.2   The Supervisory Mechanism of Late-Return Inspection, Sanitation Inspection, and High-Power Electrical Appliance Inspection

3.2.1   The Supervision Mechanism

The community self-management committee is responsible for the community late return, sanitation and high-power electrical appliance inspections of higher vocational colleges. The department takes part in the inspection with the dormitory manager. The late return inspection starts at 21:30 every day. The manager and the head of the security supervision department of the self-management committee will each lead a team to check, mainly to check how many students have not returned to the dormitory after 21:30, and the whereabouts of these students, to ensure personal security of students in school. The inspection is mainly supervised by on-duty counselors, dormitory managers and cadres from the self-management committee.

Sanitation and high-power electrical appliance inspections start at 13:00 every Wednesday. There are three teams participating. Each team is sent by the department of health and the department of security supervision and the dormitory managers. The managers will lead the team to conduct the inspection. The department of health is responsible for the cleaning condition. The security supervision department is responsible for checking whether there are any high-power electrical appliances in the student dormitories. The inspections are mainly supervised by community teachers, dormitory managers and cadres of the self-management committee.

3.2.2   The Problems

There are three main problems in the inspections. First, the late-return inspection and supervision method is supervised by the dormitory managers and the self-management committee cadres. Students in the same class may be sheltered or biased, which may lead to security risks during the late-return inspection. Second, when it comes to student-score part in the sanitation inspection and supervision, the students who are classmates or friends with the students who mark may have higher scores. Third, in the high-power electrical appliance inspection and supervision mechanism, the students who are in charge of the inspection may give advance notice to the students close to them before the inspection.

3.3   The Timeliness and Existing Problems of the Class Attendance Feedback

3.3.1   The Timeliness

Classroom is a place for teachers to teach and students to learn. A classroom now is composed of units of classes, requiring students to participate in the compulsory and elective courses chose by themselves. The current class attendance principle is the combination of teacher's roll call, students' swiping card and disciplinary inspection committee reporting. Class attendance can reflect the security and stability of a class. A class with a high attendance rate will not cause too many problems in general security. On the contrary, a class with a particularly poor attendance rate can easily have security issues. The timely feedback of classroom attendance data is of great significance to the security of the students, as well as the campus.

3.3.2 Existing Problems

There are four main problems. First, Class attendance data is carried out from three lines at the same time. There may be discrepancies in the attendance data. It is not known which data shall prevail during management. It takes a certain time to verify the data again. Second, swiping card attendance is in the operation of the machine, there may be some students who ask their classmates living in the same dormitory to swipe the card for them. Third, Teachers may have too many classes and cannot remember the name and appearance of every student. Some students will ask others to show up for them. Fourth, the discipline inspection attendance is carried out by students themselves. It may happen that the discipline inspection commissioners shield the students who have a good relationship with them.

3.4 The Closeness of Parental Contact and Existing Problems

3.4.1 The Closeness of Parental Contact

A college or university is an open school. Students live together in groups. Generally, 4 students in a class live in a dormitory. In addition to class, students also have some necessary extracurricular activities to participate in. Students not just communicate and interact with their classmates and teachers. Some students may not be able to adapt to the new city and the new environment. Students will communicate with their parents and may also communicate with their relatives and friends. Sometimes teachers in charge cannot get all the information of the students. Especially in the "Internet+" era, students are getting closer and closer to the outside world, and the relationship with the teachers and classmates in real life is aloof. This requires teachers and parents to keep in close contact with students at all times and communicate and feedback timely. Parents provide timely feedback to teachers of students' psychological conditions, and teachers provide timely feedback to parents of students' performance in school. The close contact can promote the security and stability of the school [3].

3.4.2 Existing Problems

There are mainly three problems. First, In the era of "Internet+", with the rapid development of the Internet, students are becoming more and more dependent on the Internet, and there is less and less contact with their parents. Second, In the "Internet+" era, some things students can solve online without going through their parents, including the money they need. They can borrow money online. Students rely on the Internet more than their parents. Third, In the era of "Internet+", students are increasingly dependent on the Internet, their demand for classrooms is getting smaller and smaller, and the connection with teachers is also decreasing [4].

4 The Construction of Security and Stability Work in Higher Vocational Colleges in the "Internet+" Era

4.1 Broaden the Supervision Mechanism of Community Management, Promote the Stable Development of Higher Vocational Colleges

Higher vocational education has gradually developed into an important part of China's higher education. The popularization of higher education has created a good environment for the development of higher vocational colleges. However, due to the limitation of traditional community management work methods, the procedures of community management work are too cumbersome. For a long time, students have not been able to enjoy the results of related community management. This is a flaw in traditional community management work. The characteristics of emerging media can solve this flaw, benefit more students in a low-cost, efficient, and sustainable way of community management, so that all students can improve their growth together.

The current process of informatization is accelerating. The main target of community management in higher vocational colleges is the young people who have grown up with the Internet. This situation requires educator to establish a strategic vision and actively promote the informatization of community management. by managing relevant new media account of community management, the coverage of community management work gets bigger and bigger, and the depth of influence is getting stronger and stronger. By launching a series of activities on the new media account, the Party's influence can be enhanced, mainstream ideology can be promoted, students' yearning for the Party can be increased, and the role of ideological and political work in the training of vocational students can be further strengthened. If higher vocational colleges want to fully implement the concept of sustainable development and realize the innovation of community management, they must have the courage to break through the past work practices, and actively explore the use of modern emerging media and other means to innovate and improve the form of community management.

4.2 The Effectiveness of Class Attendance Needs to be Strengthened; the Timeliness of Classroom Attendance Needs to be Improved

Class attendance is based on each class. Class attendance data is carried out from three lines at the same time. There may be discrepancies in attendance data. At this time, three lines are required to verify separately their data and report the accurate data for management [5]. The swiping card attendance is operated by machine. Some students may not come to class. The school should take advantage of the "Internet+" and use fingerprints instead of campus cards to improve the effectiveness of class attendance. The phenomenon of letting students in the same dormitory swipe their cards on their behalf will gradually disappear in the "Internet+" era. Speaking of the roll call, the teacher has too many classes and may not remember the name and appearance of every student. At this time, teachers can take the advantage of "Internet+" to ask students to sign in on the internet and upload a photo of themselves on the home page [6]. The discipline inspection attendance is carried out by students themselves. It must strictly require the student cadres, and the student cadres must be required to be fair and unselfish.

4.3 Diversify the Form and Content of Management in Higher Vocational Colleges Should be Diversified, Meet the Characteristics of Higher Vocational Colleges

The security and stability workers of higher vocational colleges should apply for a WeChat official account and a Sina Weibo account. They can use the WeChat official account to daily post security and stability-related articles to students to strengthen their ideological and political awareness and construct campus daily security education. They can use Sina Weibo account to interact with students, such as posting relevant campus security and stability work systems, so that they can continue students' ideological education, and also keep students informed about the school's community, class management, campus security systems [7].

The reform of security and stability in higher vocational colleges can be carried out by changing the linkage of some or all of the five aspects of students, teachers, dormitory managers, schools, and parents [8]. Linking students' professional qualities and abilities with campus security and stability work to increase the forms of campus security and stability work. It can provide more theoretical references for schools to improve students' professional quality and security and stability work reforms. This can, in a certain sense, make up for the excessive cost of time and energy in the work process of campus security and stability, which is more in line with the development characteristics of higher vocational colleges [9].

5 Conclusion

Facing the current situation and the problems of community security and stability in higher vocational colleges, in order to improve the security and stability of higher vocational colleges, colleges should start from the supervision mechanism of community's late-return, sanitary and high-power electrical appliance inspections, the timeliness of class feedback, and the closeness of parental contact, make full use of various favorable factors in the "Internet+" era, and strive to build a safe and stable work network in the communities in vocational colleges, and promote long-term community security and stability in higher vocational colleges.

References

1. Deng, H.: About network security education of college students. Educ. Explor. **07**, 79–80 (2014)
2. Bai, T., Chen, L., Zhang, X.: A study on network security education of university students in the background of "Internet+" and big data. J. Hebei North Univ. (Soc. Sci. Edn.) **32**(05), 116 (2016)
3. Wang, H.: Network security education for college students. J. Changchun Univ. Technol. (High. Educ. Study Edn.) **34**(04), 129–130 (2013)
4. Wan, C., Xia, J.: Analysis on the urgency of strengthening security education for college students. Sci. Technol. Inf. **08**, 32 (2015)
5. Liu, Y., Xie, S., Lin, L., et al.: Analyses and countermeasures of the security and stability in Fujian colleges and universities under the new situation. J. Chongqing Coll. Electron. Eng. **23**(01), 53–59 (2104)

6. Zhao, Y.: Influencing factors and countermeasures of college students' safety problems. Educ. Vocat. **20**, 86–88 (2014)
7. Wang, J.: Research on the management of college students' safety education under the new normal. J. Saf. Sci. Technol. **16**(12), 191 (2020)
8. Fan, J.: Research on security education of college students under the background of "Internet+". Educ. Sci. Forum **46**, 117–118 (2020)
9. Mei, X.: Analysis on the present situation and realization path of safety education for college students. New Curric. Res. **09**, 121–122 (2020)

Competitiveness Evaluation of Port Based on BP Neural Network

Chongkai Zhang[✉], Jianmin Li, Zhipeng Zhang, and Wan Zhang

Navigation College, Dalian Maritime University, Dalian, Liaoning, China
zhangchongkai@yeah.net

Abstract. To solve the problem of ignoring the influence of intelligent port and green port in the existing evaluation index of port competitiveness, this study introduces some indexes such as the level of port informatization, energy consumption per unit GDP and the output value rate of science and technology input to construction index. After building the evaluation index system of the competitiveness of the port, all the data needed in this paper are collected, and the qualitative index data is determined by expert scoring, so that all the data are dimensionless, and the BP neural network model is constructed. Seven Coastal Ports in China are selected as the training test set, and MATLAB is used for simulation experiment to control the model error within the allowable range. The results show that the error between the established index system and the actual results is very small, which verifies the accuracy and operability of the model, and provides a reference for improving the competitiveness of port.

Keywords: BP neural network · Smart port · Port competitiveness

1 Introduction

With the continuous development and breakthrough of artificial intelligence, big data, cloud computing and other emerging technologies, it is the trend of the times for ports to move towards the intelligent era. Intelligent port has become an important direction of port construction in the future. For example, Qingdao port has become the first automated terminal. In this context, the accuracy of the traditional port competitiveness evaluation index which only considers economic factors has declined.

In recent years, the research mainly focuses on the green port. Therefore, based on the traditional port competitiveness and green competitiveness, this paper introduces the intelligent port competitiveness index, and uses BP neural network to study the problem. Get the impact of different indicators on the competitiveness of smart port, and then get the competitiveness evaluation model of smart port.

J. Abawajy et al. (Eds.): ATCI 2021, AISC 1398, pp. 367–372, 2021.
https://doi.org/10.1007/978-3-030-79200-8_53

2 Construction of Evaluation Index System of Smart Port Competitiveness

2.1 Construction Principles of Evaluation Index

The port competitiveness evaluation index should follow the following principles [1]:

(1) Scientific: the selected indicators should be related to the competitiveness of the port.
(2) Comparability: indicators within the same level should be relevant and can be compared.
(3) Difference: the indicators between different levels should be discrete and have great difference.
(4) Stability: the selected indicators should be able to reflect the competitiveness of the port in a long-term and stable way.

2.2 Index Construction of Port Competitiveness

There are many evaluation systems of port competitiveness. Based on the above index selection principles, this paper uses the methods of expert consultation and literature research, references relevant literature [2–6], and considers the influence of smart port to establish the evaluation index system of port competitiveness. See Table 1.

Table 1. Port competitiveness evaluation index system

Target layer	First level indicators	Second level indicators
Evaluation system of smart port competitiveness	Economic competitiveness	Cargo throughput
		Port hinterland GDP
		Route coverage
	Environmental competitiveness	Energy efficiency
		Energy consumption per unit GDP
		Environmental governance capability
	Science and technology competitiveness	Output value rate of science and technology input
		Port information level
		Education level of port personnel

(1) economic competitiveness.

① Cargo throughput: Cargo throughput is the most basic index to measure a port. The larger the cargo throughput is, the stronger the port dominates the market.

② Port hinterland GDP:The higher the GDP of port hinterland is, the more developed the port hinterland is and the larger the total amount of import and export goods is.

③ The higher the route coverage, the higher the importance of the port in shipping.

④ Energy efficiency: This index reflects the level of energy consumption and utilization effect, that is, the comprehensive index of the degree of effective utilization of energy.

⑤ Energy consumption per unit GDP: refers to the ratio of total primary energy consumption to port GDP, reflecting port energy efficiency.

⑥ Environmental governance capability: The index refers to the input needed to control a certain amount of pollutants, reflecting the capacity of the port to deal with pollution.

⑦ Output value rate of science and technology input: It refers to the ratio of profit generated by science and technology investment to total science and technology investment, which reflects the transformation ability of port science and technology achievements.

⑧ Port information level: It refers to the application level of e-commerce, Internet of things technology and cloud computing.

⑨ Education level of port personnel: The proportion of the number of people with bachelor's degree or above and the number of senior technicians or above in the total number reflects the talent reserve and scientific and technological innovation ability of the port.

3 Establishment of Intelligent Port Competitiveness Model Based on BP Neural Network

BP neural network is a typical model of artificial neural network, the key is to determine the number of layers and the number of neurons in each layer [7]. Practice shows that there is no benefit in using more than three hidden layers [8]. Therefore, this paper uses three-layer BP neural network.

The reference formula of middle hidden layer neurons is as follows [9]:

$$n_1 = \sqrt{n + m} + a \tag{1}$$

$$n_1 = \log_2 n \tag{2}$$

The n_1 is the number of hidden neurons, m is the number of output neurons, n is the number of input neurons, and a is the constant between [1, 10].

From the above two formulas, taking n = 9, m = 1, we can get that N1 should be between 4–14. It is found that when the number of neurons is 11, the prediction error is the smallest, so the number of middle layer nodes is 11.

After determining the structure of the neural network, the collected data are processed and input into the BP neural network for training to obtain the competitiveness evaluation index of the smart port. Finally, the data of the port to be evaluated are standardized and input into the model to get the competitiveness Value of the port to be evaluated.

4   Realization of BP Neural Network

4.1   Data Processing

All the quantitative data in this paper come from «China Statistical Yearbook 2019» [10]. The qualitative index data were scored by experts. After all data are standardized, Table 2 is obtained. The full score of each index is 20. The experts sum up all the indexes of the port to get the total score.

Table 2. Port competitiveness evaluation index system

Index	Sub index	Learning sample raw data						
		Weihai port	Dalian port	Yantai port	Fuzhou port	Shenzhen port	Qinhangdao port	Wenzhou port
Economic competitiveness	Cargo throughput	−1.000	0.874	1.000	0.004	0.264	0.040	−0.782
	Port hinterland GDP	−0.893	−0.574	−0.523	−0.385	1.000	−1.000	−0.606
	Route coverage	−1.000	1.000	1.000	0.000	0.000	0.000	0.000
Environmental competitiveness	Energy efficiency	−1.000	1.000	1.000	1.000	1.000	1.000	−1.000
	Energy consumption per unit GDP	−1.000	1.000	1.000	1.000	1.000	1.000	−1.000
	Environmetal governance capability	0.700	0.700	0.700	0.700	0.700	0.700	0.700
Science and technology competitiveness	Output value rate of science and technology input	−1.000	1.000	1.000	1.000	1.000	−1.000	−1.000
	Port information level	−1.000	1.000	1.000	1.000	1.000	1.000	−1.000
	Education level of port personnel	−1.000	0.000	0.000	0.000	1.000	−1.000	−1.000
Total score		−1.000	0.650	0.575	0	1.000	−0.875	−0.55

The simulation results are shown in Fig. 1, and the comparison between simulation prediction results and expert prediction is shown in Table 3.

The result shows that error between the model results and the expert evaluation score is within 1%, which proves the accuracy of the model.

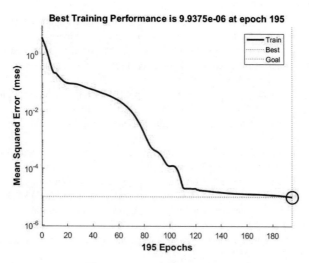

Fig. 1. The simulation results

Table 3. Comparison between simulation prediction results and expert prediction

	Weihai port	Dalian port	Yantai port	Fuzhou port	Shenzhen port	Qinhangdao port	Wenzhou port
Expert evaluation	35	101	98	75	115	40	53
Actual output	34.995	100.761	98.232	75.102	114.999	39.999	53.002
Relative error	0.014%	0.236%	0.236%	0.136%	0.001%	0.001%	0.001%

5 Conclusion

In this paper, seven ports in China are selected as test data, and using BP neural network to construct the evaluation index system model of port competitiveness. After testing, the model can accurately reflect the value of port competitiveness, which is conducive to the application of BP neural network to port evaluation, and provides a reference for improving the index system of smart port competitiveness.

References

1. Zhao, B.: Port competitiveness evaluation based on BP neural network. Dalian Maritime University (2016). (in Chinese)
2. Ou, Y., Wang, L., Huang, J., et al.: Research and application of green low carbon port evaluation index system. Water Transp. Eng. (004), 73–80 (2015). (in Chinese)

3. Li, Y.S., Huang, L.: Financial early warning model of listed companies based on BP neural network. Theory Pract. Syst. Eng. **25**(1), 12–18 (2005). (in Chinese)
4. Hao, G.: Study on prediction of urbanization level based on GA-BP neural network. In: Proceedings of the 21st International Conference on Industrial Engineering and Engineering Management 2014. Atlantis Press (2015)
5. Han, Y., Xu, Y.: Evaluation of green port competitiveness based on grey cloud clustering model. Logist. Technol. 38(03), 82–85+119 (2019). (in Chinese)
6. Zhang, X.: Construction of smart port evaluation index system. Port Sci. Technol. **169**(03), 3–5 (2020). (in Chinese)
7. Zhen, Z.: Competitiveness model of Chinese port manufacturing industry based on global value chain. J. Coast. Res. **103**(sp1), 701 (2020)
8. Hao, J., Lin, Y., Ren, G., et al.: Comprehensive benefit evaluation of conservation tillage based on BP neural network in the loess Plateau. Soil Tillage Res. **205**, 34–36 (2021)
9. Chen, J., Huang, S.: Evaluation model of green supply chain cooperation credit based on BP neural network. Neural Comput. Appl. **33**, 1–9 (2021)
10. National Bureau of statistics of the people's Republic of China. China Statistical Yearbook. China Statistics Press (2019)

Target Simulation of Bucket Reflector Based on Complex Structure

Jiaxing Hao[1](✉), Xuetian Wang[1], Hongmin Gao[1], Sen Yang[2], and Cuicui Yu[1]

[1] School of Information and Electronics, Beijing Institute of Technology, Beijing, China
[2] Department of UAV Engineering, Army Engineering University, Shijiazhuang, Hebei, China

Abstract. Radar cross-section (RCS) means that at a given frequency, the emitted electromagnetic wave illuminates the surface of the object and then reflects back to the radar or the receiving antenna signal. The change of the attitude angle of the target will greatly affect the RCS by changing the shape and size of the measured target, then the purpose of jamming the radar signal can be realized. Therefore, this paper proposes a bucket reflector design based on complex structure by changing the material structure and attitude angle, it achieves the purpose of simulating the RCS distribution curve of the horizontal azimuth angle of the target. Finally the demand target is simulated to achieve radar interference.

Keywords: Radar reflection cross-sectional area · RCS · Bucket Reflector · Composite material · Target mimulation

1 Introduction

With the rapid development of modern radar technology, the radar signals are increasingly becoming large, the ability of the radar to distinguish targets has been continuously improved, and its functions have become more powerful. The existing radars can not only accurately capture the location of the target, but can also obtain the shape, size and movement of the target. Radar jamming can be divided into active jamming and passive jamming in terms of electromagnetic radiation or not. And rain, snow, sea clutter, fog and other weather factors will interfere with the radar signal, forming noise echo. In addition, radar Angle reflector and chaff and other man-made object jammers, these are passive jamming [1–4].

Radar echo simulation is to receive the radar signals, analyze and simulate one or more targets, and then transmit the simulated echo to the radar. The existing radar target echo simulation research mainly includes two parts: target radar cross-section (RCS) calculation and target echo generation. Currently, for target RCS calculation, the electromagnetic simulation software is generally used to generate a simulated target, and the corresponding RCS value. Under the condition of scene coupling, the omnidirectional RCS distribution characteristic, the target echo generates the corresponding echo waveform through its own RCS.

In recent years, China has made significant progress in radar echo simulation. In 2010, Lei Ting et al. applied the MTD-frequency domain pairing method to the terminal

J. Abawajy et al. (Eds.): ATCI 2021, AISC 1398, pp. 373–379, 2021.
https://doi.org/10.1007/978-3-030-79200-8_54

of the LFMCW radar signal processing system, which improved the system operation rate and the measurement accuracy [5]. Liu Xian et al. proposed a single-sample echo radar signal acquisition method to achieve compression processing of sparse frequency signals and reduce the error to a resolution unit inside [6]. In the United States, a company named Sensis designed a radar echo simulation source named AN/TPS-59. However, there are problems such as low accuracy [7–9].

This paper mainly focuses on the distribution characteristics of radar target RCS by changing the structure of a certain horizontal angle position, it can show the RCS characteristic distribution of civil aircraft under the radar signal scanning, so as to achieve the purpose of target simulation.

2 Design Principle

2.1 Radar Cross-Section

Radar cross-section means that the radar emits a series of electromagnetic waves that will produce a reflected echo signal when irradiated on the surface of the object. The radar obtains the signal characteristics of the target after filtering and discriminating the signal [10]. The larger the reflected electromagnetic wave, the larger the RCS and vice versa. The RCS is not a fixed constant, but a distribution characteristic closely related to the changes in the angle.

The energy density of a radar incident plane wave is defined as:

$$\omega_i = \frac{1}{2}E^i H^i = \frac{1}{2Z_0}\left|E^i\right|^2 \tag{1}$$

where E^i and H^i are the incident electric and the magnetic field strengths, respectively, while $Z_0 = \sqrt{\mu_0/\varepsilon_0} = 377\,\Omega$ represents the free wave impedance.

The reradiated power density back to the radar is:

$$\frac{P_t G}{4\pi R^2} \cdot \frac{\sigma}{4\pi R^2} \tag{2}$$

The RCS is defined as:

$$\sigma = 4\pi R^2 \frac{|E_r|^2}{|E_i|^2} \tag{3}$$

where R is the distance between the radar and the target, E_r is the electric field strength of the echo signal, and E_i is the incident electric field strength of the radar electromagnetic wave to the target.

2.2 Attitude Angle

When the size of the target is larger than the wavelength of the emitted wave, which is called the scattering "optical" region so that the change of attitude angle can greatly affect the RCS of the target.

The target attitude angle includes the azimuth angle A_θ and the pitch angle A_ϕ.

Where $V_i(i = \theta, \phi, \varphi)$ represents the unit vector of the angle i between the target and the corresponding coordinate axis.

According to the simulation, the RCS distribution curve with the azimuth angle from $0°$ to $180°$ is shown in Fig. 1.

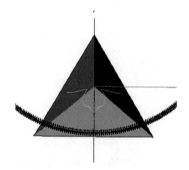

Fig. 1. 3D view of angular reverse RCS

As shown in Fig. 1, the RCS reaches its maximum at the axis of symmetry $45°$ from the edge of the triangle. The reason for the larger RCS near $-45°$ and $45°$ is the clutter caused by the edge diffraction and the specular reflection.

Next, Table 1 examines the influence of different pitch angles on the RCS value at the axis of symmetry.

Table 1. Units for magnetic properties.

Pitch angle (°)	RCS (dBsm)
85	5
75	11
65	15
58	21
55	24.5
50	22
45	17

RCS values corresponding to different pitch angles. It can be seen from Table 1 that when the pitch angle is $55°$, the RCS is consistent with the theoretically calculated value of 24.5 dBsm. Therefore, the RCS is the largest when the trihedral corner reflector takes the elevation angle of $55°$ and the azimuth angle of $45°$ as the central axis.

3 Simulation Results and Analysis

The computing environment used for simulations is configured with Win 7, I5-8500, 16G running memory and GTX 1060 6G graphics card. The radar model is Jms-1031 shipborne radar. First, the model is simulated using FEKO. A metal plate is simulated and the ability of the composite material to absorb incident electromagnetic waves is analyzed. The simulation model is shown in Fig. 2.

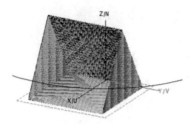

Fig. 2. FEKO simulation model

In Fig. 2, The length of the plate is 0.4m, the side is an isosceles triangle, the frequency of electromagnetic wave is set at 10GHz, and the pitch Angle of its normal line is 60° and the azimuth Angle is 45°.

The simulation results are shown in Fig. 3.

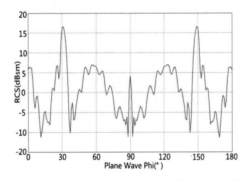

Fig. 3. Reverse simulation results of bucket angle

As shown in Fig. 3, when the horizontal azimuth of the bucket angle is 30° and 60°, the RCS reaches a maximum of 17 dBsm, which is caused by edge diffraction and mirror diffraction. At a horizontal azimuth of 90°, there is a crest with a RCS of 4 dBsm.

Next, take civil aircraft as an example to study the RCS distribution characteristics of the bucket angle reflector. As shown in Fig. 4, the RCS distribution characteristics of civil aircraft are studied.

As shown in Fig. 4, the aircraft has one high peak and two secondary peaks, which are the nose of the aircraft and two ejectors. Therefore, it is necessary to make corresponding changes to the structure of the bucket Angle reflector, as shown in Fig. 5.

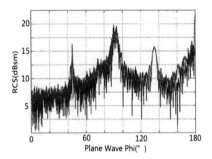

Fig. 4. The RCS distribution curve of the civil aircraft

Fig. 5. The simulation model of improved bucket angle reflector

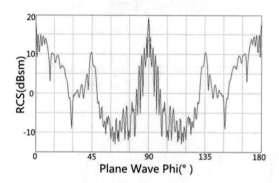

Fig. 6. Reverse simulation results of improved bucket angle

The simulation results are shown in Fig. 6.

As shown in Fig. 6, when the horizontal Angle is 90°, the RCS of the bucket Angle reflector is almost the same size as that of civil aircraft. However, the diffraction of the 0° and 180° mirrors is larger, so the material and structure need to be changed to absorb certain waves. The changed relative dielectric constant parameters are shown in Table 2.

As shown in Table 2, the decrease of the real part of the relative permittivity significantly increases the plane RCS. When the real part of the relative permittivity is 0.1, the imaginary part is 0.2, the RCS reaches the maximum value of 7.5 dBsm. Therefore,

Table 2. RCS drop values of different dielectric constants.

Real part (ε)	Imaginary part (ε)	Reduction (RCS)
4	0.3	1.5
2	0.3	4.0
1	0.3	5.5
0.5	0.3	4.0
1	0.2	7.5

the composite material chooses the relative dielectric constant $1 + i * 0.2$ to meet the requirements.

The physical object of the bucket Angle reflector is tested accordingly, as shown in Fig. 7.

Fig. 7. A physical picture of the improved bucket angle reflector

As shown in Fig. 7, the physical picture of the improved bucket angle reflector, and the RCS distribution characteristics of $0° -180°$ are measured, as shown in Fig. 8.

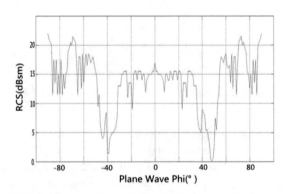

Fig. 8. The RCS distribution characteristics of the improved bucket angle reflector

As shown in Fig. 8, in the actual simulation of the improved bucket angle reflection, a peak value of 17 dBsm appears 90° away from the edge. However, the materials of 45° and 135° absorb less wave, which needs to be further improved in the future work.

4 Conclusions and Prospects

In this paper, a design scheme of bucket reflector based on complex structure is presented, and its simulator is simulated and verified in practice, so as to realize the RCS distribution characteristics of small target simulation of large passenger aircraft. The design makes full use of the influence of different attitude angle on the RCS of the bucket angle reflector, changes the material surface structure reasonably, increases or decreases the RCS in different positions with stronger applicability and flexibility.

For different target interference requirements, the design scheme in this paper can also be modified, which effectively reduces the production cost and the production time. In the following work, the problem of passive interference in the simulation should be analyzed more specifically, so as to promote the development of relevant technologies.

References

1. Zhang, C., Lin, C.: Universal radar target RCS modeling method. Mod. Defense Technol. **22**, 118–123 (2017). (in Chinese)
2. Chen, B., Li, G.: Calculation and simulation of target RCS. Electron Technol. **3**, 33–35 (2017). (in Chinese)
3. Qin, Q., Wang, X., Jiao, J.: Based on FEKO software target RCS calculation and data analysis. Appl. Electron. Technol. **44**, 108–110 (2016)
4. Zhang, Z., Li, X., Zhang, J.: Attitude based on correction program standard RCS dynamic measurement method. Syst. Eng. Electron. Technol. Surg. **41**, 1242–1248 (2019)
5. Zhu, Y.: Passive techniques for target radar cross section reduction: a comprehensive review. IEEE Trans. Aerosp. Electron. Syst. **1**, 17–25 (2010)
6. Bilal, A., Hamza, S.M., Salamat, S.: Multi-frequency analysis of Gaussian process modelling for aperiodic RCS responses of a parameterised aircraft model. IET Radar Sonar Navigat. **14**, 20–27 (2020)
7. Zhang, B., Yang, Y.: Statistical characteristics of full-polarization RCS of Ku band fixed-wing Uav. Mod. Radar **42**, 41–47 (2020)
8. Choi, W.-H., Song, W.-H., Lee, W.-J.: Equivalent pattern modeling method of periodic pattern surface and its application to the broadband radar absorbing sandwich structure. Adv. Compos. Mater **29**, 96–103 (2020)
9. Eskelinen, E., Ruuskanen, P., Rasanen, U.: Lowcost simulator for basic radar signals. IEEE Aerosp. Electron. Syst. Mag. **6**, 7–11 (1994)
10. Meng, K., Ma, W.: Study on electromagnetic scattering characteristics of floating twentyhedral angle reflector. Digit. Ocean Underwater Offense Defense **3**, 437–442 (2019)

Computer Molecular Biology Technology in the Field of Clinical Microbiological Examination

Yu Liu, Huizi Sun[✉], and Xiaoming Dong

The Second Affiliated Hospital of Qiqihar Medical University, Qiqihaer 161006, Heilongjiang, China

Abstract. The great achievements of molecular biology theory have had a great impact on human epistemology, way of thinking and social development. This research mainly discusses the application of computer molecular biology technology in the field of clinical microbiological examination. Use a high-throughput sequencing platform for actual measurement. According to the sample information, select appropriate reagents and methods to extract the total nucleic acid (total RNA) in the sample and store it for later use. Nucleic acid is randomly interrupted, that is, nucleic acid molecules are fragmented. It can be purified by ultrasonic fragmentation method, enzyme digestion method, etc. Using RNA molecules as templates, adding primers required for reverse transcription, constructing a suitable PCR reaction system, and gradually synthesizing the first and second cDNA strands. Add the sample and a series of reaction reagents to the reaction system, set the parameters, and run the sequencer for sequencing. Then conduct bioinformatics analysis,. when the virus mutation rate is 10%, all analysis procedures can calculate a relatively high coverage of the viral pathogen genome. This research contributes to the promotion of computer molecular biology.

Keywords: Computer molecular biology · Clinical microbiology testing · High-throughput sequencing · PCR reaction system

1 Introduction

Microorganisms also have various functions, which are both harmful and beneficial to humans. The types of microorganisms are complex and can adapt to almost any environment, and some are even able to withstand high temperatures and low cold. Microbiological testing has also developed rapidly. With the continuous deepening of the reform of my country's medical and health system, due to the actual needs of the development of clinical microbiological testing disciplines, the calls for improving the current unreasonable charging prices of clinical microbiological testing projects are increasing.

In order to reflect the principle of evidence-based decision-making and explore the true operating cost of microbial testing technology in a real environment, this study uses blood culture as a starting point to measure the cost range of the above-mentioned

J. Abawajy et al. (Eds.): ATCI 2021, AISC 1398, pp. 380–386, 2021.
https://doi.org/10.1007/978-3-030-79200-8_55

blood culture testing technology, and explore the problems in the development of clinical microbiological testing in our country. In order to break through the bottleneck of industry development and provide evidence-based basis for price reform decisions [1, 2]. And it will be helpful to further play the prominent role of microbiological testing in the clinical and public health fields [3, 4]. In today's society, computer technology continues to develop rapidly, and computers have become essential equipment for various scientific research institutions [5, 6]. Many inspection items of products have more or less introduced computer software to assist or replace manual operation [7, 8]. However, the microbiological inspection work that uses the microbiological detection method formulated by the Ministry of Health of the People's Republic of China still uses manual operations, and the identification results are obtained through multiple operations, and the results are manually compared to the standard to determine the results. Therefore, the issuance of the results often takes a long time and is prone to errors. Therefore, manual judgment of the results begins to be greatly challenged [9, 10].

Establish a high-throughput sequencing platform based on next-generation sequencing technology, considering different application backgrounds. Established a laboratory pathogen sequencing platform based on Iumiman's HSeq2500 sequencer and MSeq sequencer. For on-site environments or environments with limited laboratory conditions at the grassroots level, an on-site sequencing platform based on ABI's Ion torent sequencer was established. The rapid development of high-throughput sequencing technology has brought new breakthroughs to microbionomics research. Through full-omics sequencing, microbes can be studied in a comprehensive and detailed manner. Among the various functional analysis of microorganisms, the more important one is the analysis of microbial drug resistance, which has practical application significance in the fields of medicine and agriculture.

2   The Field of Clinical Microbiological Testing

2.1   Computer Molecular Biology

As an important part of the philosophy of science, biology has attracted more and more people's attention in the research of contemporary philosophy of science. Whether at home or abroad, the research on the philosophy of biology is a hot area in the philosophy of science. The establishment of the DNA double helix structure model opened the curtain of the era of molecular biology. In recent decades, the rapid development of molecular biology theory and the great achievements have had a great impact on human epistemology, thinking mode and social development. The new priority mechanism is expressed as follows.

$$Angle(a, r) = \arccos\left(\frac{a \cdot r}{|a| * |r|}\right) \tag{1}$$

The vector a is obtained by connecting each x to the starting point, and the vector r represents the search direction of the load setting information.

$$E_c = \frac{1}{L}\sqrt{\sum_{i=0}^{l-1}\left[y(k-i) - y_m(k-i)\right]^2 + y_m(k-i)} \tag{2}$$

Here, L is the evaluation time zone. E is the estimated average square error of the time domain model. The purpose of group distance selection is to calculate the density of individuals, select relatively sparse individuals, improve the diversity of individuals, and make individuals evenly dispersed. Sort the population in ascending order according to the target value of each dimension, and finally get the dense value of the individual.

$$Crowd[i]_d = Crowd[i] + \frac{Crowd\lfloor i+1 \rfloor_m - Crowd\lfloor i-1 \rfloor_m}{f_m^{max} - f_m^{min}} \quad (3)$$

Here, $Crowd[i]_d$ represents the maintenance target value of MTH.

2.2 Microbiological Inspection

Microbiological testing is an important means for patients to accurately use drugs, prevent and control nosocomial infections, regulate the use of antimicrobial drugs, and contain antimicrobial resistance. Microbiological testing plays an important role in determining the pathogenicity of microorganisms, fighting against new pathogenic microorganisms, and preventing the spread of infectious diseases. For example, the SARS epidemic has swept the world. The use of reverse transcription-polymerase chain reaction (RT-PCR) technology to detect new coronaviruses provides an important basis for the study of the pathological mechanism of SARS and provides corresponding measures for the prevention and control of infectious diseases. In our country, medical institutions at all levels have generally insufficient investment in the clinical microbiological testing specialty, and they are mostly dependent on the laboratory and other sub-specialties to maintain their operation. Moreover, the clinical microbiological testing projects have been repeatedly reduced in price and the fees are too low, which has led to the long-term. In a state of loss-making operation, it is difficult to effectively compensate the cost of pathogen culture and identification, which has become one of the important reasons that seriously hinder professional development and affect the effectiveness of clinical microbiological testing.

3   Clinical Microbiological Testing Experiment

Use a high-throughput sequencing platform for actual measurement.

(1) Nucleic acid extraction. According to the sample information, select appropriate reagents and methods to extract the total nucleic acid (total RNA) in the sample and store it for later use.
(2) Nucleic acid random interruption, that is, nucleic acid molecule fragmentation, the purpose is to break larger RNA molecules into small molecules of appropriate length to facilitate subsequent library construction and sequencing. It can be purified by ultrasonic fragmentation method, enzyme digestion method, etc.
(3) Reverse transcription. Using RNA molecules as templates, adding primers required for reverse transcription, constructing a suitable PCR reaction system, and gradually synthesizing the first and second cDNA strands.

(4) PCR amplification of cDNA library. Due to the requirements of next-generation sequencing for the initial loading volume, many experiments require PCR amplification of trace nucleic acids. Generally, whole transcriptome amplification is used to minimize amplification bias.

(5) Sequencing on the computer. The sample and a series of reaction reagents (DNA polymerase, dNTP, etc.) are added to the reaction system, the parameters are set, and the sequencer is run for sequencing.

(6) Obtain data. Then conduct bioinformatics analysis. HiSeq and MiSeq sequencing overview are shown in Table 1.

Table 1. HiSeq and MiSeq sequencing overview

Statistics project	HiSeq	MiSeq
Total number of samples sequenced	259	766
Number of sequencing	7	33
Virus sample	42	99
Bacteria sample	45	145

4 Microbiological Inspection and Analysis

4.1 Variability of the Sequencing Depth on the Whole Genome and Segment of HIN1

Table 2 shows the variability of sequencing depth on the whole genome and segments of HIN1. In order to have an intuitive quantitative observation index for the uniformity of coverage, we adopted the coefficient of variation (CV) to calculate the coefficient of variation for the entire genome of each sample and the coverage depth of each segment. It can be seen from the table that the variability of the coverage depth of the untreated samples is significantly smaller than that of the other treated samples (1.00–1.59). The degree of variability in each segment showed the same trend (Wlcoxon Iank sum test $p < 0.05$). We observed that in these segments, NA, NP, and NS segments seem to be more susceptible to experimental pretreatment methods and severe coverage bias (CV > 1) during this process.

Table 2. Sequencing depth variability in the whole genome and segments of HIN1

Treatment	Segment CV			Genome CV	P
	NA	NS	NP		
No retreatment (0.55%)	0.54	0.79	0.48	0.49	–
BD (0.55%)	0.92	1.20	1.06	1 03	0.008
8-h WTA (0.55%)	0.80	1.55	1.06	159	0.008
No retreatment (1.50%)	0.51	0.58	0.46	0.47	–
BD + 2-h WTA (1.50%)	1.08	0.63	1.18	1 05	0.008

4.2 Qualcomm's Most Sequencing Analysis of the Reactor Under Different Working Conditions

Working condition 1: The total nitrogen concentration of the inlet water is about 800 mg/L, the inlet water volume is 500 ml/d, and the submerged water inlet. Working condition 2: The total nitrogen concentration of the inlet water is about 600 mg/L, the inlet water volume is 700 ml/d, and the immersion type water inlet. High-throughput sequencing was performed on the filler samples in each working condition in the reactor, and effective sequences were obtained after quality analysis and screening. The number of filler sample sequences in Working Condition 1 and Working Condition 2 are 27516 and 2551 respectively, and the average sequence length is 464.6bp and 455.9bpo respectively. After comparing the measured sequences, perform OTU classification, and obtain 3883 in Working Condition 1. OTUs, 877 OTUs were obtained in the second working condition. O1 and O2 sample coverage (Coverage) were 92% and 77%, respectively, indicating that the amount of sequencing of the samples is reasonable and can truly reflect the microbial community structure in the reactor. Qualcomm's best sequencing of the reactor under different working conditions is shown in Fig. 1.

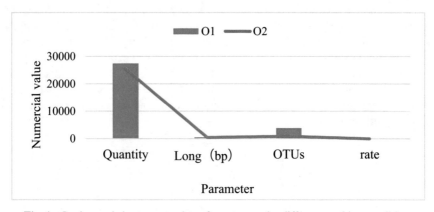

Fig. 1. Qualcomm's best sequencing of reactors under different working conditions

We chose Humanadenovirus 55 (gil256028986IGBIFJ643676.1) as the target pathogen, and set the mutation rate of the genome to 10%, 20%, 30% and 100% (i.e. completely new genome) to construct In the simulated sequencing data set, the average coverage depth of the sequencing data to the pathogen genome is also 10X, the sequencing read length is 150bp, paired-end sequencing, and the sequencing platform is Ilumina. The results show that when the mutation rate is 10%, all analysis procedures can calculate a relatively high viral pathogen genome coverage, but when the mutation rate is higher than 20%, only the P2 process can detect the target virus pathogen well. P3 and P4 analysis process, their analysis process depends on the known database. Comparing the completely new viral genome sequence with the known database will only treat them as unnecessarily matched reads, so subsequent analysis of the target virus sequence cannot be performed. That is to say, the analysis process (such as P1 and P2) that does not depend on the prior knowledge of the pathogen sequence can be suitable for the detection of novel viral pathogens. The results of human adenovirus sequencing are shown in Fig. 2.

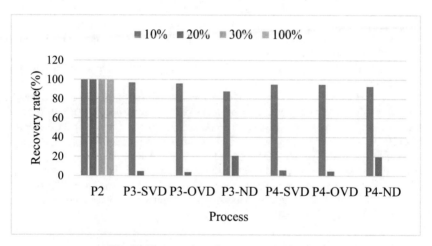

Fig. 2. Human adenovirus sequencing results

5   Conclusion

This research mainly discusses the application of computer molecular biology technology in the field of clinical microbiological examination. Use a high-throughput sequencing platform for actual measurement. According to the sample information, select appropriate reagents and methods to extract the total nucleic acid (total RNA) in the sample and store it for later use. Nucleic acid is randomly interrupted, that is, nucleic acid molecules are fragmented. It can be purified by ultrasonic fragmentation method, enzyme digestion method, etc.

Using RNA molecules as templates, adding primers required for reverse transcription, constructing a suitable PCR reaction system, and gradually synthesizing the first and second cDNA strands. Add the sample and a series of reaction reagents to the reaction system, set the parameters, and run the sequencer for sequencing. Then conduct bioinformatics analysis. This research contributes to the promotion of computer molecular biology.

References

1. Gao, S., Song, S., Cheng, J.: Incorporation of solvent effect into multi-objective evolutionary algorithm for improved protein structure prediction. IEEE/ACM Trans. Comput. Biol. Bioinf. **15**(4), 1365–1378 (2018)
2. Bhattacharjee, P., Mitra, P.: A survey of density based clustering algorithms. Front. Comput. Sci. **15**(1), 1–27 (2021)
3. Khan, Y.D., Rasool, N., Hussain, W.: Iphosy-pseaac: identify phosphotyrosine sites by incorporating sequence statistical moments into pseaac. Mol. Biol. Rep. **45**(6), 2501–2509 (2018)
4. Levin, M., Bongard, J., Lunshof, J.E.: Applications and ethics of computer-designed organisms. Nat. Rev. Mol. Cell Biol. **21**(11), 1–2 (2020)
5. Yamasaki, S., Amemiya, T., Yabuki, Y.: Togo-WF: prediction of RNA tertiary structures and RNA–RNA/protein interactions using the KNIME workflow. J. Comput. Aided Mol. Des. **33**(5), 497–507 (2019)
6. Breuer, M., Ananthanarayanan, V.: Cell scientist to watch -Vaishnavi Ananthanarayanan. J. Cell Sci. **132**(11), 1–2 (2019)
7. Acuna, V.V., Hopper, R.M., Yoder, R.J.: Computer-aided drug design for the organic chemistry laboratory using accessible molecular modeling tools. J. Chem. Educ. **97**(3), 760–763 (2020)
8. Khedr, A.M., Ibrahim, M.H.: LPB: a new decoding algorithm for improving the performance of an HMM In gene finding application. IAENG Int. J. Comput. Sci. **47**(4), 4–13 (2020)
9. Khan, T., Ahmad, R., Azad, I., Raza, S., Joshi, S., Khan, A.R.: Computer-aided drug design and virtual screening of targeted combinatorial libraries of mixed-ligand transition metal complexes of 2-butanone thiosemicarbazone. Computat. Biol. Chem. **75**(4), 178–195 (2018)
10. Zhao, Z., Zhang, W., Che, W.: An evaluation of chinese human-computer dialogue technology. Data Intell. **1**(2), 187–200 (2019)

Cloud Computing Network Security Technology Based on Big Data Analysis

Fanxing Meng[(⊠)]

Shandong Institute of Commerce and Technology, Jinan, Shandong, China
20060494@sict.edu.cn

Abstract. As a new network model, Internet big data cloud computing technology can integrate excellent computing practices in the Internet, significantly improve the efficiency of big data computing, and make full use of network resources. The development of network technology has reached a new height, and the significance and role of network data security has also been widely valued. Based on this, this paper begins to study the application of cloud computing network security technology in big data analysis. Based on this, this paper studies the cloud computing network security technology under the big data analysis. In this paper, literature analysis and statistical analysis are selected to study. In these studies, by collecting a large number of relevant information, it has laid a solid theoretical basis for the in-depth study of this paper. This paper takes the network and data security in cloud computing environment as a key research topic. According to the characteristics of cloud computing, this paper introduces the related concepts and definitions of cloud computing, and the current research results in the field of cloud computing network and cloud computing security. In view of the dynamic data operation and the integrity and consistency verification of data files in cloud computing environment, this paper introduces the concept and definition of cloud computing, this word proposes a verifiability experiment based on Merkle hash tree technology to support data dynamic operation, and proposes an access control experiment based on attribute encryption algorithm for data file access control in cloud computing environment. The results show that the hierarchical Merkle hash tree construction experiment in this paper has obvious advantages over other experiments in terms of time cost. The attribute based encryption algorithm can minimize the generation of key and the times of encrypting plaintext and decrypting ciphertext, and can effectively improve the permission change rate of massive users sharing data files in cloud computing environment.

Keywords: Big data · Cloud computing · Network security technology

1 Introduction

In recent years, with the rapid development of computer network technology, China has ushered in the era of big data. Information technology is the theme of the times, and also an important force in social and economic development [1, 2]. With the rise and development of cloud computing, information technology has risen to a new level to

© The Author(s), under exclusive license to Springer Nature Switzerland AG 2021
J. Abawajy et al. (Eds.): ATCI 2021, AISC 1398, pp. 387–394, 2021.
https://doi.org/10.1007/978-3-030-79200-8_56

achieve effective resource saving [3, 4]. Therefore, the application scope of information technology is also expanding, which is closely related to people's life [5, 6]. As a new mode of network operation, big data cloud computing can integrate excellent computer experiments into the Internet and significantly improve its efficiency. The development of network technology has reached a new level, and the role of network data security has also been paid special attention [7, 8].

In the research of cloud computing network security technology based on big data analysis, many scholars have studied it and achieved good results. For example, Khaled Salah transplanted IDS, SSL, IPS, as and other security Mboxes to cloud computing VM to build a scalable network security architecture, which can not only provide security services for cloud computing, data centers and their enterprise networks, but also make full use of the dynamic scalability of cloud computing [9]. Cloud watcher mentioned customized cloud computing network security services. According to cloud user services, security requirements are filled in and distributed to cloud watcher. Cloud watcher analyzes these requirements and forms customized security services [10].

In order to better carry out the research, this paper first expounds the main problems of network security technology, and then classifies the implementation path of network security technology, and in view of these problems, adopts the feasible way, puts forward a kind of technology based on Merkle hash tree, In order to verify the effectiveness of these technologies, we analyze the experimental data.

2   Research on Cloud Computing Network Security Technology Based on Big Data Analysis

2.1   Main Problems of Network Security Technology

(1) Environmental issues
     Due to the increasing number of network users, the scope of information and data is also expanding, but the Internet itself is open, any user can access the cloud computing system, the current user identity analysis will be affected. In addition, some criminals rent servers to access the cloud computing system, so it is difficult for the system to produce appropriate crisis, which will inevitably increase the possibility of data leakage.

(2) Problems of laws and regulations
     Some network hackers will try to steal business secrets by various means, but the Internet is confidential, it is difficult to identify these intruders. Moreover, the existing laws and regulations in the computer network security still have some shortcomings, which increases the risk of network risks.

(3) Technical issues
     Technical problems are mainly reflected in the system, when the system fails, the user's operation is inevitably limited, there are certain security risks, both information transmission and communication. At present, the system failure will inevitably lead to the reduction of protection ability, and the possibility of hacker and virus attack is higher, which is also the key factor of future network security technology. Network problems can also encourage hacker attacks. For example, the common problem of e-mail security is the main manifestation of information storage.

Although the traditional network environment has been shared and shared data has been processed, some data still run by themselves. Whether in the cloud platform of network operators or in the network environment of users, the main target of hacker attacks has become the use of network channels to plant viruses and steal important information, which is more serious.

2.2 Network Security Technology Implementation Path

(1) System protection system

System protection is a technology to protect the subsystem in the cloud computing environment, including the investigation and repair of system vulnerabilities, and the programming of system software. When users generate data in the network environment, in order to avoid attacks between virtual machines, the system will isolate data, enhance platform security, focus on abnormal conditions, monitor Trojan content, and send it to the server analysis system for detection, the detection results will be sent to the client and set the warranty.

(2) Identity authentication system

In order to ensure the security of network data communication, we need to improve and optimize the identification. Through the ID card identification technology, we can effectively identify the hacker intrusion and the use of ordinary users. In order to ensure the security of the system, we will not be cheated by the network and blocked by the network. For example, hackers often forge IP addresses, steal data in order to invade the system, and reach the target through malicious data In order to destroy the system, users who have the right to access the network can pass the real name authentication, and can determine the authorized subject of the network. With the further intrusion of hackers, administrators should regularly check the identity recognition system to identify possible security risks. Some unauthorized users should deeply analyze their data and use digital security operation procedures to protect the system in real time. For example, network operators can also use different identification methods to ensure the security of the system through a variety of channels. In the process of a user's connection, a blacklist system can be created to enhance the network security with the computing cloud, and the users with normal data flow can not be affected by the system.

2.3 Correlation Function Formula

$C = $ Encrypt(P_k, M, T): This function uses public parameters P_k and access control structure T on plain text M encryption, and get the cipher text C. Access control structure constructed by any node of T, y the value is k_y, Randomly generated for nodes $k_y \rightarrow 1$ polynomial q_y, $q_y^{(0)}$ is secret information stored for nodes. Let $q_R^{(0)} = s$, $s \in Z_p$, and it is randomly selected, among R is represents the root node. Corresponding, Other nodes $yq_y(0) = q_{father(y)}(tag(y))$, among father$(y)$ representative $y's$ father node, tag(y) represents the number of the node y. Let collection of all leaf nodes is X, It meets the requirements of authorization set collection of corresponding access control structure is

Γ The function Encrypt(P_k, M, T) is shown in (1), among, att(x) is returns the attribute information of the node x.

$$\left(\Gamma, C^{\sim} = Me(g, g)^{as}, C = g^{bs}, \forall x \in X : C'_x = T_{att(x)}^{q_y(0)}\right) \tag{1}$$

S_k = SkeyGen(M_k, A): This function uses the master key M_k, And user attribute collection A to generate user private key S_k A as the collection of properties associated with the user's private key, Is a non empty subset of the data U file property set. Select random number $\gamma \in Z_p$ Individual properties $s \in A$, random number $\gamma_s \in Z_p$ The function SkeyGen(M_k, A) is shown in formula (2).

$$\left(D = g^{(a+\gamma)/b}, \forall s \in A : D_s = g^{\gamma} T_s^{\gamma_s}, D'_s = g^{\gamma_s}\right) \tag{2}$$

3 Experimental Research on Cloud Computing Network Security Technology Based on Big Data Analysis

3.1 Experimental Study on Verifiability of Data Dynamic Operation Based on Merkle Hash Tree

Because the experiment uses the cloud computing data storage security architecture, and based on certain security assumptions (reliable communication channel, TPA trusted and CDC faithful, etc.). Therefore, user can not only trust TPA to handle the verification and audit of data files, but also the maintenance and calculation of verification information need not be done by user. The experiment makes full use of the huge computing resources and storage space of CDC in the cloud computing environment, which not only maximizes the advantages of large-scale and virtualization of cloud computing, but also saves a lot of user resources. Moreover, compared with the previous research results of scholars, the experiment can support all the basic data dynamic operation in the cloud computing environment. In the aspect of data integrity and consistency verification, the experiment uses the root node value of hierarchical Merkle hash tree as verification information, which organically integrates the advantages of the data structure in data integrity and consistency verification with the distribution characteristics of data storage in cloud computing environment. In terms of the construction of merge hash tree and the calculation of root node value, the experiment cleverly uses the LTAG tag marked with data block to store geographic location information, which not only effectively reduces the number of nodes in merge hash tree In the process of calculating the root node value of tree, due to the large amount of time consumed by the internal communication of CDC, the system does not need to use the hash value of all data blocks of the data file to reconstruct the file Merkle hash tree and calculate the root node value after the dynamic operation of the data file by user, which effectively improves the generation rate of the verification information of the data file after the dynamic operation.

3.2 Experimental Research on Access Control Based on Attribute Encryption Algorithm

The experiment first uses AES symmetric encryption algorithm to encrypt the data file. AES symmetric encryption algorithm has been proved to have high security, so the data file stored in CDC can effectively avoid the snooping of CDC. In addition, the experiment uses the attribute encryption algorithm based on cipher text strategy to encrypt the symmetric key. The attribute encryption algorithm based on cipher text policy guarantees the confidentiality and security of the algorithm through attribute set and access control tree structure. Because the experiment uses the cloud computing data storage security architecture, and based on certain security assumptions (reliable communication channel, trusted TPA and loyal CDC, etc.), the verification of data file access control permission after the introduction of tPA has better security than the traditional access control experiment in the authentication and private key redistribution after the access permission change. In addition, the introduction of the key attribute Ka and the file identification attribute RA can effectively avoid the occurrence of user's unwitting misreading of dirty data after the data file is changed. It is very suitable for the storage of sensitive data files in cloud computing environment.

In terms of time cost, user leaves a large number of encryption tasks and key generation and distribution to CDC, which saves a lot of computing resources and time, and gives full play to the huge computing power of CDC in cloud computing environment. At the same time, through the use of cloud computing data storage security architecture, the trusted third-party audit institutions are introduced to support public audit. User does not need to verify the data files stored in CDC, which saves user's valuable time and strengthens the audit ability of data files.

4   Cloud Computing Network Security Technology Based on Big Data Analysis

4.1   Research and Analysis of Data Dynamic Operation Verifiability Based on Merkle Hash Tree

The simulation experiment describes the time comparison between the hierarchical merge hash tree reconstruction and the unstructured merge hash tree reconstruction after the data file is inserted under the condition of different number of distributed nodes. The reconstruction time comparison of merge hash tree after inserting data files is shown in Table 1.

Table 1. Refactoring time comparison of insert operations

Number of nodes	Layered	Not stratified
1	1501	1501
5	1504	475
20	1507	268
50	1502	243
100	1506	277
500	1507	531

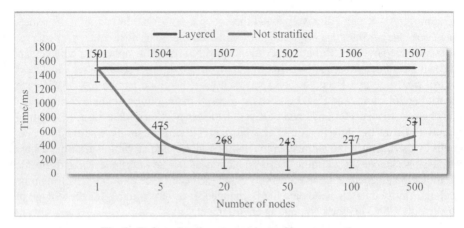

Fig. 1. Refactoring time comparison of insert operations

From the simulation results in Fig. 1, we can see that we insert the data file, and the hierarchical structure of merge hash tree obviously saves time compared with the non hierarchical structure of merge hash tree. Moreover, within a certain range of nodes, the construction time of hierarchical Merkle hash tree will decrease with the increase of nodes. Therefore, the hierarchical Merkle hash tree construction experiment adopted in this paper has obvious advantages in time cost.

4.2 Experimental Research and Analysis of Access Control Based on Attribute Encryption Algorithm

By simulating the real encryption and decryption environment, the simulation experiment records the time of key generation, plain text encryption and cipher text decryption based on attribute encryption algorithm under different number of attributes, as shown in Table 2.

Table 2. The relation between the number of attributes and time in attribute encryption algorithm

Number of attributes	Generate key	Encrypted plain text	Decrypt cipher text
5	151	784	1207
10	160	819	1298
15	172	967	1401
20	183	1155	1473
25	191	1362	1594
30	205	1578	1802

According to the simulation results in Fig. 2, the time consumption of the attribute based encryption algorithm in the process of encrypting plain text and decrypting cipher

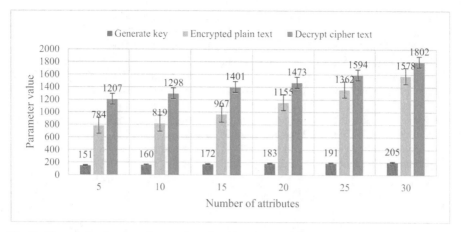

Fig. 2. The relation between the number of attributes and time in attribute encryption algorithm

text will increase greatly with the increase of the number of attributes, and the time consumption of key generation will also increase slightly with the increase of the number of attributes. Therefore, in the attribute based encryption algorithm, the number of key generation, plain text encryption and decryption can be reduced as much as possible, which can effectively improve the efficiency of applying the attribute based encryption algorithm to the access control experiment.

5   Conclusions

Although the rapid popularization and deepening of network technology provides great opportunities for people's life and work, there are still many network security problems to be solved in the big data cloud computing environment. Cloud computing and big data, as the hot spot of research and implementation in recent years, are considered as the main application architecture of next generation electronic network technology by most computer companies and internal participants. In order to solve the security problem of cloud computing network on the basis of big data analysis, this paper proposes a method based on Merkle hash Tree's access control technology and an attribute encryption algorithm technology, and its experimental research.

References

1. Zissis, D., Lekkas, D.: Addressing cloud computing security issues. Futur. Gener. Comput. Syst. **28**(3), 583–592 (2012)
2. Subashini, S., Kavitha, V.: A survey on security issues in service delivery models of cloud computing. J. Netw. Comput. Appl. **34**(1), 1–11 (2011)
3. Li, J., Li, J., Chen, X.: Identity-based encryption with outsourced revocation in cloud computing. IEEE Trans. Comput. **64**(2), 425–437 (2015)
4. Szefer, J., Lee, R.B.: Architectural support for hypervisor-secure virtualization. SIGARCH Comput. Archit. News **40**(1), 437–450 (2012)

5. Joseph, D., Stoica, I.: Modeling middle boxes. IEEE Netw. **22**(5), 20–25 (2008)
6. McKeown, N., Anderson, T., Balakrishnan, H.: Open flow: enabling innovation in campus networks. ACM SIGCOMM Comput. Commun. Rev. **38**(2), 69–74 (2008)
7. Sherry, J., Hasan, S., Scott, C.: Making middle boxes someone else problem: network processing as a cloud service. ACM SIGCOMM Comput. Commun. Rev. **42**(4), 13–24 (2012)
8. Pearce, M., Zeadally, S., Hunt, R.: Virtualization: issues, security threats, and solutions. ACM Comput. Surv. (CSUR) **45**(2), 17 (2013)
9. Mishra, A., Jain, R., Durresi, A.: Cloud computing: networking and communication challenges. IEEE Commun. Mag. **50**(9), 24–25 (2012)
10. Azodolmolky, S., Wieder, P., Yahyapour, R.: Cloud computing networking: challenges and opportunities for innovations. IEEE Commun. Mag. **51**(7), 54–62 (2013)

Balance Detection System Based on the Entropy Weight Method and Decision Tree Classification

Bocheng Liu and Boxiang Ji[(✉)]

School of Software, Nanchang University, Nanchang 330047, Jiangxi, China

Abstract. With the aging of the population, the safety problems of the elderly due to falls are particularly prominent. So far, elderly balance has not been specifically evaluated by traditional balance research and mainstream medical service platforms. Therefore, this paper studies the balance assessment system for the elderly. Based on the kinematic features extracted from the motion coordinates of the elderly, the basic information is clustered and analyzed. Then using a random forest to conduct a preliminary check on the information. Finally, according to the entropy method and the decision tree interval division method, the elderly balance is analyzed, and the final model is obtained. The accuracy can reach more than 70%. The system can evaluate elderly balance, and provide advice for the elderly to prevent falls and reduce the risk of falls.

Keywords: Balance detection · Cluster analysis · Entropy weight method · Decision tree classification

1 Introduction

Falls are a common problem when walking in elderly people [1]. Due to weakness, falls can easily have an irreversible impact on the body of the elderly, and even lead to premature death [2, 3]. So far, mainstream medical service platforms have not yet assessed the balance ability of the elderly.

This article analyzes the sample of the elderly randomly selected by the research institution from 42 human monitoring points, and performs a cluster analysis on the basic information table of the elderly. Random forest is used to conduct preliminary check on the kinematic features extracted according to the motion coordinates of the elderly. According to the entropy method and the decision tree interval division method, the final model is obtained.

Among them, K-means clustering is performed on the general information in the basic information of the elderly, which is divided into two categories. According to the dynamic state and static kinematics analysis, the final 25-dimensional feature is selected through the chi-square analysis of 40 multi-dimensional features. Before building the model, the random forest algorithm is used to cross-validate the labels and features. Finally, each elderly person is scored based on the entropy weight method, and the cart decision tree is used to divide it into multiple intervals to obtain the final model. Through the analysis of human balance ability and corresponding balance ability training, the elderly can be prevented from falling to the greatest extent.

© The Author(s), under exclusive license to Springer Nature Switzerland AG 2021
J. Abawajy et al. (Eds.): ATCI 2021, AISC 1398, pp. 395–400, 2021.
https://doi.org/10.1007/978-3-030-79200-8_57

2 Data Analysis and Preprocessing

2.1 Data Source

The data set consists of two parts, The first part is the 24 basic information data of 80 elderly people. The second part is the original data of 80 elderly people's free walking state calibrated, and the random sampling of 42 monitoring points on human body is selected. From 42 monitoring points [4], 25 can be found to represent the characteristics of body balance.

2.2 Analysis and Treatment

In mechanics, balance refers to a state of an object when the resultant force applied to the object is zero [5]. According to the basic information of the elderly, we look for factors that may affect the ability to balance: Age; Sex; Height; Weight; Fall times; BMI; Complaint;

After discarding invalid values, extract key feature values: 1) Create two new features, Staircase_cross and Force_platform, to indicate whether the stairs and the force platform cross. 2) Since the three sets of Aid by hand or not data are basically the same, one set of values is used. 3) Categorize the 4 diseases that have a large number of patients. 4) Group continuous values such as age, height, weight, BMI value, etc.

Associate the above variables with the number of falls, and select 10 features. By K-means clustering the balance divided into two categories, Class1 (balance weak) and Class2 (balance normal) [6]. According to the comparison result of eigenvalues, class2 is regarded as a class with strong balance ability. Clustering results shown in Fig. 1.

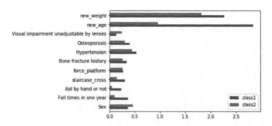

Fig. 1. Clustering results

In order to facilitate the identification of the relevant features that affect the human body balance and simplify the 42 human body monitoring points, we select some key points to describe the human skeleton [7]. Simplify the Hanavan model and divide the human body into twelve uniform rigid body links, namely the head and neck, the forearms on both sides, the upper arms on both sides, the torso, the thighs on both sides, the calves and the feet on both sides. Therefore, the calculation of the heterogeneous human heart is converted into the calculation of the center of gravity of twelve average rigid bodies. Simulate human gait movement, as shown in Fig. 2.

From this, the calculation formula for each link is obtained:

$$r_i = (1 - p)r_{pi} + r_{di} \qquad (1)$$

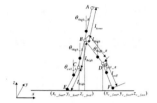

Fig. 2. Simulate human gait movement

In formula (1), r_i is the coordinate of the center of gravity of each link, r_{pi} is the position of the upper end of the joint, r_{di} is the lower end of the joint, P is the scale factor of the center of gravity of the chain link.

Filter features and divide more than 40 features into the following categories: 1) Features such as speed, acceleration, and position of the center of gravity of a special position. 2) Angle characteristics of important joint positions [8, 9]. 3) The characteristics of special moments, such as the height of the arm swing to the highest [10]. 4) Some overall characteristics, such as step length, step frequency, step width.

3 Model Building

After integrating the classification label of the data set with the data set, put it into Weka for analysis, and select the random forest algorithm with higher success rate and coverage rate. When classifying the data set in Weka, set the 10-fold cross-validation test method and obtain the calculation result.

The elderly balance model uses the 25-dimensional features extracted above as the final feature. In order to reasonably evaluate the balance ability of the elderly, 25 characteristics were weighted and scored to establish a balance ability evaluation system. The specific process: 1) Standardize the data of various indicators. 2) Calculate the 25-dimensional feature Pij of the elderly. 3) Find the information entropy of each indicator. 4) Calculate the weight of each indicator. 5) Score 76 elderly people.

The final 76 elderly balance score results are shown in Table 1.

Table 1. 76 elderly balance score.

Elder	No. 1	No. 2	No. 3	No.4	...	No. 76
Score	0.238172	0.147403	0.278230	0.309570	...	0.364201

The weight scores of 76 elderly people are ranked in ascending order. Choose two extreme values as the threshold value and divide the score table into three intervals to judge the balance. Use matlab's CART classification tree for interval selection, and the results are shown in Fig. 4.

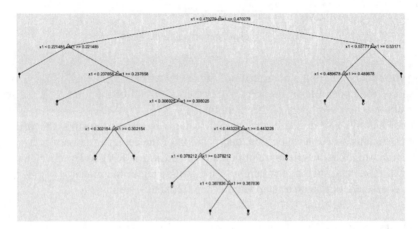

Fig. 4. Decision tree before pruning

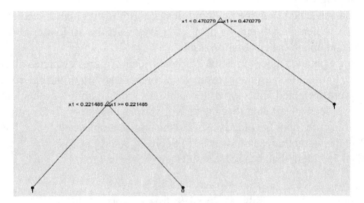

Fig. 5. Decision tree after pruning

The initial classification tree is cross-validated and then pruned to obtain the decision tree as shown in Fig. 5 above. The clustering results of the elderly information of the interval division result are further correlated, and the result of clustering and interval is shown in Table 2.

Accuracy Rate = True/False * 100% = 55/76*1 0 0% = 72.368%.

Use the random forest algorithm to combine tags and features to do a cross-validation. Finally, based on the entropy method, each elderly person is scored, and the cart decision tree is divided into intervals, and the final model is obtained. The results of stratified cross-validation and detailed accuracy by class are shown in Table 3, Table 4. After

Table 2. Result of clustering and interval

Elder number	Interval division result Y1	Cluster analysis results Y2	Y1Y2 + Y1′Y2′
No. 1	Good	Good	True
No. 2	Bad	Bad	True
No. 3	Good	Bad	False
...	...	...	...
No. 76	Good	Bad	Flase

obtaining more information about the elderly, there is a lot of room for improvement to increase the accuracy.

Table 3. Stratified Cross-validation

Category	Number	Proportion
Correctly classified Instances	57	75%
Incorrectly classified Instances	19	25%
Kappa statistic	0.4738	
Mean absolute error	0.378	
Root mean squared error	0.433	
Relative absolute error	78.98%	
Root relative squared error	88.56%	
Total number of Instances	76	

Table 4. Detailed accuracy by class

Category	TP rate	FP rate	Precision	Recall	F-Measure	MCC	ROC area	PRC area	Class
Detailed accuracy	0.804	0.333	0.787	0.804	0.796	0.474	0.771	0.806	Y
	0.667	0.196	0.69	0.667	0.678	0.474	0.771	0.685	N
Weighted avg	0.75	0.279	0.749	0.75	0.749	0.474	0.771	0.758	

4 Conclusions

This paper divides the useful information extracted into two categories through K-Means clustering. Using dynamic and static kinematics to analyze 40 multi-dimensional features, we select 25-dimensional features after performing chi-square analysis. Before building the model, cross-validate the labels and features according to the random forest algorithm, and the effect is about 75%. According to the entropy method, each elderly person is scored and divided into multiple intervals using the cart decision tree to obtain the final model. Compared with the original classification label, the accuracy rate exceeds 70%. According to the final analysis result, the overall accuracy rate reaches 0.75, and the ROC is also higher than 0.7, which has a certain degree of accuracy. The current research on balance has not been applied to the elderly, and mainstream medical platforms also have not conducted a detailed assessment of the balance ability of the elderly. Therefore, the system can evaluate elderly balance, and provide advice for the elderly to prevent falls and reduce the risk of falls. After obtaining more samples of the elderly, the system can further improve the accuracy rate, which has a lot of room for improvement.

Acknowledgement. This work was supported by Student innovation and entrepreneurship training program (no. 2020CX173).

References

1. Gill, T.A., Taylor, A.W., Pengelly, A.B.: A population-based survey of factors relating to the prevalence of falls in older people. Gerontology **51**(5), 340–345 (2005)
2. Peel, N.M.: Epidemiology of falls in older age. Can. J. Aging **30**(1), 7–49 (2011)
3. Stel, V.S., Smit, J.H., Pluijm, S.M.: Consequences of falling in older men and women and risk factors for health service use and functional decline. Age Ageing **33**(1), 58–65 (2004)
4. Pollock, A.S., Durward, B.R., Rowe, P.J.: What is balance? Clin. Rehabil. **14**(5), 402–406 (2000)
5. Meyerson, A., O'callaghan, L., Plotkin, S.: A k-median algorithm with running time independent of data size. Mach. Learn. **56**, 61–87 (2004)
6. Hawk, C., Hyland, J.K., Rupert, R.: Assessment of balance and risk for falls in a sample of community-dwelling adults aged 65 and older. Chiropract. Osteopathy **14**(1), 1–8 (2006)
7. Schimke, L.: Schimke J: Urological implications of falls in the elderly: Lower urinary tract symptoms and alpha- blocker medications. Urolog. Nurs. Off. J. Am. Urlol. Assoc. Allied **34**(5), 223–229 (2014)
8. Jahn, K., Zwergal, A., Schniepp, R.: Gait disturbances in old age: classfication, diagnosis, and treatment from a neurological perspective. Dtsch. Arztebl. Int. **107**(17), 306–315 (2010)
9. Berg, K.O., Maki, B.E., Williams, J.I.: Clinical and laboratory measures of postural balance in an elderly population. Arch. Phys. Med. Rehabil. **73**(10), 1073–1080 (1992)
10. Blanks, R.H., Fowleer, C.G., Zizz, C.A.: Postural adjustments produced by moving visual (horizontal optokinetic) patterns. J. Am. Acad. Audiol. **7**(11), 39–48 (1996)

Early Warning Mechanism of Network Public Opinion Crisis in Big Data Environment

Qing Liu[✉]

The College of Literature and Journalism of Sichuan University, Chengdu, Sichuan, China

Abstract. While the Internet brings convenience to people's production and life, it may also bring people into a crisis of public opinion. When a public crisis event occurs, the major media spread it to netizens through the Internet, which instantly arouses social attention. People express their opinions and opinions through the Internet and are easily misled by others. At this time, an online public opinion crisis may break out. It is necessary for government functional departments to take scientific and effective measures to prevent such incidents. For this reason, this article studies the online public opinion crisis early warning mechanism, which includes four mechanisms of identification, evaluation, response and coordination, which can prevent the occurrence of online public opinion crisis from the source, to improve the credibility of government departments.

Keywords: Big data · Network · Public opinion crisis · Warning mechanism

1 Introduction

At present, people can freely express their opinions and opinions on the Internet. The biggest feature of online public opinion is that there are many public opinions, unclear guidance, difficult to control, and easy to induce public opinion crisis. Although there are many factors that trigger the crisis of online public opinion, the outbreak of public opinion is also regular, which can be roughly summarized into several periods of formation, rise, change, and decline.

In real life, once there is a major event that damages the interests of the people or affects the people's livelihood, various media will follow up and report, and then arouse people's attention and discussion. This is the initial period of the formation of a crisis of online public opinion [1]. At this time, the scale of public opinion is not large, and the harm is relatively small. It is the best period for relevant departments to carry out crisis control; as more and more netizens are concerned, the enthusiasm of online public opinion is soaring, and the scope of influence continues to expand, the crisis of public opinion will be rapidly rising in a short period of time; after a period of time, as the interest of netizens decreases, and no new derivative public opinion information is exposed, the public opinion crisis will show signs of decline; once new public opinion hotspots appear or crisis events are resolved, netizens will naturally the original hot spot will gradually fade away [1]. At this time, the public opinion crisis has turned into a period of fading, but it does not mean that it will disappear forever. If a similar public opinion hot spot reappears, the public opinion crisis will break out again.

© The Author(s), under exclusive license to Springer Nature Switzerland AG 2021
J. Abawajy et al. (Eds.): ATCI 2021, AISC 1398, pp. 401–406, 2021.
https://doi.org/10.1007/978-3-030-79200-8_58

2 Significance of Establishing an Early Warning Mechanism for Network Public Opinion Crisis

In the big data environment, the Internet plays an indispensable role in people's work and life. Once an Internet public opinion incident occurs, it will inevitably have an impact on people's lives. If the Internet public opinion is allowed to spread, it will make the people feel Panic [2]. In order to eliminate the negative impact, it is necessary for relevant departments to adopt relevant online public opinion crisis early warning measures. The establishment of online public opinion crisis early warning mechanism measures will help government functional departments to better understand what netizens think, and then make more scientific decisions; early warning work usually plays a very important role in controlling the spread of online public opinion, allowing all levels to the department understands the trend of the public opinion crisis in advance, and then formulates countermeasures to prevent the crisis from worsening and cause irreparable harm; once a public opinion crisis event breaks out, it spreads quickly and is relatively dangerous, and it is often beyond the control of a department. Multiple departments need to work together to cope with it, as shown in Fig. 2. Therefore, it is very necessary to build an early warning mechanism for online public opinion.

Fig. 1. Application of big data in public opinion crisis early warning

Although it is a peaceful age, the possibility of public crisis incidents cannot be completely ruled out. After the emergence of crisis incidents, the spread of mass media will arouse heated discussions among netizens in a short period of time [2]. At this time, if the government's functional departments do not handle it properly, the people will be disappointed. It is very necessary to establish a comprehensive online public opinion crisis early warning mechanism, which will help enhance the government's credibility.

3 Analysis of the Early Warning Mechanism of Online Public Opinion Crisis

In the context of big data, the network environment is complex and changeable, and it is unrealistic to establish a single crisis early warning mechanism. It usually requires multiple mechanisms to work together to complete it. According to several development periods of the crisis of online public opinion, the early warning mechanism of online public opinion crisis constructed in this paper is a complete system including four mechanisms of identification, evaluation, response and coordination [3], as shown in Fig. 2. However, the spread of public opinion crisis hotspot information is very fast, and the early warning department has to arrange staff from time to time to improve the design of the overall mechanism to ensure that the detection task of online public opinion crisis events can be successfully completed.

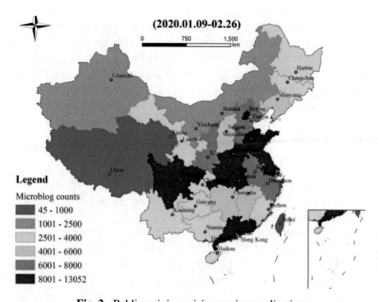

Fig. 2. Public opinion crisis warning application

1) Identification mechanism: With the occurrence of online public opinion crisis events, there will naturally be corresponding early warning mechanisms. The most common one is to screen out potentially dangerous effective information from the massive amount of online information through the identification mechanism, and then analyze the information the source and possible transmission channels of the Internet public opinion can finally give a reliable judgment on the degree of danger of the online public opinion hotspot, and provide technical support for the government functional departments to better manage and control public opinion crisis events [3].

2) Evaluation mechanism: The early warning system can process and analyze the public opinion crisis information in the past case database, and then obtain the characteristics of the information that can induce public opinion crisis. According to these characteristics, evaluate the new public opinion crisis information collected [4]. Early warning work provides accurate processing targets and improves work efficiency. There are different levels according to the degree of harm that online public opinion crisis events may cause. Among them, red represents first-level early

warning, indicating that the public opinion is very serious and the harm caused is also great. It should arouse great attention from online media and government functional departments. The establishment of an early warning evaluation mechanism is mainly to assess crisis information in advance and provide technical support for government functional departments based on the early warning level [4]. It should be pointed out that the indicators of this evaluation mechanism are not static and will be adjusted due to changes in the network environment and changes in the evolution of crisis information.

3) Response mechanism: including response decision-making mechanism and information feedback mechanism. The decision-making mechanism is the early warning system. After evaluating public opinion crisis information, government functional departments will take corresponding measures according to different levels and types of crises. This mechanism Very important in the entire early warning system. As the main body of handling public opinion crisis incidents, when taking measures, government functional departments must take the initiative of decision-making through various channels. In this process, they must also track the implementation effect and conduct evaluation and testing to ensure that they are responding to emergencies. It can quickly formulate accurate and effective early warning plans when public opinion crisis events are issued [5]. The information feedback mechanism always exists in the early warning process, and the feedback information is instant and detailed, which can play a good control effect on public opinion. This feedback information mainly includes social evaluations and opinions of the majority of netizens in the process of crisis management, which is very helpful to improve the level of early warning. The existence of this information feedback mechanism has built a bridge of communication between government functional departments and the broad masses of people [6]. The government can understand the wishes of the people at any time, and then formulate scientific and effective countermeasures.

4) Coordination mechanism: government functional departments and other departments cooperate with each other to jointly participate in the monitoring and evaluation of online public opinion crisis early warning events, timely management and control, and joint response to public crises [6]. With the evolution of public opinion, various departments play different roles. The establishment of a coordination mechanism is to allow various departments to cooperate tacitly and make the early warning work proceed in an orderly manner. In real life, there are many factors that affect the coordination mechanism, mainly including the interference of the network environment, the self-coordination consciousness of different information subjects, and the tacit understanding between them [7]. There are two ways of agreement between different subjects. One is autonomous coordination, that is, spontaneously coordinated in accordance with designated rules, in order to jointly complete the set goals, negotiate and operate with a positive attitude; the other is passive coordination and is affected by external environmental factors. As a result, the main bodies of various departments in the early warning mechanism passively form groups and can only rely on the joint cooperation between the main bodies to counter external interference.

4   Improve the Online Public Opinion Crisis Early Warning Mechanism

4.1   Use Big Data Technology to Investigate Hidden Dangers

Many examples in real life have proved that the quality of the early warning mechanism of public opinion crisis depends on the initial information and data collection and processing, because the massive data resources are affected by subjective factors in the process of artificial collection, analysis and processing, and there will inevitably be deviations and efficiency not tall. In the early warning mechanism, the prevention of crises should be considered as the first consideration [8], using big data technology to collect and analyze massive amounts of information in real time and conduct multiple monitoring and evaluations, and focus on the information that may cause harm.

4.2   Improve the Regulatory System for Crisis Early Warning by Government Functional Departments

The effect of crisis early warning is closely related to the supervisory responsibilities of each department. When a public crisis occurs, each early warning department should have a clearer division of responsibility, so that the department responsible for the function can be quickly located when responding to a specific major crisis [9]. For example, food safety-related crises should be handled by food-related units, online public opinion crises should be handled by the information office, and comprehensive crises should be handled by the National Emergency Response Office for coordination and resolution.

4.3   Encourage the Mass Media to Actively Participate in Early Warning Work

Today is the Internet age, and most of the information obtained by netizens on online public opinion depends on the dissemination of the mass media. If you want to deal with online public opinion from the root, you must be aware of the importance of mass media. Government functional departments should maintain close contact with major media and guide them to spread positive public opinion information. When a crisis breaks out, promptly transmit authoritative and accurate information to all sectors of society, so that everyone can understand the truth, control public opinion, and prevent rumors from spreading and causing social harm [10]. At the same time, the mass media should not favor any party, but maintain a neutral attitude, report objective and fair facts, and promote the development of a benign cooperative relationship between the media and government functional departments.

5   Conclusion

Government functional departments need to have a complete system to support the handling of online public opinion crisis events, and the crisis early warning mechanism is the first step and a crucial link in handling the event. A complete crisis early warning mechanism directly determines whether the society can develop in harmony and stability for a long time. In the process of handling public crises, relevant departments must cooperate closely with the government, take the initiative to assume social responsibilities, strive to incorporate big data technology into crisis early warning work, improve the responsibilities of various regulatory agencies, and encourage mass media to participate in crisis early warning work, to provide a comprehensive guarantee for China to efficiently handle the crisis of online public opinion.

References

1. Guo, Y.Y.: A review of domestic literature on online public opinion crisis early warning research. News Res. Guide **7**(23), 56–58 (2016). (in Chinese)
2. Ma, J., Hu, M., Zhang, S.: Construction and application of the evaluation model of internet public opinion crisis level. Inf. Data Work **11**(04), 36–42 (2017). (in Chinese)
3. Yan, L.: Research on the early warning mechanism of online public opinion crisis under the big data environment. Changchun: Jilin Univ. **9**(04), 112–115 (2018). (in Chinese)
4. Liu, M.J.: Research on government crisis management mechanism from the perspective of internet public opinion. Changchun: Northeast Norm. Univ. **11**(01), 87–90 (2020). (in Chinese)
5. Dong, J.F.: Analysis and research of online public opinion for public crisis early warning. Wuhan Univ. **8**(03), 17–20 (2015). (in Chinese)
6. Ma, M., Liu, D.S., Li, H.: Research on the model of network public opinion analysis system based on big data. Inf. Sci. **16**(03), 25–28 (2016). (in Chinese)
7. Huang, W.: Research on theories and methods of network public opinion dissemination and monitoring. Inf. Data Work **6**(03), 72–75 (2011). (in Chinese)
8. Chen, S.Q.: Internet public opinion analysis and research for public crisis early warning. News Res. Guide **7**(14), 223–226 (2016). (in Chinese)
9. Chu, J.W., Zhu, L.L.: Research on network public opinion early warning of emergencies based on big data analysis. Inf. Theory Pract. **10**(08), 61–64 (2017). (in Chinese)
10. Shang, H.L.: The predicament and resolution of the government governance of network public opinion in the self-media era. Adm. Forum **23**(02), 59–62 (2016). (in Chinese)

Mobile Application Behavior Recognition Based on Dual-Domain Attention and Meta-learning

Wenjun Zhang[(✉)]

School of Computer Science, Nanjing University of Posts and Telecommunications, Nanjing, Jiangsu, China

Abstract. Mobile smart devices and mobile applications carry a lot of personal information and office entertainment functions. By means of analyzing the network traffic generated by mobile applications during use, it can provide valuable information in terms of network management, privacy protection, and behavior recognition. This paper designs a recognition model based on dual-domain attention mechanism and meta-learning. First, feature extraction is performed through a deep separable convolution module, and secondly, attention is extracted from the channel and space dimensions through the attention mechanism module to enhance behavior recognition samples. At the same time, the meta-learning strategy is used for multi-task learning, so that the model can have a faster and more efficient recognition effect when facing new small sample recognition tasks. Experimental results show that compared with other small sample recognition models, the model in this paper can effectively recognize mobile application behavior.

Keywords: Mobile application behavior recognition · Deep learning · Attention mechanism · Meta-learning · Deep separable convolution · Small sample classification

1 Introduction

When users use each mobile application, a variety of network traffic will be generated. A lot of information can be obtained by analyzing these traffic. First, you can analyze the behavior of users or the behavior of users of a certain age in a certain area. Make analysis to portray the user's image to achieve better recommendations; second, it is possible to detect and prevent some malicious applications from the perspective of attackers as soon as possible, so as to avoid security issues such as privacy leaks in time; third, Which can meet the special management needs of certain scenarios or enterprises and improve the quality of network services.

Traditional mobile application traffic identification [1–4] mainly has three methods, based on port number identification, based on DPI deep packet inspection and identification, and based on machine learning. Among them, the port-based and DPI-based deep packet inspection traffic identification methods are based on rules and rely on the rules formulated by researchers for matching and identification. In addition, statistical

© The Author(s), under exclusive license to Springer Nature Switzerland AG 2021
J. Abawajy et al. (Eds.): ATCI 2021, AISC 1398, pp. 407–413, 2021.
https://doi.org/10.1007/978-3-030-79200-8_59

methods and behavior-based methods [5–7] belong to traditional machines. The learning method still requires researchers to manually select features so that the model can recognize the samples to be identified based on the established features [8]. However, with the success of deep learning [9, 10] in various fields in recent years, researchers have begun to try to use deep learning to solve the problems in traditional traffic identification methods. At the same time, considering that in actual application scenarios, mobile applications are updated at a fast speed, it is difficult to quickly accumulate a large number of samples of applications to be identified to realize the recognition of application behaviors. Therefore, the method of meta-learning multi-task training is adopted to solve the problem of recognition. The problem of the cold start of the model realizes the recognition of mobile application behavior in a small sample scenario.

2   Mobile Application Behavior Recognition Method Based on Dual-Domain Attention and Meta-learning

The classification model of mobile application behavior recognition based on dual-domain attention mechanism and meta-learning is shown in Fig. 1. After the visualization operation of traffic data, the mobile application behavior recognition task in the case of small samples is completed through deep separable convolution, extraction of dual-domain attention and training strategy module operation of meta-learning. Among them, deep separable convolution is used to replace the conventional convolution operation, in order to ensure accurate recognition and reduce the number of model parameters to achieve better generalization. By using the dual-domain attention module, the weight of useful information in the input feature graph is increased from the channel domain and spatial domain, and the useless information is suppressed to improve the recognition accuracy. With the training method of meta-learning, the dual-domain attention model is used as the base model for behavior recognition. After a lot of multi-task training, the problem of cold start of the recognition model is solved to realize mobile application behavior recognition in the case of small samples.

2.1   Feature Extraction Using Deep Separable Convolution

After preprocessing the training images, the deep separable convolution is used to replace the conventional convolution operation in the feature extraction part. The purpose of Depthwise Convolution is to improve the efficiency of computation by reducing the parameters of the model.

Deep separable convolution consists of two parts: deep convolution and point-by-point convolution. Depthwise Convolution acts on the channel level. As shown in Fig. 2(a), it is a model of per-channel Convolution. Pointtwise Convolution is similar to conventional Convolution. Figure 2(b) shows the model of point-by-point Convolution.

In fact, the use of deep separable convolution in feature extraction in this paper is not a simple way to force precision to change time. In the part of deep convolution, the data between channels does not affect each other. In addition, point-by-point convolution makes the information between channels interact, and uses the feature map of lower channels to store the feature information. In actual experiments, the Google team's

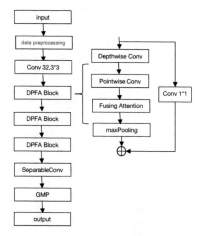

Fig. 1. This paper presents a mobile application behavior recognition model

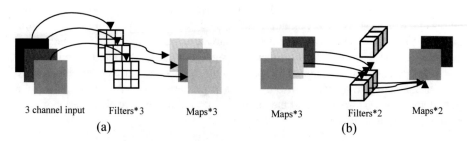

3 channel input   Filters*3   Maps*3   Maps*3   Filters*2   Maps*2

(a)   (b)

Fig. 2. Diagram of depth separable convolution

MobileNet improved speed with significantly reduced model parameters compared to VGG16 and Googlenet while maintaining nearly as good accuracy. To sum up, deep separable convolution is adopted in this paper to replace part of conventional convolution operations for behavior recognition of mobile terminal applications.

2.2 Dual-Domain Attention Extraction

In fact, the gray-scale graph of application behavior transformation also contains a lot of irrelevant information, such as the black part used to complement the length when the picture is generated, or the fixed format of traffic information. In order to avoid the interference of these irrelevant information in classification, and hope that the model can improve the effect and efficiency in classification, this paper introduces the attention module of dual-domain assembly after convolution operation.In this way, you can pay more attention to information that is helpful for classification in the learning process.The attention model used in this article is shown in Fig. 3.

Channel in a conventional dual domain attention focus is the characteristic for input figure after preprocessing based on global average pooling to channel the extraction of domain of attention, but from the perspective of frequency domain analysis, found that

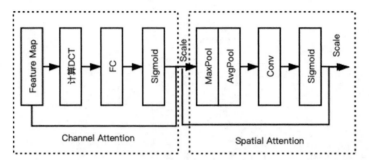

Fig. 3. Attention module

global average pooling is just a special case of the characteristics of frequency domain decomposition, thus directly do global average pooling will lose a lot of important information, in this paper, the channel of frequency domain analysis is based on DCT domain attention.

Afterwards, for the part of spatial domain attention, this article chooses to take the output of the previous channel domain attention module as input, and explore the question of where to look when further dual-domain behavior recognition for each channel. Then in this part, first perform global max pooling and global average pooling on the input feature map, and then concatenate the two results, after a convolution operation, reduce the dimensionality to a channel width feature map, after the sigmoid it can generate a spatial attention feature map, and then multiply it with the input feature map to get the final generated feature map.

2.3 Multitasking

General deep learning divides Training data into different batches, while for meta-learning, it divides Training data into different tasks. In this paper, meta-learning is used as the learner of the model, that is, the recognition model with dual-domain attention is used as the base model, and then the meta-learning method is used to train multiple tasks at the same time, and then the synthetic direction of different tasks is obtained for updating.

Then in the concrete implementation round training mission, the first round of the task needs we will learn unit parameters initialization, and need to be set in advance a batch number of tasks, and then the batch task into our recognition network training, with the data to calculate the current parameters of the query set loss under θ, the second task using the same batch job the same initial parameters, repeat the above steps in the calculation of a loss, until the end of the batch task training, calculate each loss and loss of as we study:

$$\text{Loss}(\phi) = \sum_{n=1}^{N} l^n(\theta^n)$$

In the training data, there are m different types of samples, from which 4 ($p < m$) types are randomly selected, and each type has K samples, among which 4K samples constitute the support set and 2K samples constitute the query set. The same is done for the sampling of test tasks. The sampling algorithm of meta-learning is shown in Algorithm 1.

Algorithm 1. Generate a small sample identification task from the data set
input: label: L={0,1,2,···N}
Data set: D= {(x₁,y₁),(x₂,y₂),······,(xₙ,yₙ)}，xᵢ∈Rᵈ，yᵢ∈L
D<t> represents All of the elements in D that satisfy yi=t(xi,yi)
Size of the Query set is B，The sample size of each category is K
output: Small sample identification task   T = {Sa,Q,K}
rely: RandSample(P,R) R samples are uniformly and randomly selected from the set P
1   La(la₀,···,la₃) = RandSample(L,4),Lb(lb₀, lb₁)=RandSample(L,2)
2   Generated sampling set (Sample Set)
3   for i=1 to 4:
4   Sa<i> = RandSample(D<laᵢ>,K)
5   Sa = Sa ∪ Sa<I>
6   end
7   Generate request set (Query Set)
8   Qu<0> = RandSample(D<lb₀>,B/2)
9   Qu<1> = RandSample(D<lb₁>,B/2)
10 Qu = Qu<1> ∪ Qu<2>
11 T = {Sa,Qu,K}

3   Experimental Results

Realizing the recognition of mobile application behavior in the field of small samples is still a relatively new research field [11, 12]. There are currently no research results and research data that can be directly referred to. Therefore, when conducting comparison experiments, only similar recognition models can be selected for comparison. First, the CUMMA [13] model for behavior recognition and the MAML model for small sample classification are selected for comparison experiments. The experimental results are shown in Fig. 4 below.

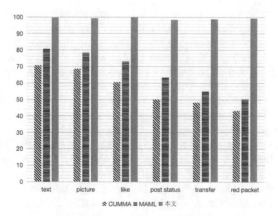

Fig. 4. Comparison of experimental results

It can be concluded that the small sample behavior recognition method proposed in this paper can have a good recognition effect in the experiment. Other comparison models and existing behavior recognition models do not have small samples in the field of behavior recognition. Both require a large number of samples for training, and only identify specific types of behavior, so the recognition model in this article is practical.

4   Conclusion

Attention mechanism was proposed based on dual domain and meta learning mobile application behavior recognition method, experiments show that mobile network data traffic information were collected in a reasonable segmentation to behavior data and translated into images, first by the depth of the separable convolution feature extraction module, secondly by attention mechanism module extracted from two dimensions of channel domain and space domain attention, strengthen the behavior recognition image texture characteristics of the sample, use learning strategies at the same time, multitasking study, makes the model can be in the face of the new task of recognition of small samples more fast and efficient recognition effect. Experimental results show that this model can be applied to mobile application behavior recognition in small sample scenarios, and has a good performance.

References

1. Malik, N., Chandramouli, J., Suresh, P., et al.: Using network traffic to verify mobile device forensic artifacts. In: The 14th IEEE Annual Consumer Communications and Networking Conference, CCNC 2017, Las Vegas, NV, USA, 8–11 January 2017, pp. 114–119 (2017)
2. Chen, Y.-C., Liao, Y., Baldi, M., et al.: OS fingerprinting and tethering detection in mobile networks. In: Proceedings of the 2014 Internet Measurement Conference, IMC 2014, Vancouver, BC, Canada, 5–7 November 2014, pp. 173–180 (2014)
3. Alan, H.F., Kaur, J.: Can android applications be identified using only TCP/IP headers of their launch time traffic, pp. 61–66 (2016)
4. Hyunchul, K., Claffy, K.C., Fomenkov, M., et al.: Internet traffic classification demystified: myths, caveats, and the best practices. In: Proceedings of ACM CoNEXT Conference-4th International Conference on Emerging Networking EXperiments and Technologies (2008)
5. Shafiq, M., Yu, X., Bashir, A.K., Chaudhry, H.N., Wang, D.: A machine learning approach for feature selection traffic classification using security analysis. J. Supercomput. **74**(10), 4867–4892 (2018). https://doi.org/10.1007/s11227-018-2263-3
6. Coull, S.E., Dyer, K.P.: Traffic analysis of encrypted messaging services. **44**(5), 5–11 (2014)
7. Fu, Y., Xiong, H., Lu, X., et al.: Service usage classification with encrypted internet traffic in mobile messaging apps. IEEE Trans. Mob. Comput. **15**(11), 2851–2864 (2016)
8. Park, K., Kim, H.: Encryption is not enough: inferring user activities on KakaoTalk with traffic analysis. In: Kim, H.-W., Choi, D. (eds.) WISA 2015. LNCS, vol. 9503, pp. 254–265. Springer, Cham (2016). https://doi.org/10.1007/978-3-319-31875-2_21
9. Wang, Q.L., Yahyavi, A., Kemme, B., et al.: I know what you did on your smartphone: inferring app usage over encrypted data traffic. In: The 2015 IEEE Conference on Communications and Network Security, Florence, Italy, 28–30 September 2015, pp. 433–441 (2015)
10. Wang, W.: Research on network traffic classification and anomaly detection methods based on deep learning. University of Science and Technology of China (2018)
11. Nan, Y., Yang, Z., Yang, M., et al.: Identifying user-input privacy in mobile applications at a large scale. IEEE Trans. Inf. Forensics Secur. **12**, 647–661 (2017)

12. Gasparis, I.: Ensuring users privacy and security on mobile devices. University of California, Riverside, USA (2017)
13. Conti, M., Mancini, L.V., Spolaor, R., et al.: Analyzing android encrypted network traffic to identify user actions. IEEE Trans. Inf. Forensics Secur. **11**, 114–125 (2016)

"Internet + Government Service" Optimization Based on Big Data

Xin Jin[1] and Yiheng Yao[2(✉)]

[1] School of Marxism, Kunming University, Kunming 650214, Yunnan, China
[2] School of Foreign Languages, Kunming University, Kunming 650214, Yunnan, China
yaoyiheng@kmu.edu.cn

Abstract. Our country's "Internet + government services" "The construction is also on the agenda and is in full swing. In this regard, it is of great significance to conduct research on "Internet + government services" from the perspective of big data. The purpose of this article is to conduct research based on the optimization of "Internet + government services" under big data. In this paper, from the information platforms such as How Net and Fang weipu, the keywords "Internet +" and "government service" are used as the search targets, and the relevant documents found in the search are studied in depth, and the related concepts of "Internet + government service" are studied. This article takes the Municipal Administrative Examination and Approval Bureau as the research object and mainly applies literature analysis method, questionnaire survey method, and adaptive clustering method to carry out research on "Internet + government service". Research shows that about 88% of those with a junior high school degree or below are not familiar with "Internet + government services". It can be seen that the level of education directly affects the acceptance of emerging things. It can be seen that the public is not familiar with "Internet + government services", and the public prefers traditional offline methods to handle business.

Keywords: Government services · Internet plus · Big data · Policy optimization suggestions

1 Introduction

With the continuous development and popularization of Internet technology, Internet information technology has always affected people's daily lives [1–4]. Based on existing research results, through the introduction of seamless government theory and public service theory, a systematic study of our country's "Internet + government service" related issues [5, 6], it is helpful to enrich theoretical discussions in related fields and help the construction of service-oriented government in the era of big data [7, 8].

In the research on the optimization of "Internet + government service" based on big data, many scholars have conducted research on it and achieved good results. For example, through investigation and research by Yun D [9], national business needs can be found in the government service hall of the national government. In this case, the government's administrative efficiency can also be quickly improved. In addition, American scholar Mamun believes that once the government government service center is established [10], those with business needs will save a lot of time to travel around. In this way, while improving the efficiency of business needs, it can also improve the government's service capabilities. Therefore, it is extremely necessary to establish a service center.

2 Optimization of "Internet + Government Service" Based on Big Data

2.1 Problems in the Construction of "Internet + Government Services"

2.1.1 Level of Development of "Internet + Government Services" Among Regions is Uneven

The overall level of Internet penetration in our country has achieved tremendous progress, and the construction of "Internet + government services" in various regions has also made some progress. However, the status quo of uneven development levels between regions still cannot be ignored. The main reasons are analyzed for the following three points: differences in economic levels between regions and insufficient funds; professional construction personnel are behind; local governments have neglected the storage and retention of network information.

2.1.2 Fragmentation of Data Information, Inter-departmental Collaboration Needs to Be Improved

When the local government builds an online government service platform, it is affected by factors such as lack of resources and insufficient professional talents. The construction of the "Internet + government service" platform is even more difficult and progress is slow. Each department of a government agency owns the data resources of their respective departments. These data are neither open to the public and enterprises, nor fully circulated among various government departments. They are just scattered, and the public and enterprises need to cross over when doing things. Many departments cannot enjoy the full range of government services, nor have they fully utilized the deeper value of these data. Only the comprehensive sharing of cross-departmental and cross-regional data can improve the construction of the "Internet + government service" platform in various places.

2.1.3 Risk of Citizens' Privacy Leakage Caused by Imperfect Legislation and Systems

Large-scale data sharing and data analysis technologies have linked together initially isolated personal privacy information, and widely used information data resources have greatly increased the value of citizens' personal privacy information. All types of personal information are collected in the social security system, which includes all types of valid data, such as personal identity, social security participation information, salary, etc. However, our country does not yet have a special law to protect personal data. Due to scattered legal regulations and unclear powers and responsibilities, it may be difficult to protect personal privacy in practice.

2.2 Existing Methods in the Construction of "Internet + Government Service"

2.2.1 Narrowing the Gap in the Construction Level of "Internet + Government Service" Between Regions

(1) The increase in special capital support and professional talent support for economically underdeveloped areas is affected by the degree of economic development. When the "Internet + government service" platform is established, some economically underdeveloped areas in the central and western regions will be subject to capital restrictions. The country needs to increase support for special funds in economically underdeveloped areas, try to cover the operation and maintenance of the entire platform while developing information technology, and supervise the flow of funds to ensure that they are used in practice, and effectively eliminate waste in the construction process. Countries and regions should strengthen the introduction of high-level professional talents, especially informatization talents, high-tech talents, and government management talents needed to build the "Internet + government service" platform. (2) Attach importance to and integrate the information and data retained by citizens. The information resources that Internet users feedback on the Internet should be fully valued. With the help of the big data era, the government's function as a citizen's information carrier should be fully utilized, and data resources can be effectively stored and further developed, to maximize the value of development data. What is easy to be overlooked in practical work is the information resources that citizens have retained in the form of paper documents. When building the "Internet + government service" platform, the information should be stored again in the form of electronic data for better maintenance.

2.2.2 Improve the Overall Structure of the "Internet + Government Service" Platform

The three levels of national platform, provincial platform, and prefecture-level platform form the "Internet + government service" platform system. Through the integration and sharing of data between levels and departments, information can be shared between the three platforms, and in terms of business cooperate with each other, supervise and evaluate, and carry out statistical analysis, etc. to realize government services in the vicinity, in the same city and long-distance.

2.2.3 Improve Relevant Legislation and System Guarantees

First of all, the foundation for the construction of online government services is to establish a safe and stable information processing platform to ensure that the infrastructure of the "Internet + government services" platform is constructed in strict accordance with regulations. Secondly, the "Internet + government service" platform must monitor in real time the dynamics of the government staff participating in the management on the network platform. Improve data recovery technology and back up information and data in a timely manner to provide long-term protection for the safe processing of data. Furthermore, the government should establish a security management system that integrates the information of various departments. Only by integrating the information technology of various departments can the huge amount of data be effectively allocated and the information security of the government and government departments can be effectively protected. In addition, through TV news, official Weibo, WeChat public account and other forms of publicity, or use the holiday time to open publicity activities in the community, such as setting up an exhibition on "Internet + government services", ask professionals to explain face to face with citizens. The operation process of online government service enables the public to quickly understand the functions and functions of the "Internet + government service" platform, so as to enhance the public's awareness of self-protection and make it proactive in maintaining personal information security.

2.3   Algorithm Based on Adaptive Clustering

Adaptive clustering is an algorithm that uses the degree of membership to determine the degree to which each data point belongs to a certain cluster. It can adaptively determine the number of cluster centers and stop clustering when the objective function converges. The algorithm process mainly includes: calculating the distance from the point to the cluster center, the membership degree of each point to the center, iteratively finding the cluster center, and calculating the objective function. Among them, the calculation operations involved in the algorithm process are defined as follows:

$$Dist_{ij} = \| Task_i - C_j \| \tag{1}$$

$Dist_{ij}$ represents the Euclidean distance between the i-th task $Task_i$ and the j-th cluster center C_j.

$$Sum(F_{ij}) = \sum_{j=1}^{k} F(i, j), (i = 1, 2, \dots, n) \tag{2}$$

$Sum(F_{ij})$ represents the sum of the membership degrees of task i to each cluster center j, and k is the number of cluster centers

$$J_m = \sum_{i=1}^{n} \sum_{j=1}^{k} F_{ij}^m (Dist_{ij})^2, (i = 1, 2 \dots, n; j = 1, 2 \dots, k) \tag{3}$$

J is used to calculate the objective function of the j-th cluster center, n is the total number of tasks, and m is the fuzzy index.

3   Experimental Optimization of "Internet + Government Service" Based on Big Data

3.1   Objects

This article will issue one-to-one questionnaires to business households who come to each window of the municipal service center to handle business, and ask them to fill in carefully to help improve the quality of government services, and set up a return questionnaire box at the exit of the hall to maximize Ensure the authenticity of the questionnaire results.

3.2   Issue Questionnaires and Return Questionnaires

The questionnaire was distributed face to face and collected, and it is expected that 350 copies will be distributed. 320 copies were actually issued, and 284 questionnaires were effectively returned, with an effective rate of 88.75%.

3.3   Algorithms Using Adaptive Clustering

Using adaptive clustering is an algorithm that uses the degree of membership to determine the degree of satisfaction of each data point, adaptively determines the number of clustering centers, and stops clustering when the objective function converges. The self-adaptive clustering algorithm is adopted to analyze and process the business households who come to each window of the municipal service center to handle business.

4   Experimental Research and Analysis of "Internet + Government Service" Optimization Based on Big Data

4.1   Familiarity of "Internet + Government Services" is Shown in Table 1.

Table 1. Familiarity of "Internet + government service"

Familiarity with "Internet + Government Service"	Number of people	Proportion (%)
Unfamiliar	150	52.82
Familiar	642	2.54
Same as e-government	32	11.27
Very familiar	38	13.38

It can be seen from Fig. 1: 52.82% of the respondents are not familiar with "Internet + government services"; it accounts for more than half of the respondents. 13.38% of the respondents are very familiar with "Internet + government services", and most of them have a college degree or above. Approximately 88% of those with a junior high school degree or below are not familiar with "Internet + government services". It can be seen that the level of education directly affects the acceptance of emerging things. It can be seen from this that the public is not familiar with "Internet + government services", and the public chooses more traditional offline methods for business. The "Internet + government service" of the Municipal Administrative Examination and Approval Bureau has not yet fully played its role, and the joint efforts of the government and the public are needed to allow more people to understand and familiarize themselves with the relevant matters handled online.

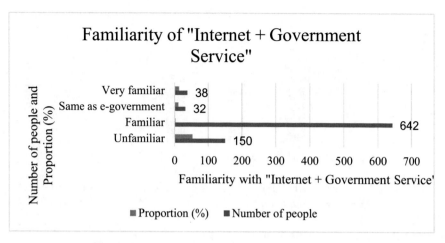

Fig. 1. Familiarity of "Internet + government service"

4.2 Satisfaction Survey

Table 2 shows the satisfaction survey of the "Internet + government service" of the Municipal Administrative Examination and Approval Bureau.

As shown in Fig. 2: 36.62% of the respondents are generally satisfied with the "Internet + government service" of the Shijiazhuang Administrative Approval Bureau; 11.97% of the respondents are satisfied with the "Internet + government service" service of the Jia zhuang Administrative Approval Bureau Very satisfied; 23.94% of the respondents are quite satisfied with the "Internet + government service" service of the Jia zhuang Administrative Approval Bureau; 19.01% of the respondents are dissatisfied with the "Internet + government service" service of the Jia zhuang Administrative Approval Bureau; there are 8.45% of the respondents who do not know the relevant situation. It can be seen that there is still room for further optimization in the construction of the "Internet + government service" system in Shijiazhuang City to better meet the needs of the public.

Table 2. Satisfaction survey

Are you satisfied with the "Internet + government services" of the Municipal Administrative Examination and Approval Bureau	Number of people	Proportion
Very satisfied	34	11.97%
Satisfied	68	23.94%
General	104	36.62%
Dissatisfied	54	19.01%
Don't know	24	8.45%

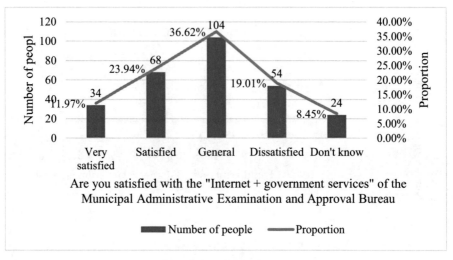

Fig. 2. Satisfaction survey

5 Conclusions

With the rapid development of Internet technology, the impact of the Internet has become greater and greater, especially when combined with many industries, it can promote qualitative changes in the industry. Therefore, local governments should pay more attention to the convenience that the Internet brings to government service. Facts have proved that the Internet + government service reform is adapted to the development requirements of the times.

References

1. Wen, X., Zhou, X.: Servitization of manufacturing industries based on cloud-based business model and the down-to-earth implementary path. Int. J. Adv. Manuf. Technol. **87**(5–8), 1491–1508 (2014). https://doi.org/10.1007/s00170-014-6348-8
2. Yang, F., Zhang, G., Li, H., et al.: Intelligent adjustment and optimization of resources based on internet big data. Design Technol. Posts Telecommun. (012), 41–45 (2018)
3. Gong, M., Cai, Q., Ma, L., Jiao, L.: Big network analytics based on nonconvex optimization. In: Emrouznejad, A. (ed.) Big Data Optimization: Recent Developments and Challenges. SBD, vol. 18, pp. 345–373. Springer, Cham (2016). https://doi.org/10.1007/978-3-319-30265-2_15
4. Salami, M., Movahedi, F., et al.: Short-term forecasting of electricity supply and demand by using the wavelet-PSO-NNs-SO technique for searching in big data of iran's electricity market. Data **3**(4), 43–43 (2018)
5. Tang, Y., Qin, G.: Research on intelligent demolition optimization based on big data platform in 4G multi-frequency network. Telecommun. Eng. Technol. Standard. **032**(009), 30–35 (2019)

6. Dias, J., Rocha, H., Ventura, T., Ferreira, B., do Carmo Lopes, M.: A heuristic based on fuzzy inference systems for multiobjective IMRT treatment planning. In: Nicosia, G., Pardalos, P., Giuffrida, G., Umeton, R. (eds.) Machine Learning, Optimization, and Big Data. LNCS, vol. 10710, pp. 255–267. Springer, Heidelberg (2018). https://doi.org/10.1007/978-3-319-72926-8_22

7. Priya, N., Geetha, G.: Dynamic programming based resource optimization in agricultural big data for crop yield maximization. J. Comput. Theor. Nanosci. **14**(9), 4464–4470 (2017)

8. Zhai, S., Duan, J., Ai, X.: Research on the third party logistics system and economic performance optimization based on big data analysis. Bol. Tecnico/Tech. Bull. **55**(11), 301–308 (2017)

9. Yun, D., Wu, C.Q., Rao, N.S.V., et al.: Advising big data transfer over dedicated connections based on profiling optimization. IEEE/ACM Trans. Netw. **27**(6), 2280–2293 (2019)

10. Al Mamun, M.A., Hannan, M.A., Hussain, A., et al.: Theoretical model and implementation of a real time intelligent bin status monitoring system using rule based decision algorithms. Expert Syst. Appl. **48**(Apr.15), 76–88 (2016)

Application Analysis of User Portrait in Library Field

Jie Dong$^{(\boxtimes)}$ and Xichen Xing

Shenyang Jianzhu University, Shenyang, Liaoning, China
xxdongjie@sjzu.edu.cn

Abstract. This paper takes the literature of user portrait application in the field of library as the research object and analyzes the collected literature through the method of visualization analysis of knowledge graph. The research shows that the current research topics of user portrait in the field of library mainly include personalization, knowledge discovery, reading promotion, reading therapy, etc., and accurate recommendation has always been a research hotspot in the field of library. The research results will help to provide reference and reference for the promotion and expansion of user portrait in the field of library and other fields in the future.

Keywords: User profiles · Academic libraries · CiteSpace · Knowledge graph

1 Introduction

With the advent of the information age, everything is inseparable from the "big data", at the same time, the term user portrait has gradually become hot. In the face of huge data in the information age, how to accurately meet the needs of users in these massive data has become the most critical problem. User portrait is a way to achieve accurate recommendation and meet the needs of users. User portrait is a labeled user model that is summarized, abstracts and mined based on user demographic data, social network relationship-ships and behavioral patterns [1].

Both domestic and foreign user portraits have been recommended in the field of library information, and the research on information filtering in foreign countries is relatively mature. Foreign countries pay more attention to data processing, domestic mainly pay attention to the application of user portrait.

In China, there are few researches on user portrait in library field using knowledge graph. This paper searches the literature from the database and analyzes it with the tool of knowledge graph to explore the research hotspots, development status and application status of user portrait in the field of library. It is expected to provide reference for the future study of user portrait in library and other fields.

© The Author(s), under exclusive license to Springer Nature Switzerland AG 2021
J. Abawajy et al. (Eds.): ATCI 2021, AISC 1398, pp. 422–428, 2021.
https://doi.org/10.1007/978-3-030-79200-8_61

2 Data Acquisition and Research Methods

2.1 Data Acquisition

In this paper, by downloading documents from CNKI (China Knowledge Infrastructure Project) and selecting the advanced retrieval method, the retrieval terms are user portrait and library, and the cross-database retrieval of international conferences, domestic conferences, newspapers and other documents is removed, a total of 185 documents are obtained, and the export format is "RefWorks".

2.2 Research Methods

Through the method of knowledge graph and clustering, this paper makes a visual analysis of these documents to study the hot spots, research status and research trends of user portrait in the field of library.

3 A Survey of User Portrait Research in Library Field

CiteSpace is the tool used for visualization analysis of knowledge graph. CiteSpace is a scientific literature analysis tool jointly developed by Dr. Chaomei Chen and Dalian University of Technology. CiteSpace version 5.7R3 was used for this article. It is through the author, institution, key words, cited analysis of literature, and then generate knowledge map, so as to discover the rules. To use CiteSpace, download the JRE (Java Runtime Envirment). CiteSpace can only handle data in the WOS format, so you first need to convert the literature in CiteSpace. Then create a new Project to generate the knowledge graph.

3.1 Analysis of Research Hotspots

After importing the literature into CiteSpace, the node type selects the keywords, and then generates the knowledge graph. Table 1 shows the key words with frequency greater than 10. The key words are the most concise words to summarize the content of the paper, or the center of the whole paper. High frequency keywords can indicate research hotspots. As can be seen from Table 1, in the field of library, library, big data, university library, smart library, precision service and digital library are the research hotspots. Count is frequency, Centrality is centrality, Year is the first proposed year, and Keywords is the keyword.

3.2 Study Author Analysis

After importing the literature into CiteSpace, the node type selects the author. The frequency of authors of these literatures can be counted. The core author of a subject area can be identified by Price's law, which is formulated as:

$$N = 0.749 \times \sqrt{n_{max}} \tag{1}$$

Table 1. Key words

Count	Centrality	Year	Key words
94	0.56	2016	User portrait
26	0.55	2016	Library
20	0.3	2016	Big data
17	0.1	2018	Academic library
15	0.32	2016	Smart library
13	0.17	2017	Precise service
11	0.14	2017	Digital library

Where, n_{max} is the maximum number of published papers by authors in the subject field, and if the number of published papers by authors is greater than N, it is the core author [2]. After formula calculation, N is 3 after rounding, and the obtained core author is shown in Table 2.

3.3 Evolution Path and Development Trend

After importing the literature into CiteSpace, select the node type and keywords. After generating the knowledge map, select Layout in the Control Panel and then TimeZone View to convert it into the map of the time context, as shown in Fig. 1.

It can be seen from the knowledge graph that user portrait, big data and smart library were widely proposed in 2016. In 2017, following the smart library, digital library continued to become the hot spot, but proposed new keywords of accurate recommendation and knowledge discovery. In 2018, university libraries became the main research object, with the emergence of reading promotion and library users. In 2019, new research contents of user portrait appeared in the library field, such as humanistic feelings, collaborative information filtering, online medical community, etc., but accurate recommendation has always been a research hotspot.

4 Research Topics

Through the keyword cluster analysis of CiteSpace, it can be found that the development of user portraits in the field of library can be divided into the following topics: personalization, knowledge discovery, reading promotion, bibliotherapy, etc. This is shown in Fig. 2.

Table 2. Core author

Count	Centrality	Year	Authors
8	0	2018	Liu Haiou
5	0	2018	Wu Zhiqin
5	0	2018	Huang Wenna
4	0	2018	Yan Sumei
4	0	2018	Zhang Yaming
3	0	2019	Bi Qiang
3	0	2016	Du jin
3	0	2016	Xiong Huixiang
3	0	2016	Li Xiaomin
3	0	2019	Zhang Han
3	0	2019	Zhang Haitao
3	0	2019	Xu Hailing
3	0	2016	Jing Ziwei

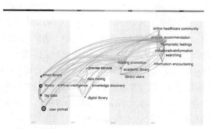

Fig. 1. Research trends

4.1 Personalized

Gong Xingang et al. [3] stated that personalized service emphasizes people-oriented, which can also be said to be one-to-one accurate service. The premise of personalized service is to collect a large amount of data from readers, and the database of readers can be established by using ontology modeling tools. Finally, the information of readers can be described as a whole, that is, user labeling. Sun Shouqiang [4] proposed the personalized service of smart library. He believed that the personalized service of smart

Fig. 2. Research topic

library was mainly realized from two aspects: reading preference and retrieval system. Using ontology modeling method, the framework is built from data acquisition layer, data processing layer, personalized service layer and so on. Li Min, Xiong Huixiang et al. [5] proposed to construct user portraits from three aspects of users' natural attributes, interest attributes and social attributes, then calculate the similarity between each aspect and books, and recommend books with the same similarity to readers.

4.2 Knowledge Discovery

Zhang Jun [6] proposed that the construction of library user portraits can be divided into three stages: data collection, behavior modeling and portrait construction. The behavior of users is modeled by machine learning method, and the demand of users is predicted by mathematical algorithm model. The techniques used in this paper include statistical method, machine learning method, semantic web technology, natural processing and cryptic meaning model. Liu Su [7] proposed that the knowledge discovery system of digital library is one of the means to realize the personalized information service of digital library. User portrait can realize the connection between knowledge point and knowledge point, knowledge point and user, user and user. When users and users are the same "note" or cluster into the same category, potential and new knowledge points can be recommended to users.

4.3 Reading Promotion

In today's environment, it is an important trend to promote reading among all the people, especially for children and college students. Liu Man [8] proposed that the reason why readers read is that they have the motivation to read. The group that constructs user portrait is university student in university library, user portrait consists of user dimension and book dimension together. It uses the ontology method to construct the conceptual model of the user portrait of college students. The reading promotion model uses clustering algorithm and collaborative filtering recommendation algorithm. The promotion types are divided into individual reading promotion and group promotion. Xiao Haiqing and Zhu Huihua [9] proposed the promotion model of participatory user portrait. Participatory user portrait promotion process is divided into internal implementation process and external implementation process. The internal implementation process refers to the process of constructing a user profile from data. The external implementation process refers to how to accurately recommend users after the user portrait is generated, including both online reading recommendation and offline reading promotion activities.

4.4 Reading Therapy

There are two kinds of bibliotherapy. One is bibliotherapy without symptoms, which is equivalent to prevention, such as some health books. Another type of bibliotherapy, which can be used for some degree of treatment or rehabilitation through reading, is an adjuvant therapy. Liu Dan et al. [10] took university library students as research objects, took Weibo and blog as data sources, calculated word similarity based on Word2vec, conducted emotional sub-screening by means of manual screening and automatic screening, and then constructed user portraits. By analyzing the user portraits of college students, we recommend books to college students with negative emotions.

Han Meihua and Zhao Jingxiu [11] took Weibo users as the data source, constructed seed words based on the Burns Depression List, extracted the comments made by users into word bags, traverse the word bags and weights, and finally calculated the depression index to form user portraits. Then, according to the depression index high and low, different books with adjunctive treatment effect are recommended.

5   Conclusion

Through the knowledge map, this paper analyzes the research hotspots, authors and research trends of user portraits in the library field in China, and expounds several topics, such as individualization, knowledge discovery, reading promotion, reading therapy and public cultural service. The results show that accurate recommendation has always been a hot topic in the field of library research, and the research content is becoming more and more novel. However, there are few researches on collaborative information filtering and public cultural services, and there are still some deficiencies. Through the analysis of this paper, the application status of user portrait in the domestic research field is combed, which can provide reference for the research of user portrait in this field. In addition, it can be compared with the application of foreign user portraits in foreign fields, and the excellent technical methods and research contents can be used for the improvement and innovation of the library field in China.

References

1. Chuanming, Y., Tian, X., Guo, Y., An, L.: User profiling based on the behaviour and content combined model. Libr. Inf. Serv. **62**(13), 54–63 (2018). (in Chinese)

2. Na, L.: Analysis on the knowledge map of domestic reading promotion reasearch. Library Work Study **39**(2), 77–86 (2018). (in Chinese)
3. Xingang, G., Lijuan, S., Zhiqian, H.: The basic database construction that takes the reader as the center. Libr. Inf. Serv. **60**(S1), 199–202 (2016). (in Chinese)
4. Shouqiang, S.: Research on personalized service of smart library based on user portrait. Libr. Study **40**(7), 60–65 (2019). (in Chinese)
5. Xiaomin, L., Xiong, H., Du, J., Jing, Z.: Research on book recommendation based on user portrait in smart library. Inf. Sci. **39**(3), 44–50 (2020). (in Chinese)
6. Jun, Z.: Research on library knowledge discovery service based on user portrait. Libr. Inf. Serv. **37**(6), 60–63 (2017). (in Chinese)
7. Liu, S.: The persona in digital library knowledge discovery system -taking Tianjin library as an example. Library Theory Practice **38**(6), 103–106 (2017). (in Chinese)
8. Man, L.: Construction of reading promotion mode of university library based on user portrait. Library Theory Practice **40**(7), 21–26 (2018). (in Chinese)
9. Haiqing, X., Huihua, Z.: Construction of precise reading promotion model of university library from the perspective of participatory user profile. Libr. Work Study **41**(6), 122–128 (2020). (in Chinese)
10. Dan, L., Xinggang, Z., Shumin, R.: User Profile-based bibliotherapy model in academic library. Chinese J. Med. Libr. Inf. Sci. **7**(27), 68–71 (2018). (in Chinese)
11. Meihua, H., Jingxiu, Z.: Research on bibliotherapy model based on User Profile- Take Depression as an Example. J. Acad. Libr. **35**(6), 105–110 (2017). (in Chinese)

The Application of AutoCAD in the Drawing of Archaeological Artifacts—Line Plot Surveying and Mapping of Standard and Complete Artifacts

Qiwang Zhao[1]([✉]) and Qianyun Lyu[2]

[1] Shanghai Urban Construction Vocational College, Shanghai, China
zachery.ok@eiwhy.com
[2] Shanghai Academy of Fine Arts, Shanghai University, Shanghai, China

Abstract. In the process of archaeological excavation and artifact sorting, the direct use of computers to draw artifact maps has not yet reached popularization. Compared with tomb maps, it is more difficult to master. It requires mappers to be skilled in engineering drawing software, archaeological typology and also the basic judgment on the characteristics of artifacts' shape and structure. This article focuses on analyzing the advantages of CAD drawing artifact maps, and taking the Jomon pottery amphora as an example, introduces the line plot operation process of relatively standard and complete artifacts by using AutoCAD software, summarizes the methods and key points of line drawing surveying and mapping, and discusses its operating principles and precautions.

Keywords: AutoCAD · Photoshop · Artifact drawing · Line plot drawing

1 Introduction

In the field of archaeology, from field investigation to data publication, it has fully entered the supporting stage of computer information technology, such as the use of GIS [1], and the multi-view 3D reconstruction of the archaeological site [2]. Relatively speaking, in the process of excavation and artifact sorting, the direct use of computers to draw tomb maps and artifact maps has not yet reached popularization. The vast majority of archaeological sites have dedicated hand-painted personnel to draw first-hand data, including maps of archaeological excavation sites and maps of unearthed artifacts [3].

There are a variety of software can draw or trace artifacts. However, in terms of the practicability of directly surveying and mapping artifacts on site, the compatibility with other 3ds Max technology [4] and mapping technology used in excavation and sorting, and the general matching of software, AutoCAD is still the best choice for archaeological drawing. It also has long-term performance development potential. The research group has also carried out long-term practice and completed related results, including archaeological excavation reports [5] and related research papers [6].

J. Abawajy et al. (Eds.): ATCI 2021, AISC 1398, pp. 429–436, 2021.
https://doi.org/10.1007/978-3-030-79200-8_62

2   CAD Drawing of Standard and Complete Artifacts

2.1   The Introduction of CAD

CAD is an engineering drawing software [7]. Its advantage lies in the accuracy and uniformity. It can be repeatedly proofread and modified. It can also use the calculation function of the software to realize the maximum recovery of the incomplete artifacts when drawing. It can provide and save all measured data. This is its greatest practical performance superior to hand-drawn or other software. It can meet the current needs of informatization data archiving and extraction, and can be used conveniently, accurately and flexibly in archaeological report sorting, writing and typesetting.

2.2   The Introduction of Standard and Complete Artifacts

Standard-type intact artifacts in the drawing refer to those that are relatively common in shape, complete with no severe deformation, no special components, or special decorations, those that are not severely damaged and do not affect the judgment of the type.

2.3   The Drawing Method

In order to give full play to the advantages of CAD drawing more accurate than direct manual surveying and drawing of artifacts, instead of direct measure and drawing, the artifact picture is generally imported into the computer first, and then be drawn. At the same time, according to the actual measurement, lofting and fixed point are used to ensure that the line plot is highly accurate in the shape and details. This method can even filter damage or deformation and restore a line plot closer to the prototype. In the drawing process, digital cameras, measuring tools, AutoCAD, and Photoshop are required [8]. The basic steps of drawing are shooting utensils, importing photos-measuring utensils, lofting photos-drawing outlines, positioning tight alignment-drawing section, marking size-exporting images, and making emblazonry. Among them, whether to loft the photos of the artifacts according to the actual size and then draw, or draw first and then loft the line drawings according to the actual size, it can be flexibly adjusted according to the drawing habits, page restrictions, and the particularity of the artifacts.

3   The Operation and Cautions of Drawing the Line Plot

Taking the Jomon pottery amphora as an example, the specific drawing operations and precautions are as follows:

3.1   Artifact Shooting

Place the amphora at a long distance for shooting [9]. The distance control is the standard to obtain a horizontal perspective and an elevation image with less perspective distortion. Try to keep the arc from the mouth to the bottom of the amphora close to straight in the front view. According to requirements, take and draw the front view (Fig. 1-①), side view, top view, bottom view, as well as detailed photos with or without the cover, the ears (Fig. 1-②), the head, and the emblazonry of the artifacts at one time.

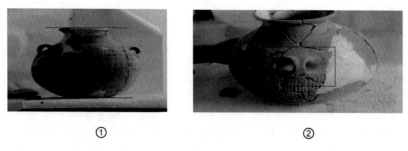

Fig. 1. Shooting the artifact

3.2 The Preparation of CAD Drawing

Fig. 2. CAD drawing interface preparation and common commands

In the specific drawing, the scope of the drawing interface can be determined and zoomed according to the actual size of the artifact, so there is no need to preset the ZOOM size. The unit is millimeter. According to the needs of displaying different content and different line types when drawing, auxiliary lines, main layers or drawing layers, filling lines, annotations, text and other layers in the layer settings can be preset. The line width and color can be preset or it can be adjusted according to the requirements when exporting the images. Dotted line type is added to the main layer in advance, and the proportion can be adjusted later as needed. Generally, the colors of auxiliary lines and labels can be selected different from the main layer. The filling line is the same color as the drawing line and the text line of the figure name and number, but the line width is about 0.1 mm thinner than the line of the main layer. Aligned label can be used, and the style can be as required. Object snapping and object snapping tracking can be used. The mode can be set as needed. Generally, endpoints, circle centers, intersections, extension lines, etc. are commonly used. Display line width and transparency can be used. (Fig. 2).

The most commonly used drawing commands in the drawing process are straight line l and polyline pl, followed by arc a, circle c, ellipse el, etc. The most common operations of line plots are move m, copy co, mirror mi, offset o, scale sc, hatch h, trim tr, extend ex, break br, etc. (Fig. 2).

The choice of the main line type in drawing line plots is often between polylines and splines. The advantage of the spline is that the lines are smooth, and the drawing speed is fast. Its linearity seems to be more suitable for drawing artifacts. But the biggest drawback of it is that it cannot fit the shape and details of the artifact accurately, especially the line cannot be positioned according to the size, fine-tuned and marked, and it is difficult to break, connect, and hatch. Polylines can be switched flexibly between straight lines and arcs, which can be accurately positioned and fit the shape of the device to the greatest extent, especially the drawing of subtle changes in the mouth, ears, feet and sudden turns. When adjusting the lines, it is rigorous and convenient, and there is no error in controlling the size and marking the size. But the biggest drawback is that if the lines are required to be smooth and beautiful, the adjustment is cumbersome, the drawing speed is slow, and the ratio should not be too large or too small. The pixels should also not be too low because the higher the definition, the more advantageous. In contrast, in order to give play to the operability of computer graphics and the accuracy and completeness of the data, it is recommended to use the polyline PLINE for drawing.

4   The Process of the Line Plot Surveying and Mapping

Measuring and recording the key size data of the artifact. Height (280 mm), caliber (128 mm), bottom diameter (90 mm), maximum abdominal diameter (275 mm).

Inserting photos of the amphora in any layer in CAD. Auxiliary layer-insert-raster image reference-select the picture-to determine the appropriate proportions inserted into the appropriate position on the screen.

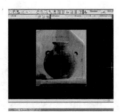

Fig. 3. Lofting the artifact image to 1:1

Lofting the picture according to actual size. Switching to the main layer or the drawing layer, and the center line drawn in the amphora picture (l) is the artifact's actual height (280). When zooming (sc), taking the actual height as the reference (r), and setting the photo to the actual size 1:1. Adding the mouth diameter line (l: 128) and the bottom diameter line (l: 90). Accurately calibrating the image position of the artifact (Fig. 3).

Surveying and mapping of the front view of the right half of the artifact. Using the polyline (pl) and the straight line (l) to fit the contour of the image of the artifact to the front view of the right half of the artifact's line plot continuously and finely (Fig. 4-①). During the drawing process, paying attention to observe the artifact, drawing the relatively complete half of the artifact, or take into account the fusion of left and right to draw half of the facade line plot, and finally flip to the right half. After the preliminary drawing, measuring the thickness of the mouth of the artifact, the height of the neck, the maximum abdominal diameter, the relative position of the ears, the height and the thickness of the ears, etc., as well as the heights and diameters of the important turning points to locate and calibrate the accuracy of the right facade line plot. After completion, closing the layer where the photo is located or deleting the photo or move the line plot to a blank space (Fig. 4-②). In the end, avoiding the anchor points and fine-tuning the arc of the polyline to make the lines coherent and smooth, and the details are delicate and smooth.

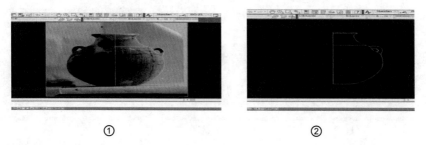

①   ②

Fig. 4. The surveying and mapping of the front view of the right half of the artifact

Drawing a cross-sectional view of the left half of the artifact. Firstly, selecting the necessary contour line (red) in the right facade line plot, and using the middle line as the mirror image (mi), without deleting the source object, forming the outer contour line in the section view and the reference line (red) for executing the offset command when drawing the section. (Fig. 5-①). Measuring the visible thickness at the edge of the artifact's mouth, the damaged and ruptured position, etc., as a basis for inferring the shape of the cross-section. Selecting the outer contour line (red) of the side wall of the artifact's body (not including the mouth, bottom, and ears) for inward offset (o). The offset distance is the measured or estimated thickness of the neck or abdomen (6 mm) to form the approximate inner contour line (red) of the side wall section view (Fig. 5-②). According to the shape judged by the cross-section of the artifact, deleting the unnecessary contour lines (green), adding the cut lines (pl) (red) of the mouth, bottom, and ears, and making some connection adjustments (Fig. 5-③). Then ensuring that the determinable outer contour line remains unchanged, and fine-tuning the section line (red) that needs to be judged, while taking into account the smoothness of the lines. The section contour line must be closed (Fig. 5-④). During the adjustment process, carefully observing and repeatedly touching the artifact. Generally, the section thickness of the mouth, bottom, and ears of the utensils will change compared with other positions of the neck and abdomen. If it is measurable, it will be drawn according to the actual measurement. If it is not measurable, it will be based on typological knowledge of common artifacts. For example, the mouth of the cross-section of the amphora becomes thicker, the joint between the ears and the body is thicker, the wall is concave, and it becomes thinner below the maximum abdominal diameter, and the bottom is concave and thinnest. Finally, switching to the hatch layer, hatching the (h) section, and adding pick points in the closed section outline. Using the 45° thin oblique line (red) in the sample, and the scale can be adjusted according to the visual effect of the drawing (Fig. 5-⑤). If the hatch fails, it is mostly because there are unclosed line segments in the hatch area, which need to be checked and corrected. If it still fails, breaking at the point (br), isolating the hatch area and then hatching it (h). Choosing to manually add the selected object. After completing the cross-sectional view, the line plot of all the main parts of the artifact can be completed, the label layer can be opened, and other required dimensions such as the maximum abdominal diameter can be added (Fig. 5-⑥).

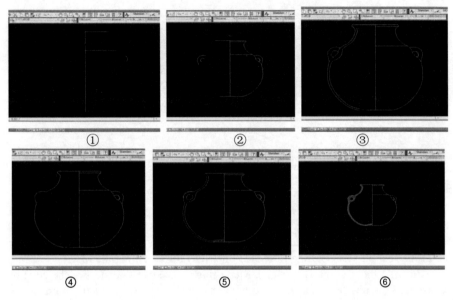

Fig. 5. Drawing steps of the left half cross-sectional view of the artifact

Drawing the front view of the artifact's ears. In addition to the front cross-sectional view, different artifacts are required to be viewed from the side and back, top or bottom according to the type and decoration requirements. The drawing method and operation steps are the same as those for the front cross-sectional view. The most common and necessary component or detail drawing is the ears. There are also animal head applique, gilding, emblazonry, etc. The use of CAD drawing mainly pays attention to the details that affect the change of the artifact type, so most of the plane decorations can be drawn and post-produced by other methods, and its drawing efficiency and aesthetics are more suitable. Take the amphora as an example to draw the front view of the ears. Inserting the raster image reference, scale (sc) the ear image to the actual size of 1:1, and use the height (red) or width of the ears as the reference (r). The ear line plot does not need to be cross-sectional, and is directly drawn with the polyline (pl) and the arc (a) on the drawing layer (Fig. 6-①). Then hiding or deleting the picture, and adjusting the position of the ear line plot to a proper distance from the artifact line plot (Fig. 6-②③). In the drawing, paying attention to the actual shape of the ears and adjusting if necessary. The lines should be smooth and round, and the height or width dimensions are always consistent with the front view of the artifact.

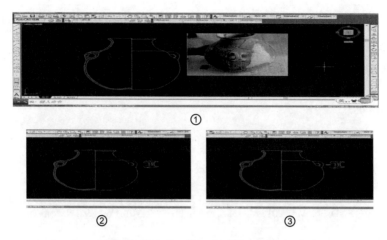

Fig. 6. Drawing the ears of the artifact

Add a plotting scale and a name table of artifacts. The scale of the CAD drawing is 1:1, so the plotting scale is also 1:1. Taking the length as needed, and putting it on the main layer with the artifact. The marking can be turned on or off according to the requirements of the drawing. Drawing the name table of the artifact, the layer can be in the main layer or the auxiliary layer, or it can be added later. Generally, the mark is used to save the collected size data (Fig. 7-①), and the figure is hidden in the report, and the scale always follows the artifact. When drawing multiple artifacts in the same CAD file, it is recommended that the artifact numbers or table names should be drawn and follow the artifacts (Fig. 7-②). The positions can be adjusted later after exporting.

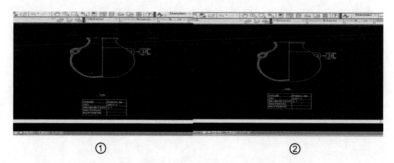

Fig. 7. The finished line plot of the artifact

5   Conclusion

At this point, the drawing of the line plot of the artifact in CAD has been completed. Saving the file for later modification or continue drawing. If it is needed to directly export the line plot of the artifact for use, CAD can directly print the drawing. But for archaeological artifacts, it is more necessary to save the line plot pictures and perform post-production, and then adjust and typeset according to different drawing requirements. Thus, the author will introduce how to save a JPG image file with clear proportions and high-resolution pixels in the coming paper.

References

1. Zhang, H.: GIS Archeology Spatial Analysis Practice Course, pp. 336–345. Peking University Press, Beijing (2016)
2. Liu, J.: Multi-view 3D Reconstruction of Archeological Sites, pp. 34–36. China Social Sciences Press, Beijing (2019)
3. Ma, H.: Archeological Artifact Drawing, pp. 101–103. Peking University Press, Beijing (2008)
4. Edited by HXSD: 3ds Max 2011 platinum manual I, pp. 04–08. Posts & Telecom Press, Beijing (2011)
5. Archaeology Major of School of History & Culture of Sichuan University, Art History & Criticism Department of Shanghai Academy of Fine Arts of Shanghai University. Xichuan Quanyangou Han Dynasty cemetery, pp. 120–122. Henan People's Publishing House, Zhengzhou (2019)
6. Zhao, Q., Pan, S.: Application of 3ds Max technology in archaeology: the 2020. In: International Conference on Machine Learning and Big Data Analytics for IoT Security and Privacy. SPIoT-2020, 02, pp. 55–58 (2020)
7. Edited by Phoenix High-tech Education: Chinese Version of AutoCAD 2016 Mechanical Drawing Basic Tutorial, pp. 57–61. Peking University Press, Beijing (2016)
8. Edited by Phoenix High-tech Education: Chinese Version of Photoshop CS6 Basic Tutorial, pp. 12–21. Peking University Press, Beijing (2016)
9. Yuan, Z.: Getting Started with Digital Photography of Cultural Relics. Cultural Relics Publishing House, pp. 92–99 (2011)

Application of Analytic Network Process in Power Grid Development-Diagnosis Management

Shiyuan Lin[1]([✉]), Yingjin Ye[1], Yaru Han[1], Yafang Zhu[1], Qingxian Li[2], and Jing Wu[2]

[1] State Grid Fujian Economic Research Institute, Fuzhou, Fujian, China
[2] Fujian Yongfu Power Engineering Co., Ltd., Fuzhou, Fujian, China

Abstract. Current power grid development-diagnosis management doesn't analyze the power grid cooperation-scale, development-speed, business-quality, operation-effectiveness and internal policy from comprehensive view. In order to improve management efficiency, this paper modified the evaluation method, built the evaluation index system based on Analytic Network Process (ANP), which took the connection and restriction among index into full consideration. And then took 5 aspect namely Scale, Structure, Safety, Effectiveness and Policy of 220 kV Grid from one province into calculation, obtained the provided the index score. Finally, concluded the advantages and disadvantages of this province 220 kV grid and provided objective and pointed suggestions to the high-quality development of power grid management as an example to other provinces.

Keywords: Power grid development management · Analytic Network Process (ANP) · Evaluation index system

1 Introduction

The power grid is one of the most important infrastructure all over the world. Providing high-quality electric energy to industry, commerce and the residents is vital factor for the economy and society development. In order to judge the grid weaknesses and management short board, power grid company carries out development-diagnosis management every year, which contains analysis about cooperation-scale, development-speed, business-quality, operation-effectiveness, internal policy and finally pointed suggestions [1–7]. However recent year, power grid development-diagnosis management concentrated on showing the data and reflecting weaknesses only from single aspect instead of considering about the link or restrict among them thereby estimating from overall situations [8–10].

J. Abawajy et al. (Eds.): ATCI 2021, AISC 1398, pp. 437–443, 2021.
https://doi.org/10.1007/978-3-030-79200-8_64

Thus this paper applicated Analytic Network Process (ANP) into Development-Diagnosis management, building up the evaluation model from specifying index system, confirming the indexes weight separately and working out the scores of different years investment effectiveness. From analyzing scores composition, clear the investment point and promoting the practicability for the power grid company. Finally stimulating the scientific development management suggestions.

2 Index System and ANP

2.1 Index System

Power grid, economy and society go hand in hand. Thus its scale and structure must fit the developmental level, which is the most important index. Secondly safe and stable operation and high-quality electricity supply are vital function of power grid. Therefore these two become another evaluation index. Meanwhile the development efficiency and quality will influence the sustainable development, which is also one of the evaluation indexes. Last but not least, as a large-scale state-owned firm, the state grid management and policy-adaptation also need taking into account. Accordingly, this paper combined specific characteristics of power grid with the structure of development-diagnosis report, specify the index system includes power grid scale, power grid structure, safety and quality, efficiency and effect, management and policy. These five belongs to control hierarchy, and contains 44 indexes at all.

2.2 Analytic Network Process (ANP)

ANP is a decision method which applies to the dependent genealogical hierarchy structure. This method used network structure to express the relationship among elements, which may have mutual effect or inter dominance, as a more effective decision method. ANP separates the elements into two parts firstly. One is the control hierarchy, which includes target and principle, and all the decision principle is independent and controlled by target index, the target index is indispensable. The other one is the network hierarchy. The index here is comprised by all the elements who demanded by the control hierarchy and they have mutual effect.

2.3 Calculation Method

Define dominance degree. Take the controlled elements into pairwise comparison to acquire the judgment matrix. The independent elements can be compared by direct dominance degree method. Otherwise, by indirect one.

Build the hypermatrix. Assume there are P_1, P_2, ...P_n in the control hierarchy, C_1, C_2, ... Cn in the network hierarchy, and Ci contains e_{i1}, e_{i2}, ..., e_{in} (i = 1, 2, ..., n). Take the Ps (s = 1, 2, ..., n) from control hierarchy as the first rule, and the e_i from Ci as the second rule, to compare the dominance degree by influence of element from Ci. Meanwhile, we can get the weight from characteristic root, and get the vector matrix W_{ij}

correspondingly. With different government subsidies, in one hurdle rate, the admission fee of one-time full charge.

$$
W_{ij} = \begin{bmatrix}
W_{i1}^{(j1)} & W_{i1}^{(j2)} & \cdots & W_{i1}^{(jn_j)} \\
W_{i2}^{(j1)} & W_{i2}^{(j2)} & \cdots & W_{i2}^{(jn_j)} \\
\vdots & \vdots & \vdots & \vdots \\
W_{in_i}^{(j1)} & W_{in_i}^{(j2)} & \cdots & W_{in_i}^{(jn_j)}
\end{bmatrix} \tag{1}
$$

Build the weighted hypermatrix. Considering about the mutual effect among elements, take the hierarchy as a whole element set and compare with other one, thereby build up the relative materiality matrix W from all sets. Each element in W is one matrix.

$$
W = \begin{bmatrix}
W_{11} & W_{12} & \cdots & W_{1n} \\
W_{21} & W_{22} & \cdots & W_{2n} \\
\vdots & \vdots & \vdots & \vdots \\
W_{n1} & W_{n2} & \cdots & W_{nn}
\end{bmatrix} \tag{2}
$$

Hypermatrix Stabilization. Calculate the limit of the matrix permutation vector, if the result is convergent and unique, the result of corresponding line from the matrix is the stable weight to the evaluation index.

$$
W^\infty = \lim_{k \to \infty} (1/N) \sum_{k=1}^{N} \overline{W}^k \tag{3}
$$

3 Calculation Example

3.1 Index System

Based on the A Province development-diagnosis report, we selected 220 kV grid as the calculating example. In the index system in Fig. 1, we can learn that cooperation scale consists of the indexes of transformer capacity, unit transformer electrical load and unit line electrical load; power grid structure only contains index of average length of line; electricity safety includes two indexes of power distribution reliability and voltage eligibility rate; operation effectiveness contains four indexes namely unit investment add load, unit investment and electricity sale, power loss rate and grid construction investment; Inner policy also includes four indexes of unit asset electricity sale, unit asset electricity revenue, EBIT and asset-liability ratio.

Among the factors, transformer capacity (C1), Unit transformer electrical load (C2), Unit Line electrical load (C3) are relevant; Unit Line electrical load (C3) and Average length of line (C4) are relevant; Average length of line (C4) and Power-loss rate (C9) are relevant; Power Distribution reliability (C5) and Power-loss rate (C9) are relevant; Unit investment add load (C7) and Grid construction investment (C10) are relevant; Unit investment add electricity sale (C8), Power-loss rate (C9), Grid construction investment

(C10) are relevant; Power-loss rate (C9), Grid construction investment (C10), EBIT (C13), asset-liability ratio (C14) are relevant; Grid construction investment (C10), Unit asset electricity sale (C11), Unit asset electricity revenue (C12) are relevant; Unit asset electricity sale (C11), Unit asset electricity revenue (C12), EBIT (C13) are relevant; Unit asset electricity revenue (C12) and EBIT (C13) are relevant; EBIT (C13) and asset-liability ratio (C14) are relevant.

3.2 Index Weight

According to the original data from the report, we chose 14 indexes to analyze the development level of A province power grid. As some data is hard to get, this diagnosis is based on acquired data to count and modify and other voltage classes data will be analyzed later. The specific data are listed in Table 1. The calculating procedure of ANP is too difficult to get the result by hand computation. Accordingly, we need to use professional calculating software—Super Decisions to calculate the weight.

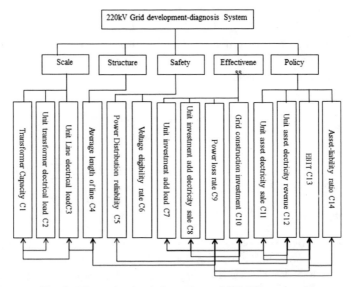

Fig. 1. The evaluation index system of 220 kV power grid

The weight results are shown in Table 2. We can learn from Table 2 that overall weight isn't positive correlation. Although regional weight of transformer capacity (C_1) and Average length of line (C_6) are very high, their overall weight is obviously low; on the contrary, the regional weights of Unit asset electricity sale (C_{11}), Unit asset electricity revenue (C_{12}), EBIT (C_{13}) aren't very high, their overall weights are top three. Hens the overall weight is based on the relevance among indexes, where higher relevance brings higher overall weight.

Table 1. 220 kV grid data from development-diagnosis report.

Index	Unit	2014	2015	2016
220 kV transformer capacity	10000 kVA	5172	5412	5802
Unit transformer electrical load	kW/kVA	0.576	0.582	0.558
Unit Line electrical load	10000 kW/km	0.259	0.261	0.253
Average length of line	km	23.5	22.1	22.1
Power distribution reliability	%	99.808	99.926	99.935
Voltage eligibility rate	%	99.701	99.996	99.997
Unit investment add load	10000 yuan	1.3	4.2	0.6
Unit investment add electricity sale	kWh/yuan	0.4	1.5	0.63
Power loss rate	%	5.65	4.75	4.75
Grid construction investment	10000 yuan	259022	176426	194201
Unit asset electricity sale	kWh/yuan	1.25	1.19	1.12
Unit asset electricity revenue	yuan/yuan	0.72	0.68	0,6
EBITDA Ratio	%	15.09	14.85	15.35
Asset-liability Ratio	%	69.7	69.49	68.2

Table 2. 220 kV grid evaluation index weight.

Index	Regional weight	Overall weight
Voltage eligibility rate	1	0.042328
Unit investment add electricity sale	0.45155	0.084524
Unit investment add load	0.18888	0.035355
Grid construction investment	0.35418	0.066297
Power loss rate	0.0054	0.00101
Average length of line	1	0.013622
Unit transformer electrical load	0.44873	0.049039
Unit Line electrical load	0.34487	0.037689
220 kV transformer capacity	0.20641	0.022557
EBITDA ratio	0.39666	**0.256867**
Unit asset electricity revenue	0.218	0.141171
Unit asset electricity sale	0.26059	0.168754

3.3 Evaluation Result

The scores of calculation based on ANP are shown in Fig. 2 that the score of 2015 is the best, followed by 2014, 2016 is the worst one. As well as the radio chart revealed, 2016 marked down by the operation-effectiveness index, which was lowered than 2015 and 2014. Meanwhile the index of grid structure was lowest too. In the later development, A province should pay more attention to the grid structure planning and design, and improve the operation-effectiveness in case of score retrogress. On the other hand, the score of 2016 safety and quality was the peak among three years, as shown in Table 3. Which reflecting cooperation attached importance to safety and quality. This laid good foundation to healthy and steady development of power grid.

Because of the restriction of index selection, this paper may not evaluate integrate data from every year. And the relevant rule-making of original data may influence a lot to the later evaluation. In the future, this method should improve the index-selecting and rule-making, to ensure the veracity and objectivity of calculating results.

Table 3. 220 kV grid evaluation result.

Year	Score
2014	90.596
2015	95.181
2016	85.711

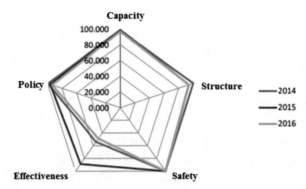

Fig. 2. 220 kV power grid development-diagnosis ratio chart

4 Conclusion

Firstly this paper comprehensive analyzed key index of A province 220 kV grid from cooperation-scale, development-speed, safety and business-quality, operation-effectiveness and internal-policy, according to development-diagnosis feature. We can obtain conclusions as followed:

Cooperation-scale, Power Grid Structure, Electricity-safety, Operation-effectiveness and Inner-policy, these five aspects consist of 14 main indexes. These 14 indexes are related to each other.

The overall weight of 14 indexes is based on the relevance among indexes instead of regional weight, where higher relevance brings higher overall weight. Especially 3 economic indexes namely Unit asset electricity sale, Unit asset electricity revenue and EBIT weigh more.

The scores of different years are relevant to all the 5 aspects. Higher scores come from better achievement and the more aspects get good scores, the higher scores will be acquired in the last.

In summary, we suggest the power grid company pay more attention to the economic indexes, which are relevant to all aspects of grid operation, and try best to improve all the aspects in order to get the higher evaluation scores, which reflects the overall operation effectiveness of the company.

References

1. Wang, Z.: Study on the diagnosis model of power grid development based on network analytic hierarchy process. J. Zhengzhou Univ. (Eng. Sci.) **39**(02), 39–43 (2018). (In Chinese)
2. Delong, C., Wei, Z.: Scientific diagnosis promotes grid management level. China Power Enterprise Manag. **06**, 53 (2020). (In Chinese)
3. Zhang, Y.: Research on power grid diagnosis, analysis and management. Sci-tech Innov. Productivity **12**, 24–25+28 (2016). (In Chinese)
4. Liu, S., Wu, D., Pan, J., Xu, L.: Deepening research and application of distribution network development diagnosis system. Rural Electrification **05**, 17–19 (2017). (In Chinese)
5. Zhao, W.: Research on the development diagnosis index system of municipal power grid. Tech. Dev. Enterprise **35**(21), 75–76+121 (2016). (In Chinese)
6. Guo, Z.: Research on power grid development diagnosis for big data. Sci. Technol. **26**(04), 83+121 (2016). (In Chinese)
7. Sheng, W., Zheng, X., Guo, X.: Frequency deviation peak calculation of sending-end network in large asynchronous interconnected power grid. J. Eng. **16**, 905–909 (2019)
8. Gencer, C., Guerpinar, D.: Analytic network process in supplier selection: a case study in an electronic firm. Appl. Math. Modelling **31**(11), 2475–2486 (2006)
9. Meade, L.M., Presley, A.: R&D project selection using the analytic network process. IEEE Trans. Eng. Manag. **49**(1), 59–66 (2002)
10. Atmaca, E., Basar, H.B.: Evaluation of power plants in Turkey using Analytic Network Process (ANP). Energy **1**, 555–563 (2002)

Time Synchronization Algorithm of Airborne Network Data

Jian Li[✉] and Kun Zhao

China Flight Test Research Institute, Xi'an 710089, China

Abstract. Accurate and reliable data is the basic requirement of flight test data processing, but the use of network test architecture brings some difficulties to data processing time synchronization. Aiming at the problems of network packet disorder and time synchronization in airborne network data processing, this paper studies the causes and characteristics of network packet data disorder. This paper analyzes and compares the advantages, disadvantages and applicability of three time synchronization algorithms based on the number of packets, the maximum reverse order and the half packet time interval in detail. Finally, the time synchronization algorithm based on the half packet time interval is selected as the airborne network data processing algorithm, and the specific implementation of the algorithm is introduced in detail.

Keywords: Network test architecture · Packet disorder · Time synchronization algorithm · Half packet time interval

1 Introduction

At present, the networked test architecture is widely used in flight test. Under the networked architecture, the parameters collected are large, the sampling rate is high, and the test equipment can use mature shelf products. In the networked architecture, the collector sends the collected data to the test system in the form of network packets, and the parameters to be processed are distributed in these network packets, When the data is transmitted on the test system, it can easily reach any device 4 and 5 connected on the switch through the switch. Because there will be a certain delay in the process of data transmission, transmission and reception, the data packets will not be arranged according to the normal sequence, and there is a phenomenon of data packet disorder [1]. If the time synchronization is not processed in the process of data processing, the resulting time of data will be disordered, and the data at other times will be treated as the current data by mistake, which will lead to the method to ensure the accuracy of model data processing, To solve the problem of out of order network packet data flight test data.

2 Causes and Characteristics of Network Packet Data Disorder

2.1 Brief Introduction of Networked Test System

The networked airborne test system includes sensor, collector, switch, recorder, data processing subsystem and telemetry transmitter. The collector collects the measured

J. Abawajy et al. (Eds.): ATCI 2021, AISC 1398, pp. 444–448, 2021.
https://doi.org/10.1007/978-3-030-79200-8_65

signals such as analog quantity, digital quantity, bus data and video data according to the set frequency, and then packages the measured signals into standard format network packet data for transmission. The network packet data is sent to the data processing subsystem, recorder and other equipment through the switch. The recorder records the received network packet data, and the telemetry transmitter sends the data to the ground in the form of PCM [2]. The data processing equipment processes the received network packet data and stores the processing results the airborne network test architecture is shown in Fig. 1.

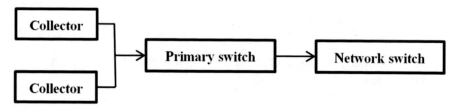

Fig. 1. Airborne network test architecture.

1) Generally, there are multiple collectors in the test system, and each collector will send multiple network data streams. The destination address and port of the data stream can be obtained through the airborne information configuration file.
2) Each data stream has its own sampling rate, and the sampling rates of each stream are not completely consistent, so the collector sends data according to the sampling rate.
3) Each network packet has its own independent key value, which is the unique identifier of the network packet.
4) Network packet data contains the code value of the parameters to be processed. Usually, the parameters to be processed are divided into different network packets.
5) Due to the complexity of the network and the uncertainty of transmission, the received network packet data may have packet loss.
6) In data processing, code value extraction and physical star transformation need to be carried out in different network data streams with different sampling rates, and the problem of time synchronization needs to be solved.

2.2 Causes and Characteristics of Network Packet Data Disorder

According to the architecture of the airborne test system, the data processing subsystem receives the network packet data sent by the collector through the switch, and then processes the data. Because the process of sending and receiving data packets will delay the data packets, and the delayed data will not be arranged in the predetermined order when processing, which leads to the disorder of data packets, so delay is the root cause of the disorder of data packets. The delay consists of three parts: transmission delay, transmission delay and reception delay. The total delay of network packet is equal to the sum of these three delays. The reasons for the delay of transmission, transmission and reception are analyzed as follows. For the delay of transmission: 1) due to the performance difference of the collector equipment [3]. The time taken by different collectors from data generation to data transmission is different: 2) for two packets with the same sampling rate of the same collector, there will be transmission delay when sending data at the same time: 1) the length of transmission path of different packets is not completely the same, For example, the path length from the data generated by the collector at the vertical tail to the data receiving device is obviously different from that generated by the collector at the fuselage. 2) The number of system nodes that thread packets pass through from generation to reception is not completely consistent. For example, some thread packets pass through the primary switch to the core switch, and some thread packets directly enter the core switch.

3   Analysis and Comparison of Three Time Synchronization Algorithms

Aiming at the problem of network packet disorder in data processing under airborne network architecture, two time synchronization processing algorithms are proposed, and the advantages and disadvantages of these algorithms are compared in detail. Finally, one of them is selected as the machine halberd data processing algorithm. The detailed description of the three algorithms is as follows.

3.1   White Processing Algorithm Based on the Number of Packets

Because the sampling rate of each network flow data packet in the airborne test system is fixed, and the sampling rate of bare problem demand is also fixed, the number of network packets to be processed per second can be obtained according to the sampling rate of network packets divided by the sampling rate of project demand, and the actual calculation time is the same.

The packet data is accessed at equal intervals according to the number of processing. The implementation of time synchronization algorithm based on half packet interval is as follows.

1) Analyze the airborne information configuration file to obtain the key value and sampling rate of each network packet;
2) Analyze the parameters to be processed, find out the key value and sampling rate of the network packet containing the parameters to be processed; assuming that there are n network packets to be processed, the sampling rate is 0.
3) According to the sampling rate s required by the project, the number of calculations per second for each packet is calculated:

$$X_i = S_i/S (i = 1, 2, \ldots, n) \tag{1}$$

4) If Xi < 1, then x = 1. This is because when the actual demand sampling rate is greater than the network packet sampling rate, each packet must be calculated. When the data is calculated, each packet is calculated once;

5) Read the network packet data and analyze the tey value and calculation times. If the key value belongs to the tey value of the parameter to be processed and the calculation times can be divided by X, the network packet will be processed; otherwise, the next network packet will not be processed and read;

6) When each network packet is processed once, the instant network packet time is regarded as the processing time, and the processing result is written into the result file.

3.2 Time Synchronization Algorithm Based on Half Packet Interval

Because the network packet data is sent to the test system according to a certain frequency, the specific network packet will not be out of order. The received network packets are arranged in order according to a specific time interval, and half of the network packet time interval is called half packet interval. The sampling rate required by the project is determined, and the time to be processed per second can be calculated according to the sampling rate required [4]. In the left and right half packet time interval of these times, there must be and only one network packet data, and the reference value at that time can be obtained by processing the network packet data. For example, if the sampling rate of a packet in the network is 16 and the sampling rate of demand parameter is 8, the half packet interval of a packet can be calculated as halfpackgap = 1000/16/2 = 15625, and the room second values of demand sampling rate are: 000, 125, 250, 325, 500, 625, 750, 875. As long as the time difference between the read network packet time points and these time points is less than half packet interval, the data packet is processed. The above time represents the sampling rate time of demand, the following time represents the real sampling rate of network packet data, the gray part represents the half packet interval of the attachment at the time of demand sampling rate, and the network packet time is processed in the gray part.

4   Conclusions

This paper first introduces the airborne networked test architecture, analyzes the causes and characteristics of network packet out of order, proposes three time synchronization algorithms for network packet out of order, analyzes and compares the contents, advantages and disadvantages of the three time synchronization algorithms in detail, and finally selects the time synchronization algorithm based on half packet interval as the airborne real-time data processing algorithm. The implementation process of the algorithm is given. The experimental results show that the algorithm can effectively solve the time synchronization problem of airborne network data processing, and give accurate processing results. The advantage of time synchronization algorithm based on maximum reverse order is that it can solve the problem of time disorder by finding the maximum reverse order time and reasonably using the disorder law; but the disadvantage is that it is more complex to implement, and the fault-tolerant ability of network data is weak. If the data packet disorder is serious, it is difficult to find the maximum reverse order time and use the disorder law.

References

1. Teng, F.: Research on Key Technologies of airborne test system based on network. Xi'an University of Electronic Science and Technology (2013)
2. Man, Y.: Research on Time Synchronization Accuracy of Networked Airborne Test System. Harbin Institute of Technology, Harbin (2014)
3. AI, B., Xing, D.: Network performance measurement in networked airborne test system. China Science and Technology Information **2016530**(z1), 45
4. Zuo, Z., Duan, X., Wang, C.: Design of new networked test system for air flight test bed. Aeronautical Sci. Technol. **27**(4), 50–53

Reservoir Storage Rainfall Dynamic Monitoring System Based on Data Mining Algorithm

Chenchen Yao, Junlong Tang$^{(\boxtimes)}$, Jinhua Liu, and Leilei Zhang

POWERCHINA HUADONG Engineering Corporation Limited, Hangzhou 311122, China
{yao_cc,tang_jl,liu_jh3,zhang_ll}@ecidi.com

Abstract. As an important water conservancy project, reservoir plays a positive role in flood interception, water supply guarantee and water resources allocation, and also provides a significant help for social and economic development and ecological civilization construction. Reservoir water storage is one of the important indicators for the healthy operation of the reservoir, which is closely related to the regional climate, humidity and precipitation. It is necessary to study the reservoir storage capacity for the rational protection and utilization of water resources and the formulation of relevant policies. With the development and maturity of data mining technology and the successful application in all walks of life, the advantages of this technology in discovering hidden rules or patterns in data are incomparable with other technologies. This paper collects and reviews a large number of relevant literature, discusses how to apply data mining technology and data analysis of reservoir storage problem.

Keywords: Data mining algorithm · Data preprocessing · Decision tree · Reservoir storage capacity

1 Introduction

In recent years, due to climate change and the increase of extreme weather, flood disaster is still the biggest natural disaster that brings loss to human society. According to the latest data from the United Nations Office for disaster reduction, the flood in 2018 alone caused property damage to 35.4 million people around the world, resulting in 2859 deaths [1]. On the other side of the flood, some areas are suffering from severe drought and lack of fresh water resources, which not only increases the pressure on domestic water for residents, but also increases the water consumption for Industry and agriculture. To this end, the governments of various countries have taken a lot of countermeasures, and the reservoir, as one of the active water conservancy projects, effectively alleviates the above situation. The reservoir has the function of intercepting flood, ensuring water supply and allocating water resources, which can not be ignored for social and economic development and ecological civilization construction. The reservoir is one of the important indicators of water storage and operation, which is closely related to the reservoir's water volume and humidity. Therefore, it is necessary to study the water storage capacity of the reservoir for the rational protection and utilization of water resources and the formulation of relevant policies.

© The Author(s), under exclusive license to Springer Nature Switzerland AG 2021
J. Abawajy et al. (Eds.): ATCI 2021, AISC 1398, pp. 449–453, 2021.
https://doi.org/10.1007/978-3-030-79200-8_66

2 Data Mining Technology

2.1 Association Rules Analysis

Association rule is a kind of relationship or mutual influence between the values of two or more data items in the data set. For example, the change of the value of one or more data items will cause the change of the value of another or more data items. Association mainly includes: simple relation Association, causality Association and time series Association. Association rule analysis is to find the association relationship hidden in the object data set, so as to mine the knowledge. The task of association rule analysis is to find those rules whose confidence is greater than or equal to the minimum confidence threshold and whose support is greater than or equal to the minimum support threshold. Association rules can also find rules in the occurrence time or sequence of events, that is, the occurrence of one or some events will cause another or some events to occur in sequence.

2.2 Neural Network

Neural network is a widely used and very important method in data mining. This method is based on the theory of neurobiology, which imitates the working mechanism and structure of human brain neural network [2]. Neural network presents many characteristics of human brain, which is composed of many interconnected neurons. It can carry out nonlinear operation and complex logic operation, and has some functions of human brain. Because the neural network is a distributed matrix structure, the neural unit has high parallelism and distribution, and can parallel calculate and store the data, which makes the neural network has strong self-adaptive, self-organizing and self-learning ability. The intelligence of neural network is mainly reflected in nonlinear mapping function, classification and recognition, associative memory, optimization calculation and knowledge processing. Based on Hebb learning rules and M model, neural network consists of feedforward neural network, feedback neural network and self-organizing neural network. This method mainly uses learning extraction method and decomposition extraction method to extract rules from neural network. The most outstanding advantage of neural network is that it can still achieve high precision prediction in the case of complex problems. However, this method also has some defects, such as: easy to be affected by excessive training; training time is long; it is not easy to be understood.

3 Classification in Data Mining

3.1 Purpose of Classification

Firstly, the data in the training sample set is analyzed. According to the characteristics of the training samples with known class labels, the description (model) of known categories is constructed. Then, the unknown data are classified by the constructed class description (model), and the class label of the unknown data is predicted. Data classification can be summarized into two steps: building a model and using a model for classification, as shown in Fig. 1.

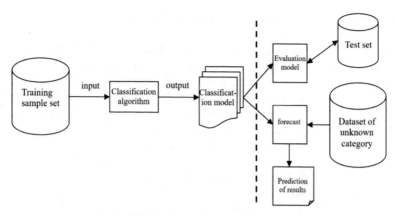

Fig. 1. Steps of classification

3.2 Classification Algorithm Based on Decision Tree

Decision tree classification algorithm is a case-based inductive learning algorithm. The algorithm can generate decision tree model based on given data set and extract rules from it directly [3]. The decision tree algorithm uses the top-down recursive method, compares the attributes of the data set by using the attribute measurement, selects the root node, and branches from the node according to the different attribute values, and the data set is divided into several sub datasets. Then recursively use the same method to divide the sub dataset, and finally reach the leaf node, each leaf node represents a category. A path from the root node to the leaf node is the prediction of the corresponding data category.

3.3 Classification Algorithm Based on Bayes

Using Bayesian algorithm for classification is to construct a statistical classifier based on Bayesian theorem, namely Bayesian classifier. The posterior probability of each pattern is obtained from the samples, and it is used to predict the probability Classification distribution of each category in the sample set. Then, the probability that the sample belongs to class c_i is:

$$P_r(C = c_i | A_1 = a_1 \wedge \Lambda \wedge A_k = a_k) \tag{1}$$

The probability that the sample belongs to which class is the highest, it belongs to which class. Bayes theorem is as follows:

$$P_r(C = c_i \backslash A_1 = a_1 \wedge \Lambda \wedge A_k = a_k) = \frac{P_r(A_1 = a_1 \wedge \Lambda \wedge A_k \backslash C = c_i)}{P_r(A_1 = a_1 \wedge \Lambda \wedge A_k = a_k)} P_r(C = c_i) \tag{2}$$

In the formula, $P_r(C = c_i)$ is the prior probability of each class, which can be obtained from the training sample set.

4   Estimation Model of Reservoir Storage Capacity

The reservoir water storage model is a mathematical model established by relevant researchers in order to quickly and accurately estimate the reservoir storage capacity. The reservoir can be generally divided into lake type reservoir and river type reservoir. The characteristics of lake type reservoir are that the water area is wide, the water flow is gentle, and the water surface of each part is basically kept at the same height; the characteristics of river type reservoir are that the reservoir body is narrow and long, the water flow area is small, and the water flow is fast, and there is an obvious water level difference between the inflow area and the outlet area, As a result, the whole reservoir will frequently submerge the surrounding fluctuating areas [4].

According to the above two characteristics, the lake type reservoir can be regarded as the static storage reservoir, and the river type reservoir can be regarded as the reservoir with the sum of static and dynamic storage capacity. Among them, the estimation of water storage capacity of river type reservoir is more complex, and the early method is difficult to accurately estimate the dynamic storage capacity of river type reservoir, and the key of research is whether the dynamic storage capacity of river type reservoir can be estimated more accurately. On the basis of literature and data, this paper studies the mainstream and classic reservoir storage estimation models, including hydrological formula method, data assimilation method and accumulation method. There are five models: method, mean value method and integration method. In practical application, the above models can be used either alone or in combination.

The application of data assimilation method in reservoir water storage estimation is to combine water level, water area and water storage with relevant mathematical models to achieve reservoir water storage estimation. The common models include polynomial model, exponential model and power function model.

5   Conclusion

In view of the current situation and shortcomings of reservoir water storage estimation, this paper proposes to use data mining algorithm to extract reservoir parameters. After verifying the accuracy of the extracted water area, the quantitative function relationship of water area difference of data mining algorithm is constructed by using data assimilation technology, Finally, a reservoir water storage estimation system based on data mining algorithm is designed and implemented.

Acknowledgements. State's Key Project of Research and Development Plan (2018YFC1508006, 2019YFC1510700).

References

1. Guojun, M., Lijuan, D., Shi, W.: Principle and Algorithm of Data Mining. Tsinghua University Press, Beijing (2005)
2. Huizhong, W., Anqun, P.: Research status and development trend of data mining. Ind. Mining Automation **2**, 33–36 (2011)

3. Zhang, L., Zhang, G., Li, F.: Study on remote sensing analysis method of Laolongkou reservoir capacity based on landsa8 and DEM. Sci. Technol. Innov. Appl. **252**(32), 126–127 (2018)
4. Yonghai, T., Zhihui, Z., Yu, S.: Design of flood regulation system of Qinghe reservoir based on web. J. Water Resources Water Eng. **21**(1), 168–170 (2010)

Numerical Calculation of Surface Plasmon Polariton Lens Based on 3D Model of Cloud System

Min Li, Hairong Wang, and Jingmei Zhao[✉]

College of Optoelectronic Engineering, Yunnan Open University,
Kunming 650223, Yunnan, China

Abstract. A new design scheme of surface plasmon polariton lens is proposed. In this scheme, dielectric gratings are placed on the outer surface of two subwavelength apertures to realize the effective convergence of human beam. The results show that the effective refractive index can be effectively controlled by adjusting the width and dielectric constant of the sub wavelength apertures, so as to realize the control of the propagation properties of the surface plasmon polaritons in the sub wavelength metal plate waveguide structure. The finite-difference time-domain (FDTD) method combined with perfectly matched layer (PML) boundary conditions is used to simulate the light field distribution in the structure. The influence of grating period number on the imaging characteristics is discussed, and the physical mechanism of nano focusing effect is deeply understood. The results show that the focal length and focal spot increase with the increase of the number of surface gratings. When the number of gratings increases from 5 to 11, the focal length increases from 1.715pm to 2.325pm, and the focal spot size increases from 0.615pm to 1.715pm. This structure may be used as nanofocusing devices in future integrated optical circuits.

Keywords: Metal waveguide · Surface plasmon element lens · Subwavelength hole · Time domain finite difference method

1 Introduction

Surface plasmon polaritons (SPPs) are electromagnetic waves propagating on the interface between metal and dielectric [1]. They are generated by the interaction between the external light field and the metal surface electrons. They exist in the form of a evanescent field on the metal dielectric interface. The intensity of SPPs decreases exponentially with the increase of the distance from the interface. The energy is highly concentrated and adheres to the metal surface. This quasi two-dimensional property makes it possible to achieve optical transmission beyond diffraction limit in sub wavelength metal microstructure materials. In 1998, Ebbesen et al. found that the transmittance of metal films with periodic holes would be greatly enhanced when some specific wavelengths passed vertically through the metal films with periodic holes. Since then, scholars have carried out a lot of theoretical and experimental studies on various nanostructures, such as nano rectangular metal strips, nano chains, nano channels and metal point dielectric structures. The most effective way to control the periodic propagation of light is to use a kind of nonmetal structure with periodic attraction.

In this paper, a periodic grating structure is placed on the outer surface of the sub-wavelength metal hole to achieve effective convergence of the incident beam. The time evolution of the light field distribution in the structure is numerically calculated by using the finite difference time domain (FDTD) method combined with the perfectly matched layer (PML) boundary conditions. The effects of the aperture size and grating period number on the imaging characteristics are discussed, In order to obtain a deeper understanding of the physical mechanism of nano focusing effect.

2   Structural Design

The structure consists of two subwavelength holes on a metal film with thickness of H, and then placing dielectric gratings on the outer surface. The width of the two holes is, ω_1, ω_2 and the width and thickness of the surface grating are a and b respectively. The width and thickness of the middle dielectric film are c and d, and the period of the outer surface grating is Λ (this paper only considers the normal incidence of TM polarized electromagnetic wave). Considering the dispersion of metal dielectric constant with wavelength and the inherent ohmic loss of metal, the dielectric constant of metal is described by Drude model:

$$\varepsilon_m = 1 - \frac{\omega_p^2}{\omega(\omega + j\gamma)} \tag{1}$$

Where p is the plasma resonance frequency of the electron, y is the parameter describing the metal loss, and ω is the circular frequency of the incident electromagnetic wave. Figure 1 shows the relationship between n_{eff} and ε_d when V is 40, 60 and 80 nm respectively.

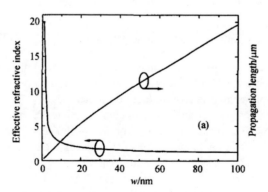

Fig. 1. Relationship between effective refractive index and aperture width and dielectric constant

3 Basic Theory of Surface Plasma

3.1 Drude Model of Metals

Metal optical properties can be described by complex permittivity. Choosing different models has a great influence on Simulation and calculation of metal optical properties. Drude model of metal materials can not only accurately describe the complex permittivity and explain the optical properties of metals, but also use the electric displacement vector as the intermediate variable, The problem caused by the negative real part of complex permittivity of metal materials is solved, and the convergence stability of Maxwell equations is guaranteed [2].

The metal electron is regarded as an ideal gas in free motion. In the process of continuous motion, the electron collides with the nucleus, lattice defect or impurity in the crystal and scatters around. The average collision interval is set as relaxation time τ, which is usually in the order of 10–14 s under room temperature. This time has nothing to do with the velocity and position of the electron, and the electron still moves freely after collision, These interactions are reflected by the collision frequency. The probability of electron collision per unit time is γ, which means that the free particle damping coefficient is assumed

$$\overrightarrow{P_{(t+dt)}} = \frac{dt}{\tau}\vec{f}dt + (1 - \frac{dt}{\tau})(\overrightarrow{p_t} + \vec{f}dt) \tag{2}$$

The formula is divided into two parts: the initial average momentum of the colliding electron is zero, the momentum at t + dt is the impulse $\vec{f}$ under the action of external force $\vec{f}$, and the momentum of the electron without collision is the momentum change caused by the original momentum and external force.

Under high vacuum condition, the dielectric constant of metal material can be obtained by measuring the reflection and transmission characteristics of metal surface with ultra-high cleanliness. In the actual numerical simulation, the parameters of metal affected by light frequency and single photon energy will change constantly, which requires the calculation of metal refractive index in different frequency bands.

3.2 Excitation Modes of Surface Plasmon Waves

Some special structures are needed to achieve wave vector matching in SPPs excitation. The common excitation methods are prism coupling, grating coupling and near-field coupling [3]. The following briefly introduces each method.

It can be divided into two types: Kretschmann coupling structure and oto coupling structure, which cover the surface of prism with metal film, and change the light wave momentum through reflection coupling to meet the dispersion relationship.

If the refractive index of prism is n, the wave vector of incident light wave is k, and the incident angle is θ, when the angle of incident light is greater than the total reflection angle, the light will have total reflection on the prism and metal surface and produce evanescent wave, then the wave vector of the evanescent wave sine. When the thickness of the metal film is not too thick (usually about tens of nanometers), SPPs will be generated at the metal air interface after the wave vector matching conditions are satisfied. If the thickness of the metal film increases until there is no evanescent wave at the metal dielectric interface, that is, the wave vector matching cannot be achieved, and SPPs cannot be generated.

3.3 Properties of Surface Plasmon Waves

Surface plasmon wave propagates along the surface of conductor. When the structure of metal surface is changed, the properties of surface plasmon, dispersion relation, excitation mode and coupling effect will change greatly. The manipulation of light propagation can be realized by the interaction between SPPs and the propagation of light field.

1. Transmission enhancement effect
   This phenomenon can be observed by etching the subwavelength single hole or small hole array structure on the metal film. The excitation of SPPs can enhance the electromagnetic field and greatly improve the diffraction of light, which is the transmission enhancement effect of SPPs [4]. At present, this phenomenon has been observed in optical frequency band, microwave band and terahertz band. At the same time, transmission enhancement phenomenon is also observed in one-dimensional and two-dimensional cases. The transmission enhancement effect is briefly introduced by using metal subwavelength small hole array (tdmsha) structure.
2. Beam collimation effect
   It is also known as beam effect. In periodic slot array structure, when SPPs excitation conditions are satisfied, the electromagnetic wave generated has a strong angle limitation, and the beam divergence angle is only $\pm 3°$.
   This kind of effect can produce new light source with high directivity, which has a wide application prospect in the fields of magneto-optical storage, nano lithography and spectral excitation.
3. Beam focusing
   The periodic fluctuation structure around the through hole can make the beam propagate tens or even hundreds of microns, and keep constant in this distance. When the aperiodic undulating structure is used, the beam size can be reduced to less than $1 \mu m$ or even to nanometer level. Because SPPs has the characteristics of convergence in the region with low phase velocity and group velocity, it can change the simultaneous interpreting of the geometric scale of metal structures in different transmission regions, modulate the propagation speed of SPPs, and achieve beam convergence.

4 Conclusion

A new surface plasmon polariton lens is designed. By placing dielectric gratings on the outer surface of two subwavelength apertures, the effective convergence of human beam is realized. The dispersion relationship of surface plasmon polaritons in waveguides is studied by genetic algorithm. It is found that the effective refractive index can be effectively controlled by adjusting the width and dielectric constant of sub wavelength apertures, so as to realize the control of propagation characteristics of surface plasmon polaritons in subwavelength metal plate waveguide structure. The finite difference time domain (FDTD) method combined with perfectly matched layer (PML) boundary conditions is used to numerically calculate the time evolution of the light field distribution in the designed structure. The intuitive physical image of the light field propagation process is given. The influence of grating period number on imaging characteristics is discussed. The results show that the focal length and focal spot size increase with the increase of the number of surface gratings. Because of the strong edge diffraction effect of traditional dielectric lens, it is not suitable for the optical system of wavelength and sub wavelength. The structure designed in this paper does not have this defect, and the focal length is only several times of the wavelength, which is beyond the traditional lens. This structure can be applied to the focusing element of integrated optical system.

References

1. Wang, Y.: Several Novel Effects in Surface Plasmon Splitters and Subwavelength Metal Arrays. Library of Harbin Institute of Technology, Heilongjiang (2010)
2. Yongyi, C., Cunzhu, T., Li, Q., Lijun, W., Jinlong, Z.: Progress in surface plasmon polariton nanolaser technology and its application. China Opt. **05**, 453–463 (2012)
3. Yankun, C., Weihua, H., Xiaoming, L., Du, Y., Yang, F.: Surface plasmon polaritons breaking through diffraction limit. Appl. Optoelectron. Technol. **04**, 39–44 (2011)
4. Qiang, W., Guoding, L.: Basic Theory of Electromagnetic Field, pp. 231–234. Tsinghua University Press, Beijing (2005)

Cloud Computing Platform for Solar Semiconductor Lighting System

Hairong Wang[✉], Min Li, and Jingmei Zhao

College of Optoelectronic Engineering, Yunnan Open University,
Kunming 650223, Yunnan, China

Abstract. Solar photovoltaic power generation as a clean and pollution-free new energy, its promotion and application development is rapid, will gradually become one of the basic human energy. White LED is a new generation of electric light source after incandescent lamp and fluorescent lamp. With the improvement of lumen efficiency, white LED is gradually developing to the field of general lighting. This paper studies and designs a stand-alone solar semiconductor lighting system based on solar photovoltaic power generation technology and white LED lighting technology

Keywords: Photovoltaic power generation · Maximum power point tracking · Battery · White light emitting diode semiconductor lighting

1 Introduction

With the development of the world economy, the demand for energy is increasing. The impact of investment on the environment is to reduce air pollution, protect the human environment and provide long-term sustainable energy. It has become a national consensus to implement the sustainable development strategy and gradually realize the energy transformation, and will continue to develop vigorously. Leila new energy is the highest in solar energy, and I am likely to use it [1].

In recent years, the global photovoltaic power generation has developed rapidly. The United States, Europe and Japan have formulated huge photovoltaic power generation development plans. In 1997, the United States and the European Union announced the "million roof photovoltaic plan", and Japan also formulated the "universal residential photovoltaic system plan" in 1997. By 2010, the United States plans to install 46 gw (including million rooftops); the European Union plans to install 67 gw (renewable energy white paper); Japan plans to install 5 GW (Japan's new sunshine program); other developing countries are expected to install 1.8 gw of solar photovoltaic modules, with an annual growth rate of 34%, Among them, Germany takes the lead in the total installed capacity of 837 mw.

J. Abawajy et al. (Eds.): ATCI 2021, AISC 1398, pp. 459–462, 2021.
https://doi.org/10.1007/978-3-030-79200-8_68

2 Principle Analysis of Independent Solar Semiconductor Lighting System

2.1 Composition of Independent Solar Semiconductor Lighting System

The stand-alone solar semiconductor lighting system is mainly composed of four parts: solar cell, battery, controller and semiconductor lighting load [2]. The purpose of adding standby power supply is to supply power directly to semiconductor lighting load by standby power supply in rainy days for a long time, so as to ensure that the semiconductor lighting system can work normally when the battery cannot supply power.

When the system works, solar energy is converted into electric energy through solar cells. However, the output power of solar cells is unstable due to the great influence of temperature and solar radiation intensity. Therefore, when the solar radiation intensity is large enough, it is necessary to use batteries to store the converted energy, In order to supply power to the semiconductor lighting load when lighting is needed, the controller is the core part of the solar semiconductor lighting system. When the system works, the controller can control the working state and manage the charging and discharging process of the battery, so that the system can work stably in different working states.

2.2 Output Characteristics of Solar Cells

The equivalent circuit model of solar cell is shown in Fig. 1, where I_{LG} is photogenerated current, R_{sh} is equivalent parallel resistance, R_s is equivalent series resistance, and R_s is equivalent resistance of load. The characteristic function of the model can be expressed by Eq. 1.

$$I = I_{LG} - I_{OS}\left\{\exp\left[\frac{q}{AKT}(V + IR_s)\right] - 1\right\} - \frac{V + IR}{R_{sh}} \tag{1}$$

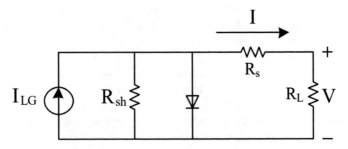

Fig. 1. Equivalent circuit model of solar cell

The output of solar cells is mainly affected by the solar temperature and solar radiation intensity, while the temperature is affected by the open circuit voltage.

3 Design of Independent Solar Semiconductor Lighting System

3.1 Overall Design Scheme of the System

The principle of independent solar semiconductor lighting system includes four main parts: solar cell, battery, controller and semiconductor lighting load, and an auxiliary part of standby power supply. The standby power supply is provided by switching power supply to convert 220 V AC mains power into low-voltage DC power supply, and only one low-voltage DC input interface is required for the standby power supply in the system design.

(1) Strong solar radiation mode: when the solar radiation intensity is strong enough (only in the daytime), that is, when the output voltage of the solar cell is not zero and the output current is greater than zero, the system decides to charge the battery with MPPT or constant voltage according to the SOC of the battery; at the same time, the battery or standby power supply is used to power the load according to the SOC of the battery.

(2) Weak solar radiation mode: when the solar radiation is weak (including night and dark day), that is, when the output current of the solar cell is zero, the system decides to supply power to the load by battery or standby power according to the SOC of the battery.

3.2 Function and Principle of Controller

The controller is the core part of the system, through which the working process of the system is managed and controlled. The main functions of the system are the management of the working state of the system, the management of the remaining capacity of the battery, the MPPT charging control of the battery, the switching control of the main power supply and the standby power supply, and the temperature compensation of the battery [3].

3.3 DC-DC Conversion Circuit

Switching DC-DC converter circuit usually has buck type, boost type and buck boost type. In this paper, the DC-DC converter in the controller is buck type converter circuit, and the DC-DC converter part of its transformation principle. Buck DC-DC converter is an important part of the controller circuit. It mainly realizes two functions: one is to cooperate with MCU to realize MPT control and constant voltage charging control, and the other is to realize overcharge protection. The circuit design is mainly about the choice of power switch and output filter capacitor.

$$L \geq \frac{V_{om}(1 - D_{cm})}{f_s|\Delta I_{Lm}|} \tag{2}$$

Where, f_s is the switching frequency of the power switch tube, D_{cm} Is the duty cycle of PWM signal at the maximum output power of DC-DC converter, ΔI_{Lm} is the peak peak ripple current value of inductance, and V_{om} is the maximum DC output voltage.

4   System Assembly and Test Analysis

The system designed in this paper is mainly composed of solar cells, valve regulated sealed lead-acid batteries, controller, standby switching power supply and LED lamps. In this paper, a simple test system is built to test the important working parameters of the system [4]. The test system adopts stp026–12/D solar cell of Wuxi Suntech; the battery adopts the valve regulated sealed lead-acid battery of Wuhan Yuao 6fm24, with the rated voltage of 12 V and the rated capacity of 24 h; the system controller is the controller designed in this paper; the LED lighting load is the LED lamp designed in this paper.

This section mainly carries out MPPT control test, battery charging test and half power discharge performance test, and makes a brief analysis. The main testing instruments used are digital multimeter, ammeter, illuminance meter and tds3052b digital fluorescent oscilloscope of Tektronix company.

In this chapter, a simple solar semiconductor lighting system is built and tested. The charging process and half power discharge performance parameters of MPPT controlled battery are tested, and the test results are analyzed. The results show that the controller can track the maximum output power point of solar cells.

5   Conclusion

As a kind of clean and pollution-free new energy, solar photovoltaic power generation is developing rapidly in the world. It is expected that the proportion of solar photovoltaic power generation in the global power generation will gradually increase and become one of the basic energy of human beings. With the improvement of LED lumen efficiency and the extension of service life, LED will be gradually used in the field of general lighting within 5–10 years. Combining solar photovoltaic power generation with semiconductor lighting, the development of 200 W level independent solar semiconductor lighting system can meet the requirements of ordinary household lighting and some special lighting fields. Based on the design of independent solar semiconductor lighting system, this paper studies the output characteristics of solar cell and the control scheme of its output, analyzes the charging and discharging control strategy of battery, designs the driving circuit of semiconductor lighting lamps, and designs the independent solar semiconductor lighting system based on the characteristics of each part.

References

1. Yang, J., Chen, Z., et al.: Prospects of solar power generation in the 21st century. J. Shanghai Electric Power Univ. **17**, 23–28 (2001)
2. Ling, L.: Opportunities and prospects of solar semiconductor lighting. New Materials Industry **11**, 38–42 (2003)
3. Guoyi, Z., Zhizhong, C.: Gallium nitride based white light emitting diode, the cornerstone of solid state lighting source. Phys. High Tech. **33**(1), 831–842 (2004)
4. Chen, W., Shen, H., Wang, D., et al.: Research on driving technology of solar semiconductor lighting. J. Lighting Eng. **16**(3), 7–10 (2005)

Evaluation System of Scientific and Technological Innovation Talents Benefit Based on Cloud Operation Management and Optimization

Yejun Wang[✉]

Shandong Xiehe University, Jinan, Shandong, China

Abstract. This paper first expounds the necessity of developing the benefit evaluation system of scientific and technological innovation talents, and then explains the evaluation model and algorithm adopted by the system. Next, it introduces the overall framework of the system and discusses the design of each functional module of the system. Finally, the prospect of the system's popularization and application is given.

Keywords: Scientific and technological innovation · Evaluation system: query statistics · Evaluation algorithm

1 Introduction

The development of the talent benefit evaluation system is based on the actual situation of the scientific and technological talents in the business technology unit. With the help of the beneficial experience gained from the current domestic and international talent benefit evaluation, the paper makes an in-depth study of the advanced theory, the rationality of the model, the versatility of the use, the friendliness of the interface and the convenience of expansion in the application of the latest achievements of information technology. We should strive to realize the scientific evaluation of talent benefits, gradually form a scientific mechanism to grasp the quality and efficiency of talent construction, and provide technical support for the transformation of scientific and technological talent team construction from quantity scale type to quality benefit type. Talent benefit evaluation system is an intelligent system to realize talent benefit evaluation and scientific and technological innovation ability evaluation. It provides a unified evaluation technology platform for each business and technical unit. It objectively reflects the overall strength and utilization efficiency of the scientific and technological personnel team of the participating units.

2 Evaluation Index System

2.1 Lay Down a Principle

The evaluation index system of this system is based on the following principles.

© The Author(s), under exclusive license to Springer Nature Switzerland AG 2021
J. Abawajy et al. (Eds.): ATCI 2021, AISC 1398, pp. 463–467, 2021.
https://doi.org/10.1007/978-3-030-79200-8_69

(1) In order to ensure the accuracy, fairness and applicability of the quantitative evaluation system, the directivity of the index system is extremely important. While highlighting the various elements of the benefit of scientific and technological innovation talents, the corresponding index system is set up to comprehensively and accurately reflect the overall situation of the scientific and technological innovation of the participating units, And play an obvious guiding role.

(2) According to the principle of incoherence, all evaluation indicators are divided into several independent categories, and each category is assigned a fixed weight[1]. This weight can be adjusted and optimized according to the situation in the use process. The internal index weight of each category has comparative significance, but the index weight belonging to different categories has no comparative significance, Thus, it avoids the difficulty of index weighting of different categories of projects caused by the unified weighting method.

(3) The universality principle allocates high weight to the projects that are easy to quantify, convenient to audit and generally recognized, and low weight to the projects that are difficult to quantify, difficult to audit and low recognition.

2.2 Index Range

The academic level index statistics the publication of natural science papers and social science papers as well as the publication of works. In natural science and social science papers, different scores are assigned according to the grades of journals or conferences in which papers are collected and published. Take natural science papers as an example, The grades of papers are ranked from high to low: Science/nature SCI/scie EI istr ISR CSCD CSCD extended library, inter national conference of core journals, provincial and Ministerial Conference of general journals. In addition, the scores of excellent doctoral dissertations and master's theses are also considered. The level of works includes science and technology monograph, social science monograph, scientific and technological translation, social science translation, natural science compilation, social science compilation and internal textbook compilation.

3 Evaluation Model and Method

3.1 Linear Evaluation Model and Method

The linear model and method are mainly used in the evaluation under the condition that the evaluation indexes are irrelevant. The evaluation results are comparable among the evaluation objects. The system adopts the weighted linear model evaluation method, and different indexes are assigned different weights. The specific description is as follows.

If the number of participants is n, then the I-type self-evaluation score is $C_i = (i = 1 \cdots 6)$. The total score of the self-test of the participating units is: $D_i = C_i/n(i = 1 \cdots 6)$. Set the weight coefficient of each type of index assigned by the superior unit's summary evaluation system is $W_i = (i = 1 \cdots 6)$ The j-th unit's I-type self-evaluation score is $F_i = W_i * D_i/M_i = (i = 1 \cdots 6)$, the total number of participating units is m, and the highest score of type I evaluation of all participating units is:

$$M_i - Max(D_{ij})(i = 1 \cdots 6, j = 1 \cdots N) \tag{1}$$

Then the final score of item I of each participating unit is:

$$F_i = W_i * D_i/M_i (i = 1 \cdots 6) \tag{2}$$

The total score of participating units is

$$F = \sum F_i (i = 1 \cdots 6) \tag{3}$$

3.2 Stu Evaluation Model and Method

Different from the former two models, the evaluation results of this model are not comparable, and mainly focus on the evaluation of the investment and contribution of the evaluation object (unit). Based on the analysis and improvement of the latest international input-output model analysis method (DEA), we established the stu (Science Technology University) model analysis method[2]. Its function is to obtain the input and output benefits of a participating unit through evaluation, and point out the aspects that need to be improved and the near-term goals that can be achieved. The analysis is as follows:

Let $X_j = (x_{1j}, x_{2j}, K, x_{mj}), Y_j = (y_{1j}, y_{2j}, K, y_{mj})$ Where, x_{ij} is the input of the ith sub index in the j-th index; y_{ij} is the contribution of the i-th sub index in the j-th index; and (x_j, y_j) can be used to represent the j-th index of the evaluation model. The input and output contribution of each sub index of each index have corresponding weight coefficients. Based on the weight coefficient, each type of index has its corresponding efficiency evaluation function, The optimal programming model can be used to solve the problem by selecting reasonable weight coefficient.

4 System Structure and Function Overview

4.1 Data Acquisition and Processing

The original data is the basis of the evaluation and the most fundamental factor to determine the accuracy of the system evaluation results. In this system, except for the information initially set by the management personnel, all other basic data should be entered by the individuals and units participating in the evaluation. The collected information is divided into personal data and unit data according to the attributes. Personal information includes: basic information, talent attributes, academic papers and Monographs on talent training, teaching and scientific research achievements, etc.; unit information includes: degree and major, technical support amount of scientific research projects, scientific research funds, invention patents, equipment and technical innovation, continuous education hours, etc. In addition, according to the requirements of function test and training, the system also provides a simulation data generator, which can automatically generate personal information and unit information of simulation randomly, and carry out simulation query statistics and evaluation after data loading.

4.2 Information Query and Statistics

In order to improve the practicability of the evaluation system, facilitate users' participation and use, and reflect the scientificalness, impartiality and accuracy of the evaluation process, the system provides various types of real-time query and statistics functions, and provides a variety of display and output modes such as tables, curves, bar charts and pie charts to facilitate users' horizontal and vertical comparison [3]. All kinds of charts can be printed as required, According to the needs of the project, results, patent unit performance, personal performance, papers published and other content can be conveniently archived. The statistical module adopts the hierarchical design method. For each level of units, the system provides corresponding query, comparison and statistical functions, so that the system can be applied to business and technical units of various scales.

4.3 Quantitative Assessment

According to the evaluation model and algorithm provided by the system, combined with the specific index system, the original information in the submitted and processed database is extracted, classified, combined, calculated, counted and summarized to obtain the evaluation score of each major category of indicators, and then weighted, and then the evaluation results obtained will be transmitted to the display module.

(1) Personal assessment. According to the personal data of the evaluation object, the evaluation model and algorithm are used to evaluate the individual sub index score and finally form the total score.

(2) Unit evaluation. Unit evaluation can be carried out at different organizational levels, such as teaching and Research Office, Ministry, Institute or technical support unit. According to the six types of evaluation items defined, the corresponding data in the database are extracted, classified and weighted, and the scores of the sub indicators are obtained respectively. The total score of the evaluation unit is obtained by using the evaluation model and algorithm.

(3) Review by the headquarters. Based on the unit evaluation, according to the evaluation model and algorithm requirements, the headquarters summary evaluation determines the extremum of each major type of index, and obtains the total.

4.4 System Settings

In order to improve the universality of the evaluation system, facilitate the use of users and maintenance of the system, the system adopts modular and object-oriented methods in the overall design to build the system. Specifically, under the condition of system authorization (different users' rights can be different), the user can set up the interface provided by the system [4]. It can easily set up the compilation sequence, evaluation index and weight system of the evaluation unit, and the system automatically adjusts the program according to the content set by the user. In the process of data acquisition and processing, all the contents and projects are set for human-computer interaction. This function is mainly completed by the system administrator before using the evaluation system.

5 Conclusion

The evaluation results of scientific and technological innovation evaluation system are conducive to the reasonable allocation of education and training funds and scientific research funds, promote the benign competition and self construction of each business and technical unit, dynamically reflect the trend of higher education, teaching reform and scientific research development of the army, facilitate the headquarters to carry out quality supervision on the business and technical units, and guide the rational allocation of talents and the correct flow of funds within the whole army. The system adopts the method of qualitative and quantitative evaluation.

Acknowledgements. Key R & D projects (Soft Science) in Shandong Province: The Research on the Influencing Factors and Mechanism of Sci-tech Talents Flow in Shandong Province, Project No: 2019RKB01362.

References

1. Internet University Corporation. Index system IEB of China University Rankings/OLJ. http://rank2003.netbig.com/cn/rnk.002.htm. Accessed 16 Jan 2004
2. Tan, L.: Establishment of modern talent evaluation method [EB/OL]. http://edu.sinacomcn/wander/2000-05-03/2563.shtml. Accessed 07 May 2000
3. Peng, P.: Several problems in the development and application of talent evaluation technology [EB/OL]. http://www.aphr.org/ta/view.asp. Accessed 21 Oct 2002
4. Yuquan, C.: Further understanding of dynamic input-output model. J. Qufu Normal Univ. **21**(5), 46–50 (1998)

Enterprise Management Simulation Training System Based on Cloud Information System Architecture

Wen-Jing Hu(✉)

Management College, Hubei Institute of Business, Wuhan 430073, Hubei, China

Abstract. Business management is a major with strong practicality, specialty and operability. The construction of a perfect simulation training teaching system can help business management students actively practice experience, expand their knowledge system, broaden their horizons, and guide students to better integrate into the practice of modern enterprise business management. This paper focuses on the current situation of business management simulation training teaching, analyzes the main problems of business management simulation training teaching, and summarizes the basic strategies of business management professional simulation training construction around the needs of enterprises for business management talents, so as to improve the effectiveness of business management simulation training teaching mode.

Keywords: Business administration · Simulation training · Teaching mode

1 Introduction

Practice teaching is of great significance to improve the professional ability of students majoring in business administration. Although colleges and universities have emphasized the establishment of standardized simulation training room in recent years, which provides a lot of human and material support for business management simulation training teaching, the scientific aspect of business management simulation training teaching needs to be improved [1]. First of all, there is still a lack of long-term simulation training update and planning mechanism, and the simulation training can not be improved and optimized around the needs of enterprises for business management talents. Secondly, the lack of professionalism of business management simulation training activities, students lack of subjective consciousness in simulation training, students' awareness of participating in simulation practice is relatively passive, which is not conducive to students' active acquisition of professional skills in simulation training activities. Third, teachers and students do not cooperate with each other enough, teachers can not pay attention to the specific problems of students, and can not apply effective simulation training mode to comprehensively improve the quality of simulation training teaching activities.

Only by keeping effective interaction between enterprises and colleges can we continuously optimize and improve the quality of business management personnel training,

J. Abawajy et al. (Eds.): ATCI 2021, AISC 1398, pp. 468–472, 2021.
https://doi.org/10.1007/978-3-030-79200-8_70

solve the shortage of business management simulation training teaching, and at the same time provide enterprises with high-quality business management talents. However, the current social competition pressure of enterprises is greater, and the tasks of business management posts are relatively heavy. Some enterprises have not been able to effectively cooperate with colleges and universities and can not provide students with rich simulated training posts. Secondly, modern enterprise office often relies on the self-developed automation and intelligent office system. For the needs of confidentiality and security, enterprises often do not want students to contact the core information of the enterprise, which makes students can not contact the core business when participating in the enterprise simulation training activities, and the training students are not familiar with and participate in the enterprise operation mechanism and management process. Thirdly, the interaction between enterprises and colleges is insufficient, and enterprises can not effectively guide the construction of campus simulation training platform, which leads to the relatively low level and efficiency of business management simulation training carried out by colleges and universities.

2 How to Improve the Teaching Method of Simulation Training for Business Administration Major

2.1 Establish Simulation Platform to Meet the Needs of Enterprises

To improve the effectiveness of business management simulation training teaching, we need to build an effective business management simulation training platform. Only by giving full play to the advantages of increasingly rich and diverse information technology platform and developing the simulation platform of enterprise simulation environment based on multimedia, can the effectiveness of practical teaching of business management specialty be improved [2]. Therefore, both ERP software and business management project practice, as well as related human resources, financial resources, quality and customer management training projects should be based on the real enterprise environment, and the relevant training laboratories should start from the actual needs of enterprises and follow the basic process of enterprise operation, For example, the basic operation methods of procurement and enterprise customer management are set up to meet the needs of training practical business management talents for enterprises. Teachers should also supervise the students' specific training and arrange the problems of professional situation for students, so as to improve the application value of the simulation training platform and better cultivate the students' practical training ability.

2.2 Creating Diversified Simulation Training Situation

To improve the effectiveness of business management simulation training activities, it is necessary to create a variety of simulation training situations, and constantly improve the pertinence of simulation training on the basis of actively guiding enterprises to participate, especially focusing on integrating the frontier problems in enterprise innovation and entrepreneurship activities into the practice activities of industrial and commercial management. First of all, optimize the teaching mode of school enterprise cooperation, focus on enriching simulation training content on the basis of enterprise employment orientation, innovate simulation training projects, and improve the effectiveness of simulation training. Secondly, enrich the simulation training channels, improve the school training mechanism, guide students to enter the enterprise, on the basis of flipping school enterprise cooperation teaching, promote students to obtain more practical training harvest. Third, explore the way of school enterprise cooperation, guide students to make full use of surrounding resources to carry out simulation training, so that students can practice more specific business processes such as marketing and enterprise management.

During normal operation, the control system adopts constant DC power control mode:

$$I_D = \frac{P_{ref}}{U_d} \tag{1}$$

Analog impedance ratio:

$$M_Z = \frac{M_V^2}{M_p} \tag{2}$$

3   The Main Mode of Perfecting Simulation Training for Business Administration Major

3.1   ERP Training Platform

ERP is the main electronic platform to cultivate the training ability of business management specialty. The baozi mode is mainly composed of physical sand table, electronic sand table and software training room [3]. It can provide students with a simulated business management environment and create a real background similar to the enterprise environment according to the needs of students to practice business management professional knowledge. It includes simulating the office environment of enterprises, setting up basic operation departments, building basic task modules such as supply chain management, production management and sales management, guiding students to be familiar with enterprise processes, participating in enterprise operation management and other activities, and integrating innovation and entrepreneurship activities to promote students to carry out simulated employment practice activities.

3.2   Task Oriented Model

Task oriented mode has the characteristics of combining cognitive and practical teaching activities. Task oriented mode emphasizes the development of students' autonomy and strives to better operate practice on the basis of students' groups. Teachers play an auxiliary role in the task-oriented business management practice. According to the needs of enterprise business management talents, teachers set up roles and divide tasks for students, put forward task problems with occupational situation for students, and guide students to collect resources independently and design solutions to problems independently with the supply of necessary information resources, And ultimately improve the students' problem-solving ability. The task oriented training mode can further promote students' integration into the market and professional environment, stimulate students' curiosity, and promote students' independent practice driven by tasks.

3.3 School Enterprise Cooperation Mode

The school enterprise cooperation mode guides students to participate in real business management practice activities in the enterprise environment. School enterprise cooperation mainly plays the role of enterprise professional business management talents, and requires double qualified talents to guide students in business and ideology, so as to promote students to accumulate practical experience in the real enterprise environment. The mode of school enterprise cooperation is developing towards standardization, specialization and diversification [4].

In order to improve the effect of simulation training, first of all, we should make clear the goal of training, so that students and teachers can have a clear idea. In this way, they can pay more attention to practical teaching and actively develop their potential under the incentive of the goal. For example, in the school enterprise cooperation simulation experiment, students carry out practical training on their posts. Although the enterprise employees are busy with their own affairs and have no time to take care of the training students, but the students know what they want to learn and want to reach that level. Then in the training, they will deliberately observe, pay attention to, and think. In this way, the effect of simulation training can be improved naturally. Secondly, guide students to formulate short-term and long-term training objectives, and make overall arrangement and planning for the whole simulation training process, which is more conducive to the achievement of students' training objectives. Thirdly, before and after the simulation training, students are organized to discuss the purpose and plan of the training before the training, and the harvest and experience after the training. This will help students to reflect and process the training, so as to improve the training effect. In addition, we can teach students how to participate in the training and how to solve the problems encountered in the process of training, so as to enhance students' practical experience and enable them to get more opportunities in practice.

4 Conclusion

The key to the simulation training teaching of Business Administration major is to start from the students' individual development needs, improve the practical teaching resources, and promote the students to actively participate in the simulation training activities, and improve the students' business management skills.

References

1. Xinrui, Z.: Discussion on the practical teaching mode of Business Administration Major in multidisciplinary universities. Lab. Res. Exploration **09**, 181–185 (2012)
2. Hong, C.: On the business management professional simulation training teaching mode. Managers **18**, 348 (2014)

3. Yang, Y.: Difficulties and Countermeasures of "simulation company" in practice training of business administration specialty. Frontier **08**, 87–89 (2013)
4. Chen, X.: Research on the transformation of newly established universities. Educ. Dev. Res. (1) (2009)

Genetic Algorithm and Cloud Computing Platform for SaO2

Zhenwu Zhou[1] and Weizheng Sun[2(✉)]

[1] College of Mechanical and Power Engineering, Zhengzhou University, Zhengzhou 445500, Henan, China
[2] College of Information and Electrical Engineering, Hunan University of Science and Technology, Xiangtan 411100, Hunan, China

Abstract. In order to solve the problem that oxygen saturation can be described during continuous monitoring. We use ARIMA model to describe the change of blood oxygen saturation of each person, and use pca-ga-bp neural network model to represent a person by inputting parameters under the new index system. The network topology is constructed by 33 training samples, and the index of the three test samples is the quantitative value of the sample number of the three test samples. The accuracy of the test model is 100%. Compared with BP neural network system, pca-ga-bp neural network model can more accurately represent a person. Then, the correlation between age and SpO2 was analyzed by data visualization, deterministic coefficient method and regression coefficient. First of all, we add the mean value and variance of 36 groups of SpO2 data to describe the average level and stability of individual SpO2, and correlate them with age, smoking history, gender and BMI to form a series model of SpO2. We get: the correlation between age and mean, variance and BMI is very small, but also affected; age and mean is negatively correlated, age and variance is positively correlated, age and BMI is positively correlated. In other words, compared with young people, the average oxygen saturation of the elderly is lower, the stability of individual oxygen saturation is lower, and they are more likely to gain weight.

Keywords: ARIMA model · pca-ga-bp neural network model · Data visualization · Determination coefficient method

1 Problem Analysis

In this paper, we establish an ARIMA model to represent the change of oxygen saturation of each person, and use the parameter value of the model as a part of the index to represent a person. Therefore, the indicators for each person included age, BMI, gender, smoking history and ARIMA model parameter values. In order to solve this problem, we use ARIMA model to describe the change of oxygen saturation of each person, and use pca-ga-bp neural network model to describe a person by inputting parameters under the new index system. However, the parameters in ARIMA model can not describe the change of individual oxygen saturation. Therefore, we added the mean value and variance of oxygen saturation data given by 36 groups of questions to describe the average level of personal oxygen saturation and the stability of personal oxygen saturation, and compared them with age, smoking history, gender and BMI to form a series model of oxygen saturation.The values of some parameters in 36 groups are shown in Table 1.

Table 1. Partial parameter values

Name of record	Parameter 1	Parameter 2
301116b	−0.172	−0.357
301116a	−0.078	−0.699
051216a	−0.198	−0.516
...	...	...
150317b	0.0878	0.996

2 Modeling and Solving

We first established ARIMA model to represent the change of oxygen saturation of each person, and used 36 sets of parameter values in the model as two indicators to represent a person. Then, principal component analysis (PCA) was carried out on all the characteristic indexes. SPSS software was used to analyze the six indexes of age, BMI, gender and smoking status in ARIMA model, and two parameter values. When four principal components were found, the cumulative contribution was less than 85%, and there was no need for principal component analysis [1]. Therefore, we take six indexes as input and incorporate them into GA-BP neural network model, which are randomly divided into 33 training samples and 3 test samples. The network topology is constructed by 33 training samples, and the indexes of three test samples are input, and then the quantitative value of the number of test samples is obtained to test the accuracy of the model. Through MATLAB software, the calculation results are shown in Table 2.

Table 2. 36 sets of parameter values.

Name of record	Original value	Predictive value	The same person
250117b	13	12.86	Right
080217a	20	19.86	Right
010317a	32	31.62	Right

However, the parameters in ARIMA model can not describe the change of individual oxygen saturation. Therefore, we first visualized the correlation between age and mean, variance, smoking history, gender and BMI. Secondly, we use the determinate coefficient to describe the fitting of the five regression curves quantitatively, and use the regression coefficient to describe the positive correlation and negative correlation. Then the correlation between age and mean, variance, smoking history, gender and BMI was analyzed quantitatively. The visualization results between age and each indicator are shown in Fig. 1.

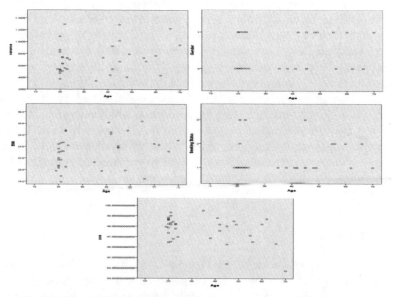

Fig. 1. Visualization of age and mean, variance, BMI, smoking history, sex

By observing Fig. 1, we can see that age has almost nothing to do with gender and smoking history. Therefore, gender and smoking history are excluded. The remaining indicators are average, variance, and BMI. Then use the coefficient of determination to quantitatively describe the fitting degree of the five groups of regression curves, and then quantitatively analyze the correlation between age and the average, variance, smoking history, gender, and BMI [2].

The coefficient of determination, also known as the goodness of fit, aims to construct a standard that can directly judge the quality of the fitting result. The value of the goodness of fit reflects the degree of interpretation of the independent variable X to the dependent variable Y. The greater the value of the coefficient of determination Larger, the higher the degree of interpretation of the independent variable X to the dependent variable Y. In other words, the coefficient of determination is the standard for how well the sample regression line fits the data. If the value of the coefficient of determination is large, it means that the regression curve fits better. If the value of the coefficient of determination is small, it indicate that the regression curve fits poorly.

The coefficient of determination is the square of the correlation coefficient. It is recorded that the fitted value is y, the mean is recorded as y, the fitted value is $\hat{Y}$, and SR is the regression sum of squares.

3   Summary

In this paper, we used data visualization, determination coefficient method and regression coefficient to analyze the correlation between age and blood oxygen saturation sequence pattern. First, we added the average value of oxygen saturation data and the variance of oxygen saturation data given by 36 groups of questions to describe the average level of personal oxygen saturation and the degree of stability of personal oxygen saturation, and correlate it with age, Smoking history, gender, and BMI together constitute the model of oxygen saturation sequence. Secondly, using a visual way to show the correlation between age and average, variance, smoking history, gender, BMI, it was found that age is almost independent of gender and smoking history. Then, the coefficient of determination was used to quantitatively describe the fit degree of the remaining three sets of regression curves, and the positive and negative correlations were described by the regression coefficients. It was found that age has little correlation with the mean, variance, and BMI, but it is also affected; age and The average value is negatively correlated, age is positively correlated with variance, and age is positively correlated with BMI; that is, compared with young people, the average level of oxygen saturation of the elderly is lower, the degree of stability of personal oxygen saturation is lower, and it is easier to grow fat [3, 4].

4   Conclusion

The purpose of this paper is to detect SpO2 and PR. In order to complete the real-time and noninvasive monitoring of physiological parameters, this paper studies the blood oxygen saturation monitor, and designs a simple, reliable and easy to operate monitor. The main contents include:

(1)   This paper introduces the basic knowledge of blood oxygen saturation, the research purpose and significance of blood oxygen saturation monitor, the development process and the research status at home and abroad.
(2)   Firstly, this paper completes the design of the oxygen saturation monitor, and introduces the system scheme, system hardware design, system software design, pulse Porter feature extraction algorithm, blood oxygen value and pulse rate value calculation method in detail; Secondly, two kinds of PC software are designed, one is LabVIEW PC based on Windows platform and the other is Android platform. These two kinds of PC are used to monitor and store PPG data, SpO2 and PR in real time.
(3)   In this paper, a fast filtering method for noise signals of different frequencies in pulse wave signal is completed, including high-frequency burr noise, low-frequency baseline drift noise, mixing slight sudden motion interference noise.

Acknowledgements. Follow-up research project on B topic "Variability of Oxygen Saturation" in the Ninth "Certification Cup" Mathematics China Mathematical Modeling International Competition 2020.

References

1. He, X., Li, J., Hao, Z., Sui, Q.: Brake friction lining wear prediction based on ARIMA model. Auto Digest **12**, 39–42 (2020)
2. Zhang, Y., Liao, Y., Wang, P., Wu, Z.: PCA-RVM model for prediction of triaxial compressive strength of sandstone under THM. Mining Res. Dev. **40**(11), 52–58 (2020)
3. Yuan, Z., et al.: A data preprocessing method for desulfurization system based on PCA and related functions. Beijing: CN111882006A, 03 November 2020
4. Yu, J., Peng, B., Hou, M., Yang, Q., Geng, D.: Research on the rotor position error prediction of permanent magnet torque motor eccentric load based on GA-BP algorithm. Mach. Tool Hydraulics **48**(10), 12–17 (2020)

Landscape Design Virtual Platform Based on iPad Operating Environment

Shihan Hu(✉)

Liao Ning Urban Construction Technical College, Shenyang 110122, China

Abstract. In order to improve the practical application effect of the virtual platform app interface for landscape design, the app interface of the virtual platform for landscape design was designed combined with iPad running environment. The processing structure module of landscape design virtual platform is improved, and the 3D virtual simulation framework of environmental landscape is optimized. Combined with the principle of neural network, the 3D landscape virtual algorithm is optimized. Finally, the effective design of APP interface of landscape design virtual platform is realized. Finally, through the experimental verification, compared with the traditional interface design effect, the design results of APP interface of landscape design virtual platform based on IAD running environment have higher accuracy and effectiveness, and fully meet the research requirements.

Keywords: iPad interface design · Running environment · Landscape design · Virtual platform · Algorithm optimization · Processing structure

1 Introduction

With the increasing demand of environmental landscape design, the virtual platform of landscape design is optimized and improved combined with iPad operating environment, and the environmental landscape design is based on the needs of the masses [1]. Because the traditional environmental landscape design usually adopts the two-dimensional plane drawing mode, it is difficult to accurately simulate the landscape environment and details of the environmental landscape, so as to ensure the rationality of the landscape design. Combined with 3D simulation technology, 3D processing is carried out in the running environment of pad to realize the virtual design of landscape environment, and 3D display of environmental landscape is carried out on virtual platform app.

In order to better simulate the complex landscape environment accurately, this paper describes and processes the landscape details based on the principle of infinite sensor network and neural network node, carries out the three-dimensional rendering of the image for the complex environment and landscape details, and effectively processes and reasonably allocates the requirements of the spatial environmental landscape design with the infinite network nodes, Using 3D virtual vision technology and panoramic information recording principle to collect, analyze and effectively transmit the environmental landscape characteristic values, and store them to the client platform for landscape environment display, so as to realize the reasonable design of the landscape design virtual platform.

2 Virtual Landscape Design Platform

2.1 3D Virtual Simulation Interface of Environmental Landscape

Design environment landscape design virtual platform app interface, need to collect environmental landscape information data, and the collected data, 3D virtual simulation in the platform. In order to ensure the effectiveness of virtual simulation, it is necessary to optimize the configuration of virtual simulation interface, and add information collection, information interaction, virtual vision and 3D simulation processing modules in the virtual simulation interface [2].

Based on the above three-dimensional virtual interface structure of environment landscape, the simulation effect of three-dimensional graphics is set and displayed, and the information is stored and sorted in the interactive platform for processing. Adding a central computer processor to the 3D virtual interface structure of environmental landscape can better define, classify, store and adjust the environmental landscape feature information. According to the re adjusted information features, the association processing of landscape detail parameters is carried out, and the random control observation points are added in the 3D virtual simulation interface, so as to better regulate and improve the urban landscape information, and modify and process the landscape position more comprehensively. According to the processing results, different landscape data interaction modes are optimized, and the landscape design platform structure is improved.

According to the collected environmental landscape information, the location of the landscape to be processed is processed, and the comprehensive virtual auxiliary processing of environmental landscape design is further carried out by using network technology and infinite network nodes, so as to improve the efficiency of environmental landscape virtual design. In the process of three-dimensional simulation of environmental landscape, it is necessary to take into account the regional landscape topography information, vegetation planting information, building appearance and building feature data to carry out the three-dimensional design and rendering of the scene, so as to realize the multi angle and multi-directional construction, compilation and restoration of the landscape information.

2.2 3D Landscape Virtual Algorithm

In the above virtual environment, the 3D landscape virtual algorithm is improved and optimized, and the interface display and design are carried out by adding environmental impact value and terrain and geomorphic feature information, so as to achieve the desired virtual simulation effect. Suppose that the transmission rate of information features in the process of landscape design is r, the common feature of platform information is z, and the transmission and receiving time of information features are t_1 and t_2, respectively. Then the numerical algorithm of building simulation code is.

$$rule : L = \cup z \cdot \lim_{0 \to \infty} \frac{t_1 - t_2}{2r} \tag{1}$$

Based on the above algorithm, the processing level of landscape 3D digital platform is further optimized, including landscape environment data acquisition module, digital control module, 3D landscape image scanning module and data information storage module. Aiming at the three-dimensional digital platform of environmental landscape and TPM three-dimensional simulation image simulation display platform, the information simulation data are effectively managed [3].

Based on the above virtual landscape management interface, the discrete elevation algorithm is introduced, and the 3D data virtual simulation algorithm is optimized to display the 3D image. The 3D image simulation calculation is carried out according to the principle of inverse distance weight interpolation algorithm. If the simulation weight value is x and the discrete value of 3D simulation is n, then the 3D space standard display algorithm is as follows:

$$L(1+x)^n = rule : L\left[1 + \frac{x}{1!} + \frac{2(n-1)x^2}{2!} + \Lambda\frac{n(n-1)x^n}{n!}\right] \qquad (2)$$

The discrete method of data coding is as follows:

$$\varpi = \sum\sum[\frac{1}{2\omega}\lim_{0\to\infty} L(1+x)^n - h]^2 - u \cdot (q+z) \qquad (3)$$

Where: ω is the weight stability information collected from the landscape environment; h is the image conversion data weight in the dimensional virtual simulation platform; u is the pixel resolution of the virtual platform display image; q is the inverse distance weighting coefficient of the 3D image in the platform interface.

According to the calculation results, the 3D virtual environment landscape is displayed and processed in the platform driver. Combined with the principle of neural network, the feature location and embedding processing of the actual landscape space are carried out, and the embedding results are output. Combined with ZigBee program, 3D virtual display driver is processed and corresponding 3D virtual interface is established. Combining java script language and C language to recompile and display the interface information to ensure that the interface is flexible and convenient. Combined with the virtual algorithm of 3D simulation, the virtual landscape is displayed.

2.3   Realization of Virtual Platform Interface Display

In order to ensure the effect of virtual interface design, the interference value of virtual display platform design is extracted, and the 3D virtual simulation interactive landscape information is mapped. The 3D model of landscape environment is constructed by 3ds MAX and other related software. In the process of model construction, the interference factors of surrounding environment and network information in the simulation process are eliminated.

$$x = \varpi\sum\sum s + x_i \qquad (4)$$

According to the above calculation results, the information data is processed, and the landscape dimension is scaled and moved. Based on the data collected from the real environment, the virtual vision transformation is carried out, and the mapping relationship of landscape environment is processed with 3ds MAX to ensure the multi angle and multi-dimensional observation and design of the landscape. Combined with the platform interface of rotation, scaling and other related functions to optimize the details, to achieve the reasonable improvement of the environmental landscape. The virtual processing of 3D information is carried out by AP I software, and the VR display of environment landscape is carried out with 3D Max, MAYA, Auto CAD and other modeling tools. In order to ensure the effectiveness of data acquisition, data processing, data modeling and other related steps, it is necessary to establish spatial data repository and index path in the process of multiple information dataset format conversion, so as to effectively store and transfer information.

In the process of environmental landscape virtual platform interface design, it is necessary to improve the landscape interface parameter modification steps, including: determine the three-dimensional coordinate information of environmental landscape, collect and scale the corresponding environmental landscape light source, transparency, modeling structure, and carry out the simulation and later modification of the solid building structure in the app interface. Planning the virtual platform of landscape processing information, clear landscape angle and scaling ratio, to ensure the best position and angle of virtual platform display, showing a good interface design effect. Finally, vegetation and task setting are added to ensure that the architectural interface display effect is more realistic, so as to improve the interface design and display process of landscape virtual platform.

3   Analysis of Experimental Results

In order to verify the design effect of APP interface of landscape design virtual platform based on iPad running environment, comparative detection was carried out, and the detection environment was set uniformly to ensure the detection effect.

CAD and 3ds MAX software are used to simulate the landscape. The information in the landscape is collected and the information of the virtual project is processed from top to bottom [4]. In the process of experimental demonstration, uz900 main frequency processor, can bus, image memory and windows 2007 are used.

Based on the above detection results, compared with the traditional virtual platform interface design results, the proposed virtual platform app interface design effect based on iPad running environment has relatively high accuracy and effectiveness even in the case of strong interference, This proves that the app interface of the virtual platform for landscape design based on iPad running environment has high practicability and fully meets the research requirements.

4   Conclusion

Combined with virtual simulation technology, the 3D virtual design of environmental landscape is carried out, and the corresponding app interface of virtual platform is optimized to improve the effectiveness and authenticity of environmental landscape design, and the accuracy of virtual interface for environmental landscape simulation and display is improved.

References

1. Hongli, Q.: Influencing factors of mobile reading service satisfaction of book platform app. Libr. J. **36**(4), 25–31 (2017)
2. Design of graphical virtual simulation experimental platform based on Internet of things. Modern Electron. Technol. **40**(1), 32–35 (2017)
3. Yanling, N., Jiening, W.: Experimental study on seismic performance of retaining wall of landscape architecture after optimization design. J. Earthquake Eng. **24**(3), 596–600 (2019)
4. Pengyu, S., Xiaowen, Y.: 3D landscape image sample optimization query simulation in the network. Comput. Simul. **34**(12), 313–317 (2017)

Cloud Based Information System Architecture in Construction Site

Jun Zhao[✉]

Department of Civil Engineering, Shandong Vocational College of Science and Technology,
Weifang 261053, Shandong, China

Abstract. Based on the research background of artificial intelligence technology, the application of artificial intelligence technology in intelligent building is studied. Firstly, the concept of artificial intelligence technology and intelligent building is analyzed. Secondly, on the basis of expounding the problems existing in the development of intelligent buildings, the paper explores the practical application of artificial intelligence technology in intelligent buildings, so as to provide reference for similar projects.

Keywords: Artificial intelligence technology · Intelligent building · Intelligent control

1 Overview of Artificial Intelligence Technology and Intelligent Building

1.1 Overview of Artificial Intelligence Technology

Artificial intelligence technology is also called a technology, which can realize intelligent simulation, extension and expansion method and technical system after it is put into use. Artificial intelligence needs to fully understand the essence of intelligent technology, so as to develop and apply advanced intelligent equipment, and can show a similar reaction with human beings [1]. Artificial intelligence technology includes robots, image recognition, language recognition and other parts. In the process of application, human thinking and thinking mode can be simulated according to the system settings. High end intelligent equipment can simulate human thinking, and even some aspects are very outstanding. Artificial intelligence technology has been developed and matured gradually, and its application scope is gradually expanding. In the future development.

1.2 Overview of Intelligent Building

Intelligent building is the application of intelligent information control technology in the building, so as to achieve the intelligent management standard of the building. According to the various functional standards required in the operation of the building, the use function can be optimized to provide a more comfortable, safe and healthy living environment for human beings. In the normal operation of intelligent building, the most

J. Abawajy et al. (Eds.): ATCI 2021, AISC 1398, pp. 483–487, 2021.
https://doi.org/10.1007/978-3-030-79200-8_73

critical is the computer system, which can detect the operation of the building at any time by using sensors and corresponding systems, and can record various data information in detail, make corresponding response through the detection results in the computer, and then complete the overall intelligent control. Many of the smart buildings developed by Toshiba and China's headquarters, such as Lloyds building in London, have been applied in many areas [2]. The intelligent stadium and gymnasium built in 2008 Beijing Olympic Games is a very important intelligent building construction project in the society at this stage.

2   Problems in the Development of Intelligent Buildings

From the analysis of the current actual situation, it is found that there is still a single function in the long-term development of intelligent buildings. Different professional construction units will set up a single intelligent function according to their own needs, without realizing the collaborative processing of multiple intelligent systems.

Although the current level of science and technology there are still some defects. In the operation of intelligent buildings, there are subsystems out of operation. Therefore, the system is prone to chaos in the work, which makes the specific operation state unable to achieve effective control, and the optimization efficiency and quality of intelligent buildings are relatively low, It has seriously hindered the long-term development of intelligent buildings. In addition, there are many defects and problems in the mutual control of each subsystem in the intelligent system, which has a negative impact on the operation and control of the whole system. Therefore, in the operation of intelligent building, technical personnel should deeply analyze the situation of each system, which leads to the workload of technical personnel is relatively large, but also makes the supervision and management difficult to carry out smoothly. In order to make the intelligent building run stably, technicians must realize the coordination and optimization of the whole system [3].

3   Application of Artificial Intelligence Technology in Intelligent Building

3.1   Expert Control System Technology

Expert control system technology is based on the current system operation needs to establish an expert database, while using the data information in the database to control the whole system. In the research and development of artificial intelligence technology related technology, the expert database can realize higher value technology, and at the same time use this technology to realize the development and application of comprehensive control system, which can ensure that expert knowledge plays its due role, It can also make the intelligent system have the thinking mode of experts, or use the knowledge system of experts to analyze the corresponding problems, effectively deal with relevant problems, and form a more perfect computer intelligent software system. The system integrates the special professional knowledge of experts, forming a more perfect expert database, and can simulate the operation of experts in the system control link, In order to better deal with the problem of complexity. The experience and knowledge of experts are introduced into the intelligent system, which has a very good effect on intelligent building control, and can deal with complex structure problems. The simplified model of the system is shown in Fig. 1.

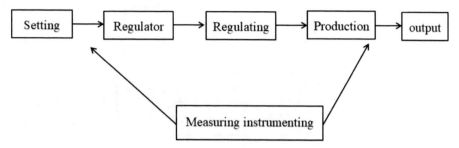

Fig. 1. Simple model of expert control system

Through the reasonable application of knowledge expression technology, the knowledge and experience of experts can be formed into a knowledge base, and the internal data can be deeply reasoned and analyzed, thus effective countermeasures can be formulated, and the control level of the whole intelligent system can be effectively improved, and the operation efficiency of the system will be higher, Such as building automation (BAS), strong current automation, safety automation (SA), fire automation (FA) and other independent information control system. Under this condition, there are obvious differences between expert system and traditional system in operation, not only through mathematical model, but also by comprehensive processing of knowledge and mathematical model, knowledge and control technology, so as to ensure that the system can realize automatic control and provide basic conditions for the operation of automatic control system in the building, It can also provide effective intelligent control and management for internal service system and user management.

3.2 Artificial Neural Network Control System

Artificial neural network control system is the core technology in the current intelligent technology, which has a huge effect, mainly reflected in the intelligent control system can learn independently, optimize the structure, and realize the recognition and processing of image and voice. Especially in the complex system control, it can better calculate and analyze, and reduce the energy consumption of the system, To ensure that a large number of data in the whole building can be processed according to the logic rules, to achieve multi-level relationship control needs, but also to achieve self-learning and adapt to a variety of environments, the system control ability it shows has a good effect. In the intelligent building project, the main function of the equipment control device is to improve the control level of the system. It can also independently understand and master the uniqueness of the building through the neural system. It can automatically adjust the follow-up control state according to the analysis results of various parameters of the system, so as to meet the requirements of different system operation, It can provide good support for the signal detection, control and maintenance of intelligent equipment in the building, and can adapt to the operation requirements of different environments. Combined with the current control requirements of equipment system in intelligent building, the composition structure of automatic fire alarm system designed in neural system is shown in Fig. 2.

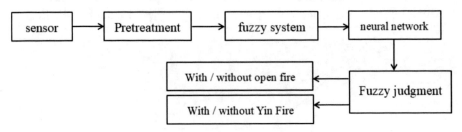

Fig. 2. Automatic fire alarm system

In addition to the above autonomous learning ability, the intelligent building system should also effectively carry out necessary simulation and simulation control according to the control requirements of the internal system. The new neural system can fully realize the dynamic learning of the internal system, which makes the operation of the model not only more complex, but also operation efficiency of the system, The hardware system of the current artificial neural network system is mainly composed of micro controller, that is, chip structure, which can completely [4].

The composition structure mode can better realize various forms of intelligent building management and control, and can learn independently according to the actual operation requirements, improve and improve its own structural performance, so that the accuracy and speed of system operation can be improved. The reasonable application of micro controller can also be considered. The system structure is optimized, and all functions can be realized and the operation is stable Sex has improved.

4   Conclusion

Based on the gradual improvement of scientific and technological level, artificial intelligence technology has been developed rapidly, and has been widely used in intelligent buildings. It can realize the intelligent control of various functions in the building. At the same time, it can also have a certain ability of learning and improving. According to the actual operation requirements, it can continuously improve people's quality of life and give full play to the advantages of intelligent control system, Actively promoting the comprehensive development of China's construction engineering field can also promote the sustainable development of human society.

References

1. Qian, X., Wei, J.: Application of artificial intelligence in modern network technology. Comput. Sci. Appl. **2019**(4), 787–7892 (2019)

2. Tian, K., Fan, C., Li, Y., et al.: Application of artificial intelligence technology in electrical automation professional certification. Artif. Intell. Robot Res. **7**(2), 63–67 (2018)
3. Zhou, Y., Guo, S., Zhang, F., et al.: Application of artificial intelligence in Hydrological Forecasting Research on water resources **8**(1), 2 (2019)
4. Yang, F.: Application Research of artificial intelligence technology in intelligent building. decoration Tiandi **2017**(9), 59 (2017)

BIM Technology Based on Cloud Computing in Urban Design

Qianyi Zhu and Rui Wang[✉]

Chongqing Jianzhu College, Chingqing 400072, China

Abstract. Information technology under the demand of early morning, through BIM , can effectively improve the work efficiency and focus on the development of big data, analyze and explore the application results of information technology in urban planning.

Keywords: Big data · Information technology · urban planning · Design

1 Application of BIM Technology

1.1 BIM Concept

BIM is a three-dimensional digital model software which combines geometry, component and construction schedule data together. Participants can get the results of calculation efficiently and share information. BIM is not only an application software, but also an excellent technology to integrate computing information and data with software, and then assist all participants to complete the task [1]. Compared with the traditional way, BIM model has great advantages.

1.2 BIM Technology Features

BIM Technology needs information from many fields and people to complete the project. Because of the huge resources and different formats, the integrated management of information is very difficult. In the traditional mode of work environment, the same kind of information will be lack of management and repeated modification and reset. The emergence of BIM Technology can ensure the same format of the model infrastructure, which can be reused many times without affecting the data. It has high utilization rate and is easy to modify. In China's various industries, the construction industry using BIM also includes a variety of construction types. It can make different types of work cooperate with each other to complete the task with high efficiency and quality. BIM can be used to integrate project data and realize unified collection of information, which facilitates the centralized management and storage of data information for project participants. This reflects the centralized characteristics of BM technology.

J. Abawajy et al. (Eds.): ATCI 2021, AISC 1398, pp. 488–492, 2021.
https://doi.org/10.1007/978-3-030-79200-8_74

1.3   Application Status in China

At present, BIM Technology is often used in urban construction. Due to the advantages of technical characteristics, the design and management of urban roads are inseparable from BIM Technology. The development of BIM Technology can be linked to many fields, and it can be used as an auxiliary tool to promote the development of BIM Technology. However, at present, the use of BIM Technology in China is still limited to urban construction, engineering design and mechanical manufacturing, etc. [2]. BIM Technology should be promoted in more fields to develop harmoniously in the environment of social market economy.

2   Application of BM in Municipal Road Design

2.1   Application Design of Topographic Map

In the application design of municipal road, the accuracy of topographic map is required to be very high. The BIM Technology can meet the requirements of topographic map data more quickly, and do a good job in the preliminary design work. BIM Technology has the characteristics of high efficiency, convenience and strong comprehensiveness. Its application in urban road design can speed up the construction speed and construction quality. In the process of using BM technology, project team members need to know the local terrain and environment in advance, and conduct accurate and detailed investigation. There are two kinds of topographic maps: one is 2D topographic map, the other is 3D topographic map. In practical application, there are many problems in two-dimensional topographic map, which bring difficulties to design and construction. The two-dimensional topographic map can not simplify the design of urban road, and it is easy to be affected by its own factors in the construction, which can not be accurately expressed in the two-dimensional topographic map. But 3D topographic maps are different. Three dimensional topographic map can directly and accurately reflect the comprehensive data required for construction. Drawing three-dimensional topographic map is a kind of application technology of BIM Technology.

2.2   How to Design Circuit

According to the on-site field survey and word search, obtain all terrain and feature information affecting road route selection, such as mountains, buildings, rivers, cables, cultural relics, etc., and carry out indoor alignment according to relevant municipal road design specifications or regulations. Put the line position into the three-dimensional topographic map system for inspection, check the rationality of the road plane line position and its conflict with important terrain and features, and evaluate the impact on it; use the driving angle to judge the driving comfort of the line position, and identify the safety problems such as sight distance of the intersection. It can also visually display the road filling and excavation, the effect of intersection with the Jiaojiao Road, and the connection with surrounding land, etc., and can report to the construction unit for decision-making.

2.3 Algorithm of Data Mining Technology from the Perspective of Big Data

Hierarchical clustering algorithm this algorithm is also an important tool in data mining, the main principle is to decompose the mining data layer by layer until the best data is obtained [3]. In order to make full use of the two classification methods, the first method is to divide the two categories into two categories according to the principle of clustering.

Compared with other bottom-up hierarchical clustering methods used in various fields, the purpose of the proposed clustering algorithm is to provide more accurate and effective clustering with the same computational complexity $O(N^2)$. The algorithm utilizes the advantages of neural network and intersection to implement hierarchical clustering. Intersection is the basic concept used in this algorithm.

$$pairs = \{(IoP_1, IoNN_1), (IoP_2, IoNN_2) \ldots (IoP_n, IoNN_n)\} \tag{1}$$

$$NN = \min \sqrt{\sum_{i=1}^{n} (x_i - y_i)^2} \tag{2}$$

3 Implementation Process of Urban Road Design

The promotion of BIM application technology in China is very good. The application research of BM theory in urban road design is still in the development stage. The government should encourage construction participants to choose BIM Technology, understand the advantages of θ IM technology in construction, and bring benefits to urban road construction. According to the situation in China and the advantages of Japan M technology, the implementation process is designed. In the traditional mode of urban road design, written or paper-based information transmission is often used, which is not only troublesome, but also easy. First, teaching materials and documents waste time and increase the use of manpower for repeated carding and calculation of drawings. 7 based on the three-dimensional road model, designers can directly convey the geometric information data in construction to structural engineers, It is convenient to design related traffic structures in the software, which can achieve the goal of multi-disciplinary joint design. Data is associated with groups in the model, making it easy to update and quickly change the design. At the same time, BIM Technology can let people who do not do data can also clearly see the design results and save the data project, which is very convenient in all aspects.

4 Application of Information Technology in Urban Planning and Design in the Era of Big Data

4.1 Application of Information Technology in Urban Planning Data Collection

Under the background of big data, the application of information technology to urban planning and design can ensure the smooth development of urban planning data collection. At present, in the process of urban planning and design data collection, there are two widely used technologies, one is global positioning technology, the other is remote sensing technology rs. In the process of using global positioning technology GPS, data collection can be carried out in a global scope. The collected data are mainly three-dimensional data, which can realize all-weather and all-round data collection under the effect of global satellites, and can provide accurate and efficient data to users. In the process of urban planning and design, through the application of this technology, time data and spatial data can be obtained to guide the improvement of urban planning and design level. By collecting real geographic information, we can provide professional support for urban planning and design. As a modern equipment, RS sensor is applied to the planning and design of alkali City, which can realize long-distance reflection of electromagnetic wave, collect relevant data and obtain scene information. Combined with the information, the actual situation of the scene is monitored and identified.

4.2 Application of Information Technology in Urban Planning Data Analysis

In the big data environment, the generation of multiple data leads to the interaction between the city and the masses. In the process of urban planning and design, it is necessary to do a good job of urban analysis, strengthen the processing of individual data of the masses, master the overall situation of the city, and ensure the rationality and standardization of urban planning and design. For example, in the process of urban traffic planning and design, the relevant traffic data such as bus swiping card and floating car need to be processed scientifically. According to the urban population distribution and the actual traffic conditions, the scientific planning of urban traffic can be realized [4]. In practice, it is necessary to integrate and research planning data with the help of geographic information technology. Geographic information technology can promote the integration of data, realize the data acquisition and processing, extend the information, and ensure the real-time and accuracy of data information. Through the application of geographic information technology, we can prevent repeated analysis of the status of the data, and provide data support for the development of urban planning and design.

5 Conclusion

In a word, under the big data environment, the planning and design direction of alkali city has changed, and strict standards have been put forward for the application of information technology. In practice, it is necessary to do a good job in the collection, preservation and management of urban planning data with the help of the application value and effectiveness of information technology, and apply global positioning technology and geographic information technology to improve the efficiency and level of urban planning and design, and escort the development of urban construction.

Acknowledgements. Chongqing Municipal Higher Education Association's Higher Education Scientific Research Project Key Project (CQGJ19A37).

Chongqing Municipal Education Commission Science and Technology Research Project (KJ1604102).

Chongqing Higher Education Teaching Reform Research Project (193485), Municipal Education Commission Science and Technology Research Project (KJQN202004303).

References

1. Yunmei, B., Xiping, Y., Shu, G., Zhiqi, J.: Application of GIS technology in urban planning. China W. Transp. (Second Half Mon.) **18**(11), 220–222 (2018)
2. Zijing, L.: Analysis of urban planning and design from the perspective of informatization. Hosshe **2018**(27), 92 (2018)
3. Lixing, Z., Xiaolan, T., Yujie, R.: The role of information processing technology in urban park green space planning and design under the digital background . J. Suzhou Univ. Sci. Technol. (Eng. Technol. Edit.) **31**(03), 41–49 (2018)
4. Shi, K.: On the application of green urban design principle in planning and design practice. Eng. Constr. Des. **2018**(15), 53–54+57 (2018)

Cyber Intelligence for AI, VR, Blockchain Applications and Innovations

Application of Robot-Assisted Percutaneous Spinal Endoscopy in the Treatment of Lumbar Fractures

Weiguo Chen[1], Xiangfu Wang[1](✉), Wanqian Zhang[2], Heng Zhao[2], Xiangyu You[2], Huanying Yang[2], Gang Zheng[2], Tingrui Sang[2], and Chao Zhang[2]

[1] Gansu Hospital of Traditional Chinese Medicine, Lanzhou 730050, Gansu, China
[2] Gansu University of Traditional Chinese Medicine, Lanzhou 730000, Gansu, China

Abstract. Objective To investigate the feasibility and safety of robotic orthopedic assisted percutaneous spine endoscopy in the treatment of lumbar fractures. Methods Orthopedic robot assisted with percutaneous spinal endoscopy to treat a patient with a lumbar fracture, Intraoperative CT scan reconstruction of the lumbar vertebra image, robot-guided pedicle screw placement, pedicle connecting rod, and reduction of the fractured vertebral body. Under the guidance of the robot navigation system, percutaneous spinal endoscopy was used to remove the loose bone mass and drive the fracture block into the vertebra. The visual analogue scale (VAS) and Oswestry disability index (ODI) of patients were compared before and after 3 days to evaluate the improvement of patients' symptoms. Results The patient successfully completed minimally invasive surgery, the operation time was about 180 min, and the intraoperative blood loss was about 80 ml. A review of CT after 3 days showed that the fractured vertebral body was ideally reset. The VAS score of lumbago was 8 points one day before surgery, 3 points three days after surgery, 48% ODI score one day before surgery, and 22% ODI score three days after urgery. Conclusions Robot-assisted percutaneous spine endoscopy is safe, minimally invasive and effective in treating lumbar fractures, and can be used as a minimally invasive urgical method for lumbar fractures.

Keywords: Orthopedic robot · Percutaneous endoscopic spine surgery · Percutaneous pedicle screw technique · Lumbar burst fracture

1   Case Report

Mr Zhang, 45 years old, pain from lumbar caused by a fall, went to our hospital after a day's limited mobility. Zhang's symptoms was: pain from lumbar, limited mobility, lower mobility in left leg, clear consciousness, in good spirit, normal appetite, unresolved bowel movements, normal urination, poor sleep. Physical examination applied: passive position, pressing and percussion pain in L3-L5 vertebrae (failed to examine spinal mobility because of his pain), normal tension in four extrimities, Trace muscle strength in left side iliopsoas and quadriceps femoris muscle, hypesthesia in front left thigh and buttock skins, other parts saw no abnormal motility. Imaging study indicated

J. Abawajy et al. (Eds.): ATCI 2021, AISC 1398, pp. 495–503, 2021.
https://doi.org/10.1007/978-3-030-79200-8_75

burst fracture in L2 vertebral body, the fracture block intruded the spinal canal and occupied about 2/3 space (Fig. 1, 2, 3). Presenting for percutaneous pedicle screw fracture reduction surgery under robot-assisted percutaneous spinal endoscopy.

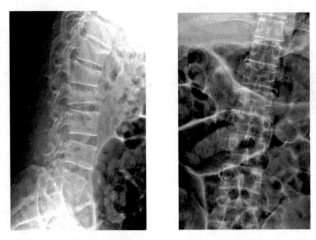

Fig. 1. X ray film showed L2 vertebral fracture

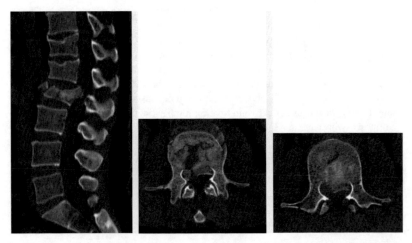

Fig. 2. CT showed a burst fracture of L2 vertebral body, with the fracture block protruding into the spinal canal

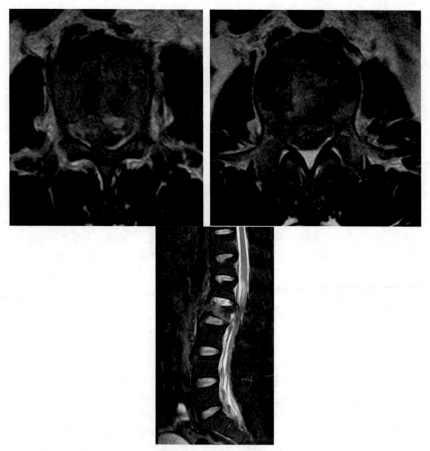

Fig. 3. Preoperative MRI showed burst fracture of L2 vertebral body and compression of dural sac and nerve root

2   Surgical Approach

Patient took prone position, lumbar excessive extend posture and was placed postural cushion after anesthesia to help fracture reduction. Scanning and positioning the injured vertebra (L2) with the help of C-arm X-ray fluoroscopy machine, then documenting it on the skin. Operating theatre sterilized routinely, laid aseptic towel. Incised the skin for 1.5 cm on T12 spinous process, make an incision to supraspinal ligament. Dissected the subperiosteal and embed the tracker. Unfolding the robotic arm and assemble the ruler, adjust the arm to let the ruler stick to the skin on the L1-L3 supraspinal ligament. Let the C-arm X-ray scan and reconstruct the lumbar image, and then transmitted the data to the robot to work out the needle insert route of L1-L3 pedicle. Install the director then robot simulate the route, dissect the skin for 1.5 cm pierce the L1, L2, L3 vertebral arch pedicle in turn and put in guide wire. 3D scan again, make sure the puncture needle is in appropriate position. Put a separate pedicle screw of 6.0*45 mm, 6.0*35 cm and 6.5*45 mm to L1, L2 and L3 vertebral arch pedicle via the guide wire. Intraoperative fluoroscopy saw vertebral arch pedicle screws were in good position. Instal pre-bending vertebral arch pedicle connection rod(with length of 90mm) that through the cut and the subcutaeous, then couple with locking nut. Performed another fluoroscopy shown the fixation was in good condition that the fracture was ideally recovered. Broking the tail of pedicle screw. Rinsing the cut. Then sew up layer-by-layer. Positioning the lamina gap between L1 and L2 vertebra under the robot's navigation and documenting it on the skin. Incised the skin for 0.7 cm, screwed the working wire and inserted spinal endoscopy. Sorted out the soft tissue and stopped bleed under the endoscopy to reveal the superior margin of L2 vertebra lamina and the posterior margin of zygapophysial joints. Use mirror power grinding drill to strip upper third L2 vertebra lamina, outward to medial wall of the pedicle and bite off ligamentum flavum, then insert endoscopy through spinal canal from the lateral spinal dura mater, fracture patterns that intrude the spinal canal can be seen under the endoscopy. Remove the loose bone mass after exploration. Screw the working wire while protecting spinal dura mater, take out the endoscopy, set fracture diver through the wire and drive the fracture block to the vertebra. Posterior border of the vertebra can be seen smooth under the endoscopy. Extend the cut to 1.5 cm, perform L-shape fracture impactor, meanwhile inserting the working wire of the endoscopy. Under the endoscopy insert fracture diver through spinal canal from the lateral spinal dura mater, drive the contralateral fracture block into the vertebral body. Under the endoscopy saw no active bleeding then take out the endoscopy. In cut leave a drain tube. Sew up layer-by-layer, applied sterile gauze. The whole surgery was a success and well anaesthetised. There was bit intraoperative blood loss. Patient back to ward safely after the surgery (Fig. 4, 5, 6).

3   Postoperative Care

Patient took recumbent position and was given antibiotic (to prevent infection), non-steroidal anti-inflammatorydrug (BSAID) and mecobalamin (to trophic nerve) after operation. Performed dorsal foot extension and straight leg rising during patient's rest. Meanwhile patient himself did quadriceps femoris relaxation excercise. Remove drain tube a day after the surgery. 3 days after the surgery patient can do out-of-bed activity with the help of stick.

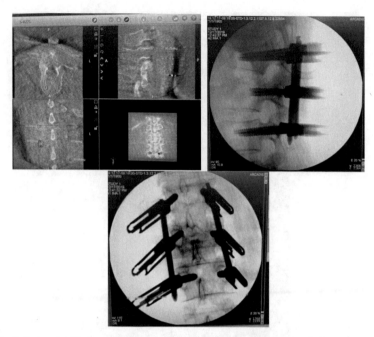

Fig. 4. Robot assisted placement of percutaneous pedicle screw and pedicle connecting rod.

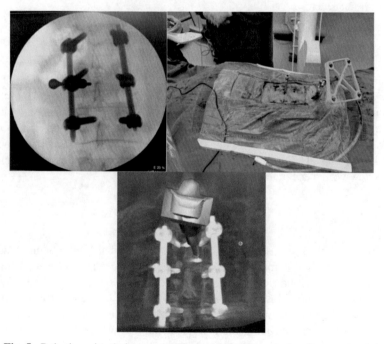

Fig. 5. Robotic assisted placement of the percutaneous spinal endoscopy system

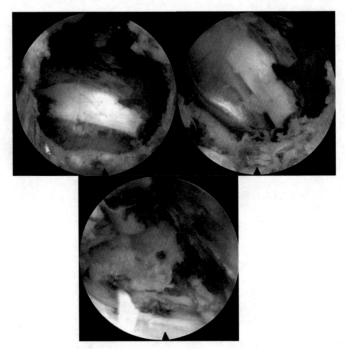

Fig. 6. The fracture was reduced under percutaneous endoscopy, and the nerve root pulsation was good after the reduction

4   Results

Evaluating the clinical efficacy after 3 days of operation, Zhang's waist has no obvious tenderness, fair muscle strength in left side iliopsoas and quadriceps femoris muscle, left thigh skin feeling are normal. No adverse reaction and surgery complication during the peroperative period. VAS was 3 and ODI was 22%. And patient was satisfied about the results of treatment. A Review of X-ray and CT after 3 days showed that the fractured vertebral body was ideally reset, the hight of vertebral body was ideally recovery, the pedicle screw was in accurately position (Fig. 7–8).

5   Discussion

As a common clinical diseases, the incidence rate of lumbar burst fractures increase by year. Traditionally treatment always performed pedicle screw open reduction and internal fixation, which has telling clinical efficacy. But it has large operation wound, great intraoperative blood loss, severe damage to the muscle close to the vertebra and bone structure, and long postoperative course [1, 2]. With the development of minimally invasive concept, percutaneous pedicle screw technique was generally performed in clinical treatment, which has advantages of minimally invasive, low level bleeding and short recovery processing [3, 4]. But X-ray fluoroscopy machine is needed to perform

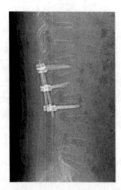
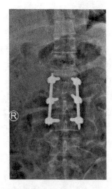

Fig. 7. Postoperative X-ray showed vertebral height recovery

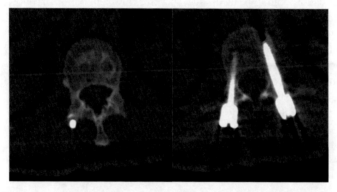

Fig. 8. Postoperative CT showed ideal decompression in the spinal canal

percutaneously. Poor percutaneous perform may result in deviation of the pedicle screw, even cause damage to nerve, spinal cord and blood vessel. However, much percutaneous perform generate large radiation to both patient and caregivers [5]. The failure rate of pedicle screw placement range from 4.9% to 37.5% according to literature description [6]. Meanwhile, patient that fracture block intruding spinal canal and whose root signs was obvious, it's impossible to reset the fracture block via percutaneous pedicle screw technique, to decompress the spinal cord and nerve root, the method has low clinical efficacy [7].

Percutaneous endoscopic spine surgery developed rapidly in recent years, its huge clinical advantage is reflected in treat prolapse of lumbar intervertebral disc, lateral recess stenosis, herniation of cervical disc and herniation of thoractic disc [8–10]. Meanwhile, scholars east and west perform the percutaneous endoscopic spine technique in more orthopedic surgery, including the treatment of cervical spondylotic myelopathy, lumbar spondyloisthesis and lumbar vertebral body posterior edge epiphyseal amputation, achieved a good clinical effect [11]. The percutaneous endoscopic spine technique was even performed in clinical treatment of spinal infectious diseases [12]. We performed the percutaneous endoscopic spine technique in treating spinal fracture, which reflected it has advantages of minimally invasive, low level bleeding, small degree of surgical injury and short recovery processing, the clinical efficacy was great. But its long operation period, comparatively more fluoroscopy steps, higher requirements for surgeon and greater risk, the accuracy of the surgery needs to be further improved [13].

With the development of science and technology in recent years, robot in orthopedics develop rapidly. The third version "Tianji" shown the characteristics of accurate, minimally invasive and rapid when assisting placing pedicle screw [14]. It using 3D imaging technology to plan the operating route and the clinical puncture route was ideal. Scholar as Kantelhardt [15] find out the accuracy of pedicle screw placement of "Tianji" was 94.5%. The performance of "Tianji" in orthopedic obviously increased the accuracy and safety of operation, reduced the frequency of fluoroscopy, shorten the operation period and lowered the threshold of surgeon [16]. Have checked related literature, the clinical report of treating lumbar burst fractures with robot-assisted percutaneous endoscopic spine technique was not yet found. This may result from high threshold and difficulty of percutaneous endoscopic spine technique, as well as low penetration of robot technology.

With orthopedics robot, the author treating lumbar burst fractures combined the percutaneous endoscopic spine technique with the percutaneous pedicle screw technique, can achieve the goal of anatomical reduction of vertebral fractures. It has advantages of high accuracy and safety, minimally invasive, less frequency of fluoroscopy, low level bleeding and fast recovery. But due to few cases observed in this group and short observation time, the statistical significance is insufficient, large sample of clinical contrast was needed to study the efficacy.

Acknowledgments. Lanzhou Talent Innovation and Entrepreneurship Project (2020-RC-54) «Relevant Research on Guiqi Tongbi Decoction in Treating Lumbar Intervertebral Disc Herniation with Residual Symptoms after Percutaneous Spine Endoscopy».

References

1. Wang, H., Zhou, Y., Li, C., et al.: Comparison of open versus percutaneous pedicle screw fixation using the sextant system in the treatment of traumatic thoracolumbar fractures. Clin. Spine Surg. **30**(3), E239–E246 (2017)
2. Vanek, P., Bradac, O., Konopkova, R., et al.: Treatment of thoracolumbar trauma by short-segment percutaneous transpedicular screw instrumentation: prospective comparative study with a minimum 2-year follow-up. J. Neurosurg. Spine **20**(2), 150–156 (2014)
3. Chung, W.H., Eu, W.C., Chiu, C.K., et al.: Minimally invasive reduction of thoracolumbar burst fracture using monoaxial percutaneous pedicle screws: surgical technique and report of radiological outcome. J. Orthop. Surg. (Hong Kong) **28**, 2309499019888877 (2020). https://doi.org/10.1177/2309499019888877
4. Sun, X.Y., Zhang, X.N., Hai, Y.: Percutaneous versus traditional and paraspinal posterior open approaches for treatment of thoracolumbar fractures without neurologic deficit: a meta analysis. Eur. Spine J. **26**(5), 1418–1431 (2017)
5. Tkatschenko, D., Kendlbacher, P., Czabanka, M., Bohner, G., Vajkoczy, P., Hecht, N.: Navigated percutaneous versus open pedicle screw implantation using intraoperative CT and robotic cone-beam CT imaging. Eur. Spine J. **29**(4), 803–812 (2019). https://doi.org/10.1007/s00586-019-06242-4

6. Faizan, A., Kiapour, A., Kiapour, A.M., et al.: Biomechanical analysis of various footprints of transforaminal lumbar interbody fusion devices. J. Spinal Disord. Tech. **27**(4), E118-127 (2014). https://doi.org/10.1097/BSD.0b013e3182a11478

7. Pannu, C.D., Farooque, K., Sharma, V.: Minimally invasive spine surgeries for treatment of thoracolumbar fractures of spine: a systematic review. J. Clin. Orthopae-dics Trauma **10**(Suppl 1), S147–S155 (2019). https://doi.org/10.1016/j.jcot.2019.04.012

8. Hoogland, T., Kvanden, B.D., Schubert, M., et al.: Endoscopic transforaminal discectomy for recurrent lumbar disc herniation: a prospective, cohort evaluation of 262 consecutive cases. Spine **33**(9), 973–978 (2008)

9. Zhang, B., Kong, Q., Yang, J., et al.: Short-term effectiveness of percutaneous endoscopic transforaminal bilateral decompression for severe central lumbar spinal stenosis. Zhongguo Xiu Fu Chong Jian Wai Ke Za Zhi **2019**(33), 1399–1405 (2019). https://doi.org/10.7507/1002-1892.201904131

10. Li, C., Tang, X., Chen, S., et al.: Clinical application of large channel endoscopic decompression in posterior cervical spine disorders. BMC Musculoskelet. Disord. **20**(1), 548 (2019). https://doi.org/10.1186/s12891-019-2920-6

11. Yang, J.S., Chu, L., Chen, C.M., et al.: Foraminoplasty at the tip or base of the superior articular process for lateral recess stenosis in percutaneous endoscopic lumbar discectomy: a multicenter, retrospective, controlled study with 2-year follow-up. Biomed. Res. Int. **2018**, 7692794 (2018) https://doi.org/10.1155/2018/7692794

12. Yang, J.S., Chu, L., Deng, R., et al.: Treatment of single-level thoracic tuberculosis by percutaneous endoscopic débridement and allograft via the transforaminal approach combined with percutaneous pedicle screw fixation: a multicenter study with a median follow-up of 36 months. World Neurosurg. **122**, e1472–e1481 (2019). https://doi.org/10.1016/j.wneu.2018.11.087

13. Yang, J.S., Wang, X.F., Zhao, K., et al.: Posterior unlocking of facet joints under endoscopy followed by anterior decompression, reduction, and fixation of old subaxial cervical facet dislocations: a technical note. World Neurosurg. **130**, 179–186 (2019). https://doi.org/10.1016/j.wneu.2019.06.239

14. Keric, N., Doenitz, C., Haj, A., et al.: Evaluation of robot-guided minimally invasive implantation of 2067 pedicle screws. Neurosurg. Focus **42**(5), E11 (2017)

15. Kantelhardt, S.R., Martinez, R., Baerwinkel, S., et al.: Perioperative course and accuracy of screw positioning in conventional, open robotic-guided and percutaneous robotic-guided, pedicle screw placement. Eur. Spine J. **20**(6), 860–868 (2011)

16. Roser, F., Tatagiba, M., Maier, G.: Spinal robotics: Current applications and future perspectives. Neurosurg. **72**(Suppl1), 12–18 (2013)

Application of Robot-Assisted Percutaneous Endoscopic Spine Surgery for Thoracolumbar Tractures

Wanqian Zhang[1], Xiangfu Wang[2]([✉]), Huihui Zheng[1], Weiguo Chen[2], Xiangyu You[1], Huanying Yang[1], Gang Zheng[1], Tingrui Sang[1], and Chao Zhang[1]

[1] Gansu University of Traditional Chinese Medicine, Lanzhou 730000, Gansu, China
[2] Gansu Provincial Hospital of Traditional Chinese Medicine, Lanzhou 730050, Gansu, China

Abstract. Thoracolumbar fracture is one of the most common types of traumatic spinal fracture. At present, its surgical treatment is still a severe challenge and a thorny problem faced by clinicians. In recent years, percutaneous endoscopic technology has made great progress and development. The combination of modern minimally invasive concept and endoscopic technology has been gradually applied to the treatment of thoracolumbar fractures, which has the advantages of small trauma, fast recovery, fewer complications and other traditional surgery. With the development of science and technology, the third generation of "Tianji" robot technology has made rapid development in the field of orthopedics. There are few reports on robot assisted endoscopic treatment of thoracolumbar fractures, which may be related to the long learning curve and technical difficulty. This article reviews the application of robot assisted percutaneous endoscopy in thoracolumbar fractures.

Keywords: Robot assisted · Percutaneous spinal endoscopy · Thoracolumbar fractures · Minimally invasive surgery

Thoracolumbar fracture is one of the most common types of traumatic spinal fracture. Its junction (T_{10}-L_2) is a special transitional area. Anatomical characteristics make this section vulnerable to high energy trauma. The surgical treatment is a serious challenge at present. The classification of fracture, the general condition and complications of the patients should be considered before operation. Traditional open surgery is limited to relatively complex indications, and has some disadvantages, such as postoperative pain, blood loss and hospitalization. At present, the technology of percutaneous endoscopy assisted by robot is becoming more and more popular in the treatment of thoracolumbar fracture, and it has been proved to be safe and effective.

1 Classification of Thoracolumbar Spine Fractures

Thoracolumbar fractures were first described by Holdsworth and further classified by different students. According to the mechanism of injury, some scholars [1] classified thoracolumbar injuries in different ways: buckling damage and stretch injury [2]. In

J. Abawajy et al. (Eds.): ATCI 2021, AISC 1398, pp. 504–510, 2021.
https://doi.org/10.1007/978-3-030-79200-8_76

1983, Denis [3] put forward the concept of "three pillars" for the first time. According to Denis' three column concept, thoracolumbar fractures include anterior and middle column injuries, and bone fragments contract into the spinal canal, which is the imaging manifestation. The typical clinical symptoms include acute low back pain, motor limitation and nerve injury, such as motor and sensory changes and progressive kyphosis. Here, the stable and unstable thoracolumbar injuries are distinguished, and four basic injury mechanisms are introduced, which are characterized by specific morphology: compression, flexion, extension and flexion rotation injuries. In the 1990s, AO Group [4] considered that Denis's classification was not suitable for the classification of injury types due to its limitations. Therefore, a more logical concept needs to be developed. AO classification is based on Whitesides to propose three basic functions of stabilizing the spine: axial compression force, axial traction tension and torsion force, which rotate around the longitudinal axis respectively. These mechanisms include compression injury, stretch injury and torsion injury, which represent the shortening, lengthening and rotation injury of spine. Their analysis clearly showed that L1 fracture was the dominant injury at the thoracolumbar junction, followed by T12, L2 and L3, and other vertebral body injuries had the same probability.

2 Surgical Treatment

2.1 Choice of Treatment Methods

The treatment principle of thoracolumbar spine fracture is spinal stability repair, kyphosis correction, spinal canal nerve decompression. So far, the ideal treatment of thoracolumbar fractures is still a problem. Although conservative treatment has achieved satisfactory clinical results, most patients can get effective rehabilitation through non-surgical treatment. However, the difference between non-surgical treatment and surgical treatment is still controversial in the optimal treatment of thoracolumbar fractures. At present, there is no consensus in the industry, and it still needs to be confirmed by high-quality randomized controlled trials. For thoracolumbar fractures, the compression effect of bone fragments on the spinal canal can cause neurological damage, the incidence of which is 15%–30% reported abroad. There is a certain correlation between the degree of spinal damage and the accompanying neurological deficit [5]. Neurological damage is one of the absolute indications for immediate surgical treatment of thoracolumbar fractures.

2.2 The Purpose of Operation and the Choice of Operation Method

The purpose of thoracolumbar fracture surgery is to restore the height of the vertebral body, correct kyphosis, fully decompress the spinal canal and promote the recovery of neurological function. Surgical approaches include anterior, posterior or combined anterior and posterior approaches. But traditional open surgery has many disadvantages, such as more bleeding, high incidence of complications, postoperative tissue ischemia, muscle atrophy, long hospital stays and so on [6]. With the development of minimally invasive concept, more and more minimally invasive surgery is applied to the treatment of thoracolumbar fractures, in order to achieve safe and effective clinical efficacy [7].

At present, the options of thoracolumbar surgery for severe neurological injury include posterior pedicle screw fixation combined with posterior decompression, anterior decompression combined with anterior fixation, anterior decompression combined with posterior fixation. Anterior approach has the advantages of direct decompression and anterior column support. However, when large bone fragments in the spinal canal are removed by this method, there is a risk of secondary nerve injury. In addition, the anterior approach is in a disadvantageous position due to the large trauma and high incidence of complications. Compared with anterior approach, the risk of secondary nerve injury related to surgery is higher. On the contrary, minimally invasive percutaneous internal fixation technique can achieve the same fixation effect as open posterior approach, and has fewer complications related to the approach [8]. In thoracolumbar fractures, the injury of nerve function is often caused by the compression of the ventral spinal cord. Compared with posterior approach, anterior approach can provide more direct decompression and is more conducive to the recovery of neurological function. However, compared with anterior and posterior surgery, combined anterior and posterior surgery has no greater advantage in neurological function recovery than combined anterior surgery.

3 Basic Research of Robot Assisted Percutaneous Spinal Endoscopy in the Treatment of Thoracolumbar Fractures

In recent 20 years, percutaneous spinal endoscopy has been widely used in clinic. In 1997, Yeung et al. Proposed that the spinal system (YESS) is a minimally invasive approach, which can be used for the treatment of conventional lumbar diseases [9]. Hoogland et al. [10] developed the spinal system (TESSYS) in early 2000, which has been widely used in spinal surgery, including the treatment of lumbar spinal stenosis and cervical spondylosis [11]. It has been proved that percutaneous spinal endoscopy can remove the fragments of spinal canal fracture, and combined with percutaneous internal fixation can replace the traditional surgery in the treatment of thoracolumbar fractures. Successful endoscopic surgery requires a solid knowledge of local anatomy of the spine, and the ability to judge the relevant tissue structure in the operation without direct vision. In addition, it is necessary to have the ability to process the diseased tissue through the narrow working channel. It is necessary to select the appropriate instruments to complete the operation under different visual field conditions, and the minimally invasive surgical approach can safely avoid the above areas.

In recent years, with the development of science and technology, robot technology has made rapid development in the field of orthopedics. The third-generation robot "Tianji" embodies the characteristics of precision, minimally invasive and rapid in the auxiliary placement of pedicle screws. It uses three-dimensional imaging to plan the surgical path and obtain the ideal clinical puncture path, Kantelhardt et al. [12] found that the accuracy of "Tianji" pedicle screw placement was 94.5%. The application of "Tianji" in orthopedic surgery significantly improved the accuracy and safety of surgical operation, reduced the number of intraoperative fluoroscopy, shortened the operation time, and reduced the learning curve. According to the relevant literature at home and abroad, there is no clinical report on the treatment of lumbar burst fracture with percutaneous spinal endoscopy combined with orthopedic surgical robot, which may be related to the long learning curve of percutaneous spinal endoscopy, great technical difficulty and less application of robot technology.

4 The Evolution of Surgical Concept

The purpose of the treatment of spinal fracture is not only to restore the spinal sequence and fully decompress the spinal canal, but also to reconstruct the stability of the spine and create conditions for the recovery of neurological function. Traditional surgical treatment mostly adopts open reduction and pedicle screw internal fixation, which has significant clinical effect. However, it has large trauma, more bleeding during the operation, more damage to the paravertebral muscle and bone structure, and slow postoperative recovery.

With the development of minimally invasive concept, minimally invasive percutaneous screw internal fixation technology is widely used in the clinical treatment of thoracolumbar fractures, and reflects the clinical advantages of small trauma, less bleeding, fast recovery and so on [13], however, it needs percutaneous operation under X-ray fluoroscopy, less fluoroscopy times may affect the position of pedicle screw, and even damage nerves, spinal cord or blood vessels. More fluoroscopy times during operation will cause greater radiation to patients and medical staff, it is reported that [14] the failure rate of pedicle screw implantation is 4.9%–37.5%; meanwhile, for patients with obvious root symptoms, percutaneous pedicle screw technique is not able to reduce the fracture block and reduce the spinal cord and nerve root. The clinical effect is poor. Percutaneous endoscopic technique is the fastest developing minimally invasive surgery in recent years, which has shown great clinical advantages in the treatment of cervical, thoracic and lumbar disc herniation and lateral recess stenosis; at the same time, scholars at home and abroad have applied this technology to more orthopedic surgery, including the clinical treatment of cervical spondylosis, lumbar spondylolisthesis and vertebral posterior epiphyseal dismemberment, and achieved good clinical curative effect, and even applied it to the clinical treatment of spinal infectious diseases; we apply this technique to the treatment of spinal fracture, which shows the clinical advantages of small trauma, less bleeding and quick recovery after operation, and the clinical effect is significant; however, the operation time is long, there are more perspective during operation, higher requirements for the surgical technique and higher risk of operation. The accuracy of operation needs to be further improved.

5 Controversy in Operation

5.1 Decompression

In theory, spinal canal decompression should be completed as soon as the condition permits, because the long interval between two operations may affect the recovery of neurological function. Therefore, in the future, according to the patient's condition, we can consider shortening the time interval between endoscopic decompression and percutaneous minimally invasive internal fixation, or even simultaneously. Secondly, this technique is not without limitations in the treatment of thoracolumbar fractures. First of all, clinicians need to master two kinds of operation; second, the relationship between nerve recovery and decompression time is still uncertain, a little carelessness may miss the best period of nerve decompression; finally, it is confirmed that the indication of this minimally invasive operation is thoracolumbar fracture with nerve function defect, but it still needs further practice and research to prove.

5.2 Fusion

Whether intraoperative fusion is necessary has always been controversial. Some studies in favor of non-fusion have shown that internal fixation alone can reduce blood loss and operation time, maintain better segmental movement and avoid complications. On the other hand, fixed fusion can reduce the risk of implant failure and improve X-ray parameters. Therefore, there is still controversy about the necessity of internal fixation with fusion in vertebral fractures. Therefore, it is still necessary to systematically analyze the existing literature to evaluate the efficiency, security and potential advantages of fusion compared with non-fusion. A scholar [15] Meta analyzed 445 patients with thoracolumbar fractures, and compared the safety and effectiveness of fusion and non-fusion, indicating that non fusion can obtain similar and satisfactory clinical results with fusion. Some studies have shown that extensive tissue dissection is needed in fusion technology. In addition, bone graft fusion consumes time and leads to vascular injury. Therefore, fusion is a more complex and invasive procedure than non-fusion. A prospective randomized study conducted by Wang et al. [16] showed that the non-fusion group had significantly better radiological parameters than the fusion group. On the contrary, a study conducted by Huang et al. [17] revealed that fusion is desirable and can obtain a better final result in the correction of kyphosis. In theory, it's easy to understand that fusion limits spinal movement. However, this result should be interpreted with caution, and more sample studies are needed to confirm this finding. In conclusion, due to the limitations of the current study, more long-term follow-up and large sample RCT are needed to further prove the safety and effectiveness of fusion and non-fusion in the treatment of thoracolumbar fractures.

6   Foreground and Prospects

Lumbar spine fracture is a common trauma, and its surgical treatment is a complex decision-making process, which is still a huge challenge. Traditional open surgery is often limited to relatively complex indications, with the disadvantages of postoperative pain, blood loss and hospitalization. With the continuous development of artificial intelligence technology, robot assisted spinal endoscopy technology combined with percutaneous internal fixation technology is applied in the clinical treatment of lumbar burst fracture, which can achieve the purpose of anatomical reduction of vertebral fracture with exact curative effect. Like all minimally invasive surgery, it has the advantages of accurate operation, high safety, small trauma and fast postoperative recovery. However, at present, there are few clinical cases, short observation time and insufficient statistical significance, so a large sample of clinical control study is needed to observe the clinical efficacy.

Acknowledgments. Lanzhou talent innovation and Entrepreneurship Project (2020-RC-54) "related research on Guiqi Tongbi formula in the treatment of residual symptoms after percutaneous spinal endoscopic surgery for lumbar disc herniation".

Gansu Provincial Natural Science Foundation Project (18JR3RA067) "study on the mechanism of Huoxue Bushen prescription selectively regulating osteoporotic fracture based on Wnt10b/β-catenin signaling pathway".

References

1. Ferguson, R.L., Allen, B.L., Jr.: A mechanistic classification of thoracolumbar spine fractures. Clin. Orthop. Relat. Res. **189**, 77–88 (1984)
2. Lopez, A.J., Scheer, J.K., Smith, Z.A., et al.: Management of flexion distraction injuries to the thoracolumbar spine. J. Clin. Neurosci. **22**(12), 1853–1856 (2015)
3. Denis, F.: The three column spine and its significance in the classification of acute thoracolumbar spinal injuries. Spine (Phila Pa 1976) **8**(8), 817–31 (1983)
4. Magerl, F., Aebi, M., Gertzbein, S.D., et al.: A comprehensive classification of thoracic and lumbar injuries. Eur. Spine J. **3**(4), 184–201 (1994)
5. Aono, H., Tobimatsu, H., Ariga, K., et al.: Surgical outcomes of temporary short-segment instrumentation without augmentation for thoracolumbar burst fractures. Injury **47**, 1337–1344 (2016)
6. Li, X., Zhang, J., Tang, H., et al.: Comparison between posterior shortsegment instrumentation combined with lateral-approach interbody fusion and traditional wide-open anterior-posterior surgery for the treatment of thoracolumbar fractures. Medicine **94**, e1946 (2015)
7. Kalra, R.R., Schmidt, M.H.: The role of a miniopen thoracoscopic-assisted approach in the management of burst fractures involving the thoracolumbar junction. Neurosurg. Clin. North Am. **28**, 139–145 (2017)
8. Defifino, H.L.A., Costa, H.R.T., Nunes, A.A., et al.: Open versus minimally invasive percutaneous surgery for surgical treatment of thoracolumbar spine fractures- a multicenter randomized controlled trial: study protocol. BMC Musculoskelet. Disord. **20**, 397 (2019)
9. Yeung, A.T., Tsou, P.M.: Posterolateral endoscopic excision for lumbar disc herniation: surgical technique, outcome, and complications in 307 consecutive cases. Spine **27**, 722–731 (2002)
10. Hoogland, T., Schubert, M., Miklitz, B., et al.: Transforaminal posterolateral endoscopic discectomy with or without the combination of a low-dose chymopapain: a prospective randomized study in 280 consecutive cases. Spine **31**, E890–E897 (2006)
11. Ohmori, K., Ono, K., Hori, T.: Outcomes of full-endoscopic posterior cervical foraminotomy for cervical radiculopathy caused by bony stenosis of the intervertebral foramen. Mini-invasive Surg. **1**, 63–68 (2017)
12. Kantelhardt, S.R., Martinez, R., Baerwinkel, S., et al.: Perioperative course and accuracy of screw positioning in conventional, open robotic-guided and percutaneous robotic-guided, pedicle screw placement. Eur. Spine J. **20**(6), 860–868 (2011)
13. Chung, W.H., Eu, W.C., Chiu, C.K., et al.: Minimally invasive reduction of thoracolumbar burst fracture using monoaxial percutaneous pedicle screws: surgical technique and report of radiological outcome. J. Orthop. Surg. (Hong Kong) **28**(1), 2309499019888977 (2020)
14. Faizan, A., Kiapour, A., Kiapour, A.M., et al.: Biomechanical analysis of various footprints of transforaminal lumbar interbody fusion devices. J. Spinal Disord. Tech. **27**(4), E118–E127 (2014)

15. Lan, T., Chen, Y., Hu, S.Y., et al.: Is fusion superior to non-fusion for the treatment of thoracolumbar burst fracture? A systematic review and meta-analysis. J Orthop. Sci. **22**(5), 828–833 (2017)
16. Wang, S.T., Ma, H.L., Liu, C.L., et al.: Is fusion necessary for surgically treated burst fractures of the thoracolumbar and lumbar spine? A prospective, randomized study. Spine **31**(23), 1724e31 (2006)
17. Hwang, J.H., Modi, H.N., Yang, J.H., et al.: Short segment pedicle screw fifixation for unstable T11-L2 fractures: with or without fusion? A three-year follow-up study. Acta Orthop. Belg. **75**(6), 822e7 (2009)

Development and Application of Artificial Intelligence Technology to Unmanned Driving Under the Background of Wireless Communication

Zhenxing Bian$^{(\boxtimes)}$

Software Engineering Department, Shandong Polytechnic College,
Jining 272000, Shandong, China

Abstract. With the continuous development of China's modern economy and the popularization of wireless communication technology, wireless communication technology continues to develop in the direction of digitization with today's information age as the background. The rapid development of artificial intelligence technology has penetrated into all areas of social life, and the cross-development of artificial intelligence and automobile research has revolutionized the transportation industry today. The development of unmanned driving technology provides sufficient development space for artificial intelligence. Unmanned driving technology is a new direction for the development of the automotive industry, and its goal is to solve people's high requirements for the safety, comfort and reliability of car driving. Artificial intelligence technology is widely used in natural language processing, speech recognition, sensor fusion and other fields, achieving epoch-making development in the field of unmanned driving. Automation companies, network companies and large car manufacturers are actively exploring the use of artificial intelligence technology to achieve the ultimate unmanned driving. This article elaborates and analyzes the artificial intelligence technology in unmanned driving under the background of wireless communication, and points out the future development of unmanned driving and the application of artificial intelligence technology in unmanned driving.

Keywords: Wireless communication · Artificial intelligence technology · Driverless · Automobile industry development

1 Introduction

Driverless cars refer to the use of modern technology to realize cars that do not require drivers to drive. Because there is no driver to drive, they use the on-board system of the car to perceive the environment around the vehicle, based on the perception of pedestrians, roads, obstacles and traffic lights, so as to obtain the status of the road, and then use computer technology to plan the route cleverly, quickly and accurately, and control the vehicle to reach the specified destination [1]. The creation of system perception, information processing, and execution of instructions is the key to unmanned driving

© The Author(s), under exclusive license to Springer Nature Switzerland AG 2021
J. Abawajy et al. (Eds.): ATCI 2021, AISC 1398, pp. 511–518, 2021.
https://doi.org/10.1007/978-3-030-79200-8_77

technology, so it must be applied to artificial intelligence technologies such as deep learning, image processing, and data processing [2]. As a new driving mode, unmanned driving technology has changed the traditional driving experience, truly liberated the driver's hands, and greatly improved the efficiency and safety of the transportation system [3]. At the same time, it can also improve social benefits and better protect personal safety [4].

From the perspective of the advantages of wireless communication, due to its small dependence on physical lines, it basically does not consume too much money, so it has been welcomed by different industries [5]. As far as the wireless communication equipment of unmanned vehicles is concerned, due to the normal and safe operation of unmanned vehicles, some immediate situations often occur, which are deeply affected by the surrounding environment [6]. Coupled with equipment blocking and current interference, wireless communication disconnects more frequently. In this context, there are higher technical requirements for the layout of outlets and the selection of antennas [7].

The current automatic driving can be roughly divided into two types. One is currently very popular unmanned driving that belongs to the L4 stage of the autonomous driving category. Unmanned driving emphasizes the automatic driving of vehicles, providing a comfortable driving experience and reducing labor costs. Representative examples are the driverless vehicles of Baidu and Google; the other type is ADAS (full name Advanced Driver Assistance System), which has a long history of development and entered the layout of automobile factories in the 1970s [8]. According to the regulations of the Highway Safety Administration, the development of auto-driving cars can be divided into 5 stages (respectively: the first stage is the driver's complete driving; the second stage has more than one automatic control function; the third stage can be carry out a variety of tasks autonomously; the fourth stage requires human intervention in automatic driving with cars as the main body; the fifth stage is fully automatic driving.). The current driverless development technology is in the third stage of development [9]. We started to promote the development of driverless cars in 2009, and in 2015 the first fully driverless car was opened on public roads in history. The self-driving car produced in cooperation with the car manufacturer in November 2016 has already experienced actual road operation. The actual experience process is to use L4 driverless technology on a 3.16 km road. Driverless cars have realized signal light recognition, pedestrian avoidance, overtaking and other situations that will be encountered on actual road driving [10]. In other words, the current pioneers in the field of unmanned driving are mainly concentrated in network companies, and network companies have mastered the core technologies of voice recognition, image processing, and voice recognition required by unmanned driving technology.

2   Method

2.1   Teaching and Learning

Teaching learning is an important field of robotics. Foreign researchers' research on teaching learning can be traced back to the 1990s and is widely used in the field of robotics. In many technologies, it is difficult to obtain the decision-making mode of human experts, so learning this mode is also an important means to solve substantive decision-making problems.

Compared with indirect teaching learning, the idea of direct teaching learning is very simple, assuming we can get a teaching path network composed of experts:

$$\{\tau_1, \tau_2, \tau_3, \tau_4, \ldots, \tau_m\} \tag{1}$$

Any one of the trajectories τ_i contains a series of action-state pairs:

$$\tau_i = \left\{ s_1^i, a_1^i, s_2^i, a_2^i, \ldots, s_{n_i}^i, a_{n_i}^i, s_{n_i+1}^i \right\} \tag{2}$$

Extract all the state-action pairs to get a data set:

$$D = \left\{ \left(s_1^i, a_1^i \right), \left(s_2^i, a_2^i \right), \ldots, \left(s_N^i, a_N^i \right) \right\} \tag{3}$$

Among them:

$$N = \sum_{i=1}^{m} n_i \tag{4}$$

2.2 Reinforcement Learning

Reinforcement learning is a type of machine learning, and its basic theory can be traced back to the 1990s. Reinforcement learning is an iterative process. It continues to interact with the environment, obtain rewards, update strategies, and then continue to interact with the environment. For the value function, two definition methods are usually used: cumulative step-T prizes and cumulative discount prizes:

$$V_T(s) = E_\pi \left[\frac{1}{T} \cdot \sum_{t=1}^{T} r_t | s_0 = s \right] \tag{5}$$

$$V_\gamma(s) = E_\pi \left[\sum_{t=0}^{+\infty} \gamma^t \cdot r_{t+1} | s_0 = s \right] \tag{6}$$

There are two ways to define the action-value function:

$$Q_T(s, a) = E_\pi \left[\frac{1}{T} \cdot \sum_{t=1}^{T} r_t | s_0 = s, a_0 = a \right] \tag{7}$$

$$Q_\gamma(s, a) = E_\pi \left[\sum_{t=0}^{+\infty} \gamma^t \cdot r_{t+1} | s_0 = s, a_0 = a \right] \tag{8}$$

Commonly used algorithms based on the action value function iteration are the SARSA algorithm and the Q learning algorithm. These two algorithms use action value functions for iteration, but the time difference error used in strategy evaluation is different:

Q learning uses the following methods to update the Q value:

$$\begin{cases} \delta_t = r_{t+1} + \gamma \cdot max_{a'} Q\left(s_{t+1}, a' \right) - Q(s_t, a_t) \\ Q' = Q(s_t, a_t) + a_t \cdot \delta_t \end{cases} \tag{9}$$

SARSA is updated in the following ways:

$$\begin{cases} \delta_t = r_{t+1} + \gamma \cdot Q(s_{t+1}, a_{t+1}) - Q(s_t, a_t) \\ \qquad Q' = Q(s_t, a_t) + a_t \cdot \delta_t \end{cases} \tag{10}$$

Among them, Q and Q' respectively represent the action-value function before and after the update, a'represents any action that can be selected in the state, and the learning rate (step size) in the current state.

In fact, the strategy adopted by Q-learning in each policy evaluation is different from the strategy executed when interacting with the environment, so this kind of algorithm belongs to the "off-policy" algorithm. As for the SARSA algorithm, each strategy evaluation and execution strategy is the same, so it belongs to the "on-policy" algorithm.

3 Experiment

3.1 Subject

In order to verify the reliability of artificial intelligence technology in the decision-making system of unmanned vehicles under the background of wireless communication, so that the vehicle can realize the autonomous driving function on structured roads, the actual road environment is selected as the actual vehicle test site. The test environment includes standards. The actual traffic scenes such as lane lines, U-Turn intersections, obstacles, stop lines, and intersections.

3.2 Experimental Design

3.2.1 Experimental Purpose

① Whether the planning path result of the global path planning module is correct. In the urban road network, the starting point and the end point are arbitrarily given, and the task file is given to complete the determination of the driving section.
② Whether the state transition of the driving behavior of the intelligent decision system is correct.
③ The rationality of the state selection of the intelligent decision system under the premise of safety and driving efficiency.
④ The safety of the path planning results of the intelligent decision-making system, and the executability of the planning path under different task states and dynamic environments.

3.2.2 Experimental Program

Before the start of the experiment, you need to define the road network file of the regional road. The road network file contains information such as the road section number, the location of the road feature point and the intersection feature point; the European distance between the two feature points on the road is about 200m, and each section of the road, they are all connected with other roads and are marked with the keyword "connect" in the road network file. At the same time, in order to verify the real performance of the global planning module of the decision-making system, experts are invited to formulate driving routes and make MDF files, which will be issued 30 min before the experiment. At the same time, in order to verify the ability of the behavior decision module to handle complex scenarios, under the guarantee of safety, two real and controllable driving scenarios in the urban environment are designed, and the decision results of the unmanned vehicle are compared with those of the skilled driver. To compare and analyze the performance of the entire behavioral decision-making module, the road length of the entire scene is about 2 km.

3.2.3 Experimental Scenario

Usually when driving a car, you will encounter such a situation. When the vehicle is running normally, the vehicle in front suddenly decelerates and brakes in order to avoid pedestrians or the vehicle fails. This is a real emergency scenario. In this situation, the driver will have two strategies to deal with this emergency: the first is to slow down, and the second is to adjust the direction to try to bypass the obstacle. This is a typical emergency situation. If the unmanned vehicle adopts inappropriate behavior, there is a danger of collision.

4   Results

During the driving process, record the perception map, vehicle behavior decision results, speed and vehicle trajectory information (Fig. 1).

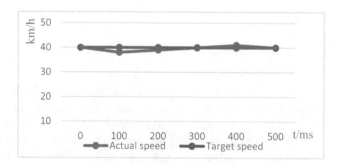

Fig. 1. Record result of target speed and actual speed

Based on the established decision matrix, since the current vehicle is on the first lane, after logical inspection, the right lane change plan is deleted. From the perception map, the corresponding value of each attribute is extracted, and the driving behavior decision matrix A is constructed (Table 1).

After model calculation, the objective weight of each attribute of the behavior decision matrix is:

$$(0.067, 0.128, 0.213, 0.380)$$

Table 1. Decision matrix A

A	f1	f2	f3	f4
x1 (Lane keeping)	5.57	1.93	41.2	1.93
x2 (Left lane change)	1.88	5.62	1.88	12.18
x3 (Accelerated)	5.57	1.93	41.01	1.93
x4 (Deceleration)	5.57	1.93	41.93	1.93

The combined weight vector determined based on the entropy weight method and the AHP analytic hierarchy process is:

$$(0.05, \ 0.08, \ 0.15, \ 0.23)$$

The relative closeness G between each plan and the ideal plan is (Fig. 2):

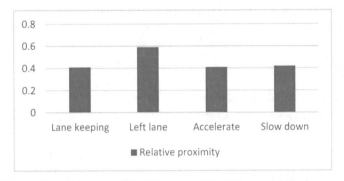

Fig. 2. The relative closeness of the candidate solution to the ideal solution

Therefore, the unmanned vehicle enters the left lane changing state after the vehicle in front suddenly decelerates, and it changes lanes smoothly without colliding with other vehicles during driving. In the process of subjective weight assignment, the safety index assignment is larger. The driving attributes related to driving safety are the distance to the left and right curbs of the road, the distance to obstacles in the left and right lanes, and the driving safety factor of vehicles in the lane. After changing lanes, because the upper limit speed of the target lane is higher, the vehicle can drive to the end faster, so its comprehensive evaluation is higher. Scheme 1 and Scheme 4 can complete the driving task safely, but the driving safety of scheme 1 is lower. After the vehicle in front decelerates, the distance between the vehicle and the vehicle in front decreases, which affects its driving safety. Although the fourth scheme has a higher safety index value in the lane, it reduces the speed of the vehicle, affects the efficiency of driving, and sacrifices time in exchange for safety.

In view of the above traffic scenarios, please have experienced drivers make subjective judgments on the decision results of test drivers or unmanned vehicles, and give an evaluation on the degree of satisfaction with the decision results. The comment set is (unsatisfied, less satisfied, general, Relatively satisfied, satisfied), and replaced by corresponding numbers (1, 2, 3, 4, 5) respectively as a score for the degree of satisfaction with the decision result. The evaluation of satisfaction is mainly based on the safety of decision-making results and driving efficiency. The safety factor mainly reflects the degree of safety of the vehicle during the driving process subjectively judged by the driving expert. The driving efficiency factor mainly reflects the speed characteristics of the unmanned vehicle. Related to behaviors such as deceleration, acceleration, and lane keeping. The data in Table 2 reflects the satisfaction degree of the driving expert group for driving safety and real-time speed in scenario 1.

Table 2. Expert scoring results

Expert number	(1)	(2)	(3)	(4)	(5)	(6)	(7)
Has a driver	5	4	4	4	5	4	4
Unmanned vehicle	4	3	4	4	4	4	4

From the data in the table, we can see that there is no significant difference between the decision-making results of the driverless vehicle and the driver in the scenario. Using the single factor analysis method to compare with human drivers, it can be proved that the unmanned vehicle decision-making results are equivalent to the driver's decision-making results in the experimental scene. The design method of this decision-making system meets the driving requirements of unmanned vehicles on urban roads.

5   Conclusion

As a new thing, driverless cars increase the intelligence of driving comfort, decision-making efficiency and control that people face on the road. The comfort and problems it brings to us coexist for a certain period of time, and there are some contradictions that are currently difficult to resolve quickly. However, with the further development and improvement of artificial intelligence technology, and the country's high investment and high attention to unmanned driving technology, unmanned driving technology will become more mature. Unmanned driving technology benefits from the application and promotion of artificial intelligence technology. It can provide great help for people's travel, and even make people apply for a car driving license older. The technology of driverless programs is not yet fully mature. To move towards true industrialization and commercialization, more efforts are needed in the application of artificial intelligence technology, but driverless cars will surely accelerate toward practicality and synchronization.

References

1. Varshney, A., Sharma, V.: A comparative study of microwave rectangular waveguide-to-microstrip line transition for millimeterwave, wireless communications and radar applications. Microw. Rev. **26**(2), 21–37 (2021)

2. Prakash, V.D., Vedachalam, N., Ramesh, R., et al.: Assessment of the effectiveness of the subsea optical wireless communication system in the Arabian sea using field data. Mar. Technol. Soc. J. **53**(1), 9–19 (2019)
3. Hong, X., et al.: Discrete multitone transmission for underwater optical wireless communication system using probabilistic constellation shaping to approach channel capacity limit. Opt. Lett. **44**(3), 558–561 (2019)
4. Chen, W.L., Lin, Y.B., Ng, F.L., et al.: Ricetalk: rice blast detection using internet of things and artificial intelligence technologies. IEEE Internet Things J. **7**(2), 1001–1010 (2020)
5. Concepcion, R.I.S.: Advancement in computer vision, artificial intelligence and wireless technology: a crop phenotyping perspective. Int. J. Adv. Sci. Technol. **29**(6), 7050–7065 (2020)
6. Khairy, D., Abougalala, R.A., Areed, M.F., et al.: Educational robotics based on artificial intelligence and context-awareness technology: a framework. J. Theor. Appl. Inf. Technol. **98**(1817–3195), 2227–2239 (2020)
7. Perry, T.S.: Here comes driverless ride sharing: cruise unveils the origin, a fully autonomous SUV designed for app-controlled urban transportation - [spectral lines]. IEEE Spectr. **57**(3), 4 (2020)
8. Wu, J., Ji, Y., Sun, X., et al.: Price regulation mechanism of travelers' travel mode choice in the driverless transportation network. J. Adv. Transp. **2020**(3), 1–9 (2020)
9. Veeramani, C., Sawada, Y., Yoshino, N.: From import substitution to integration into global production networks: the case of the Indian automobile industry. Asian Dev. Rev. **36**(2), 72–99 (2019)
10. Soni, Y., Gupta, R.C.: Optimization in logistics for supply chain management of an automobile industry using Fuzzy DEMATEL matrix method. J. Sustain. Dev. Transp. Logist. **5**(2), 29–36 (2020)

Key Technologies of Autonomous Driving System of Rice and Wheat Agricultural Robots Based on High Precision Temporal and Spatial Information

Wei He[1,3], Lunning Zhang[2,3(✉)], Liankai Song[3], and Guoxin Yu[2]

[1] Academy for Engineering and Technology, Fudan University (FAET),
Shanghai 200433, China
[2] School of Economics and Business, Xinjiang Agricultural University, Urumqi 830052,
Xinjiang, China
[3] Shanghai Huace Navigation Technology Co., Ltd., Shanghai 201702, China
wei_he@huacenav.com

Abstract. The application of agriculture robots in wheat and rice provides solutions to the problem of low efficiency in the traditional mode of agricultural production, and the robot guiding system plays a key role in guaranteeing the functions of the rice and wheat agricultural robots, it is necessary to study the key technology. Based on the breakthrough in the key technology, the agriculture robots have better applicability, and its application values are given full play. Based on the application of agricultural robot, this paper analyzes the key technologies of high-precision spatiotemporal information based automatic driving system so as to clarify the development direction of the key technologies of automatic driving system.

Keywords: High-precision spatio-temporal information · Rice and wheat agricultural robots · Automatic driving system · The key technology

In recent years, modern agriculture has achieved rapid progress, and the development of biotechnology, network communication technology, navigation technology gained momentum in development. The application of technology has created more favorable conditions in precision agriculture [1]. It is under the joint action of a variety of new technologies that rice and wheat agricultural robots can be developed and applied, creating conditions for the implementation of precision agriculture. The application of automatic driving technology has effectively improved the performance of rice and wheat agricultural robots. The application of automatic driving system should give full play of its roles in high-precision spatiotemporal information, which should be taken as the breakthrough in key technology to ensure that the system has more powerful functions [2].

© The Author(s), under exclusive license to Springer Nature Switzerland AG 2021
J. Abawajy et al. (Eds.): ATCI 2021, AISC 1398, pp. 519–525, 2021.
https://doi.org/10.1007/978-3-030-79200-8_78

1 Background of Developing Automatic Navigation and Driving System for Rice and Wheat Agricultural Robots

The application of rice and wheat agricultural robots creates conditions for efficient planting. As the core component of robot development, the navigation automatic driving system is the product of the integration of computer technology, electronic technology, control technology, hydraulic technology and navigation technology [3]. Through the Beidou/GNSS/INS combination system, the automatic driving system of rice and wheat agricultural robot can send the positioning and attitude information to the controller in real time. The controller can control the driving direction of the rice and wheat agricultural machine according to the coordinates determined by GNSS positioning, the INS attitude and the wheel positioning. The application of this technology can effectively promote the development of high-precision and low-cost GNSS/INS integrated navigation technology, promote the development of automation technology in China, and facilitate the development of low-cost INS sensors. The market prospect is very broad.

GNSS/INS integrated navigation achieved by low cost IMU module is the trend of mainstream technology of global research and development. The system selects the appropriate low-cost IMU (Inertial Measurement Unit) device, combined with GNSS speed information, to determine the final high-precision position and attitude information, which is in line with the mainstream trend of the current integrated navigation field, which also expands the positioning means of rice and wheat agricultural automatic driving system [4].

2 Key Technologies Used in the Rice and Wheat Agricultural Robot Robot Guiding System Based on High-Precision Spatio-Temporal Information

2.1 Establishment of High-Precision Space Network

2.1.1 Determination of Networking Mode

In the specific application, there are many ways of networking, such as acentric contribution, which can tip the traditional point-to-point mode into a flat mode to meet the needs of communication in waters. The transmission mode of different nodes in the system can also be flexibly configured, or any node can be selected as the benchmark network to realize the communication between different nodes. Therefore, the command center is easy to change to facilitate the dispatch of agricultural robot ships on the sea to provide communication services, and the support scheme in communication security can be more flexible [5].

In terms of data support, the nodes are equivalent. Different nodes can be linked with the surrounding nodes, and the links can be selected based on routing information for transmission. Adjacent nodes can communicate with each other. Based on the dynamic routing protocol, different nodes can be selected to dynamically form a network according to the actual situation. Each node can be set as a relay station to ensure the coverage effect of signals in the network. With the support of protocol, the dynamic change of network topology can adapt to the demand of mobile communication of agricultural robot ships. Networking mode to ensure the network destruction resistance, any node failure will not affect the network.

During communication transmission, the modulation mode can be adjusted automatically according to the received signal strength. When conditions are available and the received signal meets the requirements, the encoding mode modulated is high order, such as 16QAM3/4, with the help of efficient encoding mode, the need of high-speed data transmission is guaranteed. As the application of high bandwidth services, a large number of data transmission can be quickly realized. Under the condition of weak signal, low order modulation is adopted, such as BPSK1/2 is selected, and the transmission bandwidth can guarantee the needs in voice and data transmission.

The system is made with the adoption of TOD synchronization and jump synchronization to ensure the synchronization of no time difference. After networking, any choice can set up group head mode. Group mode and distributed "whole network time benchmark" can ensure the highest time rating. For other nodes, the time level will decrease with the increase of distance hops. Nodes can broadcast to specific areas periodically, and the time benchmark and response range can be extended after networking. If the node that is not connected to the network obtains the information of the neighboring nodes, it will automatically complete TOD information synchronization with the node already connected to the network, and it can also automatically update TOD and RTC. According to the new TOD information, the frequency hopping sequence can be realized. In this way, the time in the network can be synchronized at any time, and can be synchronously mapped to the network. With the implementation of higher level networking, node time can be automatically updated periodically.

2.1.2 Implementation of Networking Mode

The network access adopts TDMA mode, duplex TDD. The frame can be subdivided into fixed allocation frame and dynamic allocation frame to ensure the access effect. The fixed frame allocation is based on the network scale and the configuration of network nodes. For multiple robots, dynamic frame allocation is adopted, and dynamic distribution is realized through signaling. The fixed allocation frame adopts two different methods, and the sub-frame is controlled by network or data. Dynamic frame allocation is a key function module in network access.

The data subframe is more flexible and can be better applied to the transmission of small flow data [6]. Combined with the characteristics of data sub-frames, the dynamic distribution frame after organization ensures the realization of large flow services, such as the transmission of maritime measurement and control data. In the node, the expression of the dynamic allocation frame is defined as common, and different nodes communicate with each other through sending tasks. As for the control of sub-frames, interactive signaling is used to ensure the negotiated use of bandwidth. In order to save the wireless bandwidth, the routing method can be adopted on demand in the movement of agricultural robot.

The routing protocol is based on standard AOD, and the QoS transformation is implemented for routing discovery, which can better meet the needs of maritime microwave transmission. The AODV protocol takes into account the control of routing nodes. After a network is formed, different nodes broadcast requests to routing, and packet nodes will set node routing according to the instructions given by the broadcast information, and "reverse routing" can be realized. After the network is put into use, the reverse path node should also take into account the needs of the broadcast source node, and send the information to the neighboring nodes through the nodes, and spread it to the neighboring nodes. In networking, if necessary, the route should be added, that is, the establishment of "forward route".

In actual networking, the forward path sends information to neighboring nodes based on the RREP source node. Forward and reverse paths are established by RREP. According to the requirements of the protocol, the receiving neighbor packets be made with consideration of the grouping conditions, decide the search range of the route and judge whether to add items in combination with the grouping results. The source node should be made to complete the broadcast in accordance with the standard protocol according to the packet type during the information transmission. If the intermediate node failed to guarantee that the grouping will be applied, it will not forward such requests and will automatically exit the search.

The realization of node should be made with consideration in how to guarantee the grouping success rate of source node. To ensure that the node sends out the information to get the reply, the source node can be made to choose the most reliable communication effect path to realize the transmission service. If the packet failed to guarantee the reliability of the transmission path, the source node will adjust the quality requirements according to the characteristics of the business class. The path is still not established based on the minimum quality criteria, the service is suspended, and the bandwidth is used to complete other business. After networking, it is necessary to maintain the routes of active nodes periodically and detect new paths. If the new path has better transmission quality, it will be switched to the new path, which is more suitable for the fast movement of the ship [7].

2.2 Control Algorithm

The main components of the system include navigation controller, communication module, hydraulic sensor, hydraulic controller and navigation light target, etc. The working principle of the system is as follows: Firstly, the walking line is set on the navigation light target, the navigation mode (straight line or curve) is set, and the precise positioning information and attitude information are sent to the controller in real time through the Beidou/GNSS/INS integrated system. When the controller receives the information, the direction sensor sends the motion direction of the wheel to the controller in real time. The controller sends instructions to the hydraulic control valve in real time according to the coordinates of GNSS positioning, INS attitude information and wheel rotation. By controlling the flow and flow direction of hydraulic oil in the steering hydraulic system, the driving direction of the rice and wheat farm machine is controlled to ensure that the rice and wheat farm machine runs according to the route set by the navigation light target [8].

In the research process, the parameter configuration and interface, data flow and format, protocol and standard among all subsystems, as well as the specific tasks and index allocation of each subsystem will be determined. The control algorithm is the core technology to realize the automatic navigation of rice and wheat farm machinery, which mainly includes steering control and path tracking control of front wheel of rice and wheat farm machinery. In the process of automatic driving, the vehicle sensors (such as RTK-Beidou, speed sensor, etc.) are used to acquire the motion parameters of the rice and wheat farm machine in real time, and then the actual position and heading information of the rice and wheat farm machine are compared with the pre-defined path to calculate the lateral deviation and heading deviation. The navigation decision controller takes the lateral deviation and course deviation signals as input to calculate the expected front wheel rotation angle through the built-in control algorithm and transmits it to the lower computer. The lower computer controller controls the front wheel steering of the rice and wheat farm machine to track the expected front wheel rotation angle, so as to reduce the lateral deviation and course deviation, and thus achieving automatic navigation [9].

2.3 High-Precision and Low-Cost GNSS/INS Integrated Navigation and Positioning Technology

2.3.1 Application of Integrated Navigation

In the whole design process of GNSS/INS integrated rice and wheat agricultural machine automatic driving system, integrated navigation is one of the key parts, so the quality in its design is directly related to the success of the system. Because INS is fully autonomous and flexible, it can provide navigation in a variety of modes. Currently, INS can be widely used in many different fields, but the influence of errors should be taken into account in the application of INS. With the accumulation of time, the error will increase, which is a difficult problem to be solved in the application of INS. At present, the Beidou Navigation Satellite System has more complete functions, which can provide users with all-weather, all-time, continuous and accurate three-dimensional position information, and simultaneously guarantee the synchronization of three-dimensional velocity and time information. However, Beidou has poor reliability under dynamic conditions, and is vulnerable to external interference, and it is difficult to ensure high frequency of data sampling. Since Beidou and INS have their own advantages and their complementarity can be achieved to establish a Beidou/INS integrated navigation system [10].

To realize automatic navigation of rice and wheat farm machine, control algorithm must be solved. As the core technology, it includes the control of front wheel steering and the tracking control of planting path. In order to realize the automatic driving of rice and wheat farm machine, it is necessary to dynamically obtain various parameters of the farm machine in operation with the help of vehicle-mounted sensors (such as Beidou, sensors, etc.), compare the actual position of the farm machine with the pre-defined path of the course, and then calculate the existing lateral deviation and course deviation. The system has a navigation decision controller. After inputting lateral deviation and course deviation signals, the front wheel rotation angle can be determined by the built-in control algorithm and the information can be transmitted to the lower computer. The lower computer controller can control the front wheel steering of agricultural machinery, and track the change of the front wheel angle, and eliminate the lateral deviation and course deviation to ensure the accuracy of automatic navigation.

2.3.2 Determination of Walking Paths

There are many kinds of tracking control methods of agricultural machinery walking path, such as linear model, optimal control, PID control, neural network, fuzzy logic, pure tracking model and so on. The linear model has the characteristics of simple algorithm and easy to optimize the control parameters, but it is difficult to eliminate the steady-state error in time. PID control is the most common control method, reflecting the advantages in simple algorithm and strong robustness. Steady error is easy to be eliminated, but the defect is that the control parameters are difficult to be optimized. The optimal control method can obtain the optimal control parameters directly, but the model parameters should be optimized with kinematics in application to ensure the adaptability of curve tracking. Fuzzy control does not need the high speed of model parameters for the controlled object, so it shows strong robustness, but the tracking error is large, so it is difficult to realize fast correction. Neural network has good adaptability under nonlinear motion condition, but its application is based on high quality training and the comparison support of verification samples, so its generalization ability is weak. The pure tracking model is derived from geometric angle, and the choice of forward-looking distance will affect the control accuracy.

When the rice and wheat agricultural machinery run in the farm area, it will incline to a certain extent under the influence of ground undulation. In the automatic driving system, the kinematics model with the center of the rear axle as the control point, the kinematics model with the center of the front axle as the control point, and the dynamics model with the center of mass as the control point of the rice-wheat agricultural machine tracking straight line driving were established. The optimal navigation control was analyzed for vehicle kinematics and dynamics models, and a method of weighting the feedback control parameters of the two models was proposed. The fuzzy navigation control rules of rice and wheat agricultural machine automatic driving were studied. Aiming at the problems to be solved by fuzzy control in conventional mode, the complex fuzzy-PID parallel connection was used to solve the problems, and the asymmetry of front wheel rotation angle control caused by simplification of rice and wheat agricultural machine model was compensated. On the basis of measuring the front wheel angle with the front wheel angle sensor, the driving mode of the rice and wheat farm machine was judged by the pressure sensor installed on the navigation valve group.

3 Conclusion

The research on autonomous driving system of rice and wheat agricultural robots based on high-precision spatiotemporal information has created favorable conditions for efficient and intelligent agricultural production. In the development of automatic driving system, it is necessary to solve the problems of key technologies in application path to ensure the reliability and stability of the system, and to ensure that the function of automatic driving can be better realized.

References

1. Wang Jian, X., Guoyan, C.J., et al.: Introduction to Unmanned Driving Technology, p. 100. Tsinghua University Press, Beijing (2018)

2. Lijun, Z., Qinzhi, T., Pingtao, W., et al.: Structure and Principle of Electric Vehicle, pp. 149–150. Peking University Press, Beijing (2012)
3. Chen, J., Zhang, J., Gao, Y., et al.: Automobile Structure (vol. 1). The 3rd Edition. Beijing: China Machine Press, 13–15 (2018)
4. GB 7258–2017, Technical Conditions for Operation Safety of Motor Vehicles. Beijing: China Standards Press, 2017:21
5. Chunyan, Y.: Discussion on automation of agricultural machinery. West China Dev. **6**, 188 (2010)
6. Li, Y., Zhao, Z., Gao, J.: Design and research of agricultural navigation and positioning system based on integration of GPS-SINS. J. Agric. Mech. Res. **3**, 16–22 (2014)
7. Wei, S.: Research on Guoton mechanical navigation system based on GPS and inertial navigation. Northwest Agriculture & Forestry University (2013)
8. Gong, Y.: Research on path tracking and control of unpiloted vehicles. Nanjing University of Science and Technology (2014)
9. Jun, L., Cunjie, S., Jiefa, Y.: Hardware design of vehicle sensing and intelligent terminal. Microcomput. Appl. **F31**, 31–33 (2012)
10. Liu, H.: Research on INS/GPS integrated navigation positioning and attitude determination based on MEMS inertial component. Shenyang T University (2013)

Protection Method of Network Data Privacy Security Issues Based on Blockchain Technology

Huikui Zhou[1](✉) and Mudan Gu[2]

[1] School of Art Media and Computer, Jiangxi Tourism and Commerce Vocational College,
Nanchang, Jiangxi, China
[2] School of Information Engineering, Jiangxi Modern Polytechnic College,
Nanchang, Jiangxi, China

Abstract. As a new model of Internet-based computers, blockchain technology has always been a hot spot for scholars and the industry. Although blockchain technology has been developed for many years, data security and reliable computing are still the main challenges facing the current blockchain technology applications. In order to solve this problem, many researchers have conducted a lot of research on it and proposed many models, including data integrity verification and secure multi-part calculation. However, most of these solutions face problems such as high computational complexity or lack of scalability. This paper proposes a data release method based on differential privacy protection for the problems of the release of static data sets and the release of dynamic data, and uses the built platform for experimental verification. Aiming at the privacy leakage problem faced by the multi-level and fine-grained search of big data, this paper proposes an incremental data indexing strategy based on conceptual lattice granular deduction, and builds a complete data retrieval service system that supports differentiation on the cloud platform architecture, it also supports practical search modes such as multi-keyword ranking search, similarity search and fuzzy search, and systematic research on the protection methods of network data privacy security issues under blockchain technology. The experimental results surface: Compared with other methods, the algorithm in this paper has greatly reduced information loss under the same privacy requirements, and the algorithm has high execution efficiency. This shows that the results of this study can provide ideas for the research of big data security and privacy protection, and have certain reference significance.

Keywords: Blockchain technology · Network data privacy · Security issue protection · Big data security

1 Introduction

With the continuous development of technology, while using the computing and storage services provided by blockchain technology, users also put forward higher requirements for the security of data stored in the cloud and the credibility of outsourced computing.

J. Abawajy et al. (Eds.): ATCI 2021, AISC 1398, pp. 526–533, 2021.
https://doi.org/10.1007/978-3-030-79200-8_79

This is also the main obstacles restricting the further development of blockchain technology [1, 2]. Unlike the traditional computing model where users completely control data calculation and storage locally, blockchain technology needs to centralize the user's data and physical servers in the cloud service provider management, and the user only retains some control over the leased virtual machine permission [3, 4].

Apply blockchain technology to the field of cloud computing and combine the respective advantages of blockchain and cloud computing to construct remote data integrity verification and secure multi-party computing solutions based on blockchain technology. This solution can solve the problems of data security and computing credibility in cloud computing from a technical level, and apply blockchain technology to cloud computing to provide users with safe and efficient data verification and secure multi-party computing services [5, 6]. Despite years of research and development of blockchain technology, related technologies and application solutions are becoming more and more perfect, it still faces two major technical difficulties: data security and trusted computing, which has also become a major obstacle to the further development and application of blockchain technology [7]. All transaction data in a typical public chain is open and transparent to all access nodes. The simplest form of privacy is to use a kind of pseudo-anonymity, but under this simple mechanism of pseudo-anonymity, through big data analysis, gather combining class analysis with certain methods of attacking the network, it is easy to obtain his personal information. Users in public and transparent transaction documents [8]. Therefore, data privacy research aims to study better data privacy technologies, provided that the availability of big data should be improved as much as possible to prevent intruders from using big data technology to extract data privacy information and cause data leakage [9, 10].

This article applies blockchain technology to cloud computing data integrity protection and secure multi-party computing solutions, combining the respective advantages of blockchain and cloud computing, and constructs an improvement plan for key cloud computing technologies based on blockchain technology. The data integrity protection scheme based on the blockchain and the secure multi-party computing protocol based on the blockchain are respectively proposed, using the consensus consensus mechanism of the blockchain to improve on the basis of the existing algorithms. Compared with the existing fuzzy search strategy, in terms of privacy protection, two types of inference analysis attacks caused by redundant data transmission are completely blocked.

2 Overview of UAV and Fuzzy Control Theory

2.1 Security Analysis of Mimic Blockchain

(1) Security analysis of mimic smart contracts
    In the mimic smart contract architecture, the condition for an attacker to successfully attack is that more than k of the n smart contract calculation results are the same and different from the normal output. We define the attack success rate as P_s, n languages that implement smart contracts as $\{L_1, L_2, \cdots, L_n\}$, n code obfuscation modes as $\{O_1, O_2, \cdots, O_n\}$, the probability of successfully attacking n languages that implement smart contracts as $\{P_1^L, P_2^L, \cdots, P_n^L\}$, and the success probability

of attacking n code obfuscation modes as $\{P_1^O, P_2^O, \cdots, P_n^O\}$, select the set of k code obfuscation modes from n code obfuscation modes in lexicographic order as $\{O^1, O^2, \cdots, O^{c_n^k}\}$, we get:

$$P_s = \sum_{j=1}^{c_n^k} \prod_{i=1}^{k} (P_i^{L_j}, P_i^{O_j}) \tag{1}$$

When fault tolerance is not considered, the security of a mimic smart contract is better than that of a single smart contract, then:

$$P_s = \prod_{i=1}^{n} (P_i^L, P_i^O) \tag{2}$$

$$P_s \le \min_{1 \le i \le n} (P_i^L, P_i^O) \tag{3}$$

(2) Security analysis of mimic signature algorithm

To attack the mimic signature algorithm system, three different signature algorithms must be successfully attacked simultaneously. And the cryptosystems that we designed for the three signature algorithms are not the same, so there is no correlation between attacks on these three signature algorithms. Define the probability of successfully attacking the mimic signature algorithm as P_r. The probabilities of successfully attacking the three signature algorithms are P_1^a, P_2^a, and P_3^a respectively, then:

$$P_r = P_1^a \cdot P_2^a \cdot P_3^a \tag{4}$$

The attacker must successfully attack the three signature algorithms at one time. Once one or two signature algorithms are successfully attacked, the system will replace the successfully attacked algorithm with an alternate signature algorithm. Therefore, the attacker must also successfully attack the new three signatures at the same time.

(3) Security analysis of mimic consensus process

In the mimic consensus process, we define that the attacker's successful attack means that the attacker's gain is greater than the attacker's consumption. We define the attacker's successful attack on B $PoW(1)$, $PoW(2)$, and PoS as I_1, I_2, and I_3, respectively. The consumption is E_1, E_2 and E_3. Then there are:

$$P_c = P_1^c \cdot (1 - P_2^c) \cdot (1 - P_3^c) \cdot (1 - P_4^c) \cdot (1 - P_5^c) \cdot (1 - P_6^c) \cdot (1 - P_7^c) \tag{5}$$

2.2 Big Data Security Technology

(1) Data access control technology

When the amount of user data is huge, it is difficult to store the data directly locally. As a new computing model, cloud computing provides technical support for data intensification, scale and specialization. In order to ensure the privacy of data, it is necessary to audit data access rights. How to achieve effective access control has become a challenge for cloud storage data security.

(2) Searchable data encryption

When the user stores the data in the cloud storage environment, the user loses control over the data. The data owner only knows the content of the data, but cannot know the storage location and security of the data with certainty. In order to ensure data security and user privacy data, the safest solution is to encrypt all data and then store it on a cloud server. Searchable encryption can be divided into symmetric and unpaired searchable encryption from the perspective of construction. The main advantage of symmetric searchable encryption is that it is more efficient. Most of the primitives of the symmetric searchable mode are based on block ciphers and pseudo-random functions.

(3) Big data desensitization technology

The security technology involved in cloud computing can certainly solve some data security problems, but it still faces the possibility of being compromised, and data privacy still faces the risk of leakage. If the sensitive data in the data is deprived or deformed, and then combined with the security technology of the cloud computing platform, the security of the data will be better guaranteed. Because the data contains a lot of sensitive information, and the use of data is essential to cause the leakage of sensitive information. Therefore, the research on big data desensitization technology and desensitization system is very important. The research of big data desensitization technology provides technical support for improving the confidentiality of data and the security of data use, thereby reducing the risk of data leakage.

3 Experimental Design

3.1 Construction of Data Processing Platform

The data processing platform consists of four parts: data collection, data storage, data processing, and statistical release results display. The collected data is transferred to a distributed storage system; then, the parallel operations of data: classification, clustering, data privacy protection and training of massive data are mainly processed by the Hadoop platform and the Spark platform. Data access control, encryption and decryption operations are also handled by the data processing platform; finally, the calculation results are published and displayed.

3.2 Test Subject

The data set used in the experiment comes from the Transactions mall transaction data provided by the "Kaggle: The Home of Data Science" website, including product type, brand, transaction date, purchase volume, and transaction amount. The description of the experimental data set is shown in Table 1.

Table 1. Experimental dataset description

Name	Size/Gb	Records	Number of Attribute Types	Proportion of Abnormal Data/%
Transactions	20.3	33625478	850	2.6

3.3 Experimental Method

The service party obtains computing services by sending computing requests to the blockchain, and provides corresponding computing functions, related data reference information, and certain remuneration. Access to these data needs to be authorized in advance from the data holder. After receiving the calculation request, the blockchain verifies whether the server is authorized by each data holder by executing the corresponding smart contract. After all nodes in the computing node set obtain data partitions, they start safe multi-party calculation of data based on a specific function. After the calculation is completed, the calculation result is checked. There are two kinds of calculation results, success or failure. The usability of the algorithm is mainly measured by two evaluation standards: total error and absolute error:

$$AE = C_i' - true \tag{6}$$

Among them, C_i' represents the released data after adding Laplace noise, and *true* represents the true value of the packet.

4 Analysis and Discussion of Experimental Results

4.1 Based on Data Availability and Privacy Measurement

As can be seen from Fig. 1, From the overall error formula, SSE(b) = 299.0, SSE(c) = 299.0, SSE(d) ≈ 19.47. S-GS grouping is to introduce the concept of difference set on the basis of GS method to deal with outliers. If there is an outlier, the point is split out, and the grouping is determined according to the difference set. In this experiment, SSE(b) = SSE(c), and SSE(d) < SSE(c), so the method described in this paper performs better in the overall error estimation.

This experiment also considers the impact of privacy budget on usability, and compares it with the GS method and the S-GS method. In the experiment, the privacy budget parameters are 0.01, 0.1, and 1, and the results are shown in Table 2.

It can be seen from Fig. 2 that the simulation test for the correctness of the data file does not include the client generating a set of homomorphic tags. It only considers the process from the client sending the request to the client obtaining the verification result of the server evidence, ignoring the data transmission delay, etc. Factors, this process can be simplified into two processes: node search and verification calculation. Considering the change of privacy budget, the trend of absolute error is obvious, and the absolute error of the algorithm decreases as the privacy budget increases.

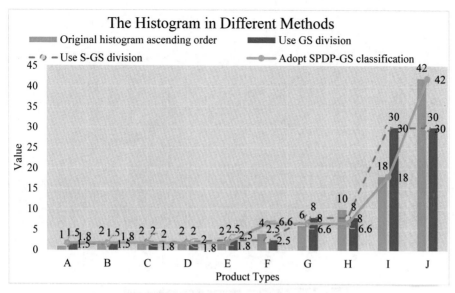

Fig. 1. The histogram in different methods

Table 2. Absolute error data table under different epsilon values

Epsilon	GS	S-GS	SPDP-GS
0.5	348.65	332.37	327.45
0.75	270.38	264.19	247.63
1	223.54	200.35	183.17
1.5	169.32	165.14	143.26

4.2 Relationship Between the Risk of Privacy Leakage and the Dimensions of Released Data

This set of experiments verifies and analyzes the relationship between the risk of privacy leakage and the dimensions of published data. The experiment uses Adultdataset as the experimental data set, and experimentally verifies the attributes of Age and Education-num in the data set. Among them, the privacy leakage risk method is used to measure the risk of privacy leakage, and the result is shown in Fig. 3.

It can be seen from Fig. 3 that after using the differential privacy protection method to perform privacy processing on the data to be released, the privacy leakage risk of data Age and Education-Num is lower than that of data D without any protection measures. And as the privacy budget increases, the risk of data privacy leakage gradually increases. This also further verifies that the privacy budget represents the privacy protection level of the algorithm. The smaller the value of the privacy budget, the higher the privacy protection level. On the contrary, the greater the risk of privacy leakage.

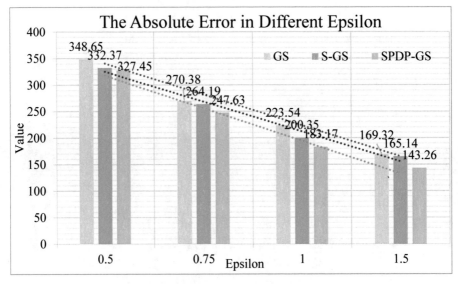

Fig. 2. The absolute error in different epsilon

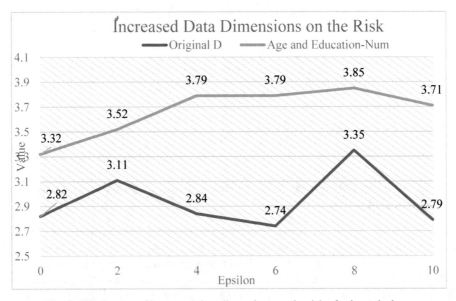

Fig. 3. The impact of increased data dimensions on the risk of privacy leakage

5 Conclusions

This article introduces the secure multi-party computing solution based on the blockchain, expounds the concepts, principles, common implementation algorithms, application scenarios and problems faced by secure multi-party computing technology, and then introduces the secure multi-party computing in the blockchain system and the application status and prospects of cloud computing. Then it focuses on the analysis of the secure and scalable multi-party computing protocol of the basic blockchain and the analysis of the security and privacy assurance of the scheme, and compares it with the GS method and the S-GS method through experiments. The results show that compared with other methods, the algorithm in this paper greatly reduces the information loss under the same privacy requirements, and the algorithm has high execution efficiency, while ensuring that the published data has better usability.

Acknowledgement. This research was financially supported by Teaching Reform Research Project of Jiangxi's Universities (Grant NO.JXJG-20-50-9) and Science and Technology Research Project of Jiangxi Provincial Department of Education (under Grant no.GJJ191100).

References

1. Sikorski, J.J., Haughton, J., Kraft, M.: Blockchain technology in the chemical industry: machine-to-machine electricity market. Appl. Energy **195**(JUN.1), 234–246 (2017)
2. Yeoh, P.: Regulatory issues in blockchain technology. J. Financ. Regul. Compliance **25**(2), 196–208 (2017)
3. Miraz, M.H., Ali, M.: Applications of blockchain technology beyond cryptocurrency. Ann. Emerg. Technol. Comput. **2**(1), 1–6 (2018)
4. Sun, J., Yan, J., Zhang, K.Z.K.: Blockchain-based sharing services: what blockchain technology can contribute to smart cities. Financ. Innov. **2**(1), 1–9 (2016). https://doi.org/10.1186/s40854-016-0040-y
5. Beck, R., Avital, M., Rossi, M., Thatcher, J.B.: Blockchain technology in business and information systems research. Bus. Inf. Syst. Eng. **59**(6), 381–384 (2017). https://doi.org/10.1007/s12599-017-0505-1
6. Sun, H., Luo, H., Sun, Y., et al.: Reversible Image Mosaic Technology Providing Large Protection Size and High Image Quality. **42**(3-4), 259–293 (2018)
7. Zhang, Y., Luo, X., Luo, H.: A multi-step attack-correlation method with privacy protection. J. Commun. Inf. Netw. **1**(4), 133–142 (2016). https://doi.org/10.1007/BF03391586
8. Cao, J., Yu, P., Xiang, X., et al.: Anti-quantum fast authentication and data transmission scheme for massive devices in 5g NB-IOT system. IEEE Internet Things J. **6**(6), 9794–9805 (2019)
9. Zou, Y., He, W., Zhang, L., Ni, J., Chen, Q.: Research on privacy protection of large-scale network data aggregation process. Int. J. Wireless Inf. Netw. **26**(3), 193–200 (2019). https://doi.org/10.1007/s10776-019-00443-w
10. Mao, B., Kawamoto, Y., Kato, N.: AI-based joint optimization of qos and security for 6g energy harvesting internet of things. IEEE Internet Things J. **7**(8), 7032–7042 (2020)

Analyses on the Monitoring Technology of CNC Machiner Under Visual System

Gan Lu[(✉)]

Chuzhou Vocational and Technical College, Chuzhou 239000, Anhui, China
lugan@chzc.edu.cn

Abstract. With the development of industrial technology in China, China's machinery manufacturing technology is developing in the direction of digitalization and automation, which plays an important role in China's manufacturing industry. Numerically-controlled machine tools can develop certain mapping relations between workpiece and machine coordinate system by measurement of computer vision system and identifying the position of workpiece in the airport and then the movement trajectory of the cutters can be concluded for processing of the workpiece. Firstly, the paper discusses the basic overview of the visual system CNC parts processing and detection technology, and then the composition of the visual inspection system of CNC parts processing, and the specific application of the visual system of CNC parts processing and monitoring technology are discussed, hoping to improve the processing technology of CNC parts in China to provide some reference.

Keywords: Visual system · CNC parts · Detection technology · Composition

1 Introduction

From 2010 to 2012, 5.038 million vehicles were sold, accounting for 45% of the country's total passenger vehicle sales. In the process of processing and manufacturing of product parts, the geometric size and quality of parts determine one of the key factors of whether industrial equipment is qualified or not. Therefore, detection technology has become one of the foundations of modern manufacturing industry and is the key element to ensure product quality [1]. Therefore, in the process of NC manufacturing of auto parts, it is necessary to strengthen the measurement of parts. This paper mainly discusses the non-contact measurement of NC parts in processing.

2 The Characteristics of Visual System Cnc Parts Processing and Testing Technology

In the processing of auto parts in China, the traditional measurement, also known as contact measurement, was mainly conducted with the use of calipers, micrometers, gauges, CMM and other tools to directly and contact the parts. The traditional calipers and micrometers are simple to use for measurement with fast speed, but in the process of measurement, the measurement results vary because of the different hand force of the person for measurement with low accuracy of the measurement results. In the process of metric measurement, specific measuring equipment needs to be selected, and workpiece detection needs to be carried out in a specific environment, which does not meet the development needs of modern industry, so it has been eliminated by many automobile manufacturers [2].

Therefore, at present, non-contact measurement is used in the processing of auto parts. Laser on-line measurement, ultrasonic on-line measurement and other technologies are used in the application of this technology. Machine vision measurement is a kind of contact measurement, visual system is capable of retracting intuitive images of workpiece on the numerical control machine by using computer software and automatically conducting processing analysis through the combination of image processing technology, image recognition technology, neural network detection technology and artificial intelligence technology, so the technology in the application has the characteristics of convenience and fast measurement speed, which, at the same time, has gradually become a new kind of detection technology [3]. The application of this technology can help CNC production tools to determine their own path and trajectory, so as to control CNC machine tools to process parts that meet the requirements of commercial standards, thereby improving the working efficiency of the machine tools, and reducing the production cost and product development cycle in workpiece production [4].

3   The Composition of Visual Monitoring System in NC Parts Machining

3.1   CCD Cameras in the Visual System

Fig. 1. CCD camera

As shown in Fig. 1, CCD camera is mainly an image sensor composed of charge-coupled devices, which are rich in a large number of highly sensitive semiconductor materials, so as to realize the orderly transformation between light and charge. In the process of application, CCD camera can convert light into digital signal through the system of AD converter chip and then the signal data is transmitted to the camera's internal flash memory card through the compression of digital signal. Later on, the signal values are transmitted to computer by cable or wireless transmission. The computer can process the images taken purposefully according to the requirements of NC machine tool on parts processing. At the same time, the CCD camera used in the visual system also has millions of photosensitive units. As long as these photosensitive units are illuminated by light in the work, they will automatically form charge reaction, and then the charge reaction will be intuitively displayed on the picture components.

CCD camera can form a relatively complete picture by concentrating the electrical signals formed by all the photosensitive units in one. The picture is also known as the original picture. After the computer processes the picture, the contour image of the workpiece can be presented. In the process of measuring the workpiece with the visual system, the actual length or diameter of the measured workpiece can be obtained only by measuring the pixel number contained in the workpiece side length or diameter in the image, and then dividing the pixel number by the constant value of the measurement ratio between the two [5].

3.2 The Lighting System in the Visual System

The vision system used in the monitoring technology of CNC parts processing is slightly different from the working principle of other sensors. The system relies heavily on light, so the system can be used with good lighting conditions in the working process, so as to ensure that the system produces a clear imaging and reduce the complexity of the system imaging work. Therefore lighting system in the visual system can be directly the main factors influencing the measurement precision of workpiece. In the measurement of the work piece, the selection and matching of the light source and lighting scheme should be made with obvious differences as far as possible, thus highlighting the contrast between the workpiece and the background. At the same time, it is necessary to ensure that the workpiece in the imaging process will not affect the quality of the lighting system on the imaging itself.

Therefore, the vision system should follow the principle of reflected light and transmission light in the process of imaging with full consideration in the relative combination between the optical lens and the light source. In addition, the optical lens, illumination brightness, spectral characteristics, uniformity and other aspects of the lighting system need to be adjusted according to the reality. Therefore, in the process of workpiece image acquisition, different light source forms need to be designed and selected according to the specific shape of the workpiece to ensure that the object can achieve the best imaging effect [6].

3.3 Image Acquisition Card in the Visual System

Image acquisition card in the visual system in the workpiece measurement process mainly assumes the role of the bridge and the data information can be converted into digital information that the computer can process by this device, so the image acquisition card production and research and development is a high content of computer technology research field. The selection of image acquisition card should be consistent with the selection of CCD camera. The image acquisition card usually has the following modules, including image information reception and A-D transformation module, bus interface module, camera control input and output interface module, etc. [7].

4 The Realization of CNC Parts Processing Monitoring Technology Under the Vision System

4.1 Noise Removal in Digital Images

In the process of machining parts by CNC machine tools, the original images obtained by the imaging equipment in the visual system cannot be directly used due to the mechanical vibration, positive and negative charge collision, human operation interference. Therefore, after taking the original picture, it is necessary to use special equipment to de-noise the original picture. In the processing process, it is necessary to carry out feature extraction, noise smoothing and filtering, geometric correction and other work on the target object in the picture.

In the experiment, it is found that there are a lot of noise in the original image obtained by the CCD camera. These noises mainly come from the chaotic signal value of dark current generated by the electric signal of the CCD camera under the condition of no light signal. The generated signal value will collide with the light signal received by the photosensitive area, which will cause some particle noise in the original image. Under the integral effect of the device taken by the original image, the shotcrete noise will form dark signal images in the photosensitive components. These dark signal images will cause the field of view of the image in the original imaging system to be limited, resulting in the blur of the edge of the target object in the image. Therefore, in the process of machining and monitoring CNC parts, filtering and denoising of the original images collected by the visual system is needed to meet the measurement requirements of the computer software in the later visual system for the target objects in the images [8].

4.2 Processing of Digital Image Workpiece

In the process of processing the digital image of the workpiece, the main purpose is to improve the quality of the original image pixel, so that a better sensory effect can be developed for operators. Vision system is required to restore and enhance the image with poor qualit. During the process, the automatic identification software in a computer system will automatically identify the objects in the images taken by the camera. Based on the target object pixel outline confirmed and the image enhancement, decoding and encoding, compression, reconstruction for the target objects in the images, the workpiece image preprocessing is finished.

Then the image is further processed. The method used in this link is to use median filter for nonlinear signal processing. Median filter in the process of handling will automatically develop sliding window with technical point, and the grey value of sliding window will replace center grey value in the process of signal processing. The median filter serves as the moving window with movement from left to right, from top to bottom. Bydoing so, the pixel module with current pixel as the center will be generated. In general, the use of median filtering will not affect the operation process of step function and slope function in the process of image processing, and can also effectively eliminate the double-half impulse and monopulse interference generated in the process of digital image processing [2].

4.3 Segmentation of Digital Parts Image

The segmentation of image of NC parts extracted by visual system, will involve the process of processing image target feature and target parameters automatically, this will make the original image change for digital image structure is more compact and abstract, so CNC parts work of image segmentation is always the focus of CNC parts processing monitoring technology.

In the process of segmentation of CNC parts, and according to the principle of image segmentation combined with the definition, the concept of the collection, so can make the whole area into several sub region is not empty, image segmentation and each subregion are overlapping, the pixels in each area have different features, in addition set segmentation criteria should also applies to all areas and image. In the process of image segmentation, the threshold method is used for image gray segmentation. A certain gray value threshold can be determined with the help of gray value, and regions with gray value less than and gray value greater than can be classified with the help of gray value threshold [9].

4.4 Edge Detection of NC Parts Image

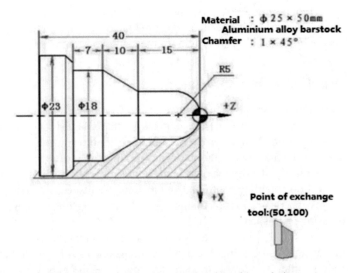

Fig. 2. Part specifications in edge detection techniques

Borderline for CNC parts image is the key to NC parts processing detection technology and the image of the object boundary detection can be conducted according to the image pixels and object edge state of adjacent area to determine whether the object pixels in the boundary of the object above, then all pixel points with characteristics are marked, and set up the edge points.

The edge points of CNC parts can be outlined by edge lines, and then the basic outline of CNC parts can be formed. In the process of dividing the edge of CNC parts, two types of roof shape and step shape can be used, and the computer data processing system can depict the edge points according to the derivative of gray value. As shown in Fig. 2, the edge detection technology is the main link, the CNC parts in image processing in the process can be extracted according to the edge of the line of the target and background region boundary defined characteristics, according to the different properties of boundary divided by feature area, you can judge each regional characteristics of grey value and texture characteristics, to achieve accurate detection for the CNC parts, then USES the method of cutting blades.

5 Conclusions

CNC machine tools can use the visual system to complete the automatic monitoring of the workpiece processing process in the process of processing the workpiece. In the application process, visual system can significantly improve the processing accuracy of CNC machine tools on the workpiece, and reduce the human resources in the processing of the workpiece. Therefore, the current research on the visual system monitoring technology has a significant practical significance for the CNC workpiece processing, and is of great help to promote the development process of modern industrialization in China.

Acknowledgements. This paper was sponsored and supported by the 2019 key research project of the college, -- Research on On-line Measurement of CNC Parts Machining Based on Vision System (NO: YJZ-2018-15).

References

1. Li, T., Liu, N.: Study on circle locating technology based on machine vision %.research on machine vision based circle positioning technology. Comput. Eng. Appl. **048**(009), 153–156 (2019)
2. Fan, X., Lu, D., Wang, J.: Research and development of NC system for coiling plate based on machine vision. Mach. Electron. **000**(005), 49–51 (2008)
3. Qin, B., Wang, H., Sun, C., et al.: Research on application of open CNC system in laser welding machine. Mach. Des. Manuf. **000**(002), 94–96 (2011)
4. Jiang, P., Yang, X., Xi, J.: Research and system development of conductive slip ring vision measurement technology in machine. Mod. Mach. Tool Autom. Manufact. Tech. **000**(003), 74–76, 79 (2017)
5. Wang, B., Ru, H., Wang, Y., et al.: Development and application of laser machining machine tool for complex shaped hole based on linuxcnc system and ethercat bus. Mod. Mach. Tool Autom. Manuf. Tech. **000**(012), 10–14 (2019)

6. Liu, M., Huang, Y., Zhang, D.: Innovative design of human-machine interaction system for CNC machine tools based on performance incentive mechanism. China Mech. Eng. **30**(21), 2554–2559 (2019)
7. Dong, F., Zheng, Z., Yuan, S., et al.: Research and development of conductive sliding ring surface micro-groove position detection system based on machine vision. Manuf. Technol. Mach. Tool **670**(04), 138–142 (2018)
8. Wu, F., Shi, F.: Research and development of automatic NC programming system based on image data. China Mech. Eng. **2015**(17), 1541–1545 (2015)
9. Mao, F., Mo, J., Wang, Q., et al.: Machine tool self-locating based on machine vision in die free progressive forming equipment. Forg. Stamp. Technol. **06**, 33–36 (2015)

Athlete's Temperature Characteristics Based on Artificial Intelligence Perspective Adapt to Infrared Thermal Image

Wumei Li[✉]

Department of Physical Education, Shandong Management University, Jinan, Shandong, China

Abstract. The application of artificial intelligence technology in the field of sports is becoming more and more matured. Because of this, the body load of athletes is getting heavier. In order to further understand the characteristics of acupoints and the mechanism of the influence of sports training on the acupoints of the human body, the research content of this article is the combination of artificial intelligence technology and sports training adapted to the analysis of the temperature characteristics of acupoints by infrared thermal imaging cameras. This article uses the characteristics of acupoint temperature and infrared thermal imaging camera to analyze the acupoint temperature. The results show that the average temperature of Neiguan and Xinshu points is higher, while the average temperature of acupoints is lower. The temperature of Xinyu is lower than that of Neiguan, but the temperature of acupuncture points is still higher than that of non-acupuncture points, and the high temperature of acupuncture points is very consistent in different acupoint regions. It can be seen that infrared thermal imaging technology can detect the thermal radiation generated by the human body through a unique imaging technology, and determine whether the distribution of the human body's heat source is normal. Through the chromatographic technique from outside to inside to analyze the characteristics of human acupoint temperature.

Keywords: Artificial intelligence · Sports training · Infrared thermal imaging

1 Introduction

The goal of sports training is to pursue the limits of the human body. Sports athletes strengthen their physical fitness and improve their speed and strength through day-to-day training. In addition, thanks to the development of science, athletes can use high-tech equipment to monitor the condition of the body and improve training efficiency.

Although relevant researchers at home and abroad have made many theoretical contributions in this field since the invention of the thermal imaging camera, they have been used less in clinical practice and less in sports.However, with the development of infrared thermal imaging technology, more and more fields can use this technology to assist diagnosis [1–3]. The thermal imaging camera has the advantages of zero radiation and low cost, and is extremely suitable for clinical and sports teams. When the human body is in a normal resting state, the external temperature of the human body is always

J. Abawajy et al. (Eds.): ATCI 2021, AISC 1398, pp. 541–547, 2021.
https://doi.org/10.1007/978-3-030-79200-8_81

about the steady state of the body [4]. However, when the human body is in an abnormal state, the temperature outside the human body will be different. The principle of infrared thermal imaging applied to the field of human motion is to detect infrared radiation, so as to reflect the change of human body surface temperature [5]. The image reflects the temperature difference by the color difference to show the temperature level, thereby judging the quality of the human body environment [6]. During or after exercise, due to muscle activity, the body's metabolic rate increases, resulting in an increase in body temperature [7]. During the test, the subject should remain quiet for a certain period of time before the temperature measurement. Because when people are in different environmental temperatures, or different emotional states, different physical states, these conditions will cause differences in body temperature.

Lan observed obvious infrared radiation trajectories along the central part of the back of the GV in healthy volunteers, and the skin temperature of GV 4 and its bilateral control points increased significantly (P < 0.001). The skin temperature of GV 4 was significantly higher than the left and right control points (P < 0.001). After moxibustion stimulation, the average skin temperature of GV 9, GV 7, GV 4 and GV 3 was increased (P < 0.01) compared with before moxibustion. Lan's research shows that moxibustion Mingmen (GV 4) can induce infrared radiation trajectories along the central part of the back covering GV and increase the skin temperature of GV points [8]. Martello also evaluated the IRT, and Martello fed 144 Nellore steers with 70-day high-concentration diets to evaluate the performance of the farm. Martello inspected nine animals classified as high RFI and measured the rectal temperature (RT), respiratory rate (RF), and IRT of the anterior, eye, eye area, cheek, hind abdomen, ribs, buttocks, and anterior portion. Chudecka determined the relationship between AN women 's body mass index, body composition (especially subcutaneous and VFM), and average surface temperature (Tmean) values. Nabenishi used four non-pregnant adult ewes in the experiment. The BST, vaginal temperature (VT) and ambient temperature (AT) of the upper neck were measured every 10 min for 7 days, and correlation analysis was performed [9, 10].

2   Research Objects, Models and Steps

2.1   Test Subject

A medical school animal experiment center provides forty-five eight-week-old healthy adult white rabbits weighing between 210 g and 230 g. It is divided into cages according to national standards, with eight animals per cage, free drinking water, and solid mixed feed for rabbits. The room temperature is between 19 and 25 °C. The light changes with nature, and all animals have not moved before the experiment. After seven days of conventional adaptive feeding, the rabbit adapted to jogging at a speed of ten meters per minute on an adjustable treadmill, with a jogging distance of no more than one hundred meters for a total of two days. Two days later, run at a speed of 10 m per minute for three minutes, at a speed of 15 m per minute for three minutes, and then rest for two to three days to screen out rabbits that are not well-adjusted during this time. Randomly divided into five groups with eight animals in each group. There was no significant difference in the weight of animals in each group.

2.2 Experimental Models and Procedures

All three groups of animals performed continuous horizontal running, and the exercise intensity was determined according to the maximum oxygen uptake of Bedford. Exercise intensity is 60% to 70% of the maximum oxygen uptake, which is moderate intensity exercise. Training once a day, 8 am to 12 am; the slope is zero degrees, training four weeks a week, every Friday days, with 15m / rain, 22m / rain, 27m / rain, 31m / rain increasing speed, time to exercise to fatigue During the exercise, use a brush to stimulate the tail of the rabbit, and the rabbit keeps moving. If someone can't exercise continuously, you can rest for two to four minutes in the middle, but the total rest time is no more than ten minutes. The main test equipment of this article is a ZR-2010 medical infrared camera produced by a medical technology company. Connect the equipment to the computer, and keep the distance between the exercise group and the control group at five meters from the lens of the ZR-2010 medical infrared camera, keep the indoor relative air still, without direct sunlight and direct light, and keep the indoor temperature at two About 16 degrees Celsius; In addition, test the animals separately, wait for ten to fifteen minutes before the body surface temperature and room temperature reach equilibrium, test; during the test and waiting period, maintain emotional stability and avoid contact with the body surface to avoid impact Test results: Start the computer, click to run the software program, collect infrared thermal images of the acupuncture points of the exercise group and the control group, and use the analysis system that comes with the instrument to measure the infrared temperature of each acupoint of the exercise group and the control group under the same conditions.

3 Analysis of Acupoint Temperature Characteristics

3.1 Analysis of the Average Body Weight and Exercise of Each Group of Animals During the Experiment

The changes in rabbit body weight are shown in Table 1.

Table 1. Changes in rat body weight

Week	A	B	C	D	E
Before the experiment	218.29 ± 31.85	223.58 ± 31.23	229.33 ± 24.56	210.35 ± 14.89	226.51 ± 12.05
1	230.73 ± 29.15	226.24 ± 31.45	221.33 ± 21.74	201.13 ± 15.24	219.35 ± 13.24
2	247.34 ± 30.13	237.46 ± 24.56	219.41 ± 24.52	210.55 ± 13.41	225.42 ± 11.45
3	251.66 ± 20.37	249.95 ± 23.71	231.56 ± 22.46	220.73 ± 14.06	240.55 ± 14.51
4	268.91 ± 24.23	262.45 ± 23.73	237.32 ± 21.93	228.61 ± 11.83	252.45 ± 13.72

During the experiment, the weight of rabbits in each group was recorded. It can be seen from Table 1 that after four weeks of feeding, the weight of the rats in the exercise group and the non-exercise group in the first week decreased, but there was no significant difference (P > 0.05); the weight of the rats in the quiet control group and the quiet sham acupuncture group increased (P > 0.05); In the second week, the weight of the rats in the exercise group and the sham acupuncture group continued to decrease (P > 0.05), the weight of the remaining three groups of rats increased, while the weight of the rats in the quiet control group increased significantly (P > 0.05); At the third week, the weight of rats in each group showed an upward trend, and there was no significant difference between the two groups; at the fourth week, the weight of all rats increased, and there was a significant difference between the non-exercise group and the exercise group, and the rest group after exercise (P < 0.05); the difference between the quiet control group and the non-exercise group was significant (P < 0.05), and there was a significant difference between the rest group after exercise (P < 0.05). The impact of the rabbit's exhaustion time on the treadmill is shown in Fig. 1.

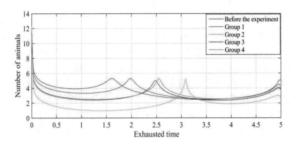

Fig. 1. The effect of the rabbit running on the treadmill until it is exhausted

It can be seen from Fig. 1 that there is no significant difference between the quiet group and the exercise control group in running to exhaustion time (P > 0.05); compared with the exercise group and the post-exercise rest group, there is a significant difference in running to exhaustion time (P < 0.05); Compared with the quiet control group, the exercise group compared with the control group and non-exercise group, the exercise group and the rest group from running to exhaustion time was also significantly extended (P < 0.05), the difference is significant.

3.2   Single Larget Tracking Experiment Result and Analysis

At present, the widely used method for evaluating the performance index of the target tracking algorithm is to evaluate it through multiple quantitative indicators. In this paper, the following two indicators will be used to evaluate the performance of the tracking algorithm: tracking accuracy and center error. An infrared image sequence of a group of rabbit targets in motion and a group of resting rabbit targets is selected as the experimental object, and compared with the classical particle filter tracking algorithm and mean-shift tracking algorithm. The tracking results of different coordinate targets in this method are shown in Fig. 2.

It can be seen that due to the randomness and unevenness of the particle distribution, the tracking result deviates from the center position and fluctuates greatly, and the center error is large. However, the tracking center of the algorithm is always near the target

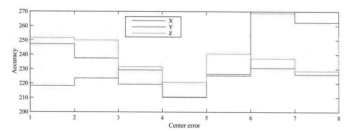

Fig. 2. Target tracking results with different coordinates

center. The X-direction deviation, Y-direction deviation and center error are small, the accuracy is high, the tracking result is more accurate, the entire tracking process is relatively stable, and there is no obvious deviation from the target.

3.3 Comparative Analysis of the Average Temperature Value of Acupoints and Non-acupoints

In 45 rabbits, the comparison of the temperature values of the different hole areas and the control points of the open side is shown in Table 2.

Table 2. Comparison of temperature values between different cave areas and side opening control points

Group	Acupoint area	Caveside area
Right Neiguan	36.43 ± 0.64	35.18 ± 0.71
Left Neiguan	36.47 ± 0.55	35.31 ± 0.73
Right heart	36.23 ± 0.42	35.43 ± 0.35

The average temperature in the right Neiguan acupoint area is (36.42 ± 0.65) °C, the temperature near the acupoint is (35.18 ± 0.73) °C; the average temperature in the left Neiguan acupoint area is (36.47 ± 0.35) °C, and the temperature near the acupoint is (35.31 ± 0.72) °C; The average temperature in the right Xinyu acupoint area is (36.22 ± 0.41) °C, and the side point is (35.42 ± 0.35) °C. The experimental results show that the average temperature of the acupoint area is higher than that of the non-acupoint area, and the temperature of each acupoint area is statistically different from the temperature of the side opening door. Figure 3 shows the comparison results of the temperature values of the different cave areas and the side-open control points.

Under constant internal and external environmental conditions, the skin temperature is mainly determined by the skin and subcutaneous tissue structure. The main factors affecting skin temperature are the distribution of blood vessels, the size of blood flow and the metabolic state of skin tissue. The results show that the average temperatures of the two points in Neiguan and Shinjuku are higher than the control points. Although the temperature of Xinyu is lower than that of Neiguan, it is still higher than that of

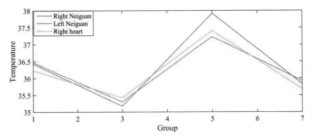

Fig. 3. Comparison results of temperature values between different cave areas and side opening control points

non-acupuncture points. The high temperature of acupuncture points is very consistent in different acupoints.

3.4 Comparison of Infrared Temperature Difference Between Acupoints Before and After Exercise in Control Group

Comparison of infrared temperature difference of acupuncture points before and after exercise in the control group is shown in Fig. 4.

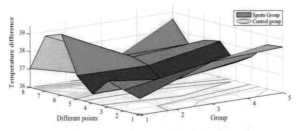

Fig. 4. Comparison of infrared temperature difference between acupoints before and after exercise in control group

No matter in the exercise group or the control group, the infrared thermal image of the two groups of rabbits before exercise generally showed that the acupoint temperature was lower, the temperature near the acupoint was higher, and the temperature difference was larger. After exercise, the temperature difference of the acupoints of most rabbits has narrowed to varying degrees, and even the acupoint temperature is higher than the temperature near the acupoint. Of course, there are some rabbits whose acupoint temperature has improved after exercise, but there is no obvious change in acupoint temperature difference before and after exercise. It can be seen that infrared thermal imaging technology can very sensitively receive the thermal radiation generated by the human body (less than 0.05 degrees Celsius), and through special imaging technology and surface to internal chromatographic technology, determine whether the distribution of human heat sources is normal. Doctors can use infrared thermal imaging technology to analyze the shape, trend, and thermal difference of abnormal heat sources to understand the overall state of the subject.

4 Conclusions

Exercise, disease and other factors have an effect on the local blood flow and tissue metabolic rate, which can lead to abnormal body temperature. Surface skin temperature reflects autonomic nerve function, tissue thermal conductivity (fat thermal conductivity is poor) and physiological compression. Therefore, when human organs or tissues are affected, the temperature of the skin surface will abnormally increase or decrease. When any of the above factors are abnormal, it will first be displayed on the body surface temperature or skin temperature. Because the thermal imaging camera is used to compare the body temperature with the body temperature, and the temperature difference statistical method is used, the difference between the subjects due to the difference in body temperature is eliminated.

There are still some deficiencies in this article. Due to the influence of the ambient temperature, the control of the laboratory environment is very important. The laboratory is affected by factors such as airflow, room temperature, and direct sunlight. When the tester performs the test, it must be ensured that the tester is in a completely quiet state, and the test can only be performed after the tester has stopped training for a period of time. The intensity of daily training is different, which may also lead to different physical conditions.

Acknowledgements. This work was supported by University-Industry Collaborative Education Program (NO.201902206043).

References

1. Jeličić, M., Uljević, O., Zenić, N.: Pulmonary function in prepubescent boys: the influence of passive smoking and sports training. Montenegrin J. Sports Sci. Med. **6**(1), 65–72 (2017)
2. Keogh, J.W.L., Winwood, P.W.: The epidemiology of injuries across the weight-training sports. Sports Med. **47**(3), 1–23 (2016)
3. Eren, U., Yilmaz, C.Y., Lu, K., et al.: A study on the sports training and instruction in turkey: an overview of higher education programs in the area of sports. World Appl. Sci. J. **34**(2), 250–255 (2016)
4. Liang, B., Liu, J.: Design and development of sports training system based on image processing technology. Clust. Comput. **22**(2), 3699–3706 (2018). https://doi.org/10.1007/s10586-018-2220-1
5. Kobelkova, I.V., Martinchik, A.N., Keshabyants, E.E., et al.: An analysis of the diet of members of the Russian national men's water polo team during the sports training camps. Vopr. Pitan. **88**(2), 50–57 (2019)
6. Hartmann, J.: Infrared thermal imaging. Phys. Unserer Zeit **49**(5), 255 (2018)
7. Sousa, E., Vardasca, R., Sérgio, T., et al.: A review on the application of medical infrared thermal imaging in hands. Infrared Phys. Technol. **85**(2017), 315–323 (2017)
8. Yining, Z., Haochun, Z., Rui, M., et al.: Evaluation of infrared thermal imaging system detection distance in different cloud and rain conditions. J. Appl. Opt. **37**(2), 288–296 (2016)
9. Liu, X.L., Fu, B.R., Xu, L.W., et al.: Automatic assessment of facial nerve function based on infrared thermal imaging. Guang pu xue yu guang pu fen xi = Guang pu **36**(5), 1445–1450 (2016)
10. Cong, H., Jin, H.: Research on undetected overheat fault of the gis bus bar contacts based on infrared thermal imaging. J. Electr. Eng. Technol. **14**(2), 839–848 (2019). https://doi.org/10.1007/s42835-018-00054-0

Artificial Intelligence-Based Tennis Match Technique and Tactics Evaluation System

Shanshan Yin[✉]

Department of Physical Education, Shandong Management University, Jinan, Shandong, China

Abstract. The technical and tactical evaluation system of tennis matches has important reference value for improving the competitive ability of tennis players. With the continuous advancement of science and technology, artificial intelligence technology is becoming more and more mature. This article explores the use of artificial intelligence methods to diagnose and evaluate tennis matches through the use of literature research methods, software engineering research methods, and mathematical modeling research methods, to scientifically evaluate the technical and tactical level of tennis players, to provide scientific and technological support for coaches and athletes to arrange training reasonably and effectively regulate the use of game skills and tactics.

Keywords: Artificial intelligence · Tennis match · Evaluation

1 Introduction

The tennis event started late in China, but the sports results achieved are very significant, achieving a substantial breakthrough in the Olympic gold medal strategy. Conducting relevant research on tennis matches and summarizing the experience of training and preparation are of very important reference significance not only for the project itself, but also for the development of all confrontational events.

Whether a tennis player can win in the game is restricted by multiple factors, such as physical, psychological, technical and tactical, and environment, etc. However, due to the complexity of the tennis game's technical and tactical system and the flexibility of tactics, the quality of its performance is determined In the "Research and Implementation of the Winning Factors of the Net-to-Net Confrontation Project" study, it is listed as the main factor for winning the game. In the national tennis team's usual training and previous preparations, it is also regarded as the most important the content of [1]. The research on tennis match techniques and tactics has realized the transformation from qualitative to quantitative and then to the study of the combination of the two. The methods have also evolved from relying solely on manual statistics in the past to relying on modern calculation methods and computer technology assistance. Computer simulation diagnosis of ball games provides new research ideas for technical and tactical diagnosis of tennis games. Some studies try to combine artificial neural network methods with computer simulation diagnosis of ball games. However, the generality of technical and tactical indicators restricts its training, and the application in game practice [2].

© The Author(s), under exclusive license to Springer Nature Switzerland AG 2021
J. Abawajy et al. (Eds.): ATCI 2021, AISC 1398, pp. 548–555, 2021.
https://doi.org/10.1007/978-3-030-79200-8_82

As an advanced machine learning method, artificial intelligence methods have been widely used in data mining, decision support and other fields. Their advantages in game technology and tactics diagnosis have begun to emerge, but there are few integrated system studies of diagnosis, evaluation and prediction. This article looks forward to an in-depth discussion on this research direction in order to provide decision support for tennis matches.

2 Research Ideas on the Evaluation System of Tennis Match Techniques

The research of tennis match technique and tactics diagnosis develops abstract and concrete in two directions. The former research is to quickly, concisely and efficiently evaluate the game. The requirement is to select the index system as simple as possible. However, this kind of research requires concise indicators and also results in the evaluation system providing less information to decision makers. The latter's research objectives are just the opposite. In order to describe the athlete's various technical performances in more detail, the evaluation system of this type of research is more specific. However, due to too many indicators of this type of research, it becomes difficult for decision makers to find the key information they need. Complexity and difficulty greatly reduce the effectiveness of tennis match evaluation. Trying to find a balance between abstract and concrete so that it has a higher guiding significance for theoretical research and game practice is the focus and difficulty of ball technical and tactical research. Related research shows that it is easy to overlook the relationship between indicators for the composition of the technical and tactical diagnosis system. Therefore, there are more studies from various aspects, but less research from the perspective of overall connections. This research attempts to combine the more important techniques in tennis matches for research, that is, to integrate technique, tactical behavior, landing and rotation to establish a diagnostic system [3, 4]. Due to the poor accuracy of video observation of tennis rotation, the evaluation results of rotation are not easy to be objective, so the diagnostic system only starts the research of the first three items. That is, the technical evaluation model is established by the index system that integrates the technology and the impact point to diagnose the technical characteristics of tennis players, and the tactical behavior and the timing of the shot are combined to establish the tactical evaluation model to study the tactical characteristics of tennis players.

3 The Connotation and Principles of Artificial Intelligence-Based Tennis Match Technical and Tactical Evaluation

Tennis game technical and tactical research mainly focuses on diagnosis and evaluation. The former mainly analyzes and diagnoses the various tactical characteristics of athletes in each game, such as advantages and disadvantages, through game observation and mathematical statistics, for training and formulating game strategies tactics provide reference while the latter is to evaluate the comprehensive strength of athletes' techniques and tactics, and to make certain predictions of athletes' competitive state. The significance of tennis game technical and tactical evaluation lies in judging the training effect by analyzing the competitive strength of both sides, setting the starting point for formulating training plans, providing reference for the formation of the competition, and ranking the strength of players.

There are three-stage index evaluation method and ten-index evaluation method that are widely used in the research of tennis match technical and tactical evaluation. Most of the later studies are the development and continuation of the former two, especially for the application and extension of the three-stage index evaluation method [5]. The two methods are mainly used in the technical and tactical evaluation of offensive players. The basic idea of tennis technical and tactical evaluation has been discussed above, and the core of the research is to reasonably determine the technical and tactical index system and its weight. The above research shows that the proposed technical system and tactical system can describe the technical and tactical level and strength of the tennis game in a more comprehensive way when performing technical and tactical diagnosis of tennis matches. The competitive efficiency values of various technical and tactical indicators have a high degree of distinction between the technical and tactical characteristics of different players, and can be used in the evaluation of technical and tactics, that is, the competitive efficiency value is used to replace the weight of the technical and tactical indicators. It should be pointed out that the above method is only suitable for the individual assessment of tennis players' skills and tactics, and a comparative analysis of a certain player's own game is used. To evaluate and compare the technical and tactical levels of different tennis players, two methods can be used: First, a model that is applicable to all players needs to be established. Second, the basis of technical and tactical diagnosis is based on the competitive efficiency value. Sort the indicators and select the most important indicators to sum to determine the evaluation value. Due to the large differences in the technical and tactical characteristics of different players, and the large disparity in technical and tactical levels, it is very difficult to establish a technical and tactical model that is applicable to all players. The second method is based on the results of technical and tactical diagnosis, and comparison after sorting is also in line with the principle of technical and tactical evaluation. In summary, the technical and tactical evaluation of tennis players is carried out in two situations. In the technical and tactical evaluation of the game, the scoring rate of each technical and tactical index is used as the evaluation index, and the utilization rate of each technical and tactical index is used as the evaluation index reference indicator.

4   Application of Artificial Intelligence-Based Tennis Match Technique and Tactics Evaluation System

4.1   Evaluation System Based on Artificial Intelligence

Artificial Intelligence is a technology that simulates the neurons of the human brain for abstract processing and realizes information processing by establishing models [6]. With the continuous deepening of neural network research, it has become more and more widely used in the fields of intelligent machines, intelligent recognition, intelligent prediction and automatic control. Real-time recognition of human body data based on neural networks using mobile sensors can provide data support for intelligent evaluation systems to make scientific decisions [7, 8].

Radio frequency identification technology (RFID) is a kind of communication technology, also called radio frequency identification or electronic tag [9]. It recognizes specific targets and reads data mainly through radio signals. Commonly used technologies mainly include passive, low frequency, ultra-high frequency and high frequency. This technology is mainly used for data function positioning in tennis training. As well as athletes' registration [10]. The sensor is a kind of signal conversion device, which is mainly composed of a conversion original and a sensitive element. The sensor has the function of sensing external signals and detecting, including heat, light and humidity, and at the same time transmitting the acquired signals to other organs and devices.

4.2 Decision Tree Classification Algorithm

The human motion data collected by sensors is based on the reference system of each sensor. Since human movement cannot be described in accordance with the sensor's reference system, such source data cannot be used directly, so it is necessary to classify and process the data through related technologies. Decision tree algorithm is a common machine learning algorithm. When doing machine learning, the decision tree uses a tree structure to resolve the problem one by one. Decision trees make selection judgments at each node, which is similar to the process of human selection when facing some decision problems.

Information entropy, information gain, conditional entropy, and information gain ratio are commonly used in decision trees as the basis for division. Information entropy indicates whether the sample subset is single, that is, the smaller the value, the single the type of data in the set, which means that the division is more ideal. Information entropy is defined as:

$$Eet(D) = -\sum_{k=1}^{|v|} p_k \log_2 p_k \tag{1}$$

Assuming that there are V possible values for the discrete attribute, a the information gain can be calculated according to the weights assigned to the branch nodes according to the number of samples:

$$Gain\,(D, a) = Ent(D) - \sum_{v=1}^{V} \frac{|D^V|}{|D|} Ent\left(D^V\right) \tag{2}$$

According to the definition of information gain, attributes that account for the majority of numbers have inherent advantages, and therefore make the division effect worse. The decision tree algorithm is different from the above, it uses the gain rate as the partition index to minimize the impact. Its definition is:

$$Gain\_ratio\,(D, a) = \frac{Gain\,(D, a)}{IV\,(a)} \tag{3}$$

among them:

$$IV(a) = -\sum_{v=1}^{V} \frac{|D^v|}{|D|} \log_2 \frac{|D^v|}{|D|} \tag{4}$$

This is called the "intrinsic value" of the attribute. At the same time, the decision tree is divided by the Gini index. The Gini value can represent the unity of data, and its definition satisfies:

$$Gini\_index(D, a) = \sum_{v=1}^{V} \frac{|D^v|}{|D|} Giin\left(D^v\right) \tag{5}$$

$$Gini\left(D, a\right) = \sum_{k=1}^{|v|} \sum_{k \neq k} p_k p_k = 1 - \sum_{k=1}^{|y|} p_k^2 \tag{6}$$

When the decision tree is divided, it will not only encounter discrete values, but also divide continuous values. When conducting experiments, sampling is often used to discretize continuous values for continuous variables. For continuous variables, discrete values are essentially processed. Therefore, dealing with the division of continuous variables is essentially dealing with the problem of a large number of possible discrete values of attributes. As samples are sampled, there will be some missing attributes in some samples. If only the samples with values on these attributes are divided, it will lead to a waste of sample sets.

Set each sample x to be given a weight w_x and define:

$$\rho = \frac{\sum x \in \tilde{D}^{wx}}{\sum x \in D^{wx}} \tag{7}$$

$$\tilde{P}_k = \frac{\sum x \in D_k^{wx}}{\sum x \in \tilde{D}^{wx}} (1 \leq k \leq |y|) \tag{8}$$

$$\tilde{r}_k = \frac{\sum x \in \tilde{D}_v^{wx}}{\sum x \in \tilde{D}^{wx}} (1 \leq v \leq |V|) \tag{9}$$

According to formula (7) to formula (9), the promotion of information gain can be defined to satisfy:

$$Gain\left(D, a\right) = \rho \times Gain(\tilde{D}, a) = \rho \times \left(Ent\left(\tilde{D}\right) - \sum_{v=1}^{V} \tilde{r}_v Ent\left(\tilde{D}^v\right)\right) \tag{10}$$

$$Ent(\tilde{D}) = -\sum_{k=1}^{|y|} \tilde{p}_k \log_2 \tilde{p}_k \tag{11}$$

Since the classification boundary formed by the decision tree is characterized by being parallel to the coordinate axis, when the boundary of the attribute value to be classified is more complicated, the decision tree must be divided into multiple segments due to the axis-parallel attribute to achieve a good fitting effect. It will increase the number of decision tree layers, which will undoubtedly lead to increased training and prediction costs. Therefore, if the function combination value of multiple attributes can be used for division, it will not only be parallel to the coordinate axis in the multi-dimensional division plane, which is to find a suitable linear classifier for the node.

4.3 Wavelet Transform

When identifying the specific pattern of data, it is necessary to understand the detailed characteristics of the data, such as the order and quantity of various frequencies, in order to better extract the signal characteristics and construct the recognition model. It is necessary to conduct a comprehensive time-frequency analysis of the signal, so it is necessary to introduce Wavelet analysis.

Wavelet analysis decomposes the signal through wavelet transform, and analyzes the signal according to the prominent local features of the signal by analyzing the local time domain or frequency domain. This solves the problem that the traditional fourier transform cannot be analyzed from the global The shortcomings of capturing the detailed characteristics of the signal, while solving the problem that the signal can not be adaptively changed with the change of the signal after the fixed window size decomposition.

Let the expression of a basic function be $\psi(t)$, let:

$$\psi_{ab}(t) = \frac{1}{\sqrt{a}}\psi\left(\frac{t-b}{a}\right) \tag{12}$$

Among them, a, b are constants, and $\psi_{ab}(t)$ is the basic function obtained through translation and expansion. Given a square-integrable signal $x(t) \in L^2(R)$, the wavelet transform of $x(t)$ is defined as:

$$ET_x(a, b) = \frac{1}{\sqrt{a}} \int_{-\infty}^{+\infty} x(t)\psi^*\left(\frac{t-b}{a}\right)dt \tag{13}$$

Among them, a, b represents the scale factor and time shift respectively, and $\psi(t)$ is the basic wavelet. Let the Fourier transform of $x(t)$ be $X(\Omega)$, then the frequency domain expression of wavelet transform is:

$$WT_x(a, b) = \frac{1}{2\pi}(X(\Omega), \psi_{a,b}(\Omega)) \tag{14}$$

A low-pass filter can filter out high-frequency noise to achieve denoising of human body motion data. However, this method cannot filter out the sensor jitter noise that is similar to the human body motion frequency, because the human body motion frequency is not fixed at a certain low frequency frequency, but has a different frequency in each motion. Figure 1 shows the steps of wavelet threshold denoising.

It can be seen from Fig. 1 that the basic steps are: wavelet transform the signal, use the threshold function to process each layer of wavelet separately, and get the signal after

Fig. 1. Wavelet threshold denoising steps

wavelet reconstruction to remove noise. In the denoising process, by using the threshold function, the noise below the threshold can be effectively filtered out. If the selected threshold is too large, part of the effective information of the signal will be filtered out, resulting in serious signal distortion; if the selected threshold is too small, the denoising will be incomplete. Therefore, the choice of the threshold is related to the quality of the filtering effect.

For the joint distribution of multi-dimensional independent normal random variables, when the data tends to infinite dimensionality, the optimal threshold is calculated according to the maximum and minimum limits of the estimated data. The constraints are:

$$\lambda = \sigma_k \sqrt{2 \ln k} \tag{15}$$

Where σ_k represents the standard deviation of noise, and k represents the signal length. ω_i represents the square value of the wavelet coefficient, and assuming that the risk vector is $\overrightarrow{R}$, the element r_i in it satisfies:

$$r_i = \left[n - 2i - (n - i)\omega_i + \sum_{k=1}^{i} \omega_k \right] / n, \ i = 1, 2, \cdots, n \tag{16}$$

On the whole, different threshold estimation methods have different denoising effects. Therefore, in actual application, it is necessary to consider the thresholds obtained by different threshold estimation methods to compare the signal denoising effects.

5   Conclusion

Based on the construction of diagnostic evaluation indicators for tennis match techniques and tactics, this article conducts a systematic study of the tennis match techniques and tactics evaluation system based on artificial intelligence methods, and conducts preliminary evaluation and prediction research on this basis. The online tennis match technique and tactics diagnosis method has been improved, the diagnosis result is clearer, the evaluation result is closer to the actual game, and the prediction result is more scientific, which can be conveniently used for the game technique and tactic diagnosis and decision support. However, this research can continue to explore in depth. According to the current technical and tactical state, predicting future trends is the direction of technical and tactical research. Other methods can be explored for tennis match prediction research. This article only discusses the preliminary technical methods in technical tactics. The application of the evaluation system can further explore the application of artificial intelligence methods in tennis match decision support.

References

1. Erlend, K., Jørgensen, Fossen T I, Ingrid Schjølberg, et al.: Experimental validation of attitude and rate-sensor bias filter using range-difference measurements. Control Eng. Pract. **73**(4), 112–123 (2018)
2. Li, X., Xiong, S., Lin, B.: Calculation of satellite's initial attitude using multi-sensor intercalibration. Guofang Keji Daxue Xuebao J. Nat. Univ. Def. Technol. **39**(1), 24–29 (2017)
3. Byeon, S.Y., Mok, S.H., Woo, H., et al.: Sensor-fault tolerant attitude determination using two-stage estimator. Adv. Space Res. **63**(11), 3632–3645 (2019)
4. Rezaei, M., Raissi, K., Mehne, H.H., et al.: Attitude determination sensor for low Earth orbit satellite based on Lorentz force. Proc. Inst. Mech. Eng. **233**(6), 2219–2230 (2019)
5. Xiong, K., Wei, C.: Multiple-model adaptive estimator for spacecraft attitude sensor calibration. Aircr. Eng. **89**(3), 457–467 (2017)
6. Shen, L., Hang, H., Zhao, J., et al.: The measuring technique of the spine shape based on attitude sensor. Chin. J. Sens. Actuators **31**(6), 841–846 (2018)
7. Zhang, Z.X.: Observation satellite attitude estimation using sensor measurement and image registration fusion. Proc. Inst. Mech. Eng. **232**(7), 1390–1402 (2018)
8. Gao, H., Wu, L., Yang, C.H., et al.: Sensor-based optimal attitude reorientation control scheme based on computational programming approach. Sens. Mater. **32**(5), 1671–1687 (2020)
9. Qiu, Z., Qin, G., Li, X.: Error compensation of star sensor imaging plane attitude detector. Nami Jishu yu Jingmi Gongcheng Nanotechnol. Precis. Eng. **15**(4), 280–285 (2017)
10. Wu, J., Zhou, Z., Li, R., et al.: Attitude determination using a single sensor observation: analytic quaternion solutions and property discussion (accepted). Ietence Measur. Technol. **11**(6), 731–739 (2017)

Digital Realization Technology of Qiannan Ethnic Pattern

Jun Yuan[✉]

The School of Computer and Information, Qiannan Normal University for Nationalities, Duyun, Guizhou, China

Abstract. There are many kinds of traditional patterns of ethnic minorities in Qiannan, with rich and colorful contents and forms. The simple and vivid ethnic patterns reflect the yearning of ethnic minorities for a better life. This paper first introduces the pattern generation technology based on pal, and then describes the pattern generation technology based on shape grammar. Shape grammar technology is a shape system that generates new patterns based on the original patterns according to the defined rules. The system can automatically generate new patterns according to the preset ideas and composition. The pattern generation technology based on pal is similar to shape grammar It is of great significance to the protection, development and utilization of national traditional pattern culture.

Keywords: Research on digital realization · Technology of qiannan ethnic · Pattern

1 Preface

Traditional patterns of ethnic minorities are created, designed and handed down by ethnic minorities in their own unique geographical and cultural environment according to the purpose of use and beautification, material, technology, economy and other conditions through artistic conception. They have distinctive national characteristics. Qiannan, located in the south of Guizhou Province, has 43 ethnic minorities, including Buyi, Miao and Shui. It is one of the 30 ethnic autonomous prefectures in China. All nationalities have created colorful national culture in their long-term production and life, which is an indispensable part of Chinese national folk culture. The traditional patterns of ethnic minorities in Qiannan include batik patterns, embroidery patterns, textile patterns and silver patterns of Miao, Buyi and Shui nationalities. The simple and vivid national patterns are not only the artistic precipitation of various nationalities' production and life for thousands of years, but also reflect their yearning for a better life. Today, with the rapid development of IT technology, how to use computer technology and multimedia technology to realize the digital generation and application of national patterns has a positive practical significance for the protection and inheritance of national culture.

J. Abawajy et al. (Eds.): ATCI 2021, AISC 1398, pp. 556–562, 2021.
https://doi.org/10.1007/978-3-030-79200-8_83

2   Pattern Generation Technology Based on PAL

Pattern assembling language Pal is a descriptive XML like pattern assembly language, which has the characteristics of simple syntax, flexible, easy to learn and use. The geometric assembly of patterns in this language is based on the common characteristics of traditional national patterns: patterns are composed of some basic parts (graphic elements), which can be mirror, zoom, rotate, ring exchange and other operations To a new pattern; [1] the composition structure of the pattern adapts to a variety of patterns, that is, changing the elements and retaining the geometric structure of the pattern can wait for a new pattern. Pattern assembly language is a language that describes the geometric structure of patterns according to specific grammar. The related sentences of the language can generate digital national patterns through the interpreter.

```
<PAL>::=<pattern parameter table><sentence table>< saving sector>
    <pattern parameter table>::={< pattern parameter define >}
        <         pattern        parameter        define        >::=define
        grid(number,number):|height :|width:
        < sentence table>::=<sentence>{< sentence >}
        <     sentence     >::=<pattern     generating     >:|<     pattern
        location>:|<assignment >
        <pattern generating >::=load<pattern_name into var>where <conditional
        expression>:|< transformation table>
        < transformation table>::=< geometric transformation >{< geometric
        transformation >}
        < geometric transformation >::=
        hzoom|vzoom|hmirror|vmirror|rotate|sheer
```

The above statements are part of pal, in which the pattern parameter table defines parameters such as the position and size of the pattern, and the sentence table defines the position and size of the pattern Table contains the generation, location and assignment statements of patterns, which are used to load patterns from the pattern library to a specific location, and can be assigned to a variable. Pattern transformation statements are used to enlarge, reduce, mirror and cut patterns. [2] The following is a simple example to illustrate how pal language generates digital ethnic patterns.

```
Pic_flower{
Load.patternset.flower.f1.petal to petal1;   // Loading petal petal primitives
from pattern library
Gen petal2 from petal1 vmirror;
Gen petal3 from petal1 hmirror; // From petal petal1, petal2 and petal3 are
generated and mirror operation is performed
Gen petal3 from petal3 rotate 35; // Petal petal3 was rotated
Load.patternset.flower.f1.leaf to leaf1; // Loading leaf f primitives from pattern
library
Gen leaf2 from leaf1 hzoom 1.5 vzoom 1.3;
Gen leaf3 from leaf1 vmirror; // Generate leaf2 and 3 from leaf1 and mirror
them
Petal1.x=20; Petal1.y=35;
Paint petal1 ; // Position the petal1 primitives and draw them on the canvas
Petal2.x=22; Petal1.y=35;
Paint petal2 ; // Position the petal2 primitives and draw them on the canvas
Petal3.x=24; Petal1.y=31;
Paint petal3 ; // Position the petal3 primitives and draw them on the canvas
leaf1.x=16; leaf 1.y=26;
Paint leaf 1 ; // Position the leaf1 and draw them on the canvas
leaf 2.x=36; leaf2.y=32;
Paint leaf 2 ; // Position the leaf2 and draw them on the canvas
leaf 3.x=18; leaf 3.y=28;
Paint leaf 3 ;} // Position the leaf3 and draw them on the canvas
```

The above is part of the sample code of pattern assembly language (PAL). Through the analysis of PAL statements, it can be seen that pal is a simple graphic operation tool language, which can realize geometric operations such as zooming, mirroring and rotation of patterns. On the basis of basic primitives and rule base, it can realize the digital generation of national patterns, and the system speed is fast, which can be used for the inheritance and development of traditional national patterns And utilization, as shown in Fig. 1.

Fig. 1. Pal pattern example

3   Pattern Generation Technology Based on Shape Grammar

In the process of creating national patterns, the design and creation of new patterns is not an easy thing. First of all, the creators should have the corresponding artistic accomplishment, and at the same time, they should have a more comprehensive understanding and grasp of the national traditional patterns. Only in this way can the patterns created by hand have both novelty and nationality, and be accepted by the people in ethnic areas, which is the best way to create national patterns by hand It is difficult to adapt to the development and utilization of national culture because of its low efficiency. With the help of modern computer and other information tools, not only the digital protection of traditional culture can be realized, but also the digital development and utilization can be realized. For traditional ethnic patterns, the use of shape grammar is an effective way to realize the rapid design and development of traditional ethnic patterns. Shape grammar is the most popular method of computer aided graphic design [3], George stiny and James Gips In 1971, the content of shape grammar was described in an article for the first time. Since then, shape method has been widely used in industrial design, clothing and other fields. This method belongs to the "parametric automatic generation technology" of pattern. The pattern generated by the change of parameters changes correspondingly, so as to realize the rapid generation of pattern. Through careful study and analysis of traditional patterns of ethnic minorities, we can find that patterns are composed of several basic components according to different rules. [4] These basic components are called primitives, which are the basic elements of patterns, as shown in Fig. 2. Mining these basic primitives and using shape method, which is a parameter automatic generation technology, can quickly design and generate new ethnic patterns, which is of great significance to traditional patterns The development and utilization of the case has positive practical significance.

Fig. 2. Example of pattern element

Shape grammars, also known as shape grammars, is a shape system that generates new patterns on the basis of original patterns/primitives according to rules. The system can automatically generate new patterns according to preset ideas and composition, as shown in Fig. 3. It is a good technical choice for the automatic and rapid generation of national patterns. Shape grammar consists of initial shape and related grammar rules. Specifically, shape grammar can be defined as a four tuple, SG = (s, l, R, I), S is the set of shapes, l is the finite set of symbols, R is the set of rules, and I is the initial set of shapes. [5] In the process of automatic generation of graphics, affine transformation and grammar reasoning can be carried out on some or all of the graphics based on the initial graphics, so as to automatically generate new graphics [6].

Shape grammar is a kind of design reasoning strategy based on the original shape and corresponding rules to generate new graphics. Shape grammar reasoning rules are mainly composed of generative rules and modifiable rules. Among them, there are two types of generative rules: addition and deletion and replacement rules. Addition and deletion refers to the addition or deletion of graph curves from the original graph partially or completely; modification rules refer to the adjustment and transformation on the basis of the original graph state, so as to make the graph form produce obvious difference changes

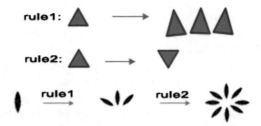

Fig. 3. Example of pattern generation rules

and form a new graph. The common graph modification is mainly affine transformation, affine transformation It belongs to linear transformation, including translation, rotation, scaling and crosscutting [7].

Translation is a relatively simple exchange operation, which means that the position of the figure in the two-dimensional coordinate system has changed, where x′, y′ represents the displacement of the figure in X and Y directions, as shown in Fig. 4.

$$S = \begin{matrix} 1 & 0 & x' \\ 0 & 1 & y' \\ 0 & 0 & 1 \end{matrix}$$

Fig. 4. Example of translation transformation and transformation matrix

Zooming is also a common exchange operation, which means that the size of the initial figure in the two-dimensional coordinate system has changed. This change can be either zooming in or zooming out. When s > 1, it is the expansion transformation, when 1 > s > 0, it is the x-reduction transformation, when SX = sy, it is the proportional zooming, and when SX ≠ sy, it is the stretch expansion, as shown in Fig. 5.

$$T = \begin{matrix} S_x & 0 & 0 \\ 0 & S_y & 0 \\ 0 & 0 & 1 \end{matrix}$$

Fig. 5. Example of scaling transformation and transformation matrix

Mirror image means that the initial figure is flipped along a certain axis in the two-dimensional coordinate system. This kind of flipping can be along the X axis or the Y axis. If KX = 1, KY = −1, the image is mirrored along the X axis; if KX = −1, KY = 1, the image is mirrored along the Y axis, as shown in Fig. 6.

Rotation refers to the new figure obtained by the initial figure rotating a certain angle counterclockwise around the coordinate origin in the two-dimensional coordinate system, where θ is the specific angle of the figure rotating around the coordinate point, as shown in Fig. 7.

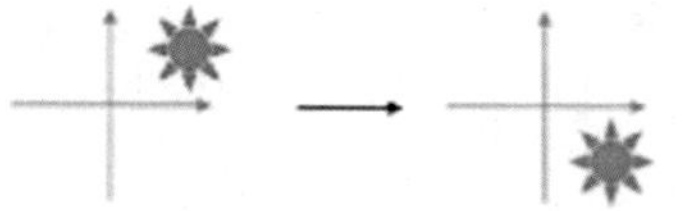

$$S = \begin{matrix} k_x & 0 & 0 \\ 0 & k_y & 0 \\ 0 & 0 & 1 \end{matrix}$$

Fig. 6. Image transformation example and transformation matrix

$$S = \begin{matrix} \cos\theta & \sin\theta & 0 \\ -\sin\theta & \cos\theta & 0 \\ 0 & 0 & 1 \end{matrix}$$

Fig. 7. Rotation transformation example and transformation matrix

Stagger is a kind of transformation in which the initial figure moves unequally along a certain coordinate direction in the two-dimensional coordinate system, resulting in the deformation of the figure. Where HX is the amount of the graph on the x-axis and hy is the amount of the graph on the y-axis, as shown in Fig. 8 [8].

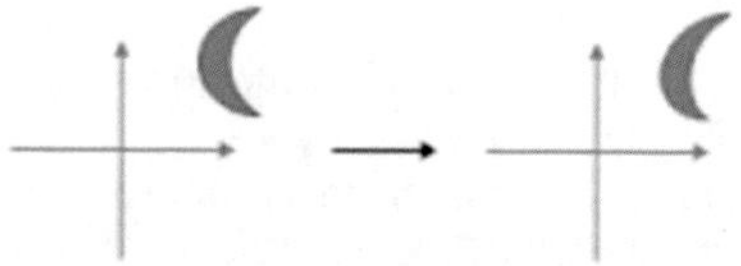

$$S = \begin{matrix} 1 & h_y & 0 \\ h_x & 1 & 0 \\ 0 & 0 & 1 \end{matrix}$$

Fig. 8. Examples and transformation matrix of stagger transformation

Shape grammar is a shape system that generates new patterns on the basis of the original patterns/primitives according to the rules. It takes the original shape as the basic element, and can automatically generate new patterns according to the preset ideas and composition and corresponding rules. It has positive practical significance for the development and utilization of traditional patterns of ethnic minorities [9].

4   Summary

There are many kinds of traditional patterns of ethnic minorities in Qiannan, with rich and colorful contents and forms. The simple and vivid ethnic patterns reflect the yearning of ethnic minorities for a better life. This paper first introduces the pattern generation technology based on pal, which is based on descriptive assembly language and has the characteristics of simple and flexible syntax. Then it describes the pattern generation technology based on shape grammar. Shape grammar is a new pattern operation and generation technology, and shape grammar is also called shape grammar, which generates new patterns on the basis of original patterns/primitives according to rules The system can automatically generate new graphics according to the pre-set ideas and composition. The sorting and application of the above two technologies has positive practical significance for the protection and inheritance of national traditional pattern culture.

Acknowledgement. This paper is part of the results of the 2018 provincial scientific research team promotion project "The Promotion Plan Project of National Culture Digital Scientific Research Innovation Team", Project number: QNYSKYTD2018005.

References

1. Lei, Z., Shifu, C.: Pal: descriptive pattern assembly language. Comput. Appl. Res. **01**, 134–136 (2001)
2. Bo, M., Zhihe, Z., Shifu, C.: Design and implementation of pattern assembly language PASL. J. Comput. Aided Des. Gr. **12**, 900–903 (2000)
3. Generative Art and algorithmic creation 06 - shape syntax. https://www.uegeek.com/181120-generative-art-creativity-06.html
4. Zhang, B., Huang, M.: Research on a new recursive algorithm of fractal tree .Comput. Inf. **15**, 216–228 (2010)
5. Zhaolin, L., Wencheng, T., Chengqi, X.: A DNA reasoning method for product design based on shape grammar. J. Southeast Univ. **07**, 704–709 (2010)
6. Jiaguang, S.: Shape grammars and shape rules. J. Comput. Sci. Technol. **1987**(04) 124–132 (1987)
7. Xinwei, Z., Jin, W.: Pattern extraction and reuse based on ontology and shape grammar. J. Zhejiang Univ. **03**, 461–471 (2018)
8. Wang, W., Yang, Y., Yang, X., Yu, S.: Research and practice of product form innovation design based on shape grammar. J. Gr. **2014**(02) 68–73 (2014)
9. Gang, Z., Beizhan, W., Yuchun, S., Jin, X.: Study on the traditional pattern retrieval method of minorities in Gansu province. SIPD1757899X0028 (2018)

Application of Artificial Intelligence Technology in Lacquerware Process Restoration

Chao Deng$^{(\boxtimes)}$ and Ting Zhang

Nanchang Vocational University, Nanchang, Jiangxi, China

Abstract. With the continuous development of science and technology, artificial intelligence has been produced and received widespread attention, and has been applied in many fields including lacquer repair. The continuous exploration of artificial intelligence in this field has made a series of discoveries. This article mainly introduces the literature reference research method, analytic hierarchy process and investigation research method. This paper uses the analytic hierarchy process to analyze the application of artificial intelligence technology in the restoration of lacquerware craftsmanship, and establishes a potential mathematical model. The model is solved by analytic hierarchy process, and the application status of artificial intelligence technology in lacquer process restoration is studied, and the model is revised using historical data to improve the accuracy of the application status research of artificial intelligence technology in lacquer process recovery. The experimental results of this paper show that the analytic hierarchy process increases the efficiency of artificial intelligence technology in the application of lacquer craft restoration by 13%, and reduces the rate of false alarms and false alarms. Finally, by comparing the application value analysis of artificial intelligence in the restoration of lacquer ware craftsmanship and the analysis of the aesthetic modern needs of lacquer craft restoration audiences, the impact of the application of artificial intelligence technology in the restoration of lacquer craftsmanship is systematically explained.

Keywords: Artificial intelligence technology · Lacquerware process restoration · Analytic hierarchy process · Literature reference research method

1 Introduction

1.1 Background and Significance

In recent years, with the development of a new generation of information and communication technology, mobile Internet is everywhere [1]. The theory of artificial intelligence, the large-scale computing power of the Internet cloud platform and continuous optimization and algorithm improvements ensure that computer systems rely entirely on deep learning to independently complete the most complex and detailed tasks, and artificial intelligence has begun to integrate. The arrival of information civilization has created rapid progress in human society [2, 3]. This is not only the result of continuous innovation in human cognition, but also an important manifestation of the realization of human

© The Author(s), under exclusive license to Springer Nature Switzerland AG 2021
J. Abawajy et al. (Eds.): ATCI 2021, AISC 1398, pp. 563–571, 2021.
https://doi.org/10.1007/978-3-030-79200-8_84

intelligence. The science and technology of computer algorithms, the Internet and artificial intelligence will inevitably become the research with the greatest attention and research potential [4]. The research and application of artificial intelligence theory has made great progress. The application of artificial intelligence in the restoration of the lacquer craft restoration industry not only enriches the restoration ability of the lacquer craft restoration industry, but also enhances the restoration ability of the lacquer craft restoration industry [5, 6]. It can be said that artificial intelligence poses many challenges to the traditional technology of lacquer craft restoration, such as the framework and concept of lacquer craft restoration model. Therefore, in view of the current situation, this topic has practical significance for in-depth thinking about the relationship between artificial intelligence and lacquer repair [7].

1.2 Related Work

Chen Jinzhou provides a method that can evaluate participatory stakeholder innovation in a complex stakeholder environment to solve essential problems [8]. Based on the principle of common value creation, he proposed an application framework that illustrates the security protection process, in which the stakeholders integrate their resources and capabilities to develop innovative wave simulation algorithms [9, 10]. In order to evaluate this evaluation framework, a number of data were collected in the study. This case represents the significance of artificial intelligence for the research and system implementation of wave simulation algorithms [11, 12]. But because the message collection process is too complicated, the data result is not very accurate.

1.3 Main Content

The innovation of this article lies in the literature reference research method, analytic hierarchy process and investigation research method. Based on the research of lacquer craft restoration under the background of artificial intelligence, the application of artificial intelligence technology in lacquer craft restoration is studied through the analytic hierarchy process. Establish a calculation method combining analytic hierarchy process and literature research method to provide research guidance for the application of artificial intelligence technology in lacquer craft restoration.

2 Application Method of Artificial Intelligence Technology in Lacquer Craft Restoration

2.1 Literature Research Method

The method of studying literature is a method of understanding scientific literature information with the help of scientific research. In the process of removing theories and cases related to varnish restoration, the author needs a lot of data and useful literature. Please consult the electronic resources and paper materials of the school library, purchase relevant books, search for relevant materials on the Internet, and collect extensive reports and statistical information related to the restoration of lacquerware craftsmanship to ensure a complete and correct understanding.

In the research and writing process, based on the defined research materials, with the help of Wanfang and CNKI databases (such as full text such as Vipu and SDOS databases), many books, master and doctoral dissertations, journals and other materials were searched, and a large number of academic achievements. The repair result of the drawer craft has played a role in the theoretical support of the writing of this article, and provided a reference for the in-depth study of the subject. Studying related theories and literature is an important learning method. Read and study the literature related to the country, understand and summarize newspapers, news, magazines, related theories and related policies.

2.2 AHP

AHP is a method that combines quality and quantity testing. This method is based on the principle of understanding things from simple to complex, and divides complex issues into several orderly and reasonable levels. By analyzing each relatively simple level layer by layer, the relative decision goals of various indicators are obtained. Through mathematical methods, the relative importance of each index is converted into weight values, and different decision-making schemes are sorted to find the best decision-making scheme.

Normalize the required feature range to classify the significance of each evaluation factor, that is, the weight value. The test shall use the following formula:

$$Q_A = \gamma_{\max} A \qquad (1)$$

The random sequence ratio of the judgment matrix is represented by EI; the general sequence index of the estimation matrix is represented by M.

$$EI = (\gamma_{\max} - M)/(M - 1) \qquad (2)$$

I represents the average random consistency index in the standard table.

The reason for adopting the analytic hierarchy process is that the analytic hierarchy process treats the research object as a system, and makes decisions according to the way of decomposition, comparison, judgment, and comprehensive thinking. It has become an important tool for system analysis developed after mechanism analysis and statistical analysis. The idea of the system is to not cut off the influence of each factor on the result, and the weight setting of each layer in the analytic hierarchy process will eventually directly or indirectly affect the result, and the degree of influence of each factor on the result is Quantitative and very clear.

This method is especially useful for systematic evaluation of unstructured characteristics and multi-objective, multi-criteria, and multi-period systematic evaluation. This method not only pursues advanced mathematics, but also does not pay attention to behavior, logic, and reasoning one-sidedly, but organically combines qualitative methods with fixed imitation methods to decompose complex systems and mathematicize people's thinking processes. Systematization, easy for people to accept, and can turn multi-objective, multi-criteria, and difficult to quantify decision-making problems into multi-level single-objective problems. After pairwise comparisons are made to determine the quantitative relationship between elements of the same level and the elements of the previous level, finally Perform simple mathematical operations. The calculation is simple, and the results obtained by the combination are simple and clear, which is easy for decision-makers to understand and master.

2.3   Investigation and Research Method

Investigation and research methods refer to methods that directly obtain relevant materials through the understanding and understanding of objective conditions and analysis of these substances. Through a large number of online materials, practical cases, etc., this article fully investigates the current situation and reasons of lacquer craft restoration in the era of artificial intelligence and new media, analyzes, synthesizes, compares and summarizes the large amount of data collected in the investigation, so as to better. There is a comprehensive and in-depth understanding of how to construct the restoration model of lacquerware in the era of new media.

3   Application Experiment of Artificial Intelligence Technology in Lacquer Craft Restoration

3.1   Experimental Design of the Application of Artificial Intelligence in the Restoration of Lacquerware Craftsmanship

Artificial intelligence has a positive effect on the restoration of lacquer craftsmanship. The application of artificial intelligence technology in the restoration of lacquer ware craftsmanship can not only help restorers to reduce work pressure, but also can effectively shorten the working time, and can also promote the development of lacquer craft restoration technology. Artificial intelligence technology can effectively help students learn the knowledge of transforming crafts from varnish. The technical form of lacquer craft restoration is different and will change with the change of the times, but the essence of lacquer craft restoration will not change. Regardless of whether they are lacquer craft restoration techniques or inherited from later, the work of lacquer craft restoration is based on human subjective thinking. After all, no matter what kind of performance artificial intelligence has achieved in the restoration of lacquer craftsmanship, one thing cannot be changed. Artificial intelligence is defined by people. The use of artificial intelligence in the restoration of lacquer craftsmanship cannot be separated from the participation and participation of people.

Artificial intelligence algorithm has outstanding advantages, it can complete the target task very well, which is why artificial intelligence is also very common in the field of paint repair. However, with the help of the application analysis of artificial intelligence, it can be found that artificial intelligence can only help people complete part of the recovery work, but cannot completely replace human thinking and analysis. This is also the main value of people in this field. Artificial intelligence can complete big data analysis in a short period of time, but cannot correctly analyze the rich thoughts and emotions of human beings, so the performance of artificial intelligence is not in place. It can be seen that artificial intelligence cannot perfectly realize the judgment and understanding of emotions at this stage, which leads to artificial intelligence in the field of lacquer repair, which is a problem that needs to be solved artificially in the recent development of artificial intelligence.

3.2 Application of Artificial Intelligence to the Satisfaction Experiment Data Collection of Lacquer Craft Restoration

The aesthetic forms used in lacquer products are mainly rhythm and rhythm, simple, coordinated contrast, relationship, rhythm and rhythm, and multiple unification. Basically, the aesthetics of lacquer can be divided into two types: the second is the internal form, which refers to creation the real and beautiful content that the reader wants to express; the second is the internal form, which is the external form that is not directly related to the content. The descriptive appearance of the interior (such as materials, lines, colors, smells, shapes, etc.). Lacquerware has certain characteristics in appearance. For example: balance, symmetry, proportion, rhythm, rhythm, change, consistency, etc.

After these unique points are combined with artificial intelligence, the success of its application in lacquer craft restoration largely depends on people's satisfaction with such restoration techniques. Therefore, this article investigates people's appraisal of artificial intelligence in lacquer ware. The satisfaction of the shape of the craft restoration, the satisfaction of the lacquer craft restoration master with the artificial intelligence technology, the work efficiency and the error rate are compared with the application of traditional technology and artificial intelligence technology in the lacquer craft restoration. The data results are shown in Table 2:

Table 1. Experimental data table of artificial intelligence applied to lacquer craft restoration

Compare items	Customer satisfaction	Employee satisfaction	Work efficiency%	Error rate%
Traditional technology	81	88	80	10
Artificial intelligence technology	85	98	95	5

It can be seen from Table 1 that there is not much difference between traditional restoration technology and artificial intelligence restoration technology in the eyes of customers. However, employees' satisfaction with artificial intelligence restoration technology is 10 points higher than that of traditional restoration technology. At the same time, work efficiency should be 15% higher and the error rate should be 5% lower. It can be seen that artificial intelligence technology is a big change for lacquer craft restoration. In order to be able to see the relationship between the two more clearly, we analyzed the data in Table 1, and the analysis results are shown in Fig. 1:

As can be seen from Fig. 1, artificial intelligence technology can improve work efficiency and reduce error rate for lacquer craft restoration, and can also increase employee satisfaction and create more benefits.

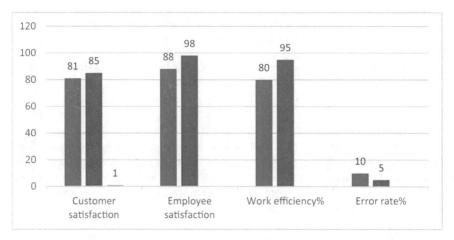

Fig. 1. Experimental data graph of artificial intelligence applied to lacquer craft restoration

4   Application Analysis of Artificial Intelligence Technology in Lacquer Craft Restoration

4.1   Application Value Analysis of Artificial Intelligence in Lacquer Craft Restoration

With the tremendous development of contemporary computer science, designers can use a large number of design programs to express their inspiration. Compared with traditional artistic creations that take months or years, computer science will help designers complete creations quickly and efficiently and mass-produce them through forms. Lacquer craft restoration restorers can use computer modeling and 3D printing technology to create products that people need; they can use a lot of colored materials in the computer to create more interesting works, or they can use a lot of art design resources in the computing network. Inspire inspiration, expand artistic thinking, and combine your own thinking with computer science thinking to achieve design creation.

In terms of the introduction of restoration work related to the development of virtual reality technology, many universities use virtual reality technology in the restoration of painted products. Virtual reality can better reflect the design work, combining restoration and technical virtual experience, environment design In order to communicate and deliver effectively, this technology can be used to make the work more realistic so that people can see the details of the artwork on the stage.

4.2   Modern Aesthetic Needs of the Audience for the Restoration of Lacquerware Technology

With the continuous development of Internet technology, three-dimensional sports expressions have gradually become a mainstream of society. By enhancing the two-dimensional image, the audience can obtain the original sensory simulated animation image. This realistic visual effect greatly enhances the audience's sense of immersion and immerses them in the composition atmosphere. Secondly, interaction with users is also very important: printing can be done with the help of HIPS material artificial intelligence technology, and the surface can be installed, shaved, and painted. After these processes are completed, the paint can be dried. Using lemon to remove the tire, melt it and take it out, not only solves the modeling problem, but also greatly reduces the recycling time of lacquerware.

Under this technology, this article conducts an experimental analysis on the recovery time and efficiency of artificial intelligence technology and traditional recovery technology. By comparing the recovery time, recovery degree, and recovery satisfaction between the two, the data results are as follows Table 2 shows:

Table 2. The modernization needs of the audience's aesthetics are lacquer craft restoration data table

Compare items	Recovery satisfaction%	Recovery time%	Degree of recovery%
Traditional restoration techniques	88%	10%	86%
Artificial intelligence technology	98%	6%	99%

It can be seen from Table 2 that in the modern aesthetic of the audience, compared with traditional technologies, artificial intelligence technology has 10% higher recovery satisfaction, 4% less recovery time, and 13% higher recovery. It can be seen that artificial intelligence technology can better meet the aesthetic needs of modern audiences. In order to be able to see the relationship between the two more clearly, we analyzed the data in Table 1, and the analysis results are shown in Fig. 2:

It can be seen from Fig. 2 that in the modern aesthetic of the audience, the restoration of lacquer craftsmanship requires the adoption of advanced technology. It is necessary to use artificial intelligence technology to improve the satisfaction of restoration, reduce the restoration time, and improve the restoration level.

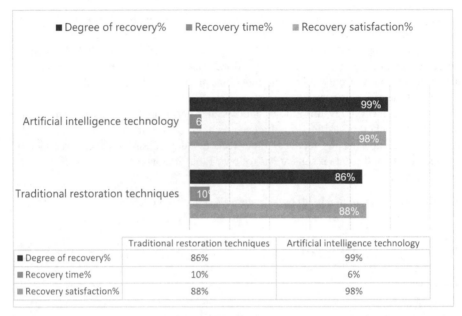

Fig. 2. The modernization needs of the audience's aesthetics are the analysis of the restoration data of the lacquerware process

5 Conclusions

Although this article has made certain research results in the literature reference research method, the analytic hierarchy process and the investigation research method, there are still many shortcomings. The application of artificial intelligence technology in the restoration of lacquer craftsmanship still has a lot of in-depth content worth studying. There are still many steps in the decision-making process that have not been involved because of space and personal ability. In addition, the actual application effect of the improved algorithm can only be compared with the traditional model from the level of theory and simulation.

References

1. Miklin-Kniefacz, S., Griesser, M., Parson, W., et al.: Searching for blood in Chinese lacquerware. Stud. Conserv. **59**(S1), S252–S253 (2014)
2. Honda, T., Lu, R., Kamiya, Y., et al.: Identification of ryukyu lacquerwares by pyrolysis-gas chromatography/mass spectrometry and 87Sr/86Sr isotope ratio. J. Anal. Appl. Pyrolysis **117**(jan), 25–29 (2016)
3. Ma, X., Shi, Y., Khanjian, H., et al.: Characterization of early imperial lacquerware from the luozhuang han tomb. China Archaeom. **59**(1), 121–132 (2017)
4. Jia, Y., Yin, L., Zhang, F., et al.: Fungal community analysis and biodeterioration of water-logged wooden lacquerware from the nanhai no. 1 shipwreck. Appl. Sci. **10**(11), 3797 (2020)

5. You, C., Han, Y., Wei, Q.: Brief analysis on the developmental and creative approach of large-scale lacquerware crafts with ethnic characteristics. Art Des. Rev. **07**(2), 115–123 (2019)
6. Zhou, Y.: Research on the application of artificial intelligence in mobile network technology%. J. Zhangjiakou Vocat. Coll. Technol. **029**(001), 51–53, 56 (2016)
7. Qian, H.: Development and application of artificial intelligence technology% development and application of artificial intelligence technology. Electr. Power Inf. Technol. **015**(009), 32–37 (2017)
8. Jinzhou, C.: The application of artificial intelligence technology in intelligent building% the application of artificial intelligence technology in intelligent building. Intell. Build. Urban Inf. **000**(003), 44–45 (2018)
9. Iqbal, M.W., Sarkar, B.: Application of preservation technology for lifetime dependent products in an integrated production system. J. Ind. Manag. Optim. **13**(5), 1–28 (2017)
10. Jiang, T., Li, L.: Application of GIS technology in geological prospecting% application of GIS technology in geological prospecting. World Nonferrous Met. **000**(009), 64, 66 (2019)
11. Shaoke, Z., Miaolei, D.: The application of RFID technology in university laboratory. Intell. Comput. Appl. **006**(005), 93–96 (2016)
12. Li, Z., Wang, J.: An optimization application of artificial intelligence technology in enterprise financial management. Bol. Tecnico Tech. Bull. **55**(11), 83–89 (2017)

Digital Protection Platform of Pingxiang Nuo Mask Based on AR Technology

Wei Yu[✉]

Jiangxi University of Applied Science, Nanchang 330100, Jiangxi, China

Abstract. In recent years, with the development of modern information technology and the increasing popularity of the Internet, digital technology continues to penetrate into people's work and life, and has also been applied in the field of cultural heritage protection. Among them, augmented reality technology has the advantages of virtual and real fusion, real-time interaction and three-dimensional registration, and has a promising development prospect. This paper mainly studies the design of digital protection platform of Pingxiang Nuo mask based on AR technology. Comprehensive theoretical research and practical application, this paper applies augmented reality in the process of digital cultural heritage, cultural heritage from the angle of interdisciplinary spread of digital display and pingxiang nuo masks AR experience summary strategy analysis and the design principle, and carries on the concrete application practice based on local cultural resources, provide the public with abundant realistic multicultural experience, conducive to the sustainable development of the cultural heritage, with its national conditions for building under the new situation characteristic of cultural heritage digital development mode to provide beneficial reference and practice of interdisciplinary thinking.

Keywords: AR technology · Digital platform · Pingxiang nuo mask · SLAM space positioning

1 Introduction

Pingxiang Nuo mask is the intangible cultural heritage of China, which is the spiritual and cultural wealth created by ancient Chinese ancestors. However, with the rapid development of modern society, the traditional culture of Pingxiang Nuo mask is also facing the change of The Times and a new development crisis. The phenomenon of "good but not popular, young people are reluctant to learn" is becoming more and more serious. Due to the impact of foreign culture and the change brought by the development of modern mechanization technology, Pingxiang Nuo mask, as a traditional characteristic handcraft, is now facing problems such as difficult inheritance, weak public awareness and social marginalization. If we do not rescue and protect it, with the passage of time and the passing of the inheritor, Pingxiang Nuo mask, a precious intangible cultural heritage, will die out. As Chinese children, it is our bounden mission and obligation to touch history, inherit culture and draw strength from cultural heritage. With the rapid

J. Abawajy et al. (Eds.): ATCI 2021, AISC 1398, pp. 572–580, 2021.
https://doi.org/10.1007/978-3-030-79200-8_85

development of information technology and the acceleration of modernization, the digital technology represented by augmented reality and virtual reality can free the display and dissemination of cultural heritage from the constraints of time and space, and let the world see the infinite possibilities of the combination of science and technology with cultural heritage. Among them, augmented reality, through the superposition and integration of physical scenes and virtual scenes, can enable people to have a kind of super-sensory experience. Therefore, this technology, also known as mixed reality, has a good development prospect in the digital protection and display of cultural heritage [1]. However, in real life, the research on the combination of the two is still in the exploratory stage, and further efforts are needed. How to expand the display space of cultural heritage on the basis of enhancing the effectiveness of its interpretation information dissemination, improve the utilization rate and development degree of cultural heritage, and at the same time to achieve the general significance of education to the public, is the field of cultural development needs to pay special attention to the issue.

The digital protection projects of intangible cultural heritage are also in full swing in various countries around the world. Most of the research and work on the digital protection of intangible cultural heritage abroad were carried out relatively early in developed countries. Samet believes that relevant institutions should adopt an effective cooperation mechanism, and it is very important to give full play to the advantages of each institution to promote the digital protection of intangible cultural heritage [2]. Fei et al. combined VR technology with traditional manual intangible cultural heritage, and discussed the specific role of VR technology in the digitization of intangible cultural heritage [3].

Taking augmented reality as the technical core, this paper analyzes its advantages in cultural heritage digitization by comparing with traditional methods, and summarizes corresponding digital display and communication strategies of cultural heritage by combining with Pingxiang Nuo mask, forming relevant theoretical methods and practical paths, and creating excellent cultural creative brands. By analyzing the application design of digital technologies such as augmented reality in the field of cultural and cultural exhibitions, the paper explores interactive ways that meet the needs of modern consumers, and provides the public with a rich and vivid multi-cultural experience in fashion and interesting display ways and interactive forms.

2 Digital Platform Based on AR Technology

2.1 AR Technology Overview

Augmented reality and basic work first through a camera sensor for real-world images and video information, analyze it in the computer and preprocessing, and then calculated according to the virtual camera to identify real environment object position and posture assessment, thus rendering in the form of corresponding space perspective of virtual objects and the coordinate transformation, synthesis of real environment with virtual objects through a display device output, after the final users see in display device has been enhanced after the scene content [4].

Display technology is the cornerstone to maintain the normal operation of AR system. Currently, headmounted displays (HMD) and non-mounted displays are mainly used in the field of augmented reality according to the different ways of wearing. The head-mounted display includes two types of video perspective type and optical perspective type according to the specific implementation principle. The non-head-mounted display is mainly divided into hand-held and projection type.

From the perspective of wear and use, the head-mounted display and the virtual reality head display are the same. Compared with the non-head-mounted display, the head-mounted display has a stronger sense of immersion and is less constrained by the environment. However, wearing and operating for a long time will cause users' physical discomfort and fatigue. Nowadays, the rapid development of mobile devices such as smart phones has greatly expanded the hardware and software devices carrying AR, thus derifying the sub-category of mobile AR, which makes mobile handheld AR devices represented by smart phones more flexible in use experience.

Tracking 3 d registration registration is AR system essential technology, on the one hand is to the user operation real-time space tracking camera positioning, identify will be enhanced feature points and 3 d coordinates of the object, on the other hand is the virtual objects in the real environment coordinate space location and add registration, and match the reality environment coordinates. Commonly used methods include tracking registration technology based on tracker, visual tracking registration technology based on computer vision system and algorithm, and composite tracking registration technology integrating tracker and vision [5]. At present, visual tracking registration technology is the mainstream technology, which can be realized in three ways: two-dimensional image localization, three-dimensional three-dimensional localization and three-dimensional space understanding based on SLAM.

AR experience quality is so intertwined with the use of human-computer interaction technology, human-computer interaction with the characteristics of multimodal integration in augmented reality scenarios, the input device is no longer confined to the keyboard and mouse, the user can through voice recognition, hand and body posture, the head tracking, eye tracking and face recognition and so on a variety of ways to control the virtual objects and operations, improve the user's engagement, to create a more natural and smooth interaction experience.

2.2   AR Platform Design Scheme

At present, mobile AR has a wide application range and high flexibility due to the improvement of technology. Using mobile phones in visiting scenes can provide users with personalized service content and meet diversified needs [6]. At the same time, considering the operating environment requirements of AR technology, mobile phone application is selected as the product carrier when defining the product form. In addition, the construction of user usage scenarios and vision statements and the transformation of user demand definition into product design function points can help define the product framework, clarify the overall design direction of the product and promote the design process. It is easier for users to understand and participate in the experience designed around the use scenario. Therefore, the user needs summarized in the previous chapter are classified and summarized according to the time flow of the museum visit experience, and then transformed into the corresponding functional elements.

In order to guide users to complete the overall operation, a complete exhibition experience process centered on user activities is designed. By simulating the user's usage process, we can more clearly understand the primary and secondary order and usage frequency between functional elements, delete unnecessary functions and merge similar functions. It is the premise of information architecture visualization to construct the main interaction scheme for the purpose of designing all-around interactive immersive experience.

Information architecture is the framework and organization principle of an application. Drawing information architecture diagram can quickly sort out the entire product operation process and analyze whether the information presentation is reasonable and friendly to users. The reasonability of its classification directly affects users' use efficiency and satisfaction [7]. Considering that the application usage scenarios can be divided into online and offline scenarios, the main application functions can be divided into experience function and display function according to situational characteristics. Deep breadth of information architecture based on choosing the way of combination of depth and breadth, to shorten the path of the user access to information, the homepage with breadth-first, covers the experience and show a total of seven major function entrance, at the same time to reduce the entry occupy usable floor area, the phone's screen will show the class three functions and Settings under the fold to the "more" entrance, within the functions with depth-first, have more space to display the current function.

Related calculation formula

$$f = \frac{a-b}{b} \tag{1}$$

$$e = \sqrt{a^2 - b^2}/a \tag{2}$$

$$e' = \sqrt{a^2 - b^2}/b \tag{3}$$

2.3 Design and Implementation of AR Module

(1) Image Based Recognition

The virtual display function of 3D model of cultural relics is mainly realized in the integrated development environment of Unity and Vuforia. Unity is a game development engine for global creators, which supports 3D rendering and provides a very friendly operation tool for AR and VR application development experience [8]. Vuforia is an AR SDK focusing on image recognition, with good cross-platform and compatibility.

Firstly, an identification map of Pingxiang Nuo mask is made in Sketch or Photoshop, and the image size is no more than 2 MB. The identification map is uploaded to the official website of Vuforia for preprocessing, and the star rating of identification accuracy is obtained by the platform, with five stars as the best. The purpose of preprocessing is to establish a camera to capture the target database. The more obvious the feature points of the recognition map are, the higher the success rate of AR recognition will be.

Secondly, add the Vuforia resource pack to the Unity virtual environment, create a new AR Camera in the scene to call the computer Camera; Import the AR identification map and the 3D cultural relic model, adjust the position and size of the model and other relevant parameters; Add audio components to the model and import corresponding commentary audio; Run the tests several times and adjust to your satisfaction.

(2)   Spatial Localization Based on SLAM

At present, the mainstream AR development platforms based on SLAM spatial positioning technology include the AR Core launched by Google and the AR Kit launched by Apple. In this project, AR Kit, a mobile terminal augmented reality platform designed for IOS system, was selected to carry out function development in combination with UN cross-platform development tool AR Foundation.

The main operation steps are as follows: import AR Foundation and AR Kit Plugin resource pack from the Unity database to build AR environment; Import the artifact model and texture map into the scene and adjust the position and size. A prompt diagram of the base of the model was made with the plane software and added as a subset of the heritage model in Unity, which is easy to guide the user's actual operation and provide visual feedback for different gestures. The AR point cloud effect is added to the scene to facilitate the camera to capture the feature points of the plane during the three-dimensional space identification calculation and plane scanning, and at the same time to remind the user of the operational range; Using C# programming language to code, the user clicks on the screen to display the cultural relic model and touch the screen with one or two fingers to operate the model zooming, moving, rotating and other interactive functions; Finally, the whole AR scene is packaged into mobile phones through Microsoft development kit Visual Studio to generate a test AR application [9, 10].

3   Platform Testing

3.1   AR System Test

In order to evaluate the usability, functionality and acceptability of the overall interactive system of AR application, and to verify the user experience satisfaction and communication effect based on the digital display and communication strategy of cultural heritage, two evaluation items were carried out for target users, namely task test and subjective evaluation scale filling. The evaluation indicators are based on the usability evaluation dimensions of learnability, efficiency, memorability, fault tolerance and satisfaction proposed by Nielsen, and combined with three indicators of cognition, psychology, attitude and behavior in the study of communication effects in communication studies.

When it comes to user recruitment, Nielsen points out that five users are enough to get 85% of the feedback, and six to eight users get 90% of the feedback. In this test, 8 representative users who are interested in museums, aged between 20 and 39, can use mobile phones skillfully, and have certain understanding of or experience with AR.

3.2 Digital Platform Testing

In order to test whether the performance of the platform meets the design requirements, special performance tests are needed after the completion of functional tests. Since the performance requirements of the platform are reflected in the login of users to load from the home page, it is necessary to load the login information of users to carry out relevant tests. During the testing, a special testing tool was written to simulate the login and load of multi-threads for testing.

According to several indicators of performance design, the tests mainly carry out three tests: concurrent authentication response, login loading response test and concurrent loading user page response test. The user page calls the returned data of other application interfaces asynchronously, which is influenced by the application itself and is not considered in the performance test. To exclude network factors, the test environment was set to a wired network.

4 Test Results of Digital Platform for Intangible Cultural Heritage

4.1 AR Module Subjective Evaluation

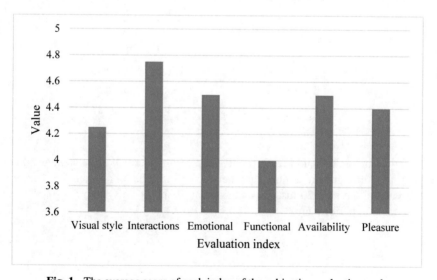

Fig. 1. The average score of each index of the subjective evaluation scale

As shown in Fig. 1, sorting out the task test results shows that users complete various tasks well. Most users can complete all tasks independently, while some users also complete tasks smoothly after simple prompts from the lead tester. In terms of application specific operation, through the observation in the process of the task 2 test users wrong operation frequency is higher, generally appear in the switch plan tour and AR navigation function of real user feedback these two function icon design exist ambiguity, could not immediately determine which one is AR live-action navigation function, moreover also observed users in the process of task 4, common to click "more" button to display the hidden menu rather than the right screen. Based on the above observations, the design of ambiguous ICONS should be optimized during subsequent iterations of the app to make it easier for users to understand, and the gesture-guided design of various types should be improved before users first use them.

4.2 Platform Concurrency Test

This test scheme is designed to simulate the number of logged in users of 2200, considering the allowance of 10%, and detect the sound time of each user to revisit the page. The tool simulates 2200 threads logging in, waits for a refresh signal, performs a page refresh at the same time, and then takes the HTML bottom test markup from the returned data stream to determine the load time of the static portion of the page. A total of 5 rounds of testing were conducted, with the maximum, minimum and average values recorded in each round.

Table 1. Load the user page concurrently in response to the test case

Test times	Min value	Max value	Average value	Conclusion
1	31	217	162	Standard
2	47	223	176	Standard
3	32	221	151	Standard
4	41	206	153	Standard
5	45	218	158	Standard

As shown in Table 1 and Fig. 2, the five test results were all up to standard. In the five tests, the minimum value fluctuated between 30 ms and 45 ms, and the maximum value fluctuated around 220 ms. This shows that the platform system designed in this paper is stable in performance.

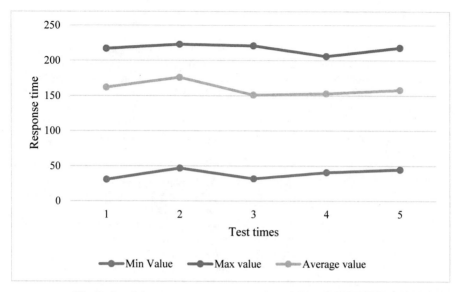

Fig. 2. Load the user page concurrently in response to the test case

5 Conclusions

In today's economic globalization and information, western values take the opportunity to infiltrate, traditional culture has suffered a strong impact, the development of cultural heritage digital work is urgent. This paper will be represented by AR technology of digital technology and the integration of cultural heritage demonstrated communication, integrated design, summed up the communication and computer technology, such as knowledge as well as cultural heritage museum of the digital display communication strategy and AR experience design principle, pingxiang nuo masks, for example, around the user research in museum AR application design practice and test assessment, designed to broaden the cultural heritage of exhibition space at the same time to strengthen the effectiveness of information communication, attract the public attention to ethnic culture and strengthen the identity. Design the AR application of Pingxiang Nuo mask on mobile terminal based on augmented reality technology display. It meets users' basic needs in visual style, interface operation and function display, and flexibly uses Unity, Vuforia, Spark AR Studio and other software to realize relevant augmented reality core functions, truly optimizing users' actual interactive experience and spiritual pleasure. The overall satisfaction of the application was analyzed through the usability task test and subjective evaluation scale to verify the effectiveness of the digital display and communication strategy of cultural heritage based on augmented reality.

Acknowledgements. Project: Science and Technology Project of Jiangxi Provincial Department of Education in 2020, No.: GJJ203010 Design and Research of Pingxiang Nuo Mask Digital Protection Platform Based on AR Technology.

References

1. Kristoffer, F., David, S.: The supply chain has no clothes: technology adoption of blockchain for supply chain transparency. Logistics **2**(1), 2–3 (2018)
2. Apte, S., Petrovsky, N., et al.: Will blockchain technology revolutionize excipient supply chain management? J. Excipients Food Chem. **7**(3), 76–78 (2016)
3. Mahyuni, L.P., Adrian, R., Darma, G.S., et al.: Mapping the potentials of blockchain in improving supply chain performance. Cogent Bus. Manag. **7**(1), 1–18 (2020)
4. King, L.: Turvo launches collaborative supply chain management platform. Air Cargo World **107**(4), 8 (2017)
5. Yazdani, M., Zarate, P., Coulibaly, A., et al.: A group decision making support system in logistics and supply chain management. Expert Syst. Appl. **88**(Dec), 376–392 (2017)
6. Li, Z., Liu, G., Liu, L., et al.: IoT-based tracking and tracing platform for prepackaged food supply chain. Ind. Manag. Data Syst. **117**(9), 1906–1916 (2017)
7. Zhu, Y., Riad, K., Guo, R., Gan, G., Feng, R.: New instant confirmation mechanism based on interactive incontestable signature in consortium blockchain. Front. Comput. Sci. **13**(6), 1182–1197 (2018). https://doi.org/10.1007/s11704-017-6338-8
8. Zhou, N., Zhang, X., Wang, S.: Timestamp reassignment: taming transaction abort for serializable snapshot isolation. Front. Comput. Sci. **13**(6), 1282–1295 (2018). https://doi.org/10.1007/s11704-018-7018-z
9. Wei, W., Amorim, P., et al.: Tactical production and distribution planning with dependency issues on the production process. Omega Int. J. Manag. Sci. **67**(Mar), 99–114 (2017)
10. Iansiti, M., Lakhani, K.R.: The truth about blockchain. Harvard Bus. Rev. **95**(1), 118–127 (2017)

Discussion on Application of Artificial Intelligence in the Construction of the Model for Senile Intertrochanteric Fracture After Hidden Hemorrhage During Treatment

Yi Liu and Difeng Ke[✉]

Shunde Hospital of Southern Medical University, Foshan 528300, Guangdong Province, China

Abstract. Objective: To explore the role of artificial intelligence in the model construction for elderly intertrochanteric fracture after hidden blood loss during treatment.Methods:120 senior people with femoral intertrochanteric fracture were diagnosed with PFNA. They were divided into the group with hemorrhage (n = 35 cases) and non-blood loss group (n = 85 cases) according to whether hidden blood loss occurred after surgery. Single factor and multi-factor logistic analysis were conducted by reference to the medical records of the two groups. Artificial intelligence R software was used to establish the nomogram model for the elderly with latent hemorrhage after femoral intertrochanteric fracture. ROC curve was drawn to analyze the predictive efficacy of the model for the latent hemorrhage after intertrochanteric fracture for senior people. Results: Univariate and multivariate Logistic analysis demonstrated that the occurence of occult blood after operation in senior people with intertrochanteric fracture was statistically significant with fracture type, internal fixation method, methods in anesthesia and administration of anticoagulant drug ($P < 0.05$).The results of the model showed that the score for fracture type was 63 points. The score for anesthesia method was 37.6 points. The score for internal fixation was 71.5 points. The results of ROC curve showed that the C-index of the histogram model for predicting latent hemorrhage after intertrochanteric fracture was 0.816, which showed good differentiation. The area under the curve was 0.832, and the 95%CI was 0.757–0.899, which showed good prediction efficiency. Conclusion: There are many influencing factors for latent hemorrhage after intertrochanteric fracture for the senior. The prediction model of hidden blood loss based on artificial intelligence has good diagnostic sensitivity and is worthy of popularization and application.

Keywords: Artificial intelligence · Femoral intertrochanteric fracture in the elderly · Hidden blood loss · Influencing factors · Prediction model · The sensitivity

Femoral intertrochanteric fracture generally occurred among the elderly, which is more likely related to the occurrence of osteoporosis in patients. The trochanteric fracture of femur is one of the more serious parts of the human body bone loss, making the incidence of femoral intertrochanteric fracture for senior people being increasing. PFNA enjoys characteristics in simple operation, surgical trauma, and high fracture healing

© The Author(s), under exclusive license to Springer Nature Switzerland AG 2021
J. Abawajy et al. (Eds.): ATCI 2021, AISC 1398, pp. 581–587, 2021.
https://doi.org/10.1007/978-3-030-79200-8_86

rate [1]. Previous studies have shown that PFNA is less invasive and has higher safety for senior people with intertrochanteric fracture [2]. However, some patients showed anemia of varying degrees after surgery, which may be due to the fact that doctors in perioperative period paid more attention to the dominant hemorrhage and drainage volume in postoperative period, ignoring blood exuding in tissue space, and increasing the incidence of hidden blood loss [3]. Previous studies have shown that strengthening the construction of the prevention model for patients with hidden blood loss after surgery is of great significance to improve the success rate of surgery and promote the recovery of patients. Therefore, this study takes senior people with intertrochanteric fracture as object for exploration of role of artificial intelligence in the treatment of elderly patients with intertrochanteric fracture and the construction of hidden blood loss model, which is demonstrated as follows.

1 Materials and Methods

1.1 Clinical Data

From March 2018 to April 2020, 120 senior people with femoral intertrochanteric fracture were selected for PFNA treatment. Patients were subdivided into blood loss group (n = 35 cases) and non-blood loss group (n = 85 cases) based on whether hidden blood loss occurred after surgery. In the blood loss group, there were 19 males and 16 females, aged (61–85), with an average age of (73.71 ± 6.53).The fracture types were stable in 26 cases and unstable in 9 cases. There were 85 patients in the non-blood loss group, including 43 males and 42 females, aged (62–84), with an average age of (73.51 ± 6.49).The fracture types: 41 stable cases and 44 unstable cases.

1.2 Inclusion and Exclusion Criteria

Inclusion criteria: (1) Be qualified with diagnostic criteria for femoral intertrochanteric fracture [4] with the age $\geq$ 60 years old; (2) Conform to the indications for surgical treatment of PFNA; (3) The patient met the diagnostic criteria for recessive blood loss after surgery, and had complete baseline data and follow-up data. Exclusion criteria: (1) Patients with infectious diseases that are difficult to control and complicated with mental abnormalities; (2) Abnormal cognitive function and severe abnormal liver and kidney function.

1.3 Methods

A. Calculation method of latent hemorrhage after intertrochanteric fracture of femur for senior people. Blood tests were performed to determine the HCT level during perioperative period. The amount of latent hemorrhage was counted based on the formula: Latent hemorrhage volume = total hemorrhage - dominant hemorrhage volume + transfusion volume. In the formula, total hemorrhage volume = blood volume × (HCT before operation-HCT after operation)/average HCT. Blood volume = K1 × H + K2 × M + K3.In the formula, H refers to height and M is weight. For male patients, K1 = 0.367,

K2 = 0.032, K3 = 0.604; For female patients, K1 = 0.356, K2 = 0.033, K3 = 0.183; B. Analysis on the influencing factors of hidden blood loss after intertrochanteric fracture of femur in the elderly. Medical records of the two groups were reviewed, and gender, age, operative time, fracture type, internal fixation method, anesthesia method, complications, anticoagulant medication, etc., were recorded. Univariate and multivariate Logistic analyses were completed. C. Artificial intelligence model construction. Artificial intelligence R software was used to develop nomogram model to predict latent blood loss after intertrochanteric fracture of femur in the elderly. The ROC curve was drawn to analyze the predictive efficacy of the histogram model for latent hemorrhage after intertrochanteric fracture for senior people [5].

1.4 Statistical Analysis

SPSS18.0 software was used for processing. The statistical data were represented by χ^2 test and n (%), while the measurement data were represented by t test and $(\bar{x} \pm s)$. The it was statistically significant in difference as $P < 0.05$.

2 Results

2.1 Univariate Analysis on Influence of Latent Blood Loss after Intertrochanteric Fracture of Femur in Elderly

All the enrolled patients were performed with the operation, and 35 cases (29.17%) had recessive blood loss after the operation. The univariate results showed that the incidence of occult blood after operation in elderly patients with intertrochanteric fracture had no statistical significance with gender, age, operation time and complications ($P > 0.05$).It had statistical significance with fracture type, internal fixation method, anesthesia method and anticoagulant drug administration ($P < 0.05$), as displayed in Table 1.

Table 1. Univariate analysis on the influence of hidden blood loss after intertrochanteric fracture for senior people

Univariate analysis		Number of cases (n = 120)	Blood loss group (n = 35)	Non-blood loss group (n = 85)	χ^2	P
Gender	Male	65	22	43	0.691	0.413
	Female	55	13	42		
Age	≥80	61	19	52	1.094	0.538
	<80	59	16	43		

(*continued*)

Table 1. (*continued*)

Univariate analysis		Number of cases (n = 120)	Blood loss group (n = 35)	Non-blood loss group (n = 85)	χ^2	P
Duration of operation	≥3 h	48	15	33	0.693	0.663
	<3 h	72	20	52		
Fracture type	Stable	67	26	41	1.213	0.491
	In-stable	53	9	44		
Internal fixation method	Internal fixation of the hip	67	31	36	5.691	0.033
	The external fixation of hip	53	4	49		
Anesthesia	General anesthesia	78	28	50	8.956	0.016
	Intraspinal anesthesia	42	7	35		
Complication	Yes	36	10	26	1.201	0.848
	No	84	25	59		
Taking anticoagulants	Yes	81	17	64	7.091	0.021
	No	39	18	21		

2.2 Multivariate Logistic Analysis on the Influence of Hidden Blood Loss

Multivariate Logistic analysis results showed that the incidence of postoperative occult blood for senior people with intertrochanteric fracture was statistically significant with fracture type, internal fixation method, anesthesia method and anticoagulant medication (P < 0.05), as demonstrated in Table 2.

Table 2. Multivariate Logistic analysis on the influencing factors of latent hemorrhage

Multivariate	B value	S.E value	Wald value	P	OR	95%CI
Fracture type	1.448	0.311	7.395	<0.0001	6.392	6.112–7.986
Internal fixation method	1.241	0.298	5.351	<0.0001	5.124	4.525–6.491
Anesthesia method	1.068	0.241	6.998	<0.0001	7.301	6.493–7.856
Anticoagulant medication	1.332	0.285	5.236	<0.0001	5.356	4.645–5.867

2.3   To Construct the Nomogram Based Prediction Model of Hidden Blood Loss by Artificial Intelligence

Artificial intelligence R software was used to establish a model of the histogram to predict the latent hemorrhage. Based on the model, the score for fracture type was 63 points. The score for anesthesia method was 37.6 points. The internal fixation method scored 71.5 points, as shown in Fig. 1.

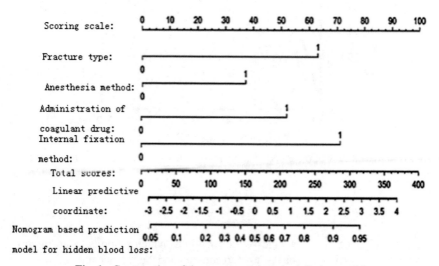

Fig. 1. Construction of the nomogram based prediction model

Note: The corresponding points of different variables can be found in the nomogram the total score can be obtained by adding them together to obtain the prediction model for blood loss after femoral intertrochanteric fracture in the elderly.

2.4   Prediction Efficacy of the Nomogram Model for Latent Hemorrhage After Intertrochanteric Fracture in the Elderly

The results of ROC curve showed that the C-index of the nomogram model for predicting hidden blood loss after intertrochanteric fracture for seniors was 0.816, which showed good differentiation. The area under the curve was 0.832, and the 95%CI was 0.7577–0.899, showing good performance in prediction, as shown in Fig. 2.

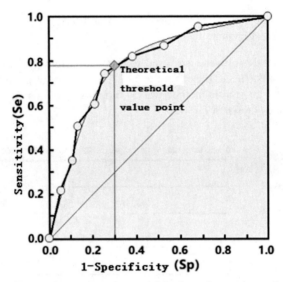

Fig. 2. Predictive efficacy of the nomogram model for latent hemorrhage after intertrochanteric fracture for seniors

3   Discussions

Artificial intelligence is a new intervention method, which belongs to an important branch of computer science. It can be used to comprehend the essence of intelligence and produce new intelligent machines that has similar response similar to human intelligence [6]. In this study, a prediction model of latent hemorrhage after intertrochanteric fracture for seniors was constructed by artificial intelligence. The results demonstrated that the fracture type scored 63 points. The score of anesthesia method was 37.6 points. The score of internal fixation was 71.5 points. The results of ROC curve showed that the C-index of the nomogram model for predicting latent hemorrhage after intertrochanteric fracture for seniors was 0.816, which showed good differentiation. The area under the curve was 0.832, and 95%CI was 0.7577–0.899, showing good predictive efficiency, indicating the possibility of constructing a prediction model of latent hemorrhage after intertrochanteric fracture for seniors with the help of artificial intelligence software, which could obtain high predictive sensitivity and develop intervention measures for the prevention of postoperative hidden blood loss in patients.

In conclusion, there are many influencing factors for latent hemorrhage after intertrochanteric fracture for seniors. The prediction model of hidden blood loss based on artificial intelligence has good diagnostic sensitivity and is worthy of popularization and application.

Acknowledgement. Special gratitude goes the research on the clinical study of tranexamic acid in reducing blood loss in senile intertrochanteric fracture with PFNA in perioperative period (Project Number: 2017AB003513) for the support of this study.

References

1. Song, C., Zhou, R., Huang, Y.: Clinical analysis of proximal femoral anti-rotation intramedullary nails in the treatment of senile femoral intertrochanteric fractures with weak lateral wall. Chin. J. Bone Joint **9**(9), 73–76 (2020)
2. Dou, H., Huang, W., Hua, J., et al.: Perioperative latent blood loss in elderly patients with unstable intertrochanteric fracture treated with FHR and its clinical significance. Chin. J. Med. Guide **21**(4), 197–201 (2019)
3. Zheng, H., Zhang, Y., Wang, H., et al.: Comparison of perioperative hidden blood loss for intertrochanteric fractures in the elderly by different intramedullary fixations: a randomized controlled study protocol. Medicine **99**(48), e21666 (2020)
4. Li, J., Ling, J., Li, C., Huang, G.: Effects of modified Shiquan Dabu decoction on postoperative latent blood loss and blood glucose levels in elderly patients with intertrochanteric fracture. Shaanxi J. Tradit. Chin. Med. **41**(9), 80–83 (2020)
5. Tian, Y., Dong, S., Liu, S., et al.: Analysis of latent blood loss after combined external fixator in elderly patients with high risk intertrochanteric fracture. Natl. Med. J. China **1**(5), 373–377 (2020)
6. Zhu, J., Zhang, X., Chen, Y., et al.: Analysis of related factors of PFNA in the treatment of senile intertrochanteric fracture blood loss. Chin. J. Bone Joint Injury **34**(4), 44–46 (2019)

Auxiliary Role of Artificial Intelligence Technology in Landscape Architecture Design

Yinglin Xiao[✉]

Nanchang Vocational University, Nanchang, Jiangxi, China
yinglinxiao@nvu.edu.cn

Abstract. With the advancement of computer science and technology, architects have an unprecedented range of mechanical applications, and the practice of robots in construction is creating a new building code system. The roles of designers and systems are not yet clear, and discussions on this potential new system are needed. The purpose of this article is to study the auxiliary role of artificial intelligence technology in landscape architecture design. It combines machine learning, big data and intelligent cloud display technology to integrate various advanced algorithms. Internet equipment allows architects and developers to perform routine analysis, planning and design tasks before construction, and accelerate architects through related tasks, support land acquisition such as conception and related analysis, and deepen design and construction after docking. Artificial intelligence technology The use of technology in artificial intelligence can save architects' labor and planning time costs, reduce design time by 25%, and effectively improve architects' work efficiency.

Keywords: Artificial intelligence technology · Landscape architecture design · Collaborative design · Simulation experiment

1 Introduction

1.1 Background and Significance

Artificial intelligence technology is a branch of computer science intelligent machine design, research and application. Therefore, machines can perform human intelligent actions. Research in related fields includes robotics, language recognition, image recognition, natural language processing and expert systems [1]. Since the birth of artificial intelligence, theory and technology have become more and more mature and more widely used. One of the main goals of this research is to be able to handle certain complex operations that usually require human intelligence [2]. Compared with traditional design methods, artificial intelligence technology is not only useful for modifying graphics, copying, and storing and managing blueprints, but also drawing computer graphics to more effectively use existing graphics resources to coordinate division of labor and teamwork, which can greatly improve work efficiency [3].

J. Abawajy et al. (Eds.): ATCI 2021, AISC 1398, pp. 588–595, 2021.
https://doi.org/10.1007/978-3-030-79200-8_87

1.2 Related Work

Yu D Z believes that in the most common traditional architectural design process, people are the center and computers are auxiliary equipment [4]. Architects and planners use computer drawing tools to manually express their design concepts as 2D or 3D blueprints. Developers can use these blueprints to communicate and take actions, even in the contractor's plan approval department and even the most common house dealer in the industry [5, 6]. In this process, the architect not only needs to be responsible for the existing painting, aesthetics, abstraction, design and other technologies, but also needs to have the ability to calculate communication and expression, integration and return on investment [7]. Wu C believes that when the basic drawing technology becomes more and more similar to the basic configuration, the construction project requires a lot of time and labor costs for the first time, but it also needs to be able to steal space [8]. Today, artificial intelligence technology has become an important part of Japan's strategic emerging industries and has been widely used in all fields. In the development of intelligent buildings, problems such as difficulty in system integration, low scalability, and difficulty in coordination between systems [9, 10]. The advancement of artificial intelligence technology provides new ideas for solving these problems. The application of artificial intelligence technology in intelligent buildings will accelerate the development of intelligent buildings [11]. But their research is not completely accurate and incomplete.

1.3 Main Content

In this article, the artificial intelligence technology-assisted architectural design is the direction of discussion and research, and the climatic characteristics and requirements of buildings in my country's hot summer and cold winter climate regions are discussed. The advanced artificial intelligence technology assists public buildings in the planning stage. Methods of analyzing and comparing different schemes. And through simulation analysis, discover and explain the advantages and disadvantages of different design strategies, in order to achieve the purpose of auxiliary design improvement. The innovation of this article is that first, this article analyzes the historical process and characteristics of artificial intelligence technology in the auxiliary role of landscape architecture design, and combines the climate characteristics of our country's hot summer and cold winter areas, and analyzes how to determine the use of artificial intelligence in the design plan stage The starting method of technology-assisted building performance design. Then, by analyzing the historical process of computer-aided architectural design, it is clear how the design aids meet the early requirements of architectural design, and the goal of simulation analysis at this stage is not to verify the design, but to compare different design strategies. Support the architect's ecological and sustainable design ideas. Finally, analyze the main characteristics of passive buildings in hot summer and cold winter areas, and analyze the methods and processes of passive energy-saving building design assisted by artificial intelligence technology in more detail.

2 Methods on the Auxiliary Role of Artificial Intelligence Technology in Landscape Architecture Design

2.1 Theoretical Research Method

(1) Read the literature extensively, summarize artificial intelligence technology theory, practical application research results and interfaces, deficiencies and existing problems, find out the key points and difficulties of this article, and provide theoretical basis for the research.
(2) Information collation, summary analysis, application scope, value, solutions, and problems of artificial intelligence technology in actual engineering, and a more comprehensive summary of the theory and application of artificial intelligence technology in architectural engineering structure design.
(3) Analyze the current artificial intelligence technology software platforms commonly used in the market, and the application status, pros and cons, and information transmission problems of various platforms in structural design, and summarize the artificial intelligence technology solutions for various platforms.

2.2 Actual Engineering Research Method

(1) Based on the artificial intelligence technology core modeling software Revit, explore the structure design process, drawing, and structure design methods based on artificial intelligence technology technology, compare with the traditional structure design process, and sort out the differences in the two design processes Differences and pros and cons.
(2) Taking Lianyungang Roquette corn deep processing by-product packaging room project as an example, using ETABS software and traditional PKPM software to carry out structural design (the whole process) respectively, and compare: structural design process and mode, information collection and transmission, structural analysis results, in terms of construction drawings and other aspects, the use of artificial intelligence technology to solve the key technical problems of structural design compares the difference between traditional design methods and structural design based on artificial intelligence technology, reflects the application value of structural design based on BM, and finds out current technical problems. The actual research algorithm:

$$S(x) = S_0 + S_1(x - x_k) + S_2(x - x_k)^2 + S_3(x - x_k)^3 \tag{1}$$

Among them, $x \in [x_k, x_{k+1}]$, S_0, S_1, S_2, S_3 is the coefficient. At this time, a cubic polynomial can be used to determine the function value corresponding to the interpolation point t on the interval.

2.3 Support Vector Machine Regression Algorithm

The energy management system in the intelligent building measures the electricity consumption and heat consumption of the entire building, the measurement of the heat supply, and the electricity consumption of each subsystem and the equipment of the subsystem, and realizes the classification and time-sharing measurement. As an important part of the energy management system, energy consumption prediction can help property managers grasp the trend of building energy consumption and provide support for decision-making. A regression support vector machine (Support Vector Machine for Regression, SVR) is used to predict and analyze the energy consumption of an experimental building. SVR regression function means:

$$f(x) = \omega \cdot \varphi(x) + b \tag{2}$$

$\varphi(x)$ represents a nonlinear mapping function, ω represents the weight vector, and b represents the threshold.

3 Experiment on the Auxiliary Role of Artificial Intelligence Technology in Landscape Architecture Design

3.1 Collaborative Design Experiment Based on Artificial Intelligence Technology

Structural engineering is a variety of design priorities in the existing 2D design process, and software tools that match the profession are usually used to complete the main design content. Each department uses it as an AutoCAD tool to merge the 2D engineering drawing of the main design content and finally the design result of each department into a 2D engineering drawing to generate the final design result of the project.However, this is not an effective way to exchange information in the process of designing another profession, and it is impossible to effectively negotiate design information changes between professions with other professional designers. This ignores the impact of major changes on other professions. Software integrated with some PKPM series software has appeared, but other professions can send and share some design data between professions through software. But in fact, the use of this integrated design software cannot meet the general design requirements of complete and rapid information provision and dissemination. The emergence of artificial intelligence technology has changed the way professional designers work together in the traditional collaborative design process. All project participants, including designers, builders and owners, use artificial intelligence models to complete a common design method for organic integration projects. Since the artificial intelligence model is a parametric information model, there is a logical match between the model and the final 2D architectural drawing. When the model information changes, the architectural drawing information also changes. As a result, reduced design changes will cover the graphics and the information will be incorrect. The difference between artificial intelligence 3D architectural design and traditional 3D architectural design is shown in Table 1:

Table 1. The difference between artificial intelligence architectural design and traditional 3D architectural design

	Traditional 3D architectural design	Artificial intelligence architectural design
Design function	Renderings and virtual reality	Rendering, virtual reality, design, analysis, management
Parametric model	No	Have
Coordination efficiency	Lower	Higher
Relationship with 2D design drawings	2D Design accessories	Logical correspondence

Comparing the existing collaborative design process of 2D and 3D buildings, we can see that artificial intelligence-based collaborative design has many advantages over traditional collaborative design methods. In the traditional design process, each design major must complete drawings once a week, and general construction majors provide preliminary plans. Therefore, the structure major and the equipment major carry out the arrangement of structural components and the installation of equipment according to the architectural drawings.

3.2 Simulation Experiment

Using Autodesk revitmep software to create artificial intelligence models for water supply and sewage treatment systems. The BM model in the main building is connected to the Revit-Mep software through a general design platform for professionals and used as a carrier in the building to complete the creation of artificial intelligence technology models for professional plumbing systems. Use the interference check function of Reviit software and Navisworks software to perform interference check and layout optimization between pipelines and structures and buildings in the artificial intelligence model. At the same time, the parameterized drive function of the software is used to generate statistical content of materials, and to control the actual design and maintenance process according to design simulation.

The simulation data is shown in Fig. 1:

Through the above data, we can see the superiority of artificial intelligence technology. In this way, architects can help architects make their design work more pure. This is consistent with the conclusion of architectural thinking and frees you from tedious work. The advancement and development of science and technology enable machines to design better buildings than humans due to lack of final completion. This motivates architects to improve and maintain self-awareness. It can also be regarded as an incentive.

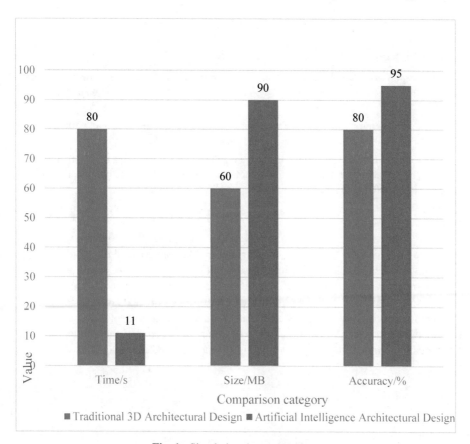

Fig. 1. Simulation data comparison

4 Simulation Experiment Data Analysis

The architectural model constructed by the traditional design method is a two-dimensional plane model composed of the most basic geometric elements such as points and lines. There is only plane information of each component in the model. The doors, windows, beams, columns and other components in the building information model constructed by artificial intelligence technology technology all contain three-dimensional information parameters. The artificial intelligence technology model composed of these basic pieces with three-dimensional attributes stores a large amount of design information, these information are stored in the huge database of the model in the form of parameters. The entire design process based on artificial intelligence technology is itself a parametric design process, and the model components are modified by modifying the parameters of the basic components in the model. The update of these modification information is interrelated. When a certain component information is modified, the artificial intelligence technology model will automatically update the modification information and directly reflect it in the design results. The analysis comparison is shown in Table 2:

594   Y. Xiao

Table 2. Comparison of artificial intelligence technologies

	The advantages of artificial intelligence in architectural design
1	It is possible to control the construction progress and estimate the construction cost during the actual construction process of the project through construction simulation and material statistics
2	Able to update information in time with changes in project design, so that all participants in the project can fully, timely and clearly understand the progress of the project
3	Able to share information, enable the design results to be presented in a timely manner through a collaborative design platform among various disciplines, and improve the quality of decision-making, thereby improving the overall project quality

The relevance of the information modified by the artificial intelligence technology model provides a basis for information sharing for collaborative design, avoids possible omissions and deficiencies in the information interaction process of traditional design methods relying on drawings, and effectively improves the accuracy of information transmission. The mesh-based simplification algorithm can be executed faster, and the amount of point cloud data can be significantly reduced, the number of triangular patches generated by subsequent reconstruction work can be reduced, and the efficiency of 3D display can be improved. The point cloud is rationalized and improved and suitable for

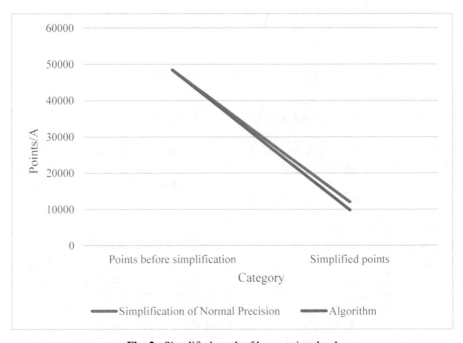

Fig. 2. Simplified result of horse point cloud

the initial stage. The simplified experimental results of this article for words and rabbit point clouds are shown in Fig. 2:

It can be seen from the comparison results that the method of simplifying the accuracy of the normal and the method of simplifying the curvature scanning can fully retain the contour and area information of the point cloud model in the geometric features of the point cloud model, and will solve the vector for multiple calculations. According to the grid in this article, the execution time can be increased by replacing other points in the grid with the center of the grid, depending on the proximity of the point cloud position in the space, and find the closest adjacency bounding box method. Parts that undergo major local changes will lose functional information, but it is useful for preliminary simplification of the point cloud.

5 Conclusions

In general, artificial intelligence technology has been widely used in the construction of construction projects in recent years. Artificial intelligence technology plays a very important role in engineering deployment because it has many advantages, including excellent visibility and adjustment capabilities, as well as the future-oriented advantages of balancing detection and simulation. It can be seen from the above that the application of artificial intelligence technology in construction engineering has become the trend of the times. Practice has proved that the application of artificial intelligence technology not only reduces the cost of construction companies, but also improves the company's business level and skills, thereby ensuring the overall quality and promoting the sustainable development of construction projects.

References

1. Jeavons, A.: What Is artificial intelligence? Res. World **2017**(65), 75 (2017)
2. Hassabis, D., Kumaran, D., Summerfield, C., et al.: Neuroscience-inspired artificial intelligence. Neuron **95**(2), 245–258 (2017)
3. Lu, H., Li, Y., Chen, M., et al.: Brain intelligence: go beyond artificial intelligence. Mob. Netw. Appl. **23**(7553), 368–375 (2017). https://doi.org/10.1007/s11036-017-0932-8
4. Li, Z.Z., Yu, D.Z.: Application prospect of artificial intelligence technology in vestibular disorders. J. Clin. Otorhinolaryngol. **33**(9), 895–897 (2019)
5. Meng, Y., Wu, N., Li, Y., et al.: BIM impact assessment of landscape architecture design. J. Phys. Conf. Ser. **1533**(4), 042093 (2020). (6pp)
6. Duzenli, Y., Alpak, E.M., et al.: The effects of model making on design and learning in landscape architecture education. Eur. J. Educ. Res. **2017**(70), 121–134 (2017)
7. Zhao, X.: Application of 3D CAD in landscape architecture design and optimization of hierarchical details. Comput.-Aided Des. Appl. **18**(S1), 120–132 (2020)
8. Li, P., Wu, C., Zheng, J., et al.: Consumer-centered collaborative design of fashion clothing brands: a communication and organizational structure study. J. Text. Inst. Proc. Abs. **111**(3), 405–415 (2020)
9. Wu, Y., Zhou, Z.: Study on the collaborative design architecture of tractor performance prototype. IOP Conf. Ser.: Mater. Sci. Eng. **892**(1), 012038 (2020)
10. Brian, H.: Discrete-event systems simulation. J. Oper. Res. Soc. **36**(5), 455–456 (2017)
11. Armbruster, D., Ringhofer, C., Thatcher, A.: A kinetic model for an agent based market simulation. Netw. Heterogen. Media **10**(3), 527–542 (2017)

Application of Computer Environment Simulation Technology in Explosion Proof Equipment Experiment

Ruotong Shi[1,2(✉)] and Weibin Zhang[1,2]

[1] State Key Laboratory of Coal Mine Safety Technology, China Coal Technology and Engineering Group Shenyang Research Institute, Fushun 113122, Liaoning, China
[2] CCTEG Shenyang Research Institute, Shenyang, Liaoning, China

Abstract. With the continuous development of science and technology, the explosion-proof technology is more and more perfect, but the explosion still exists. In view of the above situation, it is urgent to prevent the explosion and eliminate the hidden danger. At the same time, in order to avoid accidents in the research experiment, the computer environment simulation technology can be well applied. Usually, explosion-proof electrical equipment realizes explosion-proof safety through flameproof enclosure. The explosion-proof principle of flameproof electrical equipment is that the live parts of electrical equipment are placed in a special shell, which can isolate the sparks and arcs that may be generated by the electrical parts in the shell from the explosive mixture outside the shell. However, in the waste treatment of these electrical appliances, the shell will inevitably be mixed with various gases. In this paper, the combustion and explosion phenomena of explosion-proof electrical appliances in the process of explosion are studied and analyzed, and the computational fluid dynamics software FLUENT is used for simulation analysis, so as to realize the explosion-proof and explosion suppression function of explosion-proof electrical appliances. Based on the theory of gas explosion, this paper analyzes the mechanism of gas explosion in the use of electrical appliances. Assuming that the explosion source is a single methane gas, the explosion characteristic parameters such as explosion limit, explosion temperature and explosion pressure are theoretically estimated. The explosion limit is about 5.2%–14.3%, the explosion temperature is about 2883k, and the explosion pressure is about 0.87 mpa.

Keywords: Electrical enclosure explosion-proof · Numerical simulation · Explosion-proof equipment · Simulation experiment

1 Introduction

In recent years, due to the rapid development of science and technology, people's living standards have been greatly improved. The requirements for living standards are not only to meet the food and clothing, but also the safety problems in life have become the focus of public attention. There are all kinds of inflammable and explosive materials in

J. Abawajy et al. (Eds.): ATCI 2021, AISC 1398, pp. 596–604, 2021.
https://doi.org/10.1007/978-3-030-79200-8_88

people's life, especially household appliances, industrial appliances and so on. There are countless examples of explosion and even more serious consequences caused by improper operation of these explosives and their own quality problems. At present, the research of explosion protection has become an important research direction [1].

In general, the quality requirements of explosion-proof electrical appliances are higher. In order to ensure the more reliable operation of explosion-proof electrical appliances, it is particularly important to inspect them regularly. For the inspection personnel of explosion-proof electrical appliances, it is necessary to optimize the original inspection methods based on the operation characteristics of explosion-proof electrical appliances, so as to improve the inspection level of explosion-proof electrical appliances and reduce the occurrence of explosion. By analyzing the hazard sources in the inspection of explosion-proof electrical appliances and taking appropriate safety protection measures, the safety of explosion-proof electrical appliances can be improved and the probability of explosion of explosion-proof electrical appliances can be reduced. China's population base is relatively large, the daily use of electrical appliances is more frequent than other countries, and the scrap rate of electrical appliances is also higher, because the structure of electrical components control part of explosion-proof electrical appliances is more complex. There are many electrical junction boxes in explosion-proof electrical appliances. If there is no explosion-proof device, the safety of explosion-proof electrical appliances will be reduced. In addition, due to the serious aging of explosion-proof electrical circuit system and the protruding of flame-retardant rubber sealing ring, the internal circuit of explosion-proof electrical apparatus is easy to break when there is a fire outside, which affects the reliable operation of explosion-proof electrical apparatus. More lines, to a certain extent, increase the difficulty of explosion-proof electrical inspection. During the operation of explosion-proof electrical appliances, if the wire layout is unreasonable, it is easy to cause serious explosion phenomenon, affecting people's life and property safety [2].

In this paper, the explosion of electrical appliances and explosion-proof electrical appliances in daily life are studied in many aspects, and the related explosion-proof technology is described, focusing on the research and analysis of combustible gas explosion. Taking the explosion of electrical appliances in daily life as an example, the explosion-proof detection test is carried out, and the explosion-proof suppression technology of crushing equipment is studied, which can not only reduce the cost of electrical appliances The explosion probability in daily use can also greatly improve the safety of electrical appliances in the explosion, which can also provide a reference for explosion-proof test methods in other fields.

2 Related Concepts

2.1 Explosion Proof Technology

At present, the general explosion-proof technology mainly consists of explosion pressure relief, inert material explosion suppression and flame retardant quenching explosion suppression. Explosion pressure relief refers to the process that when explosion occurs in a closed container, the explosion relief device installed on the container is opened to release high temperature and high pressure gas to the external space, so as to limit the pressure inside the container to a safe range and prevent disasters. The common explosion suppression materials are inert gas, water mist and powder materials. Inert gas inerting is to inject inert gases such as nitrogen, carbon dioxide and halogenated hydrocarbons to dilute the oxygen content in the mixed gas, destroy the combustion triangle and prevent explosion [3].

2.2 Basic Forms of Combustible Gas Explosion

Gas combustion and explosion can be generally divided into the following four forms: isobaric combustion, deflagration, detonation and isochoric combustion. Before and after combustion, the pressure inside the vessel remains the same as the ambient pressure, the damage power is small, and the flame propagation speed is generally less than 1 m/s. Deflagration leads to the combustion gas cannot be discharged in time, it will produce pressure wave, the speed of pressure wave is greater than the propagation speed of flame front, so it is also called precursor shock wave. Deflagration wave is composed of precursor shock wave and combustion wave front, which constitutes a typical two wave three zone structure. Zone 0 is the initial state of combustible gas mixture, zone 1 is the state of gas swept by precursor shock wave, and zone 2 is the state of combustion products after combustion wave front passes through. In the reaction process, the explosion wave propagates at subsonic speed, and the unreacted medium is disturbed by the shock wave, which is an unstable combustion [4].

Detonation is the highest form of gas combustion. The reaction process and energy release speed are very fast, which will cause great damage. In the process of deflagration, if the obstacles or constraints are not released all the time, the pressure wave will continue to increase, the flame propagation speed will continue to increase until it catches up with the precursor shock wave, and the flame front and the precursor shock wave will merge into a supersonic shock wave with chemical reaction zone. In the deflagration process, the flame is the dominant factor, and the shock wave is the dominant factor in the detonation process [5].

Assuming that methane combustion is a one-step irreversible reaction, the chemical reaction equation is as follows:

$$CH_4 + 2O_2 = CO_2 + 2H_2O(\Delta H = -890.3KJ/mol) \tag{1}$$

2.3 Combustible Gas Explosion Mechanism

(1) Chain reaction theory

In the chain reaction, when the combustible gas molecules absorb the external energy, the molecular chain breaks and generates free radicals. Free radicals have great chemical reaction activity. They continuously collide with reactants to release energy and generate new free radicals, which make the combustion or explosion reaction continue. The chain reaction process can be divided into the following three stages: 1) chain initiation: after the combustible molecules absorb the external energy (greater than the reaction activation energy), the molecular bonds are destroyed and decomposed to produce free radicals. 2) Chain transfer (chain transfer): the highly active free radicals generated in the previous stage continuously collide with reactant molecules and free radicals to generate new free radicals and release a lot of energy. 3) Chain termination: the free radicals are consumed to form stable compounds or destroyed by collision with the reactor wall, resulting in the disappearance of the reaction carrier and the end of the chain reaction [6].

The chain reaction process of methane combustion is as follows.

$$CH_4 + O_2 \rightarrow CH_3 + HO_2 \tag{2}$$

(2) Thermal explosion theory

When the gas is ignited locally, the combustion reaction is carried out in a certain space, and the exothermic reaction is difficult to dissipate. The accumulation of heat leads to the constant rise of temperature, which in turn accelerates the reaction, and finally leads to explosion. This kind of explosion caused by thermal effect is called thermal explosion [7].

Taking methane combustion as an example, a small range of mixed gas near the ignition center is ignited, resulting in a rapid rise in temperature, forming a spherical high-temperature heat source. There is a great temperature gradient between the heat source and the unburned gas nearby, and the heat of the heat source will be transferred to the unburned gas quickly, resulting in the decrease of the temperature in the burned zone and the rapid rise of the temperature in the unburned zone, resulting in the chemical reaction. When the heat production rate of the system is higher than the heat dissipation rate, heat accumulation will occur in the system, which is called heat imbalance. The thermal imbalance makes the flame propagate continuously, forming a spherical outward propagating combustion wave front and finally forming an explosion [8].

Under the condition of heat balance, the heat production rate of the system is equal to the heat release rate, at the same time, the heat and heat dissipation per unit temperature change also reach the balance

$$\begin{cases} q_G = q_L \\ \dfrac{dq_G}{dT} = \dfrac{dq_L}{dT} \end{cases} \tag{3}$$

Heat release per unit time of the system:

$$q_G = Qv \tag{4}$$

The reaction rate V has the following relationship with the molar concentration of reactants

$$v = k[CH_4][O_2] \tag{5}$$

2.4 Numerical Simulation

CFD uses discrete methods such as finite difference method, finite element method and finite volume method to replace the continuous physical field (such as temperature field, velocity field, etc.) with the set of variables at finite discrete points. The nonlinear partial differential equations of flow field are transformed into algebraic equations at discrete points. By solving the series of simultaneous discrete control equations, the solution at discrete nodes is used as the field The approximate value of the variable.

Before using FLUENT software to analyze the flow field, it is necessary to establish the corresponding geometric model and mesh the model according to the requirements of simulation. Using the mesh generation module of ANSYS Workbench, the poor quality of mesh will even directly affect the convergence of simulation [9].

Structured grid has many advantages, such as fast generation speed, good quality of grid, good adaptability to boundary layer and so on. However, the disadvantage of structured grid is that it is difficult to deal with the complex shape model, and cannot realize the local refinement of the grid [10].

3   Experimental Simulation Design

3.1   Basic Assumptions

The process of electrical explosion is a complex three-dimensional turbulent flow process with high temperature and high pressure, involving physical and chemical reactions.

(1)   The explosion reaction is one-step irreversible combustion reaction, and the explosion source is a single methane gas, which is evenly distributed in the box;
(2)   The explosive gas is an ideal gas with uniform distribution, and its specific heat capacity is only related to temperature;
(3)   There is no heat exchange between the explosion reaction and the outside, the wall thickness of the box is ignored, and there is no slip on the wall adiabatic.

3.2   Experimental System

The control system controls the solenoid valve, vacuum pump, ignition energy generator and other devices to realize the operation of air intake, exhaust and ignition. In this paper, the probabilistic closed function transport model is used as the turbulent combustion model. Based on the molecular dynamics and conservation equations of random motion, the model has been widely used in premixed and non-premixed combustion.

4   Analysis of Simulation Test Results

4.1   Explosion Simulation

The experimental results show that the explosion limit of methane is 4.9%–5.8%. Because methane gas cannot react completely, the maximum explosion pressure has great influence on the reaction. When the concentration of methane is 10%, the maximum explosion pressure is about 0.784 MPa.According to the turbulence model and combustion model selected in the second chapter, for a spherical vessel with a radius of

Table 1. Comparison of calculated, experimental and simulated explosive parameters

Parameter	Calculated value	Experiment value	Analog value
Explosion limit (%)	5.2/14.3	4.9/15.8	–
Explosive pressure (MPa)	0.87	0.784	0.807
Explosion temperature (K)	2883	–	2780

0.17 m, the methane at stoichiometric concentration is calculated. The numerical simulation of the explosion process of the body under the condition of central ignition is carried out. The comparison of the calculated, experimental and simulated values of the explosion parameters is shown in Table 1. It can be seen that the deviation between the calculated value, the experimental value and the simulated value is small, and the calculation and numerical simulation model can better describe the explosion process of alkane gas.

4.2 Influence of Gas Concentration

In this paper, methane gas with mass fraction of 4.5%, 5%, 5.5% and 6% under normal temperature and pressure is selected as the initial value of simulation to study the influence of gas concentration on explosion development process. The proportion of oxygen and nitrogen meets the mass ratio of the two in air. The initial composition of gas is shown in Table 2.

Table 2. Component mass fractions

Component	CH4	O2	N2
Quality scores(%)	4.5	22.2	73.3
	5	22.0	73.0
	5.5	22.0	73.5
	6	21.8	72.2

The curves of temperature development, pressure development and pressure rise rate of different concentrations of gas mixture near the wall under normal temperature and pressure obtained by numerical simulation show that when the mass fraction of methane gas increases from 4.5% to 4.5%, the pressure rise rate increases In the process of 5.5%, the inflection point of temperature rise and the time of reaching stable temperature advance with the increase of methane mass fraction, and the explosion temperature also increases with the increase of alkane mass fraction. This is mainly because with the increase of methane concentration, the chemical reaction rate and combustion rate increase, and the energy released by the reaction increases with the increase of reactants. When the mass fraction of methane increases from 5.5% to 6%, the mass

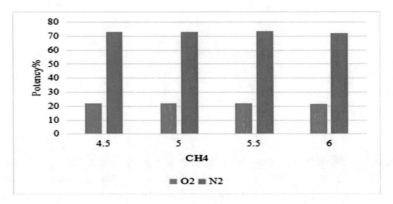

Fig. 1. Component mass fractions

fraction of oxygen decreases from 22% to 21.8%, which is not enough to support the complete combustion of methane. Because of the decrease of combustion supporting gas, the substances actually involved in the reaction will decrease with the increase of methane concentration. Therefore, with the increase of methane mass fraction, the moment of inflection point and stable temperature will be slightly delayed, and the explosion temperature will also decrease slightly. When the mass fraction of methane is 5.5% (i.e. stoichiometric concentration), the combustion explosion reaction is the most severe (Fig. 1).

4.3 Effect of Initial Concentration

It can be seen from Fig. 2 that when the stoichiometric concentration (methane mass fraction 5.5%, oxygen mass fraction 22%), the explosion temperature, maximum pressure, maximum pressure rise rate and other parameters reach the maximum value; when the methane mass fraction is lower than 5.5%, the parameters increase with the increase of methane mass fraction; when the methane mass fraction is higher than 5.5%, the explosion temperature, maximum pressure, maximum pressure rise rate and other parameters reach the maximum value 5%, the parameters decrease with the increase of methane content.

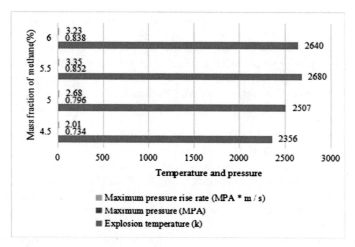

Fig. 2. Effect of initial concentration on parameters of explosion

5   Conclusion

At present, with the continuous development of science and technology and information technology, the explosion-proof technology is becoming more and more mature. However, with the deepening of science and technology, the structure of daily electrical appliances is more and more complex, and the shell material has also changed, which leads to the explosion situation of electrical appliances in daily use is more and more complex. Based on the gas explosion theory and computer simulation technology, this paper analyzes the mixed gas explosion caused by the aging of the internal circuit of electrical appliances Taking the explosion process simulation experiment as an example, the explosion-proof technology is further analyzed. In this paper, referring to the existing electrical enclosure, combined with the computational fluid dynamics software FLU-ENT, the development law of gas explosion process in electrical appliances is studied, and the passive explosion-proof and explosion suppression function of electrical appliances is realized. Based on the theory of gas explosion, the mechanism of gas explosion in crushing process is analyzed. Assuming that the explosion source is a single methane gas, the explosion characteristic parameters such as explosion limit, explosion temperature and explosion pressure are theoretically estimated. The explosion limit is about 5.2%–14.3%, the explosion temperature is about 2883k, and the explosion pressure is about 0.87 mpa.

References

1. Fotau, D., Magyari, M., Moldovan, L., et al.: Research on tests carried out on explosion proof cable entries. Environ. Eng. Manag. J. **18**(4), 811–815 (2019)
2. Brinovar, I., et al.: The classification of explosion-proof protected induction motor into adequate temperature and efficiency class. In: AIP Conference Proceedings, vol. 1866, 1, pp. 1–5 (2017)

3. Li, J., et al.: Design of the explosion-proof detection integrated system based on PGNAA technology. J. Radioanal. Nucl. Chem. **322**(3), 1719–1728 (2019)
4. Chen, M., Wang, K.: A bow-tie model for analyzing explosion and fire accidents induced by unloading operation in petrochemical enterprises. Process Saf. Prog. **38**(1), 78–86 (2019)
5. Nieto, A., Gao, Y., Fu, G.: A comparative study of gas explosion occurrences and causes in China and the United States. Int. J. Surf. Min. Reclam. Environ. **30**(4), 269–278 (2016)
6. Lauf, W.: Explosion-proof magnetic flux leakage inspection apparatus for well tubing. **12**(14), 113–114 (2016)
7. Nal, M.S., Killiolu, S.Y., Nver, B.: Comparison of explosion proof certificates of mining equipment. Madencilik **55**(4), 37–45 (2016)
8. Pasculescu, V.M., Vlasin, N.I., Florea G.D., et al.: Advances in impact resistance testing for explosion-proof electrical equipment. In: MATEC Web of Conferences, vol. 121, no. 6, pp. 110–115 (2017)
9. The quality common fault analyses of petrochemical engineering explosion-proof electrical equipment installation. (011, 135–138 (2015)
10. The safety assessment in explosives proof electrical equipment in hazardous locations based on set pair analysis, 000(002), 36–38 (2016)

Rapid Real-Time Collision Detection for Large-Scale Complex Scene Based on Virtual Reality

Sining Cheng[1(✉)], Xianjun Chen[1], and Huiyan Qu[2]

[1] Haikou College of Economics, Haikou 571127, Hainan, China
[2] School of Information Technology, Jilin Agricultural University, Changchun 130118, Jilin, China

Abstract. In the virtual environment, all kinds of geometric models are composed of basic geometric elements such as triangular patches and tetrahedrons. The key of its detection is to test the intersection of different geometric model elements. Based on the two-dimensional projection method, this paper presents a fast test method of intersection between basic geometric elements. By changing the internal data structure of the virtual object, an optimization algorithm is proposed by using the spatial decomposition method and spatial-temporal correlation. It is expected that this algorithm will sacrifice a small amount of storage space to improve the real-time performance of the algorithm. Not only that, it can also deal with the collision detection of multiple geometric models colliding at the same time. The experimental results show that this method can improve the efficiency of collision detection to a certain extent.

Keywords: Virtual reality · Collision detection · Hierarchical bounding box · Spatial location

1 Introduction

With the maturity of computer virtual technology, computing virtual reality technology is widely used in various scientific research fields. In orders to improve the experience of virtual reality, it is necessary to improve the authenticity of virtual reality. Authenticity of virtual reality can be by determining whether contact or penetration occurs between multiple objects [1, 2]. A basic law in the real world, without changing any geometric element of an object, cannot hold two identical objects in the space where only one object can be accommodated. If virtual reality technology can simulate this, can truly achieve realistic simulation of reality [3, 4]. Each geometric model consists of thousands of basic geometric elements. The geometric complexity of virtual environment greatly increases the computational complexity of collision perception. Collision detection is an unavoidable problem in the field of virtual reality, and the strict real-time and reliability of virtual environment put forward higher requirements for collision detection technology [5, 6].

© The Author(s), under exclusive license to Springer Nature Switzerland AG 2021
J. Abawajy et al. (Eds.): ATCI 2021, AISC 1398, pp. 605–610, 2021.
https://doi.org/10.1007/978-3-030-79200-8_89

Collision detection was first proposed in the field of robots, mainly to solve the problem of robot path avoidance, and then gradually developed into the field of virtual reality [7]. The most primitive and simple collision detection method depends on a large number of simple calculation methods. That is to say, double cross-test all the basic elements of the two geometric models. This method can get the correct results when the model is simple. If the degree of geometric model replication increases, this cross-test method will certainly not bear the [8, 9].

Through long-term accumulation, some mature collision detection techniques have been formed so far. There are two main types of collision detection algorithms between the two geometric models. Spatial decomposition method and hierarchical bounding box method. A core idea of the two approaches is to minimize the quantitative of the basic geometric elements required for cross-testing [10]. This paper brings the problem of collision detection into the problem of optimizing the search optimal solution. Combined with the existing hierarchical bounding box technology and particle swarm optimization algorithm, an efficient and real-time geometric model element cross detection method model is formed.

2 Algorithm

2.1 Particle Swarm Optimization Algorithm

The core idea of PSO algorithm is that a group can simulate the cooperation of individuals to find the best solution. Its advantage is that it is easy to implement and does not need a lot of data adjustment.

Update rule: initialized to a group of random solutions. Then by constantly changing the most solution. In each turnover, the particle updates by tracking two "extremum" (pbest, gbest) updates. particles to update the velocity and position by the following formula.

$$v_i = v_i + c_1 \times rand() \times (pbest_i - x_i) + c_2 \times rand() \times (gbest_i - x_i) \tag{1}$$

$$x_i = x_i + v_i \tag{2}$$

In formula (1) and (2), I = 1, 2, N, n is the total number of particles in this group. v_i is the velocity of the particle; x_i is the current position of the particle.

Rand(): a random number between (0, 1).

c_1 and c_2 is a learning factor, usually $c_1 = c_2 = 2$.

The maximum value of v_i is V_{max} (greater than 0), and if v_i is greater than W, then $v_i = V_{max}$

Formula (1) and (2) are PSO standard forms

$$v_i = \omega \times v_i + c_1 \times rand() \times (pbest_i - x_i) + c_2 \times rand() \times (gbest_i - x_i) \tag{3}$$

ω is called inertia factor, and its value is negative.

$$\omega^{(t)} = (\omega_{ini} - \omega_{gnd})(G_k + g)/G_k + \omega_{gnd} \tag{4}$$

There are also common ω for linear decreasing weights, ω the formula for variation with k is:

$$\omega = \omega_{max} - k \times \frac{\omega_{max} - \omega_{min}}{N_{iter}} \tag{5}$$

In order to avoid the particle easily exceeding the search range $[v_{imin}, v_{imax}]$, the particle velocity must be satisfied $v_{id}^k \in [x_{imin} x_{imax}]$ and the following relation exists in the iteration when the new particle velocity is generated:

$$\begin{cases} v_{imin} = \chi (x_{imax} - x_{imin}) \\ \quad v_{imax} = -v_{imin} \end{cases} \tag{6}$$

The χ is the velocity limiting coefficient and meets the $\chi \in [0.1]$. p^{best} Optimal position of individual and population g^{best}

$$P_{id}^{k+1} = \begin{cases} x_{id}^{k+1} & where\, fit\left(x_{id}^{k+1}\right) < fit(P_{id}^k) \\ P_{id}^k & others \end{cases} \tag{7}$$

$$G_d = min\left\{P_{1d}^k, P_{2d}^k, \ldots, P_{Md}^{k+1}\right\} \tag{8}$$

2.2 Update of Bounding Box Tree After Object Movement

The translation parameters along the x, y, z axes are equivalent to a translation vector $t = (x, y, z)^T$. the rotation parameters θ, φ, ψ around the X, y, z axes correspond to the transformation matrix

$$Rot(x, \theta) = \begin{matrix} 1 & 0 & 0 \\ (0 & cos\theta & -sin\theta) \\ 0 & sin\theta & cos\theta \end{matrix} \tag{9}$$

$$Rot(y, \varphi) = \begin{matrix} cos\varphi & 0 & sin\varphi \\ (\quad 0 & 1 & 0 \quad) \\ -sin\varphi & o & cos\varphi \end{matrix} \tag{10}$$

$$Rot(z, \psi) = \begin{pmatrix} cos\psi & -sin\psi & 0 \\ sin\psi & cos\psi & 0 \\ 0 & 0 & 1 \end{pmatrix} \tag{11}$$

Our default transformation order is to rotate around the z axis, the y axis, the x axis, then the synthetic rotation matrix is:

$$R = Rot(x, \theta)Rot(y, \varphi)Rot(z, \psi) \tag{12}$$

$$= \begin{matrix} cos\varphi cos\psi & -cos\psi sin\psi & sin\varphi \\ (cos\theta sin\psi + sin\theta sin\varphi cos\psi & cos\theta cos\psi - sin\theta sin\varphi cos\psi & -sin\theta cos\psi) \\ sin\psi sin\theta - cos\theta sin\varphi cos\psi & sin\theta cos\psi + cos\theta sin\varphi cos\psi & cos\theta cos\varphi \end{matrix} \tag{13}$$

3 Model Building

If the model is described by a polygon, the surface of the model should be triangulated first. The implementation steps are as follows:

(1) Calculate the covariance matrix C, which is a symmetric matrix, and its three eigenvectors are orthogonal to each other. Suppose that the coordinates of the ith triangle vertex in the model are (p^i, q^i, r^i), the area of the triangle is denoted as A^i, and the surface area of the model is the sum of the areas of the included triangles, denoted as A^H, then:

$$A^i = \frac{|(q^i - p^i) \times (r^i - p^i)|}{2} \tag{14}$$

$$A^H = \sum_{i=1}^{n} A^i \tag{15}$$

If the centroid of the ith triangle is m^i, then:

$$m^i = \frac{p^i + q^i + r^i}{3} \tag{16}$$

The centroid of the model is the weighted average of the centroids of the triangles contained in the model, denoted as m^H, and its weight is the area of the triangle

$$m^H = \frac{\sum_{i=1}^{n} A^i m^i}{A^H} \tag{17}$$

The covariance matrix is:

$$C = [c_{jk}] = \sum_{i=1}^{n} \frac{A^i}{12} \left(9m_j^i mi_k^i + p_j^i p_k^i + q_j^i q_k^i + r_j^i r_k^i\right) - m_j^H m_k^H A^H \quad 1 \le j, k \le 3 \tag{18}$$

(2) Find out the eigenvector of the covariance matrix, unit it and set the three axes of the local coordinates of the OBB; (A_x, A_y, A_z)
(3) All the vertices v^i of the model are projected on the three axes, and the maximum value ux, uy, uz and the minimum value wx, wy, wz in the three axes are calculated to determine the size of the OBB;
(4) Calculate the center of the bounding box.center

$$center = \frac{1}{2}((ux + wx)Ax + (uy + wy)Ay + (uz + wz)Az) \tag{19}$$

4 Evaluation Results

In order to compare the performance of several typical bounding boxes, this paper mainly analyzes several commonly used hierarchical bounding box methods from the aspects of construction difficulty, storage capacity, complexity and compactness of intersection test, and applicability in deformation body collision detection. The evaluation results are shown in Table 1.

Table 1. Comparison of several typical bounding boxes

	Structural difficulty	Storage	Intersection test complexity	Tightness	Calculation of bounding box update	Deformat collision suitability
A ABB	1	2	3	2	4	2
Spheres	2	1	1	4	2	4
OBB	4	3	4	3	1	3
K-DoPs	3	4	2	1	3	1

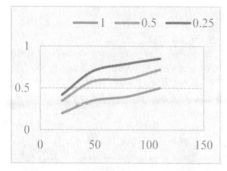

Fig. 1. Curve of search success rate in three different search spaces of Spheres model

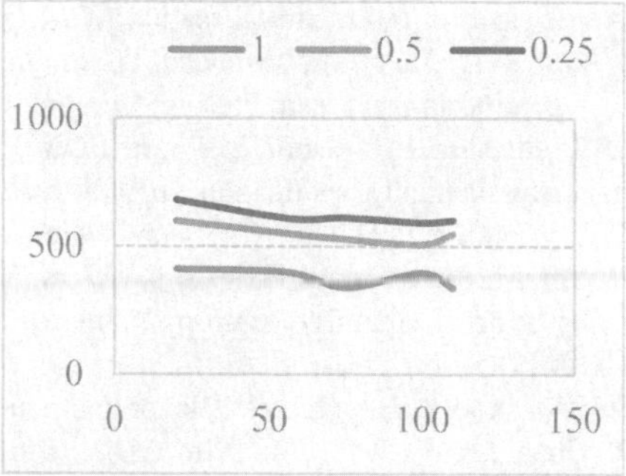

Fig. 2. Spheres time curves of three different search spaces in the model

Firstly, spheres model is used to obtain a series of sub search spaces by controlling the depth of binary tree to test the influence of search space size on search success rate and search time and obtain test data. Taking the number of contact points as the abscissa and the search success rate as the ordinate, according to the search success rate and search time consumption data under three different sizes of search space, the drawing is shown in Fig. 1 and Fig. 2. The three curves represent 1 times search space, 0.5 times search space and 0.25 times search space respectively.

5   Conclusion

This paper focuses on collision detection technology, aiming at the shortcomings of traditional algorithms, and further studies how to improve the efficiency of collision detection in complex scenes. In order to reduce the computational complexity and speed up the detection speed, different detection strategies are used according to different situations. Based on the traditional hierarchical bounding box algorithm, a novel multi object collision detection algorithm based on spatiotemporal correlation is proposed. With the increasing complexity of virtual environment, collision detection technology still needs further research.

Acknowledgments. This work was financially supported by Hainan Provincial Natural Science Foundation of China, Item number: 619QN247.

References

1. Oh, S., Hwang, S.: A GJK based real-time collision detection algorithm for moving objects. In: Wang, R., Shen, E., Gu, F. (eds.) Advances in Cognitive Neurodynamics ICCN 2008, pp. 817–820. Springer, Dordrecht (2008). https://doi.org/10.1007/978-1-4020-8387-7_142
2. Elbamby, M.S., Perfecto, C., Bennis, M., et al.: Towards low-latency and ultra-reliable virtual reality. IEEE Netw. **32**(2), 78–84 (2018)
3. Thies, J., Zollhofer, M., Stamminger, M., et al.: FaceVR: real-time facial reenactment and eye gaze control in virtual reality. ACM Trans. Graph. **37**(2), 25.1-25.15 (2018)
4. Wang, X., Tang, M., Manocha, D., et al.: Efficient BVH-based collision detection scheme with ordering and restructuring. Comput. Graph. Forum **37**(2), 227–237 (2018)
5. Raghupathi, L., Grisoni, L., Faure, F., et al.: An intestinal surery simulator: real-time collsion processing and visualization. IEEE Tans. Vis. Comput. Graph. **10**(6), 708–718 (2004)
6. Ericson, C.: Real-Time Collision Detection. CRC Press, Boca Raton (2004)
7. Mo, Y., Song, Z., Li, H., et al.: A hierarchical safety control strategy for exoskeleton robot based on maximum correntropy Kalman filter and bounding box. Robotica **37**(12), 2165–2175 (2019)
8. Esposito, D.: Hierarchical Blazor components. MSDN Mag. **34**(3), 44–46 (2019)
9. Chang, J.-W., Wang, W., Kim, M.-S.: Efficient collision detection using a dual OBB-sphere bounding volume hierarchy. Comput.-Aided Design **42**(1), 50–57 (2008)
10. Zhao, W., Cai, X.: A PSO optimization algorithm based on the solution space devision. J. Univ. (Sci. Edn.) **50**(4), 725–732 (2012)

Application of Artificial Intelligence in the Academic Search Engine

Guoqing Yue[1,2] and Shaojie Peng[3(✉)]

[1] School of Computer Engineering, Anhui Sanlian University, Hefei, Anhui, China
[2] College of Computing and Information Technologies, National University, Manila, Philippines
[3] Anhui University Library, Hefei, Anhui, China

Abstract. Based on the introduction of Google Scholar and other traditional academic search engines, this paper analyzes the comprehensive application of artificial intelligence technology in the academic search engine based on the new features of network academic resources. Under the open access environment, based on artificial intelligence, a new generation of academic search engines, accurate and quick access to high quality academic knowledge provides the conditions; this paper discusses the based on artificial intelligence of a new generation of academic search engine., Semantic Scholar Academic search in the field of Computer Science. Through technical analysis and case analysis and Research on new generations of academic search engines such as using semantic search and artificial intelligence technology, and forecasts the development trend of the academic search engines.

Keywords: Artificial intelligence · Academic search engine · Semantic search

1 Overview to Artificial Intelligence and Academic Search Engine

Artificial intelligence (AI), also known as machine intelligence, refers to intelligence manifested by man-made systems [1]. AI represents a new engineering science designed for studying and developing theories, approaches, technologies and application systems for simulating, extending and expanding human intelligence, and also forms a sub-discipline of computer science [2] (Fig. 1).

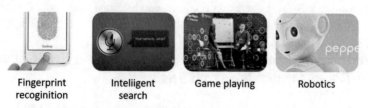

| Fingerprint recoginition | Inteliigent search | Game playing | Robotics |

Fig. 1. Latest AI applications

In the wake of application of artificial intelligence technologies in academic search engines, semantic search, individualization, intelligence, data mining and analysis, big

J. Abawajy et al. (Eds.): ATCI 2021, AISC 1398, pp. 611–616, 2021.
https://doi.org/10.1007/978-3-030-79200-8_90

data analysis and other functions have been made available in the new generation of academic search engines incorporated with artificial intelligence technologies [3].

2   Overview to Mainstream Academic Search Engine Technologies

Preliminary studies on the application of artificial intelligence in search engines have been conducted inside and outside China. Mainly, Google Scholar, Microsoft Academic (overseas representatives), Xueshu.Baidu and CNKI Scholar (domestic representatives) are available as the mainstream academic search engines [4]. Refer to Table 1 for general information.

Table 1. Overview to mainstream academic search engines

Academic search engines	Website	Release date	Main technologies
Google Scholar	https://scholar.google.com	Nov. 2004	PageRank, rank ordering based on paper citation
Microsoft Academic Search	https://academic.microsoft.com	Nov. 2009	Knowledge mapping, ordering based on relevancy and influence of objects
Xueshu.Baidu	http://xueshu.baidu.com	Jun. 2014	Ordering based on relevancy and frequency of citation
CNKI Scholar	http://scholar.cnki.net	2008	Intelligent indexing and in-depth knowledge mining

Google Scholar, the most popular free academic search engine, was released in November 2004. Google Scholar covers most published academic journals in the world and allows extensive searching of papers, books, abstracts and articles [5].

Microsoft Academic Search, an on-line free academic search engine developed by Microsoft Research Asia, provides ordering of search results based on the two factors: relevancy of words searched and influence of the objects searched in the world [6].

3   Introduction to and Technical Framework for a New Generation of AI-Based Academic Search Engine

Google Scholar is the largest academic search engine with more than 100million pieces of academic literature information. Even though, some gaps still exist in the system as a large number of literature actually does not fall into the category of academic literature from the measurement perspective of scientific researchers [7].

Listed below are two new generations of AI-based academic search engines, Semantic Scholar and wolframalpha.

3.1 Semantic Scholar – An AI-Based Academic Search (www.semanticscho lar.org)

A free academic search engine, named Semantic Scholar, was released by Allen Institute for Artificial Intelligence (AI2, a non-profit research institute) on November 2, 2015. Incorporated with massive AI technologies, working efficiency of Semantic Scholar has been greatly improved with more help from academic efforts of scientific researchers [8] (Table 2).

Table 2. Technical frame chart of Semantic Scholar.

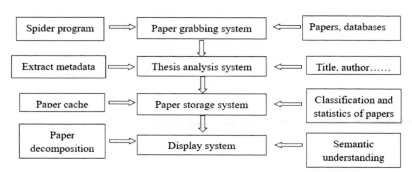

3.2 Wolfram Alpha – A Computation-Based Knowledge Search Engine

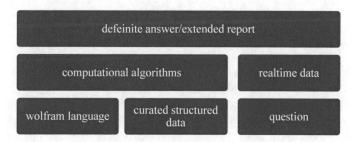

Fig. 2. Technical frame chart of Wolfram Alpha

A new generation of search engine, named Wolfram Alpha (http://www.wolframal pha.com/), was launched by Wolfram Research, Inc. in May 2009 [9]. When the users type in any question to be searched in the search bar, the search engine will directly show answers instead of plenty of web links. The users can submit command and computing requirements in a textbox, and then Wolfram Alpha starts to compute by a built-in knowledge base comprised of selected structured data and provides an answer and gets back to any related visual graphics. Wolfram Alpha is different from general semantic search which adds indexes to different answers and attempts to match the question and possible answers [9]. Wolfram Alpha allows searching, computation and analysis of all fields. Additionally, Wolfram Alpha can recognize pictures uploaded by the users (Fig. 2).

3.3 Comparative Analysis of Academic Search Engine Cases

Similarities and differences between the new generations of academic search engines were compared with traditional ones in some case studies, to analyze accuracy, timeliness and comprehensiveness of search results.

Comparison between Semantic Scholar and Google Scholar
The paper took the up-to-the-moment deep learning of artificial intelligence as an example and presents search results of both search engines as follows (Fig. 3):

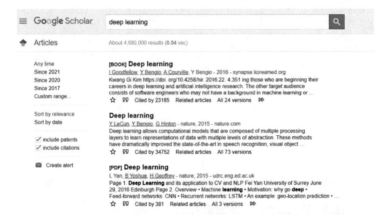

Fig. 3. Search result interface on deep learning with Google Scholar

The main search result interface of Google Scholar showed that there were 3,470,000 related literatures which were ordered as per PageRank and paper citation [10]. The research results included not only papers, but also books, internet thesis, open access resources. Some of the literatures were shown with links for complete content downloading. Author, title, journal for publication, date of publication and times of citation were also displayed. The search results could also be presented per time frame and time relevancy (Fig. 4).

It turned out that more information was presented on the search result interface on Deep Learning with Semantic Scholar, which was more helpful to the searchers. In addition to the title, author and times of citation of each academic dissertation, the Semantic Scholar can also indicate total publications of top-level dissertation, number of publications in the past years, top-level author, key words, literature review, etc. Ordering of the results is dependent on relevancy or influence factor.

By referring to the above-mentioned two research results, it was found out that Semantic Scholar search results were more relevant, higher in quality and easier to locate the review dissertation, and Semantic Scholar is also designed to present the papers as

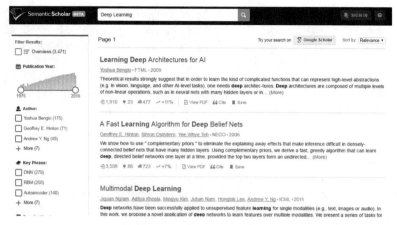

Fig. 4. Search result interface on deep learning with Semantic Scholar

per year of publication, top-level authors, etc. The user could go to Google Scholar, web link of which is available on the Semantic Scholar, for further searching should no academic result searched on Semantic Scholar is satisfactory to the user. The readers, if interested, could re-search with other key words. For this purpose, the performance on large amount of key words search by Google Scholar and Semantic Scholar was compared as follows (Table 3):

Table 3. Comparison on performance comparison of Google Scholar and Semantic Scholar

Performance Comparison	Google Scholar	Semantic Scholar
Accuracy	OK	High
Timeliness	High	High
Comprehensiveness	High	OK
Practicability	OK	High

4 Conclusions

In the context of open access, the AI-based new generations of academic search engines lay foundations for accurately and rapidly acquiring quality academic knowledge. The paper presents discussion on the AI-based new generations of academic search engines, computation-based knowledge engine wolfram alpha and www.semanticscholar.org in the computer field. It is also demonstrated in the paper that Semantic Scholar outperforms Google Scholar in accuracy and practicability through technical analysis and case study. Baidu Xueshu, Microsoft Scholar and CNKI Scholar as the rising and promising competitors are in the progress of utilizing diversified artificial intelligence and data mining technologies. In conclusion, it is the general trend to intelligentize academic search engines which also involves gradual popularization of machine translation, natural language understanding, voice search, full text understanding, graphic information extraction, as well as big data analysis.

Acknowledgements. This work was supported by 2018 Anhui Provincial Department of education outstanding young talents support program (gxyq2018128).

References

1. Russell, S., Norvig, P.: Artificial intelligence: a modern approach (2002)
2. Björk, B.-C., et al.: Open access to the scientific journal literature: situation 2009. PloS one **5**(6), e11273 (2010)
3. Ruibin, W., Guo, Y.: Study on the Baidu scholar based on the user experience. J. Mod. Inf. **4**(6), 56–62 (2017)
4. Tang, J.: Aminer: toward understanding big scholar data. In: Proceedings of the ninth ACM International Conference on Web Search and Data Mining (2016)
5. Yue, G., Peng, S.: Application mode research on knowledge sharing in the field of scientific research. Acad. J. Xi'an Univ. Posts Telecommun. **6**(3), 104–108 (2014). in Chinese
6. Thelwall, M.: Microsoft Academic automatic document searches: accuracy for journal articles and suitability for citation analysis. J. Informet. **12**(1), 1–9 (2018)
7. Zhang, H., Liu, J.: Discussion and analysis on frequently-used academic search engines. Netw. Secur. Technol. Appl. **5**(3), 157–158 (2015). in Chinese
8. Semantic Scholar search [O/L]. http://www.nature.com/news/artificial-intelligence-institute-launches-free-science-search-engine-1.18703
9. Wolfe, J.: The wolfram alpha computational knowledge engine. Math. Comput. Educ. **44**, 186 (2010)
10. Liu, Y., Song, Y.: Knowledge mapping for academic research on search engine. Doc. Inf. Knowl. **8**(6), 105–110 (2010). in Chinese

Artificial Intelligence and the Dilemma of Meaning of Life

Huaiqin Mu[(⊠)]

Xijing University, Xijing Road, Chang'an District, Xi'an, Shaanxi, China

Abstract. The meaning of life has always been an important topic for people to explore and discuss. The meaning of life is closely related to the background of the times and daily life. In the future era of artificial intelligence, great changes have taken place in the background of the times, and the meaning of life is facing difficulties. It is mainly manifested in three aspects: the change of daily life direction, the dilemma of self-transcendence of life meaning; the change of "self" essence, the dilemma of self-reflection of life meaning; the change of "self" essence, the dilemma of self-reflection of life meaning.

Keywords: AI · Meaning of life · Dilemma

A prominent feature of human nature is that people like to ask themselves who I am, where I come from, where I want to go, and what kind of life is worth pursuing. The meaning of life has always been an important topic for people to explore and discuss. The form of human existence is not only the physical existence, but also the more important spiritual existence. Therefore, human beings are not satisfied with the existence itself, but always thinking about the meaning of life and looking for the sustenance of life. The significance of life is so important, which is related to the happiness of individuals, and also affects the stability and development of a society.

What is the meaning of life? In the meaning of life, Eagleton points out that modernism regards human existence as accidental, and believes that human identity and behavior lack firm foundation; [1] let the meaning of life return to human life itself from the theoretical discussion, no longer superstitious in the definite and authoritative answer to a fixed question, but related to the actual life of an individual in a certain way – "take back the meaning of life from experts or experts, and put it back into the ordinary affairs of daily life" [1]. Giddens expressed a similar view in modernity and self-identity. He thought that the problem of "how will I live" is only to seek the answer in the trivial matters of daily life under the influence of reflexivity in modern society [2].

Obviously, different times, different social life environment, daily life trivia are also different. In primitive society, people are working hard for survival every day, and gathering and hunting are almost the whole of their daily life. In the agricultural society, people have precious and short leisure time, and the trivia of daily life is naturally enriched. But the relationship between man and nature is so close that it almost occupies the whole meaning of life. When the society enters the industrial age, the development of science reveals the essence of various mysterious phenomena. Religion is no longer

© The Author(s), under exclusive license to Springer Nature Switzerland AG 2021
J. Abawajy et al. (Eds.): ATCI 2021, AISC 1398, pp. 617–621, 2021.
https://doi.org/10.1007/978-3-030-79200-8_91

mysterious and can no longer support the sustenance of the meaning of life. Along with this, the relationship between people becomes very complex, and the inquiry of the meaning of life returns to human itself from the outside, and truly returns to the trivia of daily life. The development of society has not stopped. The era of artificial intelligence has begun to emerge, and various changes have been very obvious. In this case, the trivia of daily life that people are used to will change dramatically. With it, the meaning of life will also change, and the questioning of the meaning of life will face the dilemma of transformation.

1 The Direction of Daily Life Has Changed, and the Self Transcendence of the Meaning of Life is Facing Difficulties

Frankel, the founder of meaning therapy school, believes that the true meaning of life always points out of the self. The stronger the self-transcendence of the meaning of life, the more people can realize and achieve themselves [3]. This way of thinking, not only does not contradict with the judgment that people are looking for the meaning of life in the trivial matters of daily life, but also studies the direction of people's daily life to become a shortcut to discover the meaning of life. Whether it is the underdeveloped life or the modern life field, people are faced with two different fields: individual spiritual life and public social affairs. Whether daily life points inward to individual spiritual life or outward to public social affairs has become an important distinction in examining the meaning of life. In the era of industrialization, social division of labor is very common. Everyone is regarded as a specific part of the huge social machine. The role of the part directly affects the operation of the society. Therefore, people consciously associate individuals with the collective, and it is difficult to separate daily life from work. In order to be competent for work better, people are alienated as tools. Even if the daily life of pure leisure and entertainment activities, but also to relax themselves, in order to better play a personal role in working hours. In this state, the construction of the meaning of life is based on the other, that is, more transcendence, easier to give the sacred meaning to daily life. People live not only for themselves, but more for others to live better.

Entering the era of artificial intelligence, this state of living for others will be completely subverted. Because artificial intelligence will replace human work in a large area. People jokingly say that the future factory may only have one person and one dog. The dog looks at people and doesn't let them move machines. Of course, it is controversial to what extent artificial intelligence will replace human beings in the future. There are three kinds of conclusions. One is that artificial intelligence will replace all human work, that is, people think that artificial intelligence can complete every work better than human. At this time, if someone is willing to hire people instead of machines, the only reason is out of special emotion or nostalgia. This view is most in line with the monistic position. People who hold this view believe that everything in the world is made up of a single substance. Monism is very easy to deduce the view of the universe based on complete mechanism, which is easy to be questioned and challenged by people. The second view is that artificial intelligence will replace part of the work. People who hold this view believe that artificial intelligence will cause a large number of unemployed people, and the jobs they replace will be far more than the jobs they create. This kind of cognition is consistent with dualism, which holds that human beings have spiritual world besides material world. The material world can be simulated and replaced, while the spiritual world is mysterious and can never be replaced by artificial intelligence. The third view is that artificial intelligence will not take away any job opportunities. People who hold this view believe that human beings are different from animals and machines because they have skills that animals and machines can never replicate [4]. I prefer the second view, that is, artificial intelligence will replace part of the work, but I think the scope of the replacement will be very wide.

Just imagine such a scenario, in the future society, the vast majority of jobs are undertaken by artificial intelligence, human beings are only engaged in some jobs, and the intensity of these jobs will be greatly weakened, and a large number of surplus population will appear. Because the productivity of the whole society is highly developed, even if a small number of people are engaged in work, they can meet the needs of all human beings. Of course, the honor should be attributed to the selfless dedication of artificial intelligence. In this way, the "unemployed" population does not have to worry about doing nothing, but with the arrival of leisure time, it is not only no need to work, but also the dilemma of constructing the meaning of life. Human society no longer needs everyone to contribute, and people do not have to rely on others to help. The external direction of the meaning of life no longer exists, and the realization and achievement of itself are seriously challenged.

2   The Essence of "Self" Has Changed, and the Self Reflection of the Meaning of Life is Facing Difficulties

When the external direction does not occupy the mainstream, the self-direction will stand out. In fact, it is humanism that makes human beings get rid of the dilemma of meaningless life and groundless existence. Humanistic Religion worships human nature, expecting "human beings" to play the role of God in Christianity or Allah in Islam, or the role of natural law in Buddhism and Taoism. According to humanism, human beings must find meaning from their own inner experience. From the process of understanding the world, it is easy for people to be self-centered and radiate out like ripples. Egocentrism is human nature. As early as in ancient Greece, Socrates issued a maxim that "life without reflection is not worth living". Since ancient times, the issue of the meaning of life is closely related to the individual's self-examination and reflection. Rousseau believes that when looking for the code of conduct of life, we should understand that these codes are "in my heart, out of nature, no one can erase them. Just ask yourself what you want to do. I think good is good; I think bad is bad [5]. The subject of evaluating good and bad has changed from God to man. Cassirer put forward: "man is claimed to be an existence that constantly explores his own existence - an existence that he must examine and examine his living conditions at all times. The real value of human life lies in this kind of examination [6]. Therefore, human beings must examine their own living forms, perceive and sort out the context and direction of life, so as to have a meaningful life. In the face of the problem of the meaning of life, people must reflect on themselves, seek inspiration in the examination and reflection, and turn the examination of life into existence and meaning itself.

Obviously, the basis of self-direction lies in self-cognition. The question is what is "self"? Is "self" really certain? In the age of artificial intelligence, is the cognition of "self" still the same as before? As for what "self" is, Byron Reich believes that there are three possible choices: a brain trick, an emergent thought, or a soul [7]. The so-called brain trick mainly refers to two aspects: on the one hand, the brain obtains information from various senses; on the other hand, different parts of the brain are responsible for different tasks. The so-called emergent thinking means that the 40 trillion cells that make up the body work together and collectively acquire the physical properties that no individual thing has. The theory of soul, of course, is the individual experience of those who believe in the existence of soul. The generalization of these three kinds of "self" cognition represents people's understanding of "self" at present. In the future, what is "self" in the era of artificial intelligence may become more uncertain.

In a brief history of the future, Yuval Herali said: "in the 21st century, it is very likely that mankind will really turn to the goal of immortality" [8]. This argument sounds crazy, but it is logical. Because, for life and death, modern scientific and cultural views are totally different from religious views. They don't think that death has some metaphysical mystery. On the contrary, for modern people, death is a technical problem that we can and should solve. Especially with the development of genetic engineering, regenerative medicine and nanotechnology, this prediction is more and more optimistic. Let's make an assumption based on our current cognition. In the future, when we break the secret of the body, we will replace it wherever it is damaged. The organs of the body will not fail, but will be upgraded to more powerful super organs because of the intervention of technology, which will make the vitality more vigorous. However, it's easy to think of the thought experiment of Theseus' ship. Is the new body composed of artificial intelligence still the original "self"?

"Self" has undergone earth shaking changes. Has reflection lost its foundation? Is the meaning of life gained by reflection still valid? At this time, the self-reflection of the meaning of life is bound to face difficulties.

3   The Length of Life Has Changed, and the Ultimate Concern of the Meaning of Life is Facing Difficulties

In China, there is a saying: "it's hard to die forever". Achilles, the hero in the ancient Greek epic Iliad, went to hell after his death. He couldn't help sighing: "I would rather live in the world and be a servant of others than be king in the ghost of the dead". Poetic language portrays death as a shadowy ghost without sunshine and hope. As soon as the problem of death appears, the problem of the meaning of life immediately emerges. Up to now, a passage in Paul Kochakin's how steel is made, which is included in the middle school textbook, is still exciting, "a person's life should be spent like this On his deathbed, he was able to say, "my whole life and all my energy have been dedicated to the most magnificent cause in the world - the struggle for the liberation of mankind". Once the link between life and death is established, the meaning will come out naturally. It is said that how to live a person's life, just think about two virtues. One is the virtue of resume, that is, when we are looking for a job, how do you sell yourself on your resume? If you think of this scene, you will not waste your daily time, and will subconsciously do something worth writing to add color to your resume. The second is the virtue of eulogy, that is, if you imagine how others will deliver eulogy after you die, you will not waste your life. This is what Heidegger called "living to death".

Man is the only creature that knows that his life will come to an end. If you forget this, it's easy to live day by day, pretend that others like it, and forget the sense of urgency of life. If the problem of death is often brought to the front, people will choose what kind of life they want to live very seriously and sincerely, and every moment of life will not be easily missed. But the problem is, as mentioned earlier, with the arrival of the era of artificial intelligence, the progress of science and technology may defeat death. In this way, the meaning of life based on ultimate care will lose its support. In this way, human beings may be more anxious.

4 Concluding Remarks

On the prediction of the future, Mao Zedong made a famous conclusion. He said: "Marxists are not fortune tellers. The development and changes in the future should and can only give a general direction. They should not and cannot set the time mechanically. But the climax of the Chinese revolution that I said is coming. It is by no means an empty thing that is completely meaningless and impossible to achieve, as some people call it "possible to come". It is a ship standing on the coast and looking into the sea with the tip of the mast in sight. It is a sun shining in the east from the top of the mountain. It is a baby about to mature in her mother's womb In a sense, the future has come. Our thinking about the future is by no means groundless, but the preparation for the great possibility, even if it is an ideological experiment, the real situation must be much more complicated. Because of preparation, when the reality comes, it will not be in a hurry, and it is easy to give birth to a new meaning of life in the uncertainty of reality.

References

1. Eagleton, T.: The meaning of life, vol. 13, p. 93. Yilin publishing house, Nanjing (2012). Translated by Zhu Xinwe
2. Giddens, A.: Modernity and Self Identity: Self and Society in Late Modernity, p. 14. China Renmin University Press, Beijing (2016). Translated by Xia Lu
3. Frankl, V.: Man's Search for Meaning, pp. 110–111. Beacon Press, Boston (1992)
4. Reich, B.: Philosophy of Artificial Intelligence, vol. 5, pp. 78–115. Wenhui Publishing House, Shanghai (2020). Translated by Wang Fei
5. Rousseau, J.-J.: Emile, Paris, p. 348 (1967)
6. Cassirer: On human beings, p. 8. Shanghai Translation Publishing House, Shanghai (1985). Translated by Gan Yang
7. Reich, B.: Philosophy of Artificial Intelligence, vol. 5, p. 47. Wenhui Publishing House, Shanghai (2020). Translated by Wang Fei
8. Herali, E.: A brief history of the future, vol. 2, p. 18. CITIC Press, Beijing (2017). Translated by Lin Junhong

Cultivating Creative Talents of Digital Media Art Under the Background of Artificial Intelligence

Fei Li[✉]

School of art and Design, Yunnan Technology and Business University, Kunming, Yunnan, China

Abstract. With the rapid development of social economy and artificial intelligence informatization, the application scope of the Internet and computer technology is becoming wider and wider, including economic, political, military, education and other aspects. The major of digital media art is a classic example of the application of computer technology in education. Digital media art can be divided into two aspects: "digital media" and "art", emphasizing the integration of "technology" and "art", using computer new media design tools to cultivate composite application design talents who design and produce works of art. With the development of the times, the prospects for development are very broad, and many colleges and higher specialized universities have established related majors. These colleges and universities, as an education base that focuses on cultivating practical talents, should regard "practicality" and "innovation" as the top priority of education. Starting from the innovative training model, the professional positioning is clear and clear, and the training direction is diversified. Pay attention to the collaborative education of the integration of production and education, the construction of practical teaching in professional studios, and the integrated development of production, education and research driven by competition. This article analyzes the cultivation of innovative talents in digital media under the background of artificial intelligence, puts forward the current problems of digital media art talent training, studies and trains digital media art talents from the perspective of artificial intelligence, and conducts educational practice based on this, and finally details Explain the professional innovation of digital media art and the reform method of digital media art talent training mode under the background of artificial intelligence.

Keywords: Artificial intelligence · Digital media art · Innovative talents · Talent training

1 Introduction

Digital media is a new type of applied discipline, closely related to science and technology, mainly involving digital film and television, communication media, computer games, animation design, web design, digital illustration, cartoon comics, virtual simulation, digital audio and digital art design And other fields [1]. Quality art originates from the integration and intersection of technology, art and culture. It is an applied discipline that deeply integrates artistic creativity and science and technology. The frequency

J. Abawajy et al. (Eds.): ATCI 2021, AISC 1398, pp. 622–628, 2021.
https://doi.org/10.1007/978-3-030-79200-8_92

of information knowledge replacement has brought severe challenges to teaching staff and media creators, requiring creators to change creative ideas, concepts and methods according to new technologies, new concepts and new methods, and enhance the application value of technology [2, 3]. Therefore, strengthening the cultivation of innovative digital media art talents is not only a requirement for the development of the times, but also a requirement for the construction of laws. In terms of talent training mode, digital media art talents need to master the most cutting-edge science and technology, more need to have a higher artistic quality and creative awareness, and need to have the core qualities required for the development stage of my country's transformation [4].

The rapid development of digital media technology has made the integration of traditional newspapers, radio and television and the Internet more and more obvious. Media barriers are disappearing, the media and the communication environment have undergone tremendous changes, and the media content and communication technology are closely related. The online video broadcasting platform has created greater opportunities and space for the animation industry [5]. The traditional TV station is a closed, one-way, young and powerful broadcasting platform, and the user's choice and intimacy are not high. The network broadcast platform is a two-way, highly selective and intimate, open and win-win communication platform, with diverse content and obvious serviceability [6]. Moreover, the network platform has a large coverage and a wide audience. At present, there are 440 million computer network platform users in my country, and digital media videos cover nearly 200 million households. The user coverage ratio is second only to 68% of TV series and 44.4% of variety shows, reaching 35.2%., Mobile network user platform 1.01 billion [7].

The development of the industry requires relevant talents, and the cultivation of innovative digital media talents is urgently needed for future development [8, 9]. This kind of talent emphasizes discipline integration, cross-border training, and highlights practical ability. Judging from the current training model of colleges and universities, there are still some certainties: the curriculum model is divorced from reality, talent training is out of touch with talent needs; training goals are not clear, theory and practice are disconnected; curriculum settings are complicated and lack flexibility, and curriculum modules are not subdivided enough [10]. Emphasis on theoretical research, ignoring technical practice. This has caused a shortage of innovative and high-level digital talents in the industry. The research on course content, teaching methods and module process lacks systemicity, and the construction of the course system is not yet perfect. In response to these situations, we need to launch targeted measures to solve them.

2 Method

2.1 Establish a Sound Evaluation System

Establish an evaluation system for the quality of top-notch creative talents, conduct empirical application research on the indicator framework of digital media art training under the background of artificial intelligence, and propose solutions based on existing problems to provide scientific and effective reference materials for improving the quality of digital media art creative talents, To explore the way of digital media art creative talents, and provide practical guidance and reference for the perfect training process of digital media art creative talents.

2.2 Quantitative Data Evaluation

Most of the indicators of the talent training evaluation system are quantitative data, so objective entropy weight value methods can be used. The entropy weight value method determines the weight value according to the amount of information provided by the observation value of each indicator. The evaluation stage of the entropy method is as follows.

Step 1: If there are m individuals to be evaluated and n evaluation indicators, there is an original data matrix $X = (x_{ij})_{m \times n}$.
The x matrix is normalized to obtain the matrix y. When the j-th index x_i is a positive index, that is, the larger the better, the following changes will occur:

$$y_{ij} = \frac{x_{ij} - min(x_{ij})_i}{max(x_{ij})_i - min(x_{ij})_i} \tag{1}$$

Among them, I = 1, 2, ..., m; j = 1, 2, ..., n.
When the j-th index x_i is an inverse index, that is, the smaller the better, it is transformed as follows:

$$y_{ij} = \frac{max(x_{ij})_i - x_{ij}}{max(x_{ij})_i - min(x_{ij})_i} \tag{2}$$

Among them, I = 1, 2, ..., m; j = 1, 2, ..., n.
When the j-th index x_i is an interval index, that is, when the interval $[m_1, m_2]$ is the best and the closer to this interval, the better, make the following transformation:

$$y_{ij} = 1 - \frac{m_1 - x_{ij}}{max\left[m_1 - min(x_{ij})_i, max(x_{ij})_i - m_2\right]} min(x_{ij})_i \le x_{ij} \le m_1 \tag{3}$$

$$y_{ij} = 1 - \frac{x_{ij} - m_1}{max\left[m_1 - min(x_{ij})_i, max(x_{ij})_i - m_2\right]} m_2 \le x_{ij} \le max(x_{ij})_i \tag{4}$$

$$y_{ij} = 1, m_1 \le x_{ij} \le m_2$$

Step 2: Calculate p_{ij}, which is the proportion of the i-th index value under the j-th index:

$$p_{ij} = \frac{y_{ij}}{\sum_{i=1}^{m} y_{ij}} \tag{5}$$

Step 3: Calculate the entropy value of the jth indicator h_j:

$$h_j = -k \sum_{i=1}^{m} p_{ij} ln p_{ij} \tag{6}$$

Set $k = \frac{1}{lnm}$ to ensure that h_j is between 0 and 1, and if $p_{ij} = 0$, then $p_{ij} ln p_{ij} = 0$.
Step 4: Calculate the comprehensive evaluation value of individual I's credit status v_i;

$$v_i = \sum_{j=1}^{n} W_j p_{ij} \tag{7}$$

3 Experiment

3.1 Subject

Most applied colleges are transformed in newly established colleges. The newly-built college refers to the general undergraduate college upgraded in some junior colleges since 1998. After these colleges were upgraded to undergraduate colleges, most of them began to focus on general education, imitating the digital media art innovative talent training model of research universities to cultivate general talents, and the situation of talent training has formed a situation of high failure and low success. In recent years, with the development of artificial intelligence technology, newly-built undergraduate colleges have transformed into application-oriented undergraduate colleges, and cultivated innovative talents in digital media arts that serve social development. In order to deeply understand the talent training model of applied colleges in the transitional period and provide actual data support for this research, it is necessary to design targeted questionnaires to analyze the current application-oriented undergraduate colleges' training model of digital media art innovative talents under the background of artificial intelligence The current situation and existing problems.

3.2 Experimental Method

(1) Literature research method
   This experiment uses this method to summarize and sort out the previous literature of top innovative talents, and summarize and classify related theoretical and practical research.
(2) Model analysis method
   This study chooses this method as the theoretical basis of the study for further research. The model analysis method uses simple methods to resolve complex problems, concrete abstract problems, so as to understand the research object, organize the research process, and then analyze the problem. The model analysis method runs through the entire process of this research and is applied to the overall research.
(3) Interview method
   This study uses interviews with relevant experts in the education field to gain insights into the knowledge of experts and scholars on the quality evaluation research of the cultivation of top-notch innovative talents, find out the levels and aspects that should be paid attention to in the evaluation process, discuss the selection of indicators, and combine the existing Collect and analyze the indicators from the literature to construct the evaluation system needed for this research.
(4) Questionnaire survey method
   This study applied the questionnaire survey method twice. First, in the process of assigning weights to the indicators, education experts and teachers in related fields were selected to distribute the questionnaires, and the weights of indicators at all levels in the evaluation system were calculated based on the feedback results.

4 Results

A total of 1,000 questionnaires were sent out in this survey, 200 copies of each of the five schools of A University, B University, C College, D College, and E College. A total of 967 copies were returned, of which 940 were valid, and the effective recovery rate was 94%. Based on this, we obtain first-hand information about the current situation of digital media art creative talent training in colleges and universities, and use Excel2007 and other statistical methods to analyze and research the data obtained from the survey. According to the main content of the talent training model: guiding ideology and basic principles, basic content and realization methods, the designed survey content starts from these three perspectives, mainly from basic information, training goals and innovative awareness, education and teaching methods, methods and content, Curriculum and professional settings, supporting conditions and innovation environment and other five aspects are investigated.

The basic situation of the sample is as follows:

Table 1. Survey sample statistics of the current situation of digital media art creative talent training

Sample situation			
Gender	Male	523	55.63%
	Female	417	44.36%
Grade	Freshman	120	12.77%
	Sophomore	344	36.6%
	Junior	216	22.98%
	Senior year	260	27.66%
Subject	Natural science	632	67.23%
	Humanities and social sciences	308	32.76%

As shown in Table 1: From the gender of the collected sample, there are 417 girls, accounting for 44.36% of the total effective sample, and 523 boys, accounting for 55.63% of the total effective sample; from the collected sample grades, the freshman is 120. People, accounting for 12.77% of the total effective sample, 344 sophomores, accounting for 36.6% of the total effective sample, 216 juniors, accounting for 2.98% of the total effective sample, and 260 seniors, accounting for 27.66% of the total effective sample; from the collected In terms of sample disciplines, 632 people in the natural sciences, accounting for 67.23% of the total effective sample, and 308 in the humanities and social sciences, accounting for 32.76% of the total effective sample.

Figure 1 shows a survey of the degree of agreement among students that applied innovative talents and skilled talents are the same type of talent. The results show that: 19.57% are very agree, 28.09% are disagree, 11.91% are very disagree, 40.43% are uncertain. Analyzing the above data, it can be concluded that more than half of the students who agree with and are not sure about the options are not clear about the orientation of applied innovative talents and skilled talents, and they are not clear about their own positioning.

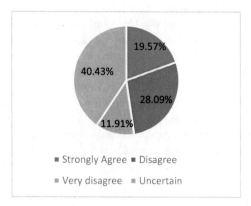

Fig. 1. Investigation on the understanding of the training orientation of applied innovative talents

Figure 2 shows a survey of which levels the colleges and universities pay more attention to in terms of talent training. The results show that: 91.06% of the focus is on the study of basic theoretical knowledge, 84.29% of the focus is on the ability to engage in related work, 20.85% of the focus is on dedication, social service, and lifelong learning, and the focus on critical spirit and innovation is the percentage. 13.19%. Analyzing the above-mentioned data can be obtained: Applied undergraduate colleges pay more attention to the cultivation of knowledge and the ability to engage in related work at the level of talent knowledge, ability and quality structure, and lack of attention to the improvement of student quality.

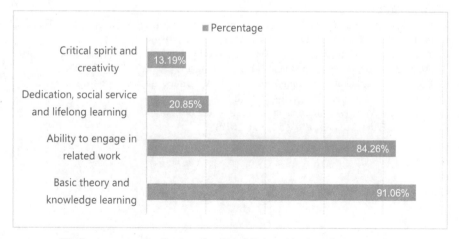

Fig 2. A survey of colleges and universities' positioning of talent training

5 Conclusion

In the process of cultivating innovative digital media art talents, colleges and universities should pay attention to the discovery of students' innovative potential, and clarify the core qualities of positive belief, continuous learning, and self-innovation that innovative talents should have in the context of innovation and transformation, so as to improve quality education With the help of concepts such as personalized education and open education, it will improve students' innovative ability, innovative consciousness and innovative thinking ability, promote students' all-round development, and establish a talent training model with a solid foundation for my country's social market economy. Our country should strengthen the training of digital media art professionals, which requires the establishment of relevant professional colleges to keep pace with the times, strengthen reforms, and build the most suitable teaching model for student development.

References

1. Mango, V.L., Sun, M., Wynn, R.T., et al.: Should we ignore, follow, or biopsy? Impact of artificial intelligence decision support on breast ultrasound lesion assessment. Am. J. Roentgenol. **214**(6), 1–8 (2020)
2. Yun, D., Xiang, Y., Liu, Z., et al.: Attitudes towards medical artificial intelligence talent cultivation: an online survey study. Ann. Transl. Med. **8**(11), 708–708 (2020)
3. Patris, K.: Influence of artificial intelligence on employment on the example of transport industry of France. Vestnik Univ. **1**(12), 71–77 (2020)
4. Liu, S., Li, T.: AVG comprehensive practice reform of digital media art based on fuzzy theory. J. Intell. Fuzzy Syst. **38**(1), 1–10 (2020)
5. Silva, D.A.E., Camargogrillo, S.V.D.: New paths for science: a contrastive discourse analysis of modifications in popularizing science through digital media. Bakhtiniana Rev. Estud. Discurso **14**(1), 54–81 (2019)
6. Burstow, S.: Screens in bed: visual art, sleep and handheld screen devices. Vis. Stud. **34**(1), 41–52 (2019)
7. Niu, W., Zhao, P.: Research on training model of innovative practical talents based on intelligent technology. World Sci. Res. J. **5**(11), 44–47 (2019)
8. Xiang, D., Gao, P., Liang, J., et al.: Teaching reform of the experiment of general chemistry principle by cultivating applied innovative talents. Univ. Chem. **35**(2), 1–5 (2020)
9. Xu, Z.: Construction and optimization of talent training quality based on data mining. Ingénierie Syst. D Inf. **25**(4), 419–425 (2020)
10. Liu, J., Liu, W.: Research on talent training model of new applied undergraduate colleges. Int. J. Inf. Educ. Technol. **9**(9), 652–660 (2019)

Content, Operation, Laboratory: Application and Innovation of Artificial Intelligence in Broadcast and TV Industry

Siwei Long and Dianyi Wu$^{(\boxtimes)}$

Department of Network and New Media, School of Advertising, Communication University of China, Beijing, China

Abstract. In recent years, with the rapid development of big data, cloud, IoT and other technologies, artificial intelligence has attracted more and more attention all over the world, and has developed rapidly in various industries. The broadcast and TV industry has been seeking media convergence, but also in the direction of intelligent transformation. Artificial intelligence technology plays an extremely important role in the transformation and development of the broadcast and TV industry. This paper introduces the application and innovation of artificial intelligence of broadcast and TV media from the perspective of content, operation, as well as the exploration of representative broadcast and TV institutions in the field of artificial intelligence laboratory. The research of artificial intelligence technology will be a major task for the industry for a long time to build a more intelligent broadcast and TV system with artificial intelligence technology, which is also a huge systematic project.

Keywords: Artificial Intelligence (AI) · Broadcast and TV Industry · Content · Operation · Laboratory

1 Introduction

In science-fiction literature and film, human beings simultaneously feel fear and allure in the presence of intelligent machines [1]. At present, the combination of artificial intelligence with various new technologies has become an important foundation for many emerging industries, and has been widely used in various fields. Vaibhav Gardé thought applications of artificial intelligence and machine learning—led by the rapid adoption of cloud computing—permeate most, if not all, marketing activities undertaken by small and large brands today [2]. While in the broadcast and television industry, Ricarda Schauerte, Stéphanie Feiereisen, Alan J. Malter believe the digital revolution is rapidly transforming the TV industry, in terms of production, distribution, and consumption [3]. Wenqing Zhang observed in order to protect the efficiency of the TV system and the steady improvement on quality, technical staff have strengthened the rational use of IT technology, and as a basis to promote the sound of television production system [4]. Similarly, the broadcast and TV industry also takes artificial intelligence as the focus

J. Abawajy et al. (Eds.): ATCI 2021, AISC 1398, pp. 629–634, 2021.
https://doi.org/10.1007/978-3-030-79200-8_93

of the current industrial layout, and is striving to establish system platforms and tools based on artificial intelligence services. Speech recognition, natural language processing, image recognition, computer vision technology and so on are all used in the process of broadcast and TV content production and operation. The new era of broadcast and TV technology comes.

2   Application of Artificial Intelligence in Broadcasting and TV Media Content

Annemarie Navar-Gill considered as streaming platforms entered the production space and became important homes for the commissioning of longform television content, they sought to build brand images as places that were both "data-driven" and characterized by work cultures of "creative freedom" [5]. The application of artificial intelligence in broadcast and TV media content is mainly realized by promoting the massive information collection, content material processing, terminal equipment adaptation and transmission security in the whole business process of media acquisition, editing, production, storage, transmission. Application of artificial intelligence has changed the working mode of media institution, improved working efficiency, and will gradually go deep into all aspects.

2.1   Content Collection

In the aspect of content collection, there are three main technology: first specific scenarios of the application of artificial intelligence, content intelligence aggregation. Through artificial intelligence learning modeling, journalists can select the required content from the massive content and gather it into hot news. Second, robot writing. Artificial intelligence technology can help to collect and analyze the Internet hot spot in real time, and automatically generate manuscripts for editing. The content of on-site interview can also be completed by intelligent editing tools such as machine writing and machine correcting. Third, intelligent material stripping and release. In the rapid process of news production, the intelligent strip splitting technology is used to strip the news materials quickly, and the news headlines are recognized at the same time.

2.2   Content Production

In the procedure of content production, artificial intelligence can be used in the following three scenarios: first, voice to text and machine translation. Second, virtual character animation. For example, produce a virtual host. Third, intelligent editing. Through the intervention of artificial intelligence technology, editors can analyze the elements, scene, emotion and other details contained in the video, and quickly generate video collections, program propaganda films and so on with the help of machine algorithm.

2.3 Content Storage

In the aspect of content storage, artificial intelligence technology has been applied to the intelligent media content asset management system. Through image recognition, speech recognition, content understanding and other artificial intelligence technologies, employees can make audio and video content written and labeled, which greatly improves the work efficiency of cataloging, saves the link of manual tagging for cataloging, and improves the accuracy of content retrieval and the reuse value of media content.

2.4 Content Distribution

In the transmission and distribution of content, artificial intelligence is mainly used in the audit process of content security. Artificial intelligence technology automatically monitors the sensitive content and illegal information in massive audio and video. In addition, broadcast and TV media institutions are also pay more and more attention to how to interact with users during the process of information distribution. Conversation robot and virtual intelligent assistant based on artificial intelligence are also used in the field of media content distribution. The service can provide users with information consultation and convenient interactive experience through natural dialogue.

3 Application of Artificial Intelligence in Broadcast and TV Operation

The proliferation of smart TVs and the surge of broadcasting contents due to the effects of Internet broadcasting and the creation of comprehensive programming channels have provided an environment in which various programs can be selected and consumed by TV viewers today [6]. Broadcast and TV industry institutions are generally facing with the challenge of how to retain users and how to increase revenue. How to do a good job in marketing products and media management, how to establish connections with users in right ways, and how to realize intelligent operation through data analysis platform, which are all need to rely on the joint promotion of big data and artificial intelligence.

Content in line with personal interests and good sensory experience is always the best element to attract users. By means of artificial intelligence and big data technology, user management, user behavior analysis, audience rating analysis, advertising effect analysis and data deep mining are carried out to provide users with more accurate personalized customized content, such as the construction of big data intelligent operation platform with thousands of people and thousands of faces, the accurate delivery of user images, and the combination of VR and AR technology means to show the interactive media content, get more user data, so as to achieve better user experience, get better communication effect. Taking NBA as an example, using artificial intelligence means, according to the characteristics of different terminal platforms and different preferences of each user, personalized recommendation of content, automatic generation of topics, collections, etc. Similarly, users can customize their own content. For example, users can choose all the content of a basketball star they like in a certain game, and all the plots of a star in a certain play.

In today's media operation centered on user experience, the form of advertising is not just simple and direct, more and more advertisements begin to be transformed into a part of the content, and the emergence of artificial intelligence has brought a major change in the form of advertising. Artificial intelligence decomposes and splits the content by video recognition technology, labels every commodity element in the content, arranges the advertising keywords in the character dialogue by semantic recognition, and presents them later with content assistance. These are not only part of the content, but also can be directly converted into advertisements. Users can buy them directly on TV. At the same time, with the maturity of VR/AR technology, users can experience the product experience personally, such as trying on the same clothes of idol stars. For users, this kind of advertisement is more about the content than the advertisement itself. The user's perception of the advertisement is decreasing, and the rejection of the advertisement is decreasing, but the conversion rate is improving, which brings certain economic benefits for the operation of the media.

4   Laboratory: R&D Base of Broadcast and TV Artificial Intelligence Technology

Curtis A. Knapp, Tom Polucci, Kay Sargent highlight workplace trends,explores what is on the horizon and how tech enterprises can optimize the places where their people work [7]. As early as the 1980s, MIT Media Lab, the first media lab in the world, was founded by MIT University of technology. It is committed to the innovation and application of media technology, promotes the change of news, the digitalization and intelligent development of media, provides the technology brain for the global media industry, and also becomes a pathfinder for the future technology of the global media industry. In recent years, the research scope of MIT Media Lab has gone beyond the narrow sense of the media industry. It redefines "media" as "technology of expression", and extends its research to computer, bioengineering, materials science, artificial intelligence and other fields. It uses innovative technology to enter the social service field of media, and explores the interaction mode, work mode and life style between media and people in the all-media era. Gwen Webber deems MIT Media Lab's new identity is because of the school is so broad in its scope and reputation [8].

After MIT, CNN, Time Warner, BBC, RTVE [9] and CCTV have set up laboratory institutions through cooperation with universities and institutions, or through independent operation. These laboratories provide space and human resources for the research and development of artificial intelligence technology for broadcast and TV.

From the perspective of the positioning of the laboratories of these institutions, whether the ARTIFICIAL INTELLIGENCE technologies developed by them serve the media business itself, or the media industry and society, there is one thing in common, that is, to focus on the actual needs of the organization, so that the technologies and tools developed by the laboratories can become catalysts and ultimately be used by the whole staff. Some of the representative projects developed by these laboratories focus on the reconstruction and empowerment of content acquisition and editing process, such as BBC News lab, which has developed various technical tools such as format conversion, grabbing, language translation and auditing to serve BBC journalists, editors and managers. Some focus on the creativity and performance of content. For example, Disney Studio lab has developed virtual reality, artificial intelligence, hybrid reality, UAV and other technologies on the creativity and performance of content, enabling Disney technicians and filmmakers to share and experiment with the emerging technology treasure house. Some focus on user research, use a variety of biological monitoring and metering equipment and technologies, simulate the environment, combine focus groups and in-depth interviews, and strive to optimize the content production, transmission, marketing and services of media from the perspective of understanding users. For example, Warner media innovation lab takes "insight, innovation and inspiration" as its motto and focuses on users.

Morgan G. Ames. examines the history of the learning theory "constructionism" and its most well-known implementation, Logo, to examine beliefs involving both "C's" in CSCW: computers and cooperation [10]. From the perspective of organizational structure, these laboratories are relatively independent in the media group, which fully reflects the characteristics of cross department cooperation of "business talents + laboratory research and technical talents", and are usually directly responsible by the higher-level leaders of the group. For example, BBC News lab is jointly established by BBC R & D department and future experience department, which emphasizes the cooperation of journalists and engineers in terms of personnel composition. Many representative projects of Disney Studio lab are completed through cooperation with Accenture interactive, Cisco, HP and other technical institutions. Turner and Warner Bros. of the original Time Warner Group have set up their own laboratories, while Warner media innovation Lab inherits them from the group level, bringing together the research operations and facilities of multiple departments. The chief marketing officer of Time Warner global media group is in charge of the lab, upgrading the lab to a new height and becoming the research center of the group's global business.

In the space setting and workflow, the laboratory space is independent, has a special office and entity display space, and reflects the characteristics of flexible and agile in personnel management and process. For example, the BBC News lab laboratory has its own space and autonomy, and exists in the form of streaming team, that is, a temporary team for a specific project. If someone comes up with a project that they want to explore, they usually separate it from their daily work, set up a project team in 3–12 months to conduct full-time research and exploration, and hand over their daily work to other people for help.

5 Conclusions

The opportunities and challenges brought by artificial intelligence to the development of broadcast and TV industry cannot be ignored. Broadcast and TV institutions need to seize the time, lay out artificial intelligence platform, carry out the integration and innovation of artificial intelligence in various fields, create more value and have broader application prospects. But at the same time, the research of artificial intelligence technology is still in the primary stage. It will be a major task for the industry for a long time to build a more intelligent broadcast and TV system with artificial intelligence technology, which is also a huge systematic project.

Acknowledgements. This work was supported by the Fundamental Research Funds for the Central Universities.

References

1. Geraci, R.M.: Robots and the sacred in science and science fiction: theological implications of Artificial Intelligence. Zygon **42**(4), 961–980 (2007)
2. Gardé, V.: Digital audience management: building and managing a robust data management platform for multi-channel targeting and personalisation throughout the customer journey. Appl. Mark. Anal. **4**(2), 126–135 (2018)
3. Schauerte, R., Feiereisen, S., Malter, A.J.: What does it take to survive in a digital world? Resource-based theory and strategic change in the TV industry. J. Cult. Econ. **PP**, 1–31 (2020)
4. Zhang, W.: Research on the integration of IT network technology and TV production and broadcasting system. IOP Conf. Ser. Earth Env. Sci. **100**(1), 012073 (2017)
5. Navar-Gill, A.: The golden ratio of algorithms to artists? Streaming services and the platformization of creativity in american television production. Soc. Media Soc. **6**(3), 56–63 (2020)
6. Jaekwang, K.: Recommendation method of smart TV programs reflecting content consumption concentration calculation. J. Eng. **2020**(13), 444–447 (2020)
7. Knapp, C.A., Polucci, T., Sargent, K., et al.: Rethinking the workplace: how the tech sector influences other industries. Corp. Real Estate J. **9**(1), 56–70 (2019)
8. Webber, G.: MIT media lab's new identity. Blueprint **33**(5), 45–51 (2011)
9. Subires Mancera, M.P.: The webdoc as a tool in the struggle for equality: en la brecha (2018) by Lab RTVE. Fonseca. J. Commun. **5**(6), 87–101 (2019)
10. Ames, M.G.: Hackers computers and cooperation. Proc. ACM Hum.-Comput. Interact. **2**(CSCW), 1–19 (2018)

Numerical Simulation Technology of Food Freezing and Refrigeration Process Based on Supply Chain

Yihan Hou[✉]

School of Energy and Architectural Engineering, Harbin University
of Commerce, Harbin 150028, Heilongjiang, China

Abstract. To simulate the change process of arbitrary physical properties with freezing time, this article first analyzes the current process of food freezing and refrigeration, and then numerically simulates this process in combination with supply chain theory to enhance the shelf life of food and save costs. The time-varying numerical simulation technology is applied to the self-developed simulator to analyze the food freezing and refrigeration process. The results show that comparing the water flooding characteristic curves of the two models, it can be seen that the freezing effect is weak due to less water production at the initial stage of production. When the water cut is lower than 30%, the curves of the conventional model and the time-varying model are exactly the same; as the freezing effect increases, under the same water content, the recovery degree of the time-varying model is higher, and this gap gradually increases as the production progresses. The final recovery factor of the time-varying model is 55%, which is 5% higher than that of the conventional model. Time-varying numerical simulation is more accurate than conventional numerical simulation in predicting waterflooding efficiency and ultimate recovery factor.

Keywords: Supply chain theory · Food freezing and refrigeration · Numerical simulation technology · Information symmetry theory

1 Introduction

Freezing is currently an important means to regulate consumer demand in different seasons and different regions. In 2019, the national aquatic products frozen products reached 64,816,800 tons. During the long-distance transportation, storage and sales of frozen food, the temperature fluctuations in the environment cause repeated freezing and thawing of frozen food. In order to improve this deterioration, new freezing methods can be used to reduce temperature fluctuations in the cold chain circulation process.

As one of the research methods of information asymmetry, supply chain theory has attracted the attention of scholars in recent years. From the perspective of supply chain theory, the conflict of interest between the principal and the agent, and the conflict of interest between multiple agents will affect the governance of the food supply chain. Yenipazarli A builds an incentive model based on supply chain theory in order

© The Author(s), under exclusive license to Springer Nature Switzerland AG 2021
J. Abawajy et al. (Eds.): ATCI 2021, AISC 1398, pp. 635–642, 2021.
https://doi.org/10.1007/978-3-030-79200-8_94

to achieve the goal of optimal output [1]. Khaksar E adopts mathematical analysis and game thinking to design a new incentive mechanism under asymmetric information conditions [2]. Liu Y explores the engineering quality government supervision agency chain and its multi-level incentive mechanism based on the dual supply chain theory [3]. Xu X confirmed through empirical research that the cumulative prospect theory can make a more reasonable explanation for the incentive and behavior patterns in the reward contract model [4]. Seth D builds an incentive model for owners and contractors based on multi-task principal-agent, so as to optimize the performance of engineering projects [5].

In this paper, a unified time-varying simulation framework of physical properties is added to the traditional numerical simulation methods, that is, the instantaneous areal flux is calculated by the area-weighted flow velocity, and the surface interpolation algorithm is used to calculate the corresponding phase infiltration under any areal flux and saturation. This article first analyzes the current process of food freezing and refrigeration, and then conducts a numerical simulation of this process in conjunction with supply chain theory to enhance the shelf life of food and save costs.

2   Supply Chain Management Methods for Food Freezing and Cold Storage

2.1   Supply Chain Theory Agency Model

Under the condition of information asymmetry, the supply chain theory is introduced into the dual principal-agent model, and the learning level, output effort level and cooperative effort level of producers and regulators in the absence of supply chain and the supply chain are compared respectively, and the numerical value is passed simulation to verify the validity of conclusions [6]. The learning level of producers and regulators is directly proportional to the incentive coefficient, negatively related to the learning cost, and has nothing to do with the supply chain and fixed compensation [7, 8]. Producers and regulators with high initial learning ability can increase consumers' expected income. The degree of effort of producers and regulators is positively correlated with the incentive coefficient and the influence coefficient of effort level, negatively correlated with output cost, and has nothing to do with fixed compensation, but fixed compensation can promote the establishment of a principal-agent relationship [9, 10]. The incentive coefficient is inversely related to the risk aversion coefficient and the degree of environmental volatility; when the supply chain coefficient is greater than the critical value, the effort level is negatively related to the supply chain coefficient, but the effort level is still greater than the effort level without supply chain behavior, and reaches a peak at the critical point [11, 12].

2.2   Numerical Simulation and Characterization Method of Time-Varying Laws of Physical Properties

In the time-varying process of food freezing and refrigeration, the degree of freezing of the water phase is an independent variable. It is generally expressed by the cumulative freezing amount of the water phase. The grid can be selected to describe the cumulative flux across the water surface, that is, through the water phase per unit area of the grid within a certain period of time. Cumulative volume, the formula is:

$$F = \sum_i \frac{Q_i}{A_i} \tag{1}$$

Where: F is the cumulative flux across the surface, Q_i Is the cumulative volume of the net water phase flowing into the grid in the i direction, A_i is the cross-sectional area in the i direction of the grid.

For a certain position in a certain injection-production relationship area, the value still depends on the grid selection, and numerical effects may occur. Therefore, this article uses an improved cumulative surface flux, the formula is:

$$k_{t1}[i] = \sum_j \cos(w_i^1, w_j^2) \tag{2}$$

$$F = \sum_t (v_i \Delta t) = \sum_t \sqrt{F_x^2 + F_y^2 + F_z^2} \Delta t \tag{3}$$

Where: v_t is the instantaneous flow velocity at time t, Δt is the time interval, Fx Is the instantaneous surface pass in the x direction at time t. The area-weighted darcy flow velocity is used to calculate the instantaneous area flux in the i direction of the grid, the formula is

$$F_i = \frac{(\sum A_i v_i)}{(\sum A_i)} \tag{4}$$

$$\ln\left(\frac{FI_{it}}{FI_{it} - 1}\right) = \alpha + \beta \ln FI_{it} - 1 + \varphi X_{it} - 1 + v_i + \tau_t \tag{5}$$

Where v_i is the velocity of DAX passing through a certain surface in I direction, and I + direction is positive. Formula (5) is to weight all faces in the I direction (one face in the corner grid may be adjacent to multiple faces). The dependent variables of the time-varying parameters are the parameters of food freezing and refrigeration, including porosity, permeability, relative permeability and capillary force curve. For porosity and permeability, the variation law of physical properties with freezing degree obtained from indoor freezing experiment is directly transformed into "cumulative surface flux physical property multiplier" relationship, which is arranged in table form for subsequent numerical simulation. For curve type parameters, such as relative permeability, capillary force, etc., the existing treatment method is to parameterize them (e.g., empirical formula of relative permeability, end value of relative permeability), and then characterize them as "cumulative surface flux characteristic parameter multiplier". In the simulation, the characteristic parameters are obtained first, and then the curves are calculated.

3 Numerical Simulation of Food Freezing and Refrigeration Process

3.1 Research Methods

This paper first analyzes the current process of food freezing and refrigeration, and then combines the supply chain theory to simulate the process, so as to strengthen the shelf life of food and save costs.

3.2 Experimental Design

The time-varying numerical simulation technology of food freezing and refrigeration is applied to the simulation of food freezing and refrigeration based on supply chain. In order to better illustrate the effectiveness of food freezing and refrigeration supply chain, the change law of decision variables without supply chain coefficient is used in numerical simulation. The fixed remuneration designed by consumers has nothing to do with the best efforts of producers and regulators, and can not stimulate the behaviors of producers and regulators. However, fixed remuneration designed by consumers can promote the construction of principal-agent relationship. Therefore, when consumers entrust producers and regulators to perform their duties, they should choose appropriate producers and regulators based on their professional competence. This just explains that in life, consumers tend to favor brand enterprises and trust the information released by mainstream media and government regulators.

Different supply chain coefficients should be given different quantities of food freezing coefficients. According to the analysis of numerical conclusions, the income of consumers, producers, and regulators increases first and then decreases as the supply chain coefficient increases. When the supply chain coefficient is small, its utility is smaller than when there is no supply chain; when the supply chain coefficient gradually increases, its utility first increases and then decreases, but it is greater than when there is no supply chain. That is, when the main body of the food supply chain has supply chain behaviors, the strengths and weaknesses are different.

4 Numerical Simulation Analysis of Food Freezing and Refrigeration Process

As shown in Fig. 1, the time-varying numerical simulation technology of food freezing and refrigeration is applied to the simulation of food freezing and refrigeration based on supply chain. The block is a block structure with saturated bottom water driven by weak gas cap and strong bottom water. With the continuous rise of supermarkets and shopping centers, the safety of fresh and cold food sold in supermarkets has gradually attracted people's attention. The main concern is packaging meat, seafood, vegetables, etc., which are not suitable for long-term storage, and have high requirements for supply and training.

In order to better illustrate the effectiveness of food freezing and refrigeration in the supply chain, the change law of decision variables without supply chain coefficient was

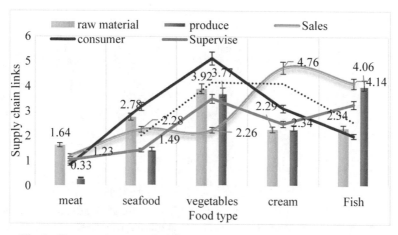

Fig. 1. Time-varying numerical simulation of food freezing and refrigeration

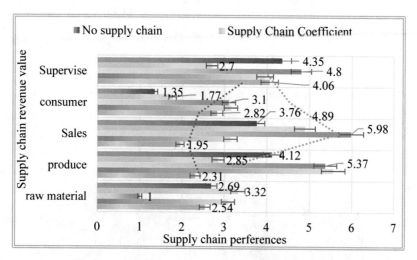

Fig. 2. Numerical simulation results of decision variables

simulated by numerical simulation. As shown in Fig. 2, the food freezing coefficient and supply chain coefficient obey the uniform distribution of expectation 0 and variance 1. There is information asymmetry in food supply chain, and the supply chain signal is difficult to be identified by the other party and cannot be accurately measured. It is easy to be affected by other variables. At the same time, the conclusion shows that the existence of supply chain has an impact on related variables, but its detailed changes are difficult to understand. Therefore, it is assumed that the supply chain coefficient obeys uniform distribution, excluding the interference of other variables, and describing the changes of related variables more accurately.

The variation of relative permeability with cumulative surface flux is shown in Table 1. The relative permeability curve of frozen food moved up slightly with freezing, while

Table 1. Change of relative permeability with cumulative surface flux

Food	Raw material	Produce	Sales	Consumer	Supervise
Meat	1.64	0.33	1.23	0.87	1.05
Seafood	2.78	1.49	2.28	3.2	1.45
Vegetables	3.92	3.77	2.26	5.13	3.52
Cream	2.29	2.34	4.76	3.13	2.51
Fish	2.34	4.06	4.14	2.03	3.29

the water phase permeability curve changed little. The results showed that the irreducible water saturation changed little, the residual oil saturation decreased, the two-phase co permeation zone increased, the isoosmotic point shifted to the right, and the food refrigeration efficiency increased. Combined with the experimental reservoir time-varying law, the time-varying numerical simulation technology is used to simulate the frozen food supply chain, and the simulation results of the time-varying model and the conventional model are compared.

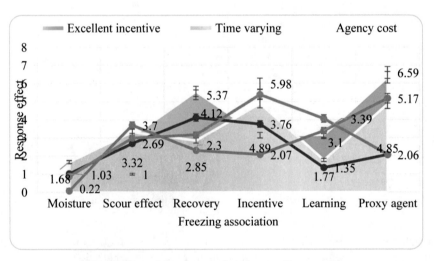

Fig. 3. Water drive characteristic curves of two models

As shown in Fig. 3, comparing the water drive characteristic curves of the two models, it can be seen that due to less water production at the initial stage of production and weak freezing effect, the curves of conventional model and time-varying model are exactly the same when the water content is lower than 30%; with the enhancement of freezing effect, the recovery degree of time-varying model is higher under the same water content, and the gap gradually increases with the production, and the final recovery of time-varying model is finally achieved. The yield was 55%, 5% higher than that of the conventional model. It can be seen from the figure that the learning ability of producers and regulators increases with the increase of incentive coefficient, which has nothing to do with their own supply chain. The incentive coefficient not only affects learning ability, but also affects the income of both principal and agent.

5   Conclusions

In the food supply chain, the propaganda of the supply chain concept can strengthen the cooperation between regulators and producers, make the interests of both parties reach an agreement, and generate a spontaneous incentive to maximize the potential of the food supply chain. Judging from the changes in the utility of consumers, producers, and regulators in the numerical conclusions, an increase in the degree of supply chain, whether it is consumers, producers, or regulators, is more effective than without a supply chain. Therefore, in the construction of the food supply chain, it is necessary to promote the concept of mutual benefit and strengthen cooperation between multiple entities, which will help achieve the overall goal of the food supply chain system. When there is a supply chain for food freezing and refrigeration, the supply chain coefficient and learning ability level of producers and regulators are taken into consideration, and the fixed remuneration only needs to set a minimum acceptable value. Producers and regulators with high initial professional competence can increase the expected income of consumers, and the optimal incentive coefficient is positively correlated with the learning ability of producers and regulators.

References

1. Yenipazarli, A., Vakharia, A.J.: Green, greener or brown: choosing the right color of the product. Ann. Oper. Res. **250**(2), 537–567 (2017)
2. Khaksar, E., Abbasnejad, T., Esmaeili, A.: The effect of green supply chain management practices on environment performance and competitive advantage: a case study of the cement industry. Technol. Econ. Dev. Econ. **22**(2), 293–308 (2016)
3. Liu, Y., Zhu, Q., Seuring, S.: Linking capabilities to green operations strategies: the moderating role of corporate environmental proactivity. Int. J. Prod. Econ. **187**, 182–195 (2017)
4. Xu, X., Luo, Q., Bai, Q.: Optimal strategy analysis of multi-level green supply chain under carbon emission constraints and carbon tradable mechanism. Ind. Eng. Manage. **21**(1), 5–9 (2016)
5. Seth, D., Rehman, M.A.A., Shrivastava, R.L.: Green manufacturing drivers and their relationships for small and medium (SME) and large industries. J. Clean. Prod. **19**(8), 1381–1405 (2018)
6. Chen, Y., Chang, T., Lin, C., et al.: The influence of proactive green innovation and reactive green innovation on green product development performance: the mediation role of green creativity. Sustainability **8**(10), 96–106 (2016)
7. Lao, K.: Research on the influence mechanism of consumer innovation on green consumption behavior. Nankai Manage. Rev. **16**(4), 106–113 (2018)
8. Liu, Y., Zhu, Q., Seuring, S.: Linking capabilities to green operations strategies: the moderating role of corporate environmental proactivity. Int. J. Prod. Econ. **18**(7), 182–195 (2017)

9. Bai, C., Tang, J.: Game analysis of green supply chain cooperation among manufacturing-sales companies. J. Syst. Eng. **32**(6), 818–828 (2017)
10. Zhu, Q., Li, X., Zhao, S.: Cost-sharing models for green product production and marketing in a food supply chain. Ind. Manage. Data Syst. **118**(4), 654–682 (2018)
11. Qu, Y., Guan, Z., Ye, T.: Research on dynamic optimization and coordination of supply chain collaborative green innovation based on CVaR guidelines. Ind. Eng. Manage. **23**(4), 78–79 (2018)
12. Hong, Z., Guo, X.: Green product supply chain contracts considering environmental responsibilities. Omega **83**(3), 155–166 (2019)

Impact and Deconstruction of Artificial Intelligence on Marriage Value

Zhen Zhang[✉]

Xijing University, Xijing Road, Chang'an District, Xi'an, Shaanxi, China

Abstract. In the future era of artificial intelligence, with the development of big data, artificial intelligence, bioengineering and other technologies, traditional marriage values will face comprehensive challenges. It is mainly reflected in the following aspects: artificial intelligence intervenes in love decision-making and challenges the concept of free love; genetic technology prolongs human life and challenges the setting of traditional marriage from one end; artificial intelligence greatly improves productivity, enriches material wealth, and the scarcity disappears, so the material symbol of love disappears; bioengineering technology fully intervenes in human reproduction and the traditional ethics of marriage will be affected It's a challenge.

Keywords: Artificial intelligence · Marriage Value · Challenge

1 Introduction

What is love? Can we have it? Can we control it? There are different opinions on this series of issues. Ever since human beings learned to think, they have been arguing endlessly and no consensus has been formed. Among many thoughts, Erich Fromm's view is representative. He thinks that love cannot be easily obtained, whether you are mature or not, the problem is whether you understand the true meaning of love. Fromm quotes a famous saying of the early Renaissance medical scientist, natural scientist and philosopher paracesas to prove his point: "a man who knows nothing has nothing to love, and a man who does nothing knows nothing. People who know nothing are worthless. A reasonable person also knows how to love, observe and discover The more you know about the nature of things, the more you love them Imagine that all the fruit ripens at the same time as the strawberry, who knows nothing about grapes. " [1] Fromm thinks that love is an art, and any art can be learned and trained, and love is no exception. Different from animals, animal love comes from instinct, while human love is rational, and human love is rational love. Why should people love? It is because people have to overcome separation. "Love is a person's ability to take the initiative, to break through the barriers that separate people, and to unite him with others. Love makes people overcome loneliness and separation, but love recognizes people's own value and maintains their own dignity. At home, there is such a state of contradiction that two people become one and still retain their dignity and personality. " [1] Fromm interprets love from the perspective of human psychology, believing that love is because human

J. Abawajy et al. (Eds.): ATCI 2021, AISC 1398, pp. 643–648, 2021.
https://doi.org/10.1007/978-3-030-79200-8_95

beings want to overcome separation and combine with each other, while maintaining personal self-esteem in the combination.

Some people think that the purpose of love is to get married, and the essence of marriage and family is sociality. "Marriage is a certain social organization of sexual relations. It must be based on the premise that both parties of marriage have certain rights and obligations recognized by the society. Any gender relationship without social approval is not marriage, even if it has a long-term nature " [2]. From this point of view, it is not difficult to see that love is an individual behavior on the surface, but from the perspective of marriage and family, its social attribute is very prominent. From the perspective of social attributes, we can more clearly distinguish between animal love and human love. "Animals courtship, but people marry, the meaning is different, is simple and clear. Courtship is biological, while marriage is social and cultural. Marriage is a kind of ceremony, a kind of combination recognized by the society, a kind of relationship that once entered, it has to bear the responsibility of recognition to the society "[3]. Engels regarded family as the epitome of society. Since its birth, family has played an important role in social production and social life with its unique function.

To sum up, no matter how we understand it, whether we start from human psychology or grasp the essence of love from sociality, we can see that what is love is closely related to the conditions of the times and social background. The concept of love listed in this paper is undoubtedly based on the establishment of human subject property and the formation of modern society. And all of these, with the continuous development of artificial intelligence, amazing changes will take place.

2   Is It Possible to Fall in Love Freely

Today, love is a personal matter, others can only suggest but can not intervene, similar views have been deeply rooted in the hearts of the people. As long as we review the history of love, or read some classical love legends and stories, we will know that love is not always decided by ourselves. In the feudal society of ancient China, we paid great attention to order and etiquette. Love should obey "parents' orders, matchmaker's words". Young men and women, in order to pursue love freedom, perform sad songs, romantic or miserable But when the right of love is really owned by individuals, it also causes a series of social problems. There are even cheaters who cheat their emotions, and they are defined as scum men and scum women. When we see similar stories, or the depression of young people after lovelorn and divorce, people will not only wonder if there is a technology involved to make people's marriage more smoothly and less risky and uncertain?

The answer is yes. When big data and artificial intelligence technology are developed enough, science and technology can provide us with scientific and accurate predictions and suggestions. In his book a brief history of the future, the Israeli philosopher Yuval herali described the following scenes. In the future, he said, many people will be happy to leave most of the decision-making process to an AI system, or at least consult when faced with important decisions. [4] For example, if a girl is faced with multiple pursuers, she may ask artificial intelligence, "John and Paul are chasing me. Who should I choose?" Artificial intelligence will help her analyze rationally: if from your inner preferences,

I know you attach great importance to appearance, of course you like handsome John more; but from the data I have from your heart to big data, your personality is not suitable. In contrast, the relationship between Paul and you will be more lasting and stable.

At this time, the girl will be more difficult to choose, is to follow the heart of love, or obey the guidance of rationality? At that time, no matter what choice she makes, her belief of love independence will be seriously challenged.

3   Will Love Be One and the Same

Exclusivity is a distinctive feature of love. In this regard, different countries, nations and cultures in the world have similar values and traditions. In Russia, behind the family, love and loyalty day, there is a beautiful love story in the period of ancient Russia, that is, the faithful love between Prince Peter of Murom and his wife fevronia. In the Orthodox Church, the couple are seen as guardians of the family and marriage. On July 8, 2008, at the initiative of Svetlana Medvedev, wife of then President Dmitry Medvedev, the whole country of Russia celebrated the day of family, love and loyalty. The anniversary originally belonging to the region was upgraded to a national holiday, with a special symbol - chamomile, and an award   love and loyalty award.

Of course, some people are wary of the private love of love. For example, bacon thinks that there may be a kind of lenient love hidden in human nature. If the love of life is not single-minded, if the love of life can not be consistent, I'm afraid that this so-called love will be given to a wider public and become benevolent and fraternal, just like some monks and friars. He thinks: "the love of husband and wife can make human beings multiply. The love of a friend can perfect the interpersonal relationship. But those wanton love will only make people degenerate. "[5] it can be seen that Bacon does not object to the loyalty of love, but proposes that there should be a higher level of love than love.

However, with the development of genetic engineering, regenerative medicine and nanotechnology, scientists and philosophers predict that people's life expectancy is expected to double to 150 years old, or even win over death and achieve immortality. In this way, our current expectation of "growing old together" may be completely defeated. You can imagine that even if you get married at the age of 40, you still have 110 years to go. Who can be sure that a marriage lasts for 110 years? The loyalty and specificity of love will surely collapse.

4   Does the Value Symbol of Love Still Exist

No matter at what stage or for what purpose, the love and marriage between men and women will inevitably lead to economic behavior or consumption behavior. In addition, all kinds of businesses add fuel to the flames, putting a series of value symbols on love. Among them, the biggest impact is diamonds. The classic advertising slogan "diamond lasts forever, one will never die" created by JWT for De Beers successfully makes diamond become the pronoun of immortality of love and establishes De Beers' leading position in the global diamond market. In fact, diamond is just one of many gemstones. It is said that diamonds are not so rare, but are monopolized by a few businesses for a long time, and insist on adopting the strategy of hunger marketing, so that the price of diamonds has been high. Thus, it fits the priceless meaning of love and acts as the value symbol of love.

As early as the 1990s, some scholars pointed out the social and economic attributes of marriage. Cai jiuzhong's (1993) [6] research shows that marriage is essentially a kind of social and economic behavior: first of all, marriage is the key link of population production and reproduction. Without marriage and sex, there is no human reproduction. Secondly, with marriage behavior, people's consumption content, structure and level will change significantly, which will affect social production, distribution and circulation. Thirdly, the four elements of marriage, including natural elements (sex), material elements (economic living conditions), spiritual elements (feelings) and social elements (traditional cultural habits), have direct and indirect relations with the development of social economy. Therefore, from the perspective of economics, we will find that marriage is a kind of transaction. From the perspective of economics, although love tends to be materialized, it reveals the simple truth that love should also be based on material. Therefore, love needs some value symbols to express.

In the future era of artificial intelligence, due to the sufficient development of science and technology and the great improvement of productivity, there will be two distinct characteristics in the society: one is the great abundance of materials, and the other is that people have enough leisure time. This seems to indicate that Communist society can really be realized. Marx believes that Communist society has at least three important marks: extremely developed productive forces, extremely rich material wealth, and free and all-round development of human beings. [7] with the rapid development of artificial intelligence, people can see the dawn of the realization of Communist society. Let's imagine that the essence of love value symbol is to reflect the value and scarcity of love, but when material is no longer scarce, can we still find the symbol matching with love? Love, which has got rid of material bondage, may be deconstructed with the disappearance of scarcity, or it may truly reflect its value itself.

Some people say that the longest love confession is company. But the current busy reality makes people feel powerless. But in the future, with the intervention of artificial intelligence, human resources will be greatly liberated, and money and leisure will no longer be a dream. The degree of production automation will be greatly improved. Machines will inevitably replace labor. There will be a close connection between machines. One machine learns, all machines learn at once. The manufacturing industry in the future will become the second agriculture: only about 20% of the labor force can produce the products needed by the whole society. [8] Some people on the Internet jokingly describe the scene of the future factory: the future factory may only have one person and one dog. The dog is watching people and does not let them move machines. Moreover, this kind of leisure is not the leisure of individuals, but the leisure of the whole human being. Working hours will be greatly shortened and spare time life will be very rich. People don't have to be ashamed of idleness, let alone fear unemployment. With leisure, people can develop their hobbies as they please, and of course, they may also devote themselves to love.

5   Do You Still Need to Worry About Eugenics

The ideal love, marriage and family are integrated, and the ultimate goal is to breed the next generation and realize the continuation of human beings. But natural sexual reproduction is like buying lottery tickets. It depends on luck. There is an interesting celebrity story about the meeting of Nobel Prize winner Anatole France and beautiful and talented dancer Isadora Duncan in 1923. At that time, it happened that the movement of fluency and eugenics in the society naturally became their conversation. Duncan said, "imagine how good it would be for a child to have my beauty and your intelligence!" while Frances replied, "but on the contrary, what if a child had my beauty and your intelligence?" This story has no clear source, it may be fictional, but it vividly reveals the lucky elements of natural reproduction. Therefore, it also determines a series of taboos in the pursuit of love, with the development of related social and ethical issues.

At the end of 2018, the world's first gene editing baby was born, which caused serious criticism all over the world. The mainstream opinion of the academic community is that human beings still can't accept editing human embryo genes for breeding purposes, no matter from the perspective of technical maturity and ethical completeness. The biggest impact on people mainly comes from the challenge to the bottom line of "who is a person". People are surprised to find that with the intervention of cybernetics, Internet, virtual reality and the development of genetic engineering and bioengineering technology, the definition of human and the way of human life will be greatly changed. At the same time, human subjectivity is endowed with artificial internality [9].

If we say that when the technology is mature enough, the ethics is complete enough, and people are fully prepared, the eugenics under the technology intervention may be very common. What we are discussing is whether the love that does not need to worry about reproduction and heredity is more extensive. Will the same-sex or even close relative love be recognized by social ethics when it is not constrained by reproduction? We can only wait and see.

6   Concluding Remarks

In a sense, the future has come, but people's psychology is not ready enough. When artificial intelligence is deeply involved in human life, love, marriage and family will be inevitably impacted and challenged. In these challenges, many traditional concepts and ethics will be completely deconstructed. It is necessary to adapt to the development of the times and seek a new way out. This paper discusses the challenges of free love, love loyalty, value symbol and sexual reproduction. There may be many similar challenges. For example, with the development of robot technology, it is highly similar to human beings. Especially, it can fully experience and perceive human emotions. Will there be love and marriage between human beings and robots? In the age of artificial intelligence, there are many uncertainties in marriage that need to be further explored.

References

1. Erich, F.: The art of love. People's Lit. Press **11**(4), 31 (2018). translated by Liu Futang

2. Xie, M.: The Origin of Marriage and Family. China Social Science Press, 181 (1983)
3. Goodall, W.J.: Family. Social Sciences Literature Press, 1 (1986). translated by Wei Zhangling
4. Euval, H.: A brief history of the future. CITIC Press, 2: 303 (2017). translated by Lin Junhong
5. Bacon, F.: Bacon on life. Yilin Publ. H. **3**, 16 (2016)
6. Cai, J.: Marriage consumption and its market significance. J. Chengdu Teach. Coll. (Lib. Arts Edit.) **1993**(2), 39–43 (1993)
7. Introduction to the basic principles of Marxism. Higher Education Press, pp. 276–283 (2018)
8. He, F.: Variables: survival strategies in the local era. Elephant Press **2021**(1), 129–130 (2021)
9. Csicsery-Ronay Jr., I.: The SF of theory: Baudrillard and Haraway. Sci. Fict. Stud. **18**(3), 388 (1991)

Construction of Art Farmland and Beautiful Countryside Based on Satellite Image Identification Technology

Hongpeng Yang(✉)

College of Fine Arts, Henan University, Kaifeng, Henan, China

Abstract. The creation of "art farmland" is one of the important ways to carry out the construction of beautiful countryside in China. Satellite image recognition technology can help the planning, inspection and adjustment of art farmland. Rural construction and farmland planners should establish close contact with local high-resolution satellite image data acquisition centers for data cooperation; local high-resolution satellite image data acquisition centers can also provide accurate data and personalized services for the project according to needs. The combination of satellite image identification technology with UAV technology and ground observation can create art farmland more efficiently and help the construction of beautiful countryside.

Keywords: Image recognition · Image content understanding · High-resolution satellite · Art farmland · Beautiful countryside

On March 22, 2013, the Ministry of agriculture and rural areas of the people's Republic of China published the document "Ministry of Agriculture Office scientific research [2013] No. 10" on its official website, entitled "opinions of the general office of the Ministry of agriculture on the establishment of beautiful countryside" [1]. This document proposes that the creation of "beautiful countryside" is the need to improve the living environment of rural people and enhance the level of building a new socialist countryside. To create a beautiful countryside suitable for living, setting up a business and playing is the comprehensive promotion of the concept, content and level of building a new countryside. Since then, China's new rural construction began not only to consider the practical needs, but also constantly enhance the aesthetic elements in practice.

Farmland is the most important part of rural areas, farmland landscape constitutes the main rural landscape. "The diversified rural development path guided by art, especially the planning, design and application of art farmland, is a new starting point for the revitalization of" aesthetic economy "in rural China in the future." [2] In the construction of beautiful countryside, how to plan and design the farmland landscape artistically and make it present a beautiful visual feeling is a subject that rural builders should carefully consider. This paper attempts to discuss the construction of art farmland under the background of the completion of China's high-resolution satellite project and the civil application of satellite image identification technology.

© The Author(s), under exclusive license to Springer Nature Switzerland AG 2021
J. Abawajy et al. (Eds.): ATCI 2021, AISC 1398, pp. 649–654, 2021.
https://doi.org/10.1007/978-3-030-79200-8_96

1 The Creation of "Art Farmland" is One of the Important Ways to Implement the Construction of Beautiful Countryside

From the perspective of environmental composition, rural areas are mainly composed of farmland, dwellings, mountains, wetlands and other natural spaces. In these elements, farmland has a crucial position, its existence is the basis of the establishment of the term "rural"; at the same time, farmland often accounts for the largest proportion of the overall rural area. Therefore, whether the farmland landscape is beautiful or not, to a considerable extent, determines whether the countryside to which the farmland belongs is beautiful or not. "Farmland, as an important landscape element in rural areas, has gradually changed from productive land to mixed land for leisure services and production, forming a recreational farmland." [3] It is an important way to turn the rural area of farmland into a beautiful one to optimize the design of farmland landscape by means of artificial intervention and beautify farmland by means of art, so as to turn ordinary farmland originally used for crop cultivation into "art farmland" for appreciation.

"Art farmland" is to plan and design agricultural plants and fields, forest areas, production facilities and other elements based on the theories of aesthetics, garden art and landscape ecology by using artistic techniques and other engineering and technical means, so as to achieve a certain artistic effect and achieve three bumper harvests of agricultural material products, agricultural art products and agricultural leisure products. The purpose of the project." [4] From the definition of "art farmland" by previous researchers, it can be seen that art farmland is elaborately planned and designed by multiple means, and can finally be appreciated; at the same time, the original function of farmland to produce agricultural material products has not been lost. Therefore, the creation of art farmland is different from the pure artistic creation such as easel painting. It is a spiritual function on the basis of meeting the basic material functions. It is also a superimposed creation. It is also restricted by the geographical features of farmland itself, the overall appearance characteristics of crops and various ancillary facilities around farmland.

Just because "art farmland" still belongs to "farmland" in terms of its core functions and basic attributes, its existence clearly indicates that the dwellings it surrounds, sets off and belongs to belong to "countryside"; and once it has aesthetic attributes, it obviously makes the village it is in add the appearance of beauty. This is a logic from the representative part to the whole, and it is also the basis for the construction of beautiful countryside from the creation of art farmland.

2 Satellite Image Recognition Technology Can Help the Planning, Inspection and Adjustment of Art Farmland

The main idea of creating art farmland is to present the overall visual aesthetic effect on a relatively large spatial scale. It is different from easel painting, which uses canvas and pigment to serve aesthetic creation freely; it is also different from bonsai and flower garden. Although it uses natural plants, its focus is on the shape of each plant or the visual effect of several plant combinations; "large space scale" and "overall visual aesthetic effect" are the focus of creating art farmland. The first consideration of art farmland is not the posture of a leaf, a flower, a branch, or a plant itself, or even the effect of several trees, clusters, or clumps of plants. It pays more attention to the overall appearance of a crop, several plots, or even several villages and several farms. Fields are different from flower beds. They are not measured in square meters, but in Mu, hectare, or even square kilometers. They also require designers not to look at them as closely as they look at the paintings on the shelves, but to stand in the distance, look down on the mountains, and enjoy themselves.

It is precisely because of this "large space scale" and "overall visual effect" of art farmland that gives space for satellite image recognition technology. The technology of satellite image identification is a kind of technology that uses satellite to obtain the ground space image in the space of hundreds of kilometers and process and identify it clearly. Although the recognition of ground objects by civil high-resolution satellites has reached the scale level of less than meters, it is mainly used to obtain large-scale images of the ground. This fundamental use attribute coincides with the idea of art farmland construction.

The basic path of using satellite image identification technology in the construction of art farmland is not complex, and the design flow Fig. 1 is as follows:

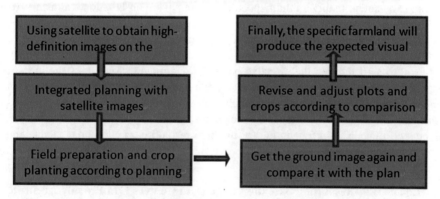

Fig. 1. Application of satellite image identification technology in artistic farmland construction

An important function of satellite image identification technology here is to obtain accurate ground image data and provide accurate information services for the planning, inspection and adjustment of art farmland. This function can not be replaced by manual measurement only on the ground, and its advantages mainly lie in three points: ① the satellite that obtains high-resolution image can quickly obtain data in a large space scale on the ground; ② the satellite that obtains high-resolution image can obtain the overall image data from the high-altitude perspective, which is more suitable for the blueprint of the overall planning; ② the satellite that obtains high-resolution image can obtain the image data from the high-altitude perspective; ③ The satellite that obtains high-resolution image obtains the data presented as an image as a whole, including not only the length and width data, but also the color, contrast, discrimination and other data information.

3   The Specific Idea of Satellite Image Identification Technology to Serve the Construction of Beautiful Countryside

3.1   Rural Construction and Farmland Planners Should Establish Close Contact with Local High-Resolution Satellite Image Data Centers for Data Cooperation

At present, China's major project of high-resolution earth observation system (referred to as the high-resolution project) has been fully implemented. Several satellites have been launched to acquire high-resolution images from "high resolution 1" to "high resolution 14", and high-resolution satellite image data centers have been set up in provincial administrative regions, and high-resolution satellite image data centers have been set up in municipal administrative regions Resolution satellite image data sub center, relying on the military and civil integration office, officially provides a variety of civil services. Taking "high resolution 6" as an example, "high resolution 6" satellite has the characteristics of high resolution, wide coverage, high quality imaging, high efficiency imaging and high localization rate. Its design life is 8 years. It is equipped with 2m panchromatic/8m multispectral high resolution camera, 16m multispectral medium resolution wide range camera, 2m panchromatic/8m multispectral camera, 90 km observation width and 16 m multispectral camera 800 km." [5] With such a configuration, the ground images obtained are enough for the planning and design of art farmland.

Therefore, decision makers and planners who are willing to build beautiful villages and want to start through the effective path of art farmland should take the initiative to establish contact with high-resolution satellite image data centers around the world, explain their intentions, communicate their ideas, and actively provide their own positioning and planning plans, so that they can obtain the data they need more efficiently and effectively It lays a solid foundation for the high-resolution satellite image data center to provide more accurate and personalized services.

3.2   Local High-Resolution Satellite Image Data Centers Can Provide Accurate Data and Personalized Services as Needed

While rural construction and farmland planners actively seek cooperation with high-resolution satellite image data centers, local high-resolution satellite image data centers should also actively cooperate with their work and further upgrade their services. Because, to provide remote sensing data support for agricultural and rural development, ecological civilization construction and other major needs is also a part of high-resolution satellite special project. At the beginning of its birth, satellite image identification technology was mainly used for military purposes. However, in the era of peace, this huge investment project is bound to release its surplus capacity to other fields, and even turn to civilian use. This is exactly what military and civilian integration offices are engaged in.

When satellite image identification technology is used for civil use, it is faced with different requirements and different choices of data accuracy and data types. Even if it also serves agriculture, the demand for monitoring agricultural disasters is not the same as that for artistic farmland construction. The former may be more sensitive to abnormal data, while the latter may have more specific requirements for the visual effect of the image. Local high-resolution satellite image data centers should provide more accurate and even more personalized services for these needs. For example, can we develop a special "aesthetic algorithm" according to some specific shapes, proportions, color gamut, etc., so as to automatically determine which farmland area is more suitable for the needs of art farmland in advance? As far as the existing technical conditions are concerned, the realization of these ideas is not difficult. The key lies in whether the high-resolution satellite image data centers in various places have the active service consciousness, and whether they deeply understand the relevant concepts of beautiful rural construction and art farmland. This is another reason why rural builders and high-resolution satellite image data centers should keep close contact and actively communicate.

3.3 Satellite Image Recognition Technology Can Be Combined with Other Technology Platforms in the Construction of Art Farmland

There is no doubt that satellite image recognition technology is indeed an advanced and efficient choice to obtain large-scale ground space images, and it is also the primary object for us to rely on in the construction of art farmland and beautiful countryside. However, any technology platform has its own advantages and disadvantages. Satellite image recognition technology should also be combined with other technologies to play a better role. The advantages of satellite image recognition technology lie in the fast and efficient acquisition of ground image and the large-scale coverage of the ground. However, because it is more suitable for acquiring large-scale images, it is limited for the acquisition of fine ground images. In other words, its ground image is difficult to achieve fine-grained presentation; although under specific instructions, the satellite can image a certain ground area with very high resolution, the fineness and coverage of imaging naturally constitute a pair of contradictions. We should make full use of the specific advantage that satellites can efficiently acquire large-scale images on the ground, and hand over more precise image acquisition to a more suitable platform for implementation, so that we can learn from each other and build a more comprehensive and efficient three-dimensional data system.

At present, the increasingly mature UAV technology is the best way to obtain the ground fine image information from the perspective of overlooking. Compared with the high-resolution satellite, UAV can fly in a very low airspace to obtain the image data in a very small scale on the ground. It is not disturbed by clouds, and can conduct very accurate control. It is more able to take concentrated and repeated photos of specific areas in a short period of time, rather than having a fixed scanning interval for specific ground shooting as satellites do. Therefore, the cooperation between satellites acquiring high-resolution images and UAV platforms helps to learn from each other.

Finally, because the images obtained by the satellite platform and UAV platform belong to the overlooking perspective, and the main viewing objects of the art farmland are villagers, visitors, tourists, etc., the main viewing perspective and sight distance include many daily visual habits such as head up, overlooking, remote viewing, close observation, etc.; therefore, in addition to the combination of satellite image recognition technology and UAV technology, the satellite image recognition technology must finally be established To the direct visual inspection of people. Of course, people on the ground can also obtain more specific and accurate images and data with the help of digital cameras, rangefinders and other tools, which constitutes a large, medium and small-scale full coverage data acquisition network with satellites and UAVs. Satellites in space, UAVs at low altitude, people on the ground Only in this way, can we build a comprehensive and three-dimensional image data system, make the art farmland have accurate and timely data support from planning, implementation to inspection and adjustment, and build our beautiful countryside more efficiently!

References

1. http://jiuban.moa.gov.cn/zwllm/tzgg/tz/201302/t20130222%5F3223999.htm
2. Duan, S.H.: Enlightenment of foreign farmland art application to the development of creative agriculture in China. World Agriculture. No. 9, pp. 74–80 (2020)
3. Sa, L., Yujun, Z.: "Low Tech" strategy of recreational farmland landscape construction. Garden. No. 11, pp. 38–41 (2016)
4. Liu, S.Y., Hou, Y.K., Xiang, Y., et al.: The Landscape Features and Planning and Design of Art Farmland. Northern Horticulture. No. 10, pp.234–238 (2009)
5. Bai, G.L., Hu, Z.: High resolution special "High resolution 6" satellite successfully launched. Spacecraft Recovery & Remote Sensing. No. 39, p. 50 (2018)

Drug Traceability System Based on RFID and Alliance Block Chain Technology

Hongjin Li[(✉)] and Xiaohua Wang

Zunyi Medical University, Zunyi, Guizhou, China
Lhj1983@scixueshu.com

Abstract. On the basis of RFID technology and alliance block chain Hyper ledger Fabric technology, combined with the existing drug production, storage, logistics, sales model, this paper puts forward a set of drug anti-counterfeiting traceability system implementation scheme, with the help of distributed books, intelligent contracts, consensus mechanism and other technologies to write drug data into the block chain to ensure that drug information cannot be tampered with, for consumers, quality inspection agencies and the State Food and Drug Administration to trace drugs. High transparency and non-tampering characteristics of alliance block chain Hyper ledger Fabric technology for pharmaceuticals Supply management provides feasible solutions. Application of RFID and block chain to establish the whole process of drug production to processing and marketing anti-counterfeiting traceability system, can identify fake and inferior drugs, in the process of drug supply, automatically record and store all kinds of related information of drugs.

Keywords: RFID technology · Alliance block chain · Drug traceability system · Intelligent contract

1 Introduction

As a special commodity for the prevention, diagnosis and treatment of diseases, drugs are closely related to the health and safety of human life. Because of the high illegal profit, many drug safety accidents have appeared in recent years, which exposes the great disadvantages of the current drug anti-counterfeiting traceability system, that is, all the data are stored in the central database, the drug traceability certification adopts the central authentication method, the data is tampered with in a series of, such as uploading, storing, querying and so on, and the drug information stored in the drug anti-counterfeiting traceability system is incomplete, so it is difficult for consumers to obtain the truly useful information from it, and it is also acceptable in the information query process Easy to cause consumer privacy information disclosure [1].

At present, all the data of the medical traceability system exist in the central database, the traceability authentication adopts the central accounting mode, and the data can be tampered with in a series of operations, such as uploading, storing, querying and so on [2]. At present, there are still three shortcomings in the anti-counterfeiting traceability system: first, the stored data information is incomplete, consumers can not get the detailed

© The Author(s), under exclusive license to Springer Nature Switzerland AG 2021
J. Abawajy et al. (Eds.): ATCI 2021, AISC 1398, pp. 655–661, 2021.
https://doi.org/10.1007/978-3-030-79200-8_97

information of the drug by querying the drug traceability code; second, the privacy of the data information can not be guaranteed. Participants and final consumers will get the same information in the process of drug circulation, which can easily lead to consumer personal information leakage, third, drug information input efficiency, low security, the current record of drug information Bar code is widely used, but bar code has low security, easy to copy, and has high requirements for bar code identification conditions, and can not identify multiple bar codes at the same time [3]. Application of RFID technology and block chain technology provides a new way to solve the problems of low efficiency, low security, easy tampering, incomplete information and privacy of information.

Block chain has many applications in traceability, anti-counterfeiting and tracking systems abroad at present, such as Figorilli S. et al. apply block chain to wood supply traceability system to track the flow of wood in the whole supply chain [4]. Sikorski J. J. et al. use block chain technology to conduct comprehensive supervision of the production, circulation and use of drugs [5]. There are also many studies in China to apply block chain to the traceability system of various products, such as domestic large-scale online shopping platforms also began to introduce block chain into the traceability of goods.

Some domestic enterprises have applied RFID to drug supply chain management, such as China Resources Pharmaceutical Commercial Group, which applies RFID technology to cold chain logistics of medicine and drugs in logistics Low temperature storage for regulatory control. RFID technology is applied in drug traceability system in this paper, focusing on the realization of more complete drug traceability system, and constructing drug information big data Internet of things, which makes drugs from production line to factory, logistics, warehousing management, sales and other links are regulated [6]. Combined with the advantages of RFID and alliance block chain, the drug traceability system of automatic recording of drug information, automatic uploading of data and data security is not easily tampered with.

2   Block Chain Technology

Block chain technology is an innovative technology which integrates the functions of database, openness, security and so on. It uses alliance block chain Hyper ledger Fabric technology to develop drug traceability system. Participants and consumers in drug production, circulation and sale can connect to this system. Each organization generates a code after registration authorization [7].

The block chain system itself is that all peer nodes in the whole network jointly maintain the same account data, the non-tampering of the block chain, and the characteristics of the distributed storage account book, which makes the data in the block chain traceable. This provides a great help for the tracking of drug information data. Hyper ledger Fabric advantages for drug data traceability are as follows:

(1) Facilitate the management of members within multiple organizations of the Alliance's block chain: Fabric member management services that allow more nodes and users to join safely and easily, while authorizing certificates and private keys to newly added nodes.

(2) To overcome the disadvantages of centralization: Fabric itself, as a decentralized distributed account book, overcomes all kinds of drawbacks of the centralization system and can avoid the risk of artificial tampering with data.

(3) Peer nodes participate in breaking information islands: the block chain system itself is that all peer nodes in the whole network maintain the same account data together, so that the drug data in the block chain are traceable [8].

3 RFID Technology

Radio frequency identification (RFID) is a technology that can realize wireless data exchange and information identification. The RFID label can be divided into low frequency, high frequency, ultra high frequency and microwave according to its working frequency. According to their respective characteristics, in the process of implementation, the drug box label working frequency of 13.56 MHz, which has strong anti-interference ability Small size, reading distance up to ten centimeters, can meet the normal use requirements.

This design uses a RFID label that meets the ISO18000-6C standard. The 860 ~ 960 UHF passive RFID tag is attached to the drug packaging, and the unique EPC code of each tag is used as the unique identification code of each box of drugs. The KLM900s module supports the ISO18000-6C/EPC C1G2 protocol and can communicate with the computer equipment through the RS232-TTL pin on the interface. This Because the module has low power consumption small volume of the module, and its reading efficiency per unit time is much higher than that of the traditional bar code, the reader can be used in various wireless RFID applications such as hand held mobile devices, goods and logistics management, warehousing management, goods anti-counterfeiting, electronic commodity monitoring and manufacturing and processing, production automation and so on.RFID applied to drug packaging are divided into two types: high frequency label on small drug packaging box of each sales unit and UHF label on packaging carton [4]. A complete RFID system consists of four parts:

(1) Labels, i.e. transponders, are mostly passive units composed of coupling elements and microchips. Each tag has a unique electronic code and is attached to the object to identify the target object, which is the information carrier of the RFID system. RFID labels on drug packaging mainly include drug regulatory code, drug name, origin, batch number, validity period and other information.

(2) A reader, usually composed of a coupling module, a transceiver module, a control module and an interface unit, is a device for reading or writing tag information, can be designed as a hand-held or stationary type, and is a RFID system information control and processing center.

(3) Antenna is mainly used to transmit RF signals between tags and readers. At present, advanced electronic printing technology has been developed which can print antenna on commodity packaging box and other substrate materials, but it is not mature and has high cost.

(4) Software systems are used to process data collected by readers. RFID data acquisition system can connect with the existing application system according to the actual needs of users.

4 Intelligent Contracts for Drug Traceability Systems

The intelligent contract system is Hyper ledger Fabric an open source block chain underlying system, which provides a rich API interface. It can develop various block chain application scenarios on it and is a block chain application for enterprise alliance. According to the specific requirements of drug traceability system, the storage and query functions of drug traceability data are designed in this system, which mainly include two parts: drug data structure and contract function interface.

Store meta data and hash values related to source data in smart contracts. By adding a Identifier structure to the intelligent contract, this method contains some additional information, that is, meta data, digital signature, time stamp, hash value, data owner. This structure is then mapped to smart contracts instead of data hash values. Therefore, the basic structure of identifiers introduces the concept of decentralization, which is defined as follows.

```
struct Identifier
{bytes dsign;
uint256 timestamp;
bytes hash;
address owner;}
```

Data Integrity Verification and Analysis Algorithm for Drug Traceability System Based on Block Chain.

(1) A node needs a secure hash function to hash the data to get the hash value of the data.
(2) The node digitally signs the hash value and obtains the digital signature. Using decentralized time stamp and unique data identification technology. Add hash values, digital signatures, and time stamps and data owners of transactions as additional information to the Identifier structure. Finally, this structure is mapped to the intelligent contract instead of the data hash value.
(3) The user stores the data in the local server and retains meta data information in the data identity for real-time data integrity verification.
(4) When the user needs to perform the integrity verification service, the hash value of the data stored in the local server is first calculated hash0, then the hash is calculated according to the preserved time stamp information, and finally a new hash value is obtained hash1, Finally, using the user's private key, the value is hash1 digitally signed and the new digital signature dsign0. is obtained
(5) The node interacts with the data integrity contracts by using digital signatures dsign0, search for data hash values stored in block chains corresponding to Identifier structures in smart contracts
(6) Compare two hash values hash and hash0, verify the integrity of the data. If the two values are the same, the source data will successfully complete the data integrity verification, indicating that the data has not been tampered with, otherwise, the source data will be tampered with or destroyed.

The data integrity verification technology based on block chain can ensure the correctness of data storage and verification, thus further ensuring the integrity of medical data. In fabric, intelligent contract is called chain code, which is essentially the business logic of how different entities or related parties in the control block chain network interact or trade with each other. Chain code encapsulates business network transactions in code. Chain code can be called to set up and fetch books or world state. The contract interface of the system is divided into two categories, that is, the drug information interface and the drug information query interface. Release of drug information mainly includes: release of drug production information, transportation information, sales information. Drug information query interface includes: query drug production information, transportation information and sales information.

5   Drug Traceability System Architecture

The functional module of the drug traceability system, as shown in Fig. 1, consists of eight modules: the RFID data acquisition module, which is used to complete the bottom data automatic collection and authentication module, which is used to manage the identity information of all users, including the registration, login and cancellation of each node user. All users who need to join the alliance chain must be checked and verified by the system node before they can join; the authority management module, which assigns read and write rights to all node users who join the alliance chain, and manages clear responsibilities uniformly; and the network communication module, which is used to transmit data information that needs to be shared between P2P network nodes through the network; Data transfer module, P2P network transmission and storage format for medical records data and validation of data information; drug information management module for publishing and querying drug details, creating and modifying drug information data and supporting system database access to nodes locally; block chain storage module for storing key information for the entire drug cycle, keeping each transaction record on the block chain, including participants, time stamps, etc.; cryptographic tool module, using PKI system and cryptography principles to provide hash algorithm, digital signature algorithm, and secret key verification, transaction management; The consensus mechanism module is the implementation system According to the algorithm of consensus mechanism, the Byzantine fault-tolerant algorithm is used in this system. All the nodes with write data permission reach a consensus according to the PBFT algorithm, and write the drug data into the newly generated block [9].

The process of constructing drug traceability system is as follows:

(1)   The Internet of things system based on RFID is built: by deploying the corresponding RFID chip on the box body, box body, bracket and transport truck, the basic information of the drug is automatically collected, the relevant information in the process of drug circulation is automatically recorded, and the unique identification ID of the RFID chip is sent to the block chain network to record the key information in the block chain network.

(2)   Fabric network environment and chain code development: the roles of the parties involved in drug traceability range in terms of authority, including manufacturers,

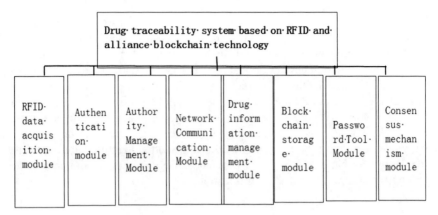

Fig. 1. Functional module diagram of the system

dealers, medical institutions, pharmacies, consumers, etc. Each member is a trusted party with access permits, and together these members form an enterprise alliance. The data in the alliance chain is maintained by all parties. The use of the alliance chain can reduce the waste of resources in the consensus process, and the maintenance of the drug traceability system is the responsibility of each participating node, whch can simplify the record books.

(3) Drug traceability information association and drug traceability: when a drug is produced by a pharmaceutical factory, a unique drug traceability code is added to the drug packaging. At the links of drug circulation and sale, the traceability code can be used to establish the data table for the primary key. Each participating node can transmit the information related to the drug in each link to the Fabric block chain according to its own role by installing different chain codes. After purchasing or using drugs, consumers can access the drug traceability platform based on RFID and block chain by using the account number certified by the organization, and enter the traceability code of the drug, which can query the drug at all stages of production, circulation and sale Details.

6 Summary

This paper establishes a cross-platform and high-security drug traceability system model based on RFID and block chain technology. The system writes drug data into the block chain with the help of RFID', distributed accounting books, intelligent contracts, consensus mechanism and other technologies to ensure that drug information cannot be tampered with. It is of practical significance to protect the safety of drug production and circulation process.

Acknowledgments. Foundation Project: Science and Technology Plan Project of Guizhou Province (Talents of Guizhou Science and Technology Cooperation Platform [2018] 5772–050);

The Project of Science and Technology Bureau in Zunyi City, Project number: Zunyi Science and Technology Cooperation HZ (2020) NO.43.

References

1. Hugo, L., Laura, A., et al.: A review of IoT sensing applications and challenges using RFID and wireless sensor networks. Sensors **20**(9), 2495 (2020)
2. Tight, K.S., Kun, Q., Xudong, M., Zhihong, L., Ziquan, D., Xin, X.: Design and implementation of medical reagent management System based on RFID. Technol. Ind. Control Comput. **33**(01), 51–53 (2020)
3. del Carmen León-Araujo, M., Gómez-, E., Acaiturri-Ayesta, M.T.: Implementation and evaluation of a RFID smart cabinet to improve traceability and the efficient consumption of high cost medical supplies in a large hospital. J. Med. Syst. **43**(6), 1–7 (2019). https://doi.org/10.1007/s10916-019-1269-6
4. Figorilli, S., Antonucci, F., Costa, C., et al.: A blockchain implementation prototype for the electronic open source traceability of wood along the whole supply chain. Sensors **18**(9), 12 (2018)
5. Sikorski, J.J., Haughton, J., Kraft, M.: Blockchain technology in the chemical industry: machine-to-machine electricity market. Appl. Energy **195**, 234–246 (2017)
6. Fengtian, L., Jingang, M., Yang, Z., Hui, S., Hui, S.: Research on drug traceability based on block chain and RFID. China Digit. Med. **15**(01), 8–10,13 (2020)
7. Xinghua, S., Jinxuan, C., Cheng, D.: Research on RFID security and privacy protection program in intelligent medical care. Sci. Technol. Manag. Res. **39**(16), 223–229 (2019)
8. Weina, W.: Real-time monitoring system of third-party pharmaceutical cold chain logistics based on RFID technology. Logistics Technol. (Equipment) **34**(10), 212–214, 221 (2015)
9. Yong, Y., Leap, W.: Current situation and prospect of block chain technology development. J. Autom. **42**(4), 481–494 (2016)

The Application of Computer Aided Technology in Interior Design

Siqi Lin[(✉)]

Shandong Management University, Jinan, Shandong, China
siqi93@eixueshu.cn

Abstract. In recent years, the development of computer information technology has been very rapid, and more and more advanced computer information technologies have gradually been applied in the education system, and have effectively promoted the development of education. Among them, computer-aided design software provides good conditions for the teaching of interior design majors. Computer-aided design software has significant characteristics and effectively changes the traditional interior design mode. This article analyzes the application of computer-aided design software in interior design.

Keywords: Computer · Assistive technology · Interior design · Application

1 Introduction

In the process of interior design, with the continuous improvement of people's design requirements, people's design ability is constantly updated, and the application of computer technology has expanded the space of interior design [1].

2 Computer-Aided Technology

Computer-aided technology includes many parts, including computer-aided design, which is commonly referred to as CAD and computer-aided manufacturing in China, which is what we know about CAM and computer-aided instruction, called CAI. First of all, computer-aided design is mainly a kind of computer calculation that uses computer technology to assist people in designing the effect of the design plan. At present, this technology is widely used in the fields of integrated circuit design, architectural project design, and vehicle design [1]. The application of computer-aided design to various industries not only shortens the design time, improves work efficiency, saves capital costs, but also improves the accuracy and quality of the design plan; secondly, computer-aided manufacturing mainly uses computer systems to the process of product processing and control is shown in Fig. 1. In the manufacturing process, input the process route and project content of the part, and the output information is the running track of the tool. In addition, the integration of computer-aided design and manufacturing technology can

J. Abawajy et al. (Eds.): ATCI 2021, AISC 1398, pp. 662–667, 2021.
https://doi.org/10.1007/978-3-030-79200-8_98

realize the automation of design and manufacturing [1]. This technology is called an integrated manufacturing system. Finally, computer-aided teaching is mainly an application of teaching using computer technology. For example, we frequently used software such as authorware, PowerPoint, Flash. Computer-assisted teaching can improve the flexibility of classroom teaching and stimulate students' interest and enthusiasm in learning. It is an indispensable teaching method in modern education systems [2].

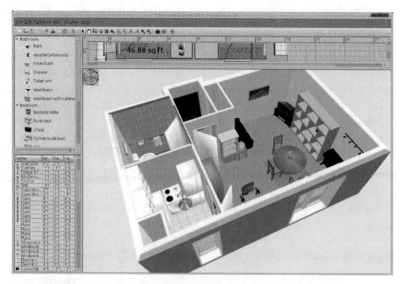

Fig. 1. Application of computer-aided technology in interior design

3 The Difference Between Computer Graphics and Traditional Graphics

Computers can express through three-dimensional graphics, so that people can directly understand the design of different positions, and in this way, the incomprehensible wrong concept of graphics can be further reduced. The computer can make the size more accurate, reduce the error to a greater extent, and the designed model is not much different from the actual object. It can be seen that the advantages of computer graphics far exceed the advantages of traditional graphics, and will not fail the work due to inaccurate graphics [2]. Moreover, the computer can form different visual experience for people, and will not exaggerate the effect through artificial means, and it can match the design effect. When using a computer, it is particularly advantageous for the replacement of the scheme. It can more intuitively control the relationship between the design object and the overall environment, which can perfectly combine the two, so that the designed model can be integrated with the environment, and the real appearance can also be reflected, through the simulation of the object under the illumination, the information of the illuminated object is transmitted, and then the object scene implementation plan is displayed later. Compared with the drawing of the plate, this method is more vivid [3]. The interior design can be modified repeatedly, and due to the development of computers, the modification is more convenient. It not only speeds up work efficiency, but also handles the inconvenience of traditional drawing modification, and improves the speed of copying and extracting, which will be the most ideal the picture is shown through the production of the designer. The computer greatly improves the output efficiency of design drawings, and can display the effects of structural drawings, plan views and three-dimensional drawings together, and can also freely select and match materials, which greatly saves manpower and time and work efficiency [3]. The computer can restore the final visual effect of the scene for people. With the continuous development of society, the technology and application of computers have been greatly improved, and technologies such as virtual roaming technology and animation design are gradually used in interior design, which greatly fills in the lack of static pictures and allows owners to the final design scene and layout are clearer in the three-dimensional space, and the picture in the mind is more quickly reflected. Moreover, it can be distinguished from the traditional design scheme, and the various parts can be broken down faster, which is also very beneficial to the observation of the organizational structure of each part [4]. The application of computer can greatly reduce the investment of manpower and space.

4 The Application of Computer-Aided Design Software in Interior Design

4.1 Design Embodiment of CAD Design Construction Drawing

Through the designer's inspection and analysis on site, the design concept can show an effect of environmental coordination, unique conception and complete layout. Then transfer the basic materials to the CAD software to draw the original floor plan, floor plan, strong and weak electricity diagram, and component elevation diagram, to show the interior design and construction drawings more perfectly [4].

4.2 3D Modeling

With the completion of the CAD design drawing, the indoor three-dimensional diagram will be displayed immediately. The 3Dmax modeling form is usually divided into three advanced methods and one basic method in form, namely: mesh, poly, nurbs, and basic modeling modifiers. Now the basic model is established and perfected, and then through specific requirements, it is changed into three modes: mesh, poly, and nurbs. When modeling, the form of modeling can be selected based on requirements [5]. First import the design structure drawing previously carried out in CAD into 3Dmax, and then complete the production of the model according to the requirements of the drawing through the design of the plan, as shown in Fig. 2.

4.3 Lighting Effect Design

For creating lights, there are differences in the types, quantities, and attributes of lights created because of the differences in the scene. Three different basic types of light sources

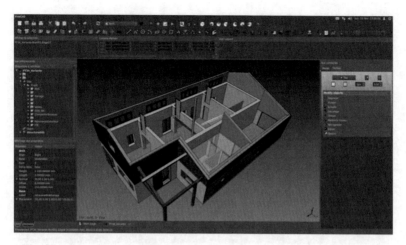

Fig. 2. Application of computer-aided technology in interior design modeling

are main light, auxiliary light, and background light. The main light becomes the most critical light to create the image of the picture. It is the light source that needs to be set first. It is usually at the three-quarter position, but it can also be changed according to specific requirements [5]. The auxiliary light can be used to fill in the areas where the main light cannot be expressed, which is called ambient light. The background light belongs to the contour light and the edge light. The edge of the object is found through the light, and the target object is separated from the background.

4.4   Cruise Animation Settings

The camera used in the scene is usually set at about 28 mm indoors, and the height is about 1.6–1.8 m. The camera can replace the person's perspective and is equivalent to the person watching. The position of the camera depends on the most exciting content in the space [6]. Generally speaking, it is best to show the most obvious content in the rendering.

4.5   The Embodiment of Vray Rendering Technology

Vray is mainly used to render some more special effects. In the material editor, it can be observed that the parameter volume exhibition of Vray special materials is different from the 3DMax built-in materials. The basic reflection, refraction, and transparency parameters are arranged in the basic parameter volume exhibition in different regions [7].

4.6   PhotoShop, PremierePro Post-processing Technology

Adjust the image sensitivity and color effects of the image screen through PhotoShop, and design the indoor surrounding environment. For the execution in PremierePro, post-processing is classified as the final step, and the video animation can be exported through

Fig. 3. Rendering of computer-aided technology in interior design

the continuous.tga file that is rendered, so that people can watch the picture effect more clearly [8] (See Fig. 3).

5   The Future Development Trend of Computer in Interior Design

5.1   Comprehensive Functions

In the future, computers will be more comprehensive in terms of various functions. The comprehensive application and development of computers in military, cultural, astronomical, and biological science research fields will require the storage capacity of the computer and the processing speed of the processor [9]. Higher and higher, full-featured computers are bound to become the main trend of future computer development.

5.2   Volume Miniaturization

With the further development and improvement of electronic technology and information technology, people have put forward higher requirements for computer performance, and hope that the computer will be powerful, easy to carry, and smaller in size [9]. For example, the intelligent wearable devices that have been gradually interested have already possessed some of the functions of computers, and it is believed that the development of computers will also develop in the direction of miniaturization in the future [10].

5.3   Intelligent

At present, the computers we use can replace part of human mental and physical labor, but their logical abilities are still relatively clumsy. Therefore, in the future, we hope that computers can develop towards intelligence. Research has given birth to a batch of comprehensive computers. Being able to specialize in the calculation of various human behaviors makes computer processing more humane and intelligent [10]. The emergence of these computer technologies is bound to comprehensively improve certain shortcomings of current computers, thereby further accelerating the development and progress of social civilization.

6 Conclusion

In summary, computer technology has an important application in the interior design process, which can further improve the design effect and meet the needs of different groups of people. For interior design, through long-term practice, the above software can roughly meet the needs of the interior design market. Through continuous optimization of content and action orientation of the engineering process, the development of the design discipline is continuously improved.

References

1. Lu, H.T.: Discussion on hand-drawing and computer expression in interior design. Inf. Technol. Inf. **3**(03), 87–88 (2010). (in Chinese)
2. Hao, Y.B.: The influence of computer technology on social science research. J. Shanxi Univ. Finance Econ. **11**(11), 54–55 (2010). (in Chinese)
3. Li, L.F., Zhang, S.: Thinking of problems in CAD software teaching of interior design specialty in higher vocational education. Inf. Technol. Inf. **12**(08), 91–94 (2017). (in Chinese)
4. Gou, Z.H.: Expression of architectural decoration engineering design effect based on computer-aided software. Autom. Instrum. **7**(11), 122–124 (2017). (in Chinese)
5. Zhang, T.: Analysis of application of computer aided design in environmental art design. Software **8**(11), 263–265 (2020). (in Chinese)
6. Chen, X.Y.: Application of computer aided design software in interior design. Inf. Technol. Inf. **12**, 77–79 (2019). (in Chinese)
7. Nie, Z.: Research on the application of computer aided design software in interior design. New Countryside (Heilongjiang) **9**(05), 162–165 (2017). (in Chinese)
8. Lin, L.M.: Research on the application of computer aided design in interior design. Art Technol. **14**(07), 194–197 (2018). (in Chinese)
9. Ye, Q.R., Xue, J.J.: Analysis of the role of auto CAD software in interior furniture design. Pop. Lit. **11**, 34–37 (2019). (in Chinese)
10. Liu, Y.T.: Study on design and application of study room based on AutoCAD and one-point perspective. Furnit. Inter. Décor. **9**(05), 40–43 (2016). (in Chinese)

New Mode of Supply Chain Management and Financing Based on Block-Chain

Jinyuan Zhang[✉] and Qun Cao

Macau University of Science and Technology, Macau, China
jinyuan_zhang@yeah.net

Abstract. Since economy is developing rapidly, supply chain finance has become a highly innovative financial model. Some characteristics of block chain technology are well adapted to the development needs of supply chain finance and bring opportunities to finance of supply chain. This paper first introduces the background and characteristics of block chain, then analyzes the main problems in the development of supply chain finance, and puts forward the optimization path of its ecosystem and block-chain technology ecosystem. By doing so, this study hopes to provide a new solution to this problem. By exploring the potential of block chain technology in the field of financing, it will contribute to the research in the field of Finance and supply chain management. This paper aims to explore the potential application of block chain in the field of financing, including the transparency of supply chain and the sustainability of contract performance. The results show that the development of market scale increases with the annual growth, and the maximum market scale in 2019 is 23.93 trillion yuan.

Keywords: Block chain · Supply chain management · Financing mode · Innovation

1 Introduction

With the development of computer technology, mufti-agent and non-high frequency transactions such as finance of supply chain are better for the application of block chain. Based on the technology, in the block chain technology, accounts receivable, bills, warehouse receipts and other assets can be digitized, and data storage is left behind, which reduces the risks of air ticket fraud, and makes the supply chain financial business very smooth [1].

As information technology is developing continuously, many experts have studied block chain financing. For example, some domestic teams have studied the GSP financial model innovation of block chain technology, applied block chain technology and finance of supply chain principles to agricultural production, and studied the rural financial intelligent platform based on block chain. Taking distributed account book, decentralization and asymmetric encryption as the core, the rural financial intelligent platform is constructed. Combined with the information digitization, the paper determines the interest connection mechanism of all parties, and proposes the scheme to solve the financing difficulty of farmers. Finally, we should establish a more fair and transparent credit system, promote the sustainable development of rural supply chain financial services, and strive to explore practical solutions to agricultural financial problems, so as to contribute wisdom to the construction of a more beautiful countryside. This paper analyzes the existing problems of GSP finance in China, and then analyzes the role and existing problems of the current financial model (including traditional financial model and Internet financial model) in the development of GSP finance, so as to build a new GSP financial model, and as a supplement to the current financial model, solve the problems in the current GSP finance in China Questions. This paper constructs a carbon trading model based on carbon finance, realizes a two-stage transaction matching model and a real auction model based on carbon finance, and compares the two models through data simulation. Design carbon finance loan mode, that is, financial institutions provide loans for enterprises with limited funds, combine carbon spot trading with futures mortgage, design a one to many bilateral matching mode, on the premise of meeting the buyer's requirements for loan term and interest, promote the carbon finance loan transaction interest rate and the buyer's loan quantity below the upper limit. Some experts have studied the supply chain financial risk control mechanism of block chain technology, analyzed four finance of supply chain in block chain by using game theory, found out the common behavior of financing parties under various financing modes, carried out static risk evaluation by using fuzzy comprehensive evaluation method and analytic hierarchy process, weighted all levels of indicators by using analytic hierarchy process, and used fuzzy comprehensive evaluation method assigns values to all levels of indicators to solve the static risk value at a certain time. The dynamic evaluation of system risk is carried out by using system dynamics, and the change of each subsystem risk with time, the relationship between the total system risk and each subsystem risk, and the change of each subsystem risk with time are obtained. Through the construction of asset sensitization alliance chain platform based on block chain database, the effect analysis after implementation shows that the application of alliance chain platform indeed improves the operation efficiency to a certain extent. Starting from the restrictive factors of traditional finance of supply chain, this paper analyzes and summarizes the main development modes of current finance of supply chain, points out its core problems, studies its feasibility, compares it with traditional finance of supply chain model, and illustrates its advantages in data sharing and trust transfer [2]. In addition, some experts have studied and established a general evolutionary game model, analyzed the evolutionary decision of traditional supply chain financial decision-makers, and discussed the cooperation between core enterprises and SMEs, and the audit supervision of SMEs and financial institutions on the loan application of core enterprises. Combined with the knowledge of global supply chain management, international settlement, international trade and other disciplines, this paper makes a detailed supply chain financing mode analysis, and objectively points out the financing difficulties faced by small and medium-sized enterprises in the current supply chain financing mode. By understanding the basic concept of block chain technology and the working principle of related technologies, this paper discusses the application and development status of block chain technology in other industries. The architecture of Supply Chain Financial Alliance Based on block chain technology and the underlying model and composition technology of Reuther block chain system suitable for finance of supply chain, innovates the new mode and operation mechanism of supply chain finance, and puts forward several business scenarios of supply chain finance from the perspective of block chain. This paper discusses the impact of block chain technology application on the model, discusses the decision-making changes and evolution path of all parties, and reveals the evolution law under the background of financial technology [3]. Although the research results of block chain financing are quite abundant, there are still some deficiencies in the research of block chain financing mode.

In order to study the new mode of block chain supply chain management and financing, this paper studies the block chain supply chain management and financing mode, and finds the supply chain mode. The results show that the block chain technology supply chain management is conducive to financing.

2 Method

2.1 Block Chain

(1) Block chain
   As a kind of distributed ledger, block chain technology establishes a shared ledger according to the structure of participating nodes, which does not need a single centralized organization maintenance [4]. The nontrade secret data of each enterprise in the chain can be stored and shared on all nodes in the chain, so that the data can be transferred in trust on the chain [5]. Consensus mechanism is not tamper, and each node can not operate data in the direction beneficial to itself. Each participant in the chain can have his own ID card and record it in the distributed ledger [6]. It helps to establish a multi-access trust system in the supply chain and provides a strong credit guarantee for the advanced finance of supply chain. Under the block chain technology, the system can limit the participating nodes in the process of transaction on the chain and require them to upload each transaction data on the chain [7]. All asset transactions are jointly endorsed by the nodes of the whole network [8, 9]. Identity information, transaction history, documents, and basic contract information are all electronically online, and complete information and records are sorted out, which is convenient for verifying the authenticity of transactions and bank support. There is no need to repeatedly consult historical data, which can greatly reduce the cost of bank participation.

(2) Supply chain management
   As a new application mode combining various computer technologies, blockchain provides solutions for transaction trust and security problems through mathematical principles [10]. The function of block chain mainly comes from its technological innovation, and finally solves the trust problem through the combination of some existing technologies [11]. The supply chain financing mode improves the limitation that the financial institutions only lend to a single enterprise in the past. It gives credit to small and medium-sized enterprises from the perspective of supply chain, increases the credit of small and medium-sized enterprises by relying on the credit of core enterprises of industrial supply chain and the operation of the whole supply chain, and alleviates the financing problems of small and medium-sized enterprises.

(3) Financing mode
Inventory financing generally occurs in enterprises with large inventory or slow inventory turnover, which means that enterprises use commodities for mortgage financing in the process of transaction to realize existing commodities in advance, so as to solve the problem of capital turnover pressure. Inventory financing generally occurs in enterprises with large inventory or slow inventory turnover, which means that enterprises use commodities for mortgage financing in the process of transaction to realize existing commodities in advance, so as to solve the problem of capital turnover pressure. In the whole supply chain, core enterprises often use their own advantages to squeeze suppliers of upstream SMEs and buy raw materials on credit. At this time, the upstream suppliers often accept the trading terms proposed by the core enterprise because of their own weakness, such as insufficient strength, small scale, limited operation ability, and unwilling to lose the high-quality customers of the core enterprise, resulting in their own losses, insufficient working capital and normal production.

2.2 Supply Chain Model

Suppose that the sales volume of core enterprises on the supply chain financing platform based on block chain technology is as shown in formula (1):

$$Q = ae = d = \delta \tag{1}$$

G represents the total revenue of the block chain supply chain financial information service platform through the core enterprises on the platform. The effectiveness of block chain supply chain financial information service platform is shown in formula (2):

$$\pi_B = aQ[g - k(a)] \tag{2}$$

Assuming that the core enterprise is a risk averse enterprise, its utility function has the characteristics of constant absolute risk aversion, so its expected return function meets the demand $E\pi_R = -e^{-p\pi_R}$. π_R Represents the total revenue of the core enterprise, ρ. The absolute risk aversion of core enterprises. Assuming that the retained income of the core enterprise is 0, the above results can be obtained, as shown in Eq. (3):

$$\pi_R = \pi_{R1} + \pi_{R2} = nQ - \frac{1}{2}fe^2 - \frac{\rho}{2}[n = +ak(a)]^2\sigma^2 \tag{3}$$

The expected return of core enterprises can be obtained by using the deterministic equivalent method, as shown in Eq. (4):

$$E\pi_R = [n + ak(a)](ae + d) \tag{4}$$

3 Experience

3.1 Experimental Object Extraction

Since the basis of block chain supply finance is the transaction between financing enterprises and core enterprises, if there are problems in the transaction between them, it will inevitably affect the financing of financing enterprises, and banks will face huge risks. Therefore, when a financing enterprise issues a financing application to a bank, the bank will review the qualification of the core enterprise and the chain financing enterprise, and review their transaction information; the financing enterprise is the demand side, and will make its information transparent to the bank, but the core enterprise does not directly participate in the transaction, and the core enterprise has its own choice on whether to share information outside the transaction, so that the bank can obtain more information. The complete information is not shared, and the enterprise has its own privacy. However, the short-term non cooperation may cause the core enterprises to encounter financing obstacles from banks in the future. Therefore, the game among core enterprises, financing enterprises and banks meets the condition of static complete information model.

3.2 Experimental Analysis

Model hypothesis: Hypothesis 1: whether the financing enterprise can fulfill the agreement on the financing date. Hypothesis 2: only core enterprises, banks and financing enterprises are considered, and other enterprises not in the scope of financing business are not considered. Hypothesis 3: financial enterprises apply to banks for financing with historical transaction information and their own business information. Hypothesis 4: all participants in the game are rational and will make decisions to maximize their own interests. In the block chain finance of supply chain, banks should supervise the financing behavior of financing enterprises to prevent the default behavior of financing enterprises. In order to complete the transaction, core enterprises will also share information with banks to prevent losses caused by transaction failure. On this basis, the game model of banks, core enterprises and financing enterprises is constructed. In the model, the behavior choices of the three participants are: financing enterprises (performance, default), core enterprises (share, do not share), banks (strict supervision, loose supervision).

4 Discussion

4.1 The Finance of Supply Chain Development

Finance of supply chain is to provide closed credit settlement, financial management and other comprehensive financial services for different nodes of the supply chain. It can help weak small and medium-sized enterprises in the supply chain to solve the financing problem, stabilize the supply and marketing channels of core enterprises, and promote and improve the capital utilization and overall efficiency of the whole supply chain. Its essence is to obtain real trade background information in time, realize risk control, help enterprises revitalize current assets and solve financing problems, which is also the main difference between enterprises and traditional financial business. As shown in Table 1.

It can be seen from the above that the market scale in 2016 is 15.93 trillion yuan, is 23.83 trillion yuan. The results are shown in Fig. 1.

It can be seen from the above that the development of market scale increases with the growth of the year, and the maximum market scale in 2019 is 23.93 trillion yuan.

Table 1. Scale and forecast of China's supply chain financial market

Particular year	2016	2017	2018	2019
Market size (trillion yuan)	15.93	18.38	19.09	23.83

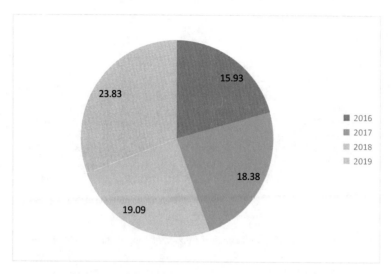

Fig. 1. Scale and forecast of China's supply chain financial market

4.2 Calculation of Relative Weight

Taking the secondary quality index A1 of substandard layer as an example, the corresponding weight is calculated. Enterprise quality judgment matrix is a fourth order matrix. By consulting the relevant literature and information, the weight of all levels of indicators under the enterprise quality is calculated. Referring to the traditional supply chain financial risk index weight, the relative importance of each index is obtained. Starting from the relative importance, the weight of each index of supply chain financial risk on the block chain is calculated (Table 2).

Table 2. Weights of various indicators of supply chain finance on block-chain

Index code	Weight value	Index code	Weight value
A	0.343	A1	0.874
B	0.536	A2	0.784
C	0.463	A3	0.832
D	0.637	A4	0.833

It can be seen from the above that the weight values of a, b, c, d are 0.343, 0.536, 0.463 and 0.637 separately. The weight values of A1, B1, C1, D1 are 0.874, 0.784, 0.832, and 0.833 separately. The results are shown in Fig. 2.

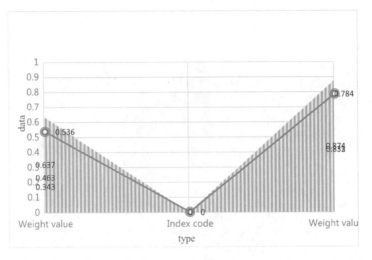

Fig. 2. Weights of various indicators of finance of supply chain on block-chain

It can be seen from the above that the weight value of A1 is higher than that of a, the weight value of B1 is higher than that of B, the weight value of C1 is higher than that of C, and the weight value of D1 is higher than that of D.

5   Conclusion

As the underlying technology and framework of bit coin, block chain is composed of credit record and credit record clearing system. It is decentralized, non-willful, anonymous and unusable data tampering. Online finance of supply chain based on big data has unique advantages in solving financing problems of small and medium-sized enterprises, but it also has new operational and regulatory risks. The application of block chain technology, through the coordinated development of block chain and online finance of supply chain, is of great significance to the risk prevention and control of online finance of supply chain. However, as an emerging technology, block chain technology is still in the exploratory stage, facing many challenges such as hardware infrastructure construction, professional talents, and block chain consensus mechanism. In order to give full play to new block chain technology in promoting finance of supply chain, finance of supply chain has become one of the hot spots in block chain technology. This paper analyzes and discusses the risks of finance of supply chain. Combined with the characteristics of block chain technology, a new method of supply chain financial risk control with block chain technology as the basis is proposed. It aims to improve the level of risk prevention and control of finance of supply chain, show its importance in supply chain risk control, and better develop supply chain finance to serve the real economy.

References

1. Azzi, R., Chamoun, R.K., Sokhn, M.: The power of a blockchain-based supply chain. Comput. Ind. Eng. **135**, 582–592 (2019)
2. Du, M., Chen, Q., Xiao, J.: Supply chain finance innovation using blockchain. IEEE Trans. Eng. Manage. **67**(4), 1–14 (2020)
3. Erik, H.: Supply chain finance and blockchain technology - the case of reverse securitisation. Foresight **20**(4), 446–447 (2018)
4. Kzlta, M.A., Cankül, D.: Yyecek Ecek Letmelernde Tedark Zncr Ve Blokzncr Teknolojs (Supply Chain and Blockchain Technology in Food and Beverage Businesses. J. Gastronomy Hosp. Travel (joghat) **3**(2), 244–259 (2020)
5. O'Leary, D.E.: Some issues in blockchain for accounting and the supply chain, with an application of distributed databases to virtual organizations. Int. J. Intell. Syst. Account. Finance Manage. **26**(3), 137–149 (2019)
6. Choi, T.M., Luo, S.: Data quality challenges for sustainable fashion supply chain operations in emerging markets: roles of blockchain, government sponsors and environment taxes. Transp. Res. **131**, 139–152 (2019)
7. Kzlta, M.A., Cankül, D.: Yyecek Ecek Letmelernde Tedark Zncr Ve Blokzncr Teknolojs (Supply Chain and Blockchain Technology in Food and Beverage Businesses). J. Gastronomy Hosp. Travel (joghat) **3**(2), 244 259 (2020)
8. Martindale, W., Hollands, T., Swainson, M.: New direction for NPD. Food Sci. Technol. **33**(1), 30–33 (2019)
9. Hazbiy, N., Ekadiyanto, F.A., Ratna, A.A.P.: Blockchain based warehouse supply chain management using hyperledger fabric and hyperledger composer. Int. J. Inf. Technol. **9**(3S), 147–151 (2020)
10. O'Dair, M., Owen, R.: Financing new creative enterprise through blockchain technology: opportunities and policy implications. Strateg. Chang. **28**(1), 9–17 (2019)
11. Alghamdi, N.S., Khan, M.A.: Energy-efficient and blockchain-enabled model for internet of things (Iot) in smart cities. Comput. Mater. Continua **66**(3), 2509–2524 (2021)

Intelligent Evaluation Method for the Credibility of Bank Digital Transformation from the Perspective of Artificial Intelligence

Xie Chen[1] and Junyi Zhang[2(✉)]

[1] School of Business, Hohai University, Nanjing, Jiangsu, China
[2] Bank of Nanjing, Nanjing, Jiangsu, China

Abstract. The application of artificial intelligence technology has become increasingly widespread and has penetrated into all areas of life. Based on artificial intelligence technology, based on the high-quality development connotation of macro-economy, enterprises and banking, this article maps the high-quality development of the three levels of macro, medium and micro to commercial banks, and explains the connotation and levels of high-quality development of commercial banks. Based on the obtained logical framework for the high-quality development of commercial banks, construct a high-quality development evaluation system for commercial banks. Research shows that the rising trend in recent years is because commercial banks have improved their own efficiency and innovation capabilities, and strengthened their shared functions with society, shareholders, and employees. Different types of commercial banks have different high-quality development status. Joint-stock banks are the best, followed by state-owned banks. City commercial banks and rural commercial banks have a tendency to differentiate, and regional banks in economically developed areas have developed better. Chinese commercial banks need to grasp the two key points of innovation and sharing, and use financial technology and sharing mechanisms as the starting point to promote the improvement of commercial banks' innovation capabilities and efficiency levels, enhance their ability to serve the real economy, and achieve high-quality development of themselves and the macro economy.

Keywords: Artificial intelligence · Commercial banks · High-quality development · Evaluation system

1 Introduction

Artificial intelligence technology has undergone three changes. As the Chinese economy shifts from high-speed growth to high-quality development, how commercial banks, as an important part of the economic system, can adjust their development strategies to high-quality development to adapt to and contribute to the high-quality development of the Chinese economy is the banking industry and important practical issues that academia are concerned about [1]. The shift of China's economy to high-quality development is manifested in the decline in economic growth, the mode of economic development has

J. Abawajy et al. (Eds.): ATCI 2021, AISC 1398, pp. 676–683, 2021.
https://doi.org/10.1007/978-3-030-79200-8_100

shifted from extensive to intensive, the structure of industry, demand, regional economy, and income distribution has been optimized, and innovation has become the main driving force for economic development. This means that the external environment faced by commercial banks has undergone major changes [2]. With the end of the era of rapid asset expansion, the banking industry has also undergone major changes: profit growth has slowed, non-performing loan ratios have risen, the asset-heavy, large-scale, and high-speed asset-liability management model has ended, and financial supervision has been continuously strengthened. Under the dual pressure of changes in the external economic environment and the unsustainable extensive development, commercial banks urgently need to adjust their development strategies and move towards high-quality development.

2   The Measurement Method and Indicator System Construction of the High-Quality Development of Commercial Banks from the Perspective of Artificial Intelligence

The rapid development of artificial intelligence technology plays a vital role in commercial banks. Based on the above definition and structure of the high-quality development of commercial banks, this article constructs a high-quality development indicator system for commercial banks.

2.1   Index Selection (See Table 1)

2.2   Evaluation Method

Existing research on the evaluation system of commercial banks mostly uses analytic hierarchy process, principal component analysis and factor analysis. The analytic hierarchy process uses expert scoring methods to determine indicator weights, and the scores of different evaluation indicators are standardized and summed with the product of their weights. Get the comprehensive evaluation score. The principal component analysis method and the factor analysis method start from the dependence relationship within the correlation matrix, group variables according to the size of the correlation, and then determine the weight according to the variance contribution rate, and then calculate the comprehensive score. The two types of methods are slightly different. This article selects the analytic hierarchy process as the evaluation method of this research from the robustness of the evaluation score and the degree of fit with reality.

Table 1. Selection table of high-quality development indicators for commercial banks

Subject	Index selection	Annotation
Safety	Non-performing loan ratio, provision coverage ratio, allocation-to-loan ratio, capital adequacy ratio, tier 1 capital adequacy ratio	Security is the basis and prerequisite of commercial banks' operations, and efforts should be made to avoid the influence of various uncertain factors to ensure stable operations and development. This paper divides security into two aspects: risk control and capital adequacy, and selects five indicators: non-performing loan ratio, provision coverage ratio, loan allocation ratio, capital adequacy ratio and Tier 1 capital adequacy ratio to measure the safety of commercial banks
fFuidity	Loan-to-deposit ratio, liquidity ratio	Liquidity is a necessary condition to meet the daily operation of commercial banks. This study selects loan-to-deposit ratio and liquidity ratio to measure the liquidity of commercial banks
Profitability	Return on net assets, return on total assets, net interest margin, net interest margin, net profit growth rate	The profit target of commercial banks is capital return, and the core is profit rate, and it needs to pay attention to the ability to continuously obtain profits. This article selects return on net assets, return on total assets, net interest margin, net interest margin and net profit growth rate to measure commercial banks
Effectiveness	Cost-to-income ratio, return on human capital, income per capita	Generally speaking, the measurement of efficiency is expressed by total factor productivity, including technical efficiency, allocation efficiency and cost efficiency, while the efficiency of commercial banks mainly includes cost efficiency, human capital efficiency and capital efficiency. Existing research uses DEA (Data Envelopment Analysis), SFA (Stochastic Frontier Analysis) and other methods of calculation, the conclusions are slightly different, but the input indicators used are basically cost, manpower and capital. This article selects the cost-to-income ratio and the rate of return on human capital for simplified analysis. And per capita income to measure the efficiency of commercial banks, covering the efficiency of manpower and cost, the efficiency of capital has been reflected in the profitability
Innovation	Proportion of intermediate business income, proportion of undergraduate employees	The traditional business of commercial banks is credit business, and the new business mainly focuses on intermediate business, and the quality of personnel is the foundation of commercial bank's innovation ability. Therefore, this article selects the proportion of intermediate business income and the proportion of undergraduate employees to measure the performance and ability
Increase	Revenue growth rate, net asset growth rate, deposit growth rate, loan growth rate	The steady growth of commercial banks is mainly manifested in the steady growth of various operating indicators without drastic fluctuations. This article selects revenue growth rate, net asset growth rate, deposit growth rate and loan growth rate to measure the robustness of commercial bank growth
Shared	Proportion of credit funds, proportion of retail loans, salary expense ratio, earnings per share, net assets per share	Commercial banks support the development of the real economy through financing and share development results with the society. This is a manifestation of the financial intermediary role of commercial banks; the role of commercial banks in promoting equal opportunities in inclusive finance is mainly reflected in the retail loan business. Therefore, this paper selects the proportion of credit funds and the proportion of retail loans as the performance of commercial banks supporting macroeconomic development. Commercial banks share development results with shareholders and employees. This article selects salary expense ratio, earnings per share and net assets per share to measure

2.3 Data Source and Sample Selection

The research data in this article is mainly derived from the public annual reports of 36 commercial banks listed on the A-share market as of September 2020, from 2010 to 2019. This article divides these 36 commercial banks into 6 large state-owned banks (Industrial and Commercial Bank, Construction Bank, Bank of China, Agricultural Bank, Bank of Communications, Postal Savings Bank), and 9 joint-stock banks (China Merchants Bank, Industrial Bank, Shanghai Pudong Development Bank) according to the nature of the bank. Bank, Minsheng Bank, China CITIC Bank, China Everbright Bank, Ping An Bank, Hua Xia Bank, Zheshang Bank), 14 city commercial banks (Bank of Beijing, Bank of Shanghai, Bank of Jiangsu, Bank of Nanjing, Bank of Ningbo, Bank of Hangzhou, Bank of Changsha, Bank of Chengdu), Bank of Guiyang, Bank of Zhengzhou, Bank of Qingdao, Bank of Suzhou, Bank of Wuxi, Bank of Xi'an) and 7 rural commercial banks (Yu Rural Commercial Bank, Zijin Bank, Changshu Bank, Jiangyin Bank, Sunong Commercial Bank, Zhangjiagang Bank, Qingdao Rural Commercial Bank) Classification research. In order to facilitate the comparison of scoring levels and interpretation of results, this paper performs normalization processing on the collected data with high-quality development, and standardizes the processing in accordance with the following formula.

$$X_{ij}^* = \frac{X_{ij} - X_{min}}{X_{max} - X_{min}} \tag{1}$$

3 Measurement of the High-Quality Development of Commercial Banks from the Perspective of Artificial Intelligence

In the past ten years, the high-quality development of listed banks in my country has declined first and then increased [3]. The main reason for the decline is that the profitability of commercial banks has fallen and the expansion speed has slowed sharply. The increase in high-quality development is mainly due to commercial banks. Banks improve their innovation capabilities and efficiency levels, and strengthen their shared functions with society, shareholders, and employees [4]. In terms of indicators, the security of my country's listed banks has declined slightly in the past decade [5]. Although major banks have strengthened their asset adequacy ratios and provision coverage ratios, the rise in non-performing loan ratios has affected the security of banks. Since the loan-to-deposit ratio is no longer included in the bank's regulatory indicators, major banks have slightly increased their loan-to-deposit ratios, but they have also strengthened their liquidity ratios, thus ensuring liquidity stability as a whole. In recent years, the profitability of my country's commercial banks has shown a downward trend, and various profit indicators are in a downward trend [6]. Although major banks have carried out scientific and refined management by controlling expenses and human cost expenditures, they have improved the efficiency of commercial banks. The expansion speed of most commercial banks has slowed down sharply, and the banking industry is also in the context of the policy of concession to enterprises, which has led to a sharp decline in the profitability of listed banks in recent years [7, 8]. In order to get rid of this situation, listed banks have vigorously developed financial technology, developed new intermediary businesses, and promoted development and transformation. The innovation capabilities of major banks have been greatly improved. In addition, under the guidance of the government's support for the real economy and inclusive financial policies, commercial banks actively develop inclusive services, strengthen the role of financial intermediary, and share development results with the society.

There are also obvious differences in the high-quality development of different types of commercial banks in the past decade. Joint-stock banks have performed relatively well in high-quality development, while rural commercial banks have fluctuated greatly, and the recent high-quality development is relatively poor. Perform classification analysis. It can be seen from Table 2 that the high-quality development of state-owned banks is relatively stable, and the scores of various first-level indicators are mostly at the midstream level. Among them, state-owned banks perform relatively well in terms of safety and liquidity, especially capital adequacy ratio and liquidity ratio. Steady, but we need to pay attention to improving the education level of employees and enhance the potential for innovation.

Table 2. 2010–2019 listed state-owned banks' high-Quality development index score table.

Years	Score	Safety	Fluidity	Profitability	Effectiveness	Innovation	Increase	Shared
2010	39.80	30.61	37.92	44.79	40.19	37.26	48.31	48.22
2011	42.46	34.16	38.91	51.12	45.97	43.90	44.39	45.30
2012	41.90	31.58	38.96	49.14	50.34	47.54	35.35	44.93
2013	41.62	27.48	37.38	44.66	50.55	60.96	31.21	47.93
2014	41.60	24.42	38.96	41.42	53.71	65.67	31.37	48.17
2015	40.75	20.89	39.02	36.33	54.53	72.60	32.58	47.40
2016	39.15	19.87	34.35	29.75	54.36	79.17	29.64	47.44
2017	38.34	21.35	28.31	23.70	50.20	84.83	23.08	54.67
2018	38.80	22.39	26.10	24.00	50.37	77.05	24.71	62.42
2019	40.13	24.03	27.61	26.46	52.89	70.41	28.92	65.12

Joint-stock banks have leveraged financial technology to vigorously develop new intermediary services and inclusive services. They have outstanding performance in innovation and sharing, and they have implemented refined management, which has significantly improved operating efficiency. However, they need to pay attention to safety and liquidity strengthen (see Table 3).

Table 3. 2010–2019 Listed joint-stock bank high-Quality development index score sheet.

Years	Score	Safety	Fluidity	Profitability	Effectiveness	Innovation	Increase	Shared
2010	38.90	23.64	41.89	45.82	44.95	47.96	38.11	40.33
2011	40.52	30.94	44.22	50.48	41.70	47.32	33.37	37.55
2012	40.24	31.96	43.94	48.12	42.99	46.04	28.12	38.66
2013	39.43	29.20	43.20	45.41	43.16	48.26	24.24	40.80
2014	38.79	27.89	41.47	43.09	44.91	46.93	24.98	42.17
2015	37.11	23.72	42.61	35.98	45.53	49.21	23.71	41.85
2016	35.59	23.13	42.83	28.51	43.68	52.04	20.21	41.21
2017	36.13	25.48	42.31	27.20	44.79	50.25	19.81	43.57
2018	37.92	28.52	46.11	27.27	46.09	49.24	19.99	45.84
2019	38.61	30.69	45.69	26.02	47.11	49.66	20.78	47.40

The high-quality development of city commercial banks and rural commercial banks has fluctuated greatly in recent years (see Table 4). Rural commercial banks in particular have fallen from the leading position in the four types of banks in 2010 to the bottom in 2019. The reason is that after the rapid decline in expansion speed and profitability, problems such as high non-performing loans, low operating efficiency, and insufficient innovation capabilities that were not visible in the previous rapid development process have gradually been exposed. In addition, the high-quality development of city commercial banks and rural commercial banks in recent years has also shown a polarized situation [9]. City commercial banks and rural commercial banks in economically developed areas have performed well. The high-quality development level of the leading city commercial banks in recent years can be comparable to that of joint-stock banks. While the high-quality development of urban commercial banks and rural commercial banks in economically backward areas fluctuates greatly [10]. This also shows the mutual promotion between the high-quality development of regional banks and the level of local economic development. Commercial banks should integrate their own high-quality development into the high-quality development of the local economy, leveraging each other, and co-exist and prosper. In terms of indicators, some city commercial banks and rural commercial banks need to control the rate of non-performing loans, improve their innovation capabilities, and strengthen their support for the powerful economy.

Table 4. 2010–2019 Listed city commercial bank high quality development index score table.

Years	Score	Safety	Fluidity	Profitability	Effectiveness	Innovation	Increase	Shared
2010	39.63	33.89	48.44	45.94	46.74	21.55	47.12	35.46
2011	41.86	36.68	47.82	54.21	48.97	24.38	43.59	35.72
2012	42.05	35.93	48.58	53.48	52.82	26.91	39.73	34.75
2013	40.96	32.33	48.39	48.40	53.23	32.17	33.85	36.65
2014	40.72	28.15	50.76	45.64	54.10	36.75	36.56	36.71
2015	39.81	27.22	52.54	38.65	54.09	41.80	36.78	34.00
2016	37.80	25.98	51.53	30.04	52.25	49.59	31.26	31.23
2017	37.11	27.69	49.12	26.87	51.04	49.23	28.35	32.45
2018	37.58	28.12	47.60	25.22	50.83	44.20	33.13	39.68
2019	38.89	29.54	47.18	26.78	53.14	41.66	30.62	45.50

4   Conclusion

Artificial intelligence technology is developing rapidly. Based on the evaluation results of high-quality development of commercial banks, my country's commercial banks need to grasp the two key points of innovation and sharing in the future transformation to high-quality development. Use financial technology as a breakthrough in technological innovation to drive business innovation and system innovation; to help the real economy, regional economy and inclusive finance as the starting point to achieve profitability and profit growth; to use a complete shareholder and employee sharing mechanism as incentives to promote improved efficiency level. Create economic value and social value more efficiently and effectively, and finally complete the strategic goal of transition to high-quality development, and then promote high-quality macroeconomic development, achieve financial prosperity, and the economy promotes a cycle of financial growth.

References

1. Shuxin, A.: Research on the Path to Promote High-quality Economic Development: A Literature Review. Contemporary Economic Management **40**(09), 11–17 (2018)
2. Xiaojing, C., Zhixin, X.: The theoretical logic and practical mechanism of China's high-quality economic development in the new era. Journal of Northwest University (Philosophy and Social Sciences Edition) **48**(06), 12–22 (2018)
3. Sujian, H., Hongjun, X., Xin, W.: On the high-quality development of state-owned enterprises. China Industrial Economy **10**, 19–41 (2018)
4. Xiaoyu, H., Kunrong, S.: Modern economic system, total factor productivity and high-quality development. Shanghai Economic Research **06**, 25–34 (2018)
5. Bei, J.: Economic research on "high-quality development." China Industrial Economy **04**, 5–18 (2018)
6. Jian, L., Lijuan, W., Fang, W.: Research on the Evaluation of High-Quality Development of Commercial Banks——The Construction and Application of "Gyro" Evaluation System. Financial Supervision Research **06**, 56–69 (2019)

7. Baohua, L., Lei, G., Lili, Z., Zhou, L., Ming, W.: Research on the Construction and Application of Evaluation System for the High-quality Development of Commercial Banks. Financ. Supervision Res. **12**, 17–33 (2018)
8. Zhiying, M., Dong, K.: From scale to value: analysis of the high-quality development model of commercial banks. Journal of Qinghai Normal University (Philosophy and Social Science Edition) **41**(05), 14–17 (2019)
9. Baoping, R.: Research on the high-quality development of China's economy. Journal of Shaanxi Normal University (Philosophy and Social Sciences Edition) **47**(03), 104 (2018)
10. Baoping, R., Fengan, W.: Judgment criteria, determinants and realization methods of China's high-quality development in the new era. Reform **04**, 5–16 (2018)

Application and Development Prospect of Virtual Reality Technology in Interior Soft Decoration Design Industry

Hongying Zhang[✉] and Yao Lu

Chongqing College of Architecture and Technology, Chongqing, China

Abstract. With the rapid development and advancement of modern technology, the working methods of designers are also changing accordingly. Among them, virtual reality (VR) technology brings new opportunities and challenges to actual design work. Traditional interior design is based on design drawings, and the designer's design concepts and design effects are all based on static renderings and three-dimensional animations. With the application and popularization of VR technology, the thinking mode and working methods of interior design have also undergone great changes. The application of VR technology makes interior design work break the traditional mode, become more vivid and more efficient. The article makes relevant research and analysis on the development prospects and application methods of VR interior design.

Keywords: Virtual reality · Interior decoration · Interior design · Design industry

1 Introduction

In recent years, VR technology has achieved rapid development in all occupations. Compared with traditional environmental art design teaching, the application of VR technology in environmental art can break the limitations of traditional design methods [1]. The interior design industry is no exception. The application of VR technology in interior design adopts advanced interactive forms. Based on computer software, VR headsets enable design content to be presented more vividly and intuitively in front of designers and customers. The rapid advancement of design work and the communication and coordination between all parties. The application of VR technology in interior design work has become an inevitable trend in the development of interior design industry [1].

2 VR Technology Overview

VR technology is also called VR technology. It can rely on computer simulation to create a three-dimensional space-folding virtual world. It can realize sensory simulation through sight, hearing, and touch, so that the operator can experience the multi-dimensional feelings brought by the technology. The application of VR technology

J. Abawajy et al. (Eds.): ATCI 2021, AISC 1398, pp. 684–690, 2021.
https://doi.org/10.1007/978-3-030-79200-8_101

in interior design is to use Internet and computer simulation methods to create a virtual model world that is highly similar to the actual space, and then cooperate with VR headsets to present the design results of interior designers to customers more three-dimensionally and intuitively [2]. By creating an immersive human-computer interaction experience for customers to show the designer's ideas more vividly and intuitively.

3   Advantages of Applying VR Technology in Interior Design

The interior decoration design is actually based on the nature of indoor use and the environment in which it is located, using scientific and technological means and the principles of architectural aesthetics to create a beautiful and comfortable indoor environment that can satisfy people's material, there are spiritual needs [2]. By designing this technique, the indoor space can better fulfill its use value, maximize its use function, and achieve the purpose required by people. Usually a house must be designed after it is built, because in this way can the use of the building space be maximized.

3.1   Conducive to Promoting Effective Communication Between Designers and Customers

The popularization of VR technology makes interior design content vividly displayed in front of designers and householders. Since two-dimensional renderings or traditional model displays require certain relevant professional knowledge to understand, and when communicating with designers, many householders hope to be able to evaluate design effects from multiple angles and in all directions [3]. VR technology uses dynamic the three-dimensional virtual scene provides customers with angles and details that cannot be displayed in static drawings, breaking the communication barriers between designers and customers, and avoiding poor communication problems caused by the owner of the household who cannot understand the drawings or cannot understand the design content., Which allows the designer and Party A to make a more comprehensive evaluation of the plan in the virtual space, and it is convenient for Party A to make some suggestions on the design, as shown in Fig. 1.

Fig. 1. VR shows the effect of indoor soft decoration

3.2   More Three-Dimensional in Terms of Design Expression

The use of VR technology to build VR models can effectively avoid visual blind spots under two-dimensional graphics and make the design content three-dimensional. It is not only convenient for designers to modify and improve the design content, but also allows designers to more efficiently manipulate the interior the space is dynamically evaluated and judged, the omissions in the design plan are corrected at any time, and the design details are optimized [4].

4   Application of VR Technology in Interior Soft Decoration Design

4.1   Optimize the Design Process

The traditional interior design mode usually relies on drawings or two-dimensional and three-dimensional imaging of the computer to show the effect of the plan. The user's participation is not high enough to feel the real effect of the space. According to the traditional "Fang, whether it is a designer or a customer, it is difficult to find the problems in the plan in the first time [5]. Usually these problems will be exposed during the construction stage, which will lead to the number of on-site rework and increase the construction cost. If you use VR Technology, can simulate the actual space scale and future decoration effects through a computer, and then cooperate with the headset VR glasses to create a strong immersive experience for users, allowing users to more intuitively feel the effects of the project after completion, and let designers [5]. In the dynamic virtual environment, discover the problems in the plan in time, make adjustments to the design ideas at any time, to make up for the defects in the design details, including the color, texture, lighting effects and other elements of the material, to avoid repeated rework during the construction phase to increase the cost and manpower cost.

4.2   Simulate Environmental Conditions

Buildings are often affected by complex natural environmental factors, such as light, temperature, wind and other natural conditions when they are used in the real environment. These conditions are also necessary considerations for designers when designing. Specifically: light lighting plays a pivotal role in the indoor environment and the lighting effect is closely related to the color temperature, illuminance, quantity, and density of the selected lamps. Traditional static renderings usually use VRAY rendering software to render lighting effects based on the designer's subjective feelings. The disadvantage is that it can only blur out the indoor light environment [6]. The specific color temperature and illuminance cannot accurately simulate the lighting effects of the real scene. Users only can imagine the effect after construction is completed through pictures. Using VR technology, it is possible to simulate lighting effects that meet the real scene based on the selection of real lamps, and through the headset, customers can truly experience the real feelings brought by different environments and different lighting effects, and the actual construction of future projects [6]. The effect is predicted and controlled to avoid the influence of the final use effect due to unreasonable lighting settings in the design stage, as shown in Fig. 2.

Fig. 2. VR simulation indoor environment effect

4.3   Establish an Integrated Network of Virtual Models and Product Sales

Another important application of VR technology in interior design is to integrate offline resources to promote the integration of program design and product sales. Specifically, through the Internet big data, the high-quality products of the major offline product suppliers in the major decoration fields are virtualized and digitized, and the data is embedded in the VR design software, so that it is convenient for the designer to call the real at any time during the plan deliberation stage [7]. According to the required program effect, a detailed list of materials (including furniture, lamps, hardware, soft decorations, etc.) can be listed in the program stage according to the product data of the product. If the customer is satisfied with the effect of the plan, he can quickly obtain the purchase link according to the generated bill of materials, which can ensure the high consistency of the actual construction effect and the plan, and build a wide range of product sales for offline home furnishing stores through the design of the plan platform, to enhance the sales and use value of offline products, and promote the economic development of the decoration field [8].

4.4   Commercial Value in Showcase

The application of VR technology in the field of home improvement can also greatly save the production cost of model rooms. In the process of model room display in the traditional home improvement industry, various styles of model rooms are designed according to different apartment types and customer tastes for consumers to visit and experience [9]. This method not only needs to spend twice the time and labor for construction work, but also makes consumers feel physically and mentally exhausted. If the virtual model room environment is constructed through the VR system, and then according to the customer's aesthetic characteristics and personal preferences, using computer technology to replace the material model of the model room at any time to meet the different preferences of customers, it can save many labor and material resources, as shown in Fig. 3. At the same time, with the help of powerful network and software, VR technology can replace a variety of interior design styles and colors in a short period, creating a diversified sensory experience for customers and increasing the transaction rate [9].

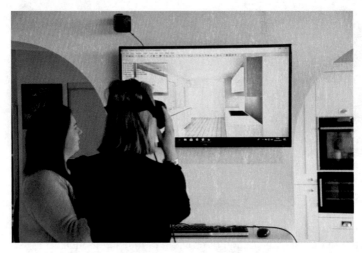

Fig. 3. VR display model room effect

5   The Development Prospects of VR in the Interior Soft Decoration Design Industry

Interior decoration design is built to meet the needs of people, that is, to reprocess existing buildings or objects to achieve beautiful and more practical purposes [8]. For example, when designing a house, some residents often feel that they have not achieved the desired effect after the design has been remodeled, and they want to remodel the design. In fact, this will waste time and financial resources, and use virtual reality. Technology can handle this problem very conveniently. When designing, first carry out the substitution of virtual reality technology, and design and operate according to the requirements of the residents. When the model appears in front of the residents, due to the fidelity of the technology, the effect from the eyes of the residents you can see the effect afterwards. If there is any dissatisfaction, you can modify it at any time. Unlike before using the technology, it takes time, energy, materials, and money [10]. Therefore, virtual reality technology should be combined with interior decoration design or Combined with each other, to make the design and development work better. In many cases, when designers design drawings according to customer requirements, they show them to customers, but they do not understand them. At this time, if you use virtual reality technology, you can make customers feel the design in a simple and clear way.

With the continuous improvement of China's economic, social development level, and the gradual deepening of the reform of the design industry at this stage, the application of VR technology in interior design has received more and more attention [11].

VR technology is the result of the continuous development of modern science and technology. The application of this advanced technological achievement in China's interior design industry has significant positive effects. At the current stage, the application of VR technology in rendering processing of interior space design renderings and the application of space design operability have received good overall feedback. Overall, the application of VR technology in China's interior design industry is not only conducive to achieving better communication and communication between target customers and professional designers, but also conducive to improving the design level of professional designers [11]. Advanced science and technology can more intuitively experience the final effect of interior design, and then ensure the final quality of interior design. For the target customer group, with the continuous development of advanced Internet technology, the target group can use the network system to feel the final design effect more intuitively and timely, so that the construction industry and customers can better meet their own needs.

6 Conclusion

The application of science and technology has made human production and life more convenient, and the rational use of science and technology can make our lives better. The application of VR technology in interior design can expand the designer's vision and improve work efficiency; it can also increase the participation of owners and promote the success rate of the design plan; at the same time, it can quickly integrate the indoor plan and offline product resources to form Integration of design and sales. In short, the birth and popularization of VR technology has promoted the transformation of design modes, created a broad scope for interior designers, and inspired designers to use more multi-dimensional observation methods and thinking modes to realize their creativity, the most efficient and accurate. The design details are carefully controlled, so that the design works can be perfectly presented in the end.

References

1. Guo, H.X., Liu, Q.: Application of virtual reality technology in interior design. Pop. Bus. Mon. **9**(02), 99–101 (2011). (in Chinese)
2. Wei, J.Y.: Discussion on the application of VR technology in architectural interior design. Juye **12**, 54–57 (2017). (in Chinese)
3. Yang, B.M.: The application of virtual reality technology in architectural planning and interior design. Electron. Today **11**, 82–85 (2011). (in Chinese)
4. Wang, M.M.: The application of virtual reality technology in environmental art design. Heilongjiang Sci. **12**, 117–119 (2018). (in Chinese)
5. Xiao, F.: The application of VR design in interior design. Home Ind. **11**(06), 57–59 (2018). (in Chinese)
6. Wang, F.T.: The use and development of virtual reality technology. Sci. Technol. Lett. **12**(06), 113–115 (2019). (in Chinese)
7. Guo, H.X.: The use of virtual reality technology in interior design. Pop. Bus. **10**, 75–76 (2018). (in Chinese)

8. Wang, X.: The development of virtual reality technology and its application exploration. Pop. Sci. Technol. **10**(05), 44–47 (2017). (in Chinese)
9. Yun, R.W.: Teaching application of virtual reality and design of simple virtual learning environment. Nanjing: Sch. Educ. Sci. Nanjing Norm. Univ. **8**(11), 120–123 (2016). (in Chinese)
10. Guo, P.C.: Application of virtual reality technology in interior decoration design. Chin. Resid. Facil. **12**, 55–58 (2018). (in Chinese)
11. Liu, S., Sun, Q.: Analysis of the application of virtual reality technology in interior decoration design. Home Ind. **11**(07), 44–46 (2018). (in Chinese)

Analysis and Construction of Visual Supply Chain for Internet Plus Fresh Agricultural Products in the New Era

Ping Yang$^{(\boxtimes)}$ and Yanran Huang

Informatiaon Management Department, Chengdu Neusoft University, Chengdu, Sichuan, China

Abstract. Consumers' requirements for the quality of fresh agricultural products have continued to improve with the improvement of living standards . Supply chain is the key to ensure the high quality of fresh agricultural products purchased by consumers finally. In the traditional fresh agricultural products supply chain, the links of production, processing, storage, transportation and distribution are relatively isolated, and the information among producers, distributors, retailers and consumers is not correct. These problems lead to the blind production of traditional fresh food supply chain, the increase of bullwhip effect in all links and the difficulty of traceability for consumers. This paper will build a visual supply chain for fresh agricultural products using the internet plus in the new era. At the same time, it will optimize the supply chain continuously and meet the consumers' requirements for high-quality fresh agricultural products gradually according to the total quality management, PDCA cycle.

Keywords: Internet plus · Fresh agricultural products · Visual supply chain · Total quality management · PDCA

1 Introduction

In the new era, the internet has changed the operation mode of traditional industries, and changed every link and every participant from production to consumption. The supply chain is closely connected with the development of the internet [1]. The scale of fresh e-commerce transactions in China exceeded 200 billion in 2018 according to relevant data. The traditional fresh agricultural product supply chain is facing great tests and challenges behind the huge transaction [2]. The supply chain of traditional fresh agricultural products is in urgent need of reconstruction because of false platform information [2], delivery timeliness, product quality and other issues. This paper uses the internet plus to visualize the logistics, the information flow and the capital flow in the fresh agricultural supply chain to meet the development of fresh agricultural products supply chain in the new era [3].

© The Author(s), under exclusive license to Springer Nature Switzerland AG 2021
J. Abawajy et al. (Eds.): ATCI 2021, AISC 1398, pp. 691–697, 2021.
https://doi.org/10.1007/978-3-030-79200-8_102

2   Problems of Traditional Fresh Agricultural Products Supply Chain

Fresh agricultural products supply chain refers to the function network chain structure of fresh agricultural products from production to consumers. Each link plays its specific function in the network chain. If any link fails, it will cause problems in the whole supply chain.

2.1   Isolated Supply Chain

Fresh agricultural products go through the production, processing, storage, transportation, distribution and other links to arrive at the hands of consumers finally. In order to ensure the quality of fresh agricultural products, we must ensure that the information of each link of the whole fresh agricultural products supply chain is smooth. In the traditional supply chain, each link is relatively isolated, as shown in Fig. 1. During the production of farmers, the demand information of consumers cannot be well obtained, analyzed and integrated. Consumers are not clear about the production, processing, storage, transportation and other information of the purchased fresh agricultural products. And it is difficult to link the production information of farmers with the demand of consumers in each link of the supply chain [4].

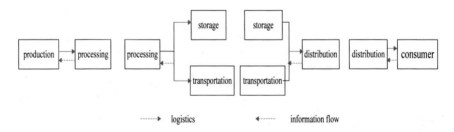

Fig. 1. Isolated supply chain

2.2   Access for Consumers to Fresh Agricultural Product Quality Information is Limited

Consumers can only judge their freshness by vision and touch for fresh agricultural products generally. However, consumers have no way to know the residual pesticides or other chemical substances. In large supermarkets or wholesale markets, the relevant quality inspection departments will inspect the quality of vegetables. But in the secondary and tertiary markets, consumers know nothing about the contents of the vegetable quality inspection report and have no access to vegetable quality information. For pollution-free, organic, green agricultural products, the state had the corresponding quality marks to indicate. But if consumers want to obtain more detailed information behind these quality marks, the channel is also cumbersome.

2.3 The Bullwhip Effect of Middleman Increases

Bullwhip effect refers to the phenomenon that when the information of the supply chain is transferred from the downstream to the upstream, the information is magnified step by step, because it is unable to share the information effectively [5]. Wholesalers, distributors and other middlemen play a key role in the supply chain. At present, The middlemen in the supply chain will enlarge the information in order to reduce their own risks, because the information in the traditional fresh agricultural products supply chain is not smooth, resulting in the increase of the bullwhip effect of middlemen. This situation makes the inventory and cost of each link of the supply chain rise, and leads to the distortion of the logistics, information flow and capital flow of the whole supply chain eventually.

2.4 Great Blindness of Producers

There are two production modes of producers, push production and pull production. Push production is that producers produce products according to experience and forecast, and push products to market. Pull production is that producers arrange production according to market demand. Obviously, pull production is more in line with the market requirements of the new era. Producers of fresh agricultural products need to arrange production plan according to market demand. However, the information in the traditional supply chain is not smooth, which makes it difficult for the downstream market demand information of the supply chain to be accurately transmitted to the upstream producers, and makes the production of fresh agricultural products producers blind. "Cheap vegetables hurt farmers" is a typical example of the blind production of fresh agricultural products.

3   Construction of Visual Vegetable Supply Chain

The visualization supply chain of vegetables is proposed based on internet plus, aiming at the existing problem of vegetable supply chain. It realizes the transparency of all aspects of supply chain and make logistics and information flow smoothly in the supply chain. It can start from the establishment of rural cooperatives, corporate operation, building information platform and new retail to realize the construction of vegetable visualization.

3.1 Establishment of Rural Cooperatives

Individual farmers have great limitations in choosing the type, quantity, production technology and equipment of products, which often lead to the problems of single variety and poor quality of products, and make the competitiveness of individual farmers in the market very weak ultimately. Therefore, it is necessary to establish rural cooperatives. Rural cooperatives promote farmers to be more scientific in production through reasonable distribution and accurate analysis. Farmers can reasonably distribute the types and quantity of vegetables produced in cooperatives through the establishment of rural cooperatives. At the same time, compared with individual production, the scale of cooperative production has expanded, and more advanced and scientific production technology and equipment can be used to enhance competition in the same industry.

3.2 Corporate Operation

One of the main reasons for the blindness of agricultural producers is that individual farmers have limited ability to obtain and deal with market information during production, and they arrange production plans only by personal experience and analysis. At the same time, individual farmers are very weak in dealing with market risks, which is also a fatal weakness of small-scale peasant economy. Although rural cooperatives can arrange some production at the micro level well, the micro level can play a better role only under the guidance of the macro level. Vegetable supply chain is a relatively complex system, involving more participants and information to be processed. Individual operation cannot cope with such a complex system, so it needs company operation. The company needs to arrange production scientifically and reasonably, control production costs and improve the quality of vegetables combined with the specific situation of rural cooperatives and integrated the market demand for the quantity, variety, price and quality of vegetables. The company and rural cooperatives can achieve a win-win situation on the premise of meeting the requirements of consumers for vegetables.

3.3 Internet Platform

Each participant in the whole supply chain has the demand to obtain the information in the supply chain. The production of cooperatives needs to be arranged according to the information transmitted from the downstream; companies need to integrate the information of all links in the supply chain to prevent the bullwhip effect from increasing in all links; consumers need to be clear about the quality of vegetables they buy, such as the growing place, the use of pesticides or fertilizers, processing environment, storage conditions, distribution time, etc. Therefore, it is very necessary to build a supply chain internet platform. This platform can open part of the shared information, and set access to some information according to the identity of the participants. This can not only ensure the visualization of the whole supply chain information, but also protect the trade secrets of all participants. The company guides the cooperative farmers to select vegetable varieties and quantity through the information platform. The cooperative farmers understand the market demand through the information platform, and better negotiate and cooperate with the company. The company and the distributors, the wholesalers, the retailers at all levels of the chain exchange information on this information platform to realize the visualization of the supply chain. Consumers can have channels to understand the production, processing, storage, transportation and distribution of purchased vegetables, so as to realize the visualization of the whole vegetable supply chain [6].

3.4 New Retail

The traditional supply chain model is being reconstructed, which brings opportunities and challenges to the participants in all aspects of the supply chain with the development of internet, big data, artificial intelligence and other technologies [7]. The supply chain links can be simplified and transparent, which makes it easier for the participants to integrate and analyze the information in the supply chain, and also makes it easier to obtain the experience data of consumers with the help of these technologies. Through online or online/offline sales [8], Companies can obtain consumers' experience information data quickly and efficiently, which is convenient for communication and coordination with cooperative farmers and the middle links of the supply chain, realizing the visualization of vegetable supply chain. Consumers can also use these technologies to understand the whole process of vegetables from production to sales intuitively, so as to select suitable products according to their own needs. The company upgrades the production, circulation and sales process of vegetables, integrates online/offline and modern logistics of vegetables deeply, and realizes the visualization of vegetable supply chain with the help of new retail finally.

The visual supply chain of vegetables is constructed after analysis, as shown in Fig. 2. Rural cooperatives, companies and other links in the supply chain can share information based on the internet information platform. The visualization of logistics and information flow in the production, processing, storage, transportation and distribution of vegetables urges all links in the vegetable supply chain to guarantee and pay attention to the quality of vegetables. Consumers can understand the information of each link in the vegetable supply chain and know the quality of vegetables they buy on the basis of the new retail model and information visualization of each link.

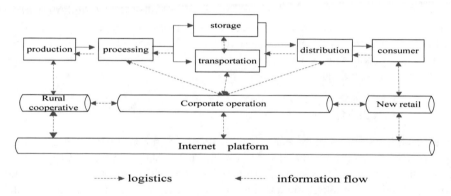

Fig. 2. Visual supply chain

4 Total Quality Management

Total quality management is to take quality as the center, full participation, taking corresponding management measures to make customers satisfied. Its characteristics are: full participation, whole process, comprehensive quality concept [9].

4.1 Vegetable Quality Management with Full Participation

From planting to the final arrival of vegetables in the hands of consumers, the personnel involved in each link will affect the quality of vegetables. It can ensure that the vegetables in the hands of consumers are of high quality only in each link of the implementation of strict quality responsibility system [10].

4.2 Whole Process Vegetable Quality Management

Vegetables have gone through the production, processing, storage, transportation, distribution and other links from the perspective of the process of vegetables reaching the hands of consumers finally, and each link is independent and interrelated. The whole process of vegetable quality management will eliminate the unqualified products in the process. At the same time, all links from the source to the final consumer form a cycle and rising process, which makes the quality of vegetables improve continuously.

4.3 Quality Concept of Total Vegetable Quality Management

The quality of total vegetable quality management is comprehensive, not only the narrow sense of vegetable quality, but also the quality of vegetable production and consumption process. Vegetable quality management should not only manage the vegetable itself, but also manage people and the whole working process, and it should improve the quality of personnel and work quality constantly.

4.4 PDCA Cycle Method

The feasibility analysis of vegetable visual supply chain construction based on total quality management can adopt PDCA cycle method, as shown in Fig. 3. PDCA refers to the four processes of plan, do, check and action, which go through a cycle and spiral up. After each cycle, the quality is improved. It is impossible to complete the complete construction of the supply chain at one time, and it needs to be gradual due to the complexity of the vegetable supply chain. First of all, it needs to start from each link in the aspect of full participation of vegetable supply chain. If it carries out full participation in vegetable quality management in each link gradually, it can realize full participation in vegetable quality management in the whole supply chain. Secondly, the quality management of the whole process of vegetable supply chain also needs a gradual improvement process. At present, each link of the supply chain is relatively disjointed. It needs rural cooperatives, companies and internet platforms to improve the quality management of the whole process of the vegetable supply chain gradually, which needs time and cost. Finally, the promotion and popularization of the concept of total vegetable quality management also need to be improved in the process of planning, implementation, inspection and action.

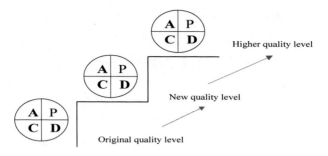

Fig. 3. PDCA cycle method

5 Conclusion

This paper constructs a visual vegetable supply chain based on internet plus. It starts from the establishment of rural cooperatives, corporate operation, building information platform and new retail to realize a visual and smooth information supply chain of vegetables from vegetable planting to consumers in order to realize the construction of visual supply chain of vegetables. The feasibility analysis of supply chain construction can use PDCA cycle method. In the actual construction of vegetable visual supply chain, technology, capital and talents are the specific practical factors. The construction should be completed step by step based on total quality management under the guidance of visual ideas.

References

1. Li, L.: Research on the innovation of e-commerce operation mode of fresh agricultural products in the internet era. Agr. Econ. **6**, 139 (2020). (in Chinese)
2. Yupeng, W.: Thinking about the quality improvement of fresh e-commerce service from the perspective of block chain. Res. Commer. Econ. **14**, 92 (2020). (in Chinese)
3. Shuwen, Z.: Research on logistics efficiency of dual channel supply chain of fresh produce under the background of internet plus. Agr. Econ. **2**, 129 (2018). (in Chinese)
4. Yangyang, L.: Big data driven fresh agricultural products supply chain model innovation and operation optimization. Res. Commer. Econ. **16**, 150 (2020). (in Chinese)
5. Liang, L.C., Xiaojin, L.: Supply chain management under the internet mode: a case study of bullwhip effect. Res. Commer. Econ. **6**, 79 (2018). (in Chinese)
6. Beilei, C.: Innovation of enterprise supply chain management mode under the background of internet economy. Res. Commer. Econ. **12**, 125–128 (2020). (in Chinese)
7. Linyi, C., Guangxue, C.: Research on packaging visualization from the perspective of supply chain. Packag. Eng. **7**, 19–20 (2018). (in Chinese)
8. Jingxiang, H.: Research on the innovation of online video marketing mode of edible fungi under the background of internet. Chinese Edible Fungi **12**, 135–138 (2019). (in Chinese)
9. Zhixiang, C.: Production and operation management, p. 245. China Machine Press, Beijing (2019).(in Chinese)
10. Tao, J.F., Zheng, D.X., Ling, Z., Yun, H.H.: Problems and countermeasures of vegetable supply chain. Value Eng. **14**, 32–35 (2015). (in Chinese)

Cloud Computing Aided Model Design of Urban and Rural Planning and Design Based on OBE Results Oriented Theory

Peiyu You$^{(\boxtimes)}$

Nanchang Institute of Technology, Nanchang, Jiangxi, China

Abstract. Hybrid teaching mode is an inevitable choice for colleges and universities to carry out teaching reform and cultivate applied talents under the background of education informatization . Based on the concept of "OBE", this paper takes the training and development course of human resource management as an example, optimizes the hybrid teaching mode through preset learning results, reverse design and positive implementation, and makes a preliminary discussion on the optimization of the hybrid teaching mode.

Keywords: College students' innovation and entrepreneurship · Project resource library · Personalized recommendation · Collaborative filtering

1 Introduction

With the wide application and gradual deepening of educational informatization, the change of educational teaching mode and learning style has become a common concern of higher education institutions, teachers and students. How to give full play to the advantages of online learning to enhance students' learning autonomy, initiative and creativity, while fully inheriting the advantages of traditional teaching, such as teachers' guidance, inspiration and monitoring role and showing personality influence, the hybrid teaching mode has become a practical choice. Based on the OBE teaching concept, aiming at the characteristics of most courses of human resource management specialty, which are applied, professional and practical, this paper designs and practices the mixed teaching mode on how to improve the teaching level and enhance the teaching effect, in order to improve the teaching level and quality of this specialty [1].

2 Design Requirements of Hybrid Teaching Mode of Human Resource Management Courses Based on OBE Concept

OBE (outcome based education) education concept, also known as results oriented education, is a kind of education concept that emphasizes students' learning results and effects as the guidance. It organizes curriculum teaching according to the structure of

© The Author(s), under exclusive license to Springer Nature Switzerland AG 2021
J. Abawajy et al. (Eds.): ATCI 2021, AISC 1398, pp. 698–702, 2021.
https://doi.org/10.1007/978-3-030-79200-8_103

professional results obtained from professional and work analysis, and attaches importance to the acquisition of professional skills. The essence of OBE concept is to focus on the analysis of students' learning output, reverse design of students' education structure and related evaluation system, so as to effectively improve the quality of teaching. Under this concept, educators need to have a clear construction of students' learning outcomes, that is, what should students' learning outcomes be and how to present them, Why do students need to achieve such results, and through reasonable curriculum structure, teaching activities, evaluation system design to effectively promote and ensure students to achieve results. Based on the OBE concept of human resource management professional mixed teaching mode design requirements.

2.1   Clear Teaching Objectives and Preset Learning Outcomes

Based on the OBE concept, teachers must analyze the professional ability required by the professional talent market in the teaching process, and complete the construction of students' preset learning outcomes accordingly. The ultimate teaching goal of human resource management specialty is to cultivate sustainable development talents who can highly integrate the theory and practice of human resource management specialty and integrate knowledge and ability [2].

2.2   Pay Attention to Learning Process and Reform Teaching Mode

Although OBE emphasizes the students' learning outcomes, it does not mean that the learning process is ignored. Under the concept of OBE, learning process and learning outcomes should show the logical relationship of reverse design and positive implementation, that is, to achieve the preset learning outcomes, we must pay great attention to the students' learning process. The design and implementation of each link of OBE teaching should be closely around the students and reflect the dominant position of students. Teachers organize teaching according to their preset learning outcomes for each student, and teachers mainly play the role of guidance, inspiration and monitoring.

3   Distribution Route Optimization Based on Hybrid Adaptive Cuckoo Algorithm

Cuckoo algorithm has good global search ability and versatility, simple parameters and easy implementation, but it has the problems of slow convergence speed and insufficient convergence accuracy. Therefore, in order to optimize the convergence speed and solution accuracy of cuckoo algorithm, this chapter proposes a hybrid adaptive cuckoo algorithm (HACs), which combines cuckoo algorithm with particle swarm optimization algorithm, adopts adaptive parameter update, introduces fitness weight factor to enhance the information exchange among individuals in the population, and in order to solve the path optimization problem, the adaptive cuckoo algorithm is used to improve the convergence speed and solution accuracy, In the discrete HACs algorithm, the local optimization operator is introduced to optimize the local path of the solution [3].

The formula of the cuckoo Levy's flight is as follows:

$$\vec{x}_i^{t+1} = \vec{x}_i^t + \alpha_{step} \oplus L(\delta) \tag{1}$$

Lewy flight random walk has better search ability in solution space and larger step size range:

$$L(\delta) \sim \mu = t^{-\delta}(1 < \delta < 3) \tag{2}$$

In order to optimize the path and eliminate the cross edges on the path, a local path optimization operator is introduced. 2-opt algorithm is a local path search algorithm. You can select a path in the current path file and change the order of the points on the path. If this change reduces the path length, the path scheme after exchange will be retained, otherwise the original path will be retained, as shown in Fig. 1.

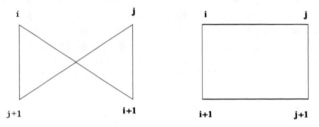

Fig.1. Example of local path optimization

4   Optimization Path of Hybrid Teaching Mode of Human Resource Management Curriculum Based on OBE Concept

OBE provides a clear goal orientation, reverse design and positive implementation idea for the practice of hybrid teaching mode. Taking the training and development course of human resource management as an example, the author puts forward the specific optimization path.

4.1   Teaching Resources Design and Development Stage

According to the teaching goal, the teaching task is made. Based on OBE's results oriented teaching philosophy, the author investigates the quality requirements of external enterprises and relevant experts for human resources training personnel, and organizes teachers and students of human resources to discuss the teaching objectives of this course, To further clarify the teaching objectives of training and development course is to enable students to systematically master the relevant technologies and methods of training demand analysis, training process implementation and training effect evaluation in the process of human resource training and development, cultivate the ability to analyze and solve problems, and provide new methods and means for human resource allocation through innovative thinking [4].

4.2 Online Learning and Classroom Teaching Implementation Stage

The implementation stage of blended teaching mode is composed of online learning and classroom teaching, which complement each other and form the advantages of blended teaching mode. Online learning is mainly reflected in the stage of classroom teaching preview and after class improvement, that is, before each class, students carry out autonomous learning according to the teaching resources released by teachers on the network platform, and put forward the problems they encounter online, and the teachers and other students participate in the discussion and answer. Teachers can record the common problems which are difficult to solve in class. This process fully reflects the dominant position of students' autonomous learning, and also lays the foundation for the improvement of classroom teaching efficiency. In the stage of after class promotion, teachers publish the key and difficult problems of classroom summary and analysis to the network platform, and arrange a certain number of exercises according to the needs of teaching content, consolidate knowledge points, and guide students to further think.

The classroom teaching activities of hybrid teaching mode are the process of face-to-face communication between teachers and students. Generally speaking, we are composed of two links. One is that teachers should form a series and in-depth excavation of the fragmentation problems in the process of students' self-learning according to the collected students' online learning situation. This paper focuses on the common problems, analyzes the causes of the problems, and then provides effective learning and research methods in classroom teaching. Second, the task driven teaching method is adopted to change the traditional teaching based on teaching. Through the task carrier and taking problem-solving as the main line, the students can make full use of the learned knowledge and carry out innovative thinking in the process of practice, so as to achieve the purpose of internalizing knowledge and acquiring relevant skills. Due to the strong practice and operability of training and development courses, teachers can set corresponding tasks for each topic module. When setting tasks, teachers are required to consider not only the students' pre class learning level and practical ability, but also the challenge of tasks and the extension of knowledge.

5 Conclusion

Paying attention to formative assessment is helpful to timely revise and improve the teaching design in the process of achieving results, and to understand students' personalized learning needs. Through the statistical analysis function of the network teaching platform, combined with the portfolio evaluation method, it grasps the students' participation in online interactive teaching activity, contribution to the class, online and classroom group discussion, personal report and other activities, so as to comprehensively evaluate the students. The evaluation of teachers can use online questionnaire survey and students' anonymity to score and give advice to teachers, so as to obtain real teaching feedback and help teachers improve teaching. On the basis of formative evaluation, developmental evaluation pays more attention to students' future growth, discovers and develops students' potential through teaching process, and helps students to know themselves, Build self-confidence so as to give full play to the educational function of evaluation. This requires teachers to use fragmented time to comprehensively evaluate the advantages and disadvantages of students and encourage their healthy development based on the values and emotional attitude, learning ability, analytical ability, communication ability and conflict resolution ability of students when they use various means to complete tasks in online learning and classroom teaching.

References

1. Guangmei, L.: Achievement oriented education theory and its application [J]. Educ. Rev. **1**, 20–21 (2007)
2. Bo, J.: OBE: results based education [J]. Foreign Educ. Res. **30**, 35–37 (2003)
3. Fupeng, T., Daoli, J.: Practical exploration of mixed teaching mode in Colleges and Universities under the information environment [J]. Res. Audio Vis. Educ. **4**, 63–65 (2005)
4. Donghuai, G., Liyan, P.: Construction and application of teaching mode based on Network Course [J]. Modern Educ. Tech. **1**, 80–83 (2013)

Rural Meteorological Video Service Platform Based on Particle Swarm Optimization Algorithm

Haimin Cheng[✉]

Hebei Meteorological Service Center, Shijiazhuang 050021, China

Abstract. Meteorological film and television in the development of popular science work, are often faced with a wide range of service groups, complex and diverse service needs. Among them, because the disaster prevention system is relatively weak, the disaster prevention knowledge is relatively lacking, and the production and life of rural groups are greatly affected by meteorological factors, therefore, the rural groups have more urgent service needs for the disaster prevention and popular science work of meteorological film and television. How to protect the safety of life and property of rural groups, help rural groups to effectively prevent disasters and avoid risks, reduce the losses caused by meteorological disasters; at the same time, provide meteorological services for the production and life of rural groups, so that rural groups can get better economic benefits, are two problems facing meteorological film and television services. According to the actual situation of Guangxi, this paper puts forward several working ideas of providing meteorological services for rural people, in order to improve the ability of meteorological film and television to serve agriculture, meet the needs of rural people for meteorological services, and promote the sustainable development of society and economy.

Keywords: Meteorological film and television · Serving agriculture ·
Integration and expansion

1 Tap Program Resources and Improve Service Quality

Guangxi's agrometeorological service products are rich in content and wide in coverage, including: agrometeorological information, grain yield forecast, agrometeorological condition forecast of major agricultural diseases and insect pests, Agroclimatic Resource Utilization, agroclimatic division, agroweather forecast and other business service fields, as well as the extraction and estimation of the area of bulk crops represented by the advantageous industries such as sucrose, etc., and their growth trend State monitoring, disaster monitoring and yield prediction, etc. Development process of main crops (early rice, late rice, sugarcane, corn, litchi, longan, banana, Mango) in all counties and cities of the region, key points of disease and pest prevention, weather conditions and impact assessment on crop growth, weather trend in the next week, production suggestions, etc. At the same time, in addition to the regular release of agrometeorological

© The Author(s), under exclusive license to Springer Nature Switzerland AG 2021
J. Abawajy et al. (Eds.): ATCI 2021, AISC 1398, pp. 703–709, 2021.
https://doi.org/10.1007/978-3-030-79200-8_104

service products in all periods of the year, the release frequency and service provision will be increased according to the service needs. Due to the high universality and great reference value of the agrometeorological service products, we can make the agrometeorological service products suitable for release to the society into a TV Meteorological program combining with weather warning and forecast information and popular science knowledge. For rural groups, by fully mining the program resources, we can improve the information utilization rate and information arrival rate, so as to do a good job in meteorological film and television services [1].

Throughout agriculture, forestry, water, land and other agriculture related departments, the meteorological department has an unparalleled advantage over other departments - that is, it has the unique service platform of TV Meteorological programs. If meteorological film and television want to do a good job of serving agriculture, it should not only rely on the meteorological department's own agrometeorological service products, but also strengthen the communication with agriculture, forestry, water, land and other agriculture related departments, open the platform of TV Meteorological Program in the way of cooperation and sharing, integrate and use the external technical force and basic resources, so as to expand the service field and increase the service hand To improve the added value of service products and promote the construction of new countryside. In practical work, we can invite experts from agriculture related departments to participate in program production, analyze and make suggestions based on various agricultural production activities and weather conditions, and even invite experts to the field to have field discussion and interactive demonstration with farmers' friends [2].

2   Basic Particle Swarm Optimization

Particle swarm optimization (PSO) is a new group intelligent evolutionary computing technology proposed by Kennedy and Eberhart in 1995. Its basic concept is derived from the study of foraging behavior of birds. Through the study of birds' foraging behavior in flight, we can simulate the collective cooperation of birds to find the optimal solution at the fastest speed. Similar to genetic algorithm, PSO is also an optimization algorithm based on "population" and "evolution". The basic solution process is: first, the system is initialized to a group of random solutions, and then according to the size of individual fitness to operate, iterative search for the optimal value [3].

It is assumed that the meteorological space is d-dimensional. The process of particle swarm optimization can be described as follows: the system is randomly initialized as a group of M particles, and the speed and position of the group are updated iteratively to search for the optimal value in the meteorological space. In each iteration, each particle tracks the individual extremum and global extremum, and adjusts its own speed by the distance between the individual's current position and the individual extremum and global extremum. Then, with this speed, the particle's position is updated. The principle of particle movement shown in Fig. 1.

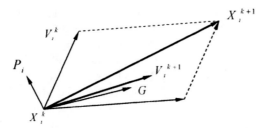

Fig. 1. Schematic diagram of particle movement

The process of particle swarm update can be expressed by the expression of particle position velocity update. Then the formula of d Dimension component update for the $k + 1$ iteration of the ith particle is as follows:

$$v_{Id}^{k+1} = \omega \cdot v_{Id}^{k+1} + c_1 r_1 \left(p_{Id}^k - x_{Id}^k \right) + c_2 r_2 \left(g_d^k - x_{Id}^k \right) \tag{1}$$

$$\begin{cases} v_{Id}^{k+1} = v_{maxd}, \text{ if } v_{Id}^{k+1} > v_{maxd} \\ v_{Id}^{k+1} == v_{maxd}, \text{ if } v_{Id}^{k+1} < -v_{maxd} \end{cases} \tag{2}$$

$$x_{Id}^{k+1} = v_{Id}^k + x_{Id}^k \tag{3}$$

Particle swarm optimization algorithm reflects the complex social behavior of groups. From the sociological point of view, formula (1), in which the first part is the acceleration of the particle's previous speed multiplied by a weight, indicating the trust of the particle in its current state, and the inertial motion is carried out according to its own speed, so the weight is called "inertia weight"; The second part is "cognitive part". The distance between the particle's current position and its optimal position indicates the thinking of the particle itself, that is, the part from its own experience in particle motion; the third part is "social part". The distance between the particle's current position and the group's optimal position indicates the information sharing and cooperation between particles, that is, the particle motion originates from the group The other part of particle experience, through cognition, imitates the movement of better companion. In the process of seeking oneness, individual particles often remember their own cognition of search space, and at the same time test the cognitive results of their peers. When the individual perceives that the cognition of the peer is better, it will adjust adaptively, so as to promote the group to move towards the common cognitive direction. The behavior of particle swarm, which depends on its own experience and peer experience, is very similar to human decision-making [4].

3   Strengthen the Sharing of Video and Audio Image Data to Improve the Timeliness and Refinement of Services

Integrate the broadcast platform of TV Meteorological programs in the whole region and normalize the service for agriculture.

At present, the broadcast platform of Guangxi TV weather program has the advantages of wide coverage and great influence: Guangxi satellite TV and Guangxi station of China weather network cover the whole country; Guangxi public, variety, urban and international channels and other provincial channels broadcast TV weather program every day; 14 prefecture level and city TV stations in the whole region also have TV weather program hosted by the host;China Meteorological channel covers more than 2.8 million digital TV users in Guangxi. The powerful release channel greatly improves the information arrival rate of TV weather program, and makes the information dissemination more efficient. By integrating the regional meteorological program broadcasting platform, relying on the public meteorological business system and basic business products, and aiming at the special needs of the rural audience such as the unique viewing time, viewing habits and preferences, the existing TV Meteorological programs are properly adjusted and improved, and the agricultural service programs are produced in the way that farmers and friends like to hear and see, with fixed time, frequency and channel broadcast We can normalize the work of serving agriculture [5].

3.1 Make Use of "Centralized Manufacturing, End Distribution" to Ensure that the Service Meets the Actual Needs

On the one hand, the "centralized production and end distribution" can be used to uniformly issue early warning and forecast, agricultural meteorological service information for the whole region, and provide reference for disaster prevention and mitigation and agricultural production. On the other hand, it can help the rural audiences in different regions and different service needs to contact the service information as quickly and conveniently as possible, and directly produce the service effect. For example, March of each year is a busy time for transplanting and seeding. Through "centralized production and end distribution", corresponding agricultural suggestions can be given according to different weather conditions of each city and local agricultural production progress, so as to improve the pertinence of meteorological services in different time and space scales through program differentiation. At the same time, through "centralized production, terminal distribution", it can also realize the rapid collection and transmission of meteorological visual products, as well as the rapid collection and transmission of video information on the scene of meteorological events. These visualization products can not only enrich the content of the program, but also enable the program director to better grasp the progress of agricultural production in various regions. The production mode of "bottom-up" and "front-line to rear" makes the service of TV weather program more practical.

Development of digital TV network on demand platform
In the on-demand platform of the digital television network, we will increase the provision of agricultural meteorological service products, and connect with the mature agricultural meteorological service products that can be released to the outside world in the meteorological guidance network, such as the ten day (monthly) agricultural meteorological report, forest fire risk meteorological grade, etc. In this way, the rural audience can learn about relevant information freely on demand through digital TV at any time of the day.

At present, Guangxi local program of China Meteorological channel is broadcasted from 6:56 a.m. to 10:56 p.m. in the 26th and 56th minutes of each hour, with 33 times of broadcasting frequency in the whole day. Through the local broadcast program in Guangxi, the audience can learn the latest meteorological information of Guangxi, including regional weather forecast, major cities in Guangxi, major tourist attractions in Guangxi, highway weather forecast, urban meteorological life index forecast, etc. In the local program of China Meteorological channel, developing the column of serving agriculture can strengthen the timeliness of service, give full play to the advantages of the professional meteorological channel in quick and effective service, improve the arrival rate of meteorological information in all parts of Guangxi, especially in remote mountainous areas, and reflect the service concept of "no time, no time".

4 Build Program Advantage Brand

4.1 Strengthen the Deep Processing of Service Products and Help Farmers to Seek Benefits and Avoid Disadvantages

Weather forecast is the first demand of farmers' friends for meteorological services, but it is not enough to only let the rural groups know the weather. We should process the weather forecast information in terms of service depth and personalization, combine the service products for agriculture, grasp the key nodes of weather and agricultural production, and help farmers' friends to arrange production scientifically, avoid natural risks and obtain better benefits according to the different needs of different crops for weather. At the same time, we can test and improve the agricultural service products through practice, so as to further improve the service level.

4.2 Build and Divide the Program Section, and Strengthen the Function and Pertinence of the Program

There are differences in geographical characteristics between the north and the south of Guangxi, uneven distribution of meteorological elements in time and space, and various demands of agricultural production for meteorological services in different regions. This determines that TV Meteorological programs must be produced and broadcasted in series for a certain service point, make overall planning for agricultural services in the whole region as a whole, and improve the service guidance of programs for different regions. In practical work, we should divide the program sections according to the knowledge structure of rural groups, strengthen the function and pertinence of the program for agriculture, and put the meteorological and agricultural terms in the TV program, "Translation" into the language of farmers' friends, so that they want to see, love to see.

4.3 Strengthen the Operability and Visualization of Service Products to Make Services More Intuitive and Specific

The meteorological department and the agricultural department have rich service products. By improving the operability of each service product, the service can be closer to the actual application. By understanding the different meanings of each meteorological data for agricultural production, the function of each data can be fully developed, the function and value of meteorological data can be improved, and the service level of the program for agriculture can be improved. On the other hand, compared with other media, TV has an unparalleled advantage, that is, the synchronization of sound and painting. Therefore, TV weather programs should make full use of the advantages of TV media and strengthen the visualization function of service products, so as to guide farmers' friends to carry out agricultural activities more intuitively and obtain better economic benefits.

5   Explore New Ways to Serve the New Countryside

5.1   Use New Media to Realize the Service Concept of "Everywhere"

If the meteorological channel can realize the service concept of "always on" meteorological service, then through the new media with portable mobile devices as the terminal and Internet as the media, we can provide meteorological film and television services to any area with network coverage, and build a three-dimensional comprehensive service system in time and space. With the increasing popularity of Internet and mobile phones in rural areas, farmers Instead of staying in front of the TV, friends can receive service messages in the fields. actively provide services in combination with meteorological 121 SMS platformIn case of major weather process or disastrous weather, meteorological film and television can also use the meteorological 121 short message platform to issue short message prompt to rural groups, list the broadcast time and content of the program, remind farmers' friends to take the initiative to watch the latest warning and forecast information, and effectively grasp the knowledge and skills of disaster prevention and avoidance.

Use the network of meteorological informants to strengthen the popularization of science for rural groups.

During this year's spring ploughing and planting, the relevant departments found that some local seedlings were weak. It is understood that farmers did not know how to raise seedlings scientifically. When soaking early rice, when the root length of rice seeds did not meet the standard, they put it into the seedling tray. Fundamentally speaking, because of the lack of science popularization, farmers' friends do not have enough awareness of scientific cultivation. Therefore, it is necessary to increase publicity and education, and through various ways, make popular science work deeply rooted in the people's hearts, so that farmers' friends do not eat in the sky, but use science and technology to arm themselves and get a good harvest. We can cooperate with agricultural departments and agricultural enterprises to print popular science manuals for meteorological informants to distribute in different areas in view of the frequent disastrous weather in Guangxi, various main cash crops and aquaculture and livestock and poultry breeding in Guangxi; and make video discs for meteorological informants to organize farmers' friends to watch in the slack season.

6 Conclusion

With the sustained and rapid development of economy, the demand of rural groups for meteorological services is becoming more and more diverse and demanding. Therefore, in the work of serving agriculture, meteorological film and television should adhere to the demand as the traction, service as the guidance, rely on the public meteorological business system and basic business products, do a good job in the service work, realize scientific disaster prevention, active disaster avoidance, and effective disaster reduction, so as to make greater contributions to the realization of the prosperity of agriculture and the prosperity of Guangxi.

References

1. Zhang, J.: My opinion on Agrometeorological program. In: Collected Papers on Meteorological Film and Television Technology, vol. 7, pp. 151–154 (2011)
2. Keng, X.: On the advantages of popular science of daily agricultural TV Meteorological programs. In: Proceedings of Meteorological Film and Television Technology, vol. 7, pp. 145–149 (2011)
3. Zhu, Y.: Discussion on the development of agricultural TV Meteorological Program. In: Collected Papers on Meteorological Film and Television Technology, vol. 6, pp. 194–196 (2009)
4. Qiu, L., et al.: Discussion on the quality of the broadcast manuscript and the improvement of the weather forecast program at prefecture level. Meteorol. Res. Appl. **1**, 110–112 (2010)
5. Liu, O., Zhong, W.: Application of meteorological information visualization in TV programs. In: Proceedings of Meteorological Film and Television Technology, vol. 6, pp. 238–242 (2009)

Optimization of Bank Credit Customer Financial Rating Model Based on Artificial Intelligence Algorithm

Li Lin[✉]

Harbin Finance University, Harbin 150030, Heilongjiang, China

Abstract. At present, financial analysis is still the main method to evaluate the risk of corporate credit customers. However, the design and modification of financial analysis model in credit practice are mostly based on the statistical data and experience of a certain section, which is not scientific and verifiable. In this paper, we try to apply the genetic algorithm model in the field of artificial intelligence to establish a set of solutions for the continuous iterative optimization of financial model parameters, and preliminarily verify its effectiveness, and discuss its possible shortcomings and development direction.

Keywords: Artificial intelligence · Genetic algorithm · Financial model · Credit customer rating · Optimization

1 Introduction

As a business risk industry, how to accurately measure the risk level and describe the risk characteristics of customers is the embodiment of its core competitiveness. Since the reform of China's banking industry in 1998 and the implementation of the commitment to the WTO to open the market to the world in 2006, all banks have made great efforts to develop their data-driven quantitative risk analysis tools. Among them, for small and micro businesses and retail customers, banks have gradually begun to explore ways to reduce their dependence on traditional credit data, and turn to big data support, including consumer behavior data and social media data into the model, so as to improve the accuracy of customer risk characterization. However, for the corporate customers who still account for more than 50% of the bank's credit business on average, because their traceable activities as a legal entity are basically limited to business operation, on the assumption that any bank can not obtain the big data of the whole industry chain upstream and downstream of the enterprise, the only quantitative data that the bank can rely on is still the financial statements of the enterprise [1].

According to traditional accounting theory, the three financial statements of an enterprise can fully reflect the business health of an enterprise in a certain period on the premise of reliable quality and no consideration of concealment or whitewashing. Therefore, the development of risk measurement technology of modern company business mostly

J. Abawajy et al. (Eds.): ATCI 2021, AISC 1398, pp. 710–714, 2021.
https://doi.org/10.1007/978-3-030-79200-8_105

focuses on improving the construction of enterprise statement index system, and processing various kinds of information presented or implied in financial statements with the help of various mathematical tools, so as to predict the risk degree of enterprises in the future. Practice has proved that this method combined with the qualitative judgment of credit personnel can reduce the error rate caused by the uneven level of business personnel and different judgment standards to a certain extent, excavate the risk factors in a standardized way, and improve the level of risk judgment. But experienced credit personnel often put forward many doubts about the pure mechanization model, such as the coefficient of each index factor can not be dynamically adjusted according to the economic environment, the coefficient optimization process can not be verified, the comparable industry indicators lag or can not be corresponding one by one, etc. To solve these problems, the introduction of modern computer science method is a promising way.

2 Overview of Artificial Neural Network

2.1 Characteristics of Artificial Neural Network

Compared with human brain and von Neumann computer, artificial neural network has the following characteristics:

(1) Massively parallel processing. Artificial neural network is fundamentally different from the current computer in structure. It has many small processing units connected with each other. The function of each processing unit is simple, but the collective and parallel activities of a large number of simple processing units are identified as expected, and the calculation speed is fast. The high parallel structure and parallel implementation ability of neural network make it have better fault tolerance and faster overall processing ability [2].
(2) Nonlinear processing. Neural network has inherent nonlinear characteristics, which is due to its ability to approximate arbitrary nonlinear mapping (transformation). This characteristic brings hope to nonlinear control problems.
(3) Self organization and self adaptability. Neural network can adapt to online operation, and can simultaneously carry out quantitative and qualitative operation. The strong adaptability and information fusion ability of neural network make the network process can input a large number of different control signals at the same time, solve the problem of complementary and redundant input information, and realize information integration and fusion processing.

2.2 Artificial Neuron Model

The structure of neural network is determined by the basic processing unit (neuron) and its interconnection method. The basic processing unit connecting mechanism structure and neurophysiology analogy are often called neurons [3].

If the dimension of input is increased by one dimension, the threshold can be included. For example.

$$u_k = \sum_{i=0}^{p} w_{ki} x_i, \ y_k = f(v_k) \tag{1}$$

A conventional S-type function can be expressed by the following formula:

$$f(v) = \frac{1}{1 + e^{-av}} \tag{2}$$

The feedforward network has a hierarchical structure, which is composed of some layers with no interconnection between neurons in the same layer. The signals from the input layer to the output layer flow through unidirectional connection; neurons are connected from one layer to the next layer, and there is no connection between neurons in the same layer; as shown in Fig. 1, in Fig. 1, the solid line indicates the actual signal flow, while the dotted line indicates the back propagation. Feedforward network can be divided into many kinds: feedforward network with input and output feedback, feedforward inner interconnection network, feedback full interconnection network and feedback local connection network.

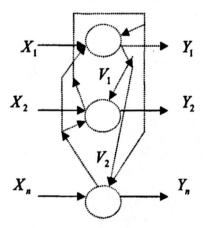

Fig. 1. Recursive feedback network

3   Financial Rating Model Case

This paper takes a bank's Financial Rating Model (hereinafter referred to as "quantitative indicators") as an example to explore its optimization path.

Firstly, it is assumed that the quantitative indicators of the model can not describe the credit status of an enterprise independently and completely, so it is tentatively determined that the quantitative indicators contribute 60% of the prediction weight in the credit rating process, and the remaining 40% of the weight is a group of descriptive problems (hereinafter referred to as "qualitative indicators") judged and scored by bank credit personnel.

Secondly, the model defines the scoring rules of quantitative indicators. The basic principle is to calculate all kinds of financial ratios, and compare them with the latest "national economic industry classification" to clearly judge the industry and scale. After comparing them with the standard value of the corresponding industry, they are classified into five grades: excellent, good, medium, poor and poor, and the corresponding comparative scores are obtained.

Then, the model weights the above scores according to the "coefficient" preset by the experts, and transforms each index into the total score of customer quantitative evaluation by the power function constructed by the weighting process.

Finally, the above quantitative and qualitative evaluation scores are weighted to get the total score of customer rating. The bank has set up a 9-level rating system for each score segment, corresponding to the statistical default probability value, so as to guide the business development.

1. Weight. The model presupposes the weight of each index in the scoring, but obviously these weights do not verify its scientificity and optimal degree, and do not change with the enterprise category and development stage. For example, in practice, the focus of index analysis of trade enterprises with light assets is significantly different from that of chemical enterprises with heavy assets, so it should not be scored according to the unified weight [4].

2. Industry foundation. The model calls the industry average, and on this basis, according to the characteristics of the industry, the industry index level is divided into five grades, the absolute value of the index is substituted into specific grades, and then the relative score is obtained by the equal difference interpolation method. This design is also unreasonable, because from a statistical point of view, each index may conform to different distribution characteristics (such as normal distribution, Poisson distribution, etc.) in the industry, and the confidence interval corresponding to the absolute value of each index varies with the distribution, so the index score relative to the industry average should not be calculated by the equal difference interpolation method.

3. The monotonicity of index influence. The model assumes that each index is monotonous in judging the quality of an enterprise. For example, "the asset liability ratio index is lower than the industry average, which means that the enterprise's solvency is better. The lower the index is, the better the solvency is. But in practice, the low "debt ratio" indicates that there may be insufficient use of leverage, which will affect the profitability and development ability of the enterprise, and eventually have a negative impact on the solvency of the enterprise. Therefore, it should not be assumed that all industry standard values are monotonic, but that they are better in a reasonable range, and the fluctuation to both sides is unfavorable, so as to build a two-sided parameter system or more reasonable.

4 Conclusion

As far as the traditional bank financial analysis and rating system is concerned, the parameters preset by experts are not scientific and verifiable enough. In this paper, by reviewing the relevant research results and summarizing the practice, it is considered that the rating model is a typical multivariate nonlinear optimization problem, and it is suitable to be improved by using artificial intelligence genetic algorithm. After that, this paper carries out the optimization experiment on the simplified bank rating model, and the results show that the genetic algorithm modeling has certain advantages over the traditional rating model, so the improvement direction may be feasible.

Acknowledgements. Optimization of Financial Rating Model for Bank Credit Customers Based on Artificial Intelligence Algorithm by the Provincial Basic Business Fee Project of Harbin Institute of Finance(2018-KYYWF-E007).

References

1. Yin, K., Liu, X.: Research on credit rating model based on commercial banks. Econ. Manage. Rev. **22** (2006)
2. Zhou, Z.: Recognition Model of Financial Statement Authenticity of Listed Companies Based on Support Vector Machine. Jilin University, Changchun (2007)
3. Liang, Y.: Application of BP neural network in bank loan classification. China Finan. Comput. News **7**, 59–62 (2005)
4. Zhou, Z., Cao, C.: Neural Network and its Application. Tsinghua University Press, Beijing (2004)

General Freshness Recognition Method Based on Electronic Nose and Improved Unsupervised Signature Projection Algorithm

Zhaojun Fan[1]([✉]) and Yongling Wang[2]

[1] Tianjin Modern Vocational Technology College, Tianjin 300350, China
[2] Tianjin Ocean Pal Carol Biotech CO., Ltd., Tianjin 300350, China

Abstract. The odor information of living hairy crab was collected by self-made electronic nose system. The dimension of multi-dimensional characteristic response of hairy crab samples was reduced by popular learning algorithm, and the low-dimensional feature vector of samples was extracted. Then, the freshness of hairy crab was identified by back-propagation neural network, and compared with the physical and chemical index volatile basic nitrogen. The results show that the accuracy of the algorithm can reach 98.1%, and the results based on the electronic nose technology and the physical and chemical indicators are basically the same. Therefore, the non-destructive identification method based on electronic nose technology is feasible.

Keywords: Hairy crab · Freshness electronic nose · Unsupervised discriminant projection algorithm · Back propagation neural network

1 Introduction

The key of aquatic product freshness recognition based on electronic nose technology lies in the later data processing, including data preprocessing, feature selection, feature extraction, classification and recognition, among which feature extraction is the core of the whole data processing. At present, there are linear dimensionality reduction methods such as principal component analysis (PCA), linear discriminant analysis (LDA) and nonlinear dimensionality reduction methods such as kernel principal component analysis (kernel) PCA.KPCA] 0 and locality preserving projection (LPP) 21.2l algorithm. PCA algorithm uses matrix operation to obtain low dimensional subspace eigenvectors from high-dimensional spatial data of samples through linear mapping, which has the advantages of low computational complexity and fast operation speed. However, it is not good for dimensionality reduction of data which is essentially non-linear distribution, while KPCA, LPP and other nonlinear dimensionality reduction algorithms do not consider the class information of samples. In addition, most of these algorithms analyze the response data of electronic nose in Euclidean space, and can not fully mine the potential information contained in high-dimensional data. There is little difference between the odor information volatilized in the storage process of the same living hairy crab, and the sensor array generally has the problems of baseline drift and poor repeatability, which makes these traditional feature extraction algorithms unable to accurately determine the freshness of hairy crab. In this experiment, the unsupervised discriminant projection 23 algorithm is combined with semi supervised machine learning method, Further optimize the performance of popular learning algorithms [1]. This semi supervised unsupervised discriminant projection (ssipp) algorithm can well mine the potential geometric structure and inherent laws of nonlinear data, and make full use of a small amount of label information to improve the performance of UDP algorithm. Therefore, the common feature extraction algorithms such as PCA KPCA can effectively identify the weak change information of sensors.

2   Materials and Methods

2.1   Material Science

In the experiment, 60 hairy crabs were purchased from Jiangsu Shajiabang special aquatic products trading Co., Ltd., and the weight of each crab was (105 ± 5) g. Quickly transport them to the laboratory with insulated barrels and ice blocks, and place them in a refrigerator with a temperature of 4 °C and a relative humidity of 90%, and label them. The odor information of 50 samples was collected by electronic nose device every day for 9 days. At the same time, according to SC/t3032-2007 "determination of volatile basic nitrogen in aquatic products", the physical and chemical indexes of the remaining 10 samples were determined nitrogen.TVB-N) Value.

2.2   Methods

The samples of hairy crabs were stored at room temperature for 2 h, and then the samples were placed in a 500 ml beaker. The container was sealed with tin foil, and the gas emitted from the samples was filled with the whole container for 40 min. At the same time, turn on the power supply of the self-made electronic nose device and preheat the sensor array for 30mn. The samples of hairy crab are crab products belonging to the same species and in the same state [2]. After preheating and headspace standing, the sampling time of electronic nose was set to be 100 s. The sampling needle is inserted into the beaker and the depth of each sampling is required to remain unchanged. The purpose is to obtain data with high repeatability. Turn on the power supply of the vacuum gas pump, and pump the gas in the top space of the beaker into the gas chamber. At the same time, the sensor array starts to sample. After the completion of sampling, the gas washing time is 120 s, and the residual gas in the pipe and gas chamber is quickly removed, and the next sample experiment is carried out.

2.3   Data Preprocessing

The original data collected by the electronic nose is the response value of 7 different gas sensors in 0–100 s, and the sampling rate is 1 point/s, so each sample can obtain 7×100 data in each measurement. Considering the measurement noise, the interference caused by the electronic nose system itself and the drift effect caused by the cross response of the sensor array, it is necessary to filter the mean value, process the baseline and eliminate the abnormal data to obtain 700 RMS. Mean filtering replaces the original value of the original sampling point by the average value in the neighborhood of the original sampling point, and realizes linear filtering to reduce the small range fluctuation of the data caused by the interference of air noise and hardware power supply. The baseline processing adopts the differential baseline processing method, that is, the untreated current sensor response data is subtracted from the previous sample, and the sensing gas is washed to the reference value before the start of the sampling stage, So the sensor drift can be compensated. Abnormal data elimination means that the abnormal data in the data obtained by the sensor is eliminated by using the 30 criterion. The abnormal data 27 caused by the gross error or temporary poisoning of the sensor will change with the adsorption process of volatile gas molecules when the signal is collected by the sensor [3]. When the adsorption is saturated, the value will be stable. If all the features of sample number are selected, the amount of calculation will become very large for multiple sensors, resulting in "dimension disaster". The information of the sensor's steady-state response is extracted from the information of the sensor's steady-state response and the transient response coefficient of the sensor. After each sampling, each sensor can obtain four eigenvalues.

3   The Characteristics of Hairy Crab were Processed

After the above sensor feature selection, each sample can obtain a 28 dimensional feature vector. According to the peak phenomenon 28, the dimension is still too high. In the case of limited test samples, the dimension of feature space needs to be reduced. In the horizontal recognition algorithm, feature extraction is used to achieve this. Considering the shortcomings of traditional feature extraction algorithms, this experiment uses popular learning algorithms as feature extraction algorithms. Popular learning algorithm is a new nonlinear feature extraction method in recent years. For the traditional feature extraction methods, this kind of algorithm is easier to deal with high-dimensional nonlinear popular structure data, with less model parameters, and the feature data after dimension reduction has the characteristics of visualization. Among the popular learning algorithms, UDP algorithm was proposed by Yang Jian et al. 3. This algorithm can obtain the correspondence between high-dimensional observation space and low-dimensional structure through unsupervised learning, and obtain the internal geometry structure by preserving the relationship between data. Its essence is a linearized manifold learning algorithm 21 based on nonlocal features. Compared with other popular learning algorithms, UDP algorithm can use the local and global characteristics of the sample to obtain the projection direction of local divergence while reducing the nonlocal divergence.

The principle of UDP algorithm assumes that the sample set in high-dimensional space $X = \{x_1, x_2, \cdots, x_N\} \subset R^D$. By looking for a projection $W \in R^{D \times d} (d < D)$, the low dimensional representation YCR of the sample is obtained, $Y \subset R^d, Y = \{y_i = W^T x_i, i = 1, 2, \cdots, N\}$ Moreover, the local divergence of the low dimensional representation should be as small as possible, and the non local divergence should be as large as possible. The adjacency matrix H = $H = (h_{ij})$ is defined. If x_1 is adjacent to

x_2, then $h_{ij} = 1$; otherwise, $h_{ij} = 0$. Then the local divergence matrix SL and nonlocal divergence matrix s of the sample set are shown in Eqs. (1) and (2).

$$S_L = \frac{1}{2 \times N \times N} \sum_{i=1}^{N} \sum_{j=1}^{N} h_{ij}(x_i - x_j)(x_i - x_j)^T \tag{1}$$

$$S_N = \frac{1}{2 \times N \times N} \sum_{i=1}^{N} \sum_{j=1}^{N} h_{ij}(x_i - x_j)(x_i - x_j)^T \tag{2}$$

SSUDP to its ability to mine the potential geometric structure and inherent laws of nonlinear data, ssidp can make full use of a small amount of label information to improve the performance of UDP algorithm.

4   Results and Analysis

TVBN value is one of the important indexes for evaluating the freshness and quality of aquatic products. This index refers to the volatile nitrogen-containing substances produced by protein decomposition of animal food due to multiple effects of enzymes and bacteria. These substances can be distilled out and then titrated with standard acid to calculate the content value. In this experiment, Kjeldahl method is used for the experiment, which is more rapid and accurate. The TVBN value of hairy crab samples was determined according to the method specified in sc_-2007. Each sample was repeatedly measured for 5 times, and the average value was taken as the TVBN value of the sample on that day. The change trend is shown in Fig. 1.

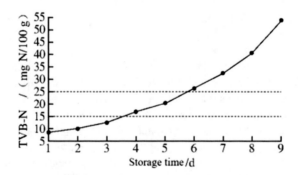

Fig. 1. Change trend of TVBN value of large usage device in stored procedure

According to the regulation of crustacean aquatic products in SC/T 3032-2007, the freshness grade of each crab sample was classified, that is, TVBN value less than 15 is fresh, greater than 15 and less than 25 is sub fresh, and greater than 25 is not fresh 30. As shown in Fig. 1, the TVN value of hairy crabs gradually increases with the increase of storage time, indicating that the longer the storage time, the worse the survival state of hairy crabs. In the first three days of cold storage, the change of TBN value was slow. At this time, the crab was in fresh state, and the change of τ VBN value increased from the fourth day, which indicated that the quality of hairy crab decreased from the fourth day to 26315 mgn100g, exceeding the standard of less than 25 mg n/100g in the second freshness. At this time, the crab was not fresh, but the sensory judgment was still fresh and there was a big error. On the 9th day, the value of TVBN increased sharply to 54602 mgn/100g [4]. It was unnecessary to study the strong odor of hairy crab.

5 Conclusion

In this paper, a method of fresh degree recognition of hairy crab based on semi super-vised UDP algorithm and electronic nose technology was studied. Firstly, the volatile odor information of hairy crab with different storage time was collected by electronic nose system, and the prediction model of crab freshness grade was established by data preprocessing, feature extraction and feature recognition. At the same time, the non-destructive testing results of the physical and chemical index of the parallel samples, tv-n, were taken as the reference standard to provide a theoretical basis for the rapid and non-destructive detection of the quality of hairy crabs. The experimental results show that the algorithm can well mine the potential geometric structure and internal laws in nonlinear data, and make full use of a small amount of label information to improve the performance of UDP algorithm. Therefore, compared with the common feature extrac-tion algorithms such as PCA and KPCA, the algorithm can effectively identify the weak change information of sensors, and can realize fast and nondestructive detection.

References

1. Du, J.: Research on Freshness Detection Technology of Chinese Mitten Crab Based on Machine Olfactory. China University of Mining and Technology, Xuzhou, pp. 1–4 (2015)
2. Huang, Z., Ding, Y., Huang, J., et al.: Comparative analysis of volatile compounds in scallop column and scallop skirt. Food Sci. **37**(4), 98–102 (2016)
3. Chen, P., Li, L., Li, L., et al.: Improvement of determination method of volatile basic nitrogen in aquatic products. Chin. J. Fish. Sci. **13**(1), 146–150 (2006)
4. Tao, B., Yan, Z., Hong, Z., et al.: Quality changes of Litopenaeus vannamei under different cold storage conditions. Fish. Sci. **28**(9), 493–497 (2009)

Optimization Simulated Annealing Algorithm for High Strength and Toughness Metal

Lijuan Zhu$^{(\boxtimes)}$

School of Materials Science and Engineering, Jilin University, Changchun 130025, Jilin, China
ljzhu@jlu.edu.cn

Abstract. The performance optimization of high alloy high strength and high toughness steel is studied by using artificial neural network. The BP algorithm is improved by combining simulated annealing algorithm, which makes the selection of network structure, determination of momentum coefficient α and learning rate more reasonable, and improves the learning efficiency of the network, The results can be used for the composition design and heat treatment process optimization of high strength and high toughness steel.

Keywords: Artificial neural network · High strength and high toughness steel · Simulated annealing algorithm · Performance optimization

1 Introduction

In recent years, with the development of materials science, computer science and artificial intelligence technology, artificial neural network, expert system and other intelligent technologies are combined with material science, The development of computer-aided material design, computational materials science and other new disciplines. Materials researchers can use computers to research, design and develop new materials. Through the existing knowledge, data and materials, according to the use conditions and performance requirements, to design a scheme that can meet the requirements. Although ultimately, it will be determined by experiment, Artificial neural network (ANN) is a system composed of a large number of highly parallel and interconnected biological neurons, which has some functions of artificial intelligence. At present, it has been widely used in material design, Artificial neural network (ANN) has unique advantages in automatically obtaining mathematical models from experimental data by self-learning: it does not need the form of formula given by people in advance, but is based on experimental data and through finite number of iterations, The neural network is especially suitable for dealing with the problems that the law is not obvious and the components and process variables are too many. Based on the experimental data, the artificial neural network method is used to establish the model, and the simulated annealing algorithm and genetic algorithm are combined to find the best composition formula and heat treatment process of high strength and high toughness steel [1].

© The Author(s), under exclusive license to Springer Nature Switzerland AG 2021
J. Abawajy et al. (Eds.): ATCI 2021, AISC 1398, pp. 720–724, 2021.
https://doi.org/10.1007/978-3-030-79200-8_107

With the development of materials science, computer science and artificial intelligence technology, artificial neural network, expert system and other intelligent technologies are combined with material science, Computer aided material design, computational material science and other new disciplines have been developed. Materials researchers can use computers to research, design and develop new materials. Based on the existing knowledge, data and materials, according to the use conditions and performance requirements, the scheme that can meet the requirements can be designed, Artificial neural network is a system with some artificial intelligence functions, which is composed of a large number of highly parallel and interconnected biological neurons. It has been widely used in material design, Artificial neural network (ANN) has unique advantages in automatic acquisition of mathematical model from experimental data by self-learning: it does not need the form of formula given in advance, but based on the experimental data, through finite number of iterative calculation, it can obtain a number reflecting the inherent law of experimental data.

2 Neural Network Modeling

In this paper, BP (back<error propagation) algorithm (error back propagation algorithm) is used to model. The input-output relationship of BP network can be regarded as a mapping relationship, which is a highly nonlinear mapping. Theoretically speaking, BP network can realize the mapping of any continuous function, The network consists of a set of nodes at different levels, such as the input layer, the output layer and the hidden layer (which can be multi-layer). The nodes of each layer output to the nodes of the next layer, as shown in Fig. 1. These output values are amplified, attenuated or suppressed due to different connection weights W, The input of each node is the weighted sum of the output values of all nodes in the previous layer. The excitation output value of each node is determined by the node input, excitation function and offset [2].

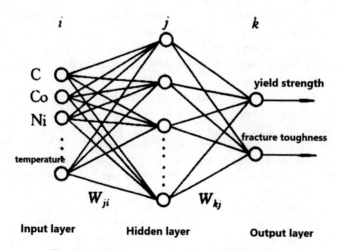

Fig. 1. Schematic diagram of BP network structure

The input value of layer J node is.

$$net_j = \sum_i W_{ji} O_i \qquad (1)$$

The output value of the node is.

$$O_j = f(net_j) \qquad (2)$$

Where f is the excitation function of the node

$$O_j = \frac{1}{1 + \exp\left[-(net_j + \theta_j)\right]} \qquad (3)$$

Parameter θ_j θ is the offset value or threshold value.

3 Material Optimization

There are many factors that affect the mechanical properties of wet formed GMT materials. Under certain preheating and molding conditions, the content of CF, the content of PP pulp, the amount of dispersant and the amount of binder affect the mechanical properties of GMT 'materials. The increase of fiber content will generally increase the mechanical properties of GMT' materials, but the increase of fiber content will reduce the dispersion effect of fiber and make it difficult to be impregnated by P, Therefore, with the increase of fiber content, it is necessary to increase the dosage of P pulp and dispersant agent; the role of PP pulp is to make the fiber evenly dispersed in CMT wet molding, and make the fiber and tree powder more evenly mixed. Because the mechanical properties of PP pulp are lower than that of PP resin, the mechanical properties of GMT materials will decrease with the increase of PP pulp content; the larger the amount of dispersant agent, the greater the amount of residual in GMT materials, which will also reduce the mechanical properties of GM materials; the role of adhesive is to bond loose sheets into integrated sheets, so as to prevent PP from moving in sheets and causing uneven distribution, It affects the impregnation of the fiber in the molding process. The results show that the mechanical properties of GMT materials decrease with the increase of binder content, and the amount of binder is related to the fiber content in GMT materials [3].

In order to obtain the ideal comprehensive mechanical properties of GM, a lot of repeated tests are needed. In order to use the obtained experimental data economically, reduce the number of experiments and improve the efficiency, people's attention has turned to the theory aided material design and prediction. Artificial neural network is an artificial intelligence system developed in recent years to simulate the biological process of human brain. At present, some progress has been made in the application of material design.

Wet forming GMT material has uniform fiber dispersion, high efficiency of fiber strength, and can be molded into complex shape product 5, but its mechanical properties are greatly affected by the forming additives. In this paper, the trained network is used to input the fiber content and the amount of additives, and the output is the mechanical properties of GMT materials.

4 Optimization of Experimental Parameters

The purpose of this study is to find out the suitable alloy element ratio and the optimum aging temperature so as to obtain the best combination of strength and toughness. Therefore, we choose the main alloy elements C, Ni, Co, Cr, Mo, Nb, Ti and aging temperature as the input mode of the network, and the output mode is the yield strength σ and fracture toughness KC, The network system is established. 35 groups of 41 experimental data are randomly selected as the learning samples of the network, and the other six groups are used to test the generalization ability of the network [4].

In order to obtain the optimal network structure and parameters, the simulated annealing algorithm is used. The network system error E is taken as the objective function, and η, α and structure parameter φ are taken as the state parameters.

$$E(\eta^*, \alpha^*, \varphi^*) = \min E(\eta_i, \alpha_i, \varphi_i) \tag{4}$$

φ_i includes the number of hidden layers and the number of nodes in each hidden layer. After repeated experiments, a four layer BP network structure of $8 \times 8 \times 6 \times 2$ was selected, with 0.62 and 0.71 α, and the system learning error $E = 0.0062$. At this time, the prediction error of six groups of inspection data is 0.0097. After the network is trained, the performance of steel is optimized by genetic algorithm, %The results show that the optimum combination of strength and toughness is c0.23, ni10.31, co1643cr2.50, mo1.37, nb0.051, ti0.070, aging temperature 486°C. The yield strength is 1786 MPa, and the fracture toughness K is 167.6 MPa.

5 Conclusion

The fiber content and additive dosage of wet forming GMT material are taken as the input of the network, and its bending strength and tensile strength are taken as the output of the network to establish a mathematical model reflecting the inherent law of experimental data, which overcomes the shortcomings of other methods, such as large amount of experiments and difficulty in obtaining the optimal solution, and optimizes the material composition, which provides an effective means for the design of GMT materials and has great application prospects.

(1) The artificial neural network model can be used to predict and optimize the properties of high alloy high strength and high toughness steel, which provides a new way for further research and development of new high strength and high toughness steel.

(2) Artificial neural network model combined with other computing technologies (such as simulated annealing algorithm, genetic algorithm, etc.) can give full play to their respective strengths and become a more functional material design system.

References

1. Wang, D., Chu, L.: Comprehensive prediction of hydrocyclone flowers by artificial neural network. Nonferrous Met. **46**(1), 29–33 (1994)

2. Application of neural adaptive algorithm in experimental design of alloy steel production. Acta Metall. Sin. **28**(11), b487–b489 (1992)
3. Lu, C.Y.: The development trend of GMI abroad (1). Glass Fiber **14**(2), 27 (1997)
4. Song, R., Zhang, Q., Zeng, M., et al.: Application of artificial neural network in the aging dynamics of 7175 high strength aluminum alloy. J. Northeast Univ. (Natural Science Board) **16**(2), 219–222 (1995)

RNN Neural Network for Recovery Characteristic System of Resistant Polymer

Qi Wang[✉]

Geological Brigade of No. 1 Oil Production Plant in Daqing Oilfield, Daqing 163000, China

Abstract. In recent years, the demand for oil field resources is increasing year by year. After the primary oil recovery, secondary oil recovery and tertiary oil recovery, the major oilfields have entered the middle and late stage of development, and the whole oilfield development is difficult to further increase. In order to meet the needs of oilfield development and create higher economic benefits, major oilfields are actively exploring tertiary oil recovery technology, and strive to use new technology to enhance oil recovery, which has broad development prospects. Based on this, this paper analyzes the relevant strategies of EOR in the tertiary oil recovery stage, which is conducive to promoting the high and stable production of China's oilfields and promoting the steady development of the oilfield industry.

Keywords: BP neural network · Salt resistant polymer and oil recovery

1 BP Neural Network

Artificial neural network (ANN) has attracted much attention in many fields because of its self-learning, self-organization, good fault tolerance and excellent nonlinear approximation ability. Exploration geophysicists have applied neural network to seismic, gravity and magnetic, electromagnetic methods and other fields, such as automatic elimination of seismic traces, automatic editing, first break pick-up 1.14.151, gravity, magnetic and electrical data inversion 161718.19, etc., and achieved good results. Multi layer feedforward neural network based on error back propagation algorithm (BP algorithm) is the most widely used neural network learning algorithm.

1.1 Network Layers

The transfer function of the hidden layer is tanh neuron, and it is an S type hiding layer. A network with B bias and at least one s layer and one linear output layer has been theoretically proved to be close to a nonlinear function with finite discontinuity. Still, the number of layers can be increased. Reduce error and improve preprocessing. Decide. This also inhibits network development and increases the network weight formation time. The model structure of neural network is shown in Fig. 1.

J. Abawajy et al. (Eds.): ATCI 2021, AISC 1398, pp. 725–729, 2021.
https://doi.org/10.1007/978-3-030-79200-8_108

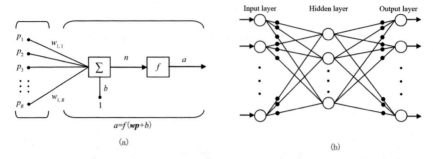

Fig. 1. Model structure of neuron (a) and three-layer BP network (b)

1.2 Number of Neurons

Depends on the actual problem to be solved. For the specific problem studied in this paper. The input vectors are the extremum ratio sequences MAG/magvz, magvz/magvzz, magvzz/magvzzz, so the input layer contains three neurons. The target is the top depth of the magnetic source, so the output layer contains one neuron. The results show that the number of neurons in the hidden layer is 15–30, which can achieve good results [1].

1.3 Neuron Transfer Function

The transfer function in BP neural model is usually a differentiable monotone increasing function. According to the specific problems in this paper, tangent sigmoid function is selected as the transfer function of hidden layer neurons, and linear function is selected as the transfer function of output layer neurons. The following function:

$$f(x) = \tanh(x) \tag{1}$$

$$f(x) = k(x) \tag{2}$$

2 Importance of Salt Resistant Polymer for Oil Recovery

At present, with the progress of oilfield development technology, tertiary oil recovery technology has been developed rapidly, and has been effectively applied in some oil-field development, which has realized the increase of oil field production and created better economic benefits. At present, the main development feature of tertiary oil recovery technology is polymer flooding. Through polymer injection, the productivity and production of oil wells are improved. In order to improve oil recovery, the following technologies can be used.

Under the polymer flooding technology, the content of polyamine in the polymer is the highest. When the polymer is mixed with water, the polymer mother liquid will be formed. When the mother liquid is aged and stirred, the polymer solution required for oilfield development will be formed. The injection of polymer solution into the reservoir is completed by injection pump [2]. The injection of polymer achieves good oil displacement effect. The viscosity of the injection agent is improved by injecting polymer into the reservoir, and the oil-water mobility is improved compared with that under this condition. This kind of oil displacement technology has the effect that traditional water flooding cannot achieve.

When a certain amount of polymer solution is injected into the reservoir, the oil-water mobility is improved, In this case, the viscosity of polymer in porous media will be further increased, the internal friction of oil flow will be increased, and the flow speed will be significantly improved, which will facilitate the production work. There is a close relationship between the concentration of polymer solution and the relative molecular weight. Polymer is easy to form high molecular weight polymer, and the oil displacement effect is more obvious. In the process of polymer flooding, there is a positive relationship between displacement effect and viscosity. When the viscosity is higher, the oil displacement effect is better, which is more conducive to oilfield development. In some oilfield development operations, ASP flooding technology is also used to assist oilfield development. Under this technology, polymer, alkali and surfactant aqueous solutions are injected into the reservoir. The injection of these solutions achieves more ideal mixed flooding effect, which is of great significance to the improvement of oil recovery. The existence of alkaline water plays an important role in oil washing in oilfield development, which effectively avoids the situation of reservoir plugging. The existence of surfactant improves the wettability of rock surface, and the permeability of reservoir rock in oilfield is better, which also helps to improve the recovery rate.

3   Dominant Percolation Channel of Reservoir After Polymer Flooding

After polymer flooding, the reservoir has undergone long-term displacement by polymer flooding and subsequent water flooding. Compared with that before polymer flooding, reservoir permeability, oil-water relative permeability and oil saturation have changed significantly. As a result of these changes, the seepage resistance of high permeability layer is obviously reduced, and that of medium and low permeability layer is obviously increased. The inter layer, intra layer and plane contradictions are further intensified on the basis of water flooding. Thus, the dominant seepage channel characterized by high permeability and low residual oil saturation is developed in the high permeability part of the reservoir, Especially after polymer flooding, the water saturation in the strong water washing section of high permeability reservoir is higher than 70%. Compared with that before polymer flooding, the relative permeability of oil phase is greatly reduced, and the relative permeability of water phase is greatly increased, resulting in the further increase of water oil fluidity ratio, resulting in the emergence of dominant seepage channels with high permeability and low residual oil, forming the low efficiency and invalid circulation of displacement fluid.

Based on the data of 20 coring wells after polymer flooding, the vertical distribution characteristics of dominant seepage channels are studied. The research results show that they are mainly distributed in the thick oil layer vertically, and the well-developed units of Pugong 2 and Pugong 3 account for 71% and 24.1% of the total dominant seepage channel thickness respectively. The thickness of dominant seepage channel in pui2 and PI3 units accounts for 23.7% and 24.9% of the unit thickness respectively. In the thick reservoir, 64.8% of the dominant seepage channels are distributed in the lower part of the composite rhythm section, 23.1% in the middle part and 121% in the upper part of the composite rhythm section, which indicates that due to the influence of gravity differentiation, it is easy to form an optimal seepage channel in the lower part of the reservoir. The profile test data show that the thickness ratio of the dominant seepage channel is only 15.9%, but the water absorption ratio is as high as 60% [3, 4]. It is necessary to take measures to control the inefficient and ineffective circulation.

The research shows that the remaining reserve potential after polymer flooding is mainly distributed in the well-developed Pu 2 and Pu 3 reservoirs, and the medium and low water washing sections and the dominant seepage channels are alternately distributed. Moreover, the internal interlayer of the thick oil layer is not developed, and the mechanical plugging and shallow profile control are ineffective, which makes the exploitation more difficult.

4 Conclusion

(1) After polymer flooding, the reservoir is scoured by polymer flooding and subsequent water flooding for a long time. Compared with that before polymer flooding, reservoir permeability, oil-water relative permeability and oil saturation have changed significantly. The combined effect of these changes results in the formation of high permeability and low residual oil saturation as the characteristics of the dominant seepage channel, especially in the strong water washing section with water saturation of more than 70%, the dominant seepage channel is more prominent, resulting in low efficiency and ineffective circulation of displacement fluid.

(2) The average oil saturation of oil reservoir after polymer flooding is 40.9%, which is 119% lower than that before polymer flooding. The cumulative thickness proportion of oil saturation greater than 30% accounts for 74.7%. After polymer flooding, there is still material basis for further enhanced oil recovery. However, the main remaining reserve potential and dominant seepage channels are vertically distributed in the well-developed Pi2 and PI3 units, The medium and low water washing sections and dominant seepage channels are distributed alternately, and the internal interlayer of thick oil layer is not developed, mechanical plugging and shallow profile control are ineffective, and the production difficulty is increased.

(3) according to the reservoir physical properties and fluid seepage characteristics after polymer flooding, an adaptive composite oil displacement method is developed, which effectively blocks the dominant seepage channel, and greatly improves the liquid absorption capacity of low permeability layer, medium permeability layer and high permeability layer, Dynamic adjustment and balanced displacement can be realized. The physical simulation results show that the oil recovery can be increased by 15.3% after polymer flooding, which is 5.1% more than that of ASP flooding, and the polymer consumption is saved by 16%.

References

1. Fu, H., Sun, L., Guo, C., Jin, Y., He, C.: Study on polymer flooding up flow in class II reservoir of high water cut sandstone reservoir. J. Pet. Chem. Eng. Univ. **33**(04), 38–42 (2020)
2. Bi, G.: Feasibility study on EOR by ASP flooding in class III reservoir. Chem. Manag. **05**, 223 (2020)
3. Hongping, W.: Exploration on the combination of tertiary infilling and tertiary oil recovery in well pattern. Chem. Eng. Equip. **12**, 35–36 (2019)
4. Jinxing, T., Tielong, C.: Experimental study on relative permeability curve of polymer flooding. Acta Pet. Sinica **18**(1), 81–84 (1997)

Real Time Thermal Error Compensation of CNC Machine Tools Based on ACOBP Algorithm

Yan Huang[✉]

Chongqing Chemical Industry Vocational College, Chongqing 401220, China

Abstract. In recent years, the compensation of communication frame loss and delay for controller area network (can) bus system is a research hotspot. For CAN bus system, it is of great significance to maximize the compensation effect under the premise of limited band competition constraint. Firstly, this paper analyzes the specific mechanism of communication frame loss and delay, At the same time, its characteristics are summarized, and how it affects the CAN bus control system is studied. Based on the two characteristics of high load sensitivity of CAN bus and the randomness of communication frame loss and delay, a distributed compensation scheme is proposed, and some common problems in the design of compensator are discussed in detail; Next, the compensation method of linear single input single output (SISC) system is studied. Finally, the simulation results show that the proposed compensation method is feasible and effective under different frame loss and delay conditions.

Keywords: Ant colony algorithm · ACOBP algorithm · Numerical control

1 The Introduction

The other error source is analyzed and measured, and a new error is created through manual intervention to offset or weaken the original error. Under the condition existing itself, various in the manufacturing machine tool, the measurement error value and error compensation model is set up, real time calculation work piece cutting point location error, will get the error information feedback input to the machine tool control system, in this way, machine run along the revised process track, and improve the machining precision of the work piece [1]. accounting for 50%–60% of the total error. A large number of statistical data show that, in the long processing process, its own and the work piece after the parts processing error (thermal error), the greater the proportion of the total processing error, the maximum can account for about 70%.

At present, the thermal error compensation at home and abroad is mainly established by the method of system identification [2], this model takes temperature, machine tool operation parameters and other parameters as input variables and thermal deformation as output variables. The methods of thermal error compensation generally include regression analysis, artificial neural network, and various theories. Neural network has a strong

© The Author(s), under exclusive license to Springer Nature Switzerland AG 2021
J. Abawajy et al. (Eds.): ATCI 2021, AISC 1398, pp. 730–735, 2021.
https://doi.org/10.1007/978-3-030-79200-8_109

self-learning function and the ability to find optimal solutions at a high speed. However, the commonly used BP neural network method still has a slow learning convergence rate, low efficiency, and is prone to fall into local optimum. Therefore, more scholars turn their attention to the algorithm research direction with strong global optimization ability. Ant colony algorithm has been developed rapidly due to its strong global search capability.

2 Principle of Thermal Error Compensation for NC Machine Tools by BP Neural Network

The main thermal error sources of machine tools are also different, so the collection points of error sources are also different. Based on prior knowledge (including system operation rules, test data or other engineering experience), network nodes of thermal error sources of machine tools are determined. All kinds of algorithms are used to get the main error sources. These error sources are used Then the calculated information is fed back to the NC controller, which makes corresponding compensation control instructions to compensate the corresponding errors. See Fig. 1.

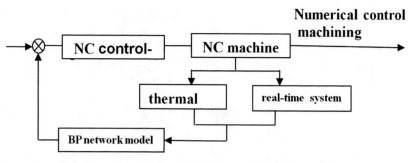

Fig. 1. Compensation system based on BP neural network

The BP neural network with more than the almost arbitrary precision. Therefore, in recent years, more and more models of NC have been established using the neural network method [3]. BP neural network or error reverse transmission multi-layer feed forward commonly used neural network at present. It is a multi-layer neural network for weight simulation training of nonlinear differentiable functions, and also a multi-layer forward feedback neural network. The most important characteristic is the forward transmission of data parameters and the reverse transmission of operational error. In the forward feedback transmission of data information, the input information is processed layer by layer through the until sent to output layer. If there is an of er and the expected value (or the actual value), the error value will enter the network in reverse for transmission, and the in the will be modified, so that the difference between the actual output value and the expected output value will gradually decrease. As the weight optimization of rule of local regional gradient descent, the slow speed of neural network training, and even oscillation and non-convergence problems in the learning process.

3   The Main Application of ACOBP Calculation Process

Ant colony algorithm can be expressed as follows: At the initial moment of the algorithm, m ants are randomly placed on n nodes, pheromone quantity on any path (i, j) is equal; let (k is a constant). Ant k (k = 1, 2 …, m) transferred from city i to city j at time t according to the heuristic information on the path. After n time, ant k completes a cycle after traversing all nodes. At this time, pheromone quantity on each path is adjusted according to the following formula:

$$\tau_{ij}(t+1) = \rho\tau_{ij}(t) + \Delta\tau_{ij} \tag{3.1}$$

$$\Delta\tau_{ij} = \sum_{k=1}^{m} \tau_{ij}^{k}(t, t+1) \tag{3.2}$$

$$\tau_{ij}^{k}(t, t+1) = \begin{cases} \frac{Q}{d_{ij}}, & \textit{If ant k goes through edge } (i,j) \textit{ in this cycle} \\ 0, & \textit{else} \end{cases} \tag{3.3}$$

The specific steps of improving BP neural network by ant colony algorithm [9] are as follows:

(1) First, a set is created for each input and output node. Each element in the set corresponds to a weight, and an initialized pheromone is set for each weight.
(2) Release m ants, each ant starts from the set I_1 and selects the next node according to the path selection rules: any, and randomly select calculated by the following formula.

$$P_j(k) = \frac{\tau(j)}{\sum\limits_{i=1}^{n} \tau(i)} \tag{3.4}$$

Select one node in each set in turn maximum number of iterations NI-max (maximum iterations), the cycle is terminated and the node label in the path is recorded.

(3) According to the weight, BP and calculate the output error sample.
(4) Modify the weight and pheromone according to the output error:
Modify global weights:

$$w_{ij}(t+1) = w_{ij}(t) + \Delta w_{ij}(t+1) + \beta\Delta w_{ij}(t) \tag{3.5}$$

Where: is the momentum term, which is generally set at 0.3–0.5.
② Modify local weights: Modify only the elements marked in step (2).

$$\lambda = \frac{e(t)}{(e(t-1) + e(t))} \tag{3.6}$$

Where: Q is a constant, which is used to represent the total number of pheromones released by ants after completing a cycle. e is the output error of each training sample of the neural network at time t.

4   Realization of Real-Time Tool ACOBP Algorithm

At present, some NC manufacturers have been equipped with real-time thermal error compensation function, but these systems are expensive and have technical limitations. Most of the ordinary domestic machine tools cannot be configured with such a high price of NC system. After the establishment system, of each point on the machine tool is detected according to the sensor in the system, which is imported into. After the predicts and calculates the compensation amount, it is then transmitted back to the numerical implement the compensation control, which is also a very practical work [4]. The thermal deformation sunlight, radiant heat of human and electronic equipment, etc. includes the cutting heat generated by the tool's interaction with the workpiece in the process of machining, movement of the working table and the movement of the supporting shaft, the loss heat of the motor change. Since the temperature and thermal expansion coefficient of each component of the machine tool are not the same, each component produces different thermal deformation and then produces relative displacement, that is, thermal error. In the actual thermal error system, the thermal deformation of the machine tool and the temperature data of some hot sensitive points are measured first, and then analyze the relationship between them and compensate.

In this paper, the thermal error compensation system based on ACOBP algorithm is applied to an HMC800A machine tool, which uses paperless recorder and laser interferometer to collect error source data and a total of 9 error sources are collected according to prior experience. Including: left motor temperature T1, right motor temperature T2, left bearing temperature T3, right bearing temperature T4, left grating temperature T5, right grating temperature T6, environment temperature T7, table temperature T8 and thermal deformation E of spindle Y-axis (collecting positioning error of Y-axis lead screw). In order to avoid the variable coupling in the error model and improve the accuracy and robustness of the thermal error model, the bayesian network was used to analyze the temperature error data of 8 collection points, find out the correlation between the thermal error and each variable, and extract the important characteristic parameters such as the thermal error compensation of the machine tool. Through analysis, four important characteristic parameters were selected from eight collection points: left motor base temperature, right bearing base temperature, right grating temperature and table temperature [5, 6].In order to eliminate the order of magnitude difference between the data in various dimensions, the 800 sets of data collected were first normalized and then grouped. One part was used for network training, and the other part was used for prediction. By comparing the predicted value of the trained BP neural network with the actual measured value, the difference obtained is our compensation value. In the thermal error compensation system, the neural network of ACOBP algorithm uses a three-layer BP network, with 4 nodes in the input layer (representing the data of 4 collection points respectively), 7 nodes in the hidden layer, and 1 node in the output layer (thermal deformation value E of the principal axis Y axis). Between the input layer and the hidden layer is a log-Sigmoid-type function, and between the output layer and the hidden layer is a tan-Sigmoid-type function. The ACOBP algorithm parameters were set as follows: the number of ants(m) was set to 30, α was set to 1, β was set to 0.5, ρ was set to 0.2, and IC-max was set to 5000. The heuristic optimization of ant colony algorithm and the advantage of global optimization are used to train the weights to be optimized, and the weights in the optimized network are accelerated, and finally the global intelligent optimization of BP neural network is realized. Figure 2 and Fig. 3 show the simulation results of general BP algorithm and ACOBP algorithm in error accuracy and error compensation accuracy in NC machine tools.

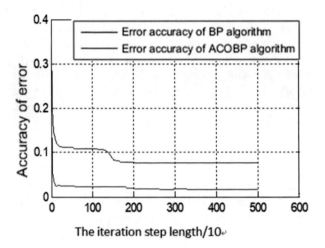

Fig. 2. The error accuracy of the two algorithms

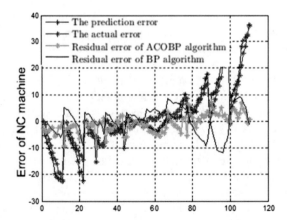

Fig. 3. Error compensation of the two algorithms

From Fig. 3, we can see that in terms of error accuracy, the ACOBP algorithm based training network can achieve a lower error accuracy in a short time and a faster speed than BP neural network. thermal ACOBP residual error control algorithm training within 6 μm, and general BP algorithm training the network of residual error sometimes have more than 10 μm. ACOBP algorithm training than normal BP algorithm training the network with higher compensation accuracy, also has better real-time characteristics in the system of l error.

5   Summary

In this paper, frame loss and delay compensation of CAN bus communication based on limited bandwidth are studied in detail. According to the mechanism of communication delay and frame loss, a distributed compensation scheme is proposed, and some common problems in the design of compensator are discussed in detail. The compensation method is applied to linear s-so system, and a feedback compensator based on controlled object model and a control compensator based on system control rate are proposed. The subsequent modeling and simulation results show that the proposed compensation method can reduce the disadvantages caused by communication frame loss and delay in a certain number of frame loss. The proposed method can effectively ensure the control system to operate with high efficiency and safety without any impact on the communication performance of CAN bus. This method also provides a reference for other similar network control systems.

References

1. Liu, H., Lin, T.: Review of time delay compensation method in real time hybrid test of building earthquake resistance. J. Seism. Eng. **42**(06), 1361–1368 (2020)
2. Zhao, G., Ye, G., Shi, Y., Yin, L., Liu, H.: Research on high adaptive grating subdivision method based on ratio linearization. Acta Metrol. Sin. **41**(07), 781–788 (2020)
3. Zheng, Y., Zhang, J., Shi, W., Anton, Sun, J.: Research on compensation of CAN bus frame loss and delay in real time control system. Comput. Meas. Control **28**(07), 107–111 + 116 (2020)
4. Yu, J., Yu, Q., Fang, A., Cheng, X.: Real time compensation technology for high efficiency grinding of large diameter silicon carbide mirror. Opt. Technol. **46**(04), 502–506 (2020)
5. Hu, Z., Wang, Z., Jin, L., Ding, G., Chen, J.: Research on real-time compensation method of non modulated two-color infrared temperature measurement error. Opt. Instrum. **42**(03), 15–20 + 56 (2020)
6. Wu, P., Dan, Q., Luo, X., Rao, X., Cai, Z.: Application of Kalman filter in real time position compensation of electric sliding table. Instrum. Technol. Sensor (06), 110–113 + 117 (2020)

Brain Activity Recognition with Deep Convolutional Neural Network

Zhengxing Yan[✉]

Shanghai Qibao Dwight High School, Shanghai 201101, China
zxyan_aiden@qibaodwight.org

Abstract. Brain-Computer Interface (BCI) can translates brain electrical fluctuations that are present in Electroencephalography (EEG) waves. Such raw signals can be quite noisy and difficult to analyze. Therefore, a robust approach is needed to achieve efficacy and efficiency. We present -based approach, a model we call as the DL-Brain, for brain activity recognition. In some sub-tasks, our model can achieve a 100% accuracy.

Keywords: Brain activity recognition · Electroencephalography (EEG) · Introduction

Brain-Computer Interface provides a pathway to connect human brains and external devices. With improvements of BCI, it can become valuable in many fields. In 1929, Hans Berger successfully extracted brain electrical waves via the method he named as Electroencephalography (EEG) in [1]. Since then, researchers started to look deeper into processing brain waves, which is now seen as a key to make BCI come true. EEG impulses are classified as the Table 1 has shown:

Table 1. Classification of EEG signals

Types of waves	Frequency band	Subject's state
Beta	12–30 Hz	Active
Alpha	8–12 Hz	Inactive
Theta	4–7 Hz	Sleeping
Delta	0–4 Hz	Deep sleeping

The concept of EEG-based brain activity recognition was also drawn out. Such concept has high expectations. However, there are problems to be solved: Average accuracy stays at a low level. Higher accuracy is needed; Researches mostly focus on intra-subject recognition, limiting their generalizability. In this paper, we propose an algorithm which: Combines CNN with Multilayer perceptron (MLP) to enhance robustness and efficiency.

J. Abawajy et al. (Eds.): ATCI 2021, AISC 1398, pp. 736–742, 2021.
https://doi.org/10.1007/978-3-030-79200-8_110

1   The Proposed Methodology

In this section, we present our model, which operates in an end-to-end manner.

1.1   The Architecture of DL-Brain

The structure of the model is shown in Fig. 1 and Fig. 2.

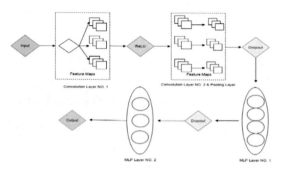

Fig. 1. Prediction process of DL-brain

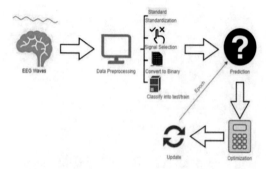

Fig. 2. Workflow of DL-brain

Below we will give explanations on components used.

CNN
In 1989, LeCun et al. proposed the Convolutional Neural Network (CNN) [2]. CNN stands out since it captures the more obvious parts of a label.

1.2   Convolution Layer

The Convolution Layer enlarges special features via kernels. The mathematical representation of this process with only one stride where f denotes the input, As shown in Fig. 3. h denotes the kernel, a denotes rows of output matrix, and b denotes columns of output matrix is:

$$G[a, b] = (f * h)[a, b] = \Sigma \Sigma h[j, k] f[a - j, b - k] \tag{1}$$

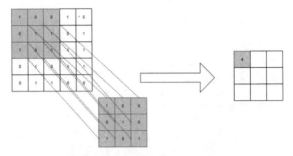

Fig. 3. Example of 2-D convolution

1.3 Pooling Layer

The pooling layer is used to downsample as shown in Fig. 4. Max-pooling function identifies obvious features via filters. The dimension of output matrices can be mathematically represented as the dimension of the filter is p * q:

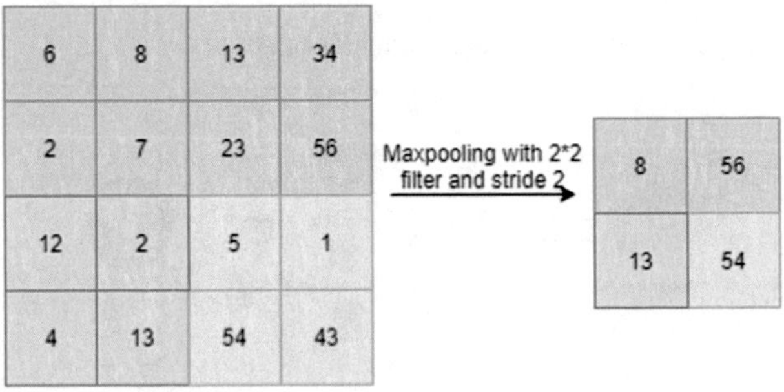

6	8	13	34
2	7	23	56
12	2	5	1
4	13	54	43

Maxpooling with 2*2 filter and stride 2

8	56
13	54

Fig. 4. An example of 2-D max-pooling

Firstly, two variables need to defined [3]. And the output should be: $y_{ij} = max(x_{i+r,j+s})$.

There are limitations when the input matrix's size is m * n: $i \leq m - p j \leq n - q$. Therefore, the dimension can be deducted significantly.

1.4 MLP

MLP was invented by Frank Rosenblatt as shown in Fig. 5. If we use bk to denote the bias, and use Wk to denote weights, we can get the following mathematical model:

1.5 ReLU

In neural networks, we use activation functions to determine which data is qualified to be inputted into the next layer of nodes as shown in Fig. 6. We adapted the Rectified Linear Unit (ReLU) function invented by Nair et al. [4]. The way it works is:

$$f(x) = max\{0, x\} \tag{2}$$

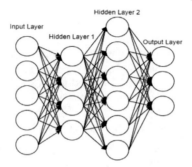

Fig. 5. Concept of MLP

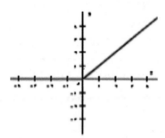

Fig. 6. The ReLU function

2 Model Optimization

There are always differences between true labels and predicted ones. The goal of optimization is to eliminate such difference.

2.1 Loss Function

For this model, we chose the Cross-Entropy Loss. The mathematical representation is the following when x denotes the features of the input, $\hat{y}$ denotes the predicted label and y denotes the actual label.

For the above possibility, we wish to make it as high as possible, so we need to add a negative sign to get a minimum value. Hence, the function should be

2.1.1 Gradient Descent

In this paper, we utilized the Stochastic Gradient Descent invented by Robbins et al. The update rule for learning parameters where α means the learning rate is:

$$\theta_j = \frac{\theta_j \partial L_{CE}(\theta)}{-a\partial\theta} \tag{3}$$

2.1.2 Experiment

Datasets

In this experiment, the EEG files we used was "EEG Motor Movement/Imagery Dataset" which was obtained from PhysioNet [5]. In their experiment, 109 healthy volunteers were asked to perform different tasks. As shown in Table 2 below:

Table 2. Testing Instructions

Task 1 Eyes open
Task 2 Eyes closed
Task 3 Clenches a fist in the direction of the icon
Task 4 Imagines clenching a fist in the direction of the icon
Task 5 Clenches fists when the icon appears on top; Closes feet when the icon appears at bottom
Task 6 Imagines clenching fists when the icon appears on top; Imagines closing feet when the icon appears at bottom

For our model, we only tested on task 3–6. A variety of annotations were also added artificially:

T0 means the subject is at a state of resting;
T1 means the subject is at the state of intending to:
Clench the left fist (correlated to Task 3 and 4);
Clench both fists (correlated to Task 5 and 6);
T2 means the subject is at the state of intending to:
Clench the right fist (correlated to Task 3 and 4);
Put both feet together (correlated to Task 5 and 6).

2.1.3 Data Preprocessing

In order to make the model fit into raw instances, few changes were applied.

3   Read and Selection

We firstly decided to adopt a band-pass filter proposed by Hans Berger which allows us to extract signals within a frequency band, which we set as 7.-30. Then, we defined a dictionary with a key named correspondingly to T1 and T2. Then, we copied the data from 0 to 3 s from the epochs so that we could include fewer samples [6]. Afterwards, we turned the values into one vector.

3.1   Standardization and Convert

After this, we used another function- "train test split" -which randomly split the processed data to training and testing sets. We designated the size of testing data as 0.1, which means 0.9 would be for training.

3.2   Implementation Details

For basic parameters, the batch size was 64, the number of epochs was 750, and the Learning Rate of SGD was 0.0001.

For the models, the parameters are set as follow:

In the first CNN layer, the number of input channels was 64 and that of output ones was 32, with a kernel size of 10, and stride of 5; In the second CNN layer, the number of input channels was 32 and that of output ones was 16, with a kernel size of 5, and stride of 3 [7]. In the first dropout layer, the dropout rate was 0.5.

In the first MLP layer, the input feature was 16 * 15, and the output one was 16. In the second dropout layer, the dropout rate was 0.2; In the second MLP layer, the input feature was 16, the output one was 3.

3.3   Baseline

Long-Short Term Memory (LSTM)
LSTM was invented by Hochreiter et al. The comparison will be shown as follows: "Results Analysis".

Results Analysis
For the following graphs, each reflects results when DL-Brain ran on sub-tasks with colored lines representing numbers of subjects.

4   Conclusion

Throughout this paper, we focused on recognizing brain signals. As tables shown in Section Results Analysis, the model had the best accuracy of 100%. Such rates can help to achieve more possibilities. For future works, we will look deeper into complicated predictions. We need to keep digging into the theme of brain activity recognition to achieve a comprehensive BCI technology.

References

1. Xiao, R., Tang, L., Zhang, Y.: Remote sensing image classification based on deep neural network. Sci.-Tech. Inf. **18**(1), 22–23 (2020)
2. Zhao, Q., Sun, L., Yuan, Y.: Image recognition of UAV based on neural network. Chin. Sci. Pap. **14**(11), 1229–1233 (2019)
3. Hou, Y., Qing, Y., Quan, J., Wang, H.: Overview of deep learning development. Ship Electro. Eng. **37**(4), 5–9111 (2017)
4. Molin, K., Li, Z., Jing, Z., Hui, Z.: Research progress of UAV image target detection technology. Measur. Control Technol. **10**(4), 1–15 (2020)
5. Hu, Y., Luo, D., Hua, K., et al.: Review and discussion on deep learning. J. Intell. Syst. **14**(1), 1–19 (2019)
6. Zhan, W., Ramatov, I., Cui, W., et al.: Overview of deep learning target detection algorithms based on candidate regions. J. Yangtze Univ. (Nat. Sci. Ed.) **18**(5), 38–45 (2019)
7. Zhang, J.: Ship Detection in Satellite Remote Sensing Image Based on Deep Learning. Harbin Institute of Technology, Harbin (2019)

Machine Learning in Error Types of Machine Translation

Xueling Zhang[(✉)]

Nanchang Institute of Technology, Nanchang, Jiangxi, China

Abstract. This studio is a performance of Javanese in Singapore and Chinese, but expected to be a hot situation, found no small review of the performance of Javanese. The qualitative analysis method is adopted to analyze the translation problems and errors, so as to make the research more practical.

Keywords: Adaptation theory · Hobent

1 Introduction

With the success of the 2008 Olympic Games and the further development of tourism, according to CNKI statistics, from 2000 to 2006, there were 49 articles about pubhc AGNs translation. The different scenic spots studied in this paper are totally different from those in other provinces. It has its own unique characteristics. Lushan has been famous all over the world since ancient times [1]. It was the first natural scenic spot in Jiangxi Province in 2007, and Sanqingshan got another natural scenic spot in 2011. As a revolutionary Province, its revolutionary culture also attracts many tourists from all over the world. One of the most famous is the Memoirs of the revolutionary martyrs in Nanchang and Jiangxi on August 1. Therefore, under the guidance of adaptation theory, the translation of public signs in scenic spots of Jiangxi Province can be used for reference by other provinces in China.

2 Introduction to Adaptation Theory

2.1 Main Theories

The percentage of Chinese voices in music shows that they believe in reproduction and family activities. tow. Varicose veins were the connection he felt in the ACA cavity. In 1987, Verschueren, Jef advanced the adaptation theory firstly after several years of modification, enrichment and elaboration. Verschueren presented his new approach on pragmatics—adaptation Theory. Language Use as a Continuous Process of Choice-making.

In order to further expound the choice-making. Secondly, speakers also adopt strategies as well as choose forms. Third the processes in which speakers make choices may show any degree of consciousnessFourth, both speakers and listeners make choices

J. Abawajy et al. (Eds.): ATCI 2021, AISC 1398, pp. 743–747, 2021.
https://doi.org/10.1007/978-3-030-79200-8_111

in producing an in interpreting anutterance. It means language use is connected with other aspects of human life. So language isn't independent or unchanged. It varies with different circumstances. It relies on many factors such as cognition, culture and customsFinally, choices will arouse the speaker's related alternatives.

3 Semantic Ontology Model and Error Elimination Mapping in English Translation

With the development and maturity of automatic translation software, the accuracy of machine translation software translation and calibration is required to be higher. In machine translation environment, semantic analysis needs to be combined with the context features of automatic translation software [2]. According to the semantic similarity, automatic translation and calibration are carried out to improve the accuracy of automatic translation semantic allocation. Under the condition of semantic heterogeneity, the automatic calibration of machine translation is mainly realized through the concept analysis of semantic similarity. The relevance and semantic similarity features of automatic translation text eliminate the semantic heterogeneity according to the similar semantics, realize the knowledge structure diagram analysis in the process of English translation, and construct the semantic concept tree. Improve the ability to eliminate translation errors. The research on the methods of eliminating errors in machine translation is of great significance to improve the intelligent level of machine translation. The research on the methods of eliminating errors in machine translation has attracted much attention.

This paper analyzes the characteristics of the associated vocabulary information in the ontology model of translation error elimination in English conversion, and establishes a language ontology document model based on deep learning. The mapping mechanism of context ontology in machine translation is constructed by using the inclusion mapping method of subject words and prepositions. In the process of English language conversion, the mapping model of semantic ontology is constructed by using the adaptive learning method of machine language, as shown in Fig. 1.

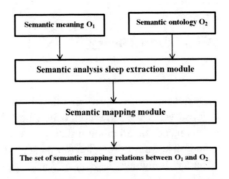

Fig. 1. Semantic ontology model of error elimination in English language conversion

Based on the above semantic ontology model, a dictionary database is constructed. Based on the large-scale ontology diffusion mapping method, a dictionary database for eliminating translation errors in the process of English language conversion is established. In order to eliminate translation errors in English conversion, a comprehensive evaluation fuzzy decision function such as formula (1) and formula (2) is proposed.

$$(\zeta_{ij}, \eta_{ij}) = \Delta \left(\frac{\min_i \min_j \Delta^{-1} + \max_i \min_j}{\Delta^{-1} d} \right) \tag{1}$$

$$(\zeta_{ij}^-, \eta_{ij}^-) = \Delta \left(\frac{\min_i \min_j \Delta^{-1} d + \max_i \max_j}{\Delta^{-1} d + \max_i \max_j} \right) \tag{2}$$

4   An Analysis of the Problems in the Translation of Public Signs

Throughout the theory, the translation is a process of choice-making and adaptation so it is with the translation of public signs. Based on Adaptation Theory, we will discuss the C-e translation of public signs mainly from two aspects: the linguistic structure and the extra-linguistic content. Each of them can be further divided into several categories which are related to each other to some extent. Translation is a process of dynamic choice and use of language. The translation of cosmetic advertisements is a process of constant choice and adaptation, which also follows the theory of adaptation. In the translation practice of cosmetic advertisements, translators should take full account of the linguistic features and purposes of cosmetic advertisements, the linguistic features of source and target languages, and national culture, and flexibly use domestication and foreignization, X: the make-up of make-up artists (Max Factor) translation: Max factor is a professional cosmetics brand in the United States, which is deeply loved by female consumers. Its advertising language translation mainly adopts literal translation and foreignization translation strategies, and the translation is similar to the original form, Focusing on the functional equivalence between the original and the translated text highlights the brand's characteristic of professionalism. At the same time, it is conducive to enhancing the affinity and influence of advertising language, which is in line with the aesthetic, cultural and value orientation of target language consumers, so as to form consumers' psychological, emotional and cultural recognition of the brand and stimulate consumers' purchase desire.

4.1   Errors in Pronunciation

Pronunciation is the most fundamental component as well as an indispensable section of the structure in a language. What makes a language different from other is itspronuciation It's obvious that English and Chinese belong to two different languages since they are spoken completely differently (Luo Xuanmin 2006). So they also have different sound system and writing system. The study of the writing system is corresponding to the investigation of the sound system. We will probe into the non-equivalence with pronunciation rules from two facets: spelling system and pinyin system.

To a large degree, the spelling errors result from the lack of the translator's responsibility. It's very common in the C-e translation of public signs in Jiangxi scenic spots.

Some people may think those errors are so trivial that they can be ignored. But infact, it is there trivial errors that reduce the attractiveness of Jiangxi in the eye of foreignvisitors. More importantly, they confuse the foreign visitors to some degree [3]. From the following examples, we can make an inference that these errors are not directly related the competence of the translator. They could have been avoided if the translator had paid more attention to it.

4.2 The Extra-linguistic Problems

In the previous section, we discuss the difference linguistic aspect from four different parts. Besides the linguistic section, there are more dissimilarities of extra-linguistic context between Chinese and English in the process of or errors are salient while the extra-linguistic context should be considered more deliberately. In this section, we will analyze the effect of the adaptation theory on the extra-linguistic context in the process of the translation. Three facets will be analyzed under the guidance of Adaptation Theory: the physical world, the social world and the mental world.

According to verschueren's adaptation Theory correlates of adaptability means the linguistic choices have to be adaptable to the context in the process of language use Verschueren divided the context into communicative context and linguistic contest [4]. The communicative context includes language users, physical world, social world and the mental world. Language users occupy the central position in the communicative context because the physical world, the social world and the mental world are activated by the language users to meet the communicative function of language. Note that language users are not limited to the speaker and the listeners, but also involve other people related to the conversation.

5 Conclusion

This paper discusses English Chinese translation from the perspective of coherence theory, which can not only explain the coherence theory. According to this theory, the choice of language is based on the reason of language. Translation into Chinese is a kind of concise expression to choose or adapt to different feelings. On the basis of a comprehensive analysis of English, we find that English is very different from English. Some effective measures are put forward to improve the quality of translation.

Acknowledgements. Study on the C-E Translation of City Publicity Materials from the Perspective of Linguistic Adaptation Theory (Social Science "13th Five-Year Plan" Project of Jiangxi Province (2016)) Project Number: 16YY16.

References

1. Qi, W.: Translation of cosmetic advertisements from the perspective of relevance theory. J. Changchun Inst. Educ. 27(10), 56 (2011)
2. He natural. Three Pragmatic Theories: Relevance Theory, Adaptation Theory and Memetics, pp. 87–92. Shanghai Education Press, Shanghai (2007)

3. Yi, S.: Translation of cosmetic brands from the perspective of adaptation theory. Jiannan Lit. (04), 118–120 (2012)
4. Guo, D.: Functional theory of translation strategies for English Cosmetic Advertisements. Mod. Chin. (Lang. Res. Ed.) **11**, 132–134 (2017)

Judgment Method of Landscape Design Rationality Based on Remote Sensing Image Recognition

Shihan Hu[✉]

Liaoning Urban Construction Technical College, Shenyang 110122, China

Abstract. Aiming at the problems of large error and low efficiency in traditional judgment methods, a method of landscape design rationality judgment based on remote image recognition is proposed. The rationality of landscape design of remote image recognition is represented by weight. The two-dimensional attributes existing in the landscape design of remote image recognition are extracted, and the rationality is identified. A fast robust recognition algorithm is introduced to analyze the landscape design rationality weight of remote image recognition, And the rationality of landscape design of remote image recognition can be effectively judged by weight comparison. The experimental results show that the improved judgment method can effectively identify the rationality of the design, and ensure the validity of the data and the accuracy of the calculation.

Keywords: Remote image recognition · Landscape design rationality · Weight comparison · Judgment error · Robust recognition · Two-dimensional attribute

1 Introduction

In the process of landscape design based on remote image recognition, due to the use of remote images, there will be some visual errors. Therefore, it is very important to judge the rationality of landscape design of remote image recognition. The rationality of landscape design of remote image recognition includes: the safety of landscape architecture structure and the error of construction process, as well as whether the designed landscape style has enough technical support. The landscape design rationality of remote image recognition can affect the overall construction process of the structure, and is also an important data index for personnel deployment and construction preparation [1]. At present, the judgment methods of landscape design rationality of remote image recognition generally include the following: scale trade-off method, ultrasonic contrast method, photographic image method, etc., but the accuracy of judgment is very low for the landscape with multi geometry and multi occlusion. In view of the rationality of landscape design in the process of remote image recognition, this paper proposes an effective method to judge the rationality of landscape design based on remote image recognition, which is verified by experimental data.

© The Author(s), under exclusive license to Springer Nature Switzerland AG 2021
J. Abawajy et al. (Eds.): ATCI 2021, AISC 1398, pp. 748–752, 2021.
https://doi.org/10.1007/978-3-030-79200-8_112

2   Weighted Representation of Landscape Design Rationality Based on Remote Image Recognition

2.1   Landscape Design Rationality Extraction Based on Remote Image Recognition

In this paper, the rationality judgment method of landscape design based on remote image recognition is designed, which uses weight to express, but it needs to extract relevant data before it can be used for effective representation. Firstly, the effective data of image is confirmed, and the process is as follows:

$$R^2_{YIU} = \frac{\left(N \sum A_i B_i - \sum A_i \sum B_i\right)^2}{\left(N \sum A_i^2 - \left(\sum A_i\right)^2\right)\left(N \sum B_i^2 - \left(\sum B_i\right)^2\right)} \tag{1}$$

Where: R^2_{YIU} is the data confirmation process of the two-dimensional reference image; A_i is the discrimination value of the reference image; B_i is the extracted value of the reference image.

After the edge data extraction process, the SVM function can extract the color difference, shadow, node and light perception of the design landscape. The process of data extraction using SVM function is as follows.

Color difference data extraction of SVM linear function:

$$K(a, b) = x \times y \tag{2}$$

Shadow data extraction of SVM polynomial and function:

$$K(a, b) = \left[(x \times y) + 1\right]^d \tag{3}$$

Light sensing data extraction based on SVM kernel function:

$$K(a, b) = \tan\left[K(x \times y) + m\right] \tag{4}$$

Where: K(a, b) is the unified expression method of the extracted data; $x \times y$ is the picture frame of the image, and the picture of each frame is extracted through certain data changes, so it can be effectively extracted only by limiting different frames. After the above process, the data extraction is completed.

2.2   Recording Rational Representation of Remote Image Recognition

In this paper, the rationality judgment method of landscape design based on remote image recognition is designed [2]. After the effective extraction of image data, the weighted representation of data can be carried out. Firstly, the amount data is weighted. The process is as follows:

$$Sum(\lambda, t) = -\sum P_t(\lambda, i, j) \log_2 P_t(\lambda, i, j) \tag{5}$$

Effective expression value, amount data mainly includes: shape, line, overall color and other data.

3   Rationality Judgment of Landscape Design Based on Remote Image Recognition

3.1   A Fast Robust Recognition Algorithm is Introduced

The method of landscape design rationality judgment based on remote image recognition is designed in this paper. The data of remote image is represented by weights. The fast robust recognition algorithm is introduced to judge the rationality of landscape design of remote image recognition by reference of this paper. However, in order to ensure the validity of the data and the accuracy of the calculation, the weighted data need to be preprocessed once to use the fast robust identification algorithm.

3.2   Rationality Judgment of Landscape Design Based on Remote Image Recognition

In this paper, the landscape design rationality judgment method based on remote image recognition is designed. After the pretreatment of the above process, the effective identification can be carried out. The first choice of fast robust identification algorithm is to express and confirm the robust coefficient, and the process is as follows:

$$r = R/(R + G + B)$$
$$g = G(R + G + B)$$
$$b = B/(R + G + B) \tag{6}$$

Where: r, g, b are the upper limit value, lower limit value and uplink transmission value of the robustness coefficient; R, G, B are the scalar used, the running scalar and the judging scalar respectively.

4   Analysis of Simulation Experiment

4.1   Parameter Setting

In order to ensure the effectiveness of the landscape design rationality judgment method based on remote image recognition, the parameters are set, and the weighted amount data is within the range of [15.5, 18.5].

In this paper, the experimental remote image data selection process is designed to separate different data effectively, so as to observe the effectiveness of the design more intuitively. The data distribution is shown in Fig. 1.

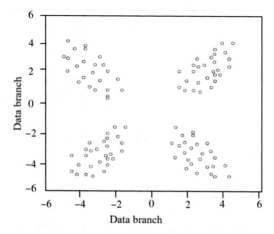

Fig. 1. Data distribution

4.2 Results Comparative Analysis

According to the results of Fig. 2, the rationality judgment method of landscape design based on remote image recognition designed in this paper is obviously higher than the traditional method in the discrimination efficiency, and the discrimination time is about 1/2 of the traditional method [3].

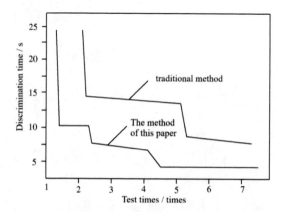

Fig. 2. Comparison test results

When using traditional methods to analyze the distribution rationality, it is difficult to accurately extract the image feature points, which has the problems of large judgment error and low speed [4]. Therefore, a 3D image simulation judgment method is proposed to analyze the rationality of landscape distribution. This method can extract and analyze the landscape design data, reduce the judgment error to a great extent, and greatly improve the speed of judgment through simple judgment steps. Through the experiment, the judgment results are verified, and it is concluded that the method has high efficiency and high precision judgment ability, which provides an important theoretical basis for the realization of the rationality of landscape distribution.

5 Conclusion

This paper proposes a method of landscape design rationality judgment based on remote image recognition, and effectively judges the rationality of remote image recognition landscape design through weight comparison. The experimental results show that the accuracy of the improved method is higher.

References

1. Wu, J., Zu, X., Xue, H.: Direction determination of text image based on feature character recognition. Comput. Eng. Des. **8**, 2260–2263 (2015)
2. Dong, W.: Remote visualization of three-dimensional image characteristics of illegal felling trees in Zijin Mountain Area. Sci. Technol. Bull. **32**(10), 148–151 (2016)
3. Peng, Z., Jiang, W., Luo, A., et al.: Remote video monitoring system based on image intelligent recognition technology. J. Guilin Univ. Electron. Sci. Technol. **24**(5), 23–26 (2004)
4. Li, X., Guo, Y., Xiao, H.: Analysis of railway landscape greening design. Microcomput. Appl. **6**(8), 49–50 (1997)

Environment Parameter Setting Control System for Sheep House Based on Wireless Network

Zhe Feng, Wen Zhong, and Rongxin Zang[✉]

Life Science and Engineering College, Northwest Minzu University, Lanzhou 730000, China

Abstract. In order to realize the intensive, digital and intelligent management of sheep raising industry and ensure the real-time collection of environmental parameters of sheep house, taking the breeding environment of sheep house in animal husbandry as the research object, using Zige short-range wireless technology and GPS technology, the real-time acquisition of temperature, humidity, CO3 concentration and ammonia concentration in sheep house environment is realized, it can be remotely or automatically controlled ventilation, heating and other equipment to ensure that the environment in the sheep house is suitable for sheep's healthy growth. The system can realize the stable acquisition of sheep shed environmental parameters. When the environmental parameters exceed the set threshold, it can timely alarm and start the corresponding control equipment, so that the environmental parameters of the sheep shed change within the set range. The operation interface of the system is simple, friendly and stable. It can alarm in time, and improve the automation level of sheep house management to a certain extent.

Keywords: GPRS · ZigBee · Environmental parameters · Sheep house

1 Introduction

In recent years, in order to restore the ecology, our country has implemented the policy of changing grain to feed, rotational grazing on grassland and banning grazing. Therefore, intensive facility sheep raising has been greatly developed. Due to the large number and high density of sheep and sheep raised in facilities, it is urgent to establish an intelligent management system of sheep house which can improve the breeding efficiency and reduce the labor intensity of personnel. Compared with foreign countries, China's sheep industry started late, there is a big gap, there are many problems, most of which are traditional extensive type". Limited by natural and environmental conditions, sheep raising facilities are simple and poor, and the ability to resist natural disasters is very weak. The development of science and technology, especially the rapid development of remote wireless technology, MCU technology and sensor technology, provides strong technical support for data acquisition and control of sheep house environment. Based on this, a sheep shed environmental parameters acquisition and control system is designed to provide theoretical basis and experimental basis for the research of sheep shed environmental control and parameter early warning [1].

© The Author(s), under exclusive license to Springer Nature Switzerland AG 2021
J. Abawajy et al. (Eds.): ATCI 2021, AISC 1398, pp. 753–757, 2021.
https://doi.org/10.1007/978-3-030-79200-8_113

2   Overall Structure of the System

The environmental parameters acquisition and control system of sheep house based on wireless network integrates sensor communication technology and computer technology. By setting up suitable environmental parameters for sheep's healthy growth, intelligent software and hardware platform is built to realize automatic monitoring and control of temperature, humidity, CO_2, ammonia and other factors in sheep house [2].

The system uses low-power microprocessor stc12c4052ad to realize the control and communication with each module. The sheep house environment data acquisition system is mainly composed of data management center, GPRS terminal, data acquisition module, sensor and actuator. The temperature, humidity, ammonia concentration, Co, concentration and other environmental parameters in different areas of the sheep house were detected by sensors to understand the growth status of sheep. The actuator can adjust and control the environmental parameters in real time to make the sheep grow in a suitable environment. The system structure is shown in Fig. 1. The connection of remote sheep house environmental data acquisition system is shown in Fig. 2.

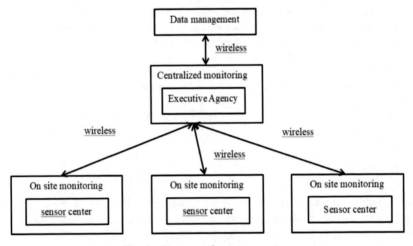

Fig. 1. Diagram of structure system

2.1   Structure Design of Data Acquisition Module

The main function of the data acquisition module is to realize the accurate measurement of temperature, humidity, ammonia concentration, Co, concentration and other parameters in the sheep house. Therefore, the temperature and humidity sensor adopts SHT11 new sensor produced by Swiss Sensirion company. The temperature range is - 40.0–123.8 °C, the temperature measurement accuracy can reach 0.4 °C, the humidity range is 0–100% RH, and the humidity measurement accuracy can reach 3% RH. Ammonia concentration sensor is composed of NH/CR-200 ammonia concentration sensor and fad-8209 ammonia concentration transmitter. The module has the advantages of high

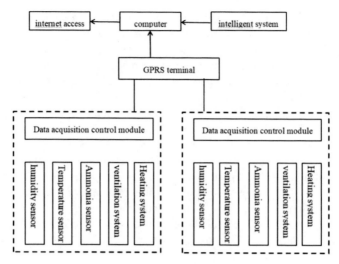

Fig. 2. Diagram of data acquisition system

measurement accuracy, good stability, strong anti-interference ability and long service life. The CO concentration sensor is mh-z14 gas sensor, which has wide linear range, high precision and low power consumption [3]. The front end of the acquisition circuit in the data acquisition module needs to be connected with various transmitters, and the standard signal, namely 4–20 mA, is sent from the transmitter. As long as the front end is connected with different types of transmitters, the acquisition circuit can transmit the collected voltage to the A/D conversion circuit. In the process of A/D conversion, the sampling circuit converts the current signal from the transmitter to the voltage signal through the sampling resistance. In order to improve the acquisition accuracy, the sampling circuit needs to use high-precision resistance.

2.2 Structure Design of GPRS Terminal Module

The data transmission terminal is the key part of the whole data acquisition system to realize remote data transmission. It is the bridge between the remote data management center and the data acquisition module. The received data is transmitted to the GPRS module through RS232 serial communication. GPRS module must be used for remote data transmission. The GPRS module used in the data transmission terminal is sim3oc of Siemens. Sim3oc module has the advantages of high reliability, complete function module and low price.

3 Software Design of Sheep House Environmental Parameters Acquisition and Control System

In order to ensure the stable and reliable operation of each part of the sheep house environmental parameters acquisition and control system, the corresponding software program must be compiled. The software of sheep house environment data acquisition system is divided into two parts: embedded bottom software and PC upper computer software [4]. The embedded bottom software includes data transmission terminal and embedded code running in MCU in data acquisition module. PC upper computer software includes data transmission terminal parameter configuration software and environment data acquisition software.

The system software consists of database unit, function unit, user management unit and remote control unit. According to the different data, the database unit establishes the corresponding data table records, and realizes the functions of adding data and searching data; The function unit includes real-time data display, parameter initialization, equipment control, historical report query and system help: the user management unit gives different levels of authority to users at different levels, which can realize a user to manage one or more information and automation equipment, and the remote control unit controls the wireless sensor network of the sheep house, Inform the system to take out the transmitted data and save it to the database. The software composition is shown in Fig. 3.

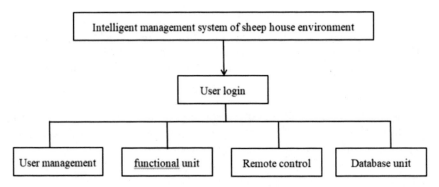

Fig. 3. Diagram of software structure

4   Test Result

The monitoring and early warning system is installed in the sheep farm. After debugging, the system runs stably, and the environmental parameter data of sheep house are collected according to the predetermined requirements. When the environmental parameters exceed the set threshold value, it can timely alarm and start the environmental parameter adjustment facilities. The user interface of the system is friendly, which effectively improves the efficiency and automation level of sheep farm management.

5   Conclusion

The intelligent facility breeding system based on GPRS Remote data transmission technology and ZigBee short distance wireless communication technology has been successfully applied. The constructed sheep house wireless monitoring and intelligent control system regulates and controls the environmental parameters in the sheep shed, which meets the requirements of sheep growth environment and effectively improves the intensive production level of animal husbandry. In this system, ZigBee short distance wireless transmission technology is used for communication between data transmission terminal and data acquisition module, and G; PRS remote data transmission technology is used for communication between data transmission terminal and remote data management center, which avoids the trouble caused by a large number of wiring caused by wired data transmission. After reliable operation, the system can accurately measure the air temperature, humidity, ammonia concentration, CO_2 concentration and other parameters in the sheep house. At the same time, the alarm range of environmental parameters can be set according to different requirements, and the system can be set to manual control or automatic control state. The development cost of the system is low, and its universality is strong, Therefore, it has broad development and application space in intelligent facility aquaculture environment control.

Acknowledgements. This work "Design and environmental monitoring of the new warm shed sheep house in Northwest alpine pastoral area" was supported by "the Fundamental Research Funds for the Central Universities" (No. 31920170026).

References

1. Boya: Current situation and development trend of animal husbandry in China. Agric. Technol. Equip. (11), 8–9 (2011)
2. Zheng, S.: Discussion on the present situation and improvement measures of comprehensive utilization of industrialized breeding facilities. J. Ningde Teach. Coll. (Nat. Sci. Ed.) **205**(3), 272–276
3. Gao, Z., Peng, K.: Current situation and sustainable development measures of animal husbandry in China. Ecol. Econ. (Acad. Ed.) **2**, 25–260 (2008)
4. Liu, Y., Zhu, L.: Wireless environmental monitoring system and application layer communication protocol. Instrum. Technol. Sens., 92–95 (2013)

Cyber Intelligence for Big Data

Impact of Big Data on Nursing Work and Application Prospects

Ke Bai[(⊠)]

Affiliated Hospital Nanyang Institute of Technology, Nanyang 473004, Henan, China
xing**yu@nyist.edu.cn

Abstract. The development of big data technology has brought tremendous changes to people's lives, affecting people's food, clothing, housing and transportation, especially for the medical industry is a brand new change. In traditional nursing work, there are problems such as low work efficiency, easy to miss patient information during handover, and long handover time. However, with the assistance of big data technology, these problems can be solved well and the original nursing care can be changed. The shortcomings in the work provide a new nursing work model. Based on this, this article proposes a research based on the impact of big data on nursing work and its application prospects, using literature data method, questionnaire survey method and experimental analysis method to explore the impact on nursing work on the basis of big data theory. This article designs an experimental study on its impact on nursing work based on big data, and explores its impact from two different perspectives: patients and nursing staff. The application of big data can improve the efficiency of nursing work, especially the time efficiency of statistics and drawing increased by 76.20% and 72.02% respectively. And the overall satisfaction of patients with it reached 85.3%, and the satisfaction of nursing professionals with the application of this technology reached 93.6%. In general, the application of big data has brought a positive impact on nursing work, and the mature application of this technology in the future will better promote the modernization and informatization of nursing work.

Keywords: Big data · Nursing · Informatization · Application prospects

1 Introduction

The development of big data has brought great convenience to our lives, and it has also changed our traditional nursing work methods. The application of big data can collect information through the Internet of Things, such as sensors, electronic tags, etc., and then use cloud computing for data mining and information parallel processing. After information mining, use distributed storage or non-relational database storage cloud platforms for data Storage, and finally the mobile Internet of Things can be used to publish information.

In the process of medical care, the information of each patient is different, and ordinary medical records cannot provide complete information, which causes trouble

© The Author(s), under exclusive license to Springer Nature Switzerland AG 2021
J. Abawajy et al. (Eds.): ATCI 2021, AISC 1398, pp. 761–769, 2021.
https://doi.org/10.1007/978-3-030-79200-8_114

for nursing workers and reduces work efficiency. However, big data has the effect of storing large amount of data, low density, and great value. Therefore, the application of big data technology in nursing work has completely subverted the previous working methods, improved work efficiency, and provided great convenience to nursing workers, and the application prospect is very good.

Black BS mainly studies the nursing work of Alzheimer's disease, assesses their cognitive function when selected, and reviews their admission records to identify all health problems/diseases and painkillers within 6 months before the selection usage of. Study the pain problem of the elderly from many aspects. However, his work did not carry out meaningful innovation and discussion on nursing work [1]. Presti comprehensively explored the application of flipped classrooms in nursing education. The literature search results produced 94 articles, 13 of which met the standards of classroom teaching methods in nursing teaching. The identified topics include theoretical foundations, strategies for implementing the flipped classroom, and student satisfaction and results with the flipped classroom approach. Conclusion: The synthesis of research results indicates that the flipped classroom approach can produce positive results. However, this method needs to be further studied to guide future implementation [2].

The innovations of this article are: (1) the combination of quantitative analysis and qualitative analysis, and adequately analyze the research data; (2) the combination of theoretical research and empirical research, and the theoretical basis of discussing big data technology, combined with nursing The actual situation of the work is empirically investigated.

2   Research Methods Based on the Impact of Big Data on Nursing Work and Application Prospects

2.1   Overview of Big Data

Big data refers to the use of traditional software tools to mine, collect, manage and process data within a specific time frame [3]. In the process of discovery and optimization, it requires a new processing mode, with strong decision-making functions and a large amount of insight, high development rate and differentiated information elements [4, 5]. Big data in the medical and health fields is mainly used for disease prevention, diagnosis and treatment, medical research and development and evaluation, clinical decision-making, medical quality management, disease prediction models, clinical trial analysis, individual treatment, public health monitoring, etc. [6].

Big data can be described from the following three aspects:

(1) User perspective: You can divide big data into non-customized big data and customized big data [7]. The pre-configuration of big data is a non-customized data resource, which will be able to meet the needs of users: most users' data storage, parallel computing, and interconnection are based on the dynamic configuration of applications. To complete this action, the position corresponding to the operation of these two numbers is transparent to the user.

(2) Cloud computing resources, platform management, and element hardware are the most important components of cloud resources. In the cloud center, the above hardware is included. The constructed equipment resources; the power resources of the cloud center are included in the POWER resources. The platform resources cover the establishment of the cloud center and the copyright of the software, and the resources should also include human resources to manage the cloud center [8].

(3) Service provider perspective: You can divide big data into: network resources, storage resources, and computing resources. The components of computing resources include the following: physical server clusters, physical servers, data security centers, virtual servers, server middleware, etc. [9]. The components of storage resources are as follows: image-based virtual machines, storage sharing, and so on. The components of network resources are: routers, switches, IP, VLANs, and so on. By distributing these resources in different geographic locations, providers can also implement dynamic configuration of data centers or more data center-related facilities to meet user needs [10].

2.2 Current Status of Nursing Work

With the rapid development of China's economy, people are paying more and more attention to their own health problems. From a national perspective, national health is also regarded as an important national policy, and medical policies are constantly being proposed. As the most important institution to protect human health, hospitals undoubtedly play an important role. In the process of protecting people's health in the hospital, doctors and nurses are mainly involved in the process.

In the medical and healthcare fields, the importance of nursing is gradually increasing. The so-called "three-point disease and seven-point nutrition" is enough to reflect the significance and importance of nursing work. The job of a nurse is an important means to ensure the quality of care and one of the foundations for patients to receive high-quality care. In the discussion of high-quality health examinations and nursing cases, effective nursing operations can prove the quality of patient care. Since the nursing business occupies an important position in medical activities, a complete nursing record is required. It can give full play to the knowledge level, nursing ability, actual nursing situation and effect of the nursing staff, and the patient's nursing situation is the most reproduced nursing work. Good performance.

At present, there is still a big gap between the domestic nursing management system and other advanced countries. For example, there are fewer nursing staff in China; long and unclear shift time can easily cause missed handover of patients; poor planning of the nursing system; poor reminder mechanism of the nursing system; the current nursing system cannot be paperless.

2.3 Application of Data Mining in Medicine

With the development of computer technology, the amount of data has increased dramatically. After social information, the operation of society is the operation of software, and the history of society is the history of data. The method of effectively acquiring the necessary knowledge from a large amount of data will become an important research topic.

The purpose of medical data extraction is to find knowledge, information, models, correlations, changes, etc. that may be effective. Extracting useful information from a large amount of medical data helps doctors make faster and more accurate diagnosis. The application of data mining in nursing work can realize the most comprehensive, timely and most important information connection, prevent the omission of the patient's condition, improve the efficiency of nursing staff, and achieve the best nursing effect.

Cluster analysis is also a commonly used method in data mining. Search and identify limited types of collections and provide detailed descriptions of the data. The law, distribution, and standard of hidden data can be discovered from a large amount of data. The application of cluster analysis is very extensive. From a business perspective, rationality analysis helps capitalists find a variety of potential customers from their core customer groups, create new market models, and study different consumer groups. In the field of biology, cluster analysis helps biologists use it to understand the classification, genetic analysis, and collective genetic structure of animals and plants. In addition, in cluster analysis, you can also classify documents on the Web to obtain information.

The data matrix is actually a relational table. The rows in the matrix represent a data object, and the columns in the matrix represent an attribute of the object. Suppose there are n objects in the data set, $A = (a1, a2, \ldots, an)$, and each object has m attributes, then this relational table can be expressed as an $n \times m$ matrix:

$$\begin{bmatrix} a_{11} \ \Lambda \ a_{1m} \\ M \quad O \quad M \\ a_{n1} \ \Lambda \ a_{nm} \end{bmatrix} \tag{1}$$

Among them, aij represents the j-th attribute value of the data object ai, $0 < i < n$, $0 < j < m$.

Data specification standardization:

$$A_{ij} = \frac{a_{ij} - \bar{a}_i}{\sigma_j} \tag{2}$$

Maximum value normalization:

$$A_{ij} = \frac{a_{ij}}{a_{j}\text{max}} \tag{3}$$

Mean normalization:

$$A_{ij} = a_{ij}/\bar{a}_j \tag{4}$$

The application of data mining technology in nursing work has demonstrated its theoretical foundation and application characteristics. It can mine, integrate, collect, and store the information needed by nursing staff, and finally give play to the advantages of big data and exert great Effect.

3 Research Experiments Based on the Impact of Big Data on Nursing Work and Application Prospects

This experiment is mainly based on the impact and application of big data on nursing work. Through experiments, explore the changes and impacts of the big data era to nursing work, and the prospect of using big data in future nursing work. This article divides the experimental subjects into experimental group and control group, and through comparative analysis, it is concluded that the convenient effect of big data on nursing work. For nursing staff, big data has significantly improved work efficiency, and can quickly and comprehensively integrate, analyze and process patients' nursing information to realize nursing modernization and informatization.

3.1 Experimental Research Objects

This article adopts the convenient sampling method and voluntary principle, and selects 45 inpatients from the cardio-cerebrovascular department of a Chinese hospital as the 24-h nursing work dynamic data collection source.

Inclusion criteria: (1) 18–65 years old, no gender limit; (2) Self-care, with certain cognitive ability, and proficient in language and written expression; (3) Patients without serious life-threatening diseases; (4) The clinical diagnosis of traditional Chinese medicine or western medicine belongs to patients with cardiovascular and cerebrovascular diseases; (5) The patients themselves have known disease diagnoses; their mental and memory are normal, and they voluntarily participate and can cooperate with this study. (6) The patient agreed to participate in this experiment and signed an informed consent form.

Exclusion criteria: (1) Cases that did not meet the inclusion criteria and were mistakenly included, as well as cases that met the inclusion criteria but did not undergo this experiment after being admitted, need to be excluded; (2) Subjects have poor compliance and fail to follow the prescribed protocol Cases under treatment, etc. The statistical analysis should be combined with the actual situation. If adverse reactions occur, they should be included in the statistics of adverse reactions; (3) The rejection and dropout rate should be <20%; (4) The age is less than 18 years old or more than 65 years old; (5) Yes Patients with severe heart, brain, liver, and kidney failure; (6) patients with severe malnutrition or unable to take care of themselves; (7) patients with mental illness and those who have no insight, cannot express accurately, or cannot write; (8)) Patients with infectious or infectious diseases; (9) Patients with brain trauma or skin trauma on the forehead, carbuncle, furuncle, etc.

3.2 Experimental Research Tools

Test equipment: a computer, a printer, a bottle of 75% ethanol, a sterile cotton ball, a sufficient data collection scale, a CRF form, a neutral black water-based pen, a wiring board and a mercury thermometer after unified calibration. The measurement error of the selected clinical thermometer is within ± 0.2 °C. The timer included in the receiver of the CIM instrument is synchronized with the network as the unified data collection time.

Experimental group: System used: real-time intelligent health monitoring system (hereinafter referred to as CIM), specification model: CD01.

Control group: Method one: traditional mercury thermometer, artificial counting (pulse, respiration); method two: ECG monitor.

3.3 Research Methods

This clinical research is aimed at the application research of the CIM system on the collected data in the clinical nursing application process. Two recorders were selected to complete the clinical 24-h continuous dynamic data observation data recording work in accordance with the clinical experimental procedures. Finally, apply research on the collected data and compare the efficiency with traditional nursing.

3.4 Statistical Processing Methods

First, enter the data into the EXCEL form through the double entry method, and then import it into the SPSS16.0 software for statistical analysis and description. Then, compare it with the big data cloud storage platform to analyze its accuracy, comprehensiveness, and timeliness.

4 Based on the Analysis of Big Data on Job Prospects and the Impact of Care

The research and analysis of this article are mainly to discuss the experimental results of the experimental group and the control group. By observing the authenticity of the data during the data collection, the doubts about the accuracy of the input and the link of the data flow transmission in the system operation mode, comparison. The experimental results of the two groups to highlight the advantages of the experimental group.

It can be seen from Table 1 and Fig. 1 that through data analysis and clinical consistency evaluation, these three different measurement methods can completely replace each other in terms of measurement accuracy. However, nursing work and traditional Compared with the surveying and mapping methods of methods 1 and 2, the time efficiency is increased by 76.20% and 72.02% respectively. Compared with traditional methods, big data technology has the following two outstanding advantages: Advantage 1: Intelligent automation, standard dynamic data collection in information collection, and standard collection, unified access and transmission, data sharing, and unified service are highlighted. Advantage 2: It shortens the operation time of nurses' manual measurement, reduces the error rate of the data input system, reduces the flow of vital signs measurement and drawing, and lays a solid foundation for the realization of scientific nursing management and intelligent decision-making.

Among the 45 patients studied in this article, 42 participated in the satisfaction questionnaire and 30 medical staff participated in the questionnaire. From the results of this experiment, it can be seen that Table 2 and Fig. 2 very well show that the nursing work under big data technology has been highly evaluated, and the overall satisfaction

Table 1. Experimental results

Serial number	Project	Measurement method/time	Drawing method/time	Total time
1	Traditional way 1	10.73	6.08	16.09
2	Traditional way 2	10.73	3.58	14.03
3	A new way based on big data	2.22	1.08	4

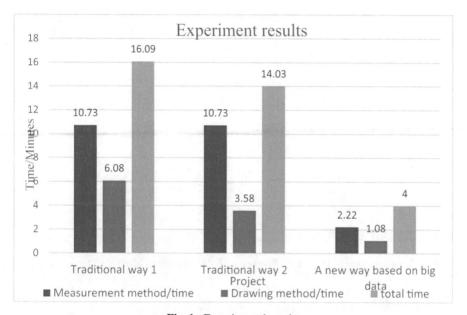

Fig. 1. Experimental results

of patients with it has reached 85.3%. Nursing professionals are the application satisfaction of this technology reached 93.6%. This shows that the development of big data technology has brought great changes and influences to nursing work, mainly to improve the work efficiency of nursing staff, to bring a more comfortable nursing environment to patients, and to fundamentally change the original. If there are defects in the nursing system, rebuild a complete and comprehensive nursing work system.

Table 2. Satisfaction of patients and nurses with the application of big data technology

Object	Very dissatisfied	Not satisfied	General	Satisfaction	Very satisfied	Total satisfaction
Patient	3.48	12.75	14.97	42.86	27.50	85.3
Nursing staff	4.45	7.98	5.59	45.72	25.81	93.6

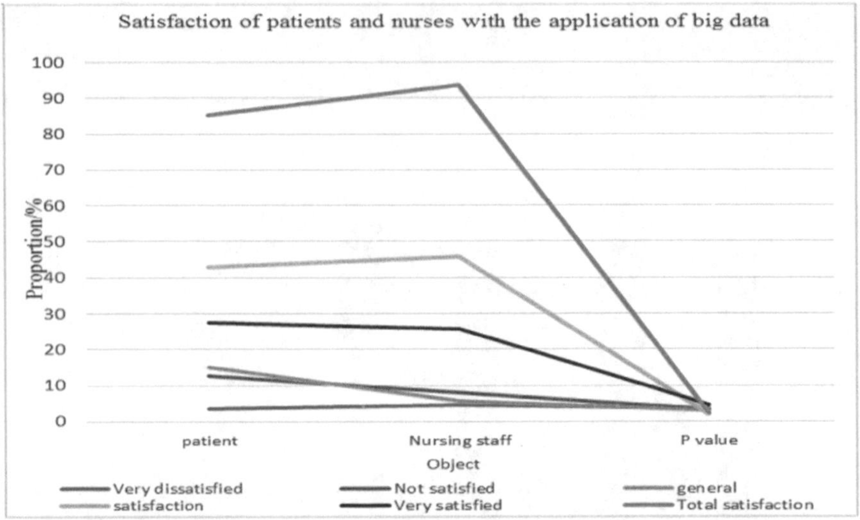

Fig. 2. Satisfaction of patients and nurses with the application of big data technology

5 Conclusion

This article is mainly based on the research on the impact of big data on nursing work and its application prospects. In the research of this article, a variety of methods such as literature data method, experimental analysis method, mathematical statistics method and so on are used to conduct a comprehensive analysis of nursing work. The innovation of this article lies in the application innovation of big data high-tech in nursing work, which is a brand-new breakthrough for nursing work. The shortcomings of this article are that the application of big data technology is not mature enough, and the measurement standards of the technology need to be improved. In the future, the application prospect of nursing work based on big data will become more and more extensive, and the nursing effect will be better.

References

1. Black, B.S., Finucane, T., Baker, A., et al.: Health problems and correlates of pain in nursing home residents with advanced dementia. Alzheimer Dis. Assoc. Disord. **20**(4), 283–290 (2016)

2. Presti, R.C.: The flipped learning approach in nursing education: a literature review. J. Nurs. Educ. **55**(5), 252–257 (2016)
3. Pytka, D., Czarkowska-Paczek, B.: Cognitive function is a prognostic factor for mortality of nursing home residents during a 3-year observational period. Dement. Geriatr. Cognit. Disord. Extra **10**(3), 163–171 (2020)
4. Morrow, M.R., Landstrom, G.: Finding a better way through discovery, theory, and nursing's pact with society. Nurs. Sci. Q. **34**(1), 33–38 (2021)
5. Mohan, D.A.: Big data analytics: recent achievements and new challenges. Int. J. Comput. Appl. Technol. Res. **5**(7), 460–464 (2016)
6. Zaharia, M., Xin, R.S., Wendell, P., et al.: Apache Spark: a unified engine for big data processing. Commun. ACM **59**(11), 56–65 (2016)
7. Obermeyer, Z., Emanuel, E.J.: Predicting the future - big data, machine learning, and clinical medicine. N. Engl. J. Med. **375**(13), 1216–1219 (2016)
8. Specht, D.: The data revolution. big data, open data, data infrastructures and their consequences. Med. Cult. Soci. **37**(7), 1110–1111 (2016)
9. Passini, R., Pigot, H., Rainville, C., et al.: Wayfinding in a nursing home for advanced dementia of the Alzheimer's type. Environ. Behav. **32**(5), 684–710 (2016)
10. Leblanc, K., Baranoski, S., Gloeckner, M.: Am. J. Nurs. **116**(11), 218–220 (2016)

Multi-dimensional Expansion of China's Economic Industry Technology in the Age of Big Data

Minglei Liu[1](✉) and Xiaoli Wang[2]

[1] School of Economics and Management, Fuzhou University of International Studies and Trade, Fuzhou 350000, Fujian, China
[2] Nanchang Institute of Science and Technology, Nanchang 330000, Jiangxi, China

Abstract. At present, with the continuous development of big data technology, global data shows a trend of rapid development. Traditional data processing methods can no longer adapt to the development and challenges of the new era. Technology needs to be constantly updated, and advanced information technology can improve the value of data. It is publicly stated that the multi-dimensional expansion of big data technology is widely used in various departments, promoting the progress and development of the industry, and China's economic development has entered a new era. The multi-level development of multi-dimensional expansion of big data technology provides new opportunities for economic statistics. Statistics departments should follow the development trend of the new era, and bring high-quality and latest multi-level technology into the work of economic departments in the new era, in order to improve work efficiency and serve the realization of high-quality development of China's economic industry. The scientific application of high data technology can fully improve the production efficiency and ensure the functional compatibility and integration of economic industry. However, the development of domestic economic industry is a kind of situation. In this case, different industries do not interfere with each other and operate independently, which seriously limits the multi-dimensional expansion of a large number of data technology. Therefore, in the era of massive data dissemination, China's economic industry should deepen the research and development of relevant technical knowledge, and help China's economic industry carry out innovation and management update. Let China's economic sectors help each other and form a huge industry group, so that China's economic development can be at the forefront of the world.

Keywords: Multi-dimensional · Big data era · Chinese economy · Industry

1 Introduction

Due to the rapid development of society and the continuous research and development of science and technology, various technological changes have begun. Among them, after the cloud and network, a large amount of data updated in the field of information technology has been popularized in all aspects of life, especially in the financial field [1]. Data based industry, looking forward to the current situation, social networks and the Internet can also grow steadily and develop rapidly. With the development of big data technology, data types of different departments can be increasingly rich, and data collection channels can also be continuously expanded [2]. In the information age, the rapid development of the Internet has enriched the channels for collecting big data, and a variety of social networks and wearable devices have become the main part of a large number of additional data resources. In addition, the development of massive data storage, computing and analysis technology provides support for high data applications [3].

The so-called big data mainly involves data integration. When the previous software was used for working time, it could not process data reasonably. On the contrary, this process is optimized in a reasonable way and improves the decision-making ability and a large amount of information resources [4]. For the multi-dimensional expansion of big data technology, it is not only to deal with a large number of data information, but also professional data processing. They are inseparable from the aid of multi-dimensional structure, and their advantage lies in their multi-dimensional structure [5]. And data, through virtualization, cloud base and other operation technology, the method of statistical analysis of big data has a very high practical effect. This method can not only optimize the business structure of enterprises, improve management efficiency, but also enhance the overall strength of enterprises. Provide enough power for the development of employees, and help enterprises focus on the fierce market competition [6].

"Real economy industry" refers to the material production industry, which uses various resources to produce goods and services to meet the needs of the people. It mainly includes primary and secondary industries and some tertiary industries [7]. In terms of national income accounting, it mainly includes agriculture and raw materials. Industry, light industry, equipment production, electricity and water supply, construction, wholesale and retail trade, hotel business and public catering, transportation and warehousing [8]. Scholars at home and abroad believe that the real economy is a concept in line with the macroeconomic virtual economy. At present, the share of virtual economy is gradually increasing [9]. The development of financial industry is more and more oriented by the rotation of virtual economy and real economy. The connection cycle in the real economy is gradually decreasing, the structure of capital distribution and service provision is gradually losing its importance, and the financial support for the real economy is gradually decreasing. The transformation and modernization of China's real economy has encountered bottlenecks in financing [10].

2 Algorithm and Method

2.1 Binary Confusion Matrix Algorithm

From the dichotomous confusion matrix, the following evaluation indexes are derived TPR (true positive rate): used to describe the proportion of the number of positive cases correctly classified in the total number of true positive cases. The true positive rate is also the recall rate. The calculation formula is as follows:

$$TPR = TP/(TP + FN) \tag{1}$$

FPR (false positive rate): used to describe the proportion of the number of false positive samples in the total number of true negative samples. The calculation formula is as follows:

$$FPR = FP/(FP + TN) \tag{2}$$

Accuracy: refers to the proportion of the number of correctly classified samples in the whole sample. The higher the accuracy is, the more accurate the prediction is. The calculation formula is as follows:

$$Accuracy = (TP + TN)/(TP + FN + FP + TN) \tag{3}$$

Precision: the proportion of the number of positive cases correctly classified to the total number of positive cases predicted. The calculation formula is as follows:

$$\Pr ecision = TP/(TP + FP) \tag{4}$$

Recall: the proportion of the number of correctly classified positive cases to the total number of true positive cases. The calculation formula is as follows:

$$\operatorname{Re} call = TP/(TP + FN) \tag{5}$$

The best expectation is that both precision and recall are high, but these two indexes are mutually exclusive. Therefore, a compromise evaluation index F1 measure is proposed. The higher the F1 measure is, the better the classification effect of positive samples is. The calculation formula is as follows:

$$F1 - Measure = (2 \times \operatorname{Re} call \times \Pr ecision)/(\operatorname{Re} call + \Pr ecision) \tag{6}$$

2.2   Establish a Echelon of High-Quality Economic Talents

Domestic big data technology research is still in its infancy, and there is still a huge gap between product integration. In order to overcome this limitation, we cannot do without the strength of excellent economic talents. Combining big data technology with economic departments puts forward higher requirements in terms of talent quality and comprehensive ability. Relevant personnel should pay attention to the real-time development of the industry, master advanced working technology, and learn to use advanced technology to promote the development of enterprises. Build and modernize the economy. Therefore, the economic industry and enterprises should pay attention to the development of economic talents, arrange regular training activities for data operators, maximize the business level, create a talent team with broad development space, absorb more complex economic talents for economic departments, and integrate high data technology and innovative technology to provide human resources. The transformation of company structure is a prerequisite for the integration of big data technology, and it is also the trend of market economy development. Due to the advanced technology, the economic departments adjust the structure according to the actual needs and the current situation of the industry, reduce the basic turnover investment and update the corporate governance system to adapt to the trend of the times.

2.3 Risk Management

Enterprises will face many risks in the process of operation, and a lot of data information will be generated in the process. The use of big data statistical method helps enterprises to find the abnormal operation in time, and make further analysis on the abnormal point, so as to confirm the problems of enterprise operation and make more accurate decisions for senior managers. Because the management risk of Finance and operation is directly related to the long-term development of the enterprise, we need to attach great importance to it. Risk management is an important part of the effective application of the data analysis technology. On the one hand, it can improve the operation structure of the enterprise, improve its safety index, and provide more accurate data for the capital operation and long-term operation of the enterprise. On the other hand, it can help enterprises find ways to solve problems, so as to solve problems faster and more effectively and reduce the impact on the sustainability of enterprise operation. Taking an enterprise as an example, when using this kind of analysis method for risk management, the focus is on overall planning. For the part with large fluctuation range of data, senior managers need to optimize and integrate according to relevant influencing factors, and carry out coordination analysis and corresponding adjustment.

3 Model Establishment

3.1 XGBoost Algorithm Model

Algorithm principle

After the training, we will get several subtrees. The process of using the model to predict the result of a sample is based on the characteristics of the sample. After the input to each tree, it will be output at a node of the tree. The output value is the result of each tree, and then the sum of all the subtree results is the final prediction value.

$$\hat{y}_i = \phi(X_i) = \sum_{k=1}^{K} f_k(X_i), \, f_k \in F \tag{7}$$

$$F = \left\{ f(x) = \omega_{q(x)} \right\} \left(q : R^m \rightarrow T, \omega \in R_T \right) \tag{8}$$

The objective function of xgboost can be expressed as:

$$Obj = \sum_{i=1}^{n} l(y_i, \hat{y}_i) + \sum_{k=1}^{K} \Omega(f_k) \tag{9}$$

$$\Omega(f) = \gamma T + \frac{1}{2}\lambda \|\omega\|^2 \tag{10}$$

Compared with other machine learning methods, xgboost uses second-order Taylor expansion to approximate the objective function of the model.

$$f(x + \Delta x) \cong f(x) + f'(x)\Delta x + \frac{1}{2}f''(x)\Delta x^2 \tag{11}$$

The above formula is the second-order Taylor expansion of the function at point X. the original objective function contains the function as a parameter and cannot be used

The traditional optimization is directly carried out in Euclidean space. As mentioned above, the xgboost model is trained in the form of addition. A function is added to each tree to make $\hat{y}_i^t$ the predicted value of the ith instance in the t-th iteration. Therefore, based on the original objective function, the objective function is transformed into the form of formula (11)

$$Obj^{(t)} = \sum_{i=1}^{n} l\left(y_i, \hat{y}_i^{(t-1)} + f_t(x_i)\right) + \Omega(f_t) \tag{12}$$

Then, the objective function of formula (12) is expanded according to the second-order Taylor expansion to get formula (13)

$$Obj^{(t)} \cong \sum_{i=1}^{n} \left[l\left(y_i, \hat{y}_i^{(t-1)}\right) + g_i f_t(x_i) + \frac{1}{2} h_i f_t^2(x_i) \right] + \Omega(f_t) \tag{13}$$

$$g_i = \partial_{\hat{y}^{(t-1)}} l\left(y_i, \hat{y}^{(t-1)}\right) \tag{14}$$

$$h_i = \partial_{\hat{y}^{(t-1)}}^2 l\left(y_i, \hat{y}^{(t-1)}\right) \tag{15}$$

From the above formula (12)–(15), we can see that the training method of xgboost can be understood as training multiple trees, and synthesizing the results of multiple trees to get the final result, which means that the calculation process between different trees will not interfere with each other, so in the calculation process, multi-core processors can be used to carry out the training. In fact, the design of xgboost is implemented in this way.

4   Evaluation Results

4.1   Investigation Results

From the data analysis in Fig. 1, we can see that there are still big problems in the current risk management, and there are not a few executives who are not satisfied with the current situation of risk management. Only a few executives are satisfied with the current

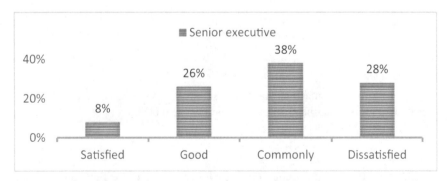

Fig. 1. Is risk management satisfactory

situation of risk management. Most of them think that the current risk management is good or not bad, but relatively satisfied or average. From the data in the Figure, it can be seen that the risk management of the economic industry can not make the majority of executives realize the advantages of multi-dimensional expansion of China's economic industry technology in the era of big data, so we should strengthen its promotion, so as to better develop China's economic industry.

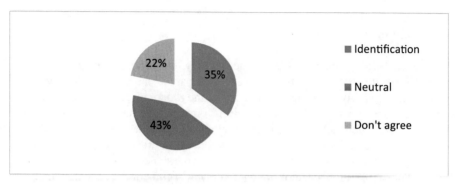

Fig. 2. Is it feasible to establish a echelon of high quality economic talents

From the data analysis in Fig. 2, it can be seen that most of the people have not expressed their views on the establishment of the echelon of high-quality economic talents. The existence of this problem has affected the development of enterprises, and these enterprises may stagnate or even regress, which will have a great impact on China's economic industry. Another quarter of the people do not agree, which is also a big problem. Therefore, to solve these problems, we should strengthen publicity, let them see the advantages of establishing a high-quality economic talent echelon, and let them identify with this matter. Only in this way can Chinese enterprises develop better, thus reflecting the good development of China's economic industry.

A survey of 100 people's opinions on risk management was conducted, in which the ratio of male to female was 1:1.

Table 1. Are male and female students satisfied with risk management

Gender	Degree		
	Satisfied	Commonly	Dissatisfied
Male	19	26	5
Female	7	23	20

According to the data of the questionnaire survey in Table 1, there is a gender difference in the degree of satisfaction with risk management between men and women. Boys are relatively satisfied with risk management than girls. Most of the people in the survey are generally satisfied with risk management. There are many girls who are not satisfied with risk management. Therefore, gender will have a very different degree of satisfaction with risk management, which is also an important factor we learned from the survey results. Therefore, in the construction of risk management, we should consider the ideas of some girls, which will be of great help to the application and development of China's economic industry technology in the era of multi-dimensional expansion of big data.

5   Conclusion

With the multi-level spread of big data technology, the economic industry urgently needs to innovate and update its resources. As a part of the system work, we must comprehensively analyze the current situation of the economic industry and the degree of mastering modern technology, and actively promote and strengthen international cooperation. Research multi-dimensional development of big data technology, and strive to improve the basic ability of economic enterprises, in order to achieve the healthy development of the economic industry. In addition to the above points, it is also important to expand the application value of big data in many aspects of the economic industry. Multi dimensional big data technology provides a new tool for economic statistics in the new era. Making full use of multi-dimensional big data technology to carry out economic statistics and scientifically evaluate the statistical results is conducive to optimizing and adjusting the structure and efficiency of economic development, and provides strong power support for the realization of high-quality economic development in the new era. If we want to achieve development in the field of economic management, we can not leave the support of multi-dimensional expansion of big data statistical analysis method. Therefore, enterprises in various industries should pay more attention to this technology and make full use of it in their own operation, so as to promote the comprehensive and stable development of enterprises and make the development of China's economic industry better.

References

1. Tang, C., Zhang, L., Zhang, Y., et al.: Bidirectional satellite communication under same frequency transmission with non-linear self-interference reduction algorithm. IET Commun. **12**(1), 52–58 (2018)
2. Yuanyu, Z.: Generation of 3rd order intermodulation products in RoF system and its use in inter-satellite communication. Int. J. Chem. Chem. Eng. **8**(1), 81–86 (2018)
3. Gladyshev, A.B., Dmitriev, D.D., Ratushnyak, V.N., et al.: Stand for measuring the directional characteristics of antennas of satellite communication systems. Spacecr. Technol. **4**(4), 201–208 (2020)
4. Kan, T., Jeong, B., Susukita, H., et al.: Experimental results of seasonal variation of shadowing by Ka-Band mobile satellite communication. Trans. Jpn Soc. Aeronaut. Space Sci. Aerosp. Technol. Jpn. **18**(6), 363–368 (2020)
5. Akpakwu, G., Silva, B., Hancke, G.P., et al.: A survey on 5G networks for the Internet of Things: communication technologies and challenges. IEEE Access **5**(12), 3619–3647 (2018)

6. Robinson, H., Al-Freih, M., Kilgore, W.: Designing with care: towards a care-centered model for online learning design. Int. J. Inf. Learn. Technol. **37**(3), 99–108 (2020)
7. Jakowluk, W., Wiercz, M.: Application-oriented experiment design for model predictive control. Bull. Pol. Acad. Sci. Tech. Sci. **68**(4), 883–891 (2020)
8. Rao, J., Vrzic, S.: Packet duplication for URLLC in 5G: architectural enhancements and performance analysis. IEEE Net. **32**(2), 32–40 (2018)
9. Gordeev, B.A., Ermolaev, A.I., Erofeev, V.I., et al.: Model-based design of magnetorheological hydromounts. Proc. Hig. Educ. Inst. Mach. Build. **10**(727), 13–25 (2020)
10. Bekimetov, A.F.: Modelling and analysis of vivaldi antenna structure design for broadband communication systems. Acta Turin Polytech. Univ. Tashkent **9**(4), 2 (2019)

Modern Digital Technology Assisted Innovative Design of Chinese Knot Button Modeling Art

Yu Zhang[✉]

Jiangxi Institute of Fashion Technology, Nanchang 330201, Jiangxi, China

Abstract. With the continuous popularization of digital technology, the digital and intelligent working mode has been realized in the field of clothing design. Designers combine the traditional design elements with modern product design by using digital technology, and integrate the traditional cultural connotation into the product. Chinese Knot Button as a representative of China's traditional art, its modeling unique, the traditional modeling elements and popular elements combined, the use of digital virtual technology to present the design effect, can better grasp the product design of the appearance of the orientation. The main methods of integrating Chinese Knot Button modeling art into product design by using digital software technology include: extracting Chinese Knot Button modeling features, using deconstructive design techniques, and using digital software for design output; Analyze the process of simulation design with PS design software.

Keywords: Digital technology · Virtual design · Chinese Knot Button modeling art · Traditional culture · Decoration design

1 Introduction

Chinese Knot Buttons as one of the elements of traditional clothing, carrying the essence of traditional Chinese culture, its beautiful shape, both functional and decorative, has a unique form of beauty, shape structure prominent cultural connotations, fully reflect people's good yearning for life and blessing [1]. In order to inherit Chinese Knot Button culture, it is necessary to imperceptibly integrate into modern people's life products, which can start from the diversified creative design of Chinese Knot Button modeling. In the design creation, design effects should be repeatedly constructed and continuous attempts should be made to achieve the combination of modern aesthetics and traditional connotation. Digital software technology is needed to carry out effective virtual design, accurately find the orientation of design, promote product research and development, and save design and research costs [2].

2 Modern Digital Technology Clothing Design Software and Features

2.1 Digital Technology Software for Modern Clothing Design

Designers use computer digital technology to collect digital materials, design the appearance of products, digital production technology, etc. Modern digital technology has become a power factor in the field of design. With the help of modern digital software technology, it can effectively improve the efficiency of design and production, realize virtual design imaging, increase the hit rate of design schemes, establish a comprehensive digital product library, save the cost of design and production links, and promote the development of fashion industry [3].

With the rapid development of digital technology, the current fashion design digital technology software is mainly used Photoshop, CorelDraw, Illustrator software output design intention, to achieve specific design scheme, intelligent clothing CAD/AM is widely used in the clothing production.

2.2 Digital Drawing Technology to Optimize Product Design Performance

Digital software image rendering technology, editing and synthesis technology, greatly enriched the way of thinking design and display of virtual design effect. From the operation tools of product style and shape design to the selection of a variety of digital resources and materials, it provides convenient operation technology for the diversity of design schemes. Digital drawing technology is a means of design expression in the field of fashion design. Photoshop software technology is more prominent in the Angle of design effect rendering production. It breaks through the traditional hand-painted effect, efficiently and accurately shows the texture of the material, and presents the designer's design inspiration with the best virtual effect rendering [4]. CorelDraw software technology highlights the structural modeling of design works, which can accurately and quickly express the structural framework. The exquisite production effect reflects the advantages of digital software technology, reduces the tedious design performance process, accurately shows the shape of the product, and enhances the appearance effect of the product design.

3 The Artistic Charm and Modeling Characteristics of Chinese Knot Button

3.1 The Modeling Implication and Artistic Charm of Chinese Knot Button

Chinese Knot Button from the knot, by the "Sakya knot", "buckle knot" itself contains beautiful, harmonious, auspicious meaning. Traditional auspicious design is often used in the modelling design of Chinese Knot Button, which expresses the implicative and show idiosyncrasy of the Chinese. It reflects people's good wishes for good luck or some kind of wish [5]. There are homophonic metonymy, symbolic meaning, implied meaning and so on. Chinese Knot Button shape is to simulate natural things as the ultimate goal, but more focus on the basis of convey the charm of natural objects, on the basis of formal beauty aesthetic requirements, highly refined, bold choices, often using technique of expression of realism, impressionistic, or deformation, make the representational China buttons modelling is no longer subject to the constraints, and abstract, concise symbols, for forms [6]. Whether from the visual hieroglyphics or from the meaning of the show its artistic beauty, bright and varied cultural colors, full of poetic and pictorial emotion, so that Chinese Knot Button show not only the form of beauty, but also a kind of artistic charm.

3.2 Selection of Chinese Knot Button Styling Themes

Chinese Knot Button in the shape of pictographic themes in people's lives are mainly popular things, often based on animals and plants, animals, auspicious words, etc., to add a kind of original ecological vitality to clothing [7]. Such as: phoenix, crane, "xi" character, "shou" character, panda button, auspicious cloud, bronze ware, swastika pattern, shape pattern, seal and so on. Symbol auspicious good meaning, contains a unique Oriental charm.

3.3 The Modeling Structure of Chinese Knot Button

Chinese Knot Button structure is divided into three parts: buckle, buckle door and buckle flower, buckle and buckle door are tied together, play a functional role, the process is relatively fixed, buckle flower is changing, arbitrary strong. According to the buckle and buckle door where the pattern, can be divided into symmetrical Chinese Knot Button and asymmetric Chinese Knot Button. Symmetry Chinese Knot Button on both sides of the figure is left and right symmetrical shape, to buckle as the center, both ends of the symmetrical shape, constitute the beauty of art; Asymmetric Chinese Knot Button on the left and right sides of the graphics have primary and secondary, weight changes, its modelling lively, three-dimensional and changeable, not subject to the balance, the constraints of examples.

3.4 Digital Representation of Chinese Knot Button Styling Elements

Chinese Knot Button modelling of digital image, can choose Coreldraw digital software to realize the Chinese Knot Button structure details, as shown in Fig. 1, through painting tool set to trace modelling is a key point, through tracing point of leverage to adjust the shape of the modelling, the designer according to trace some leverage properties can be adjusted or smooth type, symmetrical type control lever. Chinese Knot Button modeling to symmetrical type, the production of a good unilateral style modeling, you can copy, mirror shape flip, so as to obtain a complete Chinese Knot Button combination effect.

Chinese Knot Button modeling has classic representative characteristics, designers through digital drawing technology and visual imaging composite technology, can quickly obtain the design effect. In the design process to seek a new breakthrough, through the digital technology of China button modeling for digital resources classification integration, combined with modern digital software technology will be re-designed, so as to obtain a new appearance design and combination of changes [8]. In the continuous design experiments to meet the aesthetic needs of modern people, to achieve the traditional culture endowed with modern fashion products in the market effective dissemination.

Fig. 1. Stroke lever control production China buttons modelling

4   The Application of Modern Digital Technology in the Innovative Design of Chinese Knot Button Plastic Art

4.1   Changes and Reorganization of Chinese Knot Button Styling Elements

PhotoShop software can be used to achieve the effect of product design. The flower modeling of traditional Chinese Knot Button is chosen to express the yearning for natural life. The appearance features of the modeling show hierarchical characteristics, which can be used in the modeling of costume design. As shown in Fig. 2, with Chinese Knot Button modelling element in the design of sunglasses, frame edge is the key of the design adornment, using the pattern of Chinese Knot Button buttons flower modelling and the integration of picture frame edge modelling restructuring, use the pen tool path attribute, will draw the shape of the path are combined to add and subtract, forming streamline design elements, will be China's button shape feature extraction, button at the same time the door, and with glasses of the nasal bridge, combining the structure at the end of the pile head, mirror legs, have the effect of that coincide with frame model element, and the style of the entire design modelling foil.

Through the digital software pen free creation, stroke, filling, rendering, texture, layer overlay and other operation techniques, to achieve the virtual design of the imaging effect. The main body of design is to control the strength of shape change and the harmony of element recombination, making good use of digital software function is the key to assist design.

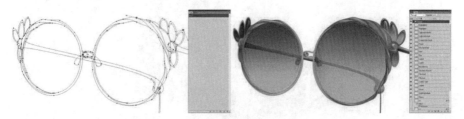

Fig. 2. Digital software to make sunglasses

4.2   Digital Technology Assisted Decoration Design and Texture Performance

Combines the design products of Chinese Knot Button modeling extension with digital printing, metal material, TPU and other modern clothing materials to form a new visual decorative art. To achieve the design effect, the combination of creative design can be carried out with the help of digital patterns and material texture materials.

(1) Firstly, basic color matching and filling are carried out for the shaped design outline. By adding layers and adjusting layer styles, basic light and shade three-dimensional shaping of the shape can be made to form multiple versions of color matching cases;
(2) Select the corresponding digital pattern or fabric material according to the design style tendency, put it in the scope of the design form, adjust the pattern position needed to choose, at the same time, the pattern can be combined and designed to change, forming a variety of versions of the pattern filling effect;
(3) After the outline and pattern decoration are established, the design scheme is basically established. The technical tools such as filter and layer style in the digital software can be used to shape the texture of the design effect drawing, increase the authenticity of the simulation design, and achieve the best effect of the design product, as shown in Fig. 3.
(4) Traditional Chinese Knot Button modeling elements can bring strong Chinese style to clothing products, widely used in modern fashion products. The use of digital software technology in the creative design of Chinese Knot Button modeling elements can quickly form a variety of design schemes, modeling combination arbitrary adjustment, prominent color pattern effect, quick modification of the scheme, renderings rendering simulation is strong. With the help of digital technology and digital resources, designers can better open up design ideas, try various schemes, better find the breakthrough point of design, and finally establish the design works that are most close to the image of design thinking [9, 10].

Fig. 3. Shows the recombination and layer effect of the digital pattern

5   The Conclusion

In the process of Chinese Knot Button inheritance and development has its constraints and bottleneck period, Chinese Knot Button open door way, the realization cost of technology skills, the inherent form of modeling, modeling style and modern people's aesthetic differences and so on. In the current clothing industry, driven by efficient, fast and strong science and technology, with the help of digital software convenient operation to create technology and rendering imaging effect, with the designer's creative inspiration and design proposition, the Chinese Knot Button plastic art for re-innovation design, in addition to the innovative design of Chinese Knot Button itself, such as modeling change, design combination form and composite effect of new materials, digital software technology can also graft its production effect into product design in other fields. The application of digital elements is fast and varied, which is not only conducive to the diversified design changes of Chinese Knot Button plastic art, but also conducive to the integration of traditional culture and contemporary pop culture, so that Chinese Knot Button have both artistic value and market value of products. Therefore, the artistic inheritance of Chinese Knot Button needs to seek new innovative points, on the basis of retaining the traditional cultural connotation and can meet the aesthetic and preferences of modern people, to find a suitable point, to regain the glory of Chinese Knot Button art.

Acknowledgements. Project No. JF-LX-201925, "Research on Inheritance and Innovation of Chinese Traditional Clothing Culture–A Case Study of Chinese Knot Button", funded by Jiangxi Institute of Fashion Technology in 2019.

References

1. Song, Y.: Research on the Creative Application of Chinese Knot Button Elements in Clothing Design. Textile Rep. **12**, 48–50 52 (2019). (in Chinese)
2. Zhou, H.: Research on Digital Representation of Traditional Pattern Inheritance and Innovation. Art Des. (Theory) **2 09**, 82–83 (2020). https://doi.org/10.16824/j.cnki.issn10082832.2020.09.025. (in Chinese)
3. Pu, Y., Zhan, H., Chen, X.: "Innovative design of digital and traditional creations: A case study of creative product design of Chu style lacquerware." Archit. Cult. **10**, 134–137 (2020). (in Chinese)
4. Ji, D.: "Digital virtual visual design analysis build." Grand View Art **35**, 72–73 (2020). (in Chinese)
5. Fang, Y.: "Brief Analysis of Design Aesthetics in Chinese Knot Button Art." Art Sci. Tech. **30**(04), 258 (2017). (in Chinese)
6. Liu, Y.: "Brief Analysis of the Innovative Application of Chinese Knot Buttons in Modern Decoration Design." Mod. Art **11**, 95 (2020). (in Chinese)
7. Shao, L.: "The research of Chinese Knot Buttons and the application of crossover design." Drama Home **16**, 124–126 (2019). (in Chinese)
8. Guo, Q., Li, C., Yan, J.: Research on the application of digital technology in modern manual bag design. In: Proceedings of 6th International Conference on Education, Language, Art and Inter-cultural Communication (ICELAIC 2019) (Advances in Social Science, Education and Humanities Research), vol. 378, pp. 614–616. Atlantis Press (2019)
9. Jiang, L., Ling, J.: The Influence Factors of Computer Fusion Media Technology on Painting Art. J. Phys. Conf. Ser. **1578**, 1 (2020)
10. Wang, Z.: Research on aesthetic design of book binding based on modern digital technology. In: Proceedings of 2019 9th International Conference on Education, Management, and Computer (ICEMC 2019). Francis Academic Press, UK, pp. 893–898 (2019)

Precise Design Research of Regional Cultural and Creative Products Under Big Data

Zhurong Wen[✉]

Xi'an FanYi University, Xi'an, Shanxi, China

Abstract. Through the rapid and accurate information processing, analysis and extraction of big data technology, the integration of big data and cultural and creative product brand design, in-depth analysis of the needs of cultural and creative product design with big data technology, mining effective regional cultural elements, and assisting cultural Creative products carry out precise optimization and upgrading in terms of design positioning, product analysis, design strategy, etc., clarify the brand strategy direction of regional elements and cultural and creative product design, promote the enhancement of brand value, and realize the conversion of new and old kinetic energy integrating regional cultural elements and the times, to promote the development of the regional economy, and strengthen the internal power support of cultural confidence.

Keywords: Big data · Regional elements · Cultural and creative product brand · Precision

1 Introduction

Today's international competition is increasingly manifested as the competition of cultural soft power, and national cultural symbols with lasting vitality and wide-ranging influence are important indicators to measure a country's cultural soft power. The cultural and creative industries play an increasingly important role in national economic development and gradually become a new type of industry that promotes national and regional economic development. Currently in China, the State Council and the State Administration of Cultural Heritage have issued policies 9 times throughout the year to encourage and guide the development of cultural and creative products. On May 16, 2016, the General Office of the State Council forwarded the "Several Opinions on Promoting the Development of Cultural and Creative Products of Cultural Heritage Units", which greatly drove the development of cultural and creative products and also clarified the importance of cultural and creative products as a new industry [1].

Big data is an efficient technology for the development of information technology. The application of big data technology to the practice of creative product design for cultural and creative products enhances the added value of traditional products through cultural and creative product design, and realizes the conversion of old and new kinetic energy that integrates traditional cultural elements with the times. It is the current society to shape the image of regional cultural and cultural and creative brands with the creativity

J. Abawajy et al. (Eds.): ATCI 2021, AISC 1398, pp. 784–789, 2021.
https://doi.org/10.1007/978-3-030-79200-8_117

of lifestyle, industrialization and fashion, to help cultural and creative design and industrial docking and high-quality development, to promote cultural heritage and innovation, and it's upsurge to create high-end brand products with regional characteristics.

2 Big Data Concepts and Characteristics

The rapid development of society has driven the rapid circulation of information technology, and big data has emerged as a product of the high-tech era. Big data, the American research structure Gartner gave this explanation: a collection of data that cannot be captured, managed and processed with conventional software tools within a certain time frame requires a new processing model to have stronger decision-making power, the massive, high growth rate and diversified information assets of insight discovery and process optimization capabilities. McKinsey Consulting's interpretation of the concept of big data is like this, "Big data is an aggregation of data whose data capacity exceeds the ability of traditional data technology to acquire, store, process and apply" [2]. Generally speaking, big data refers to data that exceeds the ability of traditional analysis and processing. Transform from the traditional sampling data collection method to all data collection methods, and summarize more accurate data results in the form of high-speed operation processing. IBM summarizes the characteristics of big data as "5V": Volume, Velocity, Variety, Value, and Veracity. Volume means massive and large-scale. It is spatial in terms of the amount of big data collection, storage, and analysis and processing. The volume of data is about one trillion times the original (quantified calculation according to the data volume: 1ZB = 240 GB = 1,099,511,627,776) [3]. Velocity means high speed, which refers to the speed at which data is processed. Variety means diversity, which means that the range of data covered is particularly wide. Value means low value density, which means that the unit value of data is very low due to the huge scale base. Veracity means authenticity, which refers to the wide and comprehensive data coverage, so the effect brought by big data is more objective, real and accurate.

Based on the advantages of big data in data collection, comprehensive analysis, and system processing, big data technology is applied to the development of regional cultural and creative product design. Scholars such as Brad Brown and Michael Chew in the United States have written "Are you ready" in 2013. For the era of 'big data'?", the first insight into the huge opportunities and challenges of big data to brand marketing [4]. Accurate analysis of the consumption behavior and consumption needs of a large number of consumers through big data forms a valuable behavior guide to provide accurate guidance for the development of cultural and creative product design. The process of big data is to mine, analyze, and organize data to form value information. The final result provides information reference for design decision makers and a theoretical basis for predicting the future development of market economy [5]. These bases have paved the way for the precise prediction of the early market research of cultural and creative products, the precise guidance of design strategies, and the precise execution of design expressions, avoiding the uncertainty of the design direction and ensuring the effectiveness of resource input.

3   Problems Faced by Regional Cultural and Creative Product Brand Design

At present, under the call of national policies, all regions are vigorously developing cultural and creative products, and cultural development has ushered in new cultural opportunities. Represented by the cultural and creative products of the Palace Museum in Beijing, sales exceeded 1 billion in 2016, and 11 museum cultural and creative exhibitions were held in 8 months. For a time, museum cultural and creative products became the focus of public attention [6]. There has been a "cultural and creative wind" in all walks of life across the country, and a series of problems have also arisen in the process of fast cultural production.

3.1   Lack of In-Depth Decomposition of Cultural Elements

The development of regional cultural and creative products should first be based on its own regional cultural characteristics, analyze the characteristics of regional cultural elements, find the differences, advantages and disadvantages between regional cultural elements and other cultural elements, and accurately refine cultural elements. When the homogeneity of the following creative products is serious, first of all, the research based on their own regional culture is not thorough enough, resulting in insufficient differentiation between brands.

3.2   Too Much Emphasis on the Inheritance of Traditional Culture

Regional culture refers to cultural resources accumulated over a certain period of time in a certain area, based on the regional environment, historical background, and humanistic conditions. Therefore, when many brands develop regional cultural and creative products, they are built based on such traditional resources. "Inheritance" is an effective way of brand regionalization. The lack of attention to changes in market demand results in regional cultural and creative products. The "local atmosphere" is too heavy to adapt to market changes.

3.3   The Applicability of Product Development is Too Old

The development of regional products is too much to follow the trend, lack of awareness of deeply digging into customer needs, and fail to update products in time for current changes in consumers' aesthetics and usage habits, and propose appropriate product design ideas, resulting in serious product homogeneity, resulting in brand damage The decline in popularity, reputation, brand sales, market share, and market coverage makes it difficult to meet consumer demand for the deep cultural connotation of cultural and creative products.

3.4   Vague Brand Management Ideas

Many cultural and creative product companies focus on products and neglect the overall product brand management. The brand itself has a cultural appeal. In today's world of information globalization and product homogeneity, when consumers choose a product, they will not only consider whether the product meets their own needs, but also examine whether the brand can show itself Taste and identity. Therefore, the construction of regional cultural and creative products is not only a process of product design, but also needs to focus on the management of the overall brand and shape the brand culture from a macro perspective, rather than simply imitating and plagiarizing.

4   Design and Construction of Regional Cultural and Creative Products Based on Big Data Service

The rapid development of information technology has led to fierce competition for brands. Design is not the designer's own performance. The success of a brand includes consumer analysis, product positioning, brand strategy and many other factors, based on large-scale data feedback provided by big data technology. The basis for guiding the design of regional cultural and creative products, the analysis and analysis of product design-related data from the three levels of behavior, material, and spirituality, while extracting effective regional cultural elements, and constructing accurate design goals and designs Performance, to provide a new model for the design of regional cultural products (as shown in the Fig. 1).

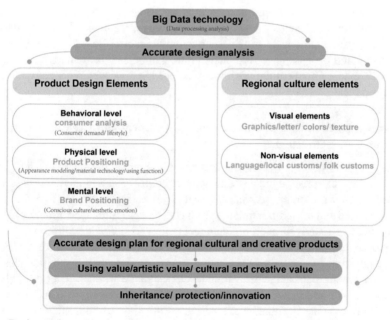

Fig. 1. Design and construction diagram of regional cultural and creative products based on big data technology

4.1 Consumer-Centric Brand Thinking

The design of traditional regional cultural and creative products takes regional culture as the core, and products as the carrier of cultural transmission. Under the background of big data, accurately design the brand thinking concept centered on consumer-centric needs, starting from the consumer's "behavior level", accurately analyzing the current consumer needs, combining consumers' lifestyle and behavior habits, and determining cultural and creative products the positioning ensures that the products are recognized by consumers, and on this basis, the regional cultural elements are integrated. At the same time, the popularization of big data technology has enabled consumers to better obtain large-scale data and information, and the consumer-centric service consciousness is more relevant in cultural and creative products. It emphasizes that "providing service providers and consumers to interact to make services more efficient, effective and needed is a comprehensive field with strong integrity and multi-disciplinary integration" [7].

4.2 Exploring the Essence of Regional Culture to Form a Differentiated Advantage

Analyzing and digging out effective regional cultural and creative elements is a breakthrough for cultural and creative products to get out of the product homogeneity dilemma. Regional culture refers to the cultural traditions that have a long history, unique characteristics, and still play a role in a specific area. It is the cultural expression of ecology, folklore, tradition, and habits in a specific area [8]. Regional culture itself is a manifestation of individual elements. Using big data technology, regional culture can be formed into orderly structure data, from which effective characteristic regional elements can be extracted, and the essence of regional culture can be analyzed and reconstructed to form big data of regional culture. Resources are used in the development of cultural and creative products.

4.3 Precise Positioning and Optimization of Cultural and Creative Brand Management Strategies

Behind a high-value brand is the highest combination of the material and spiritual attributes of an enterprise. It reflects whether an enterprise has a modern business awareness, and whether an entrepreneur has the spirit of pioneering and innovation; the brand is the most imaged and most differentiated product. The most direct performance [9]. Analyze competitive brands through big data technology, and conduct differentiated positioning for brand culture scientifically and accurately. In the cultural and creative product brand design strategy, choose products that conform to the aesthetics and lifestyles of the current consumer groups at the behavioral level, design the products in terms of physical appearance, color and material, and add spiritual culture to them in combination with regional culture on the spiritual level Inheritance, this innovation is not a simple superposition of traditional elements, patterns, colors and other cultural elements, nor is it a modern veneer of traditional culture, but a deep excavation behind cultural elements, making it more culturally connotative and increasing cultural creativity The use value, artistic value and cultural value of products also provide ideas for the inheritance and protection of regional culture and the innovative design of cultural and creative products.

5   Conclusions

The value of big data lies in the formation of new value information through analysis and extraction technology to mine the potential information contained in the data, which can provide reference for decision makers [10]. Under the strategic background of a culturally powerful country, cultural and creative products are of great significance to cultural exploration, research, inheritance and innovation, and provide an indispensable force for traditional culture to promote national strategies. Based on big data technology, it can quickly, comprehensively and truly obtain information on consumer groups, competitive brands, regional cultures, etc., and form an effective regional cultural resource data structure system, which provides a convenient way for the precise design of regional cultural and creative products. When it comes to product brand culture, the relationship between commercial value and cultural value is optimized to maximize the brand value of cultural and creative products, thereby enhancing the sense of identity with regional culture.

Acknowledgements. This work was supported by the 2020 scientific research project of the Shanxi provincial education department, and research on the design path of Shanxi cultural & creative products from the perspective of cultural confidence (No. 20JK0171).

References

1. The State Council. Several Opinions on Promoting the Development of Cultural Creative Products of Cultural Heritage Units. http://www.gov.cn/xinwen/2016-05/16/content_5073762.htm. Accessed 16 May 2016
2. Manyika, J.: Big data: the next frontier for innovation and productivity. pp. 1–4. MGI, USA (2011)
3. Jiayu, Q.: Research on the precision design of mooncake time-honored brand under the background of big data., p. 10. Guangdong University of Technology, Guangzhou (2020)
4. Brown, B., Chew, M.: Are you ready for the era of 'big data'?. pp. 3–9. McKinsey & Company, USA (2011)
5. Jin, X.: Big Data Economics, pp. 5–15. Shanghai Jiaotong University Press, Shanghai (2014)
6. Xin, Z.: Research on Cultural and Creative Product Development of Shaanxi History Museum from the Perspective of Value Chain. Xi'an University of Architecture and Technology, Xi'an (2017)
7. Guosheng, W.: Service Design and Innovation. China Construction Industry Press, Beijing (2015)
8. Baidu Encyclopedia. Regional Culture. Baidu, https://baike.baidu.com/item/regionalculture/4345010?fr=aladdin,2020:01-30
9. Yong, S.: Chinese brands make me happy and worry me. pp. 5–16. Jiefang Daily, Shanghai (2017)
10. Zhaolong, L., Honghe, G., Wei, Z.: Research on the development process of Liaoning Manchu characteristic cultural and creative products under the background of big data. Changchun: J. Jilin Univ. Arts **01**, 85–89 (2019)

Expression of Internet Altruistic Behavior in Big Data Era

Jing Lin[✉]

School of Management, Wuhan Donghu University, Wuhan 430000, Hubei, China

Abstract. The strong social impact brought by the emergency would arouse wide attention from all sectors of society, followed by the overwhelming public opinion. The emergence of big data had not only made the environment of public opinion more complex, but also increased the uncontrollability of public opinion, which had put forward new requirements for the guidance of public opinion in emergency in our country. Four forms of internet altruistic behavior in the era of big data were proposed in this paper, as well as the internet altruistic behavior that netizens tend to use. It was of great significance for improving the level of public opinion guidance in emergency and preventing the resulting public opinion crisis by using big data thinking and technology to guide the occurrence of altruistic behavior on the network.

Keywords: Internet altruistic behavior · Behavioral expression · Big data era

1 Altruistic Behavior

In the 19th century, the word "altruism" was first put forward by Auguste·Comte, an French positivist philosopher and sociologist, in the discipline of ethics [1], which meant as a selfless act without expecting anything in return. Altruistic behavior had been widely concerned by researchers and had become a common concern of many disciplines. The research on altruistic behavior had involved many fields such as biology, psychology, economics, etc., meanwhile, children and students had been selected as the mainly research objects with the research methods including observation, experiment, questionnaire and other forms. The neural basis of altruistic punishment in economic activities had been explained not only from the perspective of physiology by foreign scholars, but also from the perspective of gene-culture and evolutionism. In recent years, foreign scholars had payed more attention to the biological mechanism of altruistic behavior and the relationship between it and economic decision-making under different punishment situations. However in China, there were not many articles or researches on prosocial behavior before the 1990s [2], especially on altruistic behavior. In the future, the focus of altruistic behavior will be closely related to psychology and economics, meanwhile, prosociality, altruistic punishment and internet altruism may become the important research topics.

J. Abawajy et al. (Eds.): ATCI 2021, AISC 1398, pp. 790–795, 2021.
https://doi.org/10.1007/978-3-030-79200-8_118

2   Internet Altruistic Behavior

As a type of altruistic behavior, Internet altruistic behavior is the delayed manifestation of actual activity in the network which is not divorced from the real society. Most people who had exhibited altruistic behavior in the real society would maintain this expression method in the network, and the internet altruistic behavior was more inclined to exhibit in social communication, email and text editing [3]. In recent years, some domestic scholars had discussed the concept of internet altruistic behavior through related research. The essence of altruistic behavior in the network was the same as that in the real society, and the main difference between them lied in the environment they occurred. The former occurred in the network virtual environment which caused a certain particularity compared with the traditional altruistic behavior.

The research on network behavior under public crisis events in China mainly focused on the negative performance of internet public opinion, and the research and guidance on internet altruistic behavior was relatively weak [4]. The study on the internet altruistic behavior of public crisis events in the era of big data, especially the empirical research on guiding measures to improve the altruism behaviors, would be helpful to alleviate the psychological pressure of the public in the case of public crisis, purify network environment, make up the limitations of altruistic behavior in reality, and improve the whole quality of Internet users.

3   Expression of Internet Altruistic Behavior in Big Data Era

Since the outbreak of epidemic, people have changed the channel of expressing emotions and opinions to network due to the limitation of contact in reality. During this period, major news media had also made full use of the new media platform to carry out various interactive activities in order to improve the media exposure and the rate of interaction. Overall, the trend of netizens' preference for media interaction was from decentralized to centralized, and moved toward authoritative media gradually [5]. The main reason for this was that the number of new media platforms was large and the credibility was inconsistent.

The platform of CCTV News on Weibo was selected as a domestic authoritative media in this study, and the messages sent by netizens under the news related to the epidemic on this platform from January to December in 2020 was analyzed with the following results:

3.1   Exposure Statistics of News Related to Epidemic

The number of news related to epidemic in the platform of CCTV News on Weibo from January to December in 2020 was shown in Fig. 1 below.

As can be seen from the trend of the curve in the figure, the amount of news related to the epidemic was basically consistent with the actual situation. On this platform, epidemic-related news began to appear in late January in 2020 and reached a peak in February which had occupied almost all the topics in daily news. The theme of news reports mainly focused on the information of local epidemic numbers and the explanation of epidemic prevention and control. Due to the effective control, the proportion of relevant news had decreased from April to July, which mainly focusing on the digital notification and progress of the epidemic in foreign countries. The curve fluctuated slightly in August and since December; scattered cases had appeared in many places, which made the news data rise during this period.

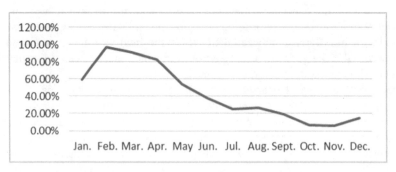

Fig. 1. Statistics on the proportion of epidemic news

3.2   Expression of Netizens' Internet Altruistic Behavior

Through the statistics and analysis of the netizens' comments in epidemic-related news, It could be clearly distinguished between internet altruistic behavior and internet aggressive behavior. Among which, the internet aggressive behavior was mainly manifested as rumors, insults, malicious comments and others. Meanwhile, it was found that internet altruistic behavior could be divided into the following types:

The first type was supportive behavior, which was specifically manifested as verbal expression of approval and support for something, such as answering questions on web pages, or helping, encouraging and supporting others in the network environment. The key words usually included "come on", "support", "agree" and so on.

The second category was directive behavior, which was specifically manifested as the views or opinions on something how to be carried out, such as volunteering to help others and providing technical support. The key words usually included "suggest", "should", "could consider" etc.

The third type was sharing behavior, which expressed the stories that happen to oneself. For example, people shared personal information in a network environment, including experiences, impressions and others. The key words usually included "how it was done in the past", "how to deal with similar situations", and so on.

The last type was reminder behavior, which was specifically manifested as reminding others to strengthen preparedness and protect themselves through warning language. For example, people might maintain the network environment, sending out warnings in time, report illegal information, and expose malicious behavior and so on. The key words usually included "vigilance", "prevention" and so on.

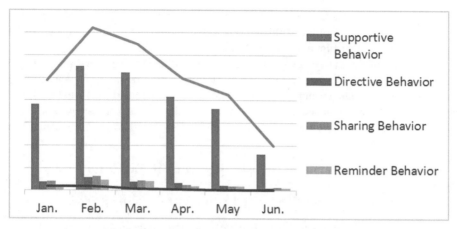

Fig. 2. The expression of internet altruistic behaviors

The analysis of netizens' responses during this period showed that the number of supportive behaviors was significantly higher than that of the others among the four types of internet altruistic behaviors (as shown in Fig. 2). In addition, the number of internet altruistic behaviors was also significantly higher than that of internet aggressive behaviors, which was closely related to the progress of online public opinion control, the improvement of the overall quality of netizens, and the favorable measures of cyberspace purification in China in recent years.

3.3 Impact of Big Data Environment on Network Behavior

According to the financial report of sina weibo in the second quarter of 2020, the number of monthly active users was as high as 523 million, among which the active users of mobile terminal had accounted for 94% and the daily active users had reached 229 million. It had become one of the important sources of information for Chinese people in their daily life. The high degree of freedom and rich information of this platform had made it easier to arouse the hot discussion among netizens.

Due to the rapid development of Internet and the undifferentiated transmission of data, the speed of information spreading had been much faster than that in the past. In addition, the popularity of Weibo, WeChat and news apps made it possible for the public to get instant information through mobile phones. In the era of big data, with the rapid development of "We Media" [6–8], the combination of network and "We Media" had become a trend, which enabled the public to participate in the discussion of events quickly to express attitudes and emotions about emergencies. It could undoubtedly promote the fermentation of public sentiment and accelerate the generation speed. On the other hand, with the advent of big data era, the public had become more and more accustomed to express their ideas and interest appeals on the network platform in face of emergencies. Due to the decentralization of information in the era of new media, these huge amounts of data had not only expanded the public's vision, but also brought confusion of authenticity. The truth that the general public wanted to know was often difficult to be published timely and effectively. Therefore, they would make empirical judgments directly based on the collected information, or blindly follow one of the opinions on the Internet, which might lead to the generation of internet public opinions and social chaos.

Therefore, in the face of crisis or uncertain environment, it is very important to use big data to control the development of internet public opinion. We should take positive behavior guidance and realize the advantages of internet altruistic behavior. By setting up a positive image and guiding positive remarks, we can effectively relieve the psychological pressure and make good use of the double-edged sword of big data.

4 Strategies of Internet Altruistic Behavior in Big Data Era

4.1 Establishing Data Thinking Guided by Internet Altruistic Behavior

The advent of big data era made the concept of traditional data analysis cannot keep up with the pace of modern society. In the face of the challenges brought by big data to public opinion guidance, we need to learn new concept and use big data to guide the Internet altruistic behavior. First of all, it is necessary to pay attention to the comprehensive analysis of massive data, as well as the deep digging the disorganized and irrelevant information to grasp the internal correlation between the information effectively, so as to carry out more targeted public opinion guidance. Secondly, data and information should be opened timely to make the public understand the truth and know how they should act next, which would be conducive to breaking the barrier and the unification and standardization of big data between regions.

4.2 Establishing Diversified Mechanism for Coordinating Guidance

In big data era, the maintenance of cyberspace required the joint participation and efforts of all relevant stakeholders, especially when some Internet users could not show the internet altruistic behavior spontaneously. It is necessary to take the government as the leading role and the network groups as the main body to encourage and guide the netizens to take the responsibility actively, and establish the public opinion guidance system for internet altruistic behavior with the joint participation of multiple subjects, so as to realize the effective response.

First of all, media should strengthen mutual cooperation to ensure the quality of news and pay attention to the language. It is necessary for them to give full play to the role of opinion leaders and create a healthy information environment to make the Internet altruistic behavior widely happened through the influence and credibility of mainstream media. Secondly, in the face of network consensus, the government and media should learn to use the theories related to the internet altruistic behavior to guide the negative opinion actively and appease the public dissatisfaction. Therefore, in addition to punishing negative public opinion or Internet aggressive behavior, it is more important to use the role of positive reinforcement to promote the generation and generalization of Internet altruistic behavior.

5   Conclusions

The emphasis on internet public opinion was disclosed for the first time in 2003 by Xinhua News Agency. The construction and management of network culture had become one of the important contents of China's cultural construction after years of development. The advantages of internet altruistic behavior had been recognized by a large number of scholars. We can correctly guide netizens' online behavior in all stages of public crisis events, relieve the psychological pressure of netizens and clean up the Internet space by using the mechanism of internet altruistic behavior. Therefore, this research had a certain reference value for network culture construction and internet public opinion management of public crisis.

Acknowledgements. This work was supported by the grants from Youth Foundation Project of Wuhan Donghu University (project number: 2020dhsk003).

References

1. Li, G., Che, J.: A review of online public opinion on public emergencies. Doc. Inf. Knowl. **2**, 111–119 (2014). (in Chinese)
2. Xi, J.: Analysis on ethical anomie in public health events – a case study of covid-19 epidemic. Radio TV J. **4**, 119–120 (2020). (in Chinese)
3. Song, L.: Review of Domestic Research on Altruistic Behavior in the Past 30 Years. J. Heilongjiang Inst. Teach. Dev. **37**(04), 89–91 (2018). (in Chinese)
4. Liu, Y., Tao, M.: Research status and trend of altruistic behavior at home and abroad. Psychol. Tech. Appl. **7**(08), 504–512 (2019). (in Chinese)
5. Jordan, J.J., Rand, D.G.: Third-party punishment as a costly signal of high continuation probabilities in repeated games. J. Theor. Biol. **12**, 421 (2017)
6. Rand, D.G., Brescoll, V.L.: Social heuristics and social roles: intuition favors altruism for women but not for men. J. Exp. Psychol. Gen. **145**(4), 389–396 (2016)
7. Xie, B.: The rational model of altruistic behavior. J. Capital Normal Univ. (Social Sciences Edition) **6**, 138–144 (2019). (in Chinese)
8. Yuan, Y.: Public crisis event communication and mainstream media responsibility. China Newspaper Ind. **6**, 10–11 (2020). (in Chinese)

Logistics Supply Chain Management Under the Background of Big Data

Ru Zhao[✉], Guoxin Gu, and Zhihui Yang

Harbin Institute of Information Technology, Harbin 150025, Heilongjiang, China

Abstract. With the rapid development of Internet, Internet of things and other technologies, data information is growing explosively, and the world has entered the era of big data. Big data promotes and optimizes the procurement in the upstream, manufacturing and production in the midstream, sales in the downstream and logistics in the supply chain activities, and promotes the supply chain enterprises to enter a higher-level competitive environment. From the perspective of logistics distribution companies, the management and control of the supply chain is an important factor in the company. Based on this point of view, this paper takes Jiangsu M company as an example to carry out a comprehensive analysis of the cost control strategy under its supply chain management. The main purpose is to find out the problem factors in the cost control of enterprises, and finally put forward relevant countermeasures and suggestions.

Keywords: Big data · Logistics enterprises · Supply chain management · Cost management

1 Introduction

With the rapid development of Internet, Internet of things, mobile computing and cloud computing, the world has gradually entered the era of data economy. All kinds of unexpected big data gradually come into people's vision. With the strengthening of data mining, analysis and application, people's perception of the surrounding environment has begun to play a great role in promoting the development of data into a new era. At the same time, according to the prediction of IDC, an international data company, and China's data output will account for 13% of the world's total, and the proportion will increase to 22% by 2020. Judging from the increasing number every day, the world has entered the era of big data [1].

With the increasing importance of big data, governments and international organizations all over the world have realized the importance of big data and raised it to the height of national strategy. At the same time, China's national strategy of big data is also rapidly implemented: 2017 in the year of the year, the Ministry of industry and information technology proposed the top level design for strengthening the standardization of big data. In the same year, Xi Jinping put forward the idea of implementing the national big data strategy and accelerating the construction of data China in the second collective learning strategy of implementing the national big data strategy. In recent

J. Abawajy et al. (Eds.): ATCI 2021, AISC 1398, pp. 796–802, 2021.
https://doi.org/10.1007/978-3-030-79200-8_119

years, the amount of data generated by the practice of end-to-end supply chain management has increased exponentially. In the current competitive environment, supply chain professionals are trying to deal with huge amounts of data. They are working on new technologies to generate, capture, organize and analyze data to provide valuable insights to the industry. Big data analysis is one of the best technologies to help them overcome their problems. As a disruptive technological change, big data gradually innovates the process and way of enterprise management, organization and decision-making [2]. Therefore, the introduction of big data into the supply chain can break the information barriers between the supply chains, realize data sharing, improve the transparency of the supply chain, and then obtain detailed commodity information. At the same time, based on the mathematical model prediction, the key business is deeply mined and analyzed to grasp its characteristics and characteristics, find the improvement opportunities and optimize them, further promote the interaction of supply and demand information, optimize the costs among the members of the supply chain, and improve the performance of the supply chain [3].

Combined with the operation characteristics of logistics enterprises and the core essence of supply chain management. Therefore, under the background of this industry, taking m logistics company as an example, this paper explores the planning of cost control in the supply chain control of the company, in order to find some ways to control the cost.

2   Related Concepts

2.1   Supply Chain Management

Now some management education institutions (such as MBA) have incorporated the management of supply into the scope of education. The reality of supply chain management is to mobilize all the resources that can be mobilized by enterprises to meet the consumption and use needs of customers to the greatest extent. In order to facilitate understanding, we can try to regard all enterprises participating in the supply chain as a cooperative alliance, and these individual enterprises are just one of these alliances. Then, the process of managing these so-called departments is the so-called supply chain management. The only difference may be that the members of the alliance will change from time to time. The reason for their change is the change of market demand [4, 5].

2.2   Integration of Logistics Supply Chain Management

Market competition always intensifies with the passage of time, and the competition will be more and fiercer. If enterprises want to form their own competitive advantage, the source of profit has to be transferred. Now, in the face of such fierce competition, sticking to the tradition has been unable to make lasting profits. We must start to transfer the source of profits and turn to an efficient enterprise logistics system as the third source of profits [5].

To pursue the degree of integration of the logistics system, the so-called integration refers to taking the enterprise's logistics system as a core, taking the product production company as the starting point, passing through the enterprises responsible for logistics, and finally delivering the products to the consumers through the sales channels. To ensure the systematicness and integrity of this process, this is the high-level form of the logistics industry, with this mode Products can be quickly transferred in the whole supply chain from production to sales, then all node enterprises in the supply chain will be profitable. If the whole society can realize logistics integration, the economic benefits of the whole society can be pushed to a higher level [6].

With the rapid development of the logistics industry, the logistics system has reached a very perfect level. The position of the logistics industry in the whole production chain is very important. It can be said that the leadership also undertakes the role of coordinating each other. The logistics service of the whole society can be contracted completely. Based on the rapid development of logistics integration, enterprises in each node of the supply chain can realize the integration of each other's functions, and form a new decision-making alliance through the complementary advantages between them. Why the United States has been able to maintain economic prosperity in a short period of more than ten years, and it will develop after in-depth analysis, which is inseparable from the fact that the United States strongly advocates the integration of logistics. Therefore, we can draw a conclusion that the efficiency of a society's production logistics is closely related to the level of logistics development [7].

2.3 Composition of Supply Chain Management Cost of Logistics Enterprises

Supply chain logistics always takes logistics activities as the top priority of the whole work. It is necessary to do a good job in the overall coordination of each link of the whole service chain. This coordination also includes the cooperation with cooperative countries, including product suppliers, middlemen, third-party service organizations and customers [8].

In other words, the operation cost of logistics enterprises can be divided into two categories, one is the cost generated by operation, and the other is the cost generated outside operation. The so-called operation cost of an enterprise refers to the cost associated with the products supplied by the enterprise, which includes the cost generated by direct labor, the cost generated by direct consumption of materials, and various expenses generated in the operation process. Correspondingly, non-operating costs represent the additional sales expenses, management expenses and other expenses of enterprises in logistics services.

Among all the costs of the enterprise's operation, all the remaining costs excluding the above costs are the operating costs. This is not an absolute concept. Affected by the circulation link, the cost will change accordingly. The cost of logistics circulation includes the cost of transportation, the cost of product processing, the cost of goods distribution, the cost of packaging, the cost of goods storage and handling [9].

3 Logistics Supply Chain Management Cost Management Practice

3.1 Object of Practice

M Logistics Co., Ltd. was founded in 2000, headquartered in Nanjing, with a registered capital of 20 million yuan. It is a large-scale international transportation enterprise that meets the national business system standards and obtains business qualification. Its subsidiary companies include m Shipping Agency Co., Ltd. and M Shipping Co., Ltd. with 280 employees At present, the enterprise is a comprehensive logistics enterprise integrating international trade and transportation, warehouse direct transportation, transportation agent, Yangtze River and coastal branch transportation, domestic and foreign shipping agent, highway container transportation, bulk cargo transportation, air transportation, import and export customs clearance and inspection.

3.2 Analysis of Practice Objects

M Logistics Company has a large number of industry elites and a strict and scientific enterprise logistics operation system. According to the different division of labor, the company has different departments, such as receiving orders, managing transportation, capital output and so on, to ensure the standardization of enterprise industrial flow.

(1) Mean square error

$$RMSE = \sqrt{\frac{\sum_{(u,i) \in T} (r_{ui} - r'_{ui})^2}{|T|}} \tag{1}$$

(2) Mean absolute error

$$MAE = \frac{\sum_{(u,i) \in T} |r_{ui} - r'_{ui}|}{T} \tag{2}$$

(3) Accuracy

$$\mathrm{Pr}\,ecision = \frac{\sum_{u \in U} |R(u) \cap T(u)|}{\sum_{u \in U} |R(u)|} \tag{3}$$

4 Data Statistics and Analysis

4.1 Cost Structure Analysis

The analysis of the cost structure of M Logistics Company is mainly based on its quarterly economic analysis report. The specific cost structure is shown in Tables 1 and 2.

Table 2 shows that the trend of direct labor costs is not stable. Although there is an upward trend, the overall trend continues to decline. The direct material costs and sales costs show a stable and fluctuating trend, and the trend is basically the same, which also verifies the fundamental characteristics of logistics enterprises. To sum up, the overall cost structure of Jiangsu m Logistics Company is dynamic and stable, and the change is not obvious, as shown in Figs. 1 and 2.

Table 1. Cost structure breakdown of M logistics companies

	2014	2015	2016	2017	2018	2019
Direct labour	269298.33	2659756.31	893719.92	839695.33	353567.06	729495.65
Direct costs	1726926.54	1214907.94	1516706.23	2501395.39	761028.57	1031692.98
Miscellaneous expenses	1342183.11	1575805.20	1673707.19	3861600.29	2524855.59	1320569.78
Indirect labour	1185479.66	1567702.95	977860.74	1226608.03	1603465.15	1448449.78
Selling expenses	123453.27	1460536.96	149675.96	1444325.38	147319.54	110919.13
General expenses	100738.04	9500.8	0	0	3783.62	0

Table 2. Share of costs of M logistics companies

	2014	2015	2016	2017	2018	2019
Direct labour	5.67%	31.33%	17.15%	8.50%	6.55%	15.72%
Direct costs	36.37%	14.31%	29.10%	25.33%	14.11%	22.23%
Miscellaneous expenses	28.27%	18.56%	32.11%	39.11%	46.81%	28.45%
Indirect lab our	24.97%	18.47%	18.76%	12.42%	29.73%	31.21%
Selling expenses	2.60%	17.21%	2.87%	14.63%	2.73%	2.39%
General expenses	2.12%	0.11%	0.00%	0.00%	0.07%	0.00%

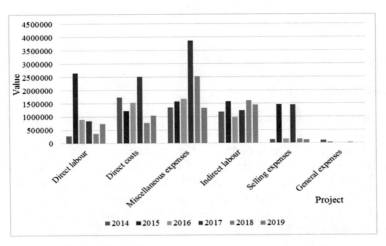

Fig. 1. M logistics company cost structure trend analysis

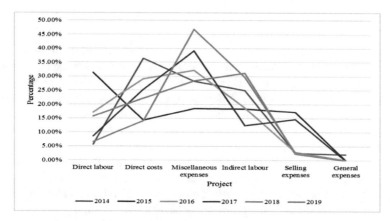

Fig. 2. Trends in the proportion of M logistics company costs

4.2 Cost Control Countermeasures of M Logistics Enterprise under the Condition of Supply Chain Management

(1) Strengthen the trust mechanism among enterprises in the supply chain

There are three types of companies cooperating with logistics companies: material supply companies; product manufacturing companies; product sales companies. To strengthen the cooperation between these companies. And it can implement the sharing of information resources and maintain the synergy. In China, the national government actively helps the logistics companies to reduce operating costs, subsidies and preferential policies for related industries, and helps to study strategies to save operating costs, so as to make the logistics companies develop in the direction of integration. In order to reduce the operation cost, we can simplify the complicated problems. For example, we can use the method of purchasing and unifying the inventory. In the development of logistics companies, mutual trust and communication with strategic partners is an important guarantee for stable development. In the cooperation between companies, the exchange of information is very important, because good exchange of information can make both sides know each other's company situation at all times, so as to carry out more effective cooperation and communication, and solve problems quickly.

The management of Logistics Company's operation cost is an important factor for enterprise development and maintaining market competitiveness. In the cost management, it can quickly improve the company's financial management technology, and make the company want to develop in a more detailed and reasonable direction. Before improving the company's operating cost management, we should have an overall analysis of its management methods, make clear the special significance of cost management in financial management, and put it in an important position of company management. In addition to the above company development strategy level of cost management methods, but also pay attention to the company's internal cost management strategy publicity, strengthen the company staff's understanding of cost management; make every staff realize the significance of cost management to the company and personal development.

5 Conclusion

Based on the situation of logistics supply chain management and the development of logistics industry under the background of big data, this paper takes m logistics company as an example. At the same time, this paper analyzes the cost control of M Logistics Company This paper analyzes the problems in the cost control of logistics companies, and then puts forward countermeasures and suggestions. In the future research process, m Logistics Company should also pay attention to strengthen the cooperation of node enterprises. Logistics companies based on the supply chain can change the service status of logistics with the help of the present form, so that they can give full play to their role in the supply chain. Therefore, we can adjust the operation of the supply chain with the help of cost and service characteristics, adjust the cooperation relationship between various companies, and constantly promote the development of supply, so as to improve the work efficiency, reduce the logistics cost, and maximize the benefits between various supply chains.

References

1. Spillan, J.E., Mcginnis, M.A., Kara, A., et al.: An empirical assessment of logistics/supply chain management in two Latin American Countries. J. Transp. Manage. **26**(2), 7–27 (2016)
2. Zhao, D.: The development trend of logistics supply chain management in Taiwan. Sci. Innov. **4**(6), 311–312 (2016)
3. Kara, A., et al.: An empirical assessment of logistics/supply chain management in two latin American countries. J. Transp. Manage. **26**(2), 7–27 (2016)
4. Wang, G., Gunasekaran, A., Ngai, E.W.T., et al.: Big data analytics in logistics and supply chain management: certain investigations for research and applications. Int. J. Prod. Econ. **176**(3), 98–110 (2016)
5. Rajkumar, C., Kavin, L., Luo, X., Stentoft, J.: Doctoral dissertations in logistics and supply chain management: a review of nordic contributions from 2009 to 2014. Logist. Res. **9**(1), 1–18 (2016). https://doi.org/10.1007/s12159-016-0132-0
6. Harrison, A., Hoek, R.V.: Valuepack: logistics management and strategy: computing through the supply chain/supply chain management: international edition/logistics & supply chain management: creating value-adding networks. Inbound Logist. **31**(7), 146–148 (2015)
7. Lundesjo, G.: Supply chain management and logistics in construction: delivering tomorrow's built environment. Int. J. Phys. Distrib. Logist. Manage. **42**(7), 1–3 (2015)
8. Peterson, M.R., Young, R.R., Gordon, G.A.: The application of supply chain management principles to emergency management logistics: an empirical study. J. Emerg. Manage. **14**(4), 245–258 (2016)
9. Zijm, H., Klumpp, M., Regattieri, A., et al.: Purchasing and supply management (Chapter 4). In: Operations, Logistics and Supply Chain Management, Lecture Notes in Logistics, pp. 45–73 (2019). https://doi.org/10.1007/978-3-319-92447-2

Digital Protection and Development of Intangible Cultural Heritage Gan Embroidery Based on Digital Technology

Lanjian Zeng[(⊠)]

Jiangxi University of Applied Science, Nanchang 330100, Jiangxi, China

Abstract. In the context of the rapid development of digital technology, the Internet and other media are used to strengthen the promotion and protection of intangible cultural heritage, and strive to promote the inheritance and development of China's intangible cultural heritage. In order to maximize the integration of intangible cultural network marketing models and intangible cultural heritage resources, the study will expand the intangible cultural heritage exchange channels through which intangible cultural heritage resources can create more economic and social benefits. Intangible cultural heritage is the quintessence of our traditional culture, the result of the efforts passed on from generation to generation, and the cultural heritage that our young people should inherit. Gan embroidery is the inheritance of Hakka culture. It is a festive costume in festivals. It is also a manifestation of the ingenuity of the Hakka people. It also interprets the profound connotation of Hakka culture with ancient production methods and beautiful patterns. This article summarizes the connotation of Gan embroidery culture, summarizes the development problems faced by the current development of Gan embroidery, and explores the feasibility of Gan embroidery art's innovative ideas and digital redesign in the digital age. The integration and development of Gan embroidery art and modern digital products can awaken the new cultural energy of Gan embroidery and promote the transmission and digital development of Gan embroidery culture.

Keywords: Digitization · Intangible cultural heritage · Gan embroidery · Traditional culture

1 Introduction

Southern Jiangxi is one of the three major Hakka settlements comparable to Fujian Province in the west and Guangdong Province in the east. After years of cultural fusion, the Hakka people based on the traditional culture of the Han nationality, formed a multi-cultural fusion with the She, Yao and other indigenous cultures-Hakka culture [1]. The Sannan, Xunwu, Anyuan and other places in southern Jiangxi are located in remote mountainous areas. There are not only famous Hakka enclosures, but also many Hakka customs and cultural relics. Some Hakka women attach great importance to married life.

Wedding embroidered shoes, embroidered pockets, embroidered hats, etc., are hand-made items worn during the child's time. These embroidered products are made of pure materials, thick and thin craftsmanship, unique shapes, exaggeration and deformation, and colorful. The embroidery patterns are very decorative, the color contrast is strong, and the Hakka style is very strong [2]. Gan embroidery mainly includes embroidered hats, embroidered cloud shoulders and bibs, embroidered pouches, embroidered shoes, embroidered aprons, etc. Gan embroidery is a symbol of women's life and sensibility, and it is a survival art that contains women's ideal colors and auspicious wishes. Regardless of the composition of the pattern and the composition of colors, they all show the attitude of supporting Hakka people's life and the characteristics of pure beauty [3, 4].

Chinese embroidery has a long history and is basically different from the four famous embroidery fabrics such as Su embroidery, which uses silk and silk as the base fabric. Gan embroidery, which is popular in Jiangxi Province, has the following characteristics. The selection of materials ranges from special summer cloth to modern living paper, and the selection of materials is very unique [5]. In particular, the fifth-stage processing inks of burnt ink, thick ink, heavy ink, light ink and clear ink, which are different from other types of embroidery, are simpler and more atmospheric. Gan embroidery has formed a complete fiber processing technology suit from its birth to the present, which provides a good technical environment for the large-scale production of Gan embroidery, and is further developed into standardization [6]. At the same time, with the development of modern large-scale machinery technology, Gan embroidery is also suitable for mass production. Gan embroidery products can accept advanced user definitions and low-cost assembly line production, in order to meet the purpose of different consumer groups [7].

Recently, an important method of inheriting intangible cultural heritage is to spread and protect intangible cultural heritage through a variety of digital new technologies [8]. Internationally, digital exchanges and the protection of intangible cultural heritage began at the end of the last century. Our country's digital communication and protection undertakings not only have an impact on advanced concepts and advanced technologies, but also should have a great impact on the future spiritual, cultural and economic development. The construction of the awakening intangible cultural heritage digital resource database is not accomplished overnight, it takes time to accumulate [9]. Digital imaging technology, digital imaging technology, three-dimensional scanning technology, virtual reality technology, augmented reality technology, documents and other new research results, through search technology and other support, long-term investment, document research to establish a distributed resource database to realize intangible cultural heritage projects. The purpose of recording, preservation, promotion, development, and inheritance of information symbols, strengthens our country's cultural soft power [10].

2 Method

2.1 Establish a Digital Museum

With the rapid development of technology, every class of life has joined the ranks of digital technology, and museum displays are no exception. Based on a variety of interactive experiences such as virtual reality and augmented reality, the digital museum is on the scene. This unprecedented display method has greatly deepened the protection and inheritance of intangible cultural heritage. Visitors can visit all museums to broaden their knowledge, broaden their horizons, understand the unique culture of the region, and use a combination of images, animation, interaction and text. The method of understanding the meaning of intangible cultural heritage greatly conforms to people's concern for culture and improves the public's cultural understanding ability. For museum managers, digital museums are "alive" cultural relics and souls. Through resource extraction and integration, the museum's operating mode is changed, the efficiency of the staff is improved, and the management and storage are convenient. Therefore, it is possible to build a digital museum with the number of characters as the theme, develop digital 3D modeling software, establish an information interaction platform, improve the number of characters data, establish a digital group business unit and other projects, carry out digital promotion of Gan embroidery culture, and attract the public. The interest of Gan embroidery culture, so as to achieve the protection and inheritance of Gan embroidery culture.

2.2 Build a Digital Service Platform for Gan Embroidery Intangible Cultural Heritage and Promote the Development of Intangible Cultural and Creative Industries

The "digital" development of Gan embroidery intangible cultural heritage should grasp the opportunity of the times, relying on the intangible cultural heritage database initially established in Jiangxi, after sorting, archiving, and publishing through a unified digital service platform. The Gan embroidery intangible cultural heritage digital service platform can provide a wide range of Gan embroidery intangible cultural heritage digital resources. It can be stationed in the intangible cultural heritage management department and the intellectual product protection department to conduct intangible cultural publicity and dissemination. It can also attract non-genetic inheritors, cultural and creative enterprises, and private individuals. Capital, etc. to cooperate in development, build a network display and sales platform for intangible cultural heritage related products and services, and develop the "digital" tertiary industry of intangible cultural heritage. Taking advantage of the increasing demand for online consumption, cultural products and services in the era of big data, the intangible cultural heritage of Gan embroidery will be transitioned from "rescue protection" to "productive protection", and the cooperation advantages of resource sharing will be brought into play. The digital resources of cultural heritage are combined with the cultural and creative industries to promote the integration of Gan embroidery intangible cultural heritage into the lives and work of modern people, enhance the vitality and competitiveness of intangible cultural and creative products, and create a characteristic cultural and creative industry development model. To develop the cultural and creative industry of Gan embroidery intangible cultural heritage, it also needs to be combined with short videos, games, film and television animation, virtual reality and other art forms. The creative industry meets the needs of modern people for culture and fashion, and makes intangible cultural heritage radiate new vitality in the contemporary era.

2.3 ElGamal Signature Algorithm

The ElGamal signature algorithm and the ElGamal signature algorithm were introduced at the same time. It is also a solution based on the discrete logarithm problem. The specific description of the algorithm is as follows:

1. Select a public key as parameter K, and a private key parameter as e, among which the largest public prime numbers a, p, calculate the transformation H(m) of the message, where:

$$k = a^e \bmod p \tag{1}$$

2. Perform signature transformation:

$$S(H(m)) = (E_1, E_2), X = a^n \bmod p \tag{2}$$

$$Y = (H(m) - eX)^{-1} \bmod (p - 1) \tag{3}$$

3. Perform signature verification: If the following formula is made to be valid, it can be regarded as a valid signature.

$$K^Y X^Y = a^H(m)(\bmod p) \tag{4}$$

Therefore, after K, X, Y can be substituted accordingly, the resulting expression is as follows:

$$K^Y X^Y = a^{eX+Yn} = a^{rmeX+H(m)-Ex} = a^{rmH(m)} \tag{5}$$

Use LSSS access structure (M, ρ) to encrypt data file M:

$$C3 + x = \beta + \rho\gamma(x) \tag{6}$$

Where ρ(x) is a permutation function, and ρ(x) maps the vector Mx to the attribute i. Encsyme(g1,g1)s(M) is a symmetric encryption scheme, the key is e(g1,g1)s.

$$C0 - Encsym\ e(g1, g1)s(M)\forall x \tag{7}$$

Finally,   the   data   owner   uploads   these ciphertexts C = {∀x, {C0, C1, x, C2, x, C3, x}; (M, ρ)} to the cloud server.

3 Experiment

3.1 Subject

From the perspective of the essence of my country's cultural heritage protection, the most common method currently is to preserve cultural heritage materials through libraries and museums in cities and regions. This method is hoped in the case of underdeveloped digital technology, but what we can see is that the cultural relics and works placed in the storage room are always objects contaminated by history, and are hardly used for cultural exchanges and activities. They have their own cultural charm, are easily affected by the actual environment, require higher maintenance fees and other shortcomings. On the whole, the construction of the "digital intangible cultural heritage" Gan embroidery intangible cultural heritage website has contributed to the inheritance of the Gan embroidery intangible cultural heritage project. At the same time, in order to build the same type of digital virtual museum, new design ideas, new thinking methods and new realization methods were proposed.

3.2 Experimental Method

This experiment analyzes the significance and value of intangible cultural heritage documentary films for cultural communication from the perspective of intangible cultural heritage documentary creation techniques through a large amount of data and documents, and comprehensively and correctly analyzes the current development status of intangible cultural heritage documentaries and their huge. The development potential and the huge benefits brought by the inheritance of Gan embroidery culture. The use of websites as a traditional cultural dissemination method has the characteristics of a large age span and a wide range of people. In order to better study the characteristics and overall framework of users, the experiment aims to target people through the "Gan embroidery intangible cultural heritage questionnaire". The collection of user information provides an information basis for the later overall planning design and model establishment.

4 Results

In order to understand people's cognition of intangible cultural heritage, the survey questions were conducted around the intangible cultural heritage of Gan embroidery. A total of 140 questionnaires were sent out and 106 questionnaires were collected. The following is the specific data analysis of the questionnaire (Table 1):

Among all the questionnaires returned, the majority are young people aged 18 to 30, and most of them are locals in Jiangxi, who have a certain understanding of local culture. According to the "Research Report on the Development of Internet Audition in China", as of December 2018, the number of Internet users in our country has reached 892 million, and the ultra-large-scale Internet usage rate provides a good user base for the digital dissemination of intangible cultural heritage (Fig. 1, Table 2).

Among the users who participated in the survey, 62.26% of users believe that it is necessary to develop and establish intangible cultural heritage websites or apps, and 70.75% of survey users believe that it is necessary to shoot intangible cultural heritage-related documentaries for people to understand intangible cultural heritage, know intangible cultural heritage. The digital display of intangible cultural heritage has huge advantages for cultural inheritance and dissemination, because digital information can be permanently retained in the cloud. However, because digitalization is the deconstruction and reshaping of the original environment of intangible cultural heritage, the problems and contradictions that will arise in it are unpredictable, resulting in a small number of people holding reservations.

Table 1. Distribution of survey personnel

Age	<18 years old	6	5.66%
	18–30 years old	86	81.13%
	30–50 years old	13	12.26%
	>50 years old	1	0.94%
Profession	Governmental agents	9	8.49%
	Corporate personnel	51	48.11%
	Freelancers	34	32.08%
	Retired people	4	3.77%
	Student	6	5.66%
	Other	2	1.89%

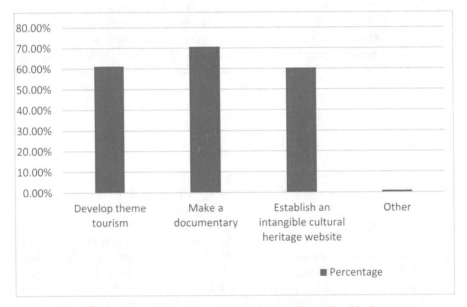

Fig. 1. Ways to protect Gan embroidery intangible cultural heritage

In order to count the number of supporters for the digital protection of Gan embroidery intangible cultural heritage, this questionnaire survey is for the general public in China. 1,500 people were surveyed. A total of about 1,500 questionnaires were distributed and 1,200 questionnaires were returned, including 982 valid questionnaires. Through a questionnaire survey, the number of supporters for the digital protection of Gan embroidery intangible heritage art (Fig. 2).

As can be seen from the statistics in the figure above, 37% are very supportive and 49% are supportive. These two categories together account for 86%. The number of

Table 2. Attitudes and views on the establishment of intangible cultural heritage websites

Is it necessary to establish an intangible heritage site	Do not know	15	14.15%
	No need	25	23.58%
	Necessary	66	62.26%
What do you care about most about intangible cultural heritage sites	Simple page	54	50.94%
	Complete target content	78	73.58%
	Attractive name icon	49	46.23%
	Simple operation	40	37.74%
	Illustrated	21	19.81%
	Audio and video combination	7	6.6%

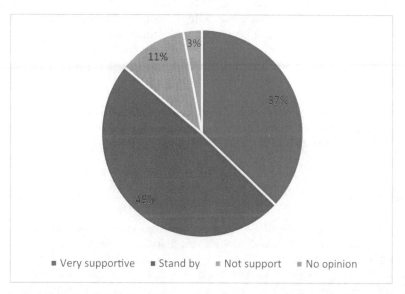

Fig. 2. Number of supporters for digital protection of Gan embroidery intangible heritage art

non-supporters accounted for 11% of the statistics, and 3% of the people did not express their opinions. The two types of people together accounted for 14%. In this challenging investigation process, a lot of gains have been gained, some results obtained are sorted and collected, and an in-depth comparison with the traditional protection status of Gan embroidery intangible heritage art. It can be concluded that in the digital protection of Gan embroidery intangible heritage art, most people attach great importance to it. There are some digitized Gan embroidery intangible cultural heritage art on the Internet. The Gan embroidery intangible cultural heritage art works that we need to travel through mountains and rivers to see can be appreciated directly by clicking the mouse. All in all, the digital protection of Gan embroidery intangible heritage art has indeed made progress, but we found that this work also has problems of one kind or another, which await the attention and solution of the people.

5   Conclusion

As an intangible culture, part of the rich folk culture of mankind, Gan embroidery exudes infinite originality and original regional spirit. It is a unique artistic treasure of the Chinese nation and a precious intangible cultural heritage. Establishing a reasonable and effective information perspective that is compatible with research in the field of culture and digital technology is the key to digital technology's intervention in the dissemination and protection of intangible cultural heritage, and it is also the promotion of general audiences, professional researchers and inheritors of intangible cultural heritage. We use digital imaging technology, digital video technology, three-dimensional scanning and other modern means to record intangible cultural heritage, disseminate intangible cultural heritage, and protect intangible cultural heritage. We hope that this age-old culture can have its place in the modern society of information explosion. We must use the new thinking, new perspectives, and new environment of the information age to give these traditional cultures modern attributes, and let these excellent cultures hidden in the folk come to life again.

Acknowledgements. Project: Culture, Art And Science Planning Project Of Jiangxi Province In 2020: "Research On Digital Protection Of Jiangxi Embroidery As Intangible Cultural Heritage", Host: Zeng Lanjian, (Project No.: YG2020155).

References

1. Rosario, S., Chavali, K.: Digitization of taxation in the changing business environment & base erosion & profit shifting (Beps) special reference to India. Eur. Sci. J. **16**(1), 61–74 (2020)
2. Cluley, R., Green, W., Owen, R.: The changing role of the marketing researcher in the age of digital technology: practitioner perspectives on the digitization of marketing research. Int. J. Market Res. **62**(1), 27–42 (2020)
3. Grant, P.: aCAdemicS: the newsletter of the SWOSU College of Arts & Sciences. **5**(5), 1 (2019)
4. Ming, Z.: Research on the transmission mode of national traditional sports culture from the perspective of intangible cultural heritage protection. Int. Core J. Eng. **6**(5), 84–87 (2020)
5. Linggih, I.N., Sudarsana, I.K.: The dynamics of rejang renteng dance in bali as an intangible cultural heritage of the world. Space Cult. India **7**(4), 45–58 (2020)
6. Vilotijevic, M.D.: The relation of music archiving and intangible cultural heritage in Serbia. Musicologist **3**(2), 171–190 (2019)
7. Ma, D., Cheng, M., Zheng, D., et al.: Development of Sichuan brocade with imitating embroidery effect based on free-floats interlacing weave. J. Text. Sci. Technol. **06**(1), 11–18 (2020)
8. Upadhayay, H., Goel, A., Gahlot, M.: Designing of Diwan cover sets inspired from Chikankari embroidery using computer aided designing. Int. J. Curr. Microbiol. Appl. Sci. **9**(1), 791–802 (2020)

9. Choi, S.: The restoration of traditional culture and cultural hybridity in Guam's Taotao Tasi. J. Mod. Engl. Drama **33**(1), 207–231 (2020)
10. Bondarenko, I., Wang, Z., Ieremenko, I., et al.: The selected architectural solution of modern museums in china in the aspect of manifestation of traditional culture. Archit. Civil Eng. Environ. **13**(2), 17–26 (2020)

Construction of Digital Protection and Knowledge Integration Platform for Intangible Cultural Heritage in the Context of Big Data

Ting Zhang[✉] and Chao Deng

Nanchang Vocational University, Nanchang, Jiangxi, China

Abstract. Traditional ways of cultural heritage protection have their own advantages, but in the display, preservation, dissemination, publicity of cultural heritage of convenience is insufficient, using digital means to protect cultural heritage has a unique advantage. This paper mainly studies the digital protection of intangible cultural heritage and the construction of knowledge integration platform under the background of big data. The intangible cultural heritage digital protection Web platform is built based on Spring Boot framework and SSM (Spring + MyBaits + SpringMVC) framework, including digital display and digital management systems. The development of the platform is carried out in accordance with the ideas and methods of software engineering. Firstly, the functional and non-functional requirements of the platform system are analyzed, and then the system is designed in general, the architecture of the system is determined, and the functional modules are divided. With the help of mature information technology, the digital protection platform of intangible cultural heritage dynamically displays the history, panorama and related dynamics of intangible cultural heritage to users. With the help of the platform, users can not only understand the relevant information of the intangible cultural heritage, but also publish the relevant information of the intangible cultural heritage on the platform for sharing, which realizes the digital protection of the intangible cultural heritage to a certain extent.

Keywords: Big data background · Intangible cultural heritage · Digital protection · MVC framework

1 Introduction

In the long history of human beings, many precious cultural heritage resources have been left, which represent different national cultures and cannot be regenerated and replaced. Due to the influence of human and natural factors, most of the cultural heritages preserved up to now have been destroyed to varying degrees, and some of them are facing extinction. Therefore, the protection of cultural heritages is urgent. At present, the protection of cultural heritage has become a worldwide research craze, and how to use high-tech means to better restore, protect and inherit cultural heritage has become a hot

J. Abawajy et al. (Eds.): ATCI 2021, AISC 1398, pp. 812–819, 2021.
https://doi.org/10.1007/978-3-030-79200-8_121

research direction of scholars in the field of cultural heritage protection and inheritance. Display and representation is the basic link of cultural heritage protection, the traditional cultural heritage demonstrated form is relatively single, give priority to with the exhibition display, historical cultural relics preserved is basically physical kept in the museum, the lost cultural relics through pictures and text form for show, this kind of display will be affected by time and space, the distance with the audience. In the 21st century with the rapid development of digital and information technology, digital technology has a unique complementary advantage in the protection of cultural heritage due to its advantages of preservation and dissemination. Digital information technology, such as virtual reality technology, can reproduce exact historical and geographic information. It can be combined with written materials and photos of cultural relics in libraries and museums, together with the explanation and consultation of some experts or scholars, which is a new method for understanding history and receiving education [1]. Therefore, with the support of mature digital technology as a supplement and substitute for the protection of cultural heritage, the digitization of cultural heritage can continue some of the disappearing cultural heritage, which is a relatively ideal and feasible solution for the protection of cultural heritage.

In the protection of cultural heritage, foreign experience is to build the corresponding digital protection platform and combined with digital modeling, digital management, virtual restoration, digital display, digital simulation, digital copy storage, technical framework, key technologies and typical platform applications and other technical means. Katiyatiya implements the architecture of digital copy, digital virtual or analog form of cultural heritage entity to realize the virtual preservation or reproduction of cultural heritage [2]. Nooren et al. discussed the application and development of digital technology in intangible cultural heritage from the perspective of digital museum. Et al. realized the digital protection of intangible cultural heritage through the 3D reconstruction method of Graph Cuts [3].

This paper selects the current mature technical means to digitize the intangible cultural heritage and establish a database, and constructs a Web network platform to manage and display these digital materials. Through the construction of the digital protection platform of intangible cultural heritage, it is expected to better realize the dynamic protection of intangible cultural heritage.

2   Digital Protection Platform for Intangible Cultural Heritage

2.1   Related Technologies

(1)   MVC Design Pattern

MVC, Model, View and Controller, is the abbreviation of Model, View and Control. It is a popular design pattern (design idea) in software development, and the classic case is JSP + Servlet + JavaBean. It decomposes the platform into three modules: M (model), V (view) and C (control), and each module performs its own duties [4].

Model layer (Model): The Model layer mainly deals with business logic and usually contains some abstract methods and objects defined by the developer. JavaBeans are part of the Model layer. The Model layer handles requests from the Controller layer and calls abstract methods to complete database operations.

View layer (View): View is mainly responsible for the interaction with users, usually used for data display and interface display, so the quality of View layer design will directly affect the user experience, JSP pages belong to the View layer. The interaction process between the view layer and the user is as follows: the view layer sends the user's request to the control layer, and the control layer calls the method of the model layer to process the data and returns it to the interface for display to the user. Data display in the view layer is controlled by the Controller layer, so code changes to the view layer do not affect the Model layer or the Controller layer.

Control layer: The control layer is mainly responsible for the connection between the model layer and the view layer. Servlets belong to the control layer. The working process of the control layer is as follows: it receives requests from the view layer, judges these requests and then calls corresponding logical methods to the corresponding model layer. When the model layer processes data and returns it to the control layer, the control layer will judge and transfer the data to the corresponding view layer according to the program [5]. The Controller layer does not involve any operation on data, but only plays the role of forwarding requests and data.

(2)   3D Panoramic Technology

3D panorama (also known as 306° panorama), also known as virtual reality panorama, is a classification of the current rapid development and rapid popularity of virtual reality technology (VR). The technology uses digital image processing technology, based on the real scene or vector modeling to achieve three-dimensional reproduction of the scene, this paper is based on the real scene of three-dimensional panoramic technology.

In general, the 3 d view is based on a panorama of the production, it is through some pictures performance good equipment (such as professional camera, fisheye lens, yuntai, tripod, etc.) to the scene of many points, then use Photoshop, software to collage of images, and so on3dmax reuse software (such as Pano2VR, Ulead College306 or 720 cloud) for dynamic panorama, finally packaging [6]. After packaging and release, users can carry out 3D panoramic virtual tour through PC or mobile phone, mainly reflected in: users can switch the panoramic perspective arbitrarily through the mouse or touch screen, and zoom the details. VR glasses can provide users with an immersive feeling, which greatly enhances user experience.

3D panoramic technology is characterized by good user experience, good interaction, simple implementation, and no restriction on terminal devices for panoramic image playback. Therefore, the solution provided by 3D panoramic technology is utilized to realize panoramic intangible cultural heritage.

(3)   Relevant Calculation Formulas

$$f = \frac{a-b}{b} \tag{1}$$

$$e = \sqrt{a^2 - b^2}/a \tag{2}$$

$$e' = \sqrt{a^2 - b^2}/b \tag{3}$$

2.2   Platform Requirements Analysis

The digitization platform of intangible cultural heritage is a Web-side digitization platform. The overall actual business requirements of the platform are as follows:

Digital information related to intangible cultural heritage published in the background should be displayed for users to browse.

It is necessary for administrators to publish digitized materials related to intangible cultural heritages to achieve the purpose of dynamic updating of the display part. This is where multiple administrators are required to show the administrative release of some of the corresponding content to ease the burden on a single administrator. Because of the need to enter the background management, we need to consider the security and authority distinction.

For the users of the forum, including ordinary users and administrators, it is necessary to distinguish the functional authority between them. For the post and comments need to verify the identity of the forum users, and the administrator needs to verify the post. For the forum users need administrators to manage them.

Therefore, the platform is divided into three modules: digital display module, digital management module and forum module.

(1)   Digital Display Module

In this module, the dynamic panorama of intangible cultural heritage is displayed. In this module, users can switch to different scenes and perspectives of intangible cultural heritage through the mouse. Equipping with VR devices allows users to feel as if they are in the scene and can visit intangible cultural heritage without leaving their homes.

(2)   Digital Management Module

In the display module, the status quo of the intangible cultural heritage is displayed in the form of pictures, so the management of the status quo of the intangible cultural heritage is mainly based on the management of the existing pictures. The management of intangible cultural heritages mainly includes adding the number of intangible cultural heritages, the introduction of intangible cultural heritages and other information, deleting some (or more) intangible cultural heritages, modifying the information of intangible cultural heritages and searching the intangible cultural heritages according to the name of intangible cultural heritages. The management of intangible cultural heritage pictures is bound to the corresponding intangible cultural heritage.

(3) Forum Module

For the forum plate management can let the forum users according to the plate classification of the post, which is conducive to the administrator of the post classification management. The management of the forum plate includes: adding new plate, including the name of the plate, the Logo of the plate and the description of the information; Modify the information of the plate; Delete unnecessary plate, when a plate is deleted, the plate theme under the plate will be deleted, for safety reasons: when there is a post under the plate, you need to delete the post before you can delete the plate; Manage the topics under the section, including adding, removing, editing topics, etc.

2.3 Platform Architecture

The design of the Web platform for digital protection of intangible cultural heritage adopts B/S mode, and is built based on the integration of Spring Boot and SSM framework. The design of the platform is constructed in a three-tier architecture.

The presentation layer is also called the User Interface layer, which provides an Interface to interact with users by displaying data and receiving User input. In the performance layer of this platform, the front-end interface is designed by using HyML/JSP, CSS and Bootstrap framework. The interaction process is roughly as follows: the user sends the request to the background through JavaScript and AJAX in the front-end page, and Spring MVC forwards the request to the corresponding Controller, who calls specific business logic to process the request, and returns to the front-end for page rendering after processing [7, 8].

The business logic layer is in the middle of the presentation layer and the data processing layer and acts as a link between the preceding and the following in the three-tier architecture. It contains the business logic interfaces of some functional modules, and calls the data access layer and is called by the presentation layer [9].

The data access layer is primarily responsible for dealing with the database, completing CUDR operations on the database, and feeding the results back to the business logic layer. The platform uses MyBatis framework in the data access layer, which can enhance the maintainability of SQL and the reusability of code [10].

3 Platform Architecture Implementation and Testing

3.1 Platform Architecture Realization

The intangible cultural heritage digital protection platform is realized based on SSM three-tier architecture. Spring Boot is adopted to simplify the configuration of Spring, and Maven technology is used to build the Spring Boot project. First of all, use Maven configuration dependency to load the JAR package needed by the project in the local warehouse, not only need the JAR package integrated with MyBatis and Spring, but also need the JAR package integrated with Spring Boot and MyBatis.

Then the configuration of the data connection pool and data source, the preparation of the Spring Boot core Boot class, the configuration of the database mapping file, etc., completes the initial architecture of the cultural heritage digital protection platform based on Spring Boot and SSM.

3.2 Platform Test Method

Platform testing is an important step before project deployment and delivery, in order to verify whether the various functions and performance of the platform meet the requirements, so as to make improvements. There are many methods to test software or platform. As the platform is divided into multiple functional modules and developed in the way of waterfall model, unit testing and black box testing are adopted for testing.

Unit testing is generally aimed at a functional module of the platform to test, is the most basic step of software testing. Unit testing is based on white-box testing to verify the correctness of a functional module code, which is typically written by a developer to check the correctness of a small, specific feature. Since SSM framework is used in this project, unit tests are completed by writing JUnit test classes.

Black box testing does not pay attention to the code, only focus on the input and output to determine whether the function meets the requirements. Inputs are not always specific and correct, and may also be wrong inputs. The correctness of platform functions is verified through input and output under various circumstances.

4 Test Results of Digital Platform for Intangible Cultural Heritage

4.1 Stress Testing

As shown in Table 1 and Fig. 1, by analyzing the pressure testing figure, when platform of traffic increase or decrease, concurrent corresponding practical platform here also stepped up, but even under the visit of 7000 times per second, the response time of the platform are also good, the actual response time is not more than 3 s, can fully meet the requirements, you can verify from the side platform can completely satisfy the user's access requirements. According to the analysis of the number of processing per second of the platform, the actual situation of processing will not change due to the increase of CPU utilization of the platform, and the performance is good.

Table 1. Stress test statistics

	60S	120S	180S	240S	300S
Virtual users	30	50	70	90	120
Response time	19.2	37.6	43.3	59.7	71.1
Number of system processing	12	11	13	12	11

4.2 CPU Resource Consumption

As shown in Table 1, combined with the platform performance test, a comprehensive test was conducted on the platform resource consumption during the test period, and the platform CPU resource consumption within 1 min was tested. It can be seen that

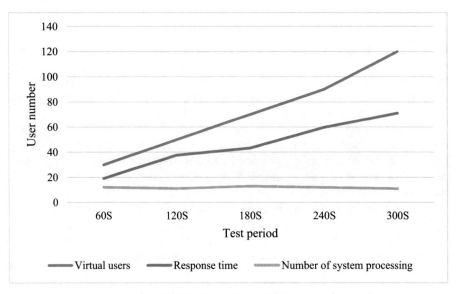

Fig. 1. Stress test statistics

the user part consumes the most CPU resources among the three parts, and the average consumption reaches 10.56%. However, the CPU consumption of user, platform and waiting reaches 30.79%, and fluctuates around 19% in the rest of the time. It shows that the platform designed and developed in this paper consumes less CPU resources and the server can meet the needs of the platform (Fig. 2).

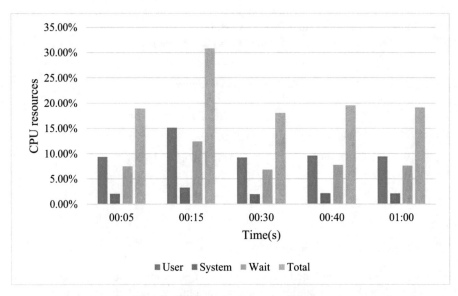

Fig. 2. CPU resource consumption test

5 Conclusions

This paper mainly describes the design and implementation of the intangible cultural heritage digital protection platform. Firstly, the source of the subject, the project background, the significance of building the intangible cultural heritage digital protection platform were introduced, and the status quo of the digital protection of cultural heritage was analyzed. Then it introduces and explains the relevant technologies used in the construction of the digitized platform of intangible cultural heritage and the reasons for its selection, including Spring Microframework, SSM framework, MVC design pattern, Maven technology, front-end technology and 3D panoramic technology. As a digital result, the intangible cultural heritage digital protection platform (WEB) provides a platform for the public to understand and spread the intangible cultural heritage culture. It is hoped that through this platform, more and more people can know about the intangible cultural heritage and become a member of the protection team of the intangible cultural heritage.

References

1. Jung, S., Yoo, J., Lee, Y.J.: A PLC platform-independent structural analysis on FBD programs for digital reactor protection systems. Ann. Nucl. Energy **103**(May), 454–469 (2017)
2. Katiyatiya, L.M., Lubisi, N.: Part B: the role of substantive equality in shaping transformative social protection in South Africa into the digital platform work. J. Poverty **4**, 1–20 (2020)
3. Nooren, P., Van Gorp, N., Van Eijk, N., et al.: Should we regulate digital platforms? a new framework for evaluating policy options. Policy Internet **10**(3), 264–301 (2018)
4. Aljamea, M., Alkandari, M.: MMVMi: a validation model for MVC and MVVM design patterns in iOS applications. IAENG Int. J. Comput. Sci. **45**(3), 377–389 (2018)
5. Song, C., Cho, E.: An integrated design method for SOA-based business modeling and software modeling. Int. J. Software Eng. Knowl. Eng. **26**(2), 347–377 (2016)
6. Tang, S.C., Shen, C.N., et al.: Pancreatic neuro-insular network in young mice revealed by 3D panoramic histology. Diabetologia: Clin. Exper. Diabetes Metabol. = Organ Euro. Assoc. Study Diabetes **61**(1), 158–167 (2018)
7. Tang, S.C., Shen, C.N., Lin, P.Y., et al.: Pancreatic neuro-insular network in young mice revealed by 3D panoramic histology. Diabetologia **61**(1), 158–167 (2018)
8. Cennamo, C., Ozalp, H., Kretschmer, T.: Platform architecture and quality tradeoffs of multihoming complements. Inf. Syst. Res. **29**(2), 461–478 (2018)
9. Pan, H., Zhongxiang, H., et al.: Research on architecture of intelligent transportation cloud platform for Guangxi expressway. AIP Conf. Proc. **1834**(1), 1–6 (2017)
10. Yoeruer H.: The role of platform architecture characteristics in flexible decision-making. Int. J. Innov. Technol. Manage. **16**(8), 1950061.1–1950061.28 (2019)

Safety and Emergency Management System for the Elderly Based on Big Data

Yongmei Tao[(✉)]

Changchun Institute of Technology, Changchun 130021, Jilin, China

Abstract. The number and structure of population are the primary factors affecting the change of global emergencies. Aging is one of the important reasons for the world economic transformation and transformation. Population aging leads to an increase in the vulnerability of the entire human system. Peace is the real need to meet the people's new expectations for a better life. Emergencies of the elderly have serious consequences. It is urgent to strengthen the construction of safety and emergency management system for the elderly. Based on the Systematic theory and the perspective of "PPRR"theory, a "five-in-one" elderly safety and emergency management system is constructed under the background of big data, including the elderly themselves, their families, communities, society and the country. To build a proactive emergency management system for the elderly to fundamentally avoid or reduce emergencies among the elderly, maintain social stability and harmony, and help to build a safer China.

Keywords: Emergency management · Safety for the elderly · System view

1 Introduction

Globalization, aging and the advent of Risk Society have put forward higher requirements for Chinese society emergency management capability [1]. Compared with the speed of economic development, the aging process of population in China is ahead of time, and the arrival of deep aging aggravates the social risks. In the 14th Five-Year Plan, "actively coping with the aging of the population" has been promoted as a national strategy. And put forward to "significantly enhance the emergency response capacity of public events, maintain social stability and security, and build a higher level of peace in China", which embodies the country's great attention to the severity and urgency of the aging society situation. In this study, a model that can effectively avoid safety risk events for the elderly at the basic level is proposed.

2 The Basic Theory

2.1 Theoretical Model of "PPRR"

American scholar Yates (2003) proposed "PPRR" theory [2], which has been widely applied in public crisis management field [3]. "PPRR" theory divided Crisis Management into four stages: prevention, preparation, response and recovery.

© The Author(s), under exclusive license to Springer Nature Switzerland AG 2021
J. Abawajy et al. (Eds.): ATCI 2021, AISC 1398, pp. 820–825, 2021.
https://doi.org/10.1007/978-3-030-79200-8_122

2.2 System Theory

System is a complex of interrelated and interacting elements. System theory regards the object as a whole composed of multi-faceted connections and multi-elements. Integrity is the basis and starting point of system theory. Once each element constitutes the system as a whole, it shows the properties and functions that the independent elements do not possess, thus form the qualitative stipulation of the new system.

2.3 Hierarchy of Needs Theory

MASLOW divides human needs from low to high into five main levels: physiological needs, security needs, belonging needs, respect needs and self-actualization needs. His hierarchy of needs reveals the relationship between human behaviors and needs. It is one of the important basic theories to study the behavior and safety risk scenarios of the elderly.

3 Safety and Emergency Management Mechanism for the Elderly Based on "PPRR" Theory

3.1 Prevention Stage

3.1.1 Safety Risks and Emergencies for the Elderly

"Preventive stage" refers to the various preventive work carried out in advance in order to eliminate the crisis or reduce the harm caused by the crisis events, which is the premise and foundation of emergency management [4]. This stage requires the application of big data technology. At this stage, the most important thing is to scientifically identify the safety risk factors of the elderly, improve the ability to predict emergencies of the elderly, and effectively carry out risk assessment and control, so as to realize the "Pass Forward" in response to emergencies [5]. Not all the safety problems of the elderly belong to emergencies, but all the safety problems of the elderly will inevitably evolve into the outbreak of emergencies if they are not identified and controlled in time (Fig. 1).

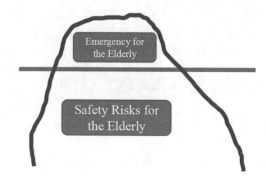

Fig. 1. Iceberg map of elderly safety risks and emergencies

3.1.2 Identification of Safety Risk Scenarios for the Elderly

Based on the framework of Maslow's hierarchy of needs theory, starting from the analysis of the needs of the elderly, comprehensively identify the types, scenarios and emergencies of risk events for the elderly. Big data technology needs to be fully utilized (Table 1).

Table 1. Elderly safety and emergency classification and main scenes

Grade	Event nature	The main event
I	Elderly emergency; Time-sensitive and dangerous	Severe and sudden disease outbreaks; Being abused; Disability and semi-disability; Electricity and water use safety emergencies; Travel and traffic safety incidents; Lost; Natural disasters and public safety emergencies in places such as The nursing home [6]; pension institutions; New coronary pneumonia and other public health events [7]
II	Very harmful, Greater urgency	Excessive medical treatment; Food safety; Cheated; Cognitive impairment; Difficult to adapt to modern technology; Hard by the pension
III	Urgent, More harmful	Daily care; The rights being invaded; The rights and interests of elderly and vulnerable female groups are violated
IV	Not easy to recognize these days, Large hidden harm	The lack of communication; Lack of hobbies; lonely; Self-worth can't be realized

3.2 Preparation Stage

It refers to all kinds of preparation work, such as early warning prediction, making emergency plan and information processing, to deal with potential risk events.

3.2.1 Make Full Use of Big Data Technology

The most important thing for monitoring and early warning of risk events in the elderly is timely acquisition of information, correct judgment and rapid response [8].

3.2.2 Improve the Mechanism for Monitoring and Early-Warning of Elderly Safety and Emergencies

Monitoring is the real-time observation and reporting of potential risk sources, emergencies and environmental parameters related to the safety of the elderly. Early warning is based on the full assessment of various potential uncertain factors and their interrelated situations, and the use of statistical analysis, logical reasoning and other technical methods, to report the information of emergency to the relevant departments [9]. The elderly safety early warning emphasizes the multi-way, full coverage, to realize the information sharing of multiple safety responsibility subjects.

3.2.3 Formulate the Safety and Emergency Management Plan for the Elderly

Emergency plan is an important content of "one plan and three systems" in Chinese emergency management. Special teams should be set up to systematically assess the emergency response capabilities of various relevant actors and to prepare emergency response plans. First of all, it is necessary to clarify the scene, type and influence degree of emergencies of the elderly. Secondly, according to the classification and classification of incidents, the emergency resources required are confirmed. Thirdly, determine the command organization and the responsibility organization, and establish the emergency rescue team combining special and part-time; Fourthly, formulate action plan; Finally, organize the elderly and their families, community personnel and other organizations to carry out training, exercise.

3.3 The Response Phase

This is the core of the elderly safety and emergency management system [10]. This stage requires the participation of multiple subjects to form a systematic advantage of multi-subject cooperation and co-governance. As shown in Fig. 2.

3.4 Recovery and Reconstruction Stage

This stage refers to restoring the normal life order of the elderly and their families, communities or institutions on the basis of summarizing experience and deficiencies after emergency treatment of the elderly, and eliminating the negative impact of the emergency on social politics, economy and public psychology.

3.4.1 Establish an Investigation and Evaluation Mechanism

The outbreak of emergencies for the elderly is mostly due to the failure of comprehensive identification or timely and effective control of the safety risk factors for the elderly. Collect, analyze and evaluate the data of the accident itself, the processing process, methods, results, the handling capacity of the emergency system and the loss situation.

3.4.2 Establish a Mechanism for Relief, Compensation and Psychological Recovery and Reconstruction

The emergency management mechanisms need to be established in order to sustain a person's basic life by providing certain material assistance and spiritual comfort. The establishment of emergency relief and compensation mechanism and psychological recovery and reconstruction mechanism of the elderly can effectively reduce the damage of accidents to the elderly themselves or their families.

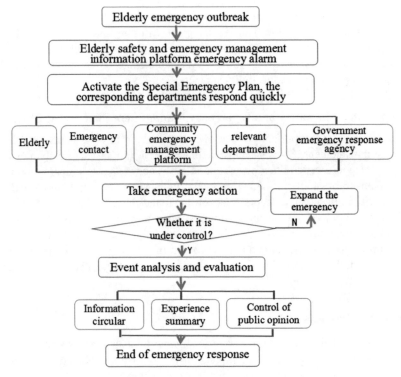

Fig. 2. Flow chart of emergency response for the elderly

3.4.3 Establish a Public Opinion Monitoring and Correction Mechanism

The emergencies of the elderly themselves have a high degree of social attention and sensitivity. The government and relevant administrative departments of the industry should establish specialized public opinion monitoring and guidance institutions for emergencies of the elderly to supervise and control the "Self-media".

3.4.4 Improve the Accountability and Improvement Mechanism

The adverse consequences caused by mistakes or dereliction of duty in risk analysis, safety management and emergency disposal of the elderly should be investigated for responsibility. The improvement and perfection mechanism combined of punishment and education should be established for the existing problems to continuously improve the safety and emergency management level of the elderly in the whole society.

4 Conclusions

Nowadays, the social risks are accumulating all over the world, and natural disasters, public health and social security emergencies are frequent. Population development and the construction of a harmonious society are not only strategic issues of global concern, but also the concentrated embodiment of socialist core values. Coping with the safety and emergency of the elderly needs the joint participation of the government and all walks of life to realize the collaborative governance of multiple subjects, so as to create a more stable and harmonious safe China while providing higher security for the elderly.

References

1. Fink, S.: Crisis management: planning for the invisibles. J. Womens Health **8**, 33–42 (1986)
2. Yates, A.: Securing critical infrastructure and the built environment. Eng. Aust. **6**, 37–48 (2003)
3. Robert, H.: Crisis Management, vol. 5, pp. 5–46. Trans. Zhongxin Publishing House (2004)
4. Tang, G.: The implementation of the community model of emergency management in the United States and its enlightenment to China. Chin. Adm. Manage. **06**, 142–146 (2017). (in Chinese)
5. Xue, L., Zhang, Q., Zhong, K.: Crisis management: challenges facing China in transition. China Soft Sci. **4**, 12–16 (2003). (in Chinese)
6. Zhang, H.: Innovative development of national emergency management system and mechanism in the new era. Acad. Front. People's Forumn **5**, 6–15 (2019). (in Chinese)
7. Kohei, H., Yusuke, H., Taichi, I., et al.: Increased incidence of hypotension in elderly patients who underwent emergency airway management: an analysis of a multi-centre prospective observational study. Int. J. Emerg. Med. **6**(1), 1–8 (2013)
8. Soon, O.K., Sun, H.B.: Emergencies in long-term care services for the elderly in korea: a mixed-methods study. Int. J. Environ. Res. Public Health **17**(1), 98–103 (2019)
9. Pallavi, B., Ashley, B., Martina, L.P., et al.: No melanoma skin cancers are more likely to be untreated in elderly patients. J. Am. Acad. Dermatol. **82**(2), 505–507 (2020)
10. Gu, Y.: Population aging: from periodic emergency to normal management. Macroecon. Manage. **10**, 58–59 + 66 (2015)

Economic Laws and Regulations for the Development of Artificial Intelligence Industry Based on Big Data

Liuhong Chen[✉]

Guangzhou Huashang Vocational College, Guangzhou 511300, Guangdong, China

Abstract. At present, with the continuous development of big data technology, the artificial intelligence industry is having an increasingly extensive and deep impact on human production and life. Out of concerns about the possible abuse of artificial intelligence technology and the possible abuse of highly intelligent robots, How to make legal adjustments to the development of the artificial intelligence industry has also begun to be concerned by the legal profession. In this regard, the purpose of this article is to study the economic laws and regulations for the development of artificial intelligence industry based on big data. This article first discusses the development status of our country's artificial intelligence industry under big data, and summarizes the main problems existing in contemporary our country's artificial intelligence industry. Then, through the method of questionnaire survey, to understand some citizens' understanding and views on the artificial intelligence industry and its related economic laws and regulations. According to the survey and research results, this article defines a new logical main line "big data technology-artificial intelligence industry-economic regulations" from the perspective of big data and based on the "technology-system" analysis framework. Next, this article conducts value evaluation and system selection at the level of economic law, and specializes in "differentiated regulation" of development plans, taxation and taxation, financing, competition, consumer protection and many other financial law systems. The experimental results show that the new framework proposed in this paper promotes the healthy development of the artificial intelligence industry and the realization of the adjustment goals of economic law, promotes the deepening of the theory of industrial law and the cross-study of "technology and economic law", and its support for the artificial intelligence industry has increased by 30%, it provides an important reference for the development of the artificial intelligence industry's economic laws and regulations.

Keywords: Big data technology · Artificial intelligence industry · Economic laws and regulations · Comparative experiments

1 Introduction

The rapid development of the information revolution has made the importance of artificial intelligence aroused widespread concern [1]. The research and application of artificial

© The Author(s), under exclusive license to Springer Nature Switzerland AG 2021
J. Abawajy et al. (Eds.): ATCI 2021, AISC 1398, pp. 826–834, 2021.
https://doi.org/10.1007/978-3-030-79200-8_123

intelligence have experienced ups and downs, and new emergence has begun in recent years [2, 3]. Based on the increasingly practical impact of artificial intelligence technology on human production and life and its positive role in promoting economic growth and social development [4, 5], and based on the vision for the future growth of smart economy and smart society, government, industry, a large number of future development plans in the fields of teaching and research, innovative products and research results have been launched [6]. Among them, artificial intelligence laws and regulations have become a research hotspot [7].

At present, the academic community has conducted a lot of discussions on key issues such as the industrial development of artificial intelligence and corresponding legal regulations [8]. In foreign countries, Lee H believes that artificial intelligence technology as a "disruptive innovation" can directly affect all aspects of production, distribution, exchange and consumption, and promote the production of technology, product innovation, forms and models, thereby leading to changes in economic structure and even the entire industrial revolution and the reconstruction of the entire economic system [9]. In China, Zhao Yinyi believes that it is necessary to attach great importance to the artificial intelligence industry to effectively promote the development of emerging intelligent industries, such as intelligent software and hardware, intelligent robots, intelligent delivery vehicles, and promote the development of artificial intelligence [10].

Based on big data, this paper studies the economic laws and regulations for the development of artificial intelligence industry. This article first starts from reality, discusses the current situation of our country's artificial intelligence industry and related economic laws and regulations, as well as the relevant opinions and suggestions of citizens. Then based on the survey, summarized the main problems existing in the economic legal system. Finally, this article puts forward a new logical main line "Big Data Technology-Artificial Intelligence Industry-Economic Laws and Regulations" based on the current situation and combined with big data and other technologies. Under the guidance of the new logical main line, the artificial intelligence industry has developed healthily, and the adjustment goals of economic law have also been achieved, which has important reference significance for the further development of economic laws and regulations for the development of artificial intelligence industry under big data.

2   Technical Research on Economic Laws and Regulations for the Development of Artificial Intelligence Industry Based on Big Data

2.1   Innovation of Artificial Intelligence Industry Regulations Based on Big Data

(1)   Range correlation function

The number of associations in the era of big data is staggering, and the associations of millions of people can basically cover a domain. Big data technology is of course associated with artificial intelligence. Under double standards, the artificial intelligence industry has new participants and opportunities. Creating tight associations is the most important thing that can happen in a virtual network environment. Now, Internet connection and offline connection occur simultaneously, which is a very important growth trend. When we use big data to integrate information from different industries, we can analyze many problems.

(2) Purification and screening function

In work, big data must be fully used, effectively explored, dynamically applied, compared with each other, and intelligently pushed. The data generated in all aspects of the artificial intelligence industry should be screened, cleaned and other scientific processing, not just Simple exchange and sharing. In fact, it is not only in the artificial intelligence industry model, but also in any field of current society, which is deeply marked by big data.

(3) Special custom function

Customization is a brand new mode of production, management and marketing for companies that produce products. The production process of customized products is the production of customized products based on the efficiency and benefits of large-scale and batch production. The final customized products are not only low in cost and large-scale production costs, but also have a short production cycle and time. Product quality and large-scale production quality of the product are almost the same, but the most important advantage is that the product is a unique customized product with strong individual characteristics. Big data can provide the artificial intelligence industry with quantitative customization momentum.

The fitness function is the standard to measure the quality of each customized product. Obviously, the cost of the small spanning tree that satisfies the degree of satisfaction of each binary code is the best standard to measure the quality of an individual. To this end, for each binary code α, first define a cost function:

$$F(a) = \sum_{e \in t} t(e) \tag{1}$$

Among them, t represents the minimum spanning tree, e represents the edge of the tree, so the fitness function Fit can be defined as:

$$Fit(a) = \max - F(a) \tag{2}$$

Among them, max is a large number that has nothing to do with the individual, and it guarantees that Fit(a) is a positive number.

(4) Deep learning function

Deep learning can perform multi-layer representation and extraction learning of data such as pictures, sounds, texts, and then understand the meaning of these data. Deep learning builds a hierarchical model structure similar to the human brain, extracts features from the bottom to the top of the data that needs to be trained, learns layer by layer, and transfers the learned features or knowledge to the next layer, making the following Neurons at higher or higher levels get a more abstract form of expression, so that the mapping from concrete and actual data signals at the bottom to abstract semantic concepts at the high level can be established well, and the way the human brain works is simulated, so that artificial intelligence and the human brain react approximate.

After layer-by-layer learning, the parameters learned in each single layer are combined into an overall model parameter, and then the parameters are fine-tuned on the entire model. After the first layer of learning, the model it produces can be written as:

$$P(v; w) = \sum_{h^1} P\left(h^1; w^1\right) P\left(v|h^1; w^1\right) \tag{3}$$

Among them

$$P\left(h^1; w^1\right) = \sum_{v} P\left(h^1, h^2; w^1\right) \tag{4}$$

If the initialization of the second layer is better, you can use the second layer

$$P\left(h^1; w^2\right) = \sum_{h^2} P\left(h^1, h^2; w^1\right) \tag{5}$$

Instead of the $P(h^1; w^1)$ in the first layer, the entire model can achieve better results. Where h^1, h^2 is the state of each layer of neurons, and w^1, w^2 is the model of each layer.

(5) Optimize function

The infiltration of big data thinking and methods can correct and improve the shortcomings of traditional methods. For example, by combining public opinion data with environmental protection data, you can discover which area haze data is fraudulent. Big data that "quantifies everything" will to a large extent change the way people perceive and understand the world. Big data improves people's understanding of the nature of things, thereby optimizing the way things are processed. Therefore, big data has undoubtedly had a great impact and influence on the original artificial intelligence industry. The application of big data in the artificial intelligence industry is not simply data concentration and accumulation, nor is it overthrowing, but the optimization and promotion of the artificial intelligence industry.

2.2 Economic Regulations for the Development of Artificial Intelligence Industry

(1) Planning promotion

In terms of planning promotion, it is necessary to examine the impact of artificial intelligence technology innovation and industrial development on the economy and society, and to respond to the system based on relevant value estimates. Among them, the choice of economic legal system is particularly important based on considering different values. In view of the government's cognitive ability and its attitude towards artificial intelligence will directly affect the formation of related systems, it is necessary to introduce or integrate a general theoretical framework to analyze the economic legislation of the "government market". This will help clarify issues related to the development of the artificial intelligence industry and economic laws and regulations.

(2) Finance and taxation promotion

In terms of fiscal and taxation promotion, as an emerging industry, the artificial intelligence industry has a huge impact on the growth of its tax system. For example, whether the artificial intelligence industry should be more inclined to finance, especially if artificial intelligence companies want to obtain funding or prioritize public procurement, various government treasury and administrative expenses, etc., all contain very important system options. If the country provides preferential institutional guarantees, it will play an important role in promoting the development of smart industries.

(3) Financial promotion

In terms of financial promotion, the artificial intelligence industry includes a lot of investment and financial issues. If it can obtain financial support through banks, capital markets and other channels during its development, and the country's financial control measures are loosened, it will affect the development of the industry. At the same time, eliminating or eliminating institutional barriers in relevant legislation is also a way to promote industrial development. For example, smart investment advisors are developing rapidly in the financial field. To prevent risks and promote the application of artificial intelligence, relevant financial legislation needs to be improved.

3 Experimental Research on Economic Laws and Regulations for the Development of Artificial Intelligence Industry Based on Big Data

3.1 Experimental Data

This article selects urban forums as the research object. In order to ensure the validity and representativeness of the research, we verify its activity. This article selects four urban areas located in different regions of our country. Considering the security of related data, the letters are replaced by: A city, B city, C city, D city. This article uses crawler technology to collect and count the forum data in four urban areas. The collected data are 425.23 M in city A, 610.12 M in city B, 556.34 M in city C, and 231.49 M in city D.

3.2 Experimental Process

This article conducts a questionnaire survey on some users in four urban areas on the forum, and retrieves the words posted and replies, counts the frequency of the words, and draws a histogram. In order to analyze the views and opinions of urban residents on the development of artificial intelligence industry and related economic laws and regulations. At the same time, this article will use the artificial intelligence industry products formed under the guidance of the two frameworks of "technology-institution" and "big data technology-artificial intelligence industry-economic regulations" for citizens to use, and then use the results to analyze this article. The effect of the proposed "big data technology-artificial intelligence industry-economic regulations" framework.

4 Experimental Analysis of Economic Laws and Regulations for the Development of Artificial Intelligence Industry Based on Big Data

4.1 Citizens' Views on the Development of Artificial Intelligence Industry and Related Economic Laws and Regulations

In order to study the current development of our country's artificial intelligence industry and the status quo of related economic laws and regulations, this paper conducts a questionnaire survey on selected representative citizens, and sorts out and counts them. According to the statistical results, a table and histogram are drawn. As shown in Table 1 and Fig. 1.

Table 1. Current state of artificial intelligence and economic regulation

	Good planning effect	Comprehensive system	Reasonable finances and taxes	Financially reasonable	Good service
Downtown A	65%	78%	55%	51%	69%
Downtown B	67%	72%	54%	49%	66%
Downtown C	70%	69%	56%	45%	70%
Downtown D	66%	77%	53%	50%	65%

It can be seen from the survey data that citizens in the four regions are not very satisfied with the current development of the artificial intelligence industry and the status quo of relevant economic laws and regulations, and the highest degree of support for various survey projects is only about 70%. This is because the industrial development of artificial intelligence and the corresponding laws and regulations have not yet been discussed from the perspective of economic law and information law, and there is an urgent need to explore modern law or postmodern law. In the micro-field technology, technological progress is often ahead of specific systems and will lead to institutional changes in related fields; in the macro-field, general or basic institutional arrangements will have a huge impact on technological innovation in specific fields. The aforementioned smart industry and the innovative development of the smart economy led by it are inseparable from corresponding policies and legal guarantees.

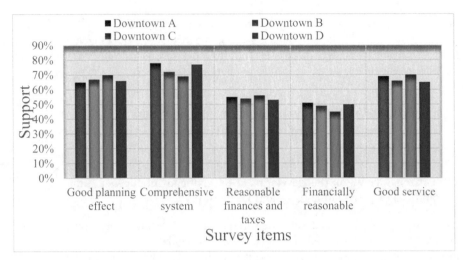

Fig. 1. Current state of artificial intelligence and economic regulation

4.2 Comparison of the Effects of the Two Regulatory Frameworks

In order to further verify the effect of the "big data technology-artificial intelligence industry-economic regulations" framework, this article will be formed under the guidance of the two frameworks of "technology-system" and "big data technology-artificial intelligence industry-economic regulations". Smart industry products are used by citizens, and then the results of the use are counted. And based on these statistical results, judge the changes recognized by the citizens for the two frameworks, visualize them, and perform curve fitting according to the mean values. As shown in Table 2 and Fig. 2.

Table 2. Comparison of the effects of two regulatory frameworks

	Initial	After 10 days	After 20 days	After 30 days	After 40 days	After 50 days	After 60 days
New regulatory framework	50%	57%	65%	71%	88%	90%	91%
Traditional regulatory framework	50%	53%	57%	65%	62%	60%	60%

It can be seen from the experimental data that, as time goes by, the support for the two frameworks is gradually similar before 33 days, but after 33 days, the gap widens. The new "big data technology-artificial intelligence" and the framework of "industry-economic regulations" are more effective. The relationship between the development of the artificial intelligence industry and economic laws and regulations is the realization

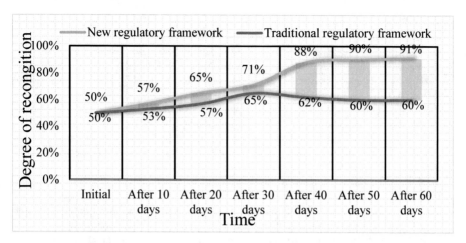

Fig. 2. Comparison of the effects of two regulatory frameworks

of the "relationship between technology and system". The progress of science and technology and the development of science and technology industries have led to changes in related economic law systems; the effective regulation of economic law has a significant impact on the progress of science and technology and the development of science and technology industries. Economic regulation means encouraging and promoting its positive influence, and restricting and prohibiting its negative influence.

5 Conclusions

Based on big data, this paper studies the economic laws and regulations for the development of artificial intelligence industry. This article first discusses the main issues of the artificial intelligence industry and related economic laws and regulations based on our country's big data. Then, through the method of questionnaire survey, to understand the understanding and opinions of some citizens. From the perspective of big data, this article proposes a new logical main line "big data technology-artificial intelligence industry-economic regulations." And this article conducts "differentiated regulation" on the level of economic law. The new framework promotes the healthy development of the artificial intelligence industry and the realization of the goal of adapting to economic laws and regulations, promotes the deepening of the theoretical regulation of economic law, and provides important reference provisions for the further development of economic laws and regulations for the development of the artificial intelligence industry.

References

1. Li, S., Hao, Z., Ding, L., et al.: Research on the application of information technology of Big Data in Chinese digital library. Libr. Manage. **40**(8/9), 518–531 (2019)
2. Yang, L.: Research on university informationization service based on big data and artificial intelligence. Exper. Technol. Manage. **035**(011), 153–156 (2018)

3. Wang, P.: Research on online learning behavior analysis model in big data environment. Eurasia J. Math. Sci. Technol. Educ. **13**(8), 5675–5684 (2017)
4. Oh, K., Kim, H.: An analysis of the influence big data analysis-based ai education on affective attitude towards artificial intelligence. J. Korean Assoc. Inf. Educ. **24**(5), 463–471 (2020)
5. Yazkova, V.: Servant or master?: catholic ethics and artificial intelligence. Sci. Anal. Herald IE RAS **17**(5), 113–120 (2020)
6. Yi, X., Wu, J.: Research on safety management of construction engineering personnel under "big data + artificial intelligence". Open J. Bus. Manage. **08**(3), 1059–1075 (2020)
7. Dong, J., Wu, Y.: Commensuration operator of urban science based on semantic network - a case study of urban design programming. IPPTA: Quar. J. Indian Pulp Paper Tech. Assoc. **30**(1), 368–373 (2018)
8. Yershova, O.L., Bazhan, L.I.: Smart city: concept, models, technologies, standardization. Stat. Ukraine **89**(2–3), 68–77 (2020)
9. Lee, H.: Trends of developing training in advanced countries and direction of the army: focusing on the LVCG training system. J. Adv. Mil. Stud. **3**(2), 1–15 (2020)
10. Zhao, Y., Ma, Y.: Research on the development of nursing under the artificial intelligence and big data. Software **040**(006), 173–175, 196 (2019)

Choice of Financing Mode for Serial Entrepreneurs in the Big Data Era

Yuyang Pan$^{(\boxtimes)}$, Heyuzi Shi, and Guoying Niu

School of Management, Wuhan University of Technology, Wuhan, Hubei, China

Abstract. The advent of the big data era has had a huge impact on the development of various industries, corporate decision-making, and personal daily life. The relevant decisions of serial entrepreneurs are bound to be affected, and the drastic changes brought about by the big data era will also bring many entrepreneurial opportunities. We select serial entrepreneurs, a special entrepreneurial group, to study the financing options of companies under their control in the era of big data. We research on the impact of big data on serial entrepreneurs' control of corporate financing, find that the immediacy and accuracy of relevant information are guaranteed and companies can get more financing opportunities. The advent of the big data era has changed the macro environment. We make some suggestions on the choice of the financing method for the serial entrepreneur to control the enterprise. We Hope to provide data support and reference opinions for entrepreneurs' financing decisions, and help companies better ride the wave of big data and achieve good development.

Keywords: Big data · Serial entrepreneurs · Financing mode

1 Introduction

The choice of corporate financing is closely related to the survival and development of the company. In 1984, S.C. Myers and others put forward the theory of financing priority order [1], which believed that corporate financing generally follows the order of endogenous financing, debt financing, and equity financing. Yu Bo and others proved that the endogenous financing, debt financing and equity financing of listed companies on the GEM are positively correlated with corporate innovation performance [2]. Endogenous financing has a more significant impact on corporate innovation performance and is a better financing method for most companies. Debt financing and equity financing are the two main financing methods of external financing [3]. Found that investments with more remote and unstable returns are more likely to use equity financing. In contrast, investments in tangible non-unique assets or assets that require a high degree of monitoring are more likely to be financed through debt. The phenomenon of Serial entrepreneurship has become more and more common. Some scholars found after investigation that 12.5% of entrepreneurs in China are serial entrepreneurs [4]. In other countries, serial entrepreneurs also account for a large proportion. In the UK, the proportion of serial

J. Abawajy et al. (Eds.): ATCI 2021, AISC 1398, pp. 835–840, 2021.
https://doi.org/10.1007/978-3-030-79200-8_124

entrepreneurs among entrepreneurs is about 19%–25%; in Germany, the economic contribution of serial entrepreneurs accounts for 18% of all entrepreneurs' economic contribution, and this figure is even close to 30% in Finland [5]. At the same time, according to academic research, the performance of companies founded by serial entrepreneurs is better than that of novice entrepreneurs [6]. It can be seen that serial entrepreneurs have gradually become the backbone of entrepreneurs, and the value and contribution of serial entrepreneurs to economic development cannot be ignored. Research on the financing options of serial entrepreneurs, a special entrepreneurial group, has more practical significance and reference value.

With the rapid development of science and technology, coupled with the huge data transmission volume brought about by the upcoming 5G communication technology, big data technology has become more and more important. Generally, "big data" is defined as a collection of data whose content cannot be captured, managed and processed with traditional database software tools within a certain period of time. It has the characteristics of large-scale, diversified categories, and rapid processing. "Big data" has quickly become a research hotspot in many fields. The advent of the big data era has had a huge impact on the development of various industries, corporate decision-making, and personal daily life. The relevant decisions of serial entrepreneurs are bound to be affected, and the drastic changes brought about by the big data era will also bring many entrepreneurial opportunities. Therefore, this article selects serial entrepreneurs, a special entrepreneurial group, to study the financing options of companies under their control in the era of big data. Hope to provide data support and reference opinions for entrepreneurs' financing decisions, and help companies better ride the wave of big data and achieve good development.

2   The Impact of Big Data on Serial Entrepreneurs' Control of Corporate Financing

2.1   The Immediacy and Accuracy of Relevant Information are Guaranteed

Scholars such as Deniz have pointed out that experienced entrepreneurs can identify more and more novel entrepreneurial opportunities [7]. It can be seen that for serial entrepreneurs, if financing is required, the timeliness and accuracy of information are very important. In the absence of information connection and integration, even with rich experience, serial entrepreneurs still need to spend a lot of time and energy to investigate, analyze and evaluate the development direction of the company, the external environment and other information, and finally determine the financing plan. Moreover, the external environment is too complicated and the information update rate is rapid. If you can't get the new information in time, it will be difficult to make correct financing decisions, thus missing the best time for development. The advent of the big data era has brought more opportunities and hopes for serial entrepreneurs. Through big data technology, serial entrepreneurs can obtain valuable information from a large amount of information, and the information obtained is more accurate and targeted, greatly reducing the degree of information asymmetry, and alleviating the current financing dilemma of enterprises. In the big data environment, serial entrepreneurs can not only get common related data, but also more valuable information, such as through data mining, web crawlers, cloud computing, and deep learning technologies. They can also analyze the news released by the media and various fragmented information to obtain horizontal comparison data between hot spots in different fields to help serial entrepreneurs make decisions that are more conducive to corporate development. In addition, big data can also track and inspect the use of financing funds, determine the degree of financing risks, enhance the ability to control the risks of financing funds, and analyze corporate financing risk analysis. At the same time, it can also help companies identify whether there is false information in finance, and then provide a basis for corporate financing decisions.

2.2 Companies Can Get More Financing Opportunities

The growth of the company, company performance, credit rating, capital structure and other levels will affect the results of financing. Companies with higher growth have greater opportunities to obtain equity financing, and companies with good business performance have more options and bargaining power in both equity financing and debt financing. Mathematical model analysis shows that serial entrepreneurs are not a random group, but the result of entrepreneurs' self-selection [8]. This shows that serial entrepreneurs are often very willing to explore new business opportunities and eager to obtain financing for development. In the era of big data, serial entrepreneurs can more completely disclose corporate information and show the charm of projects to ensure that they are understood by more investors, gain trust from more investors, and get more development opportunities. Nowadays, with the help of big data, not only its own information can be displayed more completely, but also the financing preferences of various industries can be queried in the database. For example, in the wind database, you can query financing-related data such as internal financing rates, external financing rates, and equity financing rates for various industries. Industry-related data is also more transparent, making it easier for entrepreneurs and investors to make relevant references.

2.3 The Advent of the Big Data Era has Changed the Macro Environment

In recent years, digital technologies such as the Internet, mobile Internet, and Internet of Things have been widely used in enterprise production and operation, government governance, and residents' lives. This has produced massive, high-frequency and diversified big data, which has had a great impact on all levels. At the government level, the Fourth Plenary Session of the 19th Central Committee of the Party included data for the first time in production factors and emphasized the formation of its income distribution mechanism, indicating that the government attaches great importance to the coming of the era of big data. In the current data explosion, the government statistics department is no longer the main owner and sole publisher of economic and social data. This has reduced the burden on government agencies and also brought new challenges. At the enterprise level, in the era of big data, the rise of Internet companies has gradually brought down traditional industries. Internet companies are realistic and innovative, and the integration of traditional industries and the digital economy has gradually deepened. This has given birth to many new industries, new business models, new business models and new employment models. In this context, many companies have developed a large number of self-use software in order to keep pace with the times. At the level of residents' lives, my country currently has a full range of online and offline mobile payments. At present, mobile payment is widely used in the field of consumption in my country, and it is very common to use scan code payment in online and offline shopping, catering and accommodation, life payment, travel, education and entertainment scenarios. In 2019, WeChat Pay has covered most regions, involving more than 100 industries, with an average daily transaction volume of more than 1 billion times, and connecting 50 million individual merchants and merchants. The number of Alipay users in 2019 also exceeded 1 billion. The changes in the macro environment have brought more opportunities and more challenges for serial entrepreneurs.

3   Suggestions on the Choice of the Financing Method for the Serial Entrepreneur to Control the Enterprise

In the actual business operation, there are many uncontrollable risks and uncertain obstacles. In addition to theoretical and scientific forecast guidance, the choice of corporate financing methods is also closely related to its social system, economic environment, financial system and its own conditions. Different countries, social systems, economic development levels and politics, history, culture, etc., make the capital needs and channel choices of enterprises vary greatly. The financial system selected by the government is not the same, and different financial system regulations restrict the financing methods of enterprises. The factors that enterprises need to consider when choosing financing methods are multi-dimensional, not only considering the single points of financing costs, risks assumed, and the proportion of released equity. Enterprises must also comprehensively analyze the cash flow and risk, equity structure, financial flexibility, and how to protect the equity of shareholders and creditors, while improving corporate management and maximizing corporate value. The advent of the big data era has more or less impact on all dimensions, thus affecting the financing decisions of serial entrepreneurs.

We investigated the annual reports of listed companies controlled by 229 serial entrepreneurs and analyzed their financing methods. By studying the ratio of internal financing and external financing, equity financing and debt financing that serial entrepreneurs control, we analyze the financing method preference of serial entrepreneurs in the era of big data. Through research, we found that 80.12% of serial entrepreneurs have a higher rate of endogenous financing, and only a few serial entrepreneurs use external financing as their main financing method. In equity financing and debt financing, 55.90% of serial entrepreneurs prefer to use debt financing as the main financing method. At the same time, we combine the experience of serial entrepreneurs to further analyze the two main methods of external financing for serial entrepreneurs. Experience improves the accuracy of performance forecasts [9]. Compared with novice entrepreneurs and entrepreneurs with failure experience, entrepreneurs with previous successful experience are more likely to choose the right industry and time for Serial entrepreneurship, which makes it easier to achieve success and have better performance [10]. The number of entrepreneurial entrepreneurs and pre-existing industries also have a certain impact on their decision-making. Therefore, we believe that serial entrepreneurs who have started a business for many times have multiple experience; serial entrepreneurs who have successfully started a business have successful experience; serial entrepreneurs who have controlled the enterprise and are related to the industry that currently controls the enterprise have homogeneous experience. We conducted a comprehensive analysis of this, and the results are as follows (Table 1).

Table 1. The bond financing ratio in different experience

Bond financing ratio	Number	Multiple experience	Successful experience	Homogeneous experience
0–30%	45	64.44%	17.78%	24.44%
30%–70%	139	61.15%	12.23%	18.71%
70%–100%	45	73.33%	15.56%	4.44%
Average		64.19%	13.97%	17.03%

It can be found that most serial entrepreneurs have multiple experiences, and only a few serial entrepreneurs have successful experience and homogeneous experience. Among the serial entrepreneurs whose debt financing ratio is 70%–100%, 73.33% of serial entrepreneurs have multiple experiences, which is higher than the average of 64.19%. This shows that serial entrepreneurs with multiple experiences are more inclined to adopt higher-risk financing methods to keep the asset-liability ratio at a higher level and use leverage to obtain greater profits. Simon once pointed out that entrepreneurial experience helps entrepreneurs improve their ability to develop network relationships and build social capital [11]. Therefore, it may also be that serial entrepreneurs with multiple experiences often have more developed social networks and more channels to obtain debt financing. Among the serial entrepreneurs whose debt financing ratio is 70%–100%, only 4,44% of serial entrepreneurs have homogeneous experience, which is far lower than the average 17.03%. This shows that serial entrepreneurs with homogeneous experience have a deeper understanding of the industry and are more inclined to make sound financing decisions and avoid greater risks. Therefore, in the era of big data, serial entrepreneurs need to establish self-perception systems, have a clear understanding of risks, and improve market management's cognition, social cognition, interpersonal cognition and other related content. And they need to strengthen the study of professional theoretical knowledge, make full use of big data technology, and improve the ability of information collection, measurement, evaluation, judgment, control and improve their own risk awareness. At the same time, serial entrepreneurs need to avoid blindly following the trend, making misjudgments and leading to mistakes in decision-making. Serial entrepreneurs must combine their own resources and market environmental conditions, and need to enter the market to conduct research before financing, to explore and discover profitable markets. Based on this, the concept and characteristics of the company's new products are formed, so as to smoothly enter the target market that is compatible with its own market environment. Entrepreneurs cannot decide their entrepreneurial projects based on their personal interests and experience. In addition to rigorously reviewing the project scientifically and conducting project.

Demonstrations, and not blindly operating, it is also necessary to clarify a series of favorable and unfavorable factors for the project. Entrepreneurs need to conduct a comprehensive theoretical analysis from the quality of the company's products and market, the quality of business personnel, to the company's financial status, operation and management, etc., in order to make correct financing decisions and choose the most appropriate financing method.

4 Conclusions

Big data brings more opportunities for serial entrepreneurs, but also higher risks. At present, big data is still developing, and the environment is still constantly changing. Serial entrepreneurs should correctly evaluate their own strengths, and actively seize opportunities to raise funds in a reasonable way to ensure the development of the enterprise and make a grand plan in the era of big data.

Acknowledgements. This work was supported by the Fundamental Research Funds for the Central Universities.

References

1. Myers, S.C., Majluf, N.S.: Corporate financing and investment decisions when firms have information that investors do not have. North-Holland **13**(2), 187–221 (1984)
2. Yu, B., Zhao, J.: Financing method, R&D investment and innovation performance. China Certified Public Account. **5**, 56–59 (2019). (in Chinese)
3. Grundy, B.D., Verwijmerenm P.: The external financing of investment. J. Corp. Finan. **65** (2020)
4. Anokhin, S., Grichnik, D., Hisrich, R.D.: The journey from novice to serial entrepreneurship in China and Germany: are the drivers the same? Manag. Glob. Transitions **6**(2), 117 (2008)
5. Jose, P.D.: A theory of entrepreneurship. Small Bus. Econ. **35**(4), 377–398 (2010)
6. Paik, Y.: Serial entrepreneurs and venture survival: evidence from USventure-capital-finance dsemiconductor firms. Strat. Entrep. J. **8**(3), 254–268 (2014)
7. Ucbasaran, D., Westhead, P., Wright, M.: The extent and nature of opportunity identification by experienced entrepreneurs. J. Bus. Ventur. **24**(2), 99–115 (2009)
8. Rocha, V., Carneiro, A., Varum, C.A.: Serial entrepreneurship, learning by doing and self-selection. Int. J. Indust. Organ. **40**, 91–106 (2015)
9. Cassar, G.: Industry and startup experience on entrepreneur forecast performance in new firms. J. Bus. Ventur. **29**(1), 137–151 (2014)
10. Gompers, P., Kovner, A., Lerner, J., et al.: Performance persistence in entrepreneurship. J. Finan. Econ. **96**(1), 18–32 (2010)
11. Mosey, S., Wright, M.: From human capital to social capital: a longitudinal study of technology-based academic entrepreneurs. Entrep. Theory Pract. **31**(6), 909–935 (2007)

Development and Innovation of Big Data Application in the Media Industry from the Digital Perspective: 1950–2020

Dianyi Wu, Siwei Long$^{(\boxtimes)}$, Yan Zhou, and Sucheng Chen

Department of Advertising, Communication University of China, Beijing, China
longsw@cuc.edu.cn

Abstract. Big data plays an important role in current media operation, but it must be realized that the seemingly new concept and high value of big data has actually been accumulated over decades. The author goes back to the 1950s when computers were just used and the media industry entered digital era. During the 70 years, after four stages of development, with numerous algorithms, technologies, and tools keeping mature, and media digitization process moving towards the direction of comprehensive and integrated, big data has gradually become the core force in today's media operation. The review of this development process, on one hand, demonstrates clearly the characteristics of the integration of technology and media, and the role of media in technological development, on the other hand, provides us with many new perspectives of the relationship between big data and media operation.

Keywords: Big data · Media digitalization · Media industry

1 Introduction

Where does big data come from? What is its relationship with the development of new media? Why is today's big data showing high value in new media and even broader application scenarios?

The definition of big data-'big data refers to a large amount of data that exceeds the processing capacity in the past and in turn leading to a new set of technical system adapted to this' [1] will force us to think. Big data can play its value because of the simultaneous evolution of data processing technology, and neither the development of data nor the evolution of technology is one day's work.

Today, big data has become an interdisciplinary comprehensive research field: it can handle multiple types of data such as numerical values, text, pictures, audio, video, etc., can deal with the processing needs of large amounts of data, and can describe the overall situation and appropriately predict and be able to solve problems in multiple scenarios. This involves the integration of various technologies including statistics, network science, distributed, database, data mining, natural language processing, computer graphics, etc., in fact, it has experienced a long development process.

J. Abawajy et al. (Eds.): ATCI 2021, AISC 1398, pp. 841–847, 2021.
https://doi.org/10.1007/978-3-030-79200-8_125

The current research on big data mostly started from the formal proposition of the concept of big data, which was 2008 [2], and focused on several values of big data [3]. The author, however, believes that we should go back and see that big data has existed before 2008, and its related technologies have been accumulating and forming a system under the promotion of multiple routes. After 2008, it has entered a relatively fast-growing period. Through this retrospective, we will have a deeper understanding of the connotation of big data, the technology involved, and its connection with new media

2   Data Collection and Processing

The development of big data is always closely related to the evolution of modern information technology. It may be useful to divide the development of big data into four stages: the reserve period, the embryonic period, the conceptual period, and the cultivation period.

2.1   Reserve Period: 1950–1980, Starting from the Invention and Application of Mainframes and Databases, with the Purpose of Solving the Management and Calculation Problems of Large Amounts of Data in the Fields of Business, Science, Military, and Polls

It is reasonable to say that the occurrence and development of big data can be traced back to the early stages of mainframes and databases. The reason for this is that big data points to the data and the corresponding processing system, which cannot be separated from the hardware and software required for computing. Today, a considerable proportion of data collection, management, and mining systems and ideas have to go back to the relatively early days even before the Internet and micro computers were born (Table 1).

Table 1. Milestones and cases of reserve period

Factor	Time/Event/Case
Hardware	IBM mainframe comes out
Data sources	Data from banks, governments, scientific research departments, etc.; traditional sampling methods are still very important data sources
Data management technology	Papers and products of relational database come out
Data analysis technology	The concept of artificial intelligence was proposed and explored; the concept of data mining took shape, and a variety of algorithms were proposed; NLP technology and machine translation were developed in the exploration
Data application scenarios	Banks, commercial retail, government departments, etc.
Typical institutions	IBM, ORACLE, Nielsen, etc.

After the origin of computer technology, under the impetus of the government, banks and other institutions that need a lot of computing power to deal with large amounts of data, hardware technology, data management technology, data mining and analysis technology have gradually developed. Many tools and algorithms, including relational databases [4], classic data mining algorithms [5], natural language processing algorithms [6], etc. were proposed and discussed, laying a solid foundation for future development, and are still existing and playing a role in current big data application scenarios.

At this stage, large-scale computer providers represented by IBM and large-scale relational database providers represented by Oracle, etc. provided solutions for market needs. At the same time, although the media landscape has not yet been impacted by the Internet, the data requirements (packaging analysis, market research, etc.) of the media and enterprise operations already existed. Research institutions represented by Nielsen collected and analyzed data and provided services based on sampling theory. Of course, with the increase in data volume and the complexity of calculations, after the development of computers, Nielsen also bought IBM's mainframe computers to improve efficiency.

2.2 Embryonic Period: 1980–2007, from the Birth of the Internet, the Popularization of Minicomputers, to the Publication of Google's Papers in 2007, the Rapidly Developing Internet Actually Started the Era of Big Data

The maturity and promotion of personal computers, as well as the establishment and civilian use of the Internet, actually started the era of big data. Unlike the previous stage when large amounts of data came from government, financial, scientific and other institutions, in this period data came from a wider range of ordinary users (Table 2).

Table 2. Milestones and cases of embryonic period

Factor	Time/Event/Case
Hardware	Driven by companies such as IBM, Apple, and Microsoft, personal computers came out and gradually gained popularity; virtualization technology has begun to be valued, giving the first sign of cloud computing in the future
Internet	The Internet was born, and many protocols such as TCP/IP, HTTP, Wi-Fi were developed one after another, which affected the later data collection standards in multiple dimensions
Data sources	Large-scale departmental data; personal website webpage data, user behavior data, etc.
Data management technology	Parallel computing technology, Google paper; the concept of cloud computing is proposed
Data analysis technology	Index technology, recommendation technology, etc.
Data application scenarios	In addition to the previous stage, recommendations and precise advertisements on the Internet have surfaced
Typical institutions	Fixed network operators, Google, Amazon, Baidu, Microsoft, etc.

With the maturity of the computer equipment for individual users and the standardizing of the network, the communication rules between people and between personal computers and servers were ensured, which build a solid network foundation. People were connected to each other through the Internet with computers, and people's creativity was constantly being stimulated, the number of websites and web pages was expanding, and online behaviors were constantly being recorded in the form of data.

The most representative big data institutions in the new media field at this stage were search engines and e-commerce platforms. On one hand, search engine giants such as Google and Baidu have accumulated a large number of web pages on the Internet through technologies such as crawlers. On the other hand, in order to meet people's search needs, they continue to promote the development of distributed computing [7] and natural language processing technologies [8]. E-commerce platforms represented by Amazon are also actively promoting the progress of recommendation algorithms. At the same time, these early Internet platforms actually explored the technology and products of precision marketing.

2.3 Concept Period: 2008–2018, the Concept of Big Data Was Formally Proposed Under the Circumstance of Global Expansion of Mobile Internet and then Strengthened by Cloud Computing and AI

During this period, once the concept of big data was proposed, there were many responders, and big data became a hot topic chased by capitals and scholars. At the same time, the development of the mobile Internet has added more support to big data, including the launch of Apple and Android terminals to detonate the smartphone market. The maturity of 3G and 4G and large-scale civilian use have enabled the booming of mobile services (Table 3).

Table 3. Milestones and cases of concept period

Factor	Time/Event/Case
Hardware	Apple and Android launched smart phones; GPUs were used and developed vigorously; cloud computing, virtualized and flexible computing resources were promoted
Internet	3G, 4G, Wi-Fi
Data sources	Mobile Internet data continued to explode, coming from a variety of applications; offline data collection was also valued
Data management technology	Unstructured database technology development
Data analysis technology	Artificial intelligence revived, NLP, speech recognition, especially image recognition, etc. made great progress
Data application scenarios	Programmatic buying, new retail, short video platform, etc.
Typical institutions	Apple, Alibaba, Tencent, Bytedance, etc.

Therefore, the amount and variety of data became unprecedentedly rich, and the imagination space of the combination of big data and multiple application scenarios started to make revolution in every industry. In the field of new media, programmatic adverting is one example, and the rise of personalized information platforms represented by ByteDance is another.

2.4 Cultivation Period: 2019 to Present, Digitalization Under the Context of Big Data Is Re-Emphasized, and Data-Based Applications Such as Content Technology and Marketing Technology Have Entered in-Depth Development

In the previous stage, it can be found that the main driving force should be large platforms. In fact, they realized the existence and value of big data earlier, and accumulated a large amount of data and related technologies earlier, so they can gather massive amounts of data and realize the absorption of users and customers with accurate content and advertising. In fact, it has also formed a game relationship with a large number of small and medium-sized media, various enterprises, and marketing agencies (Table 4).

Table 4. Milestones and cases of cultivation period

Factor	Time/Event/Case
Hardware	Smart homes, portable terminals, smart cars, etc., as well as more flexible computing resources
Internet	5G, Internet of Things, etc.
Data sources	More comprehensive data sources, especially individual 'biological information' has become a new data source [11]
Data management technology	Unstructured database technology development
Data analysis technology	Deep learning, etc.
Data application scenarios	Various institutions begin to attach importance to the close integration of their own data application scenarios and data; at the same time, the application of cloud computing-based tools such as marketing cloud is becoming more popular [12]
Typical institutions	Accenture Interactive, Salesforce, etc.

After the hype of the concept of big data, as well as the test and baptism of many 2C applications, whether it is media or enterprises, they have gradually realized the importance of using data to complete and optimize their business. In the context of big data, the concepts of digital transformation, 2B services, and industrial Internet are integrated. Whether it is 'smart media integration' [9] or 'enterprise data middle office', it can be regarded as a reframe of the concept of big data under this purpose. Data is owned and used by 'me' [10], and has entered the stage of intensive cultivation based on big data and industrial demand scenarios. Of course, in terms of actual operation, it must be admitted that there are still more noises at the moment, and there are many people who are greedy and adventurous and blindly pursue concepts.

3 Conclusions

Data and big data are not new terms in today's social operations, but what does it point to behind the widespread use? The connotation of "data" is a process of continuous enrichment, and it is also a process in which the data industry, information technology industry, and traditional industries enter each other. In this process, the data technology is continuously added to form a complex and three-dimensional "technology". Understanding this process will enable us to get rid of generality and truly see "data technology" clearly.

The term "big data", or "mass data", has existed for a long time in the fields of physics, biology, environmental ecology, military, finance, and communications. The development of the Internet and the information industry will push "big data". When we arrive at the stage, more and more companies and individuals are engaged in the mining and utilization of massive data. In contrast, traditional data institutions such as market research companies, as the core occupants of data resources in the era of small data (sample survey), are organically integrated with the information technology industry in the wave of big data, and are constantly upgrading their data business system. Traditional industries have a more urgent and huge demand for the application of their own business data in the grip of the Internet and the information industry. This demand drives the information industry and data industry to continuously enter new scenarios and help traditional industries build big data application architectures.

The circulation of data resources, the development of data processing science and technology, the accumulation of software and hardware resources, and the booming business driven by industry demands, let us gradually see an interactive panorama of data. At the bottom of this panorama are "data" and "data technology". The statistical science (sampling theory, etc.) that constitutes the cornerstone of the survival of traditional research institutions, the information science (entropy model, etc.) that gave birth to modern information networks, and the computer science (databases, etc.) that pervade daily life, along with the entire information industry, data industry, and traditional The development of the industry has been integrated in practice, forming an all-encompassing category of "data technology", and at the same time, it has become incarnate in all corners of the media.

References

1. Chen, M., Mao, S., Liu, Y.: Big data: a survey. Mobile Netw. Appl. **19**(2), 171–209 (2014)
2. Bill, B., Vincent, H., Ian, P., Leo, K., Helen, G., Esther, D.: Big data: the next Google. Nature **455**, 8–9 (2008)
3. Lavalle, S., Lesser, E., Shockley, R., et al.: Big data, analytics and the path from insights to valu. Mit Sloan Manage. Rev. **52**(2), 21–32 (2010)
4. Codd, E.F.: A relational model of data for large shared data banks. Commun. ACM. (6) (1970)
5. Shott, S.: Logistic regression and discriminant analysis. Publ. Am. Stat. Assoc. **73**(11), 699–705 (1978)
6. Kutas, M., Hillyard, S.A.: Reading between the lines: event-related potentials during language processing. Brain Lang. **11**(2), 354–373 (1980)
7. Michael, S., et al.: MapReduce and parallel DBMSs: friends or foes? Commun. ACM **53**(1), 64–71 (2010)

8. Saquete, E., Tomas, D., Moreda, P., et al.: Fighting post-truth using natural language processing: a review and open challenges. Exp. Syst. Appl. **141**(3),112943.1–112943.27 (2020)
9. Helles, R., Rmen, J.: Big data and explanation: reflections on the uses of big data in media and communication research. Eur. J. Commun. **35**(3), 290–300 (2020)
10. Liu, H.: Big data precision marketing and consumer behavior analysis based on fuzzy clustering and PCA mode. J. Intell. Fuzzy Syst. **2**, 1–11 (2020)
11. Han, H., Liu, W., et al.: The coming era of artificial intelligence in biological data science. BMC Bioinf. **20**(Suppl 22), 712 (2019)
12. Sheth, J.: New areas of research in marketing strategy, consumer behavior, and marketing analytics: the future is bright. J. Mark. Theory Pract. **2**, 1–10 (2021)

Tourism Development and Residents' Happiness Index Based on Big Data Analysis

Qi Zhou[✉]

Guangzhou Huashang Vocational College, Guangzhou 511300, Guangdong, China

Abstract. With the advent of the Industry 4.0 era and the advancement of our country's urbanization process, the flow of information, capital and population in cities will accelerate. The strategic goals of urban planning and construction will increase the competitiveness of the city and increase its attractiveness. Influenced by humanistic concepts, cities have gradually become a complex of life, work and leisure. On the basis of abundant material resources, the growth of spiritual and cultural demand has encouraged the "tourism" of urban residents. Tourism and entertainment have begun to integrate into urban space and industry, which also affects the happiness index of residents. This article mainly adopts the literature reference research method and model setting method, uses big data technology to study tourism development and residents' happiness index, establishes a mathematical model, uses model setting method to solve, and analyzes tourism development and residents' happiness index research, and at the same time use historical data to revise the model to improve the accuracy of the evaluation of tourism development and residents' happiness index. The experimental results show that the model setting method increases the research efficiency of tourism development and residents' happiness index by 43%, and reduces the rate of false positives and false positives. Finally, by comparing the mechanism analysis of tourism development affecting residents' happiness and the analysis of urban residents' happiness index as a tourist destination, it illustrates the influence of big data technology on tourism development and residents' happiness index research.

Keywords: Big data · Tourism development · Residents' happiness index · Literature reference research method

1 Introduction

1.1 Background and Significance

The basic purpose of development is to continuously improve lives and ultimately make people feel happier. In the traditional development-oriented approach, most studies have shown that with the rapid economic growth, the subjective well-being or satisfaction of Chinese people has not improved. The happiness of residents is not only related to personal characteristics, but also closely related to various external conditions that affect the quality of life [1, 2].

J. Abawajy et al. (Eds.): ATCI 2021, AISC 1398, pp. 848–855, 2021.
https://doi.org/10.1007/978-3-030-79200-8_126

Tourism, as one of the emerging industries, not only increases the income of the population in the tourist destination, but also improves the happiness of the local residents in the tourist destination [3, 4]. In recent years, the country has made full use of the rich local tourism resources in accordance with the development strategy of developing grand tourism, greatly improving the tourism competitiveness of tourist destinations and increasing the happiness index of local residents [5, 6].

1.2 Related Work

Jie and Zhao provide a method that can evaluate related innovations with participation in a complex related environment to solve essential problems [7, 8]. Based on the principle of common value creation, he proposed a research framework. This framework illustrates the research process of big data technology in tourism development and residents' happiness index. In this process, stakeholders integrate their resources and capabilities to develop innovative model setting methods [9, 10]. In order to evaluate this research framework, Jie and Zhao collected multiple data in their research. This case also represents the significance of big data technology in tourism development and residents' happiness index research and system implementation [11, 12]. Because of the relatively few data variables in his research, the results are not convincing.

1.3 Main Content

The innovation of this article lies in the literature reference research method and model setting method. Based on the research of tourism development and residents' happiness index under the background of big data, the relationship between tourism development and residents' happiness index is studied through the model setting method. The establishment of a model setting method combined with a calculation method of the literature research method provides guidance for studying the relationship between tourism development and the happiness index of residents in the context of big data.

2 Method of Tourism Development and Residents' Happiness Index Based on Big Data Analysis

2.1 Literature Reference Research Method

In the existing literature, there are two main factors that affect residents' well-being: one is the characteristics of individuals and families, and the other is external environmental factors that affect residents' emotions. Although people's research on residents' happiness includes many factors, the current literature has not yet fully included the research on residents' happiness through economic development methods. Faced with the ever-improving environment and resource constraints, the challenge of accelerating changes in economic development is imminent. The development of tourism does not unilaterally seek economic growth or protect the ecological environment, but the organic unity of economic and social development and ecological environment improvement. Therefore, the impact of tourism development on residents' well-being not only has different effects on economic growth, but also has different impacts on the ecological environment. There is no empirical study to verify that tourism development has an impact on residents' well-being.

2.2 Model Setting Method

This paper uses panel data to test the impact of residents' happiness index on the development of tourism. The model is as follows:

$$Tour_{it} = \eta live_{it} + \sum \delta_j X_{it} + \xi_{it} \tag{1}$$

Among them, $Tour_{it}$ is the tourism development level of i place t period, $live_{it}$ is the residents' happiness level of i place t period, and X_{it} are other control variables that affect the level of tourism development. Because residents' happiness includes many related indicators, such as economic foundation, service industry development, urban infrastructure and natural environment.

In order to study the impact of tourism development on residents' happiness, this article created a model as follows:

$$live_{it} = \gamma Tour_{it} + \mu_{it} \tag{2}$$

Among them, $Tour_{it}$ is the tourism development level of i place t period, and $live_{it}$ is the residents' happiness level of i place t period. Because residents' well-being is a multi-dimensional concept, the following content combines several indicators that affect residents' well-being and comprehensively measures them through factor analysis. Therefore, the control variables are no longer defined in Eq. (2).

3 Experiment on Tourism Development and Residents' Happiness Index Based on Big Data Analysis

3.1 Empirical Research Design

The development of urban traffic depends on the range of tourist attractions and the comfort of traffic in the area. In recent years, high-speed rail has been developed through the construction of high-level tourist attractions. The improvement of public tourism services and the tourism system has stimulated the development and transformation of urban tourism. In this paper, train passengers are used as variables for external passengers. According to the distance theory, the flow of tourists expands as the distance increases. This shows that the growth of urban population will support the development of tourism, so this article uses population to show the strength of the urban tourism market.

With the rapid development of tourism, people pay more and more attention to the economic value of tourism products, especially the acceptability of regional tourism, which plays a key role in the development of tourism. Related research also shows that improving tourism reception facilities has a positive effect on promoting the development of tourism. This article is limited to the availability of data and uses the number of star-rated hotels in the city to measure the accommodation capacity of tourists. The development of tourism and the opening of foreign economies complement each other. Relevant studies have shown that the degree of economic openness has a significant positive impact on the development of tourism, and there is a connection that cannot be ignored between international trade and inbound tourism.

3.2 Collection of Tourism Development Data

In order to measure the development level of the tourism industry more comprehensively and objectively, this paper selects four original indicators that can reflect the development level of the tourism industry, including foreign exchange income from tourism, domestic tourism income, number of inbound tourists and number of domestic tourists, through improved entropy method, to obtain the comprehensive results of the development level of the city's tourism industry, the data results are shown in Table 1:

Table 1. Entropy index data table of tourism development level

Indicator name	Foreign exchange income from tourism	Domestic tourism income	Inbound tourists	Domestic tourists
Index entropy	0.942	0.941	0.944	0.939
Index difference coefficient index weight	0.056	0.057	0.054	0.059
Indicator name	0.248	0.253	0.239	0.260

It can be concluded from Table 1 that the data difference in index entropy, index difference coefficient and index weight is not large in foreign exchange income from tourism, domestic tourism income, inbound tourist arrivals and domestic tourist arrivals. The averages are 0.9145, respectively. 0.0565 and 0.25, however, the index entropy value of inbound tourist arrivals is the highest, and the index difference coefficient and index weight of domestic tourist arrivals are the highest. In order to be able to see the relationship between the experimental class and the control class, data analysis was carried out on Table 1. The analysis results are shown in Fig. 1:

It can be seen from Fig. 2 that the index entropy value, index difference coefficient and index weight trend of tourism foreign exchange income, domestic tourism income, inbound tourist arrivals and domestic tourist arrivals are relatively flat, and the data gap between them is not large, indicating that tourism Foreign exchange income, domestic tourism income, inbound tourist arrivals and domestic tourist arrivals have a relatively even impact on tourism development.

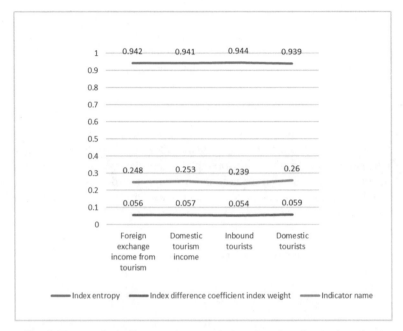

Fig. 1. Data analysis diagram of entropy index of tourism development level

4   Tourism Development and Residents' Happiness Index Based on Big Data Analysis

4.1   Mechanism Analysis of Tourism Development Affecting Residents' Happiness

The development of tourism is an organic combination of economic and social development and ecological improvement. It is a model of harmonious coexistence between man and nature. It aims to develop production, enrich life and protect ecology. The impact of tourism development plays a role through economic growth and the impact of tourism on the happiness of residents. From the perspective of the impact of economic growth, tourism and economic growth are mutual, not unilateral economic growth, but at the expense of the environment, it is based on the harmonious growth of man and nature to achieve economic growth. Regarding research on economic growth and residents' well-being, some studies based on empirical data have found that there is no obvious positive correlation between economic growth and the improvement of residents' well-being, so there is a "happiness paradox".

Regarding the research on Chinese residents' happiness, some reports believe that China's economic growth is conducive to increasing residents' happiness, but other studies have shown that there is a difference between China's economic growth and residents' happiness, for example, economic growth increases residents' wealth and income Level and improve the quality of life, but economic growth does not have much impact on improving the happiness of residents. The emergence of China's "happiness paradox" is usually related to the shortcomings of traditional growth methods. In the process of traditional economic growth, there are problems such as widening urban-rural income gap and income inequality. Insufficient government expenditures for people's livelihood and environmental degradation have had a major negative impact on the happiness of residents, and their negative impact has exceeded growth. Therefore, what is developing is not China's "happiness paradox", but the drawbacks of traditional growth methods. With the transformation of economic growth mode, economic growth under the tourism development concept is based on improving the urban and rural classes, income gap and equal opportunities, which will help to promote the happiness of residents.

4.2 Happiness Index of Urban Residents as Tourist Destinations

This article will take Huacheng as the research object and conduct an empirical analysis on the happiness index of residents living in tourist areas. As we all know, Conghua is located in the northeast of the area. This is a transitional area for collaboration and exchanges between major cities in the Pearl River Delta and surrounding areas. It has rich tourism resources and three world-class brands: "the world's rare heat source" and "China's first disease-free site", "the world's largest pumping station"; first-class Xihe National Forest Park, Shimen National Forest Park and the country's largest local race-course and other three national brands. The research object of this article is mainly the residents' happiness index of different scenic spots in Conghua City. A total of 300 questionnaires were distributed, 261 questionnaires were returned, and 215 real questionnaires were collected. The effective response rate of the questionnaire was 82.38%. The questionnaire accepts an anonymous sample survey. The survey locations include relatively densely populated scenic spots in the five cities and three streets of Conghua City. At the same time, the index system of residents' well-being in tourist destinations consists of four parts, namely material index, interpersonal index, mental index and work index. Under the four levels of index, each has a corresponding second-level index and weight to measure. Under the indicators, there are corresponding three-level indicators to be calculated according to the corresponding weights. Combining the method of measuring the happiness index of Guangdong residents and the method of measuring the happiness index of residents in tourist destinations, the data is shown in Table 2:

As can be seen from Table 2, the four values are 158.2, 168.9, 167.2, and 154.9 for the material index for measuring satisfaction, the interpersonal index for measuring value, the mental index for measuring pleasure, and the work index for measuring accomplishment. The maximum distance between these values is less than 10, and the results show that these three indicators are balanced development, that is, the satisfaction, value and enjoyment brought by the development of residents' tourism are balanced. In order to be able to see the relationship between the data tables, the data analysis of Table 2 was carried out, and the analysis results are shown in Fig. 2:

It can be seen from Fig. 2 that, on the one hand, the improvement of the residents' happiness index caused by the development of tourism is an overall improvement without structural imbalance, which reflects the integration, coordination and balance of the tourism industry. On the other hand, it also provides ideas for improving tourist destinations and increasing the happiness of residents. In this sense, improving the happiness of residents in Conghua is not only to establish a sustainable tourism industry in the city,

Table 2. Basic statistics of questionnaire survey

Basic information	Questionnaire	percentage	Resident happiness index
Education	High school and below College Undergraduate Master, PhD	16.28% 42.9% 35.35% 5.58%	158.2
Residence time	Less than 3 years 3–5 years 6–10 years Over 11 years	6.98% 26.90% 31.16% 34.00%	168.9
Marital status	Married Unmarried Divorced Widowed	17.68% 66.05% 12.09% 4.19%	167.2
Monthly income range	Below 1000 yuan 1000–2990 yuan 3000–4999 yuan 5000–10000 yuan 10,000 yuan or more	5.60% 33.49% 43.26% 13.49% 3.26%	154.9

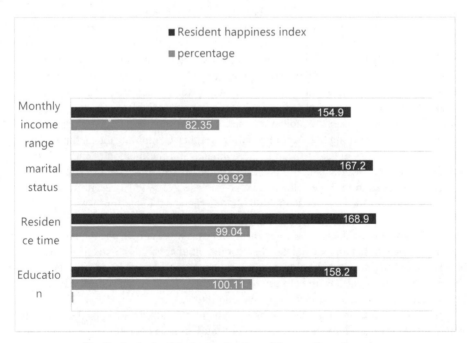

Fig. 2. Analysis of the basic situation of the questionnaire survey

but also to provide a good reference and development ideas for improving the happiness index of residents.

5 Conclusions

Although this article has achieved certain research results through the literature reference research method and the model setting method, there are still many shortcomings. There are still a lot of things worthy of in-depth research in the research of big data technology in tourism development and residents' happiness index. There are still many steps in the research process that cannot be covered because of space and personal ability. In addition, the actual application effect of the improved algorithm can only reach the level of comparison with the traditional model from the level of theory and simulation.

References

1. Kim, H.-S., Jung, L.-H., et al.: Study on the prevalence of allergic diseases based on the health behavior of multicultural families youth - the tenth Korea youth risk behavior web-based survey, 2014, centers for disease control & prevention. Korean Home Econ. Educ. Assoc. **29**(2), 41–52 (2017)
2. Gan, S., Wan, H., Wang, B., et al.: Evaluation and development strategy of Baoji intelligent tourism in the era based on big data%\"Big Data\" era of Baoji intelligent tourism evaluation and development strategy research. Henan Sci. **036**(009), 1494–1500 (2018)
3. Soukiazis, E., Ramos, S.: The structure of subjective well-being and its determinants: a micro-data study for Portugal. Soc. Indic. Res. Int. Interdiscipl. J. Quali. Life Meas. **126**(3), 1375–1399 (2016)
4. Peng, L., Liu, H., Nie, Y., et al.: The transnational happiness study with big data technology. ACM Trans. Asian Low-Resource Lang. Inf. Process. **20**(1), 1–12 (2020)
5. Zhang, Y., Padmanabhan, A., Gross, J.J., et al.: Development of human emotion circuits investigated using a big-data analytic approach: stability, reliability, and robustness. J. Neurosci. **39**(36), 7155–7172 (2019)
6. Wong, M.Y.H., Chui, W.H.: Economic development and subjective well-being: a comparative study of adolescents in Hong Kong and Macau. Child Indic. Res. **10**(1), 1–19 (2018)
7. Zhang, F.: Research on the development of tourism industry based on cluster analysis in Shanxi City% research on the development of tourism industry based on cluster analysis in Shanxi city. Jiangsu Shanglun **000**(007), 51–53 (2016)
8. Jie, Z.: Research on regional sustainable development based on evaluation of tourism competitiveness of 21 cities (states) in Sichuan Province. IOP Conf. Ser. Earth Environ. Sci. **242**(5), 52040 (2019)
9. Park, J.Y., Kim, G., Kim, C.Y., et al.: A study on tourism resource strategy of film location using social bigdata based on SNS trend analysis of Jeonju area. J. Korea Contents Assoc. **16**(11), 77–487 (2016)
10. Mao, H.: Research on the connotation and development framework of physical education teachers based on big data mining. Revista de la Facultad de Ingenieria **32**(16), 719–726 (2017)
11. Zhai, S., Duan, J., Ai, X.: Research on the third party logistics system and economic performance optimization based on big data analysis. Boletin Tecnico/Tech. Bull. **55**(11), 301–308 (2017)
12. Xu, X., Xie, F., Zhou, X.: Research on spatial and temporal characteristics of drought based on GIS using remote sensing big data. Cluster Comput. **19**(2), 757–767 (2016)

Precision Marketing Strategy of Insurance Market from the Perspective of Big Data

Ze Zhang[✉] and Qing Fan

China Mobile Information Technology Co., Ltd., Shenzhen 518000, Guangdong, China
zhangze@chinamobile.com

Abstract. In the face of the ever-changing and fierce marketing environment, enterprises are facing more and more pressure and challenges in the marketing process, and their traditional marketing methods have been unable to meet the needs of consumers. China's insurance industry has been developing for nearly 40 years since the reform and opening up, and it has also developed by leaps and bounds in recent years. At the same time, China's insurance industry presents a rapid development of diversification and scale. Firstly, based on the previous studies, this paper combs the literature on the current situation of precision marketing and big data mining at home and abroad, studies the relationship between big data mining and precision marketing on the basis of mastering certain related concepts and theories, and applies k-means algorithm to cluster the real sample data of a insurance company, and determines the minimum SSE and the most pseudo-F statistics through cyclic iteration. The optimal number of clusters can be determined by the large method. After modeling, through the analysis of each kind of sample characteristics, the user portraits are constructed, and the product preferences and precision marketing suggestions of all kinds of users are given respectively. Finally, the clustering results are divided into five categories, and the number of customers in each category presents an inverted pyramid structure, which is in line with the characteristics of the insurance company's pyramid customer group, and the marketing strategies for these five customer groups are given respectively.

Keywords: Marketing · Precision marketing · Data mining · Insurance marketing

1 Introduction

Relying on modern science and information technology, precision marketing carries out market segmentation on the premise of insight into consumers' behavior habits, and then selects one or several appropriate markets as the target market on this basis, and carries out accurate market positioning, so as to provide targeted and differentiated marketing strategies for target customers and obtain market competitive advantages [1–3]. Through decades of development, China's insurance industry shows a comprehensive and standardized development trend. At the end of 2018, China's total insurance capital exceeded 18330.8 billion yuan, an increase of 9.45% over the beginning of the year [4, 5].

© The Author(s), under exclusive license to Springer Nature Switzerland AG 2021
J. Abawajy et al. (Eds.): ATCI 2021, AISC 1398, pp. 856–865, 2021.
https://doi.org/10.1007/978-3-030-79200-8_127

This paper takes Shenzhen Branch of a insurance company as the research target, through reading the domestic and foreign literature related to the paper, on the basis of mastering a certain theory, combined with the current social development background, taking big data mining as an opportunity, through the research on the relationship between big data mining and precision marketing, finds the entry point of big data mining in precision marketing, and on this basis obtains the conclusion based on big data mining Data mining precision marketing strategy.

2 Precision Marketing Under Big Data

2.1 Premise of Precision Marketing

Precision marketing is a part of the development and evolution of marketing, which is inseparable from the essence of general marketing, that is to identify and obtain target customers, improve marketing efficiency and create profits for enterprises [4]. With the continuous change of marketing environment, marketing ideas and marketing methods are constantly upgrading. However, no matter what kind of marketing ideas or marketing strategies are adopted, they should be based on the understanding of users and be market-oriented and consumer centered. The premise of precision marketing is to have insight into consumer behavior. Only by mastering consumers' behavior habits and discovering their preferences and needs can we use personalized products and effective methods to cater to consumers' preferences and meet their needs, and truly implement the idea of taking consumers as the center [1].

2.2 Big Data Mining Methods

The task of data mining mentioned in this paper mainly includes classification and clustering. Among them, classification method is a process of training model on a given set of labeled samples, using labeled information discovery rules to build classification model, predicting unlabeled data features, and outputting its category.

Clustering method is different from classification method. It can realize the classification of data without sample label and divide the abstract set of things into several classes composed of similar objects. Clustering is first divided into several sub categories, and then to find the class features. Since clustering is an unsupervised learning algorithm, it is often regarded as a data preprocessing process [1]. In the process of clustering, we mainly follow the principle of high intra cluster similarity and low inter cluster similarity. The following is a brief introduction to the K-means clustering method used in the following.

The general steps of K-means clustering method are as follows: firstly, K observations are selected as the seeds of K class, then the distance between all observations and K seeds is calculated, and the observations are temporarily classified into the nearest seed class. Then, according to the observations in the existing class, the new seed of class center is recalculated, and then 2–3 steps are repeated until K is determined by convergence Class seeds, and finally read in all observations again to determine the class of each observation.

The objective function can be expressed as (1) [6]:

$$E = \min \sum_{j=1}^{k} \sum_{x_i \in C_j} (x_i - c_j)^2 \tag{1}$$

In the era of big data, people are no longer persistent in the pursuit of causality between things, but more to find the relationship between things. Association analysis is a method used to discover some hidden association relations, patterns or rules between things. The results of association analysis are often expressed in the form of frequent itemsets and association rules [7, 8].

3 Insurance Data Analysis Based on Big Data Mining

In order to effectively distinguish high-quality customers and general customers, and serve them with differentiated marketing strategies, this paper conducts a field survey of a insurance company, and obtains 5997 sample data, and selects eight indicators as customer segmentation criteria, including gender, age, education background, marriage, annual income, deposit amount, annual transaction amount and financial management amount.

3.1 Data Mining

As shown in Table 1, for the subsequent data calculation and analysis, this paper will process the data of each index. At the same time, in order to ensure that the index has a unified pointing effect, this paper stipulates that the larger the customer index value is, the more favorable the customer is to the business.

3.2 Empirical Analysis

K-means is a simple and easy to operate clustering method. We find the optimal number of clusters by iterating on the k-means method, and set the cluster with the smallest SSE and the largest pseudo-F statistic to win. The experimental results show that when the number of clusters is 5, the see reaches the minimum, and the pseudo-F statistic tends to be smooth. The variation trend of the two statistics with the number of clusters is shown in Figs. 1 and 2.

The initial cluster center is selected by random number, as shown in Table 2. The maximum number of iterations is set to 999, and the maximum absolute coordinate change of each iteration center is shown in Table 3 and Fig. 3.

Because there is no change or small change in the cluster center, convergence is achieved. The maximum absolute coordinates of any center are changed to. 000. The current iteration is 14. The minimum distance between the initial centers is 6.928.

Table 1. Quantification of indicators

Index	Quantification of indicators	
Age	Age under 25 years	4
	25–35 years	3
	36–50 years	2
	Age more than 50 years	1
Sexuality	Male	2
	Female	1
Educational background	Junior high school and below	1
	Senior middle school	2
	Undergraduate	3
	Graduate and above	4
Marriage	Married	2
	Unmarried	1
Annual earnings	Less than 30,000	1
	30–60 thousand	2
	60–100 thousand	3
	100–200 thousand	4
	More than 200,000	5
Amount deposited	Less than 50,000	1
	50–100 thousand	2
	100–200 thousand	3
	More than 200,000	4
Annual turnover	Less than 10,000	1
	10–200 thousand	2
	200–500 thousand	3
	More than 500,000	4
Amount of money	Less than 30,000	1
	30–60 thousand	2
	60–100 thousand	3
	100–200 thousand	4
	More than 200,000	5

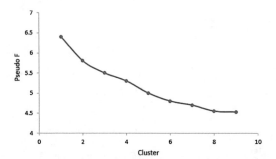

Fig. 1. Trend of pseudo-F value with clusters

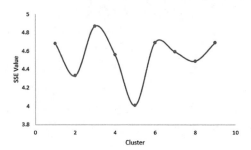

Fig. 2. Trend of SSE with clusters

Table 2. Initial cluster centres

	Cluster				
	1	2	3	4	5
Sexuality	1	2	1	2	2
Age	4	3	4	2	2
Educational background	3	3	3	2	3
Marriage	1	2	1	1	1
Annual earnings	1	2	1	4	5
Amount deposited	1	2	1	3	5
Annual turnover	2	1	2	5	4
Amount of money	3	2	3	1	5

Table 3. Iterative history

Iteration	Changes within clustering centres				
	1	2	3	4	5
1	2.708	2.719	2.816	2.799	2.749
2	0.256	0.360	0.279	0.230	0.304
3	0.189	0.216	0.211	0.182	0.212
4	0.125	0.100	0.110	0.108	0.181
5	0.098	0.057	0.113	0.108	0.172
6	0.038	0.067	0.133	0.117	0.148
7	0.072	0.060	0.138	0.079	0.122
8	0.067	0.032	0.081	0.052	0.112
9	0.048	0.017	0.065	0.022	0.047
10	0.019	0.005	0.004	0.003	0.036
11	0.005	0.007	0.000	0.000	0.015
12	0.003	0.003	0.000	0.000	0.007
13	0.003	0.000	0.000	0.000	0.005
14	0.000	0.000	0.000	0.000	0.000

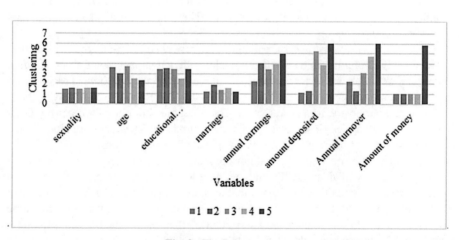

Fig. 3. Final cluster center

3.3 Data Analysis is Based on the Results of Customer Segmentation

As shown in Fig. 4. Combined with the fact that the overall income of residents in China is a pyramid distribution rather than a big pie structure of common prosperity, with the growth of the average age of each category, its total asset value is rising exponentially. The clustering results have obvious social characteristics, the fifth group of customers is the richest, the average age is the longest and the number is the least, only 73 cases;

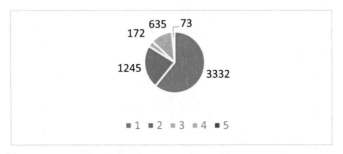

Fig. 4. Number of cases in each cluster

while the first group of samples has 3332 cases, accounting for more than 50% of the total, the average deposit is only 0.23% of the fifth group. However, the total annual turnover of the first sample is almost the same as that of the fifth sample. The above instructions for the construction of user portraits, for each type of customer precision marketing is urgent and necessary. he customer profiles of various groups are analyzed as follows.

The first type of customers are mainly under 30 years old, with bachelor's degree. Most of them are unmarried and mostly male. Their annual income ranges from 30000 to 60000. Most of them have less than 10000 deposits and about 20000 to 30000 transactions. Few people buy insurance products. Obviously, the first kind of customers have low income, low deposits, but they are keen on consumption, and their willingness to buy insurance products is lower than the other kinds. They are the initial growth customers of insurance companies.

The second kind of customers are mainly between 30 and 35 years old, with bachelor's degree, basically married and mostly male, with annual income of about 100000, deposit amount of about 30000 and transaction amount of about 50000, and almost no one buys insurance products. Most of these customers belong to the elite who have worked in the workplace for several years. Their income is fair and their consumption is not low. They buy less insurance and financial products. They should belong to the general customers of insurance companies who supply cars and houses.

The third kind of customers are about 25 years old, bachelor's degree, unmarried and mostly male, with an annual income of about 50000, but the amount of deposits is large, basically more than one million, and the annual transaction is basically more than 100000. From the above characteristics, we can see that this kind of people should be often referred to as "rich second generation", that is, high deposit and high consumption. They belong to customers with great potential.

The fourth kind of customers are mainly 35–40 years old, bachelor degree, unmarried and male, with annual income of about 150000, deposit of 500000–1000000, and annual transaction amount of 500000–1000000. These customers have higher income, deposit and transaction amount, and may be large consumer customers such as individual businesses.

The fifth kind of customers are mainly between 45 and 55 years old, with bachelor's degree, unmarried and male majority, annual income is more than one million, deposit amount is more than five million, annual transaction amount is about one million, and they will basically buy financial products, and the amount is considerable, between 500000 and one million. This kind of customers should be what people often call "diamond king five" Dedicated to consumption and financial management, they are VIP customers of insurance companies.

4   Suggestions on Precision Marketing Strategy Based on Big Data

It can be seen from the above analysis results.

(1)   Marketing strategy of the first kind of customers

The characteristics of the customer group are basically university graduates who have just worked for a short time. The current income level determines that they create less value for the insurance company. However, it is undeniable that with the increase of working hours, they will bring more value to the insurance company, so they are potential customers in the early growth stage. Because the customer accounts for the largest proportion in all the customers of the insurance company, which is the foundation of the insurance company, it is difficult to carry out targeted marketing services. Therefore, the first task of the insurance company is to create a good corporate image in the customer group, and improve the recognition of customers for the insurance company and products through some necessary promotion channels. Secondly, in response to the customer's low income but keen consumption psychology, we recommend consumer critical illness or medical insurance, because the premiums to be borne are not high, and the leverage of protection is large, which is more suitable for people with lower incomes.

(2)   The second kind of customer marketing strategy

The characteristics of the customer group are basically car loan and house loan families who have a certain time to work in the workplace, a fair income and a family. The proportion of this kind of customers in all customers of insurance companies is second only to the first kind of customers, and they are relatively stable middle-level customers in insurance companies. For this kind of customers, it is necessary to recommend automobile insurance and personal accident insurance It is suitable for married families with stable income, and recommends some low amount insurance products such as children's insurance.

(3)   The third kind of customer marketing strategy

According to the characteristics of customer groups, we call them "rich second generation" customers for the time being. In terms of income, they are similar to the first class customers, but in terms of deposit and transaction volume, they are far more than the first class customers and the second class customers. Therefore, young and high deposit customers are key customers for insurance companies, so they should buy some financial insurance products with relatively large amount and long cycle but low risk for their recommended users; on the other hand, they can recommend more property insurance products; and provide some convenient humanized services to encourage more consumption. In terms of product type, we can recommend the million medical insurance which is paid by year.

(4) The fourth kind of marketing strategy

The characteristics of customer group mostly belong to the consumption type of individual industrial and commercial households, with the characteristics of high deposit, high transaction but low financial management. According to these characteristics, we can provide different commercial insurance and other services for them according to different enterprise qualifications, and use more preferential insurance to promote the favor of such customers for the insurance company, so that their follow-up business of other merchants will be transferred to the insurance company In addition, we can recommend some stable and high amount of enterprise investment and financial insurance products to encourage them to buy consumption. In terms of insurance recommendation, we mainly recommend compensation insurance, that is, compensation according to the actual loss. In fact, the enhanced medical insurance is more in line with the actual situation of the fourth type of customers, and we can provide them with training on investment, finance and financial security.

(5) The fifth kind of marketing strategy

For insurance companies, the fifth type of customers is located at the top of the pyramid. Although the number of customers is the least, they make the greatest contribution to the insurance companies. They are VIP customers of insurance companies. For this kind of customers, insurance companies should try their best to provide one-to-one service, and arrange a professional account manager for them to serve, which can be said to be a private wealth management expert to provide professional private customized insurance products and financial products; for VIP customers, it is not only to provide private customized services, but also to let customers enjoy a kind of high-quality life clothes It's our business. In the aspect of insurance product recommendation, it focuses on high-end medical insurance and large amount life insurance. Meanwhile, it recommends savings life insurance plans for assets, and increases user stickiness and customer loyalty through annual payment.

5 Conclusion

To sum up, the Internet occupies all aspects of people's lives, and the technology of big data will have a greater and greater impact on enterprises. Fragmented media, fragmented consumers and audiences mean that using traditional marketing methods alone can no longer achieve the goal of brand communication and sales promotion, but must have a deep insight into the target population, integrate media channels, and achieve more accurate marketing Marketing, through the segmentation and mining of big data, and the comprehensive evaluation and analysis to meet the needs of customers, combined with the accurate and intelligent marketing of big data, insurance enterprises can more accurately find the needs of consumers for products through the analysis of these data, and change the traditional marketing mode, so as to improve the conversion rate of effective audience, and let the products pass through the appropriate carrier, in the right way Recommend to the right people, and achieve the goal of precision marketing.

References

1. You, Z., Si, Y.W., Zhang, D., et al.: A decision-making framework for precision marketing. Expert Syst. Appl. Int. J. **42**(7), 3357–3367 (2015)
2. Zhang, J., Wu, T., Fan, Z.: Research on precision marketing model of tourism industry based on user's mobile behavior trajectory. Mob. Inf. Syst. **2019**(4), 1–14 (2019)
3. Aldayel, M., Ykhlef, M., Al-Nafjan, A.: Deep learning for EEG-based preference classification in neuromarketing. Appl. Sci. **10**(4), 1525–1526 (2020)
4. Zhao, S., Ma, J.: Research on precision marketing data source system based on big data. Int. J. Adv. Media Commun. **7**(2), 93–100 (2017)
5. Bani-Hani, D., Khasawneh, M.: A recursive general regression neural network (R-GRNN) oracle for classification problems. Expert Syst. Appl. **135**(NOV.), 273–286 (2019)
6. Liu, H.: Big data precision marketing and consumer behavior analysis based on fuzzy clustering and PCA model. J. Intell. Fuzzy Syst. **2**, 1–11 (2020)
7. Yang, C.J., Wei, Y., et al.: Research on precision marketing of university teaching materials in big data era. Adv. Soc. Sci. Educ. Hum. **80**(3), 262–266 (2017)
8. Xiao, K., Hu, X.: Study on maritime logistics warehousing center model and precision marketing strategy optimization based on fuzzy method and neural network model. Pol. Maritime Res. **24**(s2), 13–16 (2017)

Minority Patterns in Modern Interior Design Based on Big Data

Yuhan Zhang[✉]

School of Art and Design, Yunnan Technology and Business University,
Kunming, Yunnan, China

Abstract. With the advancement of society and the increase of computer, network and digital media technology, big data technology has become an indispensable part of the modern service industry. Big data technology has been developed in all aspects of modern interior design, and the value of products and services created has gradually increased. Big data plays an important role in technology, ethnic minority patterns, and modern interior design. This article mainly introduces the application research of ethnic minority patterns in modern interior design based on big data. This paper uses the application research of ethnic minority patterns in modern interior design based on big data, adopts ethnic minority patterns for modern interior design, and reasonably analyzes the feasibility of interior design. Modern interior design is a history of thousands of years in China developed in a certain traditional regional culture. It has created the wisdom and essence of my country's traditional culture. The unique artistic spirit of the Chinese nation has been passed down in the long river of historical development. The ancestors worshipped nature and attached to all things in the world. What is reflected in the interior design is the preference for the natural color and natural texture of the materials in the selection of materials, use ethnic patterns in modern interior design. The experimental results of this paper show that the application research of ethnic minority patterns in modern interior design based on big data has increased the indoor beautification rate by 23%. Pattern application provides good modern interior design methods and approaches for analysis, discussion and summary, thereby enriching academic research results.

Keywords: Big data · Ethnic pattern · Modern interior design · Traditional culture

1 Introduction

A country is a stable society, composed of people's psychological qualities expressed in a common language, a common place, a common economic life and a common culture in history [1, 2]. On the basis of the integration of national culture and modern design technology, it provides a new impetus for the development of the minority model: at the same time understand the development trend of modern design culture, strengthen one's own cultivation culture, and improve taste. Their design ideas in modern design [3, 4], and tried to create a minority model. It is perfectly combined with modern visual design, igniting a road with ethnic minority characteristics [5, 6].

With the advancement of science and technology and the rapid development of the Internet, ethnic minority patterns have been enhanced in modern interior design. Research and exploration have been carried out based on the scientific design of semiotic thinking [7]. First, it explained the concept between the pattern and the symbol, then introduced the situation of the ethnic pattern, clarified the research ideas of the subject, and selected relevant examples for analysis and research, and then introduced the subject. Classify and interpret the visual sign language of ethnic minorities [8]. Wang believes that it is a symbolic design that summarizes their cultural consciousness and connotation from the aspects of color, shape and structure. Combining the scientific ideas of modern design with modern design examples, ethnic minority patterns have been obtained and introduced into current design symbols [9]. Xiang feels that in the design process, please follow the in-depth method of scientifically exploring ideas, explore the modern design style of ethnic patterns [10]. Finally, the pattern of ethnic minorities was adjusted, the key elements were theoretically studied, and the symbolic meaning of regeneration was explored and implemented. However, there are errors in their experimental process, which leads to inaccurate results.

The innovation of this article is to put forward a research on the modern interior design of ethnic minority patterns based on big data. The research aims at the study of ethnic minority patterns in modern interior design, and analyzes the effective countermeasures of modern interior design. The direct application of ethnic patterns in their most primitive patterns and modern design can also combine ethnic patterns with modern craftsmanship, presenting more diverse forms. The aim is to find a new path suitable for the development of modern interior design in ethnic minority patterns through this research.

2   Minority Patterns in the Big Data Environment

2.1   Ethnic Patterns

The cultural popular pattern of each ethnic minority is also different. In the early stages of human society, due to low productivity, people need some spiritual maintenance and pray for food abundance. Therefore, they should be naturally reflected on patterns, painted on utensils or knitted, embroidered and printed on clothing. The symbols of nature are gradually replaced by traditional concepts and become specific symbols. Character patterns mostly use patterns that describe life scenes, such as dancing song pattern, jumping foot pattern, dancing pattern, sedan chair pattern, eye pattern, palm pattern, etc. Religious symbols include Bodhisattva pattern, fairy pattern, human ghost pattern, continuous pattern, cloud pattern, curling wave pattern and so on. There are also some abstract patterns in daily necessities, such as shuttle pattern bamboo bridge pattern, winding board pattern, basket pattern, checkerboard pattern and other patterns.

$$X_t = \sum_{i=1}^{p} a_i x_{t-i} + \varepsilon_t \tag{1}$$

The calculation value of modern interior design algorithm is as follows;

$$X_t = \sum_{j=0}^{q} \theta_i \varepsilon_{i-j} \tag{2}$$

The application test of ethnic minority patterns should use the following formula:

$$3I_{02} = I_{A2} + I_{B2} + I_{C2} = 3U_0 j \varpi C_{02} \tag{3}$$

2.2 Ethnic Patterns Application

The process from ontological image thinking to abstract image thinking. Grasping the appearance characteristic elements of things themselves, grasping the laws of form and momentum, and adding the creator's own aesthetic consciousness to alienate things. This way of thinking radiates the original colors and beautiful artistic conception of ethnic minority patterns, and makes the patterns endless and colorful.

3 Modern Interior Design Analysis Based on Big Data Environment

3.1 Modern Interior Design

Closely integrate the colors of ethnic minorities with modern design styles, study and explore its laws of existence from the perspective of design art creation and design, apply modern scientific and technological achievements and value judgments to design, and realize the needs of humanity and aesthetics. Its characteristics are as follows: Formally, it is advocated to extract the color and modeling characteristics of ethnic minorities to form a decorative effect, advocate the principle of standardization. The national color scheme is combined with modern compositional expressionism. In the specific design, it emphasizes the overall concise design and opposes cumbersomeness.

3.2 Characteristic Culture in Modern Interior Design

At present, there are not many places where ethnic minority pattern handicrafts are well preserved. Only Guizhou and some Miao areas in Hunan still retain this traditional craftsmanship. The patterns of ethnic minorities in Hunan and Guizhou are called "mubtut" (mubtut) in Miao language. As a traditional textile printing and dyeing process in my country, it has unique ethnic characteristics. It is in a specific, closed and original cultural environment. Produced, so it has unique national characteristics. The specific results are shown in Table 1.

Table 1. Classification of upholstery fabrics

Seedlings	Carpet	Mantle	Tablecloth
Floor decoration	Wallpaper	Screen	Erbu
Wall decoration	Soft wall	Tent	Bedding and other decorations
Hanging curtain cover	Wall covering curtain	Curtain	Blanket
Seedlings	Carpet	Mantle	Tablecloth
Furniture approval category	Straw mat	Sofa covering material	Fiber art

4   Characteristic Cultural Analysis of Ethnic Minority Patterns

4.1   Ethnic Patterns of Analysis

The style is unique. This time I did a field study for my graduation thesis, and visited the Miao areas rich in ethnic patterns in Hunan and Guizhou. I learned in detail about the production process of the traditional Miao ethnic patterns and the types and characteristics of ethnic patterns. I found that the development prospects of traditional ethnic pattern crafts in the modern social environment are worrying, and realized the necessity and urgency of protecting and inheriting the pattern art of the Miao ethnic minorities in Hunan and Guizhou. The specific results are shown in Fig. 1. The essence of traditional patterns is created to create a living space that conforms to the aesthetics of modern people, and pursues the "material construction" and "spiritual construction" of furnishing design.

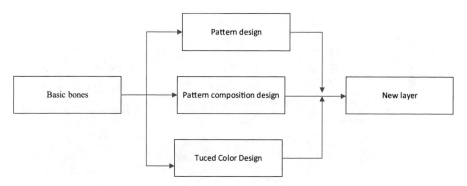

Fig. 1. Pattern design method diagram

4.2   Characteristic Cultural Analysis of Ethnic Patterns

When they open their eyes in the morning, the first thing they see is the blue sky and white clouds. Blue is a color full of deep connotation and emotional meaning, a color full of life rhythm, and a color that can evoke romantic associations. It is new, simple, stable, and quiet. It is like an emotional bridge connecting nature and people. It can lead people's hearts to the broad wilderness, and let people find people and nature from the deep green of the mountains. The specific results are shown in Table 2.

Table 2. Wall thermal energy index

Summer heat transfer coefficient	2.611
Winter external benefit coefficient	2.683
Internal heat transfer coefficient	2.251
Heat transfer attenuation	0.986
Heat transfer	0.866

The design, especially the interior decoration art design, only considers people's general practicality, and rarely or even does not design according to people's special spiritual needs or psychological demands. With the rapid development of social economy, we can almost be sure that "humanization" will be the inevitable trend of modern decoration art. Take ethnic minority pattern decoration as an example. The subject matter of ethnic minority pattern painting in interior decoration ranges from sketching landscapes, figures, geometric abstraction to the texture effect and deep style that combine modern design elements with the mutual penetration and integration, which is close to people's aesthetic psychology. The specific results are as shown in Fig. 2, therefore, the linear, L-shaped and U-shaped are typical of the traditional residences of ethnic minorities.

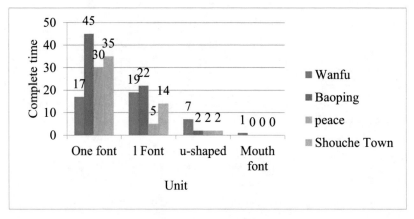

Fig. 2. Questionnaire

5 Conclusions

Although this article is aimed at the application research of ethnic minority patterns in modern interior design based on big data, there are still many shortcomings. In the intersection, collision, and fusion of traditional culture and modern civilization, the decorative arts of ethnic minorities need to form a new mode of discussion. Traditional ethnic patterns leave us with the concrete images of modern people, and we need to actively explore and continue to summarize and analyze. And once again use the rich cultural significance they contain to enlighten and expand our today's interior design and creation, and further inherit and innovate the excellent traditional culture.

References

1. Landi, D., Fitzpatrick, K., Mcglashan, H.: Models based practices in physical education: a sociocritical reflection. J. Teach. Phys. Educ. **35**(4), 400–411 (2016)
2. Mckenzie, T.L., Nader, P.R., Strikmiller, P.K., et al.: School physical education: effect of the child and adolescent trial for cardiovascular health. Prev. Med. **25**(4), 423 (2016)
3. Kirk, D.: Physical education, youth sport and lifelong participation: the importance of early learning experiences. Eur. Phys. Educ. Rev. **11**(3), 239–255 (2016)
4. Cairney, J., Hay, J., Mandigo, J., et al.: Developmental coordination disorder and reported enjoyment of physical education in children. Eur. Phys. Educ. Rev. **13**(1), 81–98 (2016)
5. Coutinho, D.A.M., Reis, S.G.N., Goncalves, B.S.V., et al.: Manipulating the number of players and targets in team sports small-sided games during physical education classes. Revista De Psicologia Del Deporte **25**(1), 169–177 (2016)
6. Ada, E.N., Zisan, K.Ç., Altiparmak, M.E., et al.: Flow experiences in physical education classes: the role of perceived motivational climate and situational motivation. Asian J. Educ. Train. 4–5 (2018)
7. Lodewyk, K.R., Muir, A.: High school females' emotions, self-efficacy, and attributions during soccer and fitness testing in physical education. Phys. Educ. **74**(2), 269–295 (2017)
8. Lander, N.J., Hanna, L., Brown, H., et al.: Physical education teachers' perspectives and experiences when teaching FMS to early adolescent girls. J. Teach. Phys. Educ. 1–16 (2017)
9. Wang, J., Shen, B., Luo, X., et al.: Validation of a teachers' achievement goal instrument for teaching physical education. J. Teach. Phys. Educ. **37**(1), 1–27 (2017)
10. Xiang, P., Ağbuğa, B., Liu, J., et al.: Relatedness need satisfaction, intrinsic motivation, and engagement in secondary school physical education. J. Teach. Phys. Educ. **36**(3), 340–352 (2017)

Qualitative and Quantitative Analysis of Financial Public Opinion Risk Based on Big Data Analysis

Xiuwen Wang[✉] and Zhen Wu

National Computer Network Emergency Response Technical Team/Coordination Center of China, Beijing, China

Abstract. Online public opinion has a strong social influence and spreads at a very fast speed. Public opinion analysis on financial public opinion reports is conducive to the relevant institutions to conveniently understand the public opinion of the event and make correct guidance and control, which is also conducive to the sustainable development of financial market. In this paper, subject model, emotion dictionary construction, public opinion analysis and other technologies are used to conduct a detailed study of financial related network public opinion, and a financial public opinion analysis model is designed. The model can effectively identify the emotional words in the text from the improved topic model, and then match the emotional words with the more comprehensive and perfect emotional dictionary in the financial field, calculate the emotional tendency value of the emotional words, so as to classify the financial text public opinion more accurately.

Keywords: Big data analysis · Financial public opinion · Financial sentiment dictionary · WHDP model

1 Introduction

The rapid development of the Internet leads to the rapid development of Internet finance and the rapid dissemination of a large amount of financial information. By analyzing the public opinion of these financial news texts, we can understand the development trend of the financial market and provide important reference value for relevant institutions [1]. In Internet finance, the news released by authoritative financial websites reflects the basic situation of China's financial market and also guides the development of China's economy. For example, a piece of good news for a company will cause a rise in the stock price and promote the economic development of the company. And the negative information can cause the share price to decline, or even shut down. It can be seen that finance has a significant impact on the economic market and has a pivotal position. With the development of the Internet and the arrival of big data, financial texts have grown exponentially. There are many hot topics in these news contents, which bring more opportunities for the financial industry to grow, but also bring the disadvantages of rapid spread of events. Network public opinion information and the credibility of

J. Abawajy et al. (Eds.): ATCI 2021, AISC 1398, pp. 872–879, 2021.
https://doi.org/10.1007/978-3-030-79200-8_129

the subject of the transaction and word-of mouth are closely related, these messages will cause the outbreak of relevant public opinion. Timely understanding and analysis of these information can enable relevant institutions to quickly grasp the market public opinion and make appropriate decisions accurately, which is conducive to the long-term development of the financial industry. At the same time, the use of public opinion data to study the financial market will enhance the growth of financial theory and lay a foundation for future research.

In recent years, scholars have carried out a large number of studies on financial public opinion analysis, and some achievements have been made. Generally, it can be divided into several categories, such as semantic based, machine learning, deep learning and hybrid method. Kenny and others based on the syntax and semantic analysis, put forward a kind of sentiment analysis engine, the analysis of financial text from the aspect of word to phrase level, and through the three kinds of evaluation method show that this method can effectively analyze the mood, also pointed out that through text messages to express emotions for predicting the change trend of the stock market is a beneficial [2]. Rosenkranz established a model to judge stock trends based on emotions in social media, and the results showed that the model combining historical prices and sentiment analysis had a more accurate effect than the model based on historical prices alone [3].

In order to solve the problem of poor classification effect of common sentiment dictionaries in financial texts, this paper constructs a sentiment dictionary in the financial field. In the experimental comparison between the basic sentiment lexicon method and the financial sentiment lexicon method, the sentiment lexicon method based on the financial sentiment lexicon method has wider coverage, more perfect, and better classification effect of public opinion.

2 Financial Public Opinion Based on Big Data

2.1 Text Preprocessing Technology

(1) Chinese Word Segmentation

Word segmentation is the process of cutting the original text into several words according to certain rules. Emotional words in the text can be obtained through word segmentation, which lays a foundation for subsequent public opinion analysis. In English writing, each word is separated by a space, there are obvious signs of division. However, Chinese text takes character as unit, and there is no distinguishing sign between each word. Some words are ambiguous, so word segmentation in Chinese text is much more complicated than that in English text [4]. There are three main categories of participles:

1) Dictionary-based method. Word segmentation is carried out according to the matching between the text and the word in the dictionary. If the text to be segmented matches the word in the dictionary, a word will be recognized. Common matching methods include forward matching method, reverse matching method and two-way matching method. However, this method is not effective for the processing of unknown words.

2) Based on the statistical method, this method believes that the more times adjacent words appear together, the greater the probability of combining words. Therefore, when the frequency of their joint occurrence is greater than a certain threshold, they can be considered as one word. However, this method relies too much on corpus and the algorithm training is complicated.

3) Comprehension-based approach, which considers the semantic relationships between words on the basis of dictionaries and uses semantic information to dis-ambiguate. However, computers cannot learn this information and need a lot of language information to support them.

In this paper, Jieba word segmentation tool is used to construct directed acyclic graph based on prefix dictionary, and the maximum probability path in the graph is found by dynamic programming to achieve word segmentation. In addition, Jieba word segmentation uses HMM model to segment unknown words, and Viterbi algorithm is used to solve the HMM prediction problem to judge the probability that a word is an unknown word.

(2) Stop Word Processing

Stoppers refer to words with high frequency but no real meaning, such as conjunc-tions, prepositions, adverbs, pronouns, modal particles, numerals and so on. In order to save space, improve the speed of information processing, reduce the time spent in public opinion analysis, more accurately express the content of the text and reduce the characteristic dimension of the text, these stops need to be removed [5]. The words that appear in the stop list after the text segmentation are the stop words, and we need to remove them by matching the stop words list.

2.2   Construction Technology of Emotional Dictionary

The tendency of public opinion in this text is reflected by sentiment words, and the accu-racy of sentiment dictionary directly affects the actual results of public opinion classifi-cation, so it is of practical significance to build a more accurate sentiment dictionary [6]. At present, the commonly used sentiment dictionaries include HowNet sentiment dictio-nary (HowNet), ntUSD, tsinghua dictionary of commendation and derogation, Chinese sentiment lexicon ontology database.

Point Mutual Information (PMI) is used to calculate the probability of the simulta-neous occurrence of two words in the text. The greater the probability, the greater the correlation [7]. The PMI values of two words word1 and word2 are calculated by the following formula

$$PMI = (word1, word2) = \log_2[\frac{p(word1, word2)}{p(word1) * p(word2)}]$$ (1)

Among them, the probability of single occurrence of word1 and word2 is represented by P (word1) and P (word2) respectively, and the probability of co-occurrence of two words is represented by P (word1,word2).

PMI(word1,word2) < 0, indicating that the two words are mutually exclusive;
PMI(word1,word2) = 0, indicating that the two words are statistically independent;
PMI(word1,word2) > 0, indicates that the two words are related. The greater the value, the greater the correlation.

Feelings tend to be some Mutual Information algorithm (Sentiment Orientation - Point Mutual Information, the SO - PMI) is the PMI thoughts into words in the emotional tendency of calculation, selected a group of strong emotional intensity of negative Nwords seed words and a set of emotional polarity strong positive seed word Pwords, calculates the unknown word in the seed word word and Nword Point of Mutual Information and word and Pword difference Point of Mutual Information, to determine whether the candidate term emotional tendencies.

$$SO - PMI(word) = \sum_{Nword \in Nwords} PMI(word, Nword) - \sum_{Pword \in Pwords} PMI(word, Pword) \quad (2)$$

SO-PMI(WORD) < 0, which is a derogatory word indicating a negative tendency;
SO-PMI(WORD) = 0, this word is neutral, indicating neutral tendency;
SO-PMI(word) > 0, this word is a positive word, indicating a positive tendency.

2.3 WHDP Theme Model

The traditional approximate inference algorithms in HDP model mainly include Gibbs sampling and variational method. Gibbs sampling assigns a topic to the current word from the topic probability distribution of each word by iteration, and the convergence rate is slow. The variational algorithm has a fast convergence speed, but each iteration of the algorithm takes a lot of time to calculate the time complexity, so the calculation complexity speed is slow. The message passing algorithm has certain advantages in increasing model accuracy and convergence speed [8, 9]. Therefore, this paper adopts message passing algorithm and Gibbs sampling to carry out approximate reasoning in WHDP model, and transforms it into an equivalent WHDP factor graph model.

According to the BP algorithm in the traditional topic model, the probabilities of K discovered topics in words can be obtained, and the probabilities of words related to undiscovered topics can be obtained according to the CRF construction, which is expressed as follows [10]:

$$\mu_{w,s}^d(k) \propto \frac{\sum_s \mu_{-w,s}^d(k) + \alpha \tau(k)}{\sum_k [\sum_s \mu_{-w,s}^d(k) + \alpha \tau(k)]} \frac{\mu_{w,-s}^d(k) + \beta}{\sum_w [\mu_{w,-s}^d(k) + \beta]} \quad (3)$$

$$\mu_{w,s}^d(K+1) \propto \frac{\alpha \tau(K+1)}{W} \quad (4)$$

3 Financial Public Opinion Analysis Model

3.1 Construction of Financial Affective Dictionary

This article in the financial website climb took a lot of financial news as corpora, the corpus after preprocessing, selection according to the word frequency from high to low from the corpus of the first 100 words, and then let the five annotation artificial remove non-financial vocabulary, again for the rest of the word emotion tagging respectively, marked with labels for: positive and negative, the final result will indicate the same vocabulary as a seed word, tagging results after different adopts the principle of the minority is subordinate to the majority of words as a seed.

After preprocessing the news corpus, a word set after word segmentation is obtained. Then, the words whose frequency is less than 5 are removed from the set and the remaining words are used as candidate words to construct the dictionary. Next, it is judged whether the word in the candidate word is in the basic sentiment dictionary. If not, the SO-PMI algorithm is used to calculate the emotional similarity between the word and the seed word. If the word is more closely related to the negative seed word, add the word to the negative emotion dictionary. On the contrary, if the word is more closely related to the positive seed word, the word is added to the positive emotion dictionary.

3.2 Public Opinion by Combining Theme Model and Financial Sentiment Dictionary

This paper will combine the WHDP model and financial sentiment dictionary to analyze the public opinion of the news text. When we calculate the emotional tendency of news text, we calculate the emotional value of the subject words with emotional tendency in the text, and then add them to get the emotional value of the whole news. The specific steps of the algorithm are as follows:

Input: financial news text to be classified, financial field sentiment dictionary;
(1) Use Jieba participles for word segmentation of the input financial text;
(2) the text after the word segmentation to stop processing;
(3) Input the preprocessed text into the WHDP/CHDP model to obtain the subject words in the text;
(4) Load the previously prepared emotional dictionary database in the financial field;
(5) Set the initial value r and make $r = 0$. Then, the subject word in the text is matched with the financial domain emotion dictionary. If the match fails, the word is skipped, and if the match succeeds, the next step is entered.
(6) If the subject word matches a word in the negative emotion dictionary in the emotion dictionary of the financial field, then the value of r is -1; if the subject word matches a word in the positive emotion dictionary in the emotion dictionary of the financial field, then the value of r is $+1$;
(7) Calculate the emotional value of each subject word in the news text according to the previous steps, and then sum up all the R values to get the final emotional value S of the whole text. If the value of S is greater than zero, it indicates that the public opinion tendency of the news text is positive. If the value of S is less than zero, it indicates that the public opinion of the news text tends to be negative. If the value of S is equal to zero, it indicates that the public opinion tendency of the news text is neutral.
Output: text classification results of public opinion.

4   Financial Public Opinion Tendency Analysis Results

This paper uses PyCharm as the editor and Python as the development language to calculate the emotional propensity values of the text introduced in the previous section. In this experiment, public opinion classification was carried out on the hotel review data set and financial data set respectively by using the basic sentiment dictionary method, the sentiment dictionary method based on the financial field, and the WHDP model combined with the sentiment dictionary method in the financial field.

4.1   Corpus of Hotel Reviews

Table 1. Hotel reviews compare results

	Accuracy	Recall rate	F value
Basic emotion dictionary	0.724	0.687	0.704
Financial sentiment dictionary	0.746	0.713	0.719
WHDP model	0.798	0.794	0.785

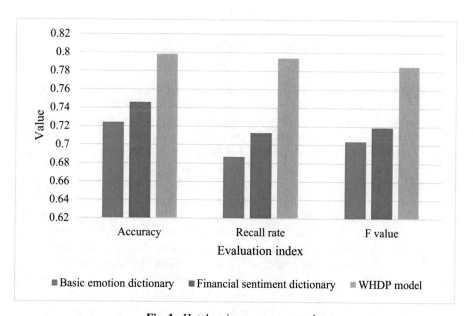

Fig. 1. Hotel reviews compare results

As shown in Table 1 and Fig. 1, you can visually see experiment based on the basic emotional lexicon in the field of general method and the methods of emotional dictionary based on the financial sector of public opinion classification results were similar, while WHDP model combined with the financial sector emotional dictionary method compared with the former two methods accuracy and recall rate is high, public opinion classification effect is better. This is because WHDP model can extract affective words from text more effectively and calculate affective tendency value more accurately.

4.2 Financial Corpus

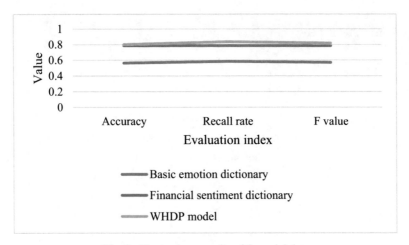

Fig. 2. Comparison results of financial data

As shown in Fig. 2, it can be clearly seen that when classifying public opinion in financial news, introducing financial sentiment words into sentiment dictionary can improve the accuracy and recall rate of the experiment, that is, the method based on sentiment dictionary in the financial field has a better effect than the method based on basic sentiment dictionary. This shows that the public opinion analysis based on sentiment dictionary depends on the selection of sentiment words to a certain extent, and the quality of sentiment dictionary determines the correctness of public opinion classification to a large extent. The correctness of public opinion classification can be better improved by constructing sentiment dictionary in the financial field. In addition, WHDP model combined with financial domain sentiment dictionary method has higher accuracy and recall rate than the previous two methods, and can more effectively classify public opinion. This is because the WHDP model considering the context semantic information in text, better able to recognize the emotional words in the text, these emotional vocabulary at the same time can get a more complete emotional dictionary in the financial sector, a more comprehensive matching, therefore, emotional dictionary WHDP model combined with the financial sector of emotions tend to value calculation accuracy is higher, the accuracy of the classification of public opinion is also higher.

5 Conclusions

This paper proposes a public opinion classification method based on WHDP model and financial sentiment dictionary. Firstly, HowNet, NtUSD, commendation and denotation dictionary and Dalian Polytechnic Sentiment Dictionary are recombined into the basic sentiment dictionary, and then the financial sentiment dictionary is constructed with SO-PMI algorithm, and then the financial sentiment dictionary and basic sentiment dictionary are merged into the financial domain sentiment dictionary. Finally, the financial text is classified by the dictionary combined with the WHDP model. Compared with the public opinion analysis experiment based on basic sentiment dictionary method and financial sentiment dictionary method, the accuracy and recall rate of this method are higher, and the results of public opinion classification are more accurate, which proves that this study is feasible for public opinion orientation analysis in the financial field.

References

1. Kristoffer, F., David, S.: The supply chain has no clothes: technology adoption of blockchain for supply chain transparency. Logistics **2**(1), 2–3 (2018)
2. Aptea, S., Petrovsky, N.: Will blockchain technology revolutionize excipient supply chain management? J. Excipients Food Chem.**7**(3), 76–78 (2016)
3. Mahyuni, L.P., Adrian, R., Darma, G.S.: Mapping the potentials of blockchain in improving supply chain performance. Cogent Bus. Manag. **7**(1), 1–18 (2020)
4. King, L.: Turvo launches collaborative supply chain management platform. Air Cargo World **107**(4), 8 (2017)
5. Yazdani, M., Zarate, P., Coulibaly, A.: A group decision making support system in logistics and supply chain management. Expert Syst. Appl. **88**(Dec.), 376–392 (2017)
6. Li, Z., Liu, G., Liu, L.: IoT-based tracking and tracing platform for prepackaged food supply chain. Ind. Manag. Data Syst. **117**(9), 1906–1916 (2017)
7. Zhu, Y., Riad, K., Guo, R.: New instant confirmation mechanism based on interactive incontestable signature in consortium blockchain. Front. Comput. Sci. China **13**(6), 1182–1197 (2019)
8. Zhou, N., Zhang, X., Wang, S.: Timestamp reassignment: taming transaction abort for serializable snapshot isolation. Front. Comput. Sci. **13**(6), 1282–1295 (2019)
9. Wei, W., Guimarães, L., Amorim, P., Almada-Lobo, B.: Tactical production and distribution planning with dependency issues on the production process. Omega Int. J. Manag. Sci. **67**(Mar.), 99–114 (2017)
10. Iansiti, M., Lakhani, K.R.: The truth about blockchain. Harv. Bus. Rev. **95**(1), 118–127 (2017)

Current Situation and Future Trends of Financial Management Development in the Era of Big Data

Zhou Yang[✉]

School of Business, Shandong Normal University, Jinan 250300, Shandong, China

Abstract. As the core of enterprise management, financial management plays an important role in improving the economic benefits and sustainable development of enterprises. This paper mainly studies the current situation analysis and future trend of financial management development in the era of big data. This paper selects five aspects of strategic planning, information system, process reengineering, operation management and personnel management as the key factors to evaluate the financial sharing ability of a company. Through the four dimensions of quality, efficiency, cost and security, it divides five first level indicators, 20 second level indicators and 51 third level indicators to quantify. According to the expert's suggestion, the analytic hierarchy process is used to calculate the weight of all levels of indicators of the standard financial sharing center and sort them. At the same time, the employees who use the service platform of the sharing center are asked to score the ability of the Financial Sharing Center. The data show that the cumulative contribution rates of the main factors are 85.157%, 85.744% and 86.225% respectively. The results show that the financial management reform is of great practical significance to improve the economic efficiency and sustainable development of enterprises.

Keywords: Big data · Financial management · Development status · Future trends

1 Introduction

With the wide application of information technology, the process of economic globalization is further accelerated, and the external environment of enterprises is more dynamic. With the advent of the era of big data, it brings massive information environment for enterprises. In this context, the management activities of enterprises are facing the huge impact and change brought by the information technology revolution.

In the era of big data, the financial department dealing with the most data is facing the problem of processing and analyzing financial big data [1]. Through the use of cloud computing and other technologies to fully mine and analyze the financial big data, the problems of information island, lack of resources and decision lag are solved, the integration of Finance and business, information resource sharing are realized, and the correlation of financial information quality is improved, thus providing new ideas for

J. Abawajy et al. (Eds.): ATCI 2021, AISC 1398, pp. 880–886, 2021.
https://doi.org/10.1007/978-3-030-79200-8_130

managers to make decisions [2, 3]. In the actual development process, its logistics, business flow, capital flow and information flow are all reflected by using the network as the carrier. In the process of four flow transportation, it presents the unstable characteristics of integration and cross line. As a result, the financial management control work also presents the characteristics of weak distribution. The control force is relatively weak and unstable, and its efficiency is relatively low, reflecting the lag [4, 5]. In financial management activities, financial data, as the core data of an enterprise, directly reflects the enterprise's operating results and status in a certain period of time [6, 7]. Under the background of big data, all kinds of accounting information are integrated through information technology, and enterprise operators form intellectual resources to serve enterprise financial management activities through the integration and sharing of multi-dimensional data [8, 9]. Financial informatization also helps to improve the efficiency of financial management, improve the accuracy of financial information, prevent misstatement, omission and false information, provide forward-looking guidance for leaders' decision-making, and have obvious benefits for bank's daily operation, cost reduction, fund-raising and other activities [10].

As an analysis method and tool, value chain analysis can be used for reference by enterprise financial management. The enterprise financial management based on value chain analysis is a more flexible and efficient financial management mode. This mode focuses on the cultivation and formation of enterprise competitive advantage and core competitiveness, aims at maximizing customer value, and ultimately serves the realization of enterprise competitive strategic objectives.

2   Big Data and Financial Management

2.1   Big Data

In the Internet, each user can create and send his own information through the man-machine interface, forming a flow of information, and transmit, process, analyze and store in the Internet through invisible flow methods. This is a kind of personal information. The user is the information exchange of the information source; other users who can establish a network connection with the information source on the Internet can receive this information through the information presentation of the information flow interface.

For two elements x and y, the correlation coefficient calculation formula between them is as follows:

$$R = \frac{\text{cov}(x, y)}{\sqrt{D(x)} * \sqrt{D(y)}} \tag{1}$$

In the formula, R is the correlation coefficient, and cov(x, y) is the covariance of x and y.

Let f(x, p) and f(x, q) be the equivalent probability distribution on χ, then the estimate of e is:

$$\hat{e} = \frac{1}{N} \sum_{i=1}^{N} I_{\{S(x_i) \geq \gamma\}} \frac{f(x, p)}{f(x, q)} \tag{2}$$

Using the Hausdorff distance formula, the solution of the optimal parameters can be derived as:

$$v^* = \arg \max_{v} E_u I_{\{S(X) \geq \gamma\}} \ln f(X; v) \tag{3}$$

2.2 Financial Management

Traditional financial work is gradually replaced by computerized system. Financial personnel must have data thinking, and change from passive data collector to data analyst, so as to excavate the business value behind the data and improve the strategic position of financial decision-making. In the enterprise, the role of financial management changes from financial information service to decision-making participant. In the era of big data, enterprises not only need financial information, but also need to integrate the financial work into the business process. They can excavate the business motivation behind the financial data, not limited to the traditional accounting work, and pay attention to participating in the enterprise strategy formulation, risk, credit and other management fields.

3   Financial Management Evaluation System Experiment

3.1 Financial Management Platform

Network platform as the basis of information transmission and sharing, based on the current situation, with the continuous development of the Internet, the environment of information exchange and resource sharing is more optimized. The functions of management software are developing and changing with the needs of enterprises. Based on the horizontal dimension, each dimension of enterprise management corresponds to the corresponding subsystem, and these subsystems are very independent, focusing on solving problems of a certain nature.

3.2 Establishment of Financial Index System

This paper selects five aspects of strategic planning, information system, process reengineering, operation management and personnel management as the key factors to evaluate the financial sharing ability of a company. Through the four dimensions of quality, efficiency, cost and security, it divides five first level indicators, 20 second level indicators and 51 third level indicators to quantify. According to the expert's suggestion, the analytic hierarchy process is used to calculate the weight of all levels of indicators of the standard financial sharing center and sort them. At the same time, the employees who use the service platform of the sharing center are asked to score the ability of the Financial Sharing Center. Considering the limitation of the traditional financial personnel, the appropriate proportion of data analysts in the financial management department should use statistical analysis, business intelligence, data analysis and other technologies to tap the potential.

4 Discussion

4.1 Analysis of Financial Management Mode

The characteristic root and contribution rate of factors are shown in Table 1. Using the method of maximum variance rotation with eigenvalue greater than 1 to extract the factors from the observation indexes of driving factors, supporting/hindering factors and innovation behavior, the main factors of 3, 4 and 3 initial variables were extracted, and their cumulative contribution rates were 85.157%, 85.744% and 86.225%, respectively. It shows that these main factors can reflect 85.157%, 85.744% and 86.225% of the information of 7 driving factors, 15 supporting/hindering factors and 5 financial management innovation behavior variables respectively. Through the analysis of the above data, it can be concluded that process reengineering is the most valued by the respondents, followed by operation management and information system. In the evaluation of process reengineering indicators, the consistency of process and business, the cost of process optimization and the degree of process continuous optimization account for a large proportion; in the evaluation of operation management indicators, the operation procedures, operation quality standards, operation procedures and the rationality of operation organization are ranked higher; in the evaluation of information system indicators, the operation management indicators are ranked higher. The results show that the system compatibility, system technology cost and system operability have great weight; in the evaluation of personnel management indicators, the operation ability of employees, employee turnover rate and employee compensation weight rank first; in the evaluation of strategic planning indicators, the degree of efficiency improvement is the most important.

Table 1. Factor characteristic roots and contribution rate

Main factor	Driving factors			Support/obstacle				Entrepreneur oriented		
	1	2	3	1	2	3	4	1	2	3
Characteristic root	2.52	2.02	1.41	4.96	3.46	2.60	1.82	1.86	1.39	1.05
Contribution rate %	36.01	28.95	20.18	33.12	23.12	17.36	12.16	37.20	27.86	21.16
Cumulative contribution rate %	36.01	64.97	85.15	33.12	56.25	73.61	85.74	37.20	65.06	86.22

4.2 Evaluation and Analysis of Financial Sharing Ability

The intra class mean difference test is shown in Fig. 1. The SIG. of the within class mean test of variables F2 and F5 is greater than 0.05, and their SIG. values are 0.158 and 0.665 respectively, which are only slightly greater than 0.05, so they have no significant effect on the discriminant results. The significant probabilities of other variables are less than 0.05, which meets the test requirements. This shows that there are significant differences between the five variables, which can be used for discriminant analysis. Although the financial sharing service center has a high degree of informatization and a low degree of correlation between various information systems at the present stage, it needs to realize data sharing through system optimization and system interface connection. Although it costs a lot in the early stage, it can reduce the cost of later maintenance and data sharing between systems in the long run, and the data can be used and analyzed in a consistent way, easier to manage. In addition, in addition to the standardized single accounting work, it saves the labor cost of the subsidiary company, and the centralized fund management and control saves the circulation and use cost of the fund. The reduction of capital cost is also reflected in the capital cost. After the financial sharing system went online, due to the financial sharing center's scheduling and management of internal funds, the interest rate of external payment decreased, and the cost of capital use was saved. In addition, by simplifying the fund processing process, financial sharing reduces many complicated manual links, optimizes the fund payment process, and shortens the payment efficiency of fund processing from 9 min to 3 min, with a payment success rate of 99.72%.

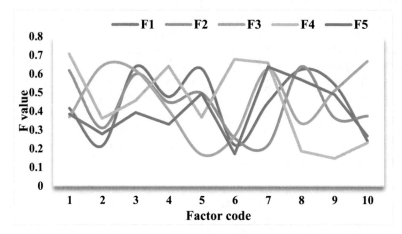

Fig. 1. Tests for differences in means within classes

The result of anomaly detection on the data stream is shown in Fig. 2. We find that the coverage is usually less than half of the total, which is likely due to the sparsity of social network links. Moreover, when k = 10, the coverage rate is basically stable at about 38%. Strengthen the management and supervision of business process. While developing the background supervision system, we should also set up the supervision and management department of the financial sharing service center. The main task of the supervision department is to manage the background supervision system, regularly check the business process operation of the staff, count the frequency of warning instructions issued by the system, and give corresponding punishment to the staff who seriously violate the rules and regulations. It is also necessary to repair the loopholes of the system regularly to avoid affecting the smooth implementation of business processing and causing unnecessary losses, so as to improve the supervision ability of the financial sharing service center. Enterprise data collects information on business activities through activity code representatives, and classifies the collected information according to different time periods, which is convenient for information demanders to understand relevant information, and can also provide effective information support for enterprise management to make decisions. At the same time of development and research, enterprises should also broaden their horizons, develop special financial early warning models of enterprises in different industries and different scales, and add them to the financial early warning system of enterprises, so as to improve their own early warning accuracy.

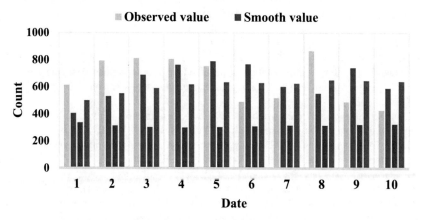

Fig. 2. Anomaly detection results on the data stream

5 Conclusions

In the financial management work, the financial management mode, as the most basic link, runs through the financial management of commercial banks. Therefore, the establishment of a scientific and perfect financial management mode is the basis of doing a good job in financial management. Commercial banks always pay attention to the operation effect of financial management mode.

Big data conditions require financial management activities to change from traditional accounting to strategic financial decision-making, and use data mining and processing to store and analyze business-related multi-dimensional data in time, so as to give full play to the role of financial support.

The new financial management mode of higher vocational colleges must be combined with the characteristics of higher vocational colleges, based on personnel training, and need support in financial budget, financial internal control risk management, financial performance evaluation system, asset accounting and other aspects.

References

1. Lemi, K., Megersa, R., Bogale, M.: The effect of financial management practice on entrepreneurial performance of micro, small and medium scale enterprises in West Oromia. J. Process Manag. New Technol. **8**(1), 50–64 (2020)

2. Junaedi, J., Handam, H.: Efforts to increase the fisherman income through financial management of the independent fisherman group in village Mattiro Bombang, Pangkep Regency. J. La Bisecoman **1**(2), 10–14 (2020)
3. Lopez, B.S., Alcaide, A.V.: Blockchain, artificial intelligence, internet of things to improve governance, financial management and control of crisis: case study COVID-19. SocioEcon. Challenges **4**(2), 78–89 (2020)
4. Catarino, G.P.D.S., Santos, L.R.D., Silva, P.V.J.D.G.: The influence of personal finance on the financial management of microenterprises in Rio De Janeiro. REMIPE - Revista de Micro e Pequenas Empresas e Empreendedorismo da Fatec-Osasco **6**(2), 312–330 (2020)
5. Kabok, J., Radii, S., Karolina, L.-K.: Application and development of financial management and control exemplified by the provincial secretariat for higher education and scientific research. Megatrend Revija **17**(2), 125–142 (2020)
6. Polyakov, D.V., Popov, A.I.: Optimization of financial management based on the theory of fuzzy sets. Vestnik Tambovskogo gosudarstvennogo tehnicheskogo universiteta **26**(1), 064–078 (2020)
7. Chen, X., Metawa, N.: Enterprise financial management information system based on cloud computing in big data environment. J. Intell. Fuzzy Syst. **39**(1), 1–10 (2020)
8. Taufiq, M., Yatminiwati, M.: Urgency of the village of financial management under the government regulations. Wiga Jurnal Penelitian Ilmu Ekonomi **10**(1), 45–59 (2020)
9. Hunde, D.A., Yohannes, D.: Financial management practices and profitability of small and medium enterprises in Kembetatembaro zone, SNNPR. J. Soc. Sci. **48**(4), 1675–1688 (2020)
10. Asandimitra, N., Kautsar, A.: The influence of financial information, financial self efficacy, and emotional intelligence to financial management behavior of female lecturer. Hum. Soc. Sci. Rev. **7**(6), 1112–1124 (2020)

Application of Transana Video Analysis Software in Teacher Case Analysis Under the Background of Big Data

Tongqing Yuan[✉]

School of Educational Science, Anhui Normal University, Wuhu 241000, Anhui, China
asdytq@mail.ahnu.edu.cn

Abstract. Under the background of big data, China attaches great importance to the integration of information technology and subject curriculum in the field of basic education. With the application and popularization of information technology in teaching, how to integrate advanced modern information technology with teaching and realize real information teaching has become a research hotspot in education and teaching. Focusing on undergraduate education, this paper designs a set of classroom teaching interactive activity information tracking method based on Transana (TSA) video analysis software, which is of great significance and practicability. For undergraduate teaching, the general form includes general lectures, seminars and experimental classes. The interaction between undergraduate students and TSA platform is higher than that between teachers and TSA platform, and the average ratio is as high as 40. The traditional teaching methods cannot meet the existing teaching objectives.

Keywords: Tsa video analysis software · Classroom recording · Big data · Information technology

1 Introduction

Against the background of big data, many information technologies provide effective auxiliary functions for teachers' teaching, including a large number of information subject tools (such as TSA video analysis software). Video analysis technology is one of the information subject tools. Some scholars and teachers at home and abroad have brought video analysis technology into the actual classroom, studied the application of video analysis technology in practical teaching, solved some teaching difficulties that can not or can not be solved by other teaching methods, and can make up for the deficiency of students' perceptual cognition by visualizing the teaching process. The emergence of video analysis technology has brought a new opportunity and breakthrough for the integration of information technology and practical teaching.

As early as the 1970s, some scholars put forward the classroom interactive behavior analysis system, classroom sociolinguistic analysis and other analytical methods [1, 2], but it is basically realized through manual analysis, in the video data such as the

J. Abawajy et al. (Eds.): ATCI 2021, AISC 1398, pp. 887–894, 2021.
https://doi.org/10.1007/978-3-030-79200-8_131

ocean is very extravagant, so people think through the machine to carry out preliminary identification and analysis, some scholars put forward that classroom teaching video through machine analysis to reduce the cognitive load of manual analysis process, and achieve automatic statistics [3, 4]; Using moving target tracking technology to model the position state information of tracking target and monitoring area, the number of people entering and leaving the monitoring area in and out of the classroom is realized. The recognition and classification of students' facial expressions and gestures are carried out by video data extraction. At present, teaching evaluation through video information is in the process of technological change. How to lighten the work burden of teaching evaluation and even realize automatic teaching evaluation requires a large number of modern information technology support [5, 6].

So far, educational researchers are constantly exploring, studying and designing various models to adapt to the development of teaching information to improve the development of education. With the development of information technology, educational researchers also apply information technology to the development of teachers. Therefore, with the help of video teaching analysis software under the background of big data, it has become an important way to improve the teaching quality and meet the teaching objectives and development needs of

2   Proposed Method

2.1   Transana Video Analysis Software

TSA platform refers to the transformation of video data into text and the analysis of text, and provides researchers with identification and organization tools, so that the video data can be refined and analyzed. A large number of videos are stored after tagging keywords. TSA platform can store a large number of digital video files; researchers intercept video segments, is not limited by the number, can be arbitrarily intercepted according to their own needs. This video analysis software is composed of a waveform generator of sound, a window to play file video, a window to record video text and a database directory tree. When running the platform, the waveform generator of the sound when analyzing the video provides the visual interface of the audio track [7, 8], and only the time code can be inserted to connect the text data with the location of the video. Thus, the video and text data play synchronously.

2.2   Big Data Era

Big data is a kind of information asset with large volume, many kinds and fast speed. It can not be collected and managed by ordinary processing tools, but must rely on virtualization technology such as cloud computing beyond a certain time and space boundary. Only through distributed deep mining can the data play its potential hidden value and play a better role in assisting people to optimize processes and other decisions. In the era of big data, the most important thing is to be able to process a large amount of data to get the potential value behind the data and expectations for the future. In other words, big data has massive data storage. The core and key of this project is data analysis and processing. In addition to relying on special processing tools and systems, when people carry out data collection, management, analysis and other steps, especially the staff of colleges and universities, they should take into account the wide range of information sources, implement specific system norms to protect privacy and attach importance to the training of specialized talents. Based on the above concept, it can be concluded that in addition to massive data information, big data needs more professional storage systems and powerful processors. Big data is not a simple accumulation of data information. It requires a certain correlation between data in order to make data more valuable for processing and analysis [9, 10]. Big data is technology, tools and methods, is the background of the times, and gradually formed scientific theory and become a way of thinking. Whether big data or artificial intelligence or other areas, Bayesian formula can answer the question:

The Bayesian theorem formula is as follows:

$$P(B_i|A) = \frac{P(B_i)P(A|B_i)}{\sum_{j=1}^{n} P(B_j)P(A|B_j)} \tag{1}$$

where P (A|B) is the possibility of A occurrence in the event of a B occurrence. A1,…An is a complete event group, that is:

$$\bigcup_{i=1}^{n} A_i = \Omega, A_i A_j = \varnothing, P\left(A_i^2\right) > 0 \tag{2}$$

When there are more than two variables, the Bayesian theorem still holds, for example:

$$P(A|B,C) = \frac{P(A)P(B|A)P(C|A,B)}{P(B)P(C|B)} \tag{3}$$

2.3 Classroom Practice

Classroom record analysis is a form of collecting data, analyzing data and studying data in class by means of modern information technology. This paper mainly analyzes the classroom teaching situation by means of TSA video analysis platform, mainly analyzes the teaching behavior of the teacher and the feedback analysis of the students' learning results, including the mutual analysis between teachers and students. This process has objectivity and comprehensiveness, and carries on the qualitative and quantitative analysis to the 6 complete teaching process, thus more pertinently analyzes and evaluates this classroom teaching process.

3 Experiments

3.1 Content of the Experiment

With the development of TSA video analysis platform, it is considered that it is necessary to provide information-based learning support for the development of undergraduate education and teaching, and to explore the influence of new teaching model on teaching effect. TSA platform can help students to learn independently, improve teaching quality and change the traditional teaching methods.

3.2   Test Methods

By collecting and reading a large number of relevant documents, we have a comprehensive and profound understanding of TSA video analysis software, big data background, classroom record and so on, and serve as a theoretical cushion. The results of classroom video observation (six lesson examples) are transformed into quantitative analysis data.

4   Discussion

4.1   Interactions Between Teachers and Technology in Classroom Teaching

Video or animation is common in the teaching resources used by undergraduate teachers. In the experiment, we selected six undergraduate class examples (microeconomics, econometrics, advanced mathematics, finance, basic introduction to Marxism, statistics) in which teachers used video or animation. Most of them were about "real work scene, working principle or operation flow, etc." Teachers used them mainly to create situations and lead to teaching themes; to demonstrate boring and understandable theoretical knowledge and principles; to present standardized and complete operation processes, etc. Human interaction with technology, that is, teacher-student interaction with technology. Taking the TSA video analysis software used by teachers as an example, the ratio of teacher-student and technical interaction in the course of classroom teaching is shown in Fig. 1:

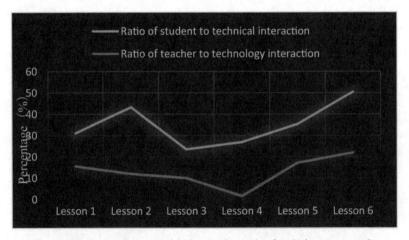

Fig. 1. Teacher/student-technical interaction ratios for six lesson examples

It can be seen from Fig. 1 that in the six lessons, the interaction ratio between students and the transana platform is higher than that between teachers and the transana platform, indicating that teachers no longer mainly use technology to interact with students in class. In other words, information technology is no longer an auxiliary teacher, but rather a tutoring student, which changes students' learning style and effectively improves traditional classroom teaching methods. In terms of teacher-technical interaction, the proportion of teacher-technical interaction behavior is very small, which can be found in combination with their classroom records, teacher and technology interaction only carried out "presentation teaching resources" and "software tool switching ", which set aside more time for students to interact with TSA platform and students' practical operation. Example 6 shows that teachers can use information technology more flexibly and conveniently by using TSA platform in classroom teaching. Teachers have more convenient information technology, and can be more proficient in the use of technology. In terms of students' interaction with technology, Among the six courses, the lowest rate of interaction between students and technology is 13.57%, and the highest is 31.30%. And from Table 1, In the six courses, the proportion of interaction between students and transana platform accounts for more than half of students' behavior, which is related to the more diverse forms of students' interaction with technology, and the integration of various information technologies and classrooms.

Table 1. Statistical tables of student-to-technical interaction rates in six lesson examples

Statistical items	Lesson 1	Lesson 2	Lesson 3	Lesson 4	Lesson 5	Lesson 6
Ratio of student behavior to technical interaction	56.00%	80.94%	62.70%	44.16%	43.27%	54.10%

4.2 Response of Students After Teaching Using Transana Platforms

During the teaching practice of using the TSA platform classroom record, through the teaching practice of six lesson examples, the students' different responses to teachers' questions are used to feedback the teaching results of using video analysis software (Fig. 2).

4.3 Use of Transana Platform Classroom Structure Statistics

By using the TSA platform to teach Microeconomics, Econometrics, Advanced Mathematics, the classroom structure statistics of these three lesson examples are shown in Table 2:

As can be seen from Table 2, in the Microeconomics class, the teacher's language is 44.04%, for about 19 min, and students' speech accounted for 28.62% of the classroom language, the cumulative time is about 12 min, effective silence occupies 17.79% of the classroom language, the cumulative time is about 8 min, 9.42% of the classroom language used TSA, the cumulative time is about 4 min. The proportion of teachers' language is lower than that of teachers' language norm 68%, the proportion of students' language is 20% higher than that of students' language norm, this shows that the teacher pays attention to the main position of the students. However, the low proportion of teachers' language indicates that teachers lack guidance to students. In econometrics class, the teacher's language is 63.02%, the cumulative time is about 25 min, the student language occupies 13.51% of the classroom language, the cumulative time is about

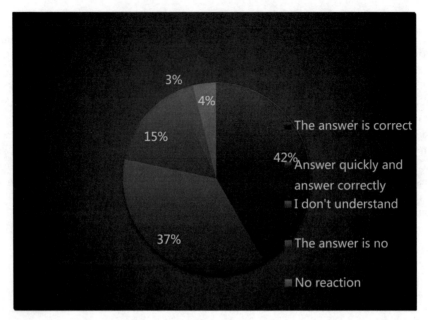

3%
4%
15%
42%
37%

The answer is correct

Answer quickly and
answer correctly

I don't understand

The answer is no

No reaction

Fig. 2. Students' responses to teacher questions (%) Unit

Table 2. Statistical unit (%) of classroom structure

Statistical items	Microeconomics	Econometrics	Higher mathematics	Average
Teacher language ratio	44.04	63.02	40.91	49.32
Pupil language ratio	28.62	13.51	23.91	22.01
Effective Silence Ratio	19.79	19.49	33.45	23.06
Proportion of TSA used	9.42	5.53	1.61	5.52

6 min, effective silence occupies 17.94% of the classroom language, the cumulative time is about 7 min, 5.53% of the classroom language used TSA, the cumulative time is about 2 min. Combined with teaching video, the teacher focuses on guiding students to think and practice, played the main role of students and the guidance of teachers. During these hours, teachers focus on explaining knowledge and asking questions, the language behavior of students is mainly to answer the teacher's questions. In addition, during this period there are students thinking, students do questions, teachers individual tutoring and so on. In advanced mathematics, the teacher's language is 40.91%, the cumulative time is about 18 min, and the student's language accounts for 23.91% of the classroom language. The cumulative time is about 10 min, effective silence occupies 33.45% of the classroom language, the cumulative time is about 15 min, 1.61% of the classroom language used TSA, the cumulative time is less than 1 min. Combined with video, it can be found that the teacher pays attention to students' practice in the teaching process, and

students have more time to do questions. In general, the teacher language and the student language occupy the main part in the three classes, each class teacher language accounts for about 20 min, the student language accounts for about 10 min, and the percentage of teacher-student language behavior accounts for more than 65% of classroom language behavior. And the proportion of teachers' language is less than that of norm 68, which shows that teachers realize that students are the main body of the classroom and pay attention to giving play to the main role of students. Secondly, the three teachers used TSA to teach to help students develop the spirit of autonomous learning and achieve the effect that each student can integrate into the classroom.

5 Conclusions

With the development of information technology, under the background of big data, the popularity of network, multimedia technology and network communication technology as the main way of information technology for learners to provide a more convenient and fast environment, conducive to the development of education, but also for the development of teachers teaching. Among them, the new media in the past scene has become the focus of teacher education.

Acknowledgements. This work was supported by Key projects of the Ministry of education in 2016 of the 13th five year plan of National Education Science, project name: Research on the application of visual analysis technology for the professional development of University Teachers (No. DCA160264)

References

1. Xu, L., Wang, J., et al.: Information Security in Big Data: Privacy and Data Mining. IEEE Access **2**(2), 1149–1176 (2017)
2. Sivarajah, U., Kamal, M.M., Irani, Z., et al.: Critical analysis of Big Data challenges and analytical methods. J. Bus. Res. **70**, 263–286 (2017)
3. Athey, S.: (Special Issue Perspective) Beyond prediction: using big data for policy problems. Science **355**(6324), 483–485
4. Xue, J.W., Xu, X.K., Zhang, F.: Big data dynamic compressive sensing system architecture and optimization algorithm for internet of things. Discr. Continuous Dyn. Syst. Ser. S **8**(6), 1401–1414 (2017)
5. Wamba, S.F., Gunasekaran, A., Akter, S., et al.: Big data analytics and firm performance: Effects of dynamic capabilities. J. Bus. Res. **70**, 356–365 (2017)
6. Tumanggor, A.M.R., Supahar, Kuswanto, H., et al.: Using four-tier diagnostic test instruments to detect physics teacher candidates' misconceptions: case of mechanical wave concepts. J. Phys. Conf. Ser. **1440**(1), 012059 (8 pp) (2020)
7. El-Dakhs, A.S., Ambreen, F., Zaheer, M., et al.: A pragmatic analysis of the speech act of criticizing in university teacher-student talk: the case of English as a Lingua Franca. Pragmatics **29**(4), 493–520 (2019)
8. Sumarni, S., Ramadhani, R., Sazaki, Y., et al.: Development of "Child Friendly ICT" textbooks to improve professional competence of teacher candidates: a case study of early childhood education program students. J. Educ. Gifted Young Sci. **7**(3), 693–708 (2019)

9. Miyakawa, T, Winsløw, C.: Paradidactic infrastructure for sharing and documenting mathematics teacher knowledge: a case study of "practice research" in Japan. J. Math. Teach. Educ. **22**(3), 281–303 (2019)
10. Kulophas, D., Kim, M . How "Thailand 4.0" principals applied leadership and technology towards teacher learning: three case studies. Kasetsart J. Soc. Sci. **41**(3), 614–619 (2020)

Enlightenment of Big Data Thinking on the Construction of Scientific Research Performance Evaluation System for Humanities Teachers in Local Universities

Ermi Zhang[1(✉)] and Wanbing Shi[2]

[1] School of Humanity and Law, Northeastern University, Shenyang 110189, Liaoning, China
[2] Special Education Research Center, Nanjing Normal University
of Special Education, Nanjing 210038, Jiangsu, China

Abstract. As a key link in university management, scientific research performance evaluation is a significant guarantee for promoting the orderly development of teachers' scientific research. However, there still exist some problems in the traditional scientific research performance evaluation of Humanities Teachers in local universities and colleges, such as the lack of independent evaluation of Teachers, the lack of specific evaluation indicators and the malposition of peer experts, Humanities Teachers' research performance evaluation and the deep integration of big data technology can build a more scientific evaluation system. This study discusses the Enlightenment of big data thinking on the construction of scientific research performance evaluation system of Humanities Teachers in local universities and colleges from three aspects: teacher self-evaluation, index selection and data collection, evaluation process and evaluation method. Through the strategic management tools such as scientific research performance evaluation model and university scientific research performance index database, the perspective evaluation of local universities and colleges is carried out, integrate the most insightful research data and results with the development status of higher education at home and abroad. On the premise of ensuring the security of data and information, let all local universities and colleges benefit from the convenience of big data, in order to gradually move towards a new era of teachers' scientific research performance evaluation which is based on big data thinking.

Keywords: Big data · Big data thinking · Humanities Teachers · Research performance evaluation

1 Introduction

The emergence of big data has created a new situation for the development of higher education informatization. Based on big data thinking, we are able to carry out various management work. For example, provide solutions for university management. As an important part of university management, the performance evaluation of University Teachers' scientific research work is an important prerequisite to ensure the quality

J. Abawajy et al. (Eds.): ATCI 2021, AISC 1398, pp. 895–902, 2021.
https://doi.org/10.1007/978-3-030-79200-8_132

of education. However, there are many problems in the construction of the scientific research performance evaluation system of Humanities Teachers in local universities. The emergence of big data thinking provides some new ideas and Countermeasures to solve these problems. Based on this, this study starts from the elements and problems of the current scientific research performance evaluation system of Humanities Teachers in local universities and colleges, and uses the concept of big data for reference, analyzes the Enlightenment of big data thinking on the construction of scientific research performance evaluation system of Humanities Teachers in local universities and colleges, so as to benefit the construction of scientific research performance evaluation system of Humanities Teachers in local universities and colleges Based on big data thinking in the future [1].

2   Overview of Big Data Thinking

Big data is an IT industry term, first proposed by NASA researchers Michael Cox and David Elsworth in 1997 to describe the huge amount of data which was generated by supercomputers in the 1990s. Researchers, data analysts and practitioners in related industries all put forward different definitions of big data. Laney, a researcher at the maita group, first proposed the definition of "big data". There are three directions for the opportunities and challenges of data growth, namely volume, namely, data size, expressed in megabytes, tables and documents; velocity, namely data, the speed of input and output, expressed in batch, near time, real-time, and flow; variability, expressed in structured, unstructured and semi-structured, and also known as "3 V" or "3vs" [2]. With the continuous in-depth study of big data by many scholars, its characteristics can be extended to "7vs", which are volume, variety, velocity, variability, veracity, complexity and low value density [3]. Generally speaking, big data must be counted by computer, and its hidden value must be deeply mined to get objective results.

Big data is an emerging research field, which provides decision-making information for people's life and work through data analysis. In recent years, experts and scholars have collected and stored more and more abundant information. Big data thinking is mainly used in business, economic management, health care and other fields to study how to effectively aggregate and associate massive data [4]. Big data has a unique way of thinking in this research field, that is, big data thinking. This way of thinking is extracted by experts and scholars in this industry from several years of practice. Fundamentally speaking, it is a set of effective methodology. Big data thinking mainly includes comprehensive thinking, fault-tolerant thinking and related thinking. universities and colleges are in an increasingly complex and competitive environment, facing increasing pressure of scientific research to satisfy the needs of national and global political, economic and cultural development. When dealing with the above pressure, university managers must develop big data thinking, because big data can effectively solve the disputes and problems in university management [5]. At present, the application of big data thinking in university scientific research management mainly focuses on teacher information management, personnel file management and scientific research achievements management. In the aspect of scientific research performance evaluation of Humanities Teachers in local universities, although experts and scholars try to combine university management with big data thinking, there is little research related to it, which also reflects the innovation of this research.

3   The Elements of the Performance Evaluation System of Humanities Teachers in Local Universities

At present, the construction of the performance evaluation system of Humanities Teachers in most local universities and colleges is inseparable from the components. It mainly starts from six components, including the evaluation subject, the object, the purpose, the standard, the evaluation system and the evaluation method. Combining with the current evaluation system of the research performance of Humanities Teachers in some local universities and colleges, and consulting the relevant existing literature, the paper summarizes the components of the evaluation system into the following six aspects:

① Evaluation subject: from the perspective of universities and colleges, the evaluation subject mainly includes the personnel department, scientific research department, discipline department and the dean of relevant colleges. For teachers of humanities, it is indispensable for them to employ experts from outside school and colleagues from inside school. If necessary, a third-party evaluation organization should be introduced. ② Evaluation object: humanities is highly creative, abstract and lagging. On the surface, the evaluation object of Humanities Teachers' scientific research performance evaluation is scientific research projects, scientific research achievements and awards. But from the deep level, we should also explore the particularity of humanities, such as the differences with other disciplines, including the differences between humanities. ③ Evaluation purpose: from the perspective of management, Humanities teachers' scientific research performance evaluation serves as a baton and plays a role in improving the overall quality of the school's scientific research; from the perspective of development, the scientific research performance evaluation of Humanities Teachers promotes the professional development of teachers, and further promotes the discipline development. ④ Evaluation criteria: the performance evaluation of Humanities Teachers in local universities and colleges are supposed to respect the basic rules and actual situation of scientific research activities, and also respond to the requirements of relevant national policies. The evaluation criteria are mainly set from three aspects of time, quantity and quality. More importantly, it is necessary to formulate special evaluation standards and indicators according to the characteristics of humanities. For example, when setting up indicators of scientific research achievements, separate indicators of art practice should be set up for art teachers in addition to papers and works. ⑤ Evaluation system: at present, the systems involved in the scientific research performance evaluation of Humanities Teachers in local universities and colleges are in connection with the evaluation cycle, evaluation frequency and feedback of evaluation results, which can be regarded as the scientific research evaluation system in the scientific research management system, that is, the system related to the scientific research project management and teachers' professional development. ⑥ Evaluation method: for Humanities Teachers, the comprehensive evaluation method of combining qualitative evaluation and quantitative evaluation is mainly used, in which qualitative evaluation accounts for a small proportion, while quantitative evaluation accounts for a large proportion, which is generally divided according to the ratio of 3:7 [6]. Among them, the evaluation subject is the initiator of the evaluation, the evaluation object is the service object, the evaluation purpose is the reason, the evaluation standard is the scale, the evaluation system is the criterion, and the evaluation method is the tool. The six elements are like a closed loop, which is connected from the beginning to the end. Through the scientific research management department, they form a continuous closed loop and a loop, and maintain a balance point through the scientific research management department, they interact and are indispensable (Fig. 1).

Fig. 1. Six elements of the performance evaluation system of Humanities Teachers in Local Universities

4   Enlightenment of Big Data Thinking on the Construction of Scientific Research Performance Evaluation System for Humanities Teachers in Local Universities

From the perspective of the future development trend of Humanities Teachers in local universities and colleges, it is indisputable to build a comprehensive scientific research performance evaluation system which is based on big data thinking, but it needs to be designed and optimized for a long time. From the current point of view, local universities should make the following three preparations for the future construction of scientific research performance evaluation system based on big data thinking.

4.1   Design and Optimize Teacher Self-evaluation Process

Through big data analysis and mining, we are able to effectively present and scientifically predict the objective law and development direction of the scientific research performance evaluation of Humanities Teachers in local universities. Different from the previous evaluation work, in the evaluation of scientific research performance of Humanities Teachers in local universities and colleges Based on big data thinking, teachers can not only conduct vertical self reflection in different time dimensions through teachers' self-evaluation, but also make a horizontal comparative analysis among teachers through teachers' mutual evaluation [7]. At the same time, as the group who knows the Humanities Teachers' scientific research work best, all teachers should participate in the establishment of scientific research performance evaluation model, track the progress of teachers' scientific research performance evaluation in real time, communicate with scientific research management departments in time, optimize the links with excellent performance, and give early warning to the links with poor performance [8]. Finally, we should improve the independent status of Humanities Teachers in scientific research performance evaluation, respect teachers' personal wishes, and truly achieve "people-oriented".

4.2 Design and Optimize Index Selection and Data Acquisition

Compared with the traditional index design and data collection of Humanities Teachers' scientific research performance evaluation in local universities, the introduction of big data thinking can design more accurate evaluation index model through more detailed data information. In the evaluation model of Humanities Teachers' scientific research, more evaluation indicators of arts research achievements can be added and data collection can be carried out. For example, the scientific research projects, scientific research achievements and awards of Arts teachers can be accurate to specific artistic practice projects, artistic practice achievements and artistic practice awards, and how to set up the evaluation index of similar local universities and colleges, so as to avoid the "one size fits all" and "simplified" closed management mode. In the scientific research work, local universities can change the traditional manual scientific research management mode by building a teacher scientific research database management platform. In addition to the statistics of scientific research projects, scientific research achievements and the number of awards, big data can supervise the scientific research funds, monitor the progress of scientific research achievements, and use the Internet for annual evaluation data statistics, online office and system maintenance, etc, Through the teacher's scientific research database management platform, the data docking of scientific research departments, personnel departments, discipline departments and colleges can be completed. On the one hand, the trouble of scientific research management personnel in filling in the scientific research project declaration form and scientific research achievements registration form can be avoided; on the other hand, the speed and efficiency of querying the relevant information of scientific research can be improved, finally, improve the quality of scientific research performance management, and provide strong data support for teachers' professional development planning (Tables 1 and 2).

4.3 Design and Optimize the Evaluation Process and Method

In the performance evaluation of Humanities Teachers in local universities and colleges, the excessive use of quantitative evaluation method is bound to cause the disgust of teachers, while the intervention of peer experts in qualitative evaluation method is easy to breed the bad phenomenon of "stressing human feelings" and "trusting relationship". In recent years, the State advocates the introduction of third-party evaluation institutions

Table 1. Quantitative standard of academic works

Work	Project category	Quantitative standard(J)			Explain
		Academic monograph	Translated and edited works Deductive works	Literary works Popular science works	
	Class A	60	45	35	Single category score calculation: $G = J \times k1 \times k2$, the authors in the works are ranked from high to low according to the number of completed words
	Class B	50	35	25	
	Class C	30	20	15	

Table 2. Example of scientific research performance data of teachers in Art College of a local liberal arts university in Liaoning Province

Serial number	Faculty number	Score of scientific research project	Academic paper score	Score of academic works	Patent score	Achievement award score
1	****	0	10	0	0	10
2	****	0	0	0	0	7.4
3	****	18.7	6	0	0	0
4	****	31.8	0	0	0	0
5	****	0	0	0	0	23
6	****	1.5	10	0	0	21.5
7	****	532	61	18.9	0	0
8	****	95.7	10	0	0	12.4
9	****	11.5	10	0	0	0
10	****	129.5	0	0	0	0
...	...	...	...	...	...	...

in the performance evaluation of university teachers, among which the official institutions or better qualified private institutions are better [9]. Taking local universities in Liaoning Province as an example, a liberal arts university in Shenyang has introduced a leading third-party evaluation institution in China since last year. This kind of institution does not rely on the experts and peers hired by the University, and completely provides professional technical support from the external perspective, so as to avoid the phenomenon of "layman evaluating expert". As we all know, improving the efficiency and level of scientific research is the central goal of scientific research management in universities and colleges. Scientific research management departments and colleges are the specific implementation units of scientific research and teacher construction in universities and colleges, which directly determine the quality and efficiency of scientific research in universities and colleges, therefore, the real-time monitoring and periodic analysis and evaluation of the scientific research performance of the college can effectively transfer the pressure of the scientific research development of the college, stimulate the scientific research vitality of the college and department, and achieve the effect of optimizing the allocation of resources, improving the level and efficiency of scientific research. In order to help local universities and colleges do a good job in this aspect, the third-party evaluation organization independently developed and designed a performance management system of colleges and departments under the guidance of big data thinking mode. The system integrates the functions of data filling, analysis and evaluation, including hundreds of data indicators reflecting the scientific research performance of colleges and departments, which can provide data support for the macro management and decision-making of the school. With the accumulation of time and the continuous expansion of underlying data, the third-party evaluation institutions will build a big data system library for each local university in line with its own scientific research situation, and build a more accurate scientific research performance evaluation model, so as to help the scientific research management departments and departments more effectively manage the scientific research performance evaluation of Humanities Teachers. Meanwhile, the data information security must be fully guaranteed [10].

Acknowledgements. This paper is the result of education project of National Social Science Foundation of China (project number: BFA150043).

References

1. Chen, S.: Reform and practice of big data driven higher education management and decision. J. Phys: Conf. Ser. **11**(1345), 1–5 (2019)
2. Klanja-Milievi, A., Ivanovi, M., Budimac, Z.: Data science in education: big data and learning analytics. Comput. Appl. Eng. Educ. **25**(6), 1066–1078 (2017). https://doi.org/10.1002/cae.21844
3. Caballero, P.D.F., Terrón, A.M., Olivencia, J.J.L.: Big data irruption in education. Pixel-Bit. Revista de Medios y Educación **57**(2), 59–90 (2020)
4. Khan, Shakir, Alqahtani, Salihah: Big data application and its impact on education. Int. J. Emerg. Technol. Learn. **17**(15), 36–46 (2020)
5. Daniel, B.: Big data and analytics in higher education: opportunities and challenges. Br. J. Educ. Technol. **46**(5), 904–920 (2015)
6. Mehandjiev, N., Layzell, P., Brereton, P.: Qualitative evaluation and research methods. Mod. Lang. J. **10**(4), 543 (2011)
7. Ohland, M.W., Loughry, M.L., Woehr, D.J.: The comprehensive scale for self- and peer-evaluation. Acad. Manage. Learn. Educ. **11**(4), 609–630 (2012)
8. Shi, Jun., Huang, Entao., Sun, Caiyun, Huang, Xia: The grey-information-entropy-based performance evaluation model. Int. Conf. Inf. Sci. Technol. **2012**(11), 435–439 (2012)

9. Stalford, C.: Social structure of third-party evaluations. New Dir. Eval. **1980**(5), 69–81 (2010)
10. Jee-wan, B., Kwang-jai, Y.: A study on constructing methods for the performance management system in Korean Universities. Focus. Univ. Adm. Organ. **2012**(8), 394–421 (2012)

Use Big Data to Analysis the Economic in China

Junming Chen[✉]

University of Miami, Coral Gables, FL, USA

Abstract. China has made remarkable progress in its economic development over the past three decades, with most of the basic indicators showing remarkably good results on the surface. This paper mainly investigated databases such as World Bank, Trading Economic, and CEIC, and compared the data with those of other countries. On the surface, although China's economy has developed rapidly in the first three decades, the potential problems and disadvantages deserve attention as well as the advantages that have contributed to its rapid development. Moreover, after analyzing the data, it can be found that the Matthew effect and the overall development slowdown in the society were actually reflected ten years ago. Therefore, if China wants to maintain its current high development rate, there are still many areas for improvement. Finally, after analyzing the data in various databases, this paper reevaluates the actual economic development trend of China in the past 30 years, its own advantages, internal problems, and future hidden dangers.

Keywords: Global · Economic growth · GDP · Saving rate · Labor · Chinese economic

1 Introduction

Shenzhen is considered a new city with a population of twenty million. But despite its large size, most people originate from other provinces in the country. This is unusual in China, as most people stay in one city or province for their entire life. Before 40 years ago, there were only several thousand people living in the small fishing village of Shenzhen: some of the original population moved to Shenzhen from Hong Kong to start new lives, others were mainly fishermen. But Shenzhen quickly became a city with 20 million people and a GDP that just surpassed Hong Kong and became one of the top 3 cities in the China by GDP ranking [1]. Since it is a newer city, the infrastructure is good, something that leads my friends to believe that Shenzhen is a much better and more modern city that other metropolises, such as Los Angeles or New York.

This is the point that I want to write this topic: why this kind of change happened, whether this change can be copied to applied to other cities inside and outside China, whether Shenzhen can maintain this rapid development rate in the future, and the potential underlying risks and flaws that people should consider in these kinds of cases.

The success of Shenzhen is just a small part of a much larger China, as the development within the country is kind of uneven. Except a few cities where change is drastic, most cities are slow-moving and tend to maintain the current situation. This is especially true for cities in inland regions, where there are few cultural and economic exchanges

J. Abawajy et al. (Eds.): ATCI 2021, AISC 1398, pp. 903–909, 2021.
https://doi.org/10.1007/978-3-030-79200-8_133

with the outside world. The rapid development of a global economy has not accelerated the economic construction and improvement in the quality of life in these cities.

2 Analysis of Global Domestic Product and Consumer Price Index

2.1 Real Global Domestic Product

The real Global Domestic Product is the index that shows the production and ability of a country, as it has already adjusted the inflation influence.

The data base after the adjusted by big data shows that from 1990 to 2010 the real Global Domestic Product of China grew stably from around 7000 to 300000 hundred million Chinese Yuan [2], which significant higher than World Bank estimates, and even reached to 500000 hundred million Chinese Yuan in last year [3]. This is an indicator that the production ability of China is still growing relative fast as a developing country. But since the cardinal number is small at first and the index value is relatively large after twenty years, the changing rate also shows that the average growing rate is decreasing. Thus, China is maintaining a relatively good growth rate from the last 20 years but is facing a problem that GDP growth is slowing down.

2.2 Real Global Domestic Product Per Capita in Purchasing Power Parity Terms

Since one way to measure the Real Global Domestic Product is by measuring the production in a country, and production is equivalent to expenditure which is equivalent to income, Real Global Domestic Product per capita can be used to measure people's average income in a country. After adjusted by purchasing power parity, we can study people's living quality changes over a given time.

This graph [3] shows people's living quality had a continuous increase from 1995 to 2020, using units of US dollar (See Fig. 1). The growth rate cannot match the growth rate of real Global Domestic Product, which may be because of population is growing too fast, or the gap between rich and poor is widening. But the overall living quality of the country increased a lot, showing how many average people can now purchase eight times the value of goods and services that they could buy in the past. This is a strong indicator that the Chinese economy has developed well over the last twenty years.

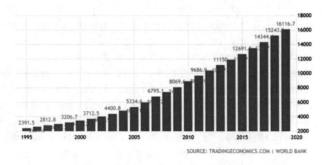

SOURCE: TRADINGECONOMICS.COM | WORLD BANK

Fig. 1. GDP per Capital in PPP terms

3   The Inequality Gap Between China and America

3.1   Real Gross Domestic Product Adjusted by Purchasing Power Parity

One way to measure the Real GDP is to measure the production in a country, as production is equivalent to expenditure which is equivalent to income. Real GDP per capita can be used to measure average income in a country. After it is adjusted by purchasing power parity, we can study people's change in living quality over a given time. Thus, this also a good standard to determine the inequality gap between two countries.

In the graph [3] (See Fig. 2), the data between the US and China is presented in US dollars, but with a different standard for the y coordinate, since the big difference in data. The left side is China's data, and the right side is America's. Based on the data, although the growth rate in real GDP per capital in terms of PPP of China is much faster, as it increased nearly ninety percent in last ten years, the actual amount is still very small compared with America's. Thus, most people in China still cannot approach the daily consumption ability of American's, or even the consumption level of American's a decade ago. China is on the losing side of the in-equality gap and needs to close the gap, which is a vicious cycle that need to be considered.

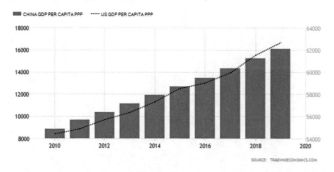

Fig. 2.   Real GDP per Capital in PPP term of China and United State in 2010–2019

3.2   The First Factor of Productivity: Human Capital

The human capital index is very stable compared with other indexes mostly because the total population and educational system is hard to have a great change; as the evidence of the HCI data base, most countries changes is less than three percentage in twenty years [4]. This may be an important factor in the constant low productivity level of China, since it is very hard to make changes in these sectors. Also, the rapidly decreasing birth rate and the Matthew effect caused by unequal resource allocation may be reasons that the human capital index has a decreasing tendency [5].

3.3   The Second Factor of Productivity: Technology

Technology is similar to physical capital in influencing productivity. Instead of concrete equipment, it can be the knowledge or an efficient process that helps workers produce more output with the same number of resources. Technology is the most important factor of productivity, as the utilization of internet plays a key role in our world today.

4 Economic Growth and Saving Rate in China

4.1 Economic Growth Rate and Average Growth Rate in China

Economic growth rate is a way for people to measure a country's economic growth and living standard growth. It uses last real GDP per capita minus the previous real GDP per capita, and then divide by the previous one to find out how much real GDP per capita increased in the given period.

In the graph [3] (See Fig. 3), which units are in percentages, people can see that China's GDP growth is around 6% for the last ten years and maintaining a decreasing tendency. For a developing country, this is still relatively good data, as the population is extremely large in China, and the growth rate is also not lower than most developing nations (the developing countries' global average economic growth rate is around 5%).

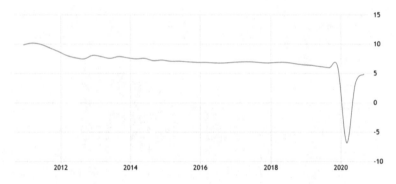

Fig. 3. GDP Annual Growth rate in 2010–2020

However, compared with the average growth rate of 8%, the real GDP per capita growth rate is actually continuously deceasing, and shows no tendency of maintaining its current level. This data shows that the previous contribution that productivity in China gave to economic growth rate is harder and harder to maintain, which is an indicator that productivity needs to be improved or other plans need to be made.

4.2 Saving Rate in China

The saving rate is also a measurement factor that influences economic growth. A high saving rate may help investment and production but have no positive contribution to consumption, further reducing the incentive of the production industry and lowering the economic growth rate. While low saving rates allow for a high-level consumption, it also can be an incentive for the investment and development of production. China always have a high saving rate, which even exceeded 53% of whole country's GDP in 2008 [6].

Even though the percentage rate of the saving over the GDP slightly lower in next decade, the amount of it is still higher than the quantity level in 2008; After analyzed data, there is a evident negative relationship between economic growth rate and the saving rate in the last 20 years [4].

4.3 Relation Between Global Domestic Product and National Saving

To sum up, the effect that saving bring to Global Domestic Product is depend on situation, saving rate may can improve the Global Domestic Product by increasing the investment and GDP growth rate will increase as follow; but saving also very probably causing negative effect on Global Domestic Product that more saving reducing the consumption, which is the main part of the Global Domestic Product, and leading horrible influence on future developing.

5 Employment and Unemployment in China

5.1 Population and Labor Force Participation Rate

Labor force participation rate and population are important indexes that contribute to a country's productivity and economics, as we have analyzed before. The larger the population, the greater the potential labor a country may have. However, if a country has a very low labor force participation rate, the large population will become a burden on the country's future development.

5.2 Employment Rate and Unemployment Rate

The employment and unemployment rate can be combined with labor force participant rate and population to analyze the situation of a country's job market and production industry.

The data shows the decrease in employment and increase in unemployment during the 1990–2020 period in China, with the units in percentages. Though the unemployment rate continuously increases, it has not changed a lot and will not cause a lot influence on society, which is around 4% to 6% [3]. However, the employment rate is dropping very quickly, which probably is one of the main reasons causing the labor force participation to also drop so quickly, which drop from 85% to 76% [4]. This means more and more workers have quit from daily jobs and do not want to go back to work. Since the retirement age has not changed for a long period of time (as discussed in the following sections) and the population has continuously grown in last fifty years, a huge problem to be considered has arisen—while the population of potential workers has not decreased, then how to stop the drastically dropping labor participation rate to maintain the productivity level and development rate of a country becomes a huge problem.

5.3 Unemployment and the Business Cycle

After comparing the GDP growth rate and the unemployment rate over the last twenty years, with the unit of percentage. During the covid virus period there is a recession and raise back quickly. Just as the common theory that the economic growth rate and unemployment keeping a negative relationship. However, these records during the recession not only show the common relationship between the unemployment rate and economic cycle, but also show that China has a relatively stable economic environment. In the recession of 2008 and 2020, the GDP growth rate did not change a lot and unemployment did not change more than 2% (from 4.4% to 4.6% during 2008–2009) [4]. Instead of the short-term serious problem happened in Covid-19 period, this stable environment may contribute to a continuous development and bring positive influence, which is a main advantage characteristic of Chinese economic development.

6 Summary

6.1 Fundamental Causes of the Prosperity Effect

China is located near many developed countries and districts with high productivity, such as Hong Kong, South Korea, and Singapore, which accelerates Chinese economic development a lot through international trading and cooperation. Also, there is no large religious groups or religions that can affect and restrict the economy, and as the data given in the Chinese saving rate analysis, China's culture lead people tend to save a lot of money. As the consequence, investment increased drastically in last half century with the support from high saving rate and the support from government, which is stimulate the economic growth and maintain the future development.

6.2 The Proximate Causes and Problems for Future Development

The economic development in China is relatively good but has already started to face some serious problems and challenges. The Real GDP growth was very good in the last twenty years, but after continuously decreasing it is now just around 5%, which is only a little bit above the average growth rate of developing countries. Besides the problem of the growth rate slowing down, according to the data of Real GDP per capita in PP terms, poverty level, and life expectancy, China has really improved citizen's life quality a lot.

However, when comparing China with America, it is clear that there is an inequality gap. China is losing in all of important factors that influence productivity in theory, especially regarding technology. Since improving productivity is the most important way to close the inequality gap, China is facing and trying to solve the problem regarding the development of human capital, physical capital, and technology [7].

References

1. Bao, Z., Lu, W.: Developing efficient circularity for construction and demolition waste management in fast emerging economies: lessons learned from Shenzhen, China. Sci. Total Environ. (2020). https://doi.org/10.1016/j.scitotenv.2020.138264

2. Feenstra, R.C., Ma, H., Neary, J.P., Prasada Rao, D.S. (2013) Who Shrunk China? Puzzles in the measurement of real GDP. Econ. J. **123**(573), 1100–1129, 1 Dec 2013. https://doi.org/10.1111/ecoj.12021

3. Tradingeconomics.com: China - Economic Indicators. (2021). https://tradingeconomics.com/china/indicators. Accessed on Nov 2020

4. Data.worldbank.org: World Bank Open Data | Data (2021). https://data.worldbank.org. Accessed on Nov 2020

5. Kūkoja, K., Kūle, L.: Different approaches in providing early childhood education services for vulnerable groups In Latvia (2018). https://du.lv/wp-content/uploads/2019/07/SZF-krajums_I_Sociologijas-aktualitates_2019_DRUKA_jauns.pdf#page=40

6. Ma, G., Yi, W.: China's high saving rate: myth and reality. Écon. Int. **2**(2), 5–39 (2010). https://doi.org/10.1016/s2110-7017(13)60028-1

7. Liu, Y., He, J., Guo, M., Yang, Q., Zhang, X.: An Overview of Big Data Industry in China (2014)

Innovation of Modern Enterprise Logistics Management Model Based on the Background of Big Data Era

Jingde Weng[✉]

Yang-En University, Quanzhou, Fujian Province, China

Abstract. In the current society with the rapid development of computer technology, the speed, type, and quantity of information dissemination in people's lives have reached unprecedented heights. Big data has gradually been integrated into people's lives and played an important role. Its inclusiveness makes it widely used in various industries. Taking logistics management as an example, the application of big data technology can significantly improve work efficiency. This not only proposes a new development direction for enterprise logistics management, but also promotes the transformation and development of enterprise logistics management in the new era. This article studies the innovation of modern enterprise logistics management mode based on the background of big data era.

Keywords: Big data · Enterprises · Logistics · Management models · Model innovation

1 Introduction

With the continuous development of the information age, it also brings many new terms. Big data is a brand-new technology that has been widely used in recent years. In modern enterprise logistics management, it is often applied to enterprises as an important part. In the process of development and transformation. For the defects and loopholes in the logistics management process, analysis with the help of big data technology can be found in time, and the loopholes can be repaired through remote monitoring technology, so as to ensure the smooth and efficient development of logistics management [1]. However, there are obvious shortcomings in the actual application of big data technology in enterprise logistics management. This article will briefly analyze the new ideas of enterprise logistics management in the era of big data.

2 Application Advantages of Big Data in Enterprise Logistics Management

2.1 Fast Information Connection

In the enterprise logistics management work, the most obvious advantage of the application of big data technology is the rapid connection between information, so as to

J. Abawajy et al. (Eds.): ATCI 2021, AISC 1398, pp. 910–915, 2021.
https://doi.org/10.1007/978-3-030-79200-8_134

avoid management problems caused by information asymmetry. In the early stage of the development of logistics management, lack of perfect technical means and more reliance on manual work, resulting in low efficiency of logistics management and very slow information connection [1]. After applying big data technology to carry out logistics management work, it not only meets the needs of modernized transformation of enterprise logistics management system, but also can monitor the actual operation process of logistics in real time. The speed of information connection has effectively improved the efficiency of enterprise logistics management. Regarding the loopholes in logistics transportation and management, managers can use big data analysis technology to find out in time and arrange for relevant technicians to repair them, so as to realize the risk control of physical management to ensure the smooth development of all work of the enterprise [2].

2.2 Increase Stickiness with Customers

In logistics management work, the work is mainly for customers, and the stickiness between customers and enterprises has naturally become the standard for measuring logistics management work. In traditional logistics management work, due to the lack of contact with customers and Information is asymmetry, resulting in low stickiness between the company and its customers and difficult for further development [2]. After the application of big data technology, companies and customers can communicate more closely, thereby speeding up the efficiency of logistics management, while increasing the stickiness with customers, understanding customer needs and developing targeted products integrating the obtained data as a basis can effectively improve the enterprise's logistics management capabilities and provide customers with better marketing services.

3 Problems in Logistics Management in the Era of Big Data

3.1 Data Security Faces Threats

Today with the highly developed information technology, big data technology is gradually being proposed and applied to all occupations, and then comes the issue of data security [3]. Although the application of big data technology in the logistics management of enterprises provides great opportunities for enterprise development great convenience, but at the same time, the issue of data security has become a major issue that has to be considered during its development. In the big data operation system, information damage, loss, distortion and other issues are the most common, so there are higher requirements for environmental safety. However, it is difficult to prevent information leakage in a man-made operation environment, which has also caused great obstacles to logistics management [3].

3.2 Inventory Management Model Is Relatively Backward

The effectiveness of the inventory management model is very important to the logistics management of enterprises. However, in terms of the current development background of logistics management in our country, the application of big data technology is still in the preliminary stage, so there is still a lack of a more complete inventory management model [4]. The efficiency and quality of enterprise logistics management cannot be effectively improved due to the lack of advanced inventory management models. In addition, business managers do not have a comprehensive grasp of big data technology, which also causes difficulties in obtaining inventory management information. Since there is no timely feedback of inventory management information to customers, which arouses customer disgust, this problem is very serious [4]. Therefore, attention should be paid to the improvement of inventory management models.

3.3 Distribution Management Issues

In the current enterprise logistics management work, the distribution management system lacks the support of advanced technology, and the distribution work has not achieved the desired effect. The lack of a complete distribution system is the key factor that causes the low distribution rate, and ultimately the actual needs of the customers are not met. The longer the customer waits for the logistics, the worse the experience is. The trust in the service quality of the enterprise decreases, and it is easier to choose other enterprises for cooperation [5]. It should also be noted that the logistics management of enterprises should pay attention to the improvement of logistics distribution status and distribution information.

3.4 Lack of Professionals

Since the logistics management industry has only become popular in recent years, the current society does not have enough talents in the logistics management field to meet market demand. The application time of big data technology is shorter. At present, there are very few professionals with strong technical ability, and enterprises lack training in related technologies, which also hinders the development of enterprise logistics management. Any industry or enterprise needs talents if it wants to develop [6]. If you do not want the logistics management development of the enterprise to be restricted in the future, then it is essential to establish a highly professional logistics management team.

4 Innovation of Modern Enterprise Logistics Management in the Era of Big Data

4.1 Speed up the Construction of Data Security Management System

Logistics management is an important part of the development of an enterprise. The logistics data information involved is the key to the enterprise. Data security must be guaranteed. The only measurement standard in the process of applying big data technology to establish a data security management system is data information. Security, so that data security risks are effectively controlled [7]. First, companies should analyze the risk factors of logistics data and formulate corresponding preventive measures to avoid unnecessary risks and prevent serious losses. Secondly, in logistics management, companies should protect information security and back up important information to reduce the risk of information loss. Finally, companies should focus on the security of the data management platform based on big data technology, and use software with higher risk resistance to effectively ensure the information security of the entire logistics management work.

4.2 Optimize Inventory Management System

To achieve the essential improvement of logistics management, one of the prerequisites is to optimize the inventory management system, rely on big data technology to establish a modern logistics management system, carry out systematic comprehensive management of inventory information, and perform inventory information management. Optimize processing and improve the safety of the environment to ensure the orderly development of inventory management [8]. Relying on big data technology, the inventory management system is constructed and the inventory management items are divided in detail, and various effective information is entered into the system, so that customers can understand the logistics inventory status by searching for the inventory information they want to know. Improve the efficiency of logistics management (As shown in Fig. 1).

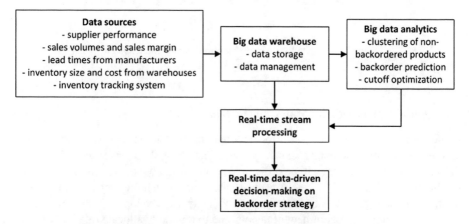

Fig. 1. Logistics inventory optimization system based on big data

4.3 Optimize the Distribution Management System

Achieving more effective internal logistics information is the key to optimizing the distribution management system. Big data technology should be used as the basis to realize the intelligence and digitization of distribution information, and provide a safer environment to carry out logistics distribution management, so that customers can Get a better experience in the distribution link, as shown in Fig. 2. First of all, in the enterprise logistics management work, logistics information should be updated in a timely manner through the distribution management system, so as to feedback the logistics status to customers in real time [9]. Secondly, real-time monitoring of logistics and distribution so as to quickly grasp all distribution routes, and achieve more reasonable logistics and distribution management.

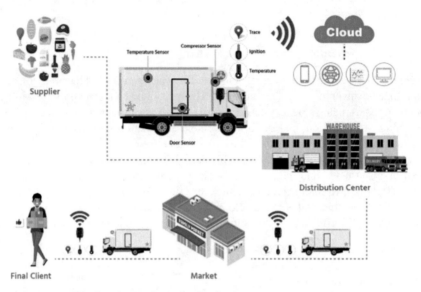

Fig. 2. Big data logistics distribution management system

4.4 Optimize the Logistics Management Team

In the context of current modernization, the main technology for constructing an enterprise's logistics management system is big data. Therefore, if no personnel really master this technology, the system will be difficult to establish and big data technology will not be effectively applied [10]. Therefore, the establishment of a more professional logistics distribution management team is necessary for enterprise development. First of all, it is to pay attention to the recruitment of talents, and to rationally allocate salaries based on the needs of professionals, so as to attract more professional talents. Secondly, the existing logistics management team must be professionally trained to improve technology and at the same time enhance its service awareness, so that it can provide customers with better services, thereby increasing the stickiness between customers and the company, and promoting the long-term stable development of the company [10].

5 Conclusion

To sum up, in the context of modernization, big data technology has gradually become synonymous with the development of computer technology. This technology is applied in more and more industries, and logistics management, as a job that transmits more information, has an impact on the application of data technology is also an urgent need at the moment. In order to realize the modern transformation of enterprise logistics management, enterprises should first identify their own shortcomings and adopt corresponding strategies for the application of big data technology.

References

1. Shao, Q.: Analysis of optimization of enterprise logistics management system based on big data. Knowl. Econ. **11**(01), 66–69 (2020). (in Chinese)
2. Jiang, J.L.: Analysis of enterprise logistics management under the background of big data. Int. Publ. Relat. **9**(01), 34–37 (2019). (in Chinese)
3. Lin, L.: Research on the logistics management mode of SMEs based on "Internet +". China Bus. Rev. **3**(14), 11–13 (2018). (in Chinese)
4. Wang, S.Z.: Analysis of the logistics management model of SMEs based on "Internet+". Mod. Bus. **19**, 96–98 (2018). (in Chinese)
5. Su, Z., Zhao, C.: Talking about the construction of enterprise logistics management system under the background of big data. Bus. News **12**(06), 18–21 (2020). (in Chinese)
6. Yu, H.Y.: Research on optimization of enterprise logistics management system based on big data. Econ. Outlook Bohai Rim **9**(04), 68–70 (2018). (in Chinese)
7. Hu, W.D., Zhou, X.H.: Research on the innovation and development strategy of logistics management mode for small and medium-sized enterprises in the "Internet +" era. J. Southwest Normal Univ. **10**(21), 42–46 (2018). (in Chinese)
8. Wei, J.H.: Research on the impact of big data application on the competitiveness of logistics enterprises. Bus. Times **7**(22), 55–58 (2018). (in Chinese)
9. Li, Y.H.: Strategic analysis of logistics enterprise upgrade management in the era of big data. Logist. Eng. Manage. **8**(05), 121–124 (2015). (in Chinese)
10. Wang, X.N.: Research on the development of smart logistics in the era of big data. Taxation **13**(06), 19–21 (2019). (in Chinese)

On the Application of Digital Humanities in the Study of Classical Literature

Jie Zhao[✉]

College of Literature and Journalism, Sichuan University, Chengdu, Sichuan Province, China

Abstract. Through a brief review of the application of digital humanities in classical literature, this paper affirms that digital humanities brings convenience to the research in the field of humanities and social sciences, it also briefly summarizes some of the problems existing in it, hoping that simple discussions can bring deeper thinking.

Keywords: Digital humanities · Big data · Classical literature · Ancient books

1 Introduction

Whether in daily life or in many fields of humanities and social sciences, under the widespread use of big data, artificial intelligence and information technology are developing rapidly, and digital content is unprecedently growing. Most of them have traces of digitization. This kind of change is both an opportunity and a challenge for enriching academic resources and consolidating discipline construction. The resulting "digital humanities" has also become one of the key topics of academic concern [1].

2 The Definition of "Digital Humanities"

Roughly speaking, digital humanities research is a new research paradigm that combines information technology and humanities in the era of big data. It is a natural projection of the level of productivity represented by the information age in the field of academic research. Of course, this trend is in the ascendant and has not yet received a unified definition. The Chinese version of Wikipedia defines digital humanities as a cross-discipline of computer computing or information technology and humanities. It is the research, teaching, publishing and other academic work of humanities carried out by new methods such as cooperation, interdisciplinary and computer computing. Looking back at the rise and development trend of digital humanities, it can be determined that, as Yang Dandan said, digital humanities combines traditional humanities and social sciences with computing and digital technology, and uses quantification and calculation to do datafication and digitization of humanities and social sciences, so ass to form new research methods, fields and categories [2].

Digital humanities, as a new type of interdisciplinary research, can be said to be brilliant in the application of classical literature. Based on the current cutting-edge research in the academic world, this article briefly expresses my opinions on the combination of digital humanities and classical literature, so as to be corrected by experts.

J. Abawajy et al. (Eds.): ATCI 2021, AISC 1398, pp. 916–921, 2021.
https://doi.org/10.1007/978-3-030-79200-8_135

3   The Application of Digital Humanities in Classical Literature

The domestic academic circles have not been exposed to the digital humanities for a long time. With the blog post The Emergence, Development and Frontier of "Digital Humanities",published by Professor Wang Xiaoguang of Wuhan University in 2009, as a sign, the relationship between the humanities community in China and the digital humanities is only over ten years. However, the research results during this period of time have confirmed that the changes brought about by the use of digital humanities in classical literature can be said to be unprecedented, not only in the integration of the resource library, but also in the influence on scholars' research thinking and research methods [3].

Here is an example of the use of digital humanities in the study of classical literature, combined with existing research results.

3.1   Ancient Literature Database

China has numerous ancient books. Those unreachable unique copies and rare books, can "fly into the homes of ordinary people" from the bookshelves of the mysterious ancient book room due to the construction of the digital literature database, and become common e-books at the hands of researchers. Those great works can also be reborn in the form of electronic data and symbols after the printing age. Readers in the shabby room can also look them up without leaving home. Among the many databases, the more eye-catching ones, such as Biography Database of Chinese Historical Figures created by the Fairbank Center for Chinese Studies at Harvard University, the Institute of History and Language of the Academia Sinica in Taiwan, and the Center for Research on Ancient Chinese History of Peking University, the open basic geographic information database Chinese Historical Geographic Information System jointly developed by Harvard University and Fudan University, Chinese Basic Ancient Books Library created by Beijing Erudition Digital Technology Research Center, Chinese Classics and Ancient Books Library launched by Zhonghua Book Company, and Complete Library in Four Branches of Literature produced by Digital Heritage Publishing Limited [4]. These resource libraries are storage, reading and retrieval systems established on the basis of digitizing and imaging traditional paper documents. The essence of books has not changed, but the preservation and presentation methods have been changed, which is more conducive to reading or retrieval. The information age is a post-printing age. The paper, silk, bamboo, wood, gold, and stone in the traditional printing age can no longer meet the requirements of preservation, dissemination, and reading of information. Although they are beautiful and rare, they must be transformed by the new era to meet the needs. In recent years, most of the electronic literature databases built in recent years have field retrieval functions of varying levels, giving the humanities new methods of analyzing and processing data, which also changes the way that researchers observe the works and present academic results. The existence of these databases not only greatly saves the search time for academic research to collect and organize documents, but also exhausts the resources in the database. More comprehensive materials and richer examples have enabled classical literature to have a broader perspective on macroscopic research.

3.2   Ancient Book Retrieval System

In classical literature research, traditional research focuses on bibliography, typology, collation and so on, so basic documentation work such as collating, cataloging, and compiling is the top priority. These tasks consume a lot of time and energy for a rigorous scholar, and it takes a long time to study intensively. However, with the rapid improvement of computer intelligence and the remarkable improvement of natural language processing technology, almost all tasks such as proofreading, punctuation, and editing can be handled by computers. For instance, the Software for Digitizing Ancient Chinese Novels produced by Mr. Zhou Wenye of Capital Normal University can compare texts of different versions of the same novel one by one, and the school notes can also be automatically generated. The Complete Poetry of Tang Dynasty Analysis System and the Complete Song Poetry Analysis System developed by Professor Li Duo of Peking University can automatically query the recurrence and interaction of Tang and Song poems in the database, and realize the automatic labeling of the rhyme of each poem, which provides great convenience for large-scale research and analysis of poetry and rhyme [4]. The systems listed above have been completed, and some are still being revised and supplemented. From the perspective of the content and type of the database, it focuses more on the study of classical poetry. Over time, with the further penetration of artificial intelligence technology into humanities research, existing databases can get smarter text analysis tools and expand the database content to other ancient books besides classical poetry.

3.3   The Application of GIS Technology

"Geographic Information System or Geo-Information system (GIS) is mainly a spatial database. It manages and analyzes the original data in chronological order and according to the spatial relationship. It visually reproduces the temporal and spatial structure of historical geographic phenomena on the computer screen through spatial analysis and thematic mapping. With the help of the comparison of original documents, it is helpful to discover the valuable information hidden under the data, discover geographical phenomena that are difficult to express by traditional methods. It can also check the correctness and rationality of the original data. These are difficult to penetrate by traditional methods. [5]" The application of this technology to the study of classical literature has improved the overall understanding of the writer group. The academic map development platform represents the full use of GIS technology. For example, in terms of classical literature, Professor Wang Zhaopeng developed the Tang and Song Literary Chronicle Map Platform, which sees geographic coordinates as the core relationship element, equally pays great attention on geography and time, and relates the four dimensions of writer, work, time and space. It shows the writer's activity scene like a panoramic view. The author believes that the most commendable innovation of this platform is the use of images and other new media to geographicalize and spatialize people and events in history, which is a moving practice of literary geography research [6]. The existence of the map platform makes the research perspective and results more three-dimensional and comprehensive. This kind of research method has been used by many scholars to carry out some surveys that require a long time-span, many characters, and require macro-control, and they have achieved gratifying results. Chronicles of writers, research on writer groups, etc. can all be used as research methods. For example, by using this platform, the poet's personal experience, associations, and regional characteristics can be seen at a glance when studying the activities and creative characteristics of the group of relegated poets in the Tang Dynasty.

The above three are some of the eye-catching results of digital humanities in classical literature research. The use of more dazzling techniques, just like a colorful garden, is worthy of careful consideration.

4   The Problems of Digital Humanities

The development of digital humanities has brought huge changes to the study of classical literature, and many of them are positive. However, while enjoying the convenience, some potential problems should not be taken lightly. Obvious problems such as, under the premise that literary literature retrieval has been so convenient and will become more and more powerful, rigorous scholars may be able to recognize the necessity of traditional reading methods and will not give up the pleasure of reading paper books. But for students, do they have the patience to finish reading a book confidently because of a problem? The new information dissemination method will definitely have a significant impact on academic research and talent training methods. Is the negative impact of it necessarily the inevitable product of technological development? This involves the transformation of humanities research under the background of the information age and other grand issues, which is not able to be discussed here. Only with regard to the practice of digital humanities in literary research, especially when the open and free Chinese database is relatively not very rich [7], the author sorts out the views of predecessors and makes a brief summary.

4.1   Resource

With the increasing maturity of digital humanities technology, the construction of digital humanities courses has gradually been put on the agenda. However, the construction of digital humanities courses is not only a technical problem, but also a resource problem. The prerequisite for solving these two problems is financial support. The development of the platform, the formation of the team, and the improvement of technology are inseparable from a large amount of funds and high-end teams. This is a bit difficult for ordinary colleges and universities, especially local ones. The use of digital humanities courses requires both liberal arts and sciences, not only profound professional knowledge, but also professional knowledge in computing and programming. Combining the current training mechanism, it is a challenge for both teachers and students. Although team building contributes to better development, technical expertise is both an advantage and a limitation. How to compatible professional technology and humanistic knowledge still needs to be explored [1].

4.2 Text

Big data relies on artificial intelligence to solve macro-level problems. Even if it is dynamic analysis, it can be solved by intelligent technology. It just takes time. Although a complete database and good dynamic analysis can replace part of manpower, in some respects it certainly can't completely replace the subjective initiative of people. The current AI technology has evolved to the point where it can automatically compile and issue news releases, but it is absolutely impossible to rely on AI technology to automatically write academic papers. Especially in classical literature, the emotions and expressions of works are more subjective or romantic. They are passionate and warm. It is not cold mechanical data analysis that can penetrate the content behind the text [8].

Being Specific to some details in ancient literature, for example, in the analysis of the details of the text, the field search provided in the literary literature database saves time and effort, and can replace the effort of close and extensive text reading to a certain extent. However, the analysis of huge search results still requires a human brain with profound knowledge to present. What's more, the complexity of the historical background and the profoundness of history and culture cannot be fully revealed through the data items retrieved. For example, the established literary literature database, even the latest tagged corpus, often fails to accurately associate the names and their style names of ancients. Take Su Shi, the great writer of the Song Dynasty, as an example. He had three style names: Zizhan, Hezhong, and Ziping. Apart from them, he had about 30 other titles. For example, "Suzi". In The First Ode for the Red Cliff, it says: "Saddened by the melancholy song, Suzi sat tight and asked the guests, 'Why was your music so disconsolate?' [9]" For another example, "Dongpo Householder". Dongpo was originally a wasteland forced to reclaim after Su Shi was demoted to Huangzhou. He built a house on it. Because of this slope, he was called "Su Dongpo", "Dongpo", and "Mr. Dongpo" and so on. Another example is "the Mad Deputy". Su Shi laughed at himself in Ding Fengbo · Two Pink Lotuses: "Persuading you to drink a few more glasses of wine to express gratitude, the drunk deputy Dongpo asked again in front of the banquet" [10]. For such problems, simple retrieval cannot solve [11].

5 Conclusion

Through a brief review of the application of digital humanities in classical literature, combined with the research results of previous scholars, it can be seen that artificial intelligence brings convenience to the humanities and social sciences such as retrieval, comprehensive data, detailed information and visualization. This paper also simply sorts out some of the macroscopic and detailed issues of digital humanities. The author hopes that the simple summary will bring deeper thinking and enable digital humanities to better play its role in the rapid development of the Internet.

References

1. Wang, J., Zhang, L.: International digital humanities research. Digit. Hum. **01**, 77–79 (2020)
2. Yang, D.: Fourth issue: the rise of "digital humanities" and the turn of literary studies. Contemp. Writers Rev. **04**, 03 (2019)

3. Zheng, Y., Duan, H.: Digitization of ancient books, digital humanities, and ancient Chinese literary research: an interview with professor Zheng Yongxiao from Chinese Academy of Social Sciences. J. Jishou Univ. (Soc. Sci. Edn.) **41**(02), 144–151 (2020)
4. Wang, Z., Zheng, Y., Liu, J.: Taking advantage of the momentum of the implement, giving new interpretation of the principle—three talks on classical literature research under the wave of "digital humanities". Lit. Art Stud. **09**, 69–73 (2019)
5. Chen, G.: Digital humanities and informationization studies for historical geography. Soc. Sci. Nanjing **03**, 136–142 (2014)
6. Wang, Z., Jiang, X.: The integration of time and space – the academic concept and value of the Tang and Song literary chronological map platform. China Three Gorges Tribune,**05**, 20–27,35 (2020)
7. Dai, A., Jiang, W., Zhao, W.: Digital humanities as a research method: the status quo and prospects of western research. Shandong Soc. Sci. **11**, 26–33 (2016)
8. Su, W., Lu, Z.: Analysis of the controversies over digital humanities methodology—Taking the distant reading project of two styles of Song Dynasty Ci-poems for example. Libr. Tribune **38** (02), 22–28,43 (2018)
9. Shi, S.: The Corpus of Su Shi, p. 06. Zhonghua Book Company, Beijing (1986)
10. Zou, T., Wang, Z.: Chronological Proofreading of Su Shi's Ci, p. 398. Zhonghua Book Company, Beijing (2002)
11. Ma, D., Ma, N.: On the name, style and literary name of Sushi clan of Meishan and the social life of scholar-officials in the Song Dynasty. J. Teach. Coll. Qingdao Univ. **23**(01), 28–36 (2006)

Analysis of the Integration of Big Data Technology and Virtual Reality

Xin Wang[(✉)] and Liang Wang

School of Art and Design, Qilu University of Technology, Jinan, Shandong, China

Abstract. Traditional data analysis has clearly seen that it cannot meet the needs of virtual reality products. Therefore, in the context of the era of big data, making full use of the huge value created by big data is an urgent and current focus. Integrate big data and virtual reality technology with each other and apply this to virtual reality actual research and development products to promote better development in the future. This article explores and analyzes the characteristics of big data, and discovers the relevance of big data and virtual reality on the basis of researching virtual reality systems, focusing on the mutual integration and common development of big data and virtual reality, and Big data in virtual reality products will better benefit human society.

Keywords: Big data · Virtual reality · Data analysis · Fusion application

1 Introduction

The rapid development of the Internet and the continuous improvement of cloud computing technology invisibly mark the arrival of the mobile cloud era. Big data gradually enters the daily life of the public, marking my country's entry into the big data era [1]. The emergence of the Internet has enabled people to understand and explore the world, and the distance from the world has been significantly shortened. In his writings, John Nessbit predicted that the production of information is as fast as automobile production, and that people are gradually being overwhelmed by information… This just shows that the rapid development of the information age is coming. Virtual reality is actually a technology that allows people to experience in virtual space, and is a three-dimensional environment simulation system that can interact with it. This technology can bring fantasy sensory experience to the experiencer. Virtual reality technology has been widely used in various industries, and related products have quickly occupied the market of various industries, showing a popular trend. The emergence of big data technology and virtual reality technology has had a greater impact on various industries [2]. Big data and virtual reality are integrated with each other, develop together, and work together continuously to effectively promote the better development of a smart society [3].

2 Big Data and Traditional Data

The era of big data is the era of human exploration of deep data. Big data is like all new things, but so far, it has not achieved its own clear meaning [4]. In the report

J. Abawajy et al. (Eds.): ATCI 2021, AISC 1398, pp. 922–927, 2021.
https://doi.org/10.1007/978-3-030-79200-8_136

elaborated by the McKinsey Institute, the concept of "big data" was first mentioned. At the same time, the report also wrote in detail: "Data has become extremely important in all industries and various related fields. Processing and production factors, as well as the increase in the types of data used by people, and the obvious increase in the amount, which implies that the new productivity will increase rapidly in the future, and there will be a larger consumption wave in the near future. "Under this premise, the amount of data is in With the continuous growth, the structure of data is becoming more and more complex. In order to solve a large number of information problems, big data technology is born [5]. Big data is the result of accumulating in accordance with the development of the times under the premise of collecting a large amount of data [6]. In my country, the application of big data is still in the initial stage of development. The Ministry of Industry and Information Technology of my country has proposed how to plan in the "Twelfth Five-Year Plan" document of the Internet of Things, and pointed out that information processing technology is included in four key technological innovation projects. Mass storage of data is one of them. In addition, the other three are information perception technology, information transmission technology, and information security technology. These technologies are also closely related to "big data" (Fig. 1).

Traditional data uses accurate and appropriate statistical methods and methods to make correct analysis of various data. The traditional level of data analysis can be done through tools such as tables to complete the relevant analysis operations, but the effective information is not extracted and combined into a conclusion. It is only a detailed study of the data itself and a process of generalization (such as Fig. 2).

By collecting and integrating relevant information, big data and traditional data can be clearly analyzed. There are similarities and differences in the following three aspects. First of all, from the comparison of analysis methods, there is no essential difference between the two. Whether it is big data analysis or traditional data analysis, it is an effective analysis and interpretation of data and indicators. Secondly, from a statistical point of view, it can be seen that there is a big difference between the two in the use of knowledge. The knowledge used in traditional data generally revolves around "whether a small amount of sampled data can be used to infer the real world". But big data mainly uses different types of overall data to design statistical schemes to obtain accurate and reliable statistical results. Finally, there are obvious differences and differences in the relationship between the two machine learning models. Most of the time, traditional data uses machine learning models as tools to assist in analyzing data.

It can be seen that big data and traditional data have obvious differences and similarities, and can dig out valuable information and be valued. The US Internet Data Center has stated that the annual growth rate of Internet data has reached 50%, and most of the world's data are from recent years. Compared with traditional data, big data has richer sources, broader coverage, more comprehensive coverage, and more professional technical means. In the era of big data, people can collect more diverse and valuable data than ever before [7].

Fig. 1. Dig data analysis   **Fig. 2.** Traditional data analysis

3   Virtual Reality and Real Reality

Compared with real reality, pseudo-reality has the direct and obvious characteristics of multi-sensitivity, immersion, interactivity and conception [8]. From the perspective of the development process alone, virtual reality has experienced three more important stages before and after, from digital reality, augmented reality to mixed reality. The mutual promotion of these three stages is the link that connects the digital world with the real world. The concepts involved mainly include virtual reality (VR), augmented reality (AR), mediated reality (MR), mixed reality (Mix R), etc. Virtual reality (VR) uses a computer to generate an environment similar to the real one, immersing users in it for simulation. It needs to be used with a headset, data gloves, location tracker, etc., to close the human vision and hearing, and make a relative isolation from the outside world, so as to guide the user to produce a real feeling in the virtual environment to achieve the goal, more It is applied to the interaction and interaction between the user and the virtual scene. In the design field, it is simulated by virtual reality technology, and the actual effect of the interior can be viewed in advance in the virtual environment, which not only saves time, but also successfully reduces manpower, material and financial resources [9]. In the military field, there are many mountains, rivers and lakes on the map. People use these statistics into data and compile them by computer. Finally, using virtual reality technology, a flat and two-dimensional map can be converted into a three-dimensional and three-dimensional terrain. Picture, more intuitive and clear display. The most obvious difference between augmented reality and virtual reality is that the data is superimposed on the user's line of sight, but at the same time the real world can still be seen without the premise of being isolated from the outside world. The scenes and characters seen by the user are half real and half virtual. However, the virtual data does not affect the user's personal perception of the external real world, and can be regarded as the visual effect of virtual embedding in reality. For example, Google's glasses, pilot's helmets, etc. The technology and application of mixed reality is to merge the real and virtual worlds to produce a new visual environment. Objects and objects are difficult to distinguish as virtual or real. The software and hardware platform integrates the five senses of sight, hearing, smell, taste and touch, and the data and instructions involved, to form an organic whole, which instantly feedbacks the external input of the operator and controls the external equipment, Adjust the output at any time to act on the operator's senses and make the experiencer feel a relatively complete human-computer interaction (Figs. 3 and 4).

Fig. 3. Virtual reality technology

Fig. 4. Augmented reality technology

Fig. 5. Hybrid reality technology

The real reality refers to the reality of this real world where human beings depend for survival, while at the same time carrying out production, labor, and interpersonal communication. The real world that actually exists is directly opposite to virtual reality, one reality and one illusion. The real reality includes the earth we live in, the environment in which we live, and the interaction between people. Compared with virtual reality, virtual reality pays more attention to people's hearts. It is the fantasy, virtual, and Unreal ideas, if they meet people's inner needs and need to be realized through virtual reality technology, it can bring people brand-new feelings and stimulation of all senses. Those who live in real and real worlds use virtual reality to simulate the world of their inner fantasy, and finally achieve the goal of dreams come true. From this, it can be seen to some extent that the future development of virtual reality may become an international and globalized comprehensive platform. The accelerated development of virtual reality technology has increased our attention to it, but it will also trigger some thinking. Whether it is from the virtual reality technology or from some input and output devices in virtual reality applications, it should be able to communicate with big data. Fusion and penetration (Fig. 5).

4   Big Data and Virtual Reality Technology

Big data technology gathers and forms a data resource library in three aspects, namely, a full range of intelligent perception, data transmission media, and cloud computing. Big data and virtual reality technology are invisible to each other.

The characteristics of big data are basically integrated and summarized into the following five items: First, big data has a huge amount of data and has data integrity. This can provide a large amount of data for virtual reality products, and can determine the value of the data to be considered and the internal information according to the size and quantity of the data, while retaining the integrity of the data. The second is high-speed data flow, which can be considered to be able to meet real-time requirements faster. This data collection for virtual reality products saves time, manpower and material resources to a large extent, and can obtain data more quickly. At the same time, in the virtual reality product experience, the delay is reduced, and it is presented in a near real-time manner. Give the experiencer a more real feeling. The third is a flexible data system, which refers to more effective management of data, which can be flexibly adapted. This lays a solid foundation for the data sorting of future virtual reality products. The fourth is rich data types, in other words, products that can help virtual reality to collect more diverse data content and types, not limited to one type. The fifth is the potential value of data. The rational use of big data in various industries is an effective way to create higher value at low cost.

In the early stage of research and development of any product, there will be a testing phase, and the same is true for virtual reality products. The test experience data related to the product can be used as the main basis for product optimization in the later stage. The correlation between hardware data and software data is also an important criterion for measuring the experience of the product. In the initial testing stage of users, there must be different feedbacks and opinions for various products. Reasonable use of big data technology can more effectively avoid problems, some invalid problems due to individual differences.

In the mid-term sales of virtual reality products, the collection of basic user information can be used as the original data for later market development. Using big data algorithms, it can more efficiently target the characteristics of target customers and major consumer groups to achieve effective sales purposes (Fig. 6).

In the later service link of virtual reality products, big data technology is used again to analyze and summarize the original user data of each link, to realize the transformation from group service to personalized service, adopt special service, collect user feedback, integrate and make appropriate changes (Fig. 7).

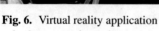
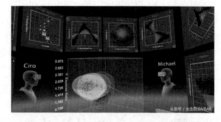

Fig. 6. Virtual reality application   **Fig. 7.** Big data and virtual reality technology

In the later supply and transportation links of the virtual reality industry, based on big data resource analysis, it can present a reasonable allocation of proportions between industries, as well as a collaborative cooperation model of resources such as prediction and prevention, and effectively improve production, warehouse storage, transportation, and post-service services. In the process, the resources are wasted. The application of virtual reality technology, from the initial stage of research and development to the output link, reduces the time cost and the occurrence of risks [10].

5 Conclusions

All in all, the era of big data has come. We need to have a deeper understanding and knowledge of big data, try to integrate big data with virtual reality, and practically apply big data to virtual reality products, so that they can develop better and faster.

References

1. Lei, G.: The application of artificial intelligence in computer network technology in the era of big data. Technol. Wind **7**, 73–74 (2021). (in Chinese)
2. Chen, J., Suli, X.: Industrial convergence application analysis of big data and virtual reality technology. J. Jilin Normal Univ. Eng. Technol. **36**(06), 83–85 (2020). (in Chinese)
3. Wei, F.: Industrial convergence application of big data and virtual reality technology. Inf. Rec. Mater. **121**(12), 149–150 (2020). (in Chinese)
4. Puxia, H.: Overview of big data technology. J. Tianjin Vocat. College **22**(12), 113–118 (2020). (in Chinese)
5. Yang, C.: Research on the application of big data technology. Comput. Knowl. Technol. **16**(35), 211–213 (2020). (in Chinese)
6. Jihui, X.: The application of big data technology in computer network information management. Textile Indus. Technol. **49**(02), 167–168 (2020). (in Chinese)
7. Peiqin, C.: Opinions on big data technology. China Secur. **4**, 46–49 (2020). (in Chinese)
8. Haiyu, W.: Research on virtual reality technology and its application. Inf. Comput. **32**(16), 23–25 (2020). (in Chinese)
9. Yan, Z.: Application and development of virtual reality technology in visual art. Technol. Mark. **27**(11), 90–91 (2020). (in Chinese)
10. You, X.: Research on the fusion of virtual reality technology and product design. Art Technol. **32**(09), 169–232 (2019). (in Chinese)

Precision Marketing Model Based on Big Data – Taking Xiaohongshu App as an Example

Daowen Ren[1,2] and Xuejun Liu[3(✉)]

[1] South China Business College, Guangdong University of Foreign Studies, Guangzhou, Guangdong, China
[2] School of Management, Guangdong Polytechnic Normal University, Guangzhou, Guangdong, China
[3] Globaltimes International Consulting Corporation, Toronto, ON L6G08, Canada
roadwen@sohu.com

Abstract. In the era of data based on big data and cloud computing, how to effectively utilize big data information technology and establish a precision marketing model in line with enterprise development among massive users and improve enterprise performance is an important topic that needs to be discussed by the academic community. Based on the big data theory, precision marketing theory and previous research literature, this paper established a precision marketing model in line with the development of enterprises. Taking Xiaohongshu App as an example, the precision marketing model was further tested and relevant suggestions were put forward. The research results of this paper hopes to have certain reference value for other enterprises.

Keywords: Big data · Little red book app · Precision marketing

1 Introduction

Based on the big data and cloud computing, the rapid development of the big data information processing technology has created a good market development environment for the enterprises. A series of data value chain based on the big data makes the enterprises no longer limited by time and space. The big data information with explosive growth rate brings a large number of active user data for the enterprises, at the same time, also make the market competition situation increasingly severe. The most obvious challenge is to shift from an overly extensive marketing model to a more effective precision marketing model. However, it is a big problem for the enterprises to obtain, integrate, analyze and reasonably use these data of massive and complicated data information into the enterprises' precision marketing to improve performance of the enterprises. Xiaohongshu App, with its explosive performance growth in recent years, has become the focus of attention from all circles. Based on the big data theory, the precision marketing theory and the review of previous research literature, this paper establishes a precision marketing model that is in line with the development of the enterprises. Taking Xiaohongshu App as an example, this paper further tests the precision marketing model and puts forward relevant suggestions. The research results of this paper expects to provide some reference value for other enterprises.

© The Author(s), under exclusive license to Springer Nature Switzerland AG 2021
J. Abawajy et al. (Eds.): ATCI 2021, AISC 1398, pp. 928–935, 2021.
https://doi.org/10.1007/978-3-030-79200-8_137

2   Review of Related Theories

2.1   Connotation of the Big Data

The big data is a data information set with a large scale that is far beyond the capability of traditional database software tools in terms of acquisition, storage, management and analysis, and is characterized by massive data scale, rapid data flow, diverse data types and low value density [1]. The big data can be acquired, managed, processed and sorted into information that can help the government agencies and the enterprises to manage and make decisions within a reasonable time [2]. Tian Juan et al. constructed the model and framework of the portrait labeling system of the enterprise, and proposed the method to process enterprise data according to the characteristics of the enterprise data [3]. The big data has the characteristics of scale, diversity, value, authenticity and high speed. At present, the big data technology has been widely used in various research fields, and provides a technical basis for the enterprises' performance optimization, process management and operational decision-making.

2.2   Concept of Precision Marketing

Philip Kotler put forward the concept of precision marketing for the first time. He believed that the marketing strategy needed by enterprises was to establish precise and quantifiable marketing communication, develop marketing communication plans with results orientation and behavior guidance, and increase investment in direct sales communication [4]. To study the quantitative problem of marketing activities, it is necessary to grasp the dynamics of marketing from different perspectives and accurately identify market opportunities, so as to improve the effect of precision marketing [5]. The purpose of precision marketing is to promote the realization of enterprise marketing, which requires sufficient influence on consumers' purchase decision, that is, to provide targeted users with services and information that can affect their purchase intention and decision within a certain period of time [6].

2.3   Literature Review of Precision Marketing Based on the Big Data

In recent years, precision marketing theory has been regarded as a product of the era of the big data, which has been continuously deepened in the need of the market development, and meanwhile inherited and developed traditional marketing theories. Through the integrated analysis of the big data, the enterprises can clearly know the demand preference and consumption tendency of the consumers, and divide consumers into different groups according to these characteristics [7]. Through searching and analyzing consumer's behavior, we can predict the purchasing pattern of the consumers, and based on this, we can subdivide the customers' data and develop differentiated marketing strategies [8]. The integration and analysis of the users' information is the fundamental to effectively realize precision marketing. The users' information obtained from the analysis can be identified so as to adopt targeted marketing methods to meet the personalized needs of the consumers and the users from multiple perspectives [9]. The arrival of the era of the big data is a double-edged sword for precision marketing. Enterprise marketing should focus on the accurate customer positioning, the customers' communication, the products and the customers' value-added services [10].

To sum up, the previous studies on the big data and the precision marketing have achieved fruitful research results and good application effects, which provide an important theoretical basis for the study of this paper. However, few scholars have systematically studied the model of the precision marketing with the big data and its application in the enterprises. Therefore, this study will build an innovative model of the precision marketing with the big data and apply it to the marketing strategies of the enterprises.

3   Construction of Precision Marketing Model Based on Big Data

3.1   Foundation of Model Build

In August 2019, the China Internet Network Information Center (CNNIC) officially released the 44th Statistical Report on Internet Development in China. As of June 2019, the number of the mobile Internet users in China has reached 847 million, and the proportion of Internet users using the mobile phones has reached 99.1%, and the number of the users using mobile online shopping has reached 622 million [11]. The huge number of Internet users provides marketing target guarantee for online precision marketing of the enterprises. In 2018, there were about 5.23 million websites and 4.49 million mobile Apps on the Chinese market. WeChat has become the most widely used social platform, with 83.4% active users, and Weibo has 42.3% active users. The users of online videos and short videos such as Douyin, Kuaishou, Tencent, Youku and iQIYI reached 648 million, accounting for 73.9% of the netizens' structure. The users' scale of the network broadcast reached 397 million, and the users utilization rate reached 47.9%. These network resources are gradually changing people's information acquisition channels and consumption behavior patterns. In addition, the highly connected and decentralized network community culture derived from the Internet era provides the marketing diffusion foundation for network community marketing, and the Internet social relations and social trust help it to carry out effective promotion of goods and information.

3.2   Precision Marketing Model Based on Big Data

Based on the big data, the enterprises' marketing general information processing needs to collect and process the big data (using the data mining techniques such as API, deep packet inspection, web crawler, cookies, site analysis, etc.)---data analysis (using the customers' analysis techniques on the big data analysis such as trigger analysis, customer portrait, semantic analysis, cluster analysis etc.)---data application (using marketing contact technology on the big data such as cooperative marketing, cross-border cooperation, cross-screen marketing, real-time marketing, etc.). Based on the theory of the big data, the data processing technology and the precision marketing theory, this paper constructs the precision marketing model from five aspects: clearing marketing goal---data collection---data analysis---users' behavior---marketing decision-making.

Clearing marketing target refers to that the marketer should determine a clear and quantifiable target with relevant interests, lead the following steps, and determine the market positioning of the enterprise through SRPT strategy segmentation.

Data collection is to collect the data available in the market on the basis of certain goals, including product support, brand loyalty and dependence, personal hobbies, satisfaction, product preferences, etc., each aspect is an important key point that cannot be ignored, which may affect the probability of success of subsequent marketing strategies. At present, the methods used data collection commonly include: keyword search, collection of consumer information, network survey, the use of emerging technology (such as GDP positioning, etc.) to obtain information, etc. If conditions permit, it is best to conduct in-depth data survey.

Data analysis is to use the big data tools for data screening and standardization. Through this process, the obtained data are normalized. In the environment of the big data, the collected data are miscellaneous and there is a large amount of information that we may not need. In this step, we should choose appropriate modeling tools to discard the data that we do not need, classify customers according to the processed data and find suitable marketing methods.

Users' behavior is divided into two parts, one is the overall behavior of the users, the other is the individual behavior of the users. The former is when a large number of the users have completed certain operations, such as registering and logging in, and the latter is when a small number of the customers have completed operations, such as browsing a particular product.

Marketing decisions-making: it meets the needs of the target consumers' groups on the basis of analyzing data, and creates the correlation between the marketers and the consumers, which is the task that the precision marketing strategy needs to complete. In this step, the enterprises need to make a clear to make different marketing decisions to different values of the customers.

4   An Example of Precision Marketing Based on Big Data – a Case Study of Xiaohongshu App

As of May 2019, Xiaohongshu App has more than 250 million registered users, with 3 billion notes exposed every day, 70% of which are UGC (user-generated content) exposed, making it the largest community e-commerce platform in the world. In the process of marketing, Xiaohongshu App uses the big data to accumulate various types of data, such as users' information, online transaction information, cross-border transaction mode, users' behavior analysis, etc., which are extremely valuable assets of Xiaohongshu App, and also the key to business competition. The users in the community platform share different content, at the same time, Xiaohongshu App aimed at the user's personalized requirements and the characteristics of marketing their products and services for the users to achieve precise matching by using big data. So the group structure of relatively stable users is maintained. Xiaohongshu App achieved the change from traditional marketing mode gradually to a more economic and efficient precision marketing model.

4.1 Determine Marketing Objectives

Xiaohongshu clear their target customers base, the customers with accurate positioning, their main users who are the young people after born in the 1990s with love of beauty, pursuit of a better life live mostly in the first and second-line cities, they have their own ideas, work principle and the pursuit of delicate, they emphasis on more personalized, can enjoy life, enjoy the fine goods, and usually choose shopping online. After determining the marketing target, Xiaohongshu defines the marketing target as acquiring, retaining and activating customers. They use a variety of methods to acquire, retain customers and activate potential users. For example, Xiaohongshu has designed a lot of small Easter eggs for users. When the network is not smooth, Xiaohongshu will prompt the users "Often on the Internet, there are always stuck".

4.2 Collection Data by the System

With the help of business intelligence system, Xiaohongshu collects all kinds of opinions raised by the users after purchase, and sends them through SMS, Email, coupons and other methods. The staff of Xiaohongshu recommend relevant products to the consumers based on their purchase history, search history and browsing history. In addition, Xiaohongshu selectively sends coupons to the users according to their portraits. By using these methods, the detailed information of the customers is sorted and classified, which is convenient for subsequent storage and management, and also lays a foundation for the next step of data analysis.

4.3 Analyze Data by Multi-channel

For Xiaohongshu App, the data collected from the different channels have slightly different data analysis methods. Xiaohongshu App analyzes the data through the multiple channels and ways, and analyzes the user's characteristics, region, education level, browser, network access provider, operating system, terminal type and other attributes, so as to prepare for the application of the big data. Through the analysis of these data, Xiaohongshu App has been further improved to better meet the needs of the users.

4.4 Clarify Consumer Behavior

Precision marketing activities always regard the consumers' demand as a starting point and the center, always emphasize that meet their individual needs. Under the guidance of the precondition of the accurate positioning and through the communication with customers, Xiaohongshu master their consumption psychology and consumption behavior, and set up a corresponding consumers' relationship management system with auxiliary use a variety of modern advanced technology, achieve effective interaction and accurately measure the users' needs. Xiaohongshu identify the consumers' preferences according to labels in the consumers' details, and then send relevant emails to the users according to their preferences. Meanwhile, Xiaohongshu can identify different consumers' consumption behaviors through the coupons and products in consumers' emails.

4.5 Accurate Brand Strategy

Xiaohongshu adopt multi-dimensional precision marketing plan to build their brand hot style. They combine the stars' products to try and share, make use of the fan effect to achieve viral diffusion, short videos, microblogs and other KOL multi-channel diffusion to strengthen the marketing effect, help the brand to obtain mass exposure, efficient flow and good reputation, and create the whole network hot style stars products. The outer ring of Xiaohongshu marketing promotion strategy has notes to brush the search page, produce phenomenal promotion; both offline and online to spread the word; good goods to share overseas. The city circle through the big data technology system help advertisers to achieve marketing effect to promote transformation, set up a big customer service team, customization, advertising creativity, media delivery strategy and effect monitoring marketing promotion.

5 Suggestions for Precision Marketing

5.1 Determine Market Positioning

To guarantee the effective development of precision marketing, the key is to find accurate market positioning. Market positioning and product and customer base has a direct relationship. Because the obvious differences between the product type, market positioning also has certain difference, to a certain extent, it still can make different between the customers groups.Because different customer groups have different shopping directions and shopping habits, various reasons will lead to different marketing content and methods. Therefore we must find the right market positioning to carry out precision marketing effectively.

5.2 Analyze Customer Data

The enterprises should collect various data through a variety of channels and conduct follow-up collation and analysis. The results through certain analysis can show what content is needed for the development of the enterprises, can reflect fully the value, and make it produce certain economic benefits. The enterprises can analyze the customer data purchased before, form the users' behavior data, and grasp accurately of the common characteristics of the customers' groups, that is, the users' portrait. There are various forms of the users' portraits, such as detailed analysis of the characteristics of the specific consumer customers to precision marketing including existing consumers groups and potential customers based on this form of data analysis.

5.3 Vigorously Promote Precision Marketing

Under the precision marketing, network advertising has the characteristics of strong interaction, high cost performance and diversified ways of expression, which can fully show that it has a very accurate classification of target users compared with traditional advertising. Through search engine marketing and network advertising marketing, advertising can be based on the click rate, forwarding rate and page views of advertising data for real-time monitoring and calculation. The enterprises should combine with the effect of advertising to optimize for the network advertising, in order to provide the consumers with better results.

5.4 Make Optimal Marketing Decisions

In the era of the big data, The enterprises should collect various needs of the users by information, so as to strengthen effectively the management of the user relations. The management of the enterprises to enhance customer relationship can maintain the harmonious relationship between the enterprises and the customers, which is conducive to the marketing decisions. The decision system must deal with various business indicators, meet different customers' needs, and clearly distinguish different marketing strategies for the customers of different values.

6 Conclusions

In the era of the big data, the enterprises frequently occur marketing difficulties, the enterprises' marketing mode needs urgently to innovate. Based on the theories of the big data and precision marketing, this paper studies the enterprises' precision marketing model based on the big data, and takes Xiaohongshu App as a case to verify the model. The research concludes that: the enterprises' marketing must be based on the precision marketing model, clear marketing objectives, mine and analyze data, track real-time the consumers' behavior and develop the optimal marketing strategy based on the big data in order to achieve the customer value maximization and improve the enterprises' performance.

Acknowledgments. This research was financially supported by the project of the normal university characteristic innovation of Guangdong province in 2017 and by the project of the Philosophy and Social Science in Guangzhou in 2018 (2018GXGJ73).

References

1. McKinsey's Global Institute Big data: The next frontier for innovation, competition, and productive (2011)
2. Ying, C.: Overview of the development course of big data. Contemp. Econ. **08**, 13–15 (2015). (in Chinese)
3. Tian, J., Dingju, Z., Wenhan, Y.: Summary of research on corporate portrait based on big data platform. Comput. Sci. **45**(S2), 58–62 (2018). (in Chinese)
4. Kotler, P.: A three-part plan for upgrading your marketing department for new challenge. Strat. Leaders. **32**(5), 4–9 (2004)
5. Paul, W., Farris, N., Bendle, T.: Marketing metrics: 50+metrics every executive should master. ECR J. **11**(2), 59–66 (2007)
6. Zabin, J., Gresh, B.: Precision Marketing. Higher Education Press, Beijing (2008)
7. Li, W., Xia, J.: Business model innovation based on "big data." China's Indus. Econ. **2013**(5), 83–95 (2013). (in Chinese)

8. Duhoy, G., Singh, S.: Finding intornational. J. Recent Technol. Eng. **2**(2), 176–179 (2013)
9. Deng, G., Weng, J.: Research on precision marketing model based on social network. Chinese Mark. **2014**(26), 8–9 (2014). (in Chinese)
10. Li, D.: On precision marketing under the background of big data. Intelligence **2016**(07), 228–230 (2016). (in Chinese)
11. China Internet Information Center: The 43th Statistical Report on Internet Development in China. Beijing (2019)

Innovation of Machinery Manufacturing Model from the Perspective of Big Data

Shuai Tao[✉]

Liaoning Jianzhu Vocational College, Liaoyang 111000, Liaoning, China

Abstract. Based on the rapid development of the current big data era, the organic integration of machine manufacturing model and big data technology is directly related to the development of China's machine manufacturing industry. Based on the above background, this article mainly analyzes the application of big data technology in machinery manufacturing. Starting from the connotation of big data technology, combined with the current situation of the machinery manufacturing industry, analyze its application in the machinery manufacturing industry, in order to better formulate the strategy of the machinery manufacturing industry. First, analyze the basic profile of the machine manufacturing company and the production characteristics of its manufacturing process, and point out the problems that the company's workshop manufacturing system faces during the production process. Secondly, on the basis of the production characteristics of the selected machinery manufacturing company and the problems faced by the workshop in the production process, a $4V^+$ model based on big data evaluation and an artificial neural network as the core of the production equipment are proposed. The integrated operation mode of the workshop manufacturing system. On this basis, in order to support the effective implementation of the integrated operation mode of the workshop, the key systems such as the information support system of the integrated operation mode have been studied, including the overall model of the distributed mechanical manufacturing energy storage aggregation model. Framework design, functional architecture and system implementation principles and steps. The experimental research results show that combined with the production status of the machinery manufacturing company, the $4V^+$ model based on big data evaluation and the CNC equipment in the integrated operation mode of the workshop manufacturing system with the artificial neural network as the core of the production equipment are selected as the implementation objects. The information support system was implemented and applied and achieved very good application effects.

Keywords: Big data technology · Mechanical manufacturing model · $4V^+$ model · Artificial neural network

1 Introduction

First of all, as far as the current development of big data technology is concerned, it can directly affect the production quality, layout and operational capabilities of machinery

J. Abawajy et al. (Eds.): ATCI 2021, AISC 1398, pp. 936–942, 2021.
https://doi.org/10.1007/978-3-030-79200-8_138

manufacturing [1]. Big data technology can not only optimize the quality inspection system of finished products, improve quality indicators and product inspection levels, but also further ensure the overall production quality of the machinery manufacturing industry [2, 3].

With the current development of science and technology in our country, automation technology has been regarded as a major research and development direction of mechanized equipment research [4, 5]. Generally speaking, the main carrier of machinery manufacturing automation is based on computers, so computer technology and its internal big data technology must be the core technology of the entire automated machinery research and development. The integration, refinement, and digitization of its data are the three main factors of the current automated machinery manufacturing technology. It is coincidence that these three factors are closely related to computer technology and big data technology, so in automated machinery and equipment If you want to innovate in manufacturing, you need to innovate its key technologies. Artificial intelligence technology is also one of the most important applications and development technologies [6]. After repeated research and application, the data processing ability of AI is much higher than the current people's independent data processing ability, because many factors need to be combined in data processing. If it is processed by humans, it will slow down the entire development. Progress and even work efficiency will be reduced [7].

The core of the development of intelligent machinery manufacturing is mechanical intelligence. In the process of realizing the development of intelligent machinery, it often involves intelligent control technology, information network technology, intelligent equipment processing, machinery intelligent detection and fault diagnosis [8, 9]. In the development process of intelligent manufacturing, relevant personnel should increase the development and use of PLC control devices and engineering control machines. After the integration of the production process, the types of equipment are simplified and replaced by more comprehensive and unified equipment, thereby reducing the difficulty of comprehensively implementing informatization. Only when the entire production line is informatized can products be digitized, and the entire industry can move toward network intelligence [10].

2 Algorithm Establishment

This paper builds a data evaluation model based on the $4V^+$ characteristics of big data. First, the definition of $4V^+$ usability evaluation is given, and finally, the big data usability evaluation model is given.

Definition 1. Let $S = <D, V>$ be a large data volume evaluation system, where D is a large data set, $\forall d\_k \in D$ is a randomly selected data sample, V_k is the volume measurement of the data sample d_k; V is the data set Volume measurement benchmark set, $\forall v\_i \in V(i = 1, ..., n)$ is a standardized data measurement benchmark. Make:

$$V_k = \begin{cases} v_0, & V_k \geq v_0 \\ v_1, & v_0 > V_k \geq v_1 \\ v_2, & v_1 > V_k \geq v_2 \\ \cdots \\ v_n, & v_{n-1} > V_k \geq v_n \end{cases} \tag{1}$$

Then the availability of the data volume V of the big data D is defined as $P_v = P\{V_k = v_k\} \times (v_k/v_0)$. Among them, $P\{V_k = v_k\}$ is the probability that the randomly selected sample data volume d_k is higher than a certain benchmark metric v_k; (v_k/v_0) is the normalized benchmark value ratio, the larger the value, the higher the availability of V.

In the above definition, the data volume measurement benchmark V can be given according to industry characteristics, or it can be automatically ranked according to user preferences. For example, the volume benchmark of the video data sample of a table tennis game can be determined according to the byte length of the file corresponding to each game: the first game-v_0, the second game-v_1, the third game-v_2,..., the seventh game ——V_6.

The definition of the diversity and availability of big data is given below.

Definition 2. Let $S = <D, M>$ be a big data data format diversity evaluation system, where D is a big data set, $\forall d\_k \in D$ is a randomly selected data sample, and M_k is a diversity measure of data sample d_k; M It is a set of data format diversity measurement benchmarks, $\forall mi \in M$ (i = 1,...,n) is a standardized data measurement benchmark. Make:

$$M_k = \begin{cases} m_0, & M_k \geq m_0 \\ m_1, & m_0 > W_k \geq m_1 \\ m_2, & m_1 > W_k \geq m_2 \\ \cdots \\ m_n, & m_{n-1} > W_k \geq m_n \end{cases} \tag{2}$$

Then the availability of the data format diversity M of the big data D is defined as $P_m = P\{M_k = m_k\} \times (m_k/m_0)$. Among them, $P\{M_k = m_k\}$ is the probability that the diversity metric d_k of randomly selected sample data is higher than a certain benchmark metric m_k: (m_k/m_0) is the normalized ratio, the larger the value, the greater the availability of M high.

Definition 3. Let $S = <D, T>$ be a big data processing urgency evaluation system, where D is a big data set, $\forall d\_k \in D$ is a randomly selected data sample, and T_k is a measure of urgency of data sample d_k; T It is the data set volume measurement benchmark set, $\forall t\_i \in T$ (i = 1,...,n) is the standardized data measurement benchmark. Make:

$$T_k = \begin{cases} t_0, & T_k \geq t_0 \\ t_1, & t_0 > T_k \geq t_1 \\ t_2, & t_1 > T_k \geq t_2 \\ \cdots \\ t_n, & t_{n-1} > T_k \geq t_n \end{cases} \tag{3}$$

Then the availability of the data processing urgency T of big data D is defined as $P_t = P\{T_k = t_k\} \times (t_k/t_0)$. Among them, $P\{T_k = t_k\}$ is the probability that the urgency measure of randomly selected sample data dk is lower than a certain benchmark measure t_k; (t_k/t_0) is the normalized ratio, the smaller the value, the higher the availability of T.

The fourth feature of big data is that the effective value of the data is relatively low. This feature is reflected in the low level of freshness of the data. For example, many technical actions in sports games have not changed much, and the difference in the data is reflected in the video content. For this reason, this article uses the difference of data content to describe and measure the availability of this feature.

2.1 Modeling Method

Distributed mechanical manufacturing energy storage aggregation model. When energy storage machinery is manufactured, it absorbs active power and SOC increases; when it releases energy, it emits active power and SOC decreases. The formula for calculating the SOC_t value at time t is:

$$SOC_t \begin{cases} SOC_0 + \left(\frac{P_{ch}\eta_{ch}}{S_{rate}} - \frac{P_{dis}}{\eta_{dis}S_{rate}} \right)\Delta t, & t = 1 \\ SOC_{t-1} + \left(\frac{P_{ch}\eta_{ch}}{S_{rate}} - \frac{P_{dis}}{\eta_{dis}S_{rate}} \right)\Delta t, & t > 1 \end{cases} \tag{4}$$

In the formula: SOC_0 is the SOC value at the initial moment of energy storage; P_{ch} and P_{dis} are the charging and discharging power between t and t−1, at most, only one is not 0; η_{ch} and η_{dis} are the charging and discharging efficiency respectively; Δt is a duration of charge and discharge; S_{rate} is the rated capacity of energy storage.

Energy storage is generally connected to the mechanical manufacturing mode through the PCS (power conversion system). While absorbing or emitting active power, it can also absorb or emit reactive power, which can provide certain reactive power support and Management role. In the actual operation process, its reactive power can be controlled according to the situation.

Constraints on energy storage.
The constraints of energy storage include SOC constraints and charging and discharging energy constraints. SOC constraints are:

$$SOC_{min} \le SOC_i \le SOC_{max} \tag{5}$$

In the formula, SOC_{min} and SOC_{max} are the minimum and maximum allowed by energy storage SOC, respectively.

The charging and discharging energy constraints are:

$$\begin{cases} 0 \le P_{ch} \le P_{ch\ max} \\ 0 \le P_{dis} \le P_{dis\ max} \end{cases} \tag{6}$$

$P_{ch\ max}$ and $P_{dis\ max}$ are the maximum charging power and maximum discharging power of energy storage, respectively.

Power distribution of distributed energy storage.
This article aggregates multiple energy storage into an equivalent centralized energy storage for dispatch by the dispatch center. In the scheduling optimization calculation, only a centralized energy storage charging and discharging variables are needed. In

order to make the power distribution of each energy storage reasonable and maintain the relative balance of SOC during the scheduling process, the rated power and SOC value of each energy storage are used to jointly determine its charging and discharging power. The specific determination method is as follows:

Recharge:

$$\begin{cases} \frac{P_{ich}}{P_{irate}f_{ch}(SOC_i)} = \frac{P_{jch}}{P_{jrate}f_{ch}(SOC_j)}, \forall i, \ j \in N \\ \sum_{i=1}^{N} P_{ich} = P_{allch} \end{cases} \tag{7}$$

Release energy:

$$\begin{cases} \frac{P_{idis}}{P_{irate}f_{dis}(SOC_i)} = \frac{P_{jdis}}{P_{jrate}f_{dis}(SOC_j)}, \forall i, \quad j \in N \\ \sum_{i=1}^{N} P_{idis} = P_{alldis} \end{cases} \tag{8}$$

Where: P_{irate} is the rated power of energy storage i; P_{allch} and P_{alldis} are the charging and discharging power of energy storage i, respectively; P_{allch} and P_{alldis} are the total charging and discharging power of dispatching demand; N is the total number of energy storage; f_ch (x) and f_dis (x) are the charge and discharge SOC functions of energy storage respectively.

In order to make the energy storage with lower SOC charge more and discharge less, the energy storage with higher SOC charge less and discharge more. In this paper, the Sigmoid function is transformed into expansion and contraction, translation, etc., as the charge and discharge SOC function of energy storage. The expression of the specific charge and discharge SOC function is:

$$\begin{cases} f_{ch}(x) = 1 - \frac{1}{1+\exp^{-20(x-0.5)}} \\ f_{dis}(x) = \frac{1}{1+\exp^{-20(x-0.5)}} \end{cases} \tag{9}$$

3   Evaluation Results and Research

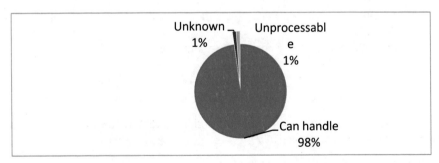

Fig. 1. The 4V$^+$ model of big data evaluation on the experimental data processing state distribution diagram A

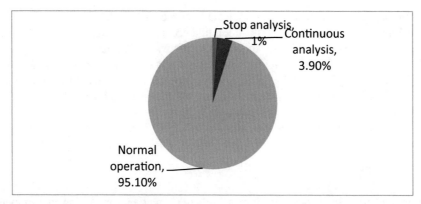

Fig. 2. The 4V⁺ model of big data evaluation on the subsequent state distribution of unanalyzed data processing B

The data shown in Figs. 1 and 2 are intelligentized in the comprehensive processing and analysis of experimental data of the 4V⁺ model evaluated by big data, and artificial intelligence processing will lead to data errors and analysis abnormalities. Situation, the consequences of this situation are more serious. Moreover, in a large number of normal operation conditions and abnormal operation conditions in the mode of mechanical manufacturing, managers mainly pay attention to abnormal operation conditions, and the rest of the normal operation status has actually been selectively forgotten. In data processing, this type of sample that is not marked and does not belong to the to-be-predicted sample plays a very important role. It is usually included in the data set for multiple risk assessments, analysis and processing; and in post-processing, this type of sample depicts the data distribution, The construction of the learning model has a great effect. Therefore, by designing experiments, this paper conducts multiple data analysis and processing on the same data set that does not have the same labeled part, uses inconsistent evaluation indicators to evaluate the performance of the model, and finally selects the best performing model to make reasonable data abnormalities prediction.

Table 1. Accuracy analysis table of the 4V⁺ model evaluated by big data

Sample	Observed	Predicted	Correct rate
During training analysis	20	19	95%
	100	98	98%
	Total proportion	117	91%
When checking and analyzing	20	18	90%
	100	97	97%
	Total proportion	115	93.5%

From the data in the accuracy analysis table of the $4V^+$ model evaluated by the big data in Table 1, it can be seen that the prediction accuracy of the accuracy model of the accuracy analysis table of the $4V^+$ model evaluated by the big data is 90% +. Among them, the first type of error in the training set and test set (normal data is regarded as abnormal data) are 16.31% and 4.920%, respectively, the second type of error (abnormal data is regarded as normal data) are 2.56% and 1.08%, respectively, and the first type of error Errors greater than the second category indicate that the model is relatively strict and the possibility of making up the numbers with false data is low.

4   Conclusion

In the current era of big data and the increasingly fierce market competition environment, facing a large number of complex and changeable information data, machinery manufacturing enterprises need to introduce mature technologies in other fields and gradually realize the application of big data technology in machinery manufacturing. In this way, it is possible to mine the truly valuable data content of the collected big data, scientifically guide the future development of machinery enterprises, and improve the competitiveness and technical level of machinery enterprises.

References

1. Dong, C., Wang, G., Sun, W.: Research on time sharing rental model of new energy sharing vehicles based on big data technology. IOP Conf. Ser. Earth. Environ. Sci. **474**(5), 052040 (2020)
2. Ghallab, H., Fahmy, H., Nasr, M.: Detection outliers on internet of things using big data technology. Egypt. Inf. J. **21**(3), 131–138 (2020)
3. Singh, N.: Big data technology: developments in current research and emerging landscape. Enterp. Inf. Syst. **2019**(9), 1–31 (2019)
4. Lei, W., Qing, C., Hongyu, G., et al.: Framework of fault trace for smart substation based on big data mining technology. Autom. Elect. Power Syst. **42**(3), 84–91 (2018)
5. Shi, J., Wang, Y., Fan, S., Ma, Q., Jin, H.: An integrated environment and cost assessment method based on LCA and LCC for mechanical product manufacturing. Int. J. Life Cycle Assess. **24**(1), 64–77 (2018). https://doi.org/10.1007/s11367-018-1497-x
6. Li, L., Wang, H.: A parts supplier selection framework of mechanical manufacturing enterprise based on D-S evidence theory. J. Algorithms Comput. Technol. **12**(4), 333–341 (2018)
7. Nie, Z., Lu, Y., Chen, T., et al.: Thermomechanical behavior of conduction-cooled high-power diode laser arrays. IEEE Trans. Comp. Packaging Manuf. Technol. **5**, 1–12 (2018)
8. Adrodegari, F., Saccani, N.: A maturity model for the servitization of product-centric companies. J. Manuf. Technol. Manag. **31**(4), 775–797 (2020)
9. Xin, G., Zhu, N., et al.: An overview of the application of artificial neural networks in lung cancer research. Zhongguo fei ai za zhi = Chin. J. Lung Cancer **22**(4), 245–249 (2019)
10. Stoffel, M., Bamer, F., Markert, B.: Artificial neural networks and intelligent finite elements in non-linear structural mechanics. Thin-Walled Struct. **131**, 102–106 (2018)

Internet Financial Innovation Under the Background of Big Data

Na Pu[✉]

Chengdu Polytechnic, Chengdu 610041, Sichuan, China

Abstract. With the rapid development of Internet technology, modern finance gradually integrates the advantages of Internet technology development, and becomes an innovative form different from traditional financial development. Taking Yuncheng Construction Bank as a case, this paper analyzes the development of Internet finance of China Construction Bank, and studies the realization and optimization of innovative business under the influence of Internet finance. Through the investigation and analysis, CCB's online transaction quota accounts for 16.7% of all the major banks in China, the total fund balance is 33.557 billion yuan, and the average daily general deposit balance is 26.585 billion yuan. The volume meets the innovation demand. At the same time, the countermeasures are formulated for the development of the bank's Internet financial innovation business in the future.

Keywords: Internet technology · Big data · Internet finance · China construction bank

1 Introduction

Since the development of the Internet, the major enterprises have begun to play the advantage of Internet resources, and driven by the "Internet plus" mode to drive the development of innovation models in various industries. Relying on Internet technology, the financial industry has established unique internet financial advantages, and formed many excellent internet financial development enterprises represented by Alibaba, which has played a typical exemplary role in the development of other industries. At the same time, it also led to the birth of Internet innovative financial business model and promoted the innovation and development of the financial industry in the Internet era. As an emerging field, Internet financial enterprises rely on modern technical means, uphold the basic principles and service concept of serving customers, create new sales means and channels, establish Internet financial market, and establish innovative Internet financial mode. Especially after 2013, the application trend of Internet finance has risen, and a series of new financial platforms, such as P2P online loan, third-party payment platform and Ali financial model, have gradually become popular and become an important part of public consumption and life. As a traditional financial intermediary, banks are the economic lifeline of a country's stability, which determines the development level of the country.

As one of the commercial banks, China Construction Bank is the pillar of China's banking system. It mainly provides funds to the government, enterprises and individuals in the form of short-term loans and long-term loans, so as to support the stable development of the national economy. With the reform of the world's financial economy and the update of modern Internet technology, China's financial market puts forward higher development requirements for traditional banks. The competition mode of financial market is gradually diversified, and the competition in financial market is more intense. Under the background of the rise of Internet finance, China Construction Bank, as a traditional bank, has been affected by its development. In the stage of actively integrating Internet finance mode, the development of Internet finance has brought development opportunities for CCB, but it also poses a certain threat to CCB as a platform of traditional banking [1, 2]. This requires China Construction Bank to combine its own advantages with Internet finance, devote itself to expanding banking business with the help of modern science and technology, and play a role in stabilizing the status of China Construction Bank industry. Therefore, under this premise, how to improve the status of traditional banks and expand traditional banking business with the help of Internet financial advantages has become the key to promote the development of CCB [3].

Internet finance is developed from business practice. This paper takes the application of Internet Finance Based on big data innovation mode as the breakthrough point, takes a branch of China Construction Bank in our city as the research object, studies the bank innovation business in the context of Internet finance, and based on the understanding of Internet Finance and Banking innovation business, it makes clear that Yuncheng Construction Bank is in the innovation industry Based on the analysis of SWOT factors of innovation business development and innovation, this paper forms the countermeasure analysis of the development of innovative business under the background of Internet of China Construction Bank, and establishes good development conditions of Yuncheng Construction Bank and Internet finance.

2   Related Concepts and Theoretical Basis

2.1   Internet Finance

"Internet finance" as the first theoretical concept put forward by Xie Ping in 2012, once elaborated, it has triggered heated discussion in the industry. Xie Ping believes that Internet finance refers to the modern financial products, with the support of Internet big data and information technology, to establish a new financial business operation mode of financing, payment and investment relying on information carriers, which realizes the combination of Internet innovation platform and traditional financial business, promotes the promotion of innovative business in the financial industry, and improves the efficiency of financial services. Internet finance is an operation mode based on the security and convenience of network technology, which can adapt to the trend of mass consumption and meet the basic requirements of current popularity for financial platform [4].

At present, the development mode of Internet finance can be divided into crowdfunding, P2P network credit, third-party payment, digital currency and other types, which determine the development pattern of Internet finance. With the help of technology platform, Internet finance can reduce transaction cost and eliminate the essence of transaction intermediary, improve service efficiency on the basis of improving information transparency, and meet the basic requirements of the public for financial services. From the perspective of basic characteristics, we should analyze the cognition of Internet finance from multiple perspectives, correctly recognize the basic characteristics of low cost, high efficiency, rapid development, weak management and high risk of Internet finance, reasonably judge the mode characteristics of Internet finance, and maximize the development advantages of Internet finance [5].

Among them, the third-party payment, as a typical application mode of Internet finance, is a platform for Internet enterprises to establish transactions with independent third-party institutions with certain strength and reputation by virtue of products. Generally speaking, the access threshold of the third-party payment is low, but the market vitality and potential are huge. Especially, as a typical way of third-party payment, online shopping attracts a large number of buyers and sellers, and drives the rapid growth of online transaction volume. The third-party payment volume of China's third-party payment increased by 296.51% in the same quarter as that of China's current third-party payment 296.9% in the third-party payment volume in China.

P2P is a kind of financial mode which mainly relies on online and offline business communication to complete the financing of both sides. According to the current situation of P2P transactions in China, most online lending companies establish credit audit with the help of project feasibility analysis of network big data, and determine the credit status of borrowers and borrowers on this basis. Based on the data of online lending home, it can be found that the loan transaction volume of P2P online lending platforms in China is huge, which has reached 468.292 billion yuan in January 2019 alone, and the development scale of P2P platforms has also established 3816. However, due to the fact that P is still the product of a new lending platform, the situation of its application platform problems continues to rise. At present, with the continuous increase of online loan problems, the number of problem operation platforms increased from 10 (2016) to 1353 (2019).

Crowdfunding is a way to raise funds for crowdfunding platform by using innovative projects to provide investors with corresponding returns. Since 2012, the crowdfunding network began to go online and set up financing platform businesses such as film and television, publishing and public welfare. By 2019, a total of 14231 crowdfunding projects were established, with a total of 732000 people supporting fund-raising. The total amount of fund-raising has reached 273 million yuan [6].

2.2 Financial Innovation

In 1912, the theory of economic development put forward the idea of innovation for the first time, and considered that innovation is the basis of the establishment of a new production function, requiring entrepreneurs to complete the re combination of various elements in the production, management and operation of enterprises. With the gradual development of innovation, innovation and financial fields began to cross. The concept of "financial innovation" appeared for the first time in the 1970s, and gradually came into the public's vision in the 1980s, and since then, the concept of financial innovation has gradually become the focus of public discussion. However, although financial innovation began to develop since then, its theoretical development and research time is still relatively short. Therefore, the systematic theoretical system of financial innovation has not yet been established. Some existing researches still focus on the understanding of financial innovation in the diversified world. After the integration of multiple financial elements, the profit maximization is realized, and then the financial situation of market reform is caused. The definition of financial innovation is based on the definition of factor restructuring, and believes that financial transmission Google includes innovative financial instruments, markets and financing methods. However, the American Dictionary of banking terms makes a special analysis of financial innovation, believing that financial innovation is a trading means created by financial enterprises, financial institutions or banks to weaken the intermediary value of money, including technological innovation, credit innovation, etc., and belongs to the originator of financial innovation [7].

From the perspective of the development of China's innovative finance, there are many problems in China's innovative financial market, such as market failure, government promotion and technology promotion. Among them, market failure refers to the low efficiency of financial market development due to the imperfect information and resource allocation conditions, and the need for financial innovation to change the current market level state, and then complete the realization of higher financial demand; government promotion refers to the current situation of financial market blank caused by the market economic state problems under the current urban-rural dual economic structure, accompanied by the economy With the promotion of reform and development, innovation of financial mode needs innovation to meet market requirements; technology promotion refers to financial innovation relying on technological innovation, social development of fast consumption requires relying on modern technical means to drive the development of new financial mode, and innovation of new financial platform to drive market demand and reduce costs [8, 9].

2.3 Relevant Calculation Formula

Stream processing:

$$Y_t = c + \sum_{i=1}^{p} \partial_i Y_{t-i} + \sum_{j=1}^{q} \beta_j X_{t-j} + \mu_t \tag{1}$$

The causal relationship between them needs to be further analyzed by Granger

$$S_t = \left[\sum \frac{(Y_{ti} - Y_t)^2}{N}\right]^{\frac{1}{2}} \tag{2}$$

$$V_t = \frac{S_t}{Y_t} \tag{3}$$

3 Internet Financial Innovation Model and Case Analysis Based on Big Data

3.1 Research Object

At present, most of the research on Internet financial business innovation of traditional banks focuses on the general description or the impact and Countermeasures of major banks, and seldom takes the designated banks in the local second tier cities as the analysis object, especially the papers focusing on the process of business innovation technology optimization under the background of the development of Internet finance The number of research cases is less. Therefore, according to the current review and analysis of relevant research content, taking the traditional research methods of relevant research as reference, this paper focuses on the influence factors of Internet Finance on Yuncheng Construction Bank, and puts forward strategies according to the Construction Bank and the actual situation, so as to establish the Internet financial innovation in line with the Construction Bank New business development mode and environmental support, to achieve a comprehensive analysis of CCB's Internet financial business.

3.2 Research Methods

(1) Literature analysis method;
(2) Inductive deduction method;
(3) Comparative analysis method;
(4) The combination of theory and practice;
(5) SWOT analysis method.

4 Case Analysis of Internet Financial Innovation Under the Background of Big Data

4.1 Online Banking Business Survey

As an innovative business carrier relying on online banking payment technology, online banking realizes the service and completion of securities, financial management, investment and other businesses with the help of Internet technology. As an extension of online financial business, which is different from traditional commercial banks, online banking effectively improves the business pressure of traditional business under the development of Internet finance. Online banking makes full use of the service advantages of 7×24 to update the latest business service information of the bank (Fig. 1).

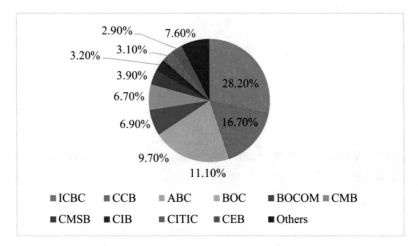

Fig. 1. Share of online banking transactions in 2019

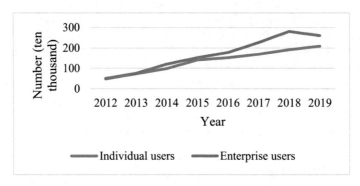

Fig. 2. Internet banking users of CCB (10000)

4.2 User Analysis

The number of online banking users of China Construction Bank is increasing year by year, and both enterprise users and individual users are in a straight-line upward trend. During the period from 2018 to 2018, however, due to factors such as the number of users and the number of users in the business sector will slow down in 2018, especially the number of users in the business sector will slow down (Fig. 2).

In terms of specific business, as shown in Table 1, the online banking services of CCB are mainly divided into the following business forms:

Table 1. Online banking business of China construction bank

Content classification	Concrete content
Information man service	Preferential information/activities, service content Business scale, branch distribution and brand image Introduction of financial information, stock exchange market, economic express, etc.
Transaction information	Personal business: account information inquiry, transfer and remittance, deposit, payment, loan, etc.
	Corporate Finance: balance inquiry, account management, online transfer business, international stage, pension, credit, trade financing, bills, etc.
Online investment	Funds, foreign exchange, precious metals, stocks, stocks, insurance, futures, bonds
Others	Online shopping mall, etc.

4.3 Analysis of the Bank's Basic Information

As a branch of China Construction Bank, a branch of China Construction Bank has opened the second development curve under the guidance of "quality and benefit development", continuously enhanced the "three capabilities", comprehensively launched the "three strategies", solidly carried out the "3 + 2" battle, adhered to the steady operation and innovative development, and achieved remarkable results (Table 2).

Table 2. Basic information analysis of banks

Poject	Total fund balance	Average daily balance of general deposits	Balance of various loans	New individual customers	New customers of the company	Gross income of intermediary business	Main business income
Business volume	33.557 billion	26.585 billion	11.841 billion	613300 households	3019 households	297 million	795 million

4.4 Innovation and Development Countermeasures

(1) Building a professional team of Internet finance professionals.

Talent, as the main structure of an enterprise, is the decisive factor to determine whether it can be sustainable development. Internet finance, as an interdisciplinary knowledge field between Internet development and financial development, needs to bring together compound talents with Internet knowledge and financial knowledge.

If China Construction Bank wants to occupy a market place in the business construction of Internet finance, it must reserve professional financial talents. For the cultivation of compound talents, we should first devote ourselves to the professional skills training of talents, which is the prerequisite for the combination of Internet talents and financial talents.

For skills training, we should pay attention to the training of Internet skills and it skills, and on this basis, carry out the teaching and training of financial knowledge. At the same time, we should pay attention to the skills training of Internet financial managers, strengthen the management concept training of enterprise employees on the basis of following the modern management mode, and focus on the formation of all-round talent reserve resources proficient in network development, mastering financial management and expanding marketing ability.

(2) Financial products and service innovation.

Financial products and services are the main components of innovative financial business. Only by implementing products and services can customers pay more attention to Internet finance, and then enhance their business enthusiasm for Internet finance. To establish the application of diversified financial products and services, we should first pay attention to the innovation of financial services. As a financial development means for CCB to meet customer demand and improve its competitiveness, product innovation program is a new banking business form established with the help of Internet technology. The development of new products and services for these businesses can increase the long-term competitive advantage of diversified products of Internet finance. Therefore, CCB should pay attention to the development of individual financial products, or increase the financial cooperation with funds, securities, insurance and other companies, and constantly expand the number of identifiable customers by enriching the categories of financial products. It can also attract a large number of scattered customers by reducing the threshold of financial management and the time of receiving accounts, systematically combine the change users with bank card business, personal loan business and personal financial management business, enrich the specific products for change users, complete the internal integration of bank financial management business and other financial systems, and improve service efficiency.

(3) Payment innovation.

With the convenience of mobile client, payment method is one of the main payment methods for big customers, which is the main application carrier of current social payment. Therefore, banks should pay attention to the innovation and practicability of payment methods as the basic principles in the innovation of diversified business.

5   Conclusion

Internet finance, as the main form of online financial business development, has gradually changed the thinking pattern of traditional financial structure, and changed the value creation and realization mode of traditional agricultural bank with new technical means, financial concepts and consumption advantages. As one of the current banking systems in China, the development of Internet finance business of China Construction Bank brings transformation opportunities for traditional banking business, but also brings development challenges. This requires traditional banks to adapt to the development environment of Internet finance, absorb its technical means and business advantages, gradually carry out the transformation of financial business, and promote the Internet finance to adapt to the current situation the basic needs of consumers.

References

1. Batu, M., et al.: Testing the effectiveness of online assignments in theory of finance. J. Educ. Bus. **93**(3), 119–127 (2018)
2. Peng, C.C.: Textbook readability and student performance in online introductory corporate finance classes. J. Educ. Online **12**(2), 35–49 (2015)
3. Kontogiannidis, P., Theriou, G., Sarigiannidis, L.: Crowdfunding exploring the factors associated with the users' intention to finance a project online. Int. J. Web Based Commun. **13**(1), 73–74 (2017)
4. Lin, Q., Peng, Y.: Incentive mechanism to prevent moral hazard in online supply chain finance. Electron. Commer. Res. **2019**, 1–28 (2019). https://doi.org/10.1007/s10660-019-09385-0
5. Yang, K., Zhang, L.: Research on credit risk evaluation of online supply chain finance with triangular fuzzy information. J. Intell. Fuzzy Syst. **37**(2), 1–8 (2019)
6. Hamdan, F., Nordin, N., Khalid, F.: Understanding the employees acceptance on online training for basic managerial finance. Creat. Educ. **10**(6), 1305–1316 (2019)
7. Hollowell, B.J.: Optimal online business finance course design. Rev. Bus. Res. **16**(3), 13–16 (2016)
8. Das, P., Ziobrowski, A., Coulson, N.E.: Online information search, market fundamentals and apartment real estate. J. Real Estate Fin. Econ. **51**(4), 480–502 (2015)
9. Gaskell, P.V., Mcgroarty, F., Tiropanis, T.: On the socialisation of finance: returns to buying online sentiment winners. J. Japan. Assoc. Chest Surg. **24**(7), 0999–1003 (2015)

Employment Environment for Overseas Students Based on Big Data

Jian Chen[✉]

Urban Vocational College of Sichuan, Chengdu, Sichuan, China

Abstract. In recent years, the number of college graduates in China has increased tenfold in the past decade . It is more and more difficult for college graduates to find a job, while the number of overseas students is also increasing, which makes it more and more difficult for students to find a job in China. This paper mainly researches and analyzes the employment environment of overseas students based on big data. Based on big data on campus, this paper analyzes and calculates students' behaviors on campus, and finally uses machine learning algorithms to build a prediction model for college students' employment. On the one hand, this research can provide decision-making suggestions for relevant departments; on the other hand to provide employment guidance for college students to provide early warning and guidance services.

Keywords: Big Data · Employment Environment · Overseas Students · Employment of International Students

1 Introduction

As a kind of higher level graduates in China's labor market, their employment process is an important part in the allocation of high-level human resources in the country, and their employment quality is related to the people's livelihood. Along with the growing number of international students, international students becomes more and more severe employment situation, employment difficulty increasing, local governments have established the employment guidance services, to promote students employment has an important role, and the network information technology as a means of promoting employment, and in employment guidance service plays an important role. More and more scholars begin to pay attention to and study the informatization construction of employment service. Therefore, as far as universities are concerned, their relevant departments store a large amount of important information and data related to students' behavior and employment. Although today, with the rapid development of information technology, universities and colleges have implemented information management and analysis on data, but they usually only adopt relatively simple and backward methods, and simply carry out statistics and display, which cannot dig out the real value of these information. Not to mention the guidance and early warning for students in school, as well as the prediction of students' future employment [1]. Therefore, how to extract effective information from the big data on campus and make use of the big data to analyze students' behaviors and employment situation on campus, dig out the interaction mechanism and the underlying rules, and then preplay and guide the overseas students on campus has become an important problem to be solved.

As for the research on the mining and application of educational data, foreign countries started earlier and have a history of many years. In the 1960s, some educators in the United States began to realize the value contained in educational data, and relevant researchers established the National Center for Educational Data Statistics with the support of the government. Choi used regression analysis and Harrod-Domar model analysis to analyze the problems of local economic growth and employment effect. The study found that the output elasticity of employment and the employment effect generated by long-term economic growth in this region were not strong [2]. Sunhee et al. established a generalized linear model and analyzed the relevant factors affecting the change of high-tech employment share in metropolitan areas. The study found that the direct factors affecting employment included science and technology policies and loan projects [3].

In this paper, we collect all kinds of data of students in school in a non-invasive way. Through the processing and analysis of these campus data, we find out the important factors that affect the graduates' employment destination, and build a model to predict the graduates' graduation destination. Compared with the traditional analysis method, this paper focuses on the dynamic change of students' multidimensional behavior over time and its impact on employment.

2 Employment of College Students Under Big Data Technology

2.1 Relevant Algorithm

(1) Common Supervised Learning Classification Algorithms.

In the era of big data, the most important thing is to conduct efficient large-scale learning in various practical applications. In the process of our study and research, big data technology and machine learning are inseparable, so to speak, the latter is the core of the former. Therefore, to study big data, machine learning is an inescapable topic. This paper studies big data on campus, which is no exception.

With the rapid development of big data technology, new requirements and conditions are constantly put forward to machine learning. Supervised learning is a type of machine learning, as well as unsupervised learning and reinforcement learning, which are beyond the scope of this article. It is to manually enter some tags when we already know the type of data set to be trained, train the data set to meet our expectations, and predict the data set of unknown tags. In general, supervised learning algorithms can be classified into two categories: regression and classification. This paper attempts to predict the employment whereabouts of students. Previously, it has divided the employment whereabouts into three labels: unemployed, self-employed and studying, so the paper studies the classification of supervised learning.

Boosting to support vector machine (SVM) has better performance in classification task than Boosting to support vector machine (SVM), which is actually a classifier ensemble based on decision tree. The randomness of the random forest (reflected in the random training samples of each tree, as well as the random selection of the characteristics of the tree) makes it difficult to overfit itself [4]. It can be used for many types of problems and is non-parametric. It doesn't have to assume the distribution of the data. It can handle different types or proportions of input attributes. In addition, random forests can operate efficiently on large data sets and process thousands of input variables at the same time without deleting variables. Even if most predictive variables are noisy, they have good predictive performance. Therefore, no variable preselection is required. In addition, random forest is a classification algorithm that can directly provide a measure of the importance of variables (related to the correlation of each variable in the classification process) [5]. These excellent features make it ideal for multi-source data, data classification, and functional selection.

(2) Deep Learning Correlation Algorithm.

Long and Short term memory (LSTM) network is a variant model of recursive neural network, which can more easily remember the data in the past for a long time, LSTM can be used as a complex nonlinear unit to build a larger deep neural network. It uses back propagation to train the model.

The LSTM model can add and delete information in the cell state through a series of "gates". There are three gates in the LSTM model, namely, the input gate, the output gate and the forgetting gate [6]. According to the test of Google, the most important in LSTM is the forgetting gate, followed by the input gate and the last is the output gate.

Enter the door. The input gate is mainly used to determine how much of the input Xt should be saved to the cell state Ct at the current moment. It has a Sigmoid function and a tanh function built in. The Sigmoid function determines which values are allowed to pass, and the value range is [0,1], where 0 means reject all and 1 means pass all. The tanH function weights the values passed to determine their level of importance. The function expressions of the two are as follows:

$$i_t = \sigma(W_i \cdot [h_{t-1}, x_t] + b_i) \tag{1}$$

$$\tilde{C}_t = tanh(W_C \cdot [h_{t-1}, x_t] + b_C) \tag{2}$$

Forget the door. The forgetting gate is mainly used to determine how many unit state CT-1 should be stored in the current unit state Ct at the current moment. It only has the built-in Sigmoid function. This is done by looking at the previous state (HT-1) and content input (Xt) and printing a value between 0 (which ignores this value) and 1 (which preserves this value) for each number in the cell state CT-1 number [7]. This allows you to selectively forget existing information, which is why it's called the forgetting gate. The function expression is as follows:

$$f_t = \sigma(W_f \cdot [h_{t-1}, x_t] + b_f) \tag{3}$$

Output the door. Output gate is mainly used to determine how much information in the unit state Ct needs to be output to the current output value HT. Similar to the input gate, it also has built-in Sigmoid function and tanH function, whose function expressions are as follows:

$$O_t = \sigma(W_O \cdot [h_{t-1}, x_t] + b_O) \tag{4}$$

$$h_t = O_t * \tanh(C_t) \tag{5}$$

2.2 Model Optimization

(1) Model Validation

After we have built the model, we need to verify the model results several times. On the one hand to avoid the occurrence of accidental results, on the other hand can verify the stability of the results. For supervised classification algorithms, we often use cross validation to verify the results. Cross-validation is a technique for evaluating a prediction model by dividing the original sample into a training model data set and a model test data set. In k-fold cross validation, we need to randomly divide the original data sample into k subsamples of uniform size. We select one of the subsamples to verify the prediction model, and the remaining K-1 subsamples will be used as the data of the training model. This process is then repeated k times, and each subsample is treated as validation data. Finally, k results are averaged (or otherwise combined) to produce a single estimate [8, 9]. The advantage of this method is that all observations are used for training and validation, and each observation is used for validation only once. For classification problems, layered K-fold cross validation is typically used, where creases are selected so that each crease contains roughly the same proportion of category labels. In this process, the value of K is very important, which directly affects the accuracy and stability of cross-validation results.

(2) Superparameter Optimization of the Model

Hyperparameter adjustment method. Hyperparameter adjustment refers to shaping the model architecture from the available space. In short, it's about finding the right superparameters to find high precision and accuracy. There are several parameter tuning techniques, but in this article, we will examine two of the most widely used parameter tuning techniques.

1) Grid search. Grid search is a technique that tends to find the right set of superparameters for a particular model. Hyperparameters are not model parameters, and it is not possible to find the best set from the training data. When we use the optimization loss function such as gradient descent, we will learn the model parameters during the training process. In this tuning technique, we only need to build one model for each combination of various hyperparameters, and then evaluate each model, with the most accurate model winning [10].
2) Random search. Random search is a method of using random numbers to find the optimal solution of function approximation, which is different from the powerful search method of grid search. The random combination of super parameters is used to find the best solution for the constructed model. It's similar to grid search, but it turns out to produce better results.

3 Model Experimental Data Processing

3.1 Student Data Collection

The data sources of this study include digital educational administration system and network center. Among them, digital educational administration system mainly contains information such as students' basic information and academic performance. The information collected by the network center is mainly about the usage of the card, including students' dining situation, medical treatment situation, shopping consumption situation, library access information and students' employment information.

3.2 Data Preprocessing

This paper focuses on the research of overseas students in the university, so it is necessary to screen the data based on personnel first, so as to eliminate the information of irrelevant personnel. For the collected data of college students, it also includes abnormal data caused by uncertain factors such as suspension of schooling, retaking and absence of examination. For various situations, we use filtering and other methods to target groups and eliminate data missing, abnormal and redundant.

Due to the low quality of the data, statistical methods should be used to compress and expand the data so as to make the data more standardized and accurate. On the other hand, in order to better show the regularity of students' behavior, it is necessary to use statistics to convert the data.

Due to the scattered sources of data, it is necessary to analyze, screen and integrate the data according to the characteristics of heterogeneous and unstructured data. Taking digital educational administration system as an example, it does not make complete statistics of all students' employment situation, but statistics of agreement employment and other employment data. Therefore, it is necessary to integrate them to obtain complete student employment information.

4 Result of the Model Experiment of International Students' Employment

4.1 Feature Importance

As shown in Fig. 1, the characteristics that have the greatest impact on the employment direction of students are the total number of visits to the library, the total amount and total number of breakfast expenses, and the variation of various behaviors (such as the change of the frequency of breakfast each week, etc.). The first three characteristics can intuitively see the specific situation of a student's behavior, which is convenient for us to investigate the causes. On the other hand, the change of behavior involves the slope of students' behavior, break point and so on. On this basis, we can explore the relationship between the change of students' behavior and the direction of employment.

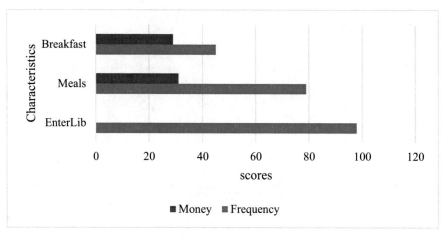

Fig. 1. Feature importance

Table 1. Go to the library several times a week

	3	6	12	18
Employment	1.8	1.5	1.9	1.8
Unemployed	3.4	4.1	5.3	3.2
Enters a higher school	5.3	4.9	6.1	5.1

4.2 Relationship Between the Number of Library Visits and Employment

As shown in Table 1, it can be intuitively seen that due to the need to prepare for the entrance examination, students of "entering higher school" go to the library the most times per week. Students of entering higher school go to the library 6.1 times per person in the 12th week, and the average number of times in the whole semester is 5.28 times. Due to internship or job hunting, students in the "employment" section went to the library the least frequently. In the sixth week, the average number was only 1.5 times, and the average number was 1.75 times in the whole semester. The "unemployed" part is the students who fail to enter college or get employed. Therefore, the number of visits to the library is between the first two. The average number of visits to the library is up to 4 times in the whole semester.

4.3 Comparison of Different Models

As shown in Table 2 and Fig. 2, by comparison, it can be found that the RF prediction model using only traditional features has an accuracy rate of only 0.612. The accuracy of the LSTM prediction model using only temporal characteristics was only 0.574. The model in this paper is constructed by combining the traditional characteristics based on statistical analysis and the time-series behavior characteristics based on deep learning.

Table 2. Compare the results of different models

	Accuracy	Presicion	Recall	F1	Auc
RF prediction model	0.607	0.643	0.612	0.668	0.685
LSTM prediction model	0.569	0.557	0.574	0.452	0.554
This model	0.714	0.735	0.712	0.729	0.758

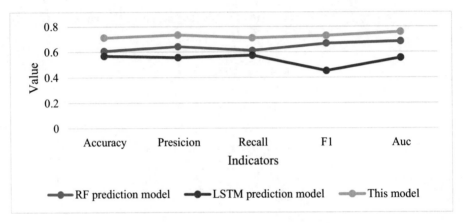

Fig. 2. Compare the results of different models

Its prediction accuracy reaches 0.712, which is much higher than the above two, and further higher than the random guess of 0.332. Therefore, it is feasible to predict students' employment through the data of students' behaviors in school.

5   Conclusions

The development of informationization puts forward new requirements for the management methods of universities, and constantly urges universities to conduct informationization, intelligentization and efficient management of university students. In the process, students' behaviors and other information on campus are collected, stored and managed, and this data is being continuously generated. This paper mainly uses the method of combining big data technology and machine learning related algorithms to study. In this paper, in the campus big data environment, the algorithm related to machine learning is used to build the employment prediction model of college students, so as to predict the employment results of college students. On the one hand, it is expected to provide employment management strategies for relevant departments of employment guidance in colleges and universities; on the other hand, it is expected to provide early warning and guidance for college students to help standardize their behaviors in school and avoid their future employment failure.

References

1. Zhang, M.: Application and research of improved AHP model in employment of college students. Open Cybern. Syst. J. **9**(1), 1212–1217 (2015)
2. Choi, J.E., Shin, D.W.: The roles of differencing and dimension reduction in machine learning forecasting of employment level using the FRED big data. Commun. Stat. Appl. Methods **26**(5), 497–506 (2019)
3. Sunhee, Y.: A study on the strategy for improvement of employment rate of using big data. Soc. Convergence Knowl. Trans. **5**(2), 49–56 (2017)
4. Deng, Z., Zhu, X., Cheng, D., Zong, M., Zhang, S.: Efficient kNN classification algorithm for big data. Neurocomputing **195**(26), 143–148 (2016)
5. Duan, M., Li, K., Liao, X., Li, K.: A parallel multiclassification algorithm for big data using an extreme learning machine. IEEE Trans. Neural Netw. Learn. Syst. **29**(6), 2337–2351 (2018)
6. Park, J., Lee, H.: Prediction of high turbidity in rivers using LSTM algorithm. J. Korean Soc. Water Wastew. **34**(1), 35–43 (2020)
7. Shi, H., Hu, S., Zhang, J.: LSTM based prediction algorithm and abnormal change detection for temperature in aerospace gyroscope shell. Int. J. Intell. Comput. Cybernet. **12**(2), 274–291 (2019)
8. Schumacher, M., Holl Nder, N., Sauerbrei, W.: Resampling and cross-validation techniques: a tool to reduce bias caused by model building. Stat. Med. **16**(24), 2813–2827 (2015)
9. Louwerse, D.J., Smilde, A.K., Kiers, H.A.L.: Cross-validation of multiway component models. J. Chemom. **13**(5), 491–510 (2015)
10. Rahmalia, D., Rohmatullah, A., Pradana, M.S.: Estimasi parameter super pairwise alignment pada Kombinasi Virus Dengue Menggunakan particle swarm optimization. TechnoCom **18**(3), 264–274 (2019)

Inclusive Finance for Intelligent Agriculture Development in Big Data

Jun Zhang$^{(\boxtimes)}$

Wuxi Taihu University, Wuxi, Jiangsu, China

Abstract. Agriculture has occupied the main economic position in our country since ancient times; the development of agriculture is of great strategic significance to the stability of the country and the improvement of people's livelihood . With the rapid development of science and technology, the agricultural field also began to adopt big data technology to develop in the direction of science and technology, convenience and high efficiency. This paper mainly discusses the role of inclusive finance in promoting agricultural development in the era of big data, Analysis of agricultural development in recent years through survey data, In order to analyze the prospects of intelligent agriculture, According to the data, agriculture has grown considerably, the outlook is promising.

Keywords: Big Data · Inclusive Finance · Intelligent Agriculture · Agricultural Development

1 Introduction

China is a country with a wide area of five thousand years of history. In the history of China, agriculture is the main industry. Agriculture, as the foundation of national economy, occupies a very important position in the whole national economic system. Throughout the development process of Chinese agriculture, although agricultural technology is constantly improving with the development of science and technology, the development model of Chinese agriculture in exchange for agriculture at the expense of resources and ecological environment has caused many serious consequences.

In recent years, with the continuous development of Internet, big data and AI and other emerging technologies, the degree of financial digitization in China has been deepening [1]. Under the background of the continuous penetration and integration of digital information science and technology into all aspects of life, an important factor in the continuous recovery of the world economy is the accelerated development of the digital economy. Intelligent agriculture is an advanced stage of agricultural development, that is, the integration of applied sensing and measurement technology, computer network and communication technology, intelligent technology and automatic control technology and other modern information technology, relying on various sensor nodes and communication networks deployed in the agricultural field to realize intelligent planting, visual management, intelligent early warning and intelligent decision-making in the field of agricultural production. The geographical area of our country is large, and the efficiency of agricultural capital allocation varies greatly in different industries and regions [2]. The exogenous development of inclusive finance is prone to inefficiency, that is, inclusive finance is mostly led by the government, not by the free market to guide the development of finance. At the same time, the development of inclusive finance has different optimization effects on the scale of social financing, financial coverage, diversity of financial services and financial use. In the process of promoting the development of inclusive finance, the influence of inclusive finance on the efficiency of agricultural capital allocation through the path of capital is also increasing, but whether this influence is positive or negative is uncertain.

This paper mainly studies the influence of inclusive financial development on the efficiency of agricultural capital allocation in China, and constructs a comprehensive response inclusive financial development index on the premise of constructing the evaluation system of inclusive financial development according to the national conditions of our country [3]. Comprehensive investigation of its agricultural capital allocation efficiency has an optimization role. By combing the scholars' research on the factors affecting the efficiency of agricultural capital allocation, this paper can infer that the influence path of inclusive financial development on the efficiency of agricultural capital allocation can make reasonable suggestions on the optimization and adjustment of the next step of inclusive financial promotion measures.

2 Relevant Concepts

2.1 Inclusive Finance

For the first time, the theme report of the 2016 G20 summit, G20 the Advanced principles of Digital inclusive Finance ("Advanced principles"), clarifies the meaning of the professional term "all actions to promote inclusive financial development through the use of digital financial services" [4]. Specifically, this is to help provide formal financial services that meet their needs through digital information technology to a wide range of groups that do not have access to financial or financial services, and in a manner that is sustainable for the service of suppliers and is itself cost-effective and accountable.

Digital inclusive finance is the result of three layers of overlapping meanings. First of all, digital inclusive finance is still finance in essence, it is not divorced from the framework of finance, on the contrary, finance is the starting point of digital inclusive finance. Secondly, the modifier of "inclusive" is to focus on the equality of finance. Inclusive Finance is committed to the welfare of the broad vulnerable groups of the society through financial services, so that the poor people can get out of the predicament, get the improvement of living standard and quality, and the small and medium-sized enterprises with financing difficulties to obtain funds to promote social equity. Finally, the addition of the word "number" brings new features to inclusive finance [5]. With the rapid development of Internet technology, many technologies such as big data and block chain can not only promote the Internet to better serve inclusive finance, but also make inclusive finance more efficient to serve the corresponding groups and improve the level of inclusive finance. The emergence of digital inclusive finance is the necessary result of the development of the times, which breaks the limitation of time and space of inclusive finance.

2.2 Relevant Formulas

The high efficiency of agricultural capital allocation indicates that agricultural capital can flow from low return area, enterprise to low return area and enterprise in time and in full.

$$\ln (K_{it}/K_{it-1}) = \alpha + \eta \ln(V_{it}/V_{it-1}) + \varepsilon_{it} \tag{1}$$

This paper uses the elasticity of total agricultural fixed assets formation (K) to agricultural output added value (V) to measure the efficiency of agricultural capital allocation.

Standardized formula

$$Z_{ij} = (X_{ij} - X_i)/S_i \tag{2}$$

Inclusive Financial Development Index

$$F = \frac{\lambda_1}{\lambda_1 + \lambda_2 + \lambda_3 + \lambda_4 + \lambda_5}F_1 + \frac{\lambda_2}{\lambda_1 + \lambda_2 + \lambda_3 + \lambda_4 + \lambda_5}F_2 + \frac{\lambda_3}{\lambda_1 + \lambda_2 + \lambda_3 + \lambda_4 + \lambda_5}$$

$$F_3 + \frac{\lambda_4}{\lambda_1 + \lambda_2 + \lambda_3 + \lambda_4 + \lambda_5}F_4 + \frac{\lambda_5}{\lambda_1 + \lambda_2 + \lambda_3 + \lambda_4 + \lambda_5}F_5 \tag{3}$$

3 Analysis of Inclusive Finance for Intelligent Agriculture Development in Big Data Era

3.1 Weight Determination

After the original data of inclusive finance are z-score standardized, the inclusive financial development index will be calculated. Although each index in the evaluation system can measure the level of inclusive financial development, the contribution of each index to the development of inclusive finance is different, and its importance can not be simply regarded as the same and directly calculated. At the same time, we can not guess the contribution of each evaluation index [6]. In order to solve the above problems, the principal component analysis method is used to determine the weight of 20 indexes in this paper.

The basic principle of principal component analysis is to reduce the dimension of the index, which will not change the contribution of the variable in the process of dimension reduction, but only change many indexes into a few indexes. It can not only guarantee the rationality of the index, but also simplify the complex problem.

3.2 Data Sources

The data of this paper mainly come from the financial development report of provinces and cities from 2011 to 2017. The data of calculating the efficiency of agricultural capital allocation mainly come from the statistics of agriculture in the National Bureau of Statistics. The remaining control variables are derived from the National Bureau of Statistics and Zhonghong database. Some provinces have no financial development reports before 2011, and some agricultural data are missing [7]. Data after 2017 have not been counted. Therefore, according to the availability of data, this paper finally chooses the panel data of provinces and cities in China from 2011 to 2017 for empirical analysis.

4 Analysis of Inclusive Finance for Intelligent Agriculture Development in Big Data Era

4.1 Agricultural Allocation Efficiency Analysis

Table 1. The efficiency of agricultural capital all location in China's provinces

Region	ICOR	Region	ICOR	Region	ICOR
Beijing	7.6504	Anhui	0.4226	Sichuan	0.3434
Tianjin	3.4739	Fujian	0.2386	Guizhou	0.0655
Hebei	1.3480	Jiangxi	0.3337	Yunnan	0.3456
Shanxi	2.5554	Shandong	−2.4746	Xizang	0.6306
Neimenggu	21.1989	Henan	−19.3878	Shanxi	0.4522
Liaoning	0.9253	Hubei	0.5072	Gansu	0.4032
Jining	−2.0456	Hunan	0.3002	Qinghai	2.2908
Heilongjiang	0.3498	Guangdong	0.1983	Ningxia	1.4713
Shanghai	0.5886	Guangxi	0.3553	Xinjiang	6.8780
Jiangsu	0.1756	Hainan	0.1397		
Zhejiang	0.2672	Chongqin	0.5244		

According to Table 1, the allocation efficiency of agricultural capital in Beijing and Tianjin is very high. Some areas have no advantages in traditional agriculture because of geographical location, area and other reasons, but their agriculture has the characteristics of high quality and high efficiency compared with other provinces, and agriculture has higher capital allocation efficiency. Qinghai, Xinjiang, Inner Mongolia has the natural resources superiority area, can build the characteristic agriculture and carries on the industrialization management to it, although the cultivated land area is small, but the pasture land area is broad, the characteristic agriculture, the agricultural tourism industry is relatively developed, therefore its agricultural capital allocation efficiency is also higher. The efficiency of agricultural capital allocation in other provinces is relatively ineffective.

4.2 Basic Agricultural Development

The total agricultural output value of the province in 2017 was 387.293 billion yuan, Up 4.5%, Production of livestock products 3.741 million tons, increasing agricultural exports by $4.06 billion, An increase of 40.2% over the same period last year. The per capita disposable income of farmers is 9862 yuan, 9.33% more than the previous year, the total grain output is 18.434 million tons, Up 1.56% year on year, High stable farmland area reached 43 million mu (Fig. 1).

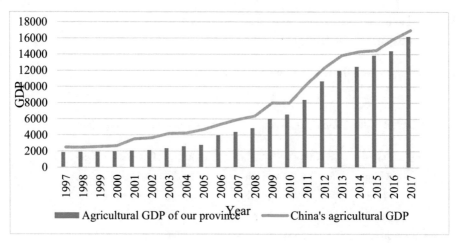

Fig. 1. Trend chart of provincial GDP and agricultural GDP (100 Million Yuan)

Table 2. Sown area of main crops (10000 HA)

Index	Foodstuff	Oilseeds	Sugar cane	Flue-cured tobacco	Vegetables	Coffee	Tea	Fruits
2013	445.32	35.76	34.24	52.55	90.08	4.94	40.06	43.17
2014	444.35	35.60	31.15	42.79	100.40	6.12	40.94	45.79
2015	445.11	35.60	31.15	42.79	100.40	6.97	42.48	50.27
2016	446.36	35.58	28.22	42.27	104.01	8.05	42.21	57.57
2017	416.92	28.88	23.99	41.25	108.48	8.34	43.79	59.43

As shown in Table 2, the trend of grain crop area in Yunnan Province is relatively stable over the years, basically maintaining at about 4.4 million hectares. The planting area of traditional oil, sugar and flue-cured tobacco crops is decreasing year by year, and the planting area of characteristic vegetables, coffee, tea and fruit is on the rise, which is precisely due to the policy of developing characteristic agriculture in our province.

4.3   Growth Rate of Production of Major Agricultural Products (Fig. 2)

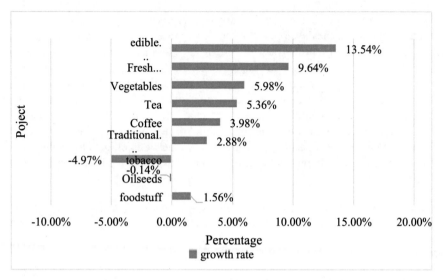

Fig. 2. Growth rate of major agricultural products

4.4   Agricultural Big Data Applications

Through the intelligent agricultural platform, producers can obtain data on plant growth factors. Massive agricultural big data has very important value.

(1) In the field of agricultural production, agricultural big data can guide farmers to grow crops more efficiently and prevent sudden disasters.

(2) In the field of agricultural economy, it can help agricultural enterprises to improve their competitiveness. With the development of science and technology, the competition of agricultural enterprises is becoming more and more fierce. In particular, the insensitive to the change of agricultural market and the slow updating of agricultural products processing technology hinder the development of agricultural enterprises and reduce the competitiveness of enterprises [8]. Agricultural big data is stored in the database, we can easily access the data to analyze and predict the changes of agricultural products. Through big data can continue to innovate technology, so that production technology continues to improve. Agricultural big data has important value in improving the competitiveness of enterprises.

(3) In terms of agricultural technological innovation, the acquisition of agricultural big data requires the continuous updating of new technologies, the formation of intelligent agricultural systems in the agricultural field and the emancipation of farmers' productivity [9]. Internet of things technology, communication technology, ZigBee technology, database technology and other continuous development to ensure the accuracy and comprehensiveness of agricultural big data.

5 Conclusions

The application degree and scope of inclusive finance are more and more extensive, which brings a turn to the development of traditional enterprises and provides technological innovation support for the development of new enterprises. The application of inclusive finance in agriculture also brings new ideas to the development of traditional agriculture [10]. With the discussion in the field of agriculture, the development model of intelligent agriculture based on big data inclusive finance is gradually formed. This paper mainly focuses on whether the development of inclusive finance in China has an optimal role in the development of intelligent agriculture. According to the investigation, the state pays more attention to agriculture and the degree of investment, and the improvement of farmers' own knowledge and quality make the domestic intelligent agriculture have a certain development. Therefore, the development prospect of intelligent agriculture is still very optimistic.

Acknowledgements. This work was supported by Jiangsu industrial circulation and agglomeration to promote rural revitalization under the background of Inclusive Finance, No. 2020SJA0903.

References

1. Arslan, A., et al.: Climate smart agriculture? Assessing the adaptation implications in Zambia. J. Agric. Econ. **66**(3), 753–780 (2015)
2. Ray, P.P.: Internet of things for smart agriculture: Technologies, practices and future direction. J. Ambient Intell. Smart Environ. **9**(4), 395–420 (2017)
3. Ahmed, N., De, D., Hussain, I.: Internet of things (IoT) for smart precision agriculture and farming in rural areas. IEEE Internet Things J. **5**(6), 4890–4899 (2019)
4. Shaw, A., Wilson, K.: The Bill and Melinda Gates Foundation and the necro-populationism of 'climate-smart' agriculture. Gend. Place Cult. **27**(3), 1–24 (2019)
5. Fallot, A.: Témoignage sur la conférence "Climate-smart agriculture 2015" (Montpellier, 16–18 mars 2015). Nat. Sci. Soc. **24**(2), 151–153 (2016)
6. Zheng, J., Xie, Z.: Research on innovative development of rural area microfinance under the background of inclusive finance. World Sci. Res. J. **6**(5), 123–129 (2020)
7. Han, J., Wang, J., Ma, X.: Effects of farmers' participation in inclusive finance on their vulnerability to poverty: evidence from Qinba Poverty-Stricken Area in China. Emerg. Mark. Financ. Trade **55**(4–6), 998–1013 (2019)
8. Guangyou, Z., et al.: Inclusive finance, human capital and regional economic growth in China. Sustainability **10**(4), 1194–1195 (2018)
9. Corrado, G., Corrado, L.: Inclusive finance for inclusive growth and development. Curr. Opin. Environ. Sustain. **24**(3), 19–23 (2017)
10. Matsui, N., Tsuboi, H.: Microcredit, inclusive finance and solidarity. SpringerBriefs Economics **13**(3), 13–25 (2015)

Convergent Operation of Traditional Media Under the Background of Big Data

Siwei Long and Dianyi Wu[✉]

Department of Network and New Media, School of Advertising, Communication University of China, Beijing, China

Abstract. This article addressed the role of big data technology in redefining traditional media operation during their convergence transformation process. Under the background of big data, traditional media has established more integrated and intelligent operation systems and used them in content, advertising and user operation. With the assistant of new data tools, traditional media has made their content operation through user portrait, content prediction, evaluation and automated production. End-to-end program advertising platforms were having been built by traditional media themselves or cooperated with new advertising technology company. Big data made it possible for DTC (direct-to-consumer) content transmission, marketing services and co-creation in social community environment.

Keywords: Media Convergence · Content Tool · Program Advertising Platform · Direct-to-Consumer

1 Introduction

The term convergence was first used by media scholar Ithiel de Sola Pool (1983) in his seminal work, The Technology of Freedom [1]. Even before the publication of Pool's book, thinkers realize that technological changes were going to affect the news media [2]. Boczkowski stressed that the new media emerge through the merger between existing socio-material infrastructures and novel technical capabilities [3]. Storsul & Krumsvik mentioned media organizations are facing disruptive changes in the industry structure, consumer habits, professional practices, and business models [4]. From digitization, two-way transmission of network, triple play and the competition of media convergence, under the background of big data, traditional media has established a more integrated and intelligent operation system, which has made breakthroughs in three aspects: Data Tool assisted content operation, programmed advertising platform, and user operation characterized by Direct-to-Consumer (DTC).

2 Data Tools Assist Content Operation

"Big data marks the beginning of a major transformation [5]." According to Pavlik, the new media have changed news reporting in four ways: first, the nature of news content is

J. Abawajy et al. (Eds.): ATCI 2021, AISC 1398, pp. 967–972, 2021.
https://doi.org/10.1007/978-3-030-79200-8_142

inexorably changing as a result of emerging new media technologies. Second, the way that journalists do their work is being retooled in the digital age. Third, the structure of the newsroom and the news industry are undergoing a fundamental transformation. Fourth, the new media are bringing about a realignment of the relationships between and among news organizations, journalists, and the public, including audiences, sources, competitors, advertisers, and governments [6]. Compaine calls for a rethinking of traditional industry-based media classifications in favor of a model characterized instead by a tripartite message-driven classification. This includes contentor substance, process (how the content is gathered and stored); and formats (how the content is displayed for consumption) [7].

In the era of big data, media content operation is changing different, media organizations and third-party tools technology companies have developed various kind of data application to help operators understand their user's requirements, optimize their content's production, customize the distribution of their content, make assessment and prediction of their communication effect, improve the revenue both from user-pay and advertising. The emergence of these tools enables the upgrade of content production, and also helps media journalists and relevant employees acquire the skills of media data analysis.

The value of data is constantly being recognized and developed, with the rapid development of computers and Internet. Data tools have been added to content production and operations as a new element. The so-called data tool refers to a tool with the functions of data collection, calculation, storage, display and analysis application, etc. It solidifies data management, data model and analysis and decision logic into a system or software as far as possible, so as to maximize the decision value of data in a more automatic, more accurate and more intelligent way.

The British Broadcasting Corporation (BBC) is not only a reference as a PSM organization in Europe but also leads innovation in the global scenario, which implements a global strategy of innovation based on the research, design and implementation of new technology [8]. On the basis of Journalism Portal, BBC has successively launched various content tools to adapt to the context of communication on social platforms and mobile terminals. Specifically, data tools provide services for the integrated content operation of traditional media in the following aspects.

User Portrait. User insight determining the user's demographic characteristics, interests, time, place, frequency and other preference can be identified with the help of data tools. What kind of specific factors promote what types of content to be read or watched by more people and more willing clicked the "like" button? To get more accurate understanding of user behavior, to get easier of making right contents.

Content Prediction. Data tools in the content space can monitor those changes in content traffic on social channels and websites. This helps editors spot topics, clues, predict upcoming hot spots, and prepare as early as possible. What kind of news or articles get more attention? What titles are more interactive? What do users really care about? What's the logic that makes users interact crazily? Etc., all of these scenarios would be answered.

Content Evaluation. Evaluation of content is enabled by data tools. With the help of data tools, real-time data of the cross-platform communication effect of a certain content, positive or negative comments of users on the content, discourse characteristics of the author on specific content topics, communication paths of topics and interactive analysis can be obtained. All the above can become the reference for the media to make the author's incentive policy and carry out some content marketing.

Automated Production. The ability of traditional media applying automated tools for content production is evolving from the manual setting in the early days to machine self-learning and deep learning nowadays. News, video, advertising, graphic or video, automate content are all making a difference in the application of data tools. First of all, automated manuscript writing robots mainly carry out programmed reports based on data, and use computer programs combined with sensors to collect event data, generate materials and complete reports instead of human beings. Narrative Science and Automated Insights, for example, can provide sports fans and stock analysts with articles that are more consistent with readers' reading habits and are filled with real-time data such as stock prices.

Lin Chao-chen finds that the new media appear in the traditional news media mainly as a source of information rather than the originator of the news agenda or a platform for expressing public opinion [9]. However, the technology is neutral. Although the core principle remains the same, technology iterates quickly. Only when technology and data productization are combined with the production system can it become a truly usable, operable, evaluable and optimized service tool to improve production efficiency and thus empower the business. Media content technology innovation is not only 3D, 4K, AR, VR, etc., things like all these kind of cool experience, shiny wrapper, but also in the form of tools, grafting to the content production, editing, distribution platform. Content data tools can effectively improve the efficiency of content operation, expand the breadth and depth of content operation, provide more convenience and services for content producers, provide content consumers with more perfect experience and feeling, throughout the content process, release the value and potential energy of technology itself, and enhance competitiveness.

3   Program Advertising Platform

Over the past 10 years, advertising technology (adtech) has shifted power from publishers to advertisers. Adtech broadly refers to a set of digital tools that give advertisers both far larger audiences and more targeted audiences than before. Google came to dominate the digital advertising market by building an end-to-end stack of software that provided all of the services that an advertiser (from a major brand to a local restaurant) would need to place an ad to a targeted group. This system was far more efficient than what publishers had historically offered (advertisers did not need to negotiate and purchase ads through human salespeople), far more targeted (there was no guess-work in the demographic that was seeing the ads), and infinitely more scalable (the market size was exponentially larger than any one publisher could offer) [10].

The advertising business of traditional media is facing great challenges from data and program, which make them have to improve significantly in data collection, intelligent response, user portrait, accurate distribution and other aspects. Therefore, Disney cooperated with Google in marketing, transferred the global trading platform of digital video advertising and display advertising business to Google Ad Manager of Google to establish an integrated and unified advertising management mechanism. In the data space, Disney has teamed up with Oracle to match consumer data and is launching Luminate, a data-driven targeted advertising product that provides advertisers with a mix of data-driven, customized and programmatic services. Brands and advertisers can find their target audiences on different platforms such as ESPN, ABC and Freeform with just one purchase.

AT&T acquired a procedural advertising company called AppNexus to create its own AD marketing platform, Xandr. Xandr now runs all of AT&T's media brands and marketing resources. On the one hand, Xandr can provide sales for media resource parties; on the other hand, it can provide brand advertisers with "one-stop" online cross-screen advertising. It completes the construction of online AD buyer and seller market, fully activates the data of AT&T in mobile, TV and streaming video business, realizes the cross-screen audience monitoring and user portrait, and provides data support and scientific decision-making reference for advertising marketing.

Comcast has worked with four different advertising providers – EffecTV, FreeWheel, Watchwith, and Adsmart – to create marketing products that have different strengths and service features. Effectv focuses on providing advertisers with accurate advertising based on geographic and demographic data, as well as insightful AD analysis and innovative integrated marketing services. Watchwith builds content analysis and video metadata editing tools linked to machine vision to gain a better understanding of the viewer's emotional state and what is happening on the couch across the screen at a particular moment in the video. More importantly, it identifies and captures the people, places, events, and behaviors that occur in a video that are relevant to viewers and advertisers, and provides contextual and addressable ads. Freewheel is better at delivering ads across screens. It provides technology that can support not only traditional tablet devices, but also cross-device, cross-website advertising services. The Adsmart platform provides advertisers with thousands of positioning approaches. Besides the traditional demographic orientation, which can provide more targeted labels about psychological insight, such as whether to be a new technology adopter orientation, moving intention orientation, whether to keep a pet orientation, financial policy orientation, etc., so as to ensure that ideal users can see their TV advertisements and better achieve the goals of marketing activities. Adsmart offers thousands of attributes and tags, in addition to the traditional demographic orientation, there are many psychological factors. Such as new technology adopters, moving intentions, pets, consumer attitudes, etc., to ensure that the ideal user can see their TV ads to achieve the marketing goals. In addition to detailed fixed attributes, tag customization is supported.

4   Direct-To-Consumer (DTC)

"To provide information to people when it wasn't available before." This is one of Jason Whittaker's conclusion about the future of traditional Journalism [11]. At present, communication operators, Internet and terminal institutions have launched media convergence services one after another, which intensifies the market competition and makes traditional media, which are already facing the challenges of transformation more pressured. In the early stage of media convergence competition, OTT (over-the-top) business model became the focus of all parties. However, when we look at it from the perspective of user operation, another trend of traditional media convergence operation emerges quietly – that is DTC (direct-to-consumer). The term DTC was put forward by Wharton Business School and applied to the retail industry [12]. It refers to the mode in which enterprises sell goods directly to consumers without the distribution of goods by retailers or middlemen. The concept was first borrowed from traditional media organizations by Disney, which first mentioned the term in its 2017 annual report, and then established the "Direct To Consumer and International Division" to provide DTC services to domestic and International users [13]. The underlying logic of DTC is the private domain operation of user data.

The BBC also recognizes the importance of DTC user operations. In the Report "The BBC's programs and services in the next Charter (2015)", it states Where broadcast had been one to everyone, today's media are one to one, one to many, many to many. It is necessary to invite content producers to jointly create content through open platforms and tools [14]. The BBC has also set up a social news team that focuses on the delivery and user interaction of BBC social network news. Traditional media seek to merge with new media to create new news content and integrate public news and user-generated content (UGC).

5 Conclusions

In the process of integration development, the focus of traditional media operation has changed from focusing on the digitization and capitalization of media content in the early stage to the production and multi-platform distribution in the medium term. At present, they have come to a more mature new stage: they begin to pay attention to the changes of user demand side, and know how to use various channels (open source channels, own channels) and resources (influencers, commercial products, activities, etc.), and establish more diversified links with them, and continuously cultivate the user value through data and tags. All these efforts ultimately nourish their own ability in content production and advertising operation.

References

1. De Sola Pool, I.: The Technology of Freedom, p. 2005. Harvard University Press, Cambridge, MA (1983)
2. Gordon, R.: The meaning and implications of convergence. In: Digital Journalism: Emerging Media and the Changing Horizons of Journalism, pp. 57–73. Rowman & Littlefield, Lanham (2003)
3. Boczkowski, P.J.: Digitizing the News: Innovation in Online Newspaper. MIT Press, Cambridge, MA, London (2004)

4. Storsul, T., Krumsvik, A.H.: What is media innovation? In: Krumsvik, A., Storsul, T. (eds.) Media Innovations. A Multidisciplinary Study of Change, pp. 13–26. Gothenburg, Nordicom (2013)
5. Viktor, M.-S.: Big Data: A Revolution That Will Transform How We Live, Work, and Think. Apple Books
6. Pavlik, J.V.: Journalism and New Media. Columbia University Press, New York (2001)
7. Compaine, B.M., Gomery, D.: Who Owns the Media? Competition and Concentration in the Mass Media Industry, 3rd edn, p. 542 (2000)
8. Fuster, M.T.Z., Aviles, J.A.G.: The role of innovation labs in advancing the relevance of Public Service Media: the cases of BBC News Labs and RTVE Lab. Commun. Soc. **33**(1), 45–61 (2020)
9. Chao-Chen, L.: Convergence of new and old media: new media representation in traditional news. Chin. J. Commun. **6**(2), 183–201 (2013)
10. Bell, E., Owen, T., Browm, P., Hauka, C., Rashidian, N.: The Platform Press: How Silicon Valley Reengineered Journalism, pp. 42–47, March 2017
11. Whittaker, J.: Tech Giants, Artificial Intelligence, and the Future of Journalism. Routledge Research in Journalism, pp. 163–164 (2019)
12. Foster, T.: Over 400 Startups Are Trying to Become the Next Warby Parker. Inside the Wild Race to Overthrow Every Consumer Category, Inc. Magazine, vol. 5 (2018). https://www.inc.com/magazine/201805/tom-foster/direct-consumer-brands-middleman-warby-parker.html
13. The Walt Disney Company: Disney 2019 Annual Report, Part I Item I Business, pp. 1–142 (2019). https://otp.tools.investis.com/clients/us/the_walt_disney_company/SEC/sec-show.aspx?FilingId=13754385&Cik=0001744489&Type=PDF&hasPdf=1
14. British Bold Creative: The BBC's Programs and Services in the Next Charter, p. 60 (2015). https://downloads.bbc.co.uk/aboutthebbc/reports/pdf/futureofthebbc2015.pdf

Multi Maneuvering Target Tracking Based on Two Point Data Association Algorithm

Hong Wang[✉], Cuijie Zhao, Nannan Zhang, Sheng Gao, and Qianqian Guo

Tianjin University of Finance and Economics Pearl River College, Tianjin 300000, China

Abstract. In the analysis of this paper, it is mainly based on two-point probabilistic data association algorithm, and on the basis of such data, it effectively combines interactive multi-model algorithm. Then a new algorithm is formed to track multi-excited targets. For the proposed technique, the traditional algorithm has high accuracy and can effectively track and locate various complex targets.

Keywords: Multi-objective · Data association · Association algorithm · Clutter

1 Introduction

In recent years, the aging population has gradually increased and the number of the elderly has increased. In order to improve physical fitness, the elderly prefer walking and other sports. If it can be calculated intelligently, it will further improve the convenience and rationality of its movement and help to improve the health status of the elderly. The application of multi-mobile target tracking technology based on two-point data association algorithm can help to achieve the above purpose. In the field of multi-objective tracking, the research center of researchers mainly lies in the problem of data association. The problem of data association is that it is necessary to design a perfect calculation method to make the target echo tracked and the specific probability of various possible target trajectories composed of redundant echo [1]. Based on this probability, as the core of the whole calculation, various types of data association algorithms can be formed one after another.

2 Multi-target Tracking Calculation

With the long research and practice, many kinds of data management association algorithms have been put forward. For example, there are many kinds of algorithms, such as nearest neighbor, full neighbor, joint probability data management association algorithm, multi-assumption tracking algorithm, intelligent theory and so on. However, the mainstream algorithm in the field is still based on Bayesian theory proposed JPDA algorithm. In the use of the algorithm, there is no need for the prior information of the target or clutter. Therefore, the algorithm is called a very efficient and objective calculation method in the field of tracking multi-target.

J. Abawajy et al. (Eds.): ATCI 2021, AISC 1398, pp. 973–976, 2021.
https://doi.org/10.1007/978-3-030-79200-8_143

However, on the other hand, in the use of the calculation method, the calculation amount will show the situation of combined explosion, which becomes an important restrictive problem in the application of the technology. Therefore, with the deepening of personnel research, various optimization measures are put forward. Based on the feasibility of JPDA technology, combined with the actual scene of using the technology at the present stage, a certain solution is put forward. In the use of the settlement method, in order to realize the simplified processing of the calculation method, the accuracy of the calculation has to be sacrificed. Therefore, the optimization of the calculation method is often suboptimal. In the actual optimization, how to ensure the calculation accuracy in the design optimization has become an important research topic in this field. In the past research, based on the two-point data association algorithm, the multi-objective data association problem is proposed, which is fully described as a three-dimensional allocation problem. Then the linear programming method is used to solve the correlation probability. For this algorithm, not only effectively improve the efficiency of calculation, but also achieve a simplified calculation method, will not reduce the accuracy of calculation, is a technical integrity and optimization. In the analysis of this paper, it is also based on this calculation method and combines the interactive multi-model algorithm to form the algorithm form of tracking maneuvering multi-target. In practical use, compared with the traditional algorithm, it has higher applicability and reliability.

3   Two-Point Data Association Algorithm

3.1   TSDA Basic Concepts of Algorithms Basic Principle of Ant Colony Algorithm

Among the proposed JPDA algorithms, when calculating the correlation probability between echo and target, it is mainly to make full use of the echo data in the sensor to calculate, which is also a two-dimensional allocation problem. Therefore, once in the actual calculation, once the sensor echo scanning process, can be based on multiple times of scanning, effectively after obtaining more information, the accuracy of the algorithm has been greatly improved. For the TSDA algorithm, it is a three-dimensional allocation problem to scan the sensor continuously twice and analyze the correlation between the echo and the target. The algorithm will use the sliding window to deal with the echo. K − 1 design, for example, has N target, and each object has completed its own trajectory, so that in the K moment, once there is an echo between the front and the next moment in the sliding window, It can be calculated. After processing the data information in the sliding window, the target motion trajectory at K time is formed. Then, in the forward shift of the window, it is necessary to remove the echo at k time to form the echo at K + 2 times. This and so on calculation method, has become the present different moment movement track. TSDA algorithm is essentially to calculate the optimal allocation mode for the echo at the time of K and K+, so that it can be allocated to N target in the optimal way.

3.2   Qualifications

TSDA algorithm has similarities with JPDA algorithm when calculating association probability, it needs to put forward two basic assumptions in calculation. First, a single echo can only act on a single target value. Secondly, it can be clearly shown in the calculation that in the TSDA algorithm used, it is the whole associated event group, and the associated time is generated when the similar function value is the largest or the cost function value is the smallest. In this way, the likelihood function can be used to calculate the probability of association. Therefore, this algorithm can be regarded as a N-P complete problem or a combinatorial optimization problem. This has become a very important solution to the current linear programming.

3.3 Wave Verification

In the echo verification phase, it is a verification process for the echo set. It can avoid some impossible correlation time in the calculation, so as to effectively reduce the amount of calculation. In the current echo verification, the commonly used algorithm is the threshold method. The specific parameters in the verification process are shown in Table 1 below.

Table 1. Specific parameters during validation

d/km distance	IMMTSDA		IMMJPDA	
	Calculation time (s)	Peak error (km)	Calculation time (s)	Peak error (km)
1.7	198	0.764	234	0.496
1.5	185	0.353	215	0.525
1.0	263	0.321	246	0.625
0.8	164	0.213	247	0.717

3.4 Performance Analysis of the Algorithm

In order to solve LP problems effectively, simplex method is often used, which can solve LP problems well. However, in the past studies, it has been found that such a calculation method presents a number-order attribute in time complexity, thus forming an exponential time algorithm. Therefore, in practical use, it can play a good role in some complex, large-scale linear programming problems. However, in the later use, many scholars have made corresponding improvements to this form method, thus putting forward various new solutions. Although this calculation method is optimized comprehensively, it always faces some computational defects. In the use of the proposed homogeneous self-dual interior point method, it converges on the theoretical level of the calculation method, and only retains the advantage of the simplex method, so the algorithm is very concise. In the concrete application, presents the polynomial time complexity. TSDA algorithm in this paper is a solution to linear programming.However, in the current TSDA algorithm, due to the introduction of the correlation probability of linear programming solution, the calculation method will be accompanied by the improvement of the target and clutter density in the actual calculation process. And constantly improve the cubic scale

[2]. However, in the JPDA algorithm, the increase of computational complexity is often accompanied by the increase of clutter density and target, and presents an exponential increase. For this reason, the use of JPDA algorithm will present the problem of combined explosion in the process of calculation. This kind of algorithm feature seriously affects the reference of the algorithm in various fields. Therefore, through comparative analysis, it can be found that the TSDA algorithm will have some advantages over JPDA algorithm. However, in terms of algorithm accuracy, more information data are obtained in TSDA algorithm, which will improve the accuracy coefficient of the algorithm [3].

4   Interactive Multi-model Two-Point Data Association

In clutter environment, in order to realize the problem of maneuvering target tracking, it has become a very important research topic in the field of multi-target tracking. In the process of putting into use, the technology can build the corresponding model of the target motion state and solve the problem of data association, which becomes a very comprehensive algorithm. At present, the combination of the proposed IMM、JPDA and other algorithms has become the key way to solve this computational problem. The research of this paper is based on this algorithm theory, successfully deduces the TSDA algorithm, and combines it with the IMM algorithm, so that we can realize the tracking calculation of excited multi-target in some tracking dense clutter environment.

Therefore, the proposed calculation method has high applicability in practical use. For example, in the process of calculating the motion of the elderly, such an algorithm can improve the overall accuracy and efficiency of the algorithm. This is the use of singularity algorithm, can not achieve the calculation effect. Therefore, in the future calculation and analysis, the use of this two-point probabilistic data association algorithm can solve many defects in the algorithm, and play a good role in various complex analysis scenarios. At the same time, in the use of the algorithm, there are still various problems, need relevant researchers to carry out more in-depth research, in order to effectively improve the effectiveness of tracking multi-motor targets.

5   Conclusion

To sum up, in the IMMTSDA algorithm proposed in this paper, the technical difficulties can be solved in the tracking multi-objective field, so as to provide more reliable and comprehensive calculation results. However, in practical application, the principle of calculation should be fully clarified, so that the calculation and analysis can be carried out more effectively and provide reliable information for researchers.

References

1. Dong, M., Ying, S.: Gait Planning and Experimental Study of Two-point Hydraulic biped Robot for Walking Southeast University, 2017.
2. Li, H., Zhang, A., Zhao, M.: Application of two-point data association algorithm in multi-target tracking. J. Military Eng. 2(05), 633–637 (2017)
3. Ye, Y., Todd, M., Mizuno, S.: AnO (nL) iteration homogenousand self-dual linear programming algorithm Ma[…]

Data Mining and Big Data Computing Platform for the Professional Identity

Jingjing Gao$^{(\boxtimes)}$

College of Education, Xi'an FanYi University, Xi'an 710105, China

Abstract. Preschool normal students are not only the important reserve talents of preschool teachers, but also in the critical period of career identity development and career preparation stage. On the one hand, their preschool teachers' professional recognition will directly affect their learning enthusiasm and learning engagement; on the other hand, it will also affect the development of their professional quality, thus affecting their career planning. Due to the imperfection of market access mechanism and the difference of preschool teachers' education level, the quality of existing preschool teachers is uneven. The low quality, low education level and low professional identity of preschool teachers are the key factors for the frequent occurrence of child abuse cases. Therefore, it is necessary to improve the professional level, quality level and professional identity of preschool teachers in time, so as to prevent the recurrence of such tragedies.

Keywords: Higher vocational education · Preschool education · Normal students · Professional identity · Strategy

1 Introduction

Preschool education is not only the beginning of lifelong learning, but also the cornerstone of national education system. Preschool education is of great significance to people's development. Research shows that preschool period is the key stage for the development of individual's social behavior, emotion, character and cognition. In addition, high-quality preschool education can not only lay a good foundation for primary school education life, It also has a significant impact on the academic performance of adolescents, and has a greater role in improving the cognitive ability of junior high school students [1]. The development of pre-school education not only meets the needs of individual survival and development, but also relates to the fundamental of the national plan. In view of the current situation that it is difficult and expensive to enter kindergartens, China has carried out a series of strategic plans, such as increasing the construction of inclusive kindergartens, enhancing the financial investment ratio of preschool education, narrowing the gap between urban and rural areas and so on. With the efforts of the party, the state and people from all walks of life, the development of preschool education in China has made a breakthrough, and has basically solved the problem of "difficulty in entering kindergarten". But it can not be ignored that in the development of preschool education today, we are faced with the problem of "good" development, and the key factor is the level of preschool teachers.

J. Abawajy et al. (Eds.): ATCI 2021, AISC 1398, pp. 977–981, 2021.
https://doi.org/10.1007/978-3-030-79200-8_144

2   Related Theory and Technology

2.1   Personalized Recommendation Theory

Content based recommendation (CBR) is an important research topic in the field of information retrieval. The main idea of the method is to assume that the items are similar in content features, and then the user evaluation is similar. According to the relevance between the attribute of the recommended item and the user's interest preference, the personalized recommendation operation is carried out for the user [2].

Keywords appear more frequently in this text, but less frequently in other texts, indicating that this keyword can effectively distinguish the types of text. The calculation of TF is shown in Formula 1.

$$tf_{i,j} = \frac{n_{i,j}}{\sum_k n_{k,j}} \tag{1}$$

In order to unify the size of the document, the normalization of the formula is usually improved. The calculation formula of IDF is as follows:

$$idf_i = \log\left(\frac{|D|}{|j : t_i \in d_j|} + 0.01\right) \tag{2}$$

Where D is the total number of files

In order to avoid the phenomenon of zero denominator, Laplace smoothing is usually used to process the formula. Through the above calculation, the total formula of integrating TF-IDF is as follows:

$$tfidf_{i,j} = tf_{i,j} \times idf_i \tag{3}$$

2.2   Collaborative Filtering Recommendation

Collaborative filtering algorithm was first proposed in 1992, and has been widely concerned by academia. In 1994, collaborative filtering technology was applied to the news recommendation system of group lens, resulting in the missing value prediction method of user item rating matrix, which was further formalized in lerlockerl and verified by experiments in Breese, which affected the development direction of recommendation system in the next decade. The core idea of collaborative filtering algorithm is to use historical behavior records to calculate the nearest neighbor set of users or items, and then predict the target user's score of unknown items according to the nearest neighbor set. The algorithm thinks that users have similar interest preferences in the past, then they will have similar interest preferences in the future. Nowadays, collaborative filtering algorithms are mainly divided into two categories: memory based collaborative filtering and model-based collaborative filtering.

3 Research Design and Implementation

3.1 Questionnaire Development Details

This paper divides the professional identity of Higher Vocational preschool education normal students into four dimensions. Professional cognition mainly includes professional characteristics cognition, professional ability cognition, professional value cognition, professional development cognition and professional social cognition. Professional emotion includes professional enthusiasm, role acceptance and environment acceptance, Professional will is the teaching attitude, the firmness of career choice and the perseverance to overcome the difficulties in professional practice. Professional behavior tendency is mainly active participation, self reflection and self-regulation. In addition, we use Likert five scale, which is divided into "very inconsistent", "not very consistent" uncertain "," relatively consistent "and" very consistent".

3.2 The Overall Situation of Preschool Teachers' Professional Identity of Preschool Normal Students

The research scale adopts the Likert five scale, which assigns the corresponding value of "completely inconsistent" and "completely consistent", so it takes 3 points as its critical value. The higher the score, the higher the recognition degree of the content of the items of Higher Vocational preschool education normal students [3].

3.3 Linking Theory with Practice to Lay a Professional Foundation

Compared with the students who are not, the students who are student leaders have more positive professional behavior and higher professional identity to preschool teachers. Therefore, it is necessary for students to combine theory with practice in the process of learning to lay a professional foundation.

First of all, we should take the initiative to participate in college practice activities, which are the source of learning theory. It is found that the professional behavior of students who are student leaders is more positive. On the one hand, students who are student leaders have higher ability and initiative. On the other hand, the practical activities organized by colleges and universities are to develop students' personal ability and class student leaders to improve their ability. Secondly, we should actively seek relevant educational practice, which is the experience of learning teachers. Educational practice is a necessary training place for normal school students, which can provide students with a career simulation, so that they can have a first career experience in the future. "Therefore, on the one hand, students should make rational use of the educational practice opportunities provided by the school, on the other hand, they should not constantly seek to increase educational practice opportunities and increase their comprehensive competitiveness. Finally, we should constantly and timely reflect on the practice harvest, which is the wealth of the learning road. The results of practice are different for each student. In order to enhance the learning effect of students in practical activities, it is necessary for students to record and reflect on practical activities in time. On the one hand, they can retain good teaching practice experience, on the other hand, they can eliminate bad teaching habits in time, so as to lay a foundation for the development of professional preschool teachers and narrow the gap with their future career.

4 Association Rule Recommendation Method Based on HITS Algorithm

The research of mobile app recommendation method is a very challenging topic in academia. At present, most of the achievements are based on the experience of traditional recommendation methods, but due to the particularity of mobile app, the traditional methods are limited in the process of application to app recommendation. Combined with the previous knowledge, this paper proposes a mobile app recommendation method based on the fusion of chain topic search algorithm and association rules, which combines the improved HITS algorithm and association rules to achieve the target user. This method not only considers the internal relationship between mobile applications, but also considers the reliability of users. Finally, experiments are carried out on a real Huawei application market data set, and the results show that the proposed method can effectively improve the accuracy of recommendation compared with other recommendation methods.

This paper proposes a mobile app recommendation method based on the fusion of chain topic search algorithm and association rules. This method combines the user's experience value with the popularity of mobile applications and association rules to mine the association of app. Experiments on real datasets show that the proposed method can effectively improve the accuracy of recommendation compared with traditional methods [4].

As we all know, in the era of rapid growth of mobile applications, it is becoming more and more difficult for users to find interesting apps in the mobile application market. In recent years, academia and business circles have begun to study the recommendation method of mobile app. Most of the existing researches focus on mining user behavior data and context aware information, and seldom consider user functional requirements, which leads to a great impact on recommendation accuracy to a certain extent. Based on this problem, this paper proposes a novel recommendation architecture, which can recommend users from the functional feature level of app. In real life, the user's preference for app may be more inclined to the function of app, so this paper focuses on the functional requirements to solve the recommendation problem in the field of app. In addition, this paper also proposes an effective method of APP function extraction and target user interest prediction. Experimental results on real datasets show that the proposed app recommendation method based on functional features is better than other comparison methods.

5 Conclusion

First of all, this paper is based on reading, combing and analyzing the literature related to preschool teachers' professional identity of Higher Vocational preschool education normal students. Using social identity theory, teachers' professional development theory and career planning theory, this paper clarifies the scientific nature of the research theoretically.

References

1. Jialu, X.: On preschool education. Qiushi **23**, 49–51 (2001)
2. Jinghui, S., Wei, R.: Case teaching and the cultivation of normal students' professional identity. Educ. Acad. Mon. **04**, 88–89 (2010)
3. Yan, Z., Hongyu, Z., Tingting, Q., Xiaohui, Z.: A study on the relationship between teachers' professional identity, learning motivation and academic achievement of free normal students. Psychol. Dev. Educ. **27**(06), 633–640 (2016)
4. Yuhua, L., Jing, S.: Empirical research on preschool teachers' professional identity. Educ. Acad. Mon. **03**, 81–84 (2011)

Data Analysis for the Hogg and Max Weber Models

Lei Shen[✉]

Accounting School, Tong Ling University, Tong Ling 244000, Anhui, China

Abstract. By adopting predictive maintenance mode and using EMS Software to analyze and summarize big data, confirm the normal performance range of system parameters, and take maintenance measures in advance to restore the degraded aircraft system to a reasonable range, so as to avoid the impact on flight safety from the aircraft with "disease" to the complete failure of the system. Predictive maintenance greatly reduces the probability of system failure and effectively ensures the safety of flights.

Keywords: Individual determinants · Social characteristics

1 Introduction

More and more studies found small business growth in various factors, such as the relevant individual characteristics of entrepreneurs and the factors that account for a large proportion. The entrepreneur acts in the introduction of new ventures. Venture capitalists tend to consider the assessment of the basic qualities of the entrepreneur, which are resilience, entrepreneurial team spirit and other individual characteristics. Social characteristics of the entrepreneur in past researches are: the entrepreneur's age, gender, family background, education levels. Considering that multiple variables introduced into the model may be highly correlated and lead to multicollinearity, resulting in the loss of significance test of variables, unreasonable economic meaning of parameter estimators and other consequences, we use SPSS software to Diagnose Multicollinearity. After the test, the standardized partial regression coefficient is obtained [1]. The larger the coefficient is, the greater the contribution of the variable to undergraduate participation is. At the same time, the test results give the hypothesis test of each variable and the corresponding p value, in which the regression coefficient of constant term has no practical significance.

2 Theoretical Hypothesis and Data Explanation

At present, the low level of College Students' innovation and entrepreneurship ability is a problem faced by colleges and universities in multi-ethnic areas, which greatly reduces the quality of innovative talents training in multi-ethnic areas. Based on the questionnaire and the findings of the project, this paper uses the "logit" as the main research object, This paper puts forward some suggestions and measures to enhance the implementation effect of the "big innovation" project in Xinjiang universities.

J. Abawajy et al. (Eds.): ATCI 2021, AISC 1398, pp. 982–985, 2021.
https://doi.org/10.1007/978-3-030-79200-8_145

3 Theoretical Hypothesis

There are some defects in using the willingness or intensity of participation (the type or level of participation in innovation and entrepreneurship projects) as the dependent variable, which can not meet the requirements of unity and comparability. Therefore, this paper selects the actual participation status (whether to participate) of Undergraduates in large-scale innovation projects as the dependent variable, based on the literature review, and combined with the actual implementation of universities in Xinjiang region, discusses the influencing factors of Undergraduates' participation in innovation and Entrepreneurship from two dimensions of personal characteristics and University supply guarantee [2]. The hypotheses are summarized as follows: (1) personal characteristics of undergraduates. Gender is generally recognized as an influencing factor in previous literature. It is generally believed that boys like to take risks and are more interested in entrepreneurship. Boys are more willing to choose entrepreneurial projects than girls. Previous literature has not mentioned the influence of individual characteristics, such as nationality, professional category, grade, student cadre or not, on the participation in the "big innovation project". In view of this, this paper takes Xinjiang as an example to make a preliminary exploration. (2) the academic situation of undergraduates. Previous studies have considered that college students' academic performance, scholarship, awards and certificates as proxy variables reflecting their academic level have a positive impact on their participation in the "big innovation project" [3]. In Xinjiang, the field research found that some students have very good academic performance, but their enthusiasm to participate in the "big innovation project" is not high. Therefore, it is worth further analyzing.

4 Understanding of "Dachuang" Project

Undergraduates' cognition of big innovation project. Academic circles generally believe that college students' participation in scientific research and entrepreneurship training has a positive effect on their studies and future planning, and the more they understand scientific research and entrepreneurship activities, the higher their participation. On the other hand, the field research in Xinjiang universities found that although some students have a high degree of understanding and cognition of the "big innovation" project, they will not actively participate in it. Therefore, it is worth to deeply analyze the related factors behind this. the related supply guarantee of "big innovation" projects in Colleges and universities. As the supplier and manager of the "big innovation" project, colleges and universities must provide corresponding education services and management system in the process of project implementation, including carrying out "big innovation" project publicity and consultation, establishing practice base, curriculum system reform, establishing tutor team and formulating management measures. Previous literature review also found that universities, as the supply managers of "big innovation" projects, have an impact on College Students' participation in "big innovation" projects through supply channels, supply intensity and supply methods.

5   Questionnaire Design, Data Sources and Samples

Descriptive statistics combined with theoretical hypothesis designed the questionnaire, and finally designed 21 variables, such as gender, nationality, whether student leaders, whether participated in scientific research projects before, participation status and so on. Other variables are divided into four order intervals according to the situation, and are treated as ordered variables. From 2015 to 2017, it is concluded that there are 9 universities in Xinjiang implementing large-scale innovative projects, and the top 4 universities with the largest number of projects are Xinjiang University, Xinjiang Agricultural University, Xinjiang University of Finance and economics and Xinjiang Normal University. The average annual proportion of "big innovation" projects of these four universities is as high as 70%. Obviously, the implementation of Da Chuang Qing project shows obvious concentration [3]. A total of 366 questionnaires were sent out, 279 were returned, and 264 were valid. The effective rate of the questionnaire was 9462. Whether the data obtained from the questionnaire is reliable or not, we must conduct reliability analysis before the questionnaire analysis, in order to test the stability of the questionnaire itself. The reliability of the questionnaire between 0.5 and 0.9 is reasonable. If the reliability coefficient is lower than 05, the survey results are not credible. The Cronbach's SA value was used to test the reliability of the scale. The overall reliability of the scale was 0.641, which was greater than 0.6. The results showed that the questionnaire was relatively reliable. correlation coefficient, F-test and Hotelling's t-squared test were adopted for further analysis. The partial correlation coefficient of each variable was less than 0.3. Both F-test and Hotelling's t-squared test passed. This study is an attempt to trace the entrepreneurial individual determinants as shown in Fig. 1.

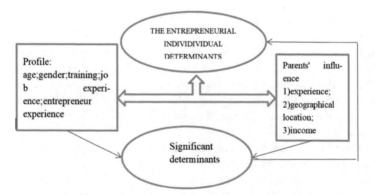

Fig. 1.   Conceptual framework of the study

The study used determinants with two components: profile and parents' influence. The study hypothesized that there is no direct relationship between and among the two components. This relationship may not be depicted by the two components.

Prob (Entrepreneur) = Y (Profile, Parents, X) The research defined the probability of entrepreneur as a function of age, gender, grades, job experience, entrepreneur training, entrepreneur experience, major, and their parents' influence.

Specifically, individual determinants are the students' social background owned by the undergraduates' themselves such as age, gender and entrepreneur experience and entrepreneur training. This is measured by the following: Entrepreneur's age and gender, entrepreneur training, entrepreneur's part time business experience, entrepreneur experience. Especially, the parents' influence from Max Weber's social theory include cultural capital, social capital and economic capital, which indicate the parent's entrepreneurial experience, parents social status, and parents' income. They could establish a competitive advantage in entrepreneur's self-employment.

The model will entail three steps: 1) The input which refers to the entrepreneurial individual determinants; 2) The process indicates how it was generated and effected entrepreneur's performance; 3) The output includes how it determined entrepreneur's performance, and how to improve it to make entrepreneur do better. it can be seen that the p value of most variables has statistical significance. Generally speaking, multicollinearity detection provides two main parameters, one is tolerance [4]. When the tolerance is less than 0.1, it indicates that there is serious multicollinearity. The other is the variance expansion factor (VF), which is the reciprocal of tolerance. Generally, it should not be greater than 5. that both tolerance and VF are within the critical value in this example, and there is no multicollinearity problem. In order to further detect multicollinearity, the eigenvalue, condition index and variance ratio are used for in-depth analysis.

6   Conclusion

From the presentation, Entrepreneurship should pay more attention to the individual studies because individuals who choose to be self-employed is not the behavior of a random process. Entrepreneurship in a large part, is subject to the individual characteristics of the entrepreneur. Mitchell (1979), Gartner (1994), Casson (1982) pointed out that the concerned individual characteristics of entrepreneurs is undoubtedly of great significance. The individual characteristics can be well explained by the occurrence of individual behavior, and the behavior of the individual has also a good predictive effect. These factors can often determine entrepreneurial capacity to carry out entrepreneurial behavior.

Acknowledgements. The innovation project of university students in Anhui (No: AH20140383095).

References

1. Qi, L., Shuai, Q.: The internal mechanism and group differences of the impact of trust on individual entrepreneurship–based on the data of China's labor dynamic survey in 2016. J. Hunan Agri. Univ. (Soc. Sci. Edn.) **21**(03), 26–34 (2020)
2. Hongwei, L., Shuo, H.: Entrepreneurial opportunity identification based on the theory of planned behavior. J. Beihua Inst. Aerosp. Technol. **30**(03), 39–41 (2020)
3. Jian, L.: Community logic, family logic and individual entrepreneurial decision: experience from Zhejiang Province. J. Wenzhou Polytech. **20**(02), 58–65 (2020)
4. Liheng, H.: Minimum wage and residents' entrepreneurial behavior: from the perspective of individual skills and entrepreneurial classification. J. Fujian Bus. Univ. **03**, 61–71 (2020)

Evaluation Index of Algorithm Performance in Building Energy Saving Optimization Design

Hai Zheng and Hongxia Yang[✉]

Yan'an University, Yanan 716000, Shaanxi, China

Abstract. Under the background of energy crisis and environmental problems, building energy consumption accounts for a large proportion in energy consumption. With the development of economy in China, the proportion of building energy consumption will continue to increase, and the situation of building energy conservation is grim. At present, there are some problems in the process of building energy-saving design, such as the lack of integrated thinking in the design process, the lag of energy-saving design in the whole process of architectural design, etc. The development of information technology provides new methods and ideas for solving problems. In this paper, based on the previous research on integrated design of building energy efficiency, based on genetic algorithm, the integrated optimization design method of building energy conservation is studied.

Keywords: Building energy saving · Integrated design · Optimal design · Genetic algorithm

1 Introduction

China is a big developing country, urban construction is still in the development period, the urbanization rate has increased to 52.6%, and the urban and rural construction area is still increasing significantly [1]. In developed countries such as Europe and the United States, the proportion of building energy consumption in terminal energy consumption is about 40%, while that in China is about 20%. With the growth of national building area the total amount of building total energy consumption will be further improved. The national energy situation is very serious. In this paper, building energy consumption refers to the energy consumed by the operation of civil buildings, that is, in office, commercial, residential, recreational, school, traffic buildings and other civil buildings, to provide users or residents with ventilation, air conditioning, heating, lighting, domestic hot water, cooking and other energy consumption to realize the building service functions.

2 Integrated Optimization Design of Building Energy Saving

2.1 Influencing Factors of Building Energy Saving Design

Building energy saving design is a systematic project. The factors that affect building energy conservation are complex and intertwined. These factors are interrelated and

J. Abawajy et al. (Eds.): ATCI 2021, AISC 1398, pp. 986–990, 2021.
https://doi.org/10.1007/978-3-030-79200-8_146

restrict each other [2]. The change of one factor will cause the change of other conditions, and then other factors need to be adjusted. At the same time, the influencing factors are gradually determined with the deepening of building energy-saving design. The feedback treatment of the early factors of energy-saving design plays a crucial role in the effect of building energy-saving. The following will analyze and summarize the influence factors of building energy-saving based on the existing literature theory with the building energy-saving theory as the tool.

2.2 Architectural Layout and Orientation

Building layout is an important part of building energy-saving planning. Through the analysis of regional climate, the form of guiding building layout is affected, and the purpose of building energy saving can be realized. The influence of wind and sunshine radiation should be considered in the design of building layout. In order to resist the harsh climate conditions, we should also make full use of the climatic conditions. In the design of the building plane and space layout, we can achieve the appropriate wind environment, good sunshine radiation and other purposes through reasonable planning and layout, and also can effectively avoid the adverse impact of natural climate. The common layout forms of buildings include determinant, peripheral, free and mixed.

2.3 Natural Ventilation Model

The driving force of wind includes wind pressure and thermal pressure. Wind pressure refers to the force of indoor and outdoor air flow caused by pressure difference when outdoor atmospheric movement acts on buildings. Hot pressure is the driving force of airflow caused by the temperature difference between indoor and outdoor air. Natural ventilation is a typical passive building energy-saving technology, which involves the fields of climate environment, building thermophysics, fluid mechanics and building environment. In a given environment, the feasibility and final effect of natural ventilation need to be analyzed by natural ventilation utilization prediction model, which is called natural ventilation potential estimation. Natural ventilation potential refers to the potential to ensure qualified indoor air quality and thermal comfort through natural ventilation, which is an important index to evaluate natural ventilation. Natural ventilation utilization rate prediction model methods include steady-state thermal balance coupled thermal comfort interval method, ventilation volume prediction based on conventional multi zone model, natural pressure difference PA hour method, multi standard evaluation method, etc. among them, natural differential pressure PA hour (PDPH) refers to the sum of the product of the effective pressure difference of building with natural ventilation higher than the minimum required pressure difference and the number of hours with this pressure difference. The model is simple and practical, and can be used in the early stage of architectural scheme. Therefore, the natural pressure difference PA hours (PDPH) in typical meteorological years is used as the evaluation index of natural ventilation potential.

It is assumed that the buildings studied are two walls with openings opposite to each other; the indoor temperature is constant at t and the temperature distribution is uniform; according to Boussinesq approximation, the air density $\rho_0(1.2 \, \text{kg/m}^3)$ is set as a constant value, and the basic calculation formula of natural ventilation rate q is as follows:

$$q = C_d A_{tot} \sqrt{\frac{2 \Delta P_{eff}}{\rho_0}} \tag{1}$$

Where:

$$A_{tot} = (A_1 + A_3) \tag{2}$$

The model is regarded as two unilateral ventilation. Therefore, according to the calculation formula of unilateral ventilation, the natural ventilation rate under the action of thermal pressure is as follows:

$$q_s = \frac{1}{3} C_d A_{tot} \sqrt{gH \frac{|\Delta T|}{T_0}} \tag{3}$$

3   Implementation and Verification Application of Building Energy Saving Integrated Optimization Design Program

3.1   Development Language and Platform

When the integrated optimization design program of building energy efficiency (iodpbee) is used to optimize the scheme, the calculation dimension is high and it will take a long time. C language with high code efficiency is the appropriate choice. C language belongs to computer high-level language. In order to complete the development of UNIX operating system, C language has the characteristics of concise and compact, strong expression ability, high quality of object code, good portability and so on. Taking CodeBlocks as the development platform, CodeBlocks is an open source, fully functional and cross platform C/C++ integrated development environment. CodeBlocks supports more than ten kinds of common compilers, takes up less hard disk space after installation, has rich personalized features, powerful functions, and is easy to learn and use [3]. CodeBlocks integrates the compiler and debugger of C/C++ editor, which makes it easy to edit, debug and compile C/C++ applications.

3.2   Generate Building Scheme Analysis

The natural pressure PA hours of real residential buildings are 6379 and 8721 respectively. Compared with the real residential buildings, the natural pressure PA hours of a building scheme generated by iodpbee design are increased by 2342, about 36.71%, and the natural ventilation potential is greater. In a word, the building scheme of iodpbee integrated optimization design has lower building energy consumption, higher average daylighting coefficient and larger natural pressure difference PA hours [4]. That is to say, the building scheme of iodpbee integrated optimization design is better than existing buildings in terms of natural lighting, natural ventilation and energy consumption, which reflects the advantages and design effect of iodpbee integrated optimization design. The design is shown in Fig. 1 below:

In this chapter, CodeBlocks is selected as the development platform to write the integrated optimization design program of building energy efficiency (IODPBEE). The optimal solution set of building scheme is generated by integrated optimization design of building energy saving as the goal of building research object by setting use conditions respectively; the fitness value of each objective function of all building schemes in the optimal solution is statistically analyzed by using IBM SPSS statistics21 to verify the correctness of iodpbee design program; Taking the basic functional requirements of a residential building in Chengdu as the design condition of architectural scheme, iodpbi is applied to the integrated optimization design of residential building scheme. A building scheme is selected from the generated optimal solution of building scheme, and the performance of building energy consumption, natural lighting and natural ventilation is compared and analyzed with existing residential buildings.

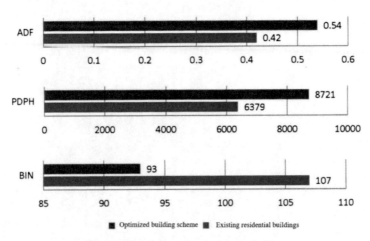

Fig. 1. Comprehensive design renderings

4   Conclusion

Building energy efficiency design research is a complex system engineering, which requires researchers to have solid theoretical knowledge and rich practical experience. The conservation based on genetic algorithm has great potential to improve the effect of building energy saving design in practice, which is limited to the accumulation of professional knowledge, the mastery of information technology and research time, There are still some deficiencies in this paper, which need to be studied in depth. The research in this paper is only a preliminary attempt to the algorithm.

Acknowledgements. 1. A Study on Indoor Thermal Environment of New Rural Residence in North Shaanxi under the Mode of Solar Energy Coupling Heating (51768073).

2. A Study on the Heat Transfer Performance of New Type Collector Heat Storagewall in North Shaanxi Area (19JK0960).

References

1. Energy saving research center of Tsinghua University: Annual Development Report of Building Energy Efficiency in China (2007). China Construction Industry Press, Beijing (2007)
2. Biao, L.: Architectural Generative Design: Research on Computer Generation Method of Architectural Design Based on Complex System. Southeast University Press, Nanjing (2012)
3. Jing, Z.: Preliminary study on the integrated design process of green buildings. Shandong University of Architecture, Jinan (2010)
4. Wei, X.: Climate zoning based on passive design strategy. Tsinghua University, Beijing (2008)

Author Index

B
Bai, Ke, 761
Bian, Zhenxing, 511
Bu, Longmin, 231

C
Cao, Qun, 141, 668
Chang, Haitao, 87
Chen, Baoxin, 57
Chen, Jian, 952
Chen, Junming, 903
Chen, Kan, 57
Chen, Liuhong, 826
Chen, Sucheng, 841
Chen, Weiguo, 495, 504
Chen, Xianjun, 605
Chen, Xie, 676
Cheng, Fei, 186
Cheng, Haimin, 703
Cheng, Sining, 605
Cui, Xiang, 322

D
Deng, Chao, 563, 812
Deng, Chunsheng, 253
Dong, Jie, 422
Dong, Li, 106
Dong, Xiaoming, 380
Dong, Ying, 106

F
Fan, Qing, 856
Fan, Zhaojun, 715
Feng, Zhe, 753

G
Gao, Hongmin, 373
Gao, Jingjing, 977
Gao, Nan, 50
Gao, Sheng, 973
Gu, Guoxin, 796
Gu, Mudan, 526
Guan, Haixing, 158
Guo, Feng, 221
Guo, Liang, 343
Guo, Qianqian, 973

H
Han, Yaru, 437
Han, Zijian, 80
Hao, Jiaxing, 373
He, Wei, 519
Hong, Zhang, 246
Hou, Yihan, 635
Hu, Shihan, 478, 748
Hu, Wen-Jing, 468
Hu, Yuehui, 236
Hua, Peng, 277
Huang, Yan, 730
Huang, Yanran, 691
Huang, Yuebing, 253

J
Ji, Boxiang, 395
Jia, Zhenghao, 322
Jiang, Huiyong, 268
Jiang, Jingfeng, 128
Jiang, Yuhua, 211, 216
Jin, Xin, 414

K
Ke, Difeng, 581
Kuang, Da, 328
Kuang, Mo, 328

L
Li, Bo, 351
Li, Fei, 622
Li, Han, 180
Li, Hongjin, 655
Li, Hongliang, 158
Li, Jian, 444
Li, Jianmin, 367
Li, Min, 454, 459
Li, Peng, 87
Li, Qingxian, 437
Li, Rongqiang, 158
Li, Shanglin, 351
Li, Shuang, 322
Li, Tingting, 151
Li, Wumei, 541
Li, Yan, 64
Li, Zhiyi, 285
Lin, Jing, 294, 790
Lin, Li, 710
Lin, Rui, 72
Lin, Shiyuan, 437
Lin, Siqi, 662
Liu, Bocheng, 395
Liu, Hualin, 343
Liu, Jing, 308
Liu, Jinhua, 449
Liu, Minglei, 770
Liu, Qing, 401
Liu, Xuejun, 928
Liu, Yi, 581
Liu, Yu, 380
Liu, Yunshan, 360
Long, Rongting, 99
Long, Siwei, 629, 841, 967
Lu, Gan, 534
Lu, Haiyang, 351
Lu, Yao, 166, 684
Lv, Ran, 200
Lyu, Qianyun, 429

M
Ma, Weiyan, 93
Mao, Yanghong, 268
Meng, Fanxing, 387
Mu, Huaiqin, 617

N
Niu, Guoying, 835

P
Pan, Yuyang, 835
Peng, Shaojie, 611
Pu, Na, 943

Q
Qi, Mingyang, 158
Qiao, Weitong, 50
Qu, Huiyan, 605

R
Ren, Daowen, 928

S
Sang, Tingrui, 495, 504
Shang, Tongfei, 308
Shen, Lei, 982
Shen, Yiting, 192
Shi, Heyuzi, 835
Shi, Ruotong, 596
Shi, Wanbing, 895
Song, Jiahan, 13
Song, Jing, 186
Song, Liankai, 519
Song, Ping, 113
Sun, Huizi, 380
Sun, Jianming, 315
Sun, Ke, 186
Sun, Lan, 206
Sun, Weizheng, 473
Sun, Yibo, 315

T
Tang, Junlong, 449
Tang, You, 87
Tao, Shuai, 936
Tao, Yongmei, 820

W
Wang, Bing, 120
Wang, Dewei, 343
Wang, Hairong, 454, 459
Wang, Hong, 973
Wang, Hongqin, 268
Wang, Jing, 211, 216
Wang, Liang, 922
Wang, Ning, 22

Wang, Qi, 725
Wang, Qianxun, 343
Wang, Rui, 488
Wang, Shanbiao, 343
Wang, Weimin, 300
Wang, Xi, 57
Wang, Xiangfu, 495, 504
Wang, Xiaohua, 655
Wang, Xiaoli, 770
Wang, Xiaozhu, 113
Wang, Xin, 922
Wang, Xiuwen, 872
Wang, Xuetian, 373
Wang, Yejun, 463
Wang, Yongling, 715
Wang, Yue, 241
Wen, Senhao, 343
Wen, Xia, 268
Wen, Zhurong, 784
Weng, Jingde, 173, 910
Wu, Di, 186
Wu, Dianyi, 629, 841, 967
Wu, Jing, 437
Wu, Rong, 260
Wu, Yujie, 206
Wu, Zhen, 872

X
Xiao, Juan, 351
Xiao, Yinglin, 588
Xin, Hua, 28
Xing, Xichen, 422
Xu, Huiting, 231
Xu, Qidan, 322
Xu, Zan, 186
Xue, Ping, 322

Y
Yan, Zhengxing, 736
Yang, Canyi, 3
Yang, Hongpeng, 649
Yang, Hongxia, 986
Yang, Huanying, 495, 504
Yang, Liqiong, 308
Yang, Ping, 691
Yang, Sen, 373
Yang, Yajun, 226
Yang, Zhihui, 796
Yang, Zhou, 880
Yangli, Ou, 241
Yao, Chenchen, 449
Yao, Yiheng, 414
Ye, Liuqing, 231
Ye, Yingjin, 437
Yin, Shanshan, 548

You, Peiyu, 698
You, Xiangyu, 495, 504
Yu, Cuicui, 373
Yu, Guoxin, 519
Yu, Haizhi, 211, 216
Yu, Hongjia, 87
Yu, Jing, 253
Yu, Wei, 572
Yu, Zhihui, 300
Yu, Ziwei, 128
Yuan, Jun, 556
Yuan, Tongqing, 887
Yue, Guoqing, 611

Z
Zang, Rongxin, 753
Zeng, Lanjian, 803
Zhang, Canhui, 231
Zhang, Chao, 495, 504
Zhang, Chongkai, 367
Zhang, Ermi, 895
Zhang, Hongying, 684
Zhang, Jinyuan, 141, 668
Zhang, Jun, 960
Zhang, Junyi, 676
Zhang, Leilei, 449
Zhang, Lunning, 519
Zhang, Nannan, 973
Zhang, Ting, 563, 812
Zhang, Wan, 367
Zhang, Wanqian, 495, 504
Zhang, Weibin, 596
Zhang, Wenjun, 407
Zhang, Xincheng, 35, 50
Zhang, Xueling, 743
Zhang, Xueqin, 334
Zhang, Yu, 778
Zhang, Yuhan, 866
Zhang, Ze, 856
Zhang, Zhen, 643
Zhang, Zhipeng, 367
Zhao, Cuijie, 973
Zhao, Dan, 231
Zhao, Heng, 495
Zhao, Jie, 916
Zhao, Jingmei, 454, 459
Zhao, Jun, 483
Zhao, Kun, 444
Zhao, Qiwang, 429
Zhao, Ru, 796
Zheng, Gang, 495, 504
Zheng, Hai, 986
Zheng, Huihui, 504
Zhong, Wen, 753

Zhong, Xia, 134
Zhou, Huikui, 526
Zhou, Liankai, 322
Zhou, Qi, 848
Zhou, Yan, 841

Zhou, Zhenwu, 473
Zhu, Lijuan, 720
Zhu, Lin, 42
Zhu, Qianyi, 488
Zhu, Yafang, 437

Printed in the United States
by Baker & Taylor Publisher Services